AN INDEX TO THE DESCRIBED LIFE HISTORIES, EARLY STAGES AND HOSTS OF THE MACROLEPIDOPTERA OF THE CONTINENTAL UNITED STATES AND CANADA

I

HARRISON MORTON TIETZ

PUBLISHED BY A. C. ALLYN
FOR
THE ALLYN MUSEUM OF ENTOMOLOGY
SARASOTA, FLORIDA
MCMLXXII

Distributed by

Entomological Reprint Specialists
P. O. Box 77971
Dockweiler Station
Los Angeles, California 90007
U.S.A.

ISBN 0-913492-01-9

INTRODUCTION

This bibliography or index as the author liked to refer to it, is the result of many years work by the late Harrison Morton Tietz (1895 - 1963) and was prepared during his tenure on the staff of the Department of Zoology and Entomology, Pennsylvania State University, State College, Pennsylvania and on the staff of the Department of Biology, Howard Payne College, Brownwood, Texas.

There has been no similar bibliography to the published hosts and life histories covering the whole of the macrolepidoptera of North American north of Mexico since that by Henry Edwards in 1889 ("Bibliographical Catalogue of the Described Transformations of North American Lepidoptera," *Bulletin of the United States National Museum,* No. 35).

Dr. Tietz began by rechecking all references in the Edwards bibliography (cited above) and then proceeded to add all the references he could find from the year 1889 through about the year 1950 when he left off his search to prepare the manuscript. The extent of the literature searched by Dr. Tietz can be seen in his list of periodicals and separate works consulted. Here are listed 226 periodicals and 127 separate works which according to Dr. Tietz were searched either in their entirety or through about the year 1950.

The nomenclature of the butterflies and moths indexed is that of the J. McDunnough 1938 check list (*Memoir of the Southern California Academy of Sciences,* volume 1), the standard reference of the period and still the latest list for most of the families of moths.

All species, subspecies and lower taxa names, including all synonyms are listed alphabetically and cross referenced.

The plant names used are mainly those found in Merritt Lyndon Fernald's rewritten and expanded eighth edition of *"Gray's Manual of Botany,"* 1950, and other similar works of that period. Dr. F. Raymond Fosberg (Department of Botany, National Museum of Natural History) aided Dr. Tietz by checking a submitted list of plant names used for the more obvious synonymy and errors in spelling.

The arrangement of references under each butterfly and moth species name is first to a list of the periodical literature in alphabetical order according to the name of the journal where the reference occurs. This is followed by a list of the separate works listed alphabetically by author. Immediately following each reference in this bibliography are listed the stages described. Here the asterisk is used after the words, egg, larva, pupa, life history and the like to show that there are illustrations in the references. The list of recorded food plants ends each species treatment.

In part I, section B, which is a list of the common names of the cited butterflies and moths together with their scientific names, the asterisk is used to denote those common names that are approved by the Entomological Society of America.

We and others could sit down and find much to criticize about the present work. As with any bibliographical work of such size and scope, errors and omissions will be found here. If we had been the authors of this work we probably would have adopted a different format. Nevertheless it is a most valuable contribution to lepidopterology and its presentation here in the format prepared by Dr. Tietz should be welcomed by all.

We know of no similar bibliography being prepared today, covering all of the macrolepidoptera. We believe that in spite of the age of this work (finished about 22 years ago) that its publication now will greatly stimulate much future work on the biology of our macrolepidoptera.

We regret that this work was not published two decades ago and we regard the fact that it has not appeared before now as a great loss to science.

The hand maiden of all research is bibliography and the only sound approach to research is to build it upon what previous workers have done. Without bibliography a student blindly gropes in the dark, ignorant of all that has been done before.

William D. Field

J. F. Gates Clarke

Department of Entomology
National Museum of Natural History

Washington, D. C.

PUBLISHER'S FOREWORD

In the preparation of this work for publication the galleys have been carefully checked against the manuscript, but no effort was made to verify or correct any errors which may exist in the original work. It is hoped that such errors, if any, will not substantively detract from the value of these volumes.

It is with pleasure that we bring to you this monumental compilation of the late Harrison Morton Tietz.

A. C. Allyn
Allyn Museum of Entomology

CONTENTS

PART I - INSECTS

SECTION A - - - WORKS CONSULTED

ENTOMOLOGICAL

A PERIODICALS, JOURNALS, BULLETINS, MEMOIRS, ETC.

ABH. NATUR. GESELL. HALLE---ABHANDLUNGEN NATURFORSCHENDEN GESSELLSCHAFT ZU HALLE, GERMANY, 1856. ALSO UNDER BURMEISTER AS SEPARATE WORK.

AGRI. GAZETTE CANADA---THE AGRICULTURAL GAZETTE OF CANADA. OTTAWA, CANADA, 1914-1924"

AGRI. PENNSYLVANIA, REPT.---REPORT ON THE AGRICULTURE OF PENNSYLVANIA, 1879, 1883, 1884 +

AMER. ACAD. ARTS. SCI., PROC.---PROCEEDINGS OF THE AMERICAN ACADEMY OF ARTS AND SCIENCE. BOSTON, MASS., 1846 +

AMER. ACAD. SCI., PROC.---SAME AS THE ABOVE ENTRY

AMER. ASSOC. ADV. SCI., PROC.---PROCEEDINGS OF THE AMERICAN ASSOCIATION FOR THE ADVANCEMENT OF SCIENCE. SALEM, MASS., 1848 +

AMER. ENT. SOC., TRANS.---TRANSACTIONS OF THE AMERICAN ENTOMOLOGICAL SOCIETY. PHILADELPHIA, PENN., 1967 + PROCEEDINGS OF THE SAME 1861-1867" ALSO UNDER SEPARATE WORKS SMITH, MONOGRAPH OF THE SPHINGIDAE.

AMER. ENT. ST. LOUIS AND NEW YORK "1868-1880" SOME OF THE EARLIER NUMBERS AS AMERICAN ENTOMOLOGIST AND BOTANIST.

AMER. JOURN. OF SCI.---AMERICAN JOURNAL OF SCIENCE. NEW HAVEN, CONN. 1818 + EARLIER ISSUES KNOWN AS SILLIMAN'S JOURNAL OF SCIENCE AND ARTS. UNDER SEPARATE WORKS SEE HARRIS--CATALOGUE OF THE NORTH AMERICAN SPHINGIDAE, SERIES I - 1819-1845; SERIES II - 1846-1870; SERIES III - 1871-1895; SERIES IV - 1896-1920; SERIES V - 1921.

AMER. MIDLAND NAT.---AMERICAN MIDLAND NATURALIST. NOTRE DAME, IND. 1909 + FIRST ISSUES KNOWN AS MIDLAND NATURALIST.

AMER. MUS. NAT. HIST., BULL.---BULLETIN OF THE AMERICAN MUSEUM OF NATURAL HISTORY 1881 +

AMER. MUS. NAT. HIST. JOURN.---JOURNAL OF THE AMERICAN MUSEUM NATURAL HISTORY. NEW YORK, N.Y., 1900-1918 LATER VOLUMES AS NATURAL HISTORY 1918 +

AMER. MUS. NAT. HIST. MEM.---MEMOIRS OF THE AMERICAN MUSEUM OF NATURAL HISTORY NEW YORK, N.Y. 1893-1930 TO BE CONTINUED AS MONOGRAPHS

AMER. MUS. NAT. HIST. NOVITATES---AMERICAN MUSEUM OF NATURAL HISTORY NOVITATES. NEW YORK, N.Y. 1921 +

AMER. NATURALIST, BOSTON, MASS. 1867 +

Amer. Philos. Soc. Proc.---Proceedings of the American Philosophical Society. Philadelphia, Penn. 1838 +

Amer. Pomological Soc., Rept. Boston, Mass. 1854

Archives Natur.---Archives fur Naturgeschichte. Wiegmann und Erichson. Berlin, Germany 1835-1911"

Argentina Soc. Cienc., Ann.---Sociedad Argentina de Ciencias Naturales, Anales. Buenos Aires, Argen. 1876 +

Arizona Agri. Exp. Sta. Bull.---University of Arizona Agricultural Experiment Station Bulletin Tucson, Ariz.

Arizona Agri. Exp. Sta. Tech. Bull.---University of Arizona Agricultural Experiment Station Technical Bulletin Tucson, Ariz.

Arkansas Agri. Exp. Sta. Bull.---Arkansas Agricultural Experiment Station, Bulletin. University of Ark.

Boston Cultivator---Boston Cultivator 1839-1876, from 1877-1915 "American Cultivator"

Boston Soc. Nat. Hist. Occas. Papers---Boston Society of Natural History Occasional Papers. See under separate works Harris Entomological Correspondence.

Boston Soc. Nat. Hist. Mem.---Boston Society of Natural History Memoirs. Boston, Mass. "1862-1942"

Boston Soc. Nat. Hist. Proc.---Boston Society of Natural History, Proceedings. Boston, Mass. 1841-1942"?

Boston Zool. Soc. Journ.-Quarterly---Journal of the Boston Zoological Society. Boston, Mass. 1881-1884"

Botanical Gazette, Crawfordsville, Ind. 1875 +

British Columbia Ent. Soc. Proc.---Proceedings of the Entomological Society of British Columbia. Victoria, B.C. 1906-1908" new series 1911 +

Brooklyn Ent. Soc. Bull.---Bulletin of the Brooklyn Entomological Society, Brooklyn, N.Y. 1878-1885, new series 1912 +

Brooklyn Hort. Soc. Rept.---Report of the Brooklyn Horticultural Society Brooklyn, N.Y.

BROOKLYN INST. ARTS & SCI. BULL.---BULLETIN OF THE BROOKLYN INSTITUTE OF ARTS AND SCIENCES. BROOKLYN, N.Y. 1906 +

BROOKLYN MUS. ARTS & SCI. BULL.---BULLETIN OF THE MUSEUM OF THE BROOKLYN INSTITUTE OF ARTS AND SCIENCES. SAME AS PRECEEDING ENTRY.

BUFFALO SOC. NAT. SCI. BULL.---BULLETIN OF THE BUFFALO SOCIETY OF NATURAL SCIENCES, BUFFALO, N.Y. 1873 +

BUTTERFLY PARK NATURE CLUB NEWS. ROSCOE, CALIF. 1929-1931"?

CALIFORNIA ACAD. SCI.BULL.---BULLETIN OF THE CALIFORNIA ACADEMY OF SCIENCES. SAN FRANCISCO, CALIF. 1884-1887"

CALIFORNIA ACAD. SCI. PROC.---PROCEEDINGS OF THE CALIFORNIA ACADEMY OF SCIENCES. SAN FRANCISCO, CALIF. SERIES I 1854-1876;

CALIFORNIA ACAD. SCI. PROC.---PROCEEDINGS OF THE CALIFORNIA ACADEMY OF SCIENCES. SAN FRANCISCO, CALIF. SERIES I 1854-1876; SERIES II 1888-1896; SERIES III, IN PARTS, AND SERIES IV 1907 +

CALIFORNIA AGRI. EXP. STA. BULL.---BULLETIN OF THE CALIFORNIA AGRICULTURAL EXPERIMENT STATION. BERKELEY, CALIF.

CALIFORNIA STATE COMM. HORT. MONTHLY BULL.---MONTHLY BULLETIN OF THE CALIFORNIA STATE COMMISSION OF HORTICULTURE. SACRAMENTO, CALIF. 1911 +

CAN. DEPT. AGRI. REPT.---REPORTS OF THE CANADIAN DEPARTMENT OF AGRICULTURE. OTTAWA, CANADA.

CAN. ENT.---CANADIAN ENTOMOLOGIST. PUBLISHED BY THE ENTOMOLOGICAL SOCIETY OF ONTARIO. LONDON, ONTARIO, CANADA 1868 +

CAN. JOURN. IND. SCI. AND ARTS---CANADIAN JOURNAL OF SCIENCE, LITERATURE, AND HISTORY (TITLE VARIES). TORONTO, CANADA 1852-1855; NEW SERIES 1856-1878" SUPERCEDED BY THE ROYAL CANADIAN INSTITUTE PROCEEDINGS 1879 +

CAN. NAT. SCI. NEWS---CANADIAN NATURAL SCIENCE NEWS. BADEN, ONTARIO, CANADA 1897"?

CAN. NAT. AND GEOL.---CANADIAN NATURALIST AND GEOLOGIST, MONTREAL, CANADA 1856-1857"?

CAN. RECORD OF SCI.---THIS REPLACES CANADIAN NATURALIST AND GEOLOGIST. MONTREAL, CANADA 1884-1916" SUSPENDED 1898, 1905-1913

CARNEGIE MUS. ANNALS---ANNALS OF THE CARNEGIE INSTITUTE MUSEUM. PITTSBURGH, PENN. 1901 +

CHICAGO ACAD. SCI. BULL.---BULLETIN OF THE CHICAGO ACADEMY OF SCIENCE. CHICAGO, ILL. 1883 +

Cincinnati Journ. Sci.---Cincinnati Quarterly Journal of Science. Cincinnati, Ohio 1874-1875"

Connecticut Agri. Exp. Sta. Bull.---Connecticut Agricultural Experiment Station Bulletins. New Haven, Conn.

Davenport Acad. Sci. Proc.---Proceedings of the Davenport Public Museum. Davenport, Iowa 1867-1914"

Denison Univ. Sci. Lab. Journ.---Denison University Bulletin---Journal of the Scientific Laboratories. Granville, Ohio 1885 +

Downing's Horticulturist---Horticulturist and Journal of Rural Art and Rural Taste. Albany, N.Y. 1846-1875"

Edge-Report on the Agriculture of Pennsylvania for the years 1879, 1883, 1884

Elliott Soc. Nat. Hist. Proc.---Proceedings of the Elliott Society of Science and Art. Charleston, S.C. 1853-1890"

Entomologica Americana. Brooklyn, N.Y. 1885 + (suspended 1891-1926)

Ent. News---Entomological News. Published by the American Entomological Society. Philadelphia, Penn. 1890 +

Ent. Soc. America Annals---Annals of the Entomological Society of America. Columbus, Ohio 1908 +

Ent. Nachtricht.---Entomologische Nachtrichten. Berlin, Germany 1875-1900"

Ent. Tidsch.---Entomologisk Tidskrift. Stockholm, Sweden 1880 +

Entomologist. London, England 1840 +

Entomologist's Intelligencer-Entomologist's Weekly Intelligencer. London, England 1856-1861"

Entomologist's Monthly Magazine. London, England 1864 +

Entomologist's Record---Entomologist's Record and Journal of Variation. London, England 1890 +

Essex Inst. Proc.---Proceedings of the Essex Institute. Salem, Mass. 1845-1868"

Everhart Mus. Bull.---Bulletin of the Everhart Museum of Natural History, Science and Art. Scranton, Penn. 1915

Field and Forest---Potomac Side Naturalist's Club. Washington, D.C. 1875-1878"

Florida Agri. Exp. Sta. Bull.---University of Florida Agricultural Experiment Station Bulletin. Lake City and Gainsville, Fla.

Florida Ent.---Florida Entomologist. Gainsville, Fla. 1917 + (Vol. 1-3 The Florida Buggist)

Geol. Surv. Terr.---United States Geological Survey of the Territories Washington, D.C.

Hovey's Mag. Hort.---The Magazine of Horticulture by C.M. Hovey. Boston, Mass. 1833-1868"?

Illinois Agri. Exp. Sta. Bull.---University of Illinois Agricultural Experiment Station Bulletins. Urbana, Ill.

Illinois Agri. Soc. Trans.---Transactions of the Illinois Agricultural Society. Springfield, Ill. 1853-1870"

Illinois Dept. Agri. Trans.---Transactions of the Illinois Department of Agriculture Springfield, Ill. 1871 +

Illinois Nat. Hist. Soc. Trans.---Illinois Natural History Society Transactions. Springfield, Ill. 1861"

Illinois, Rept. Ins. of---Report of the Insects of Illinois

Illinois State Acad. Sci. Trans.---Transactions of the Illinois State Academy of Sciences. Springfield, Ill. 1907 +

Illinois State Normal Univ. Rept.---Illinois State Normal University Report. Carbondale, Ill.

Indiana Acad. Sci. Proc.---Proceedings of the Indiana Academy of Science. Brookville and Indianapolis 1891 +

Industrialist---Kansas Industrialist. Manhattan, Kan. 1875 +

Insect Life Washington, D.C. 1888-1895"

Insecutor Inscitiae Menstruus. Washington, D.C. 1913-1926"

Iowa Academy of Science Biological Survey Publication No. 1 1936 "How to Know the Insects" by H.E. Jaques

Iowa Acad. Sci. Proc.---Proceedings of the Iowa Academy of Science. Des Moines, Iowa 1887-1893 +

Iowa Agri. Exp. Sta. Bull.---Iowa Agricultural Experiment Station Bulletins. Ames, Iowa

Iowa Hort. Soc. Trans.---Transactions of the Iowa Horticultural Society. Des Moines, Iowa

Iowa State College Research Bulletins. Ames, Iowa

Jamaica Inst. Journ.---Journal of the Jamaica Institute

Journ. Agri. Research---Journal of Agricultural Research. Washington, D.C. 1913 + Suspended 1921-1922

Journ. Econ. Ent.---Journal of Economic Entomology. Concord, N.H. 1908 +

Journ. Ent. Zool.---Journal of Entomology and Zoology Pomona College. Claremont, Calif. 1909 + Title varies - Pomona Journal of Entomology, Pomona College Journal of Entomology

Kansas Acad. Sci. Trans.---Transactions of the Kansas Academy of Sciences. Lawrence, Kan. 1868 +

Kansas Ent. Soc. Journ.---Journal of the Kansas Entomological Society. McPherson, Kan. 1928 +

Kansas Univ. Sci. Bull.---Kansas University Science Bulletin. Lawrence, Kan. 1902 + (supercedes Kan. Univ. Quarterly 1892-1901")

Kentucky Agri. Exp. Sta. Bull.---Bulletins of the Kentucky Agricultural Experiment Stations. Lexington, Ky.

Kentucky Agri. Exp. Sta. Circ.---Kentucky Agricultural Experiment Station Circulars. Lexington, Ky.

Kentucky Bur. Agri. Rept.---Kentucky Bureau of Agriculture, Labor and Statistics--Biennial Report

London Zool. Soc. Proc.---Proceedings of the London Zoological Society. London, England 1833 +

Louisiana Agri. Exp. Sta. Bull.---Bulletin of the Louisiana Agricultural Experiment Station. Baton Rouge, La.

Lycaeum Nat. Hist. N.Y. Annals---Annals of the Lycaeum of Natural History of New York, N.Y. 1823-1877" From here on as Annals of the New York Academy of Science 1877 +

Lyons (France) Linn. Soc. Annales---Annales de la Societe Linneenne de Lyon. Paris, France

Magazine of Nat. Hist.---Magazine of Natural History. London, England 1828-1836 New series 1837-1840" Later as Annals of Natural History and still later as Annals and Magazine of Natural History 1840 +

Maine Agri. Exp. Sta. Bull.---Bulletin of the Maine Agricultural Experiment Station. Onono, Maine

Maine Board Agri. Rept.---Report of the Maine Board of Agriculture. Augusta, Maine

Maine Naturalist---Journal of the Knox Academy of Arts and Science. Thomaston, Maine 1921-1930"

Marine Biol. Lab. Woods Hole---Report of the Woods Hole Marine Biological Laboratory. Woods Hole, Mass. 1930 +

Maryland Acad. Sci. Bull.---Bulletin of the Maryland Academy of Science. Baltimore, Md. 1921-1925" Superceded by its Journal 1930-1932"

Maryland Nat. Hist. Soc. Bull.---Bulletin of the Natural History Society of Maryland. Baltimore, Md. 1930 +

Maryland Nat. Hist. Soc. Proc.---Proceedings of the Maryland Natural History Society. Baltimore, Md. 1930 +

Maryland Nat. Hist. Soc. Pub.---Publication of the Maryland Natural History Society. Baltimore, Md. 1930 +

Massachusetts Agri. Exp. Sta. Bull.---Bulletin of the Massachusetts Agricultural Experiment Station. Amherst, Mass.

Massachusetts Agri. Repository Journ.---Massachusetts Agricultural Repository and Journal (The Massachusetts Society for the Promotion of Agriculture). Boston, Mass. 1796-1832

Massachusetts Agri. Soc. Proc.---Proceedings of the Massachusetts Agricultural Society. Boston, Mass.

Massachusetts Board Agri. Rept.---Report of the Massachusetts State Board of Agriculture. Boston, Mass.

Massachusetts Ploughman---Massachusetts Ploughman and New England Journal of Agriculture (New England Agricultural Society). Boston, Mass. 1840-1906"

Massachusetts Soc. Promot. Agri.---See Massachusetts Agri. Repository Journal for the year 1796

Michigan Agri. Exp. Sta. Bull.---Michigan Agricultural Experiment Station Bulletin Agricultural College and then East Lansing, Mich.

Michigan Agri. Exp. Sta. Circ.---Michigan Agricultural Experiment Station Circulars. East Lansing, Mich.

Michigan Agri. Exp. Sta. Spec. Bull.---Michigan Agricultural Experiment Station Special Bulletins. East Lansing, Mich.

Michigan Agri. Soc. Rept.---Report of the Michigan State Agricultural Society (often bound with the Report of the State Board of Agriculture). Lansing, Mich.

Michigan State Board of Agriculture Reports. Lansing, Mich.

Minnesota Acad. Nat. Sci. Bull.---Bulletin of the Minnesota Academy of Science. Minneapolis, Minn. 1875-1917"

Minnesota Agri. Exp. Sta. Bull.---Bulletin of the Minnesota Agricultural Experiment Station. University Farm and St. Paul, Minn.

Minnesota Agri. Exp. Sta. Rept.---Report of the Minnesota Agricultural Experiment Station. University Farm, Minn.

Missouri Acad. Sci. Proc.---Proceedings of the Missouri Academy of Science. Columbia, Mo. 1934 +

Missouri Agri. Exp. Sta. Bull.---Bulletin of the Missouri Agricultural Experiment Station. Columbia, Mo.

Missouri, Rept. Nox. Ins., of---Annual Report of the Noxious, Beneficial and Other Insects of the State of Missouri by C.V. Riley. Jefferson City, Mo.

Mitteilungen aus dem Roemer Museum Hildesheim Nr. 3 Jan. 1896 Die Apateliden by Grote

Montana Agri. Exp. Sta. Bull.---Bulletin of the Montana Agricultural Experiment Station. Bozeman, Mont.

Montana Univ. Bull.---Bulletin of the University of Montana. Missoula, Mont.

Mus. Comp. Zool. Bull.---Harvard Museum of Comparative Zoology Bulletin 1863 +

Nantucket Maria Mitchell Assoc. Pub.---Publications of the Nantucket Maria Mitchell Association. Nantucket, Mass. 1906 +

Nat. Acad. Sci. Mem.---Memoirs of the National Academy of Sciences. Washington, D.C. 1866 +

Naturaliste Canadien. Quebec, Canada 1868 +

Nature Study Review---by the American Nature Study Society. Ithaca, N.Y. 1905–1923" Merged into Nature Magazine, Washington, D.C. and Baltimore, Md. 1923 +

Nebraska Agri. Exp. Sta. Research Bull.---Research Bulletin of the Nebraska Agricultural Experiment Station. Lincoln, Neb.

New England Farmer, A Monthly Journal. Boston, Mass. 1848–1864; 1867–1871" Also known as The New England Farmer and Horticultural Register 1822–1846; 1865–1913"

New Hampshire Agri. Exp. Sta. Bull.---Bulletin of the New Hampshire Agricultural Experiment Station. Durham, N.H.

New Hampshire Agri. Exp. Sta. Tech. Bull.---Technical Bulletin of the New Hampshire Agricultural Experiment Station. Durham, N.H.

New Hampshire Board Agri. Rept.---Report of the New Hampshire State Board of Agriculture. Manchester, N.H.

New Jersey Agri. Exp. Sta. Bull.---Bulletin of the New Jersey Agricultural Experiment Station. New Brunswick, N.J.

New York Acad. Sci. Annals---Annals of the New York Academy of Science. New York, N.Y. 1877 + Supercedes the Annals of the Lycaeum of Natural History of New York

New York Ent. Soc. Journ.---Journal of the New York Entomological Society. New York, N.Y. 1893 +

New York State Agri. Exp. Sta. (Cornell) Bull.---New York State Agricultural Experiment Station (Cornell) Bulletin Ithaca, N.Y.

New York State Agri. Exp. Sta. (Cornell) Mem.---New York State Agricultural Experiment Station (Cornell) Memoir Ithaca, N.Y.

New York State Agri. Exp. Sta. (Geneva) Bull.---New York State Agricultural Experiment Station (Geneva) Bulletin Geneva, N.Y.

New York State Agri. Soc. Rept.---Report of the New York State Agricultural Society. Albany, N.Y. Includes Proceedings and Transactions.

New York State Cab. Nat. Hist. Rept.---New York State Cabinet of Natural History Report. Title varies sometimes Annual Report of the New York State Museum. Contains Lintner's Entomological Contributions. Albany, N.Y. 1847– 1918"

New York State Ent. Rept.---Report of the New York State Entomologist. Albany, N.Y.

New York State Mus. Bull.---Bulletin of the New York State Museum. Albany, N.Y. 1887 +

New York State Mus. Hand.---New York State Museum Handbook. Albany, N.Y. 1927 + Also see under Separate Works Saunders, A.A.

New York State Mus. Mem.---Memoirs of the New York State Museum. Albany, N.Y. 1889-1923"?

New York State, Rept. on Ins. of---Report on the Noxious, Beneficial and Other Insects of the State of New York by Asa Fitch. Albany, N.Y.

North Amer. Ent.---North American Entomologist. Buffalo, N.Y. 1879-1880"

North Dakota Agri. Exp. Sta. Bull.---North Dakota Agricultural Experiment Station Bulletin. Fargo, N.D.

Nova Acta der Ksl Leop-Carol. Deutschen Akademie der Naturforscher Halle For details see under Separate Works Gumppenberg

Nova Scotia Ent. Soc. Proc.---Proceedings of the Entomological Society of Nova Scotia (Publication of the Acadian Entomological Society). Truro, Nova Scotia 1915-1921"

Novitates Zoologica Tring Museum England. London, England 1894 +

Observer of Nature---Publication of the Kansas University Natural History Society. Lawrence, Kansas 1871-1876"

Ohio Agri. Exp. Sta. Bull.---Bulletin of the Ohio Agricultural Experiment Station. Wooster, Ohio

Ohio Agri. Exp. Sta. Rept.---Report of the Ohio Agricultural Experiment Station. Wooster, Ohio

Ohio Agri. Exp. Sta. Tech. Bull.---Technical Bulletin of the Ohio Agricultural Experiment Station. Wooster, Ohio

Ohio Farmer also known as Ohio Practical Farmer. Cleveland, Ohio 1852 +

Ohio Journ. of Science (Published by the Ohio State University Scientific Society Biological Club and the Ohio Academy of Science). Columbus, Ohio 1900 +

Ohio Acad. Sci. Rept.---Report of the Ohio State Academy of Sciences. Columbus, Ohio 1892–1902" Continued in its Proceedings 1902–1930 and further continued as the Ohio Journal of Science

Oklahoma Agri. Exp. Sta. Bull.---Bulletin of the Oklahoma Agricultural Experiment Station. Stillwater, Okla.

Oklahoma Agri. Exp. Sta. Circ.---Oklahoma Agricultural Experiment Station Circular. Stillwater, Okla.

Ontario Dept. Agri. Rept.---Report of the Ontario Department of Agriculture. Ontario, Canada

Ontario Ent. Soc. Ann. Rept.---Annual Report of the Entomological Society of Ontario. Ontario, Canada Note: The Report for any given year is printed the following year.

Oregon Agri. Exp. Sta. Bull.---Bulletin of the Oregon Agricultural Experiment Station. Corvalis, Ore.

Oregon Agri. Exp. Sta. Circ.---Oregon Agricultural Experiment Station Circular. Corvallis, Ore.

Oregon Agricultural Experiment Station Biennial Crop Pest and Horticultural Report. Corvallis, Ore.

Ottawa Naturalist---Published by the Ottawa Field Naturalists Club and known as the Canadian Field Naturalist 1887. From 1887–1919 as Ottawa Naturalist Closed 1919?

Outlook---1870–1935" New York, N.Y. Name varies: Christian Union, Outlook, and New Outlook

Pacific Coast Ent. Soc. Proc.---Proceedings of the Pacific Coast Entomological Society. San Francisco, Calif. 1901–1930" Continued as The Pan Pacific Entomologist

Pan Pacific Entomologist. San Francisco, Calif. 1924 + Published by the California Academy of Sciences

Papilio Published by the New York Entomological Club 1881–1884" Continued as Entomologica Americana by the Brooklyn Ent. Society

Peabody Acad. Sci. Rept.---Report of the Peabody Academy of Science Published by the Peabody Museum. Salem, Mass. 1867–1887 Title varies

Pennsylvania Acad. Sci. Proc.---Proceedings of the Pennsylvania Academy of Science Harrisburg, Penn. 1924 +

Pennsylvania Agri. Exp. Sta. Bull.---Bulletin of the Pennsylvania Agricultural Experiment Station State College, Penn.

Pennsylvania State Agri. Soc. Rept.---Report of the Pennsylvania State Agricultural Society 1850-1896" Continued as the Pennsylvania State Board of Agriculture Report

Phila. Acad. Nat. Sci. Journ.---Journal of the Philadelphia Academy of Natural Science Philadelphia, Penn. 1817-1842; 1847-1918" Synopsis of North American Sphingidae

Phila. Ent. Soc. Proc.---Proceedings of the Entomological Society of Philadelphia 1861-1867" Continued as the Transactions of the American Entomological Society 1867 +

Practical Entomologist Published by The American Entomological Society Philadelphia, Penn. 1865-1867"

Prairie Farmer Chicago, Ill.

Providence Park Museum Bulletin Providence, R.I.

Psyche---A Journal of Entomology 1874 + Cambridge, Mass.

Royal Agri. Soc. Eng. Journal---Journal of the Royal Agricultural Society of England London, England 1839 +

Royal Soc. Can. Trans.---Transactions of the Royal Society of Canada Ottawa, Canada 1882-1894, Series II 1895-1906; series III 1907 +

Saint Louis Acad. Sci. Trans.---Transactions of the Saint Louis Academy of Science Saint Louis, Mo. 1856 +

Science Publication of the American Association for the Advancement of Science Cambridge, Mass. and New York, N.Y. 1883 +

Sierra Club Bulletin San Francisco, Calif. 1893 +

Smithsonian Misc. Collections---Smithsonian Misscellaneous Collections Washington, D.C. 1862 +

Smithsonian Rept.---Annual Report of the Board of Regents of the Smithsonian Institution Washington, D.C. 1846 +

Smithsonian Scientific Series Washington, D.C. 1929-1932

Soc. Acclim France, Bull.---Bulletin Mensuel de la Societe Imperiale Zoologique D'Acclimatation Paris, France 1854-1863; second series 1864-1873; third series 1874-1883; fourth series 1884-1889; fifth series 1890 +

South Dakota Acad. Sci. Proc.---Proceedings of the South Dakota Academy of Science Vermillion, S.D. 1916 +

South Dakota Agri. Exp. Sta. Bull.---Bulletin of the South Dakota Agricultural Experiment Station Brookings, S.D.

Southern Calif. Acad. Sci. Bull.---Bulletin of the Southern California Academy of Science Los Angeles, Calif. 1902 +

Tennessee Agri. Exp. Sta. Bull.---Bulletin of the Tennessee Agricultural Experiment Station Knoxville, Tenn.

Texas Agricultural Experiment Station Bull.---Bulletin of the Texas Agricultural Experiment Station College Station, Texas

Trenton Nat. Hist. Soc. Journ.---Journal of the Trenton Natural History Society Trenton, N.J.

U.S.D.A. Commissioner Rept.---Report of the United States Commissioner of Agriculture Started in 1837 Up to and including 1861 combined with the Report of the Commissioner of Patents but published as a separate volume beginning with the year 1849. In 1862 separated from Patents and made a special report of the Commissioner of Agriculture

U.S.D.A. Bull.---Bulletin of the United States Department of Agriculture 1914 + Usually contributions from the Bureau of Entomology

U.S.D.A. Bur. Ent. Bull.---United States Department of Agriculture, Bureau of Entomology Bulletin Old series 1-33 1883-1895; New series 1895 + The new series bulletins are designated as Bureau of Entomology and the old series as Division of Entomology

U.S.D.A. Bur. Ent. Circ.---United States Department of Agriculture Bureau of Entomology Circulars 1895 +

U.S.D.A. Farmer's Bulletins 1889 +

U.S.D.A. Miscellaneous Publications

U.S.D.A. Div. Ent. Rept. Ent.---U.S.D.A. Division of Entomology Report of the Entomologist 1870-1873 Glover; 1880 Comstock; 1879-1886 Riley

U.S.D.A. Ent. Comm. Bull.---Bulletin of the U.S.D.A. Entomological Commission

U.S.D.A. Ent. Comm. Rept.---Report of the U.S.D.A. Entomological Commission

U.S.D.A. Division Entomology, General Publication---J.H. Comstock Report on Cotton Insects 1879

U.S.D.A. Tech. Bull.---Technical Bulletin of the United States Department of Agriculture 1927 +

U.S.D.A. Yearbook---Yearbook of the United States Department of Agriculture, 1894 +

U.S.N.M. Bull.---Bulletin of the United States National Museum (Smithsonian Institution) Washington, D.C.

U.S.N.M. Proc.---Proceedings of the United States National Museum

Utah Agru. Exp. Sta. Bull.---Bulletin of the Utah Agricultural Experiment Station. Logan, Utah

Vermont Agri. Exp. Sta. Bull.---Vermont Agricultural Experiment Station Bulletin. Burlington, Vt.

Vermont Agri. Soc. Rept.---Report of the Vermont State Agricultural Society. Montpelier, Vt.

Washington (D.C.) Biological Soc. Proc.---Proceedings of the Washington Biological Society. Washington, D.C. 1880 +

Washington (D.C.) Ent. Soc. Proc.---Proceedings of the Entomological Society of Washington. Washington, D.C. 1884 +

Wisconsin Agri. Exp. Sta. Bull.---Bulletin of the Wisconsin Agricultural Experiment Station. Madison, Wis.

Wisconsin Agri. Exp. Sta. Research Bull.---Wisconsin Agricultural Experiment Station Research Bulletin. Madison, Wis.

Wisconsin Agri. Soc. Trans.---Transactions of the Wisconsin State Agricultural Society. Madison, Wis.

Zeitschrift gesamm. Natur.---Zeitschrift die gesammten Naturwissenschaften. Place of publication varies.

Zeitschrift Mediz. Naturwiss.---Zeitschrift fur Medizinsch Naturwissenschaftliche Gesellschaft zu Jena.

Zion's Herald. Boston, Mass. 1823-1828. New Series 1829 + Title varies. May be New England Christian Herald

B Separate Works---This includes some articles which have appeared in Bulletins, Journals, etc. and also published separately.

Abbot & Smith--Lep. Ins. Georgia---The Natural History of the Rarer Lepidopterous Insects of Georgia by John Abbot and James Smith. London, England 1797

Agassiz--Lep. Lake Superior---Lake Superior, Its Physical Characteristics, Vegetation, and Animals by Louis Agassiz. Boston, Mass. 1850

Ashmead--Ins. Inj. Orange---Orange Insects or a Treatise on the Injurious and Beneficial Insects Found on the Orange Tree in Florida by W.H. Ashmead. Jacksonville, Fla. 1880

Ballard--Ins. Lives---Insect Lives, or Born in Prison by Julia P. Ballard. Cincinnati, Ohio 1879

Boisduval--Monograph Zygaenidae---Essai sur une Monographie des Zygaenides by Dr. J.A. Boisduval. Paris, France 1829

Boisduval--Spec. Gen. Lep.---Histoire Naturelle des Insectes, Species Generale des Lepidopteres Heteroceres et Sphingidae. Dr. J.A. Boisduval. Paris, France, 1874 Diurnes, 1836

Boisduval & LeConte--Hist. Lep. North America---Histoire Generale et Iconographie des Lepidopteres des Chenilles de L'Amerique Septertrionale by Dr. J.A. Boisduval et John LeConte. Paris, France, pages 1-64, 1829; 65-80, 1830; Revolution stopped the work; pp.81-196, 1833; 197-228, 1837. Text stops at page 228, plates go beyond text which does not finish notes on Alope.

Burmeister--System. Verz. Sphing.---Systematische Verzeichniss der Sphingidae Brasiliensis. See Abhandlungen Naturges. Gesell. Halle under periodicals.

Burmeister--Descr. Rep. Argent.---Atlas de la Description Physique de la Republique Argentine. Dr. H. Burmeister. Buenos Aires, Argentine. 1878-1879

Butler--Lep. Exotica---Lepidoptera Exotica or Descriptions and Illustrations of Exotic Lepidoptera by A.G. Butler. London, England, 1874.

Butler--Monograph of the Callidryas. Published one year later than Lepidoptera Exotica and is incorporated in this work.

Chenu-Demarests--Encycl. Hist. Nat. Papillons---Encyclopedie D'Histoire Naturelle des Papillons. Chenu-Demarests. Paris, France, 1878. Some parts by Chenu Lucas, others by Chenu alone.

Chenu-Lucas See Above Date 1856

Chittenden--Ins. Inj. Veg.---Insects Injurious to Vegetation by F.H. Chittenden. New York, N.Y. 1907

Comstock--Butter. Calif.---The Butterflies of California by John A. Comstock. Los Angeles, Calif., 1927.

Comstock--Guide Study Ins.---A Guide to the Study of Insects by J.H. Comstock. Ithaca, N.Y., 1895.

Comstock--How to Know Butter.---How to Know the Butterflies, a Manual of the Butterflies of the Eastern United States by J.H. Comstock. New York, N.Y., 1904.

Comstock---A Manual for the Study of Insects by J.H. Comstock. Ithaca, N.Y. 1895.

Comstock--Rept. Cotton Ins.---Report on Cotton Insects. See U.S.D.A. Division of Entomology General Publication, 1879.

Cragen--Our Ins. Friends & Foes---Our Insect Friends & Foes by Belle S. Cragen. New York, N.Y., 1899.

Crosby & Leonard--Man. Veg. Gard. Ins.---A Manual of Vegetable Garden Insects by C.R. Crosby and M.D. Leonard. New York, N.Y. 1918.

Curtis--Brit. Ent.---British Entomology by John Curtis. London, England, 1862.

Cuvier--The Animal Kingdom by Baron Cuvier. London, England, 1832.

Dewitz--Beschreibungen von Jungenstadien exotischer Lepidopteren in Nova Acta der Ksl. Leop-Carol Deutschen Akademie der Naturforscher Halle. Band xliv, Nr. 2, 1882. by H. Dewitz.

Dickerson---Moths and Butterflies by Mary C. Dickerson. New York, N.Y., 1901.

Doane etc.--Forest Insects by R.W. Doane, E.C. Van Dyke, W.J. Chamberlin, and H.E. Burke. New York, N.Y., 1936.

Duncan--Naturalist's Library by Sir William Jardine. Entomology by James Duncan. Edinburgh, Scotland, 1836-1858. See Vol. iv, vi, vii, xxii, xxvii, xxx.

Duncan--Trans. Ins.---The Transformation of Metamorphosis of Insects by P. Martin Duncan. New York, N.Y., 1873-1874.

Edge--See under periodicals Agri. Penn. Rept.

Edwards--The Butterflies of North America by W.H. Edwards. Boston, Mass. and New York, N.Y. First series 1868-1972; Second 1874-1884; Third 1887-1896.

Edwards--In Kingsley's Stand. Nat. Hist.---The Standard Natural History. John S. Kingsley. Vol. 11, Boston, Mass., 1884.

Edwards--Pacific Coast Lepidoptera by Henry Edwards. 1876. See also Proceedings of the California Academy of Science VII, paper 16, 1876.

Eliot and Soule--Caterpillars and their Moths by Ida M. Eliot and Caroline G. Soule. New York, N.Y., 1902.

Emmons--Nat. Hist. New York---The Natural History of New York, vol. v. The More Common and Injurious Species of Insects by E. Emmons, Albany, N.Y., 1854.

Essig--Inj. & Benef. Ins.--Injurious and Beneficial Insects of California by E.O. Essig. (Supplement of the Monthly Bulletin of the California State Commission of Horticulture. Berkeley, Calif. 1913 and 1915.

Essig--Ins. West. North America---Insects of Western North America by E.O. Essig. New York, N.Y., 1926.

Fernald--Butterflies of Maine by C.H. Fernald. Orono, Maine, 1884.

Fernald, C.H.--See Kingsley's Standard Natural History II, 1885. Moths.

Fernald--The Sphigidae of New England by C.H. Fernald. Orono, Maine, 1886.

Fernald--Applied Entomology by H.T. Fernald. New York, N.Y., 1926.

Fernald & Shepard--Applied Entomology by H.T. Fernald and H.H. Shepard. New York, N.Y., 1942.

Fernald, H.T. & Forbush, E.H.--Rept. on the Gypsy Moth---Report on the Gypsy Moth. Boston, Mass., 1896.

Figuier--The Insect World by Louis Figuier. London, England, 1868 and 1872.

French--Butter. East. U.S.---The Butterflies of the Eastern United States by G.H. French. Philadelphia, Penn., 1886.

Glover--Illust. North Amer. Ent.---Illustrations of North American Entomology by T. Glover. Washington, D.C., 1878.

Gosse--Can. Nat.---Canadian Naturalist. A Series of Conversations on the Natural History of Lower Canada. London, Ont., 1840 by Philip Gosse.

Gosse--Letters from Alabama Chiefly Relating to Natural History. London, Ont., 1859. by Philip Gosse.

Graham--Principles of Forest Entomology by S.A. Graham. New York, N.Y., 1939.

Grote--North Amer. Ent.---North American Entomologist by A.R. Grote. Buffalo, N.Y., 1880.

Grote, A.R.--Die Apateliden. See Mittheilungen aus dem Roemer Museum Hildesheim. Nr. 3, Jan. 1896.

Grote, A.R.--Rept. Geol. Surv. Alabama---Report on the Progress of the Geological and Agricultural Survey of Alabama. Montgomery, Ala., 1876.

Guenee--Spec. Gen. Lep.---Histoire Naturelle des Insects, Species Generale de Lepidopteres by Dr. J.A. Boisduval et A. Guenee. Paris, France. Noctuelites 1852, Deltoides et Pyralites 1854, Uranides et Phalaenites, 1857. Plates for all sections, 1858.

Gumppenberg--Systema Geometrarum Zonae Temperatioris Septentrionales by C. Freih von Gumppenberg. Halle, Germany. See also under Nova Acta der Ksl. etc. in Periodical Section Vol. 1-1887; 2-1888; 3-1890; 4-1890; 5-1892; 6-1893; 7-1895; and 8-1896.

Gundlach--Entom. Cubana---Contribucion a la Entomoligica Cubana Por el Dr. Juan Gundlach. Habana, Cuba, 1881.

Harris--Cat. North Amer. Sphing.---Catalogue of the North American Sphingidae. Reprint from American Journal of Science xxxvi, No. 2, 1839. by T.W. Harris.

Harris--Ent. Corresp.---Harris' Entomological Correspondence by Samuel Scudder in Occasional Papers of the Boston Soc. Nat. Hist. 1869.

Harris--Ins. Inj. Veg.---Report on Insects Injurious to Vegetation by T.W. Harris. Boston, Mass. First Edit. 1841; Second Edit. 1852; Third Edit., Flint Edit., 1862.

Hart--Ins. Coll. Illinois High Schools---Synopsis of Insect Collections for Distribution to Illinois High Schools, Lepidoptera by Charles A. Hart. Urbana, Ill., 1903. Illinois State Laboratory of Natural History.

Herrick--Ins. Enemies Shade Trees---Insect Enemies of Shade Trees by Glenn W. Herrick. Ithaca, N.Y., 1935.

Holland---The Butterfly Book by W.J. Holland. New York, N.Y., 1905 and 1931.

HOLLAND---THE MOTH BOOK BY W. J. HOLLAND. NEW YORK, N.Y., 1905.

HUBBARD--INS. AFFECT. ORANGE---INSECTS AFFECTING THE ORANGE BY H.G. HUBBARD. WASHINGTON, D.C., 1885.

HUMPHREY---GENERA OF BRITISH MOTHS BY H.N. HUMPHREY. LONDON, ENGLAND, 1859.

JAEGER--LIFE NORTH AMER. INS.---THE LIFE OF NORTH AMERICAN INSECTS BY B. JAEGER. NEW YORK, N.Y., 1859.

KELLOGG--AMER. INS.---AMERICAN INSECTS BY VERNON KELLOGG. NEW YORK, N.Y., 1903.

KIRBY--EUROPEAN BUTTER. & MOTHS---EUROPEAN BUTTERFLIES AND MOTHS BY W.F. KIRBY. LONDON, ENGLAND, 1882.

KOLLAR--INJ. INS.---A TREATISE ON INSECTS INJURIOUS TO GARDENERS, FORESTERS, & FARMERS BY VINCENT KOLLAR. LONDON, ENGLAND, 1840.

LINTNER--ENT. CONTRIB.---ENTOMOLOGICAL CONTRIBUTIONS BY J.A. LINTNER. SEE TWENTY-THIRD ANNUAL REPORT N.Y. STATE CABINET OF NATURAL HISTORY. ALBANY, N.Y., 1872.

LUTZ---A LOT OF INSECTS BY F.E. LUTZ. NEW YORK, N.Y., 1941.

MAYNARD--BUTTER. NEW ENGLAND---THE BUTTERFLIES OF NEW ENGLAND BY C.J. MAYNARD. NEWTONVILLE, MASS., 1886.

MAYNARD--MANUAL NORTH AMER. BUTTER.---A MANUAL OF NORTH AMERICAN BUTTERFLIES BY C.J. MAYNARD. BOSTON, MASS., 1891.

MEAD--REPT. LEP. WHEELER EXPED.---REPORT OF THE LEPIDOPTERA OF THE WHEELER EXPEDITION BY T.L. MEAD. (REPORT UPON THE GEOGRAPHICAL AND GEOLOGICAL EXPLORATIONS AND SURVEYS WEST OF THE 100TH MERIDIAN. VOL. 5 WASHINGTON, D.C. INCLUDES THE WORK OF THE YEARS 1871-1874 INCL.

MERIAN--INS. SURINAM---HISTOIRE GENERALE DES INSPECTES DE SURINAM BY MARIA SIBYLLA MERIAN. AMSTERDAM, HOLLAND, 1719.

MERIAN--METAMORPH. INS. SURINAM---METAMORPHOSIS INSECTORUM SURINAMENSIUM BY MARIA SIBYLLA MERIAN. AMSTERDAM, HOLLAND, 1705.

METCALF & FLINT--DESTR. & USEFUL INS.---DESTRUCTIVE AND USEFUL INSECTS, THEIR HABITS AND CONTROL BY C.L. METCALF AND W.P. FLINT. NEW YORK, N.Y., 1939.

MORRIS-SYNOP. LEP. NORTH AMER.---SYNOPSIS OF THE LEPIDOPTERA OF NORTH AMERICA BY J.G. MORRIS. ALSO PUBLISHED AS VOL. IV. ART. 2 SMITHSONIAN MISCELLANEIOUS COLLECTIONS, 1862.

Newman--British Moths-An Illustrated Natural History of British Moths by Edward Newman. London, England, 1869.

O'Kane--Inj. Ins.---Injurious Insects, How to Recognize and Control Them by W.C. O'Kane. New York, N.Y., 1912.

Osborn--Meadow and Pasture Insects by H. Osborn. Columbus, Ohio, 1939.

Packard--Guide to the Study of Insects by A.S. Packard. Salem, Mass., 1869.

Packard--Half Hours with Insects by A.S. Packard. Boston, Mass., 1877.

Packard--Hayden's Survey Terr.---Hayden's Geological Survey of the Territories by Hayden, 1873. Published 1874. Washington, D.C.

Packard--Ins. Inj. Forest Trees---Insects Injurious to Forest Trees by A.S. Packard, 1881. Also as Bulletin 7 U.S.D.A. Ent. Comm.

Packard--Ins. Inj. West---The Injurious Insects of the West by A.S. Packard. Salem, Mass., 1877.

Packard--Inj. Ins. New or Little Known---Injurious Insects New or Little Known by A.S. Packard. Boston, Mass., 1870.

Packard--Monograph of the Geometridae--United States Geological Survey of the Territories--F.V. Hayden vol. x, 1876.

Peairs--Insect Pests of Farm, Garden, and Orchard by L.M. Peairs. New York, N.Y., 1941.

Peale--Lepidoptera Americana by Titian Peale. Philadelphia, Penn. 1833.

Poey--Cent. Lep. Cubana---Centura de Lepidopteros in Memorias Sobre La Historia de la Isla Cuba by Felipe Poey. Habana, Cuba, 1851.

Quayle--Ins. Citrus Subtrop. Fruits---Insects of Citrus and Other Subtropical Fruits by H.J. Quayle. Ithaca, N.Y., 1941.

Ratzeburg--Die Forst Insecten--by J.T.C. Ratzeburg. Berlin, Germany, 1840.

Saunders--Butter. Alleg. State Park---The Butterflies of the Alleghany State Park by Aretas A. Saunders. New York State Museum Handbook No. 13. Albany, N.Y., 1932.

Saunders--Ins. Inj. Fruits---Insects Injurious to Fruits by Wm. Saunders. Philadelphia, Penn., 1883 and 1889.

Say--American Entomology or Descriptions of Insects of North America by Thomas Say. Philadelphia, Penn., 1824-1828.

Scudder--Butter. East. U.S. and Canada---The Butterflies of the Eastern United States and Canada. Cambridge, Mass., 1889.

Scudder--Butter. Their Struct. etc.---Butterflies, Their Structure---and Life History by Samuel Scudder. New York, N.Y., 1881.

Scudder--Char. Ins. Fauna White Mts.---Character of the Insect Fauna of the White Mountains by Samuel Scudder. Cambridge, Mass., 1863.

Scudder--Everyday Butter.---Everyday Butterflies by Samuel Scudder. Cambridge, Mass., 1899.

Scudder--Guide to Common Butter. North America and Canada---Guide to the Common Butterflies of North America and Canada by Samuel Scudder. New York, N.Y., 1899.

Scudder--Geol. New Hampshire---Ins C.H. Hitchcock's Geology of New Hampshire. Samuel Scudder. Concord, N.Y., 1874.

Sepp--Dutch Lep.---Beschouwing der wonderen Gods---nederlandsche Insecten. Jan Christian Sepp. Amsterdam, Holland, 1855.

Sepp--Surinamsche Vlinders---Surinaamsche Vlinders Papillions de Surinam by Jan Christian Sepp en Zoon. Amsterdam, Holland, 1855.

Slingerland & Crosby--Manual of Fruit Insects by M.V. Slingerland and C.R. Crosby. New York, N.Y., 1914.

Sloane--History of Jamaica---A Voyage to the Islands Madera, Barbados, Nieves, St. Christopher and Jamaica with Natural History etc. by Hans Sloane M.D. Printed by the British Museum, 1725.

Smith--The Insects of New Jersey by John B. Smith. Trenton, N.J., 1900.

Smith--Monograph of the Sphingidae. Reprint from the Transactions of the American Entomological Society xv, pp. 49-242, 1888 by John B. Smith.

Smith--Our Insect Friends and Enemies by John B. Smith, 1909.

Spuler--Die Raupen der Schmetterlinge Europas by Arnold Spuler. Stuttgart, Germany, 1904.

Stainton--Manual British Butterflies and Moths by William Stainton. London, England, Vol. I, 1857; Vol. II, 1859.

Stauffer--Report on Agriculture of Pennsylvania. 1878.

Separate Works (continued)

Stephens--Illust. Brit. Ent. Haust.---Illustrations of British Entomology Haustellate by J.F. Stephens. London, England, Vol. i, 1828; Vol. ii, 1829; Vol. iii, 1829.

Strecker--Rhop. et Heteroc.---Rhopalocera et Heterocera Indigenous and Exotic by Herman Strecker. Reading, Penn. 1873, 1876, 1877. Supplements 1898, 1899, 1900.

Stretch--Zygaen. & Bomby. North America---Illustrations of Zygaenidae and Bombycidae of North America by Robert Stretch. San Francisco, Calif., 1873.

Swainson--Zool. Illust.---Zoological Illustrations Vol. ii by William Swainson. London, England, 1821-1822.

Tenney--Manual Zool.---Natural History, A Manual of Zoology for Schools, Colleges, and General Reader by Sanborn Tenney. New York, N.Y., 1867.

Tenney--Manual of Nat. Hist.---Manual of Natural History, Same as Manual of Zoology--preceeding entry.

Vance--Life Among the Butterflies by Randolph Vance. Halderman Julius & Co. (Little Blue Book). Girad, Kansas, 1925.

Walsh--Practical Entomologist by Benjamin Walsh (Vols. 1 & 2 all published) 1865-1867. Issued by the American Entomological Society.

Weed--Life History of American Insects by C.M. Weed. New York, N.Y., 1897.

Westwood--British Butter.---The Butterflies of Great Britain by J.O. Westwood. London, England, 1855.

Westwood--In Drury's Illust. Exot. Ent.---Illustrations of Exotic Entomology by J.O. Westwood. London, England, 1837.

Westwood--Introd. Modern Class. Ins.---Introduction to the Modern Classification of Insects by J.O. Westwood. London, England, 1840.

Westwood--Introduction to Entomology by J.O. Westwood. London, England, 1839.

Wheeler--Ants, Their Structure, Development, and Behavior. Wm M. Wheeler. New York, N.Y., 1933.

Wood--Illust. Nat. Hist.---Illustrated Natural History by Rev. J.G. Wood. London, England, 1863.

Wood--Insects Abroad by Rev. J.G. Wood. London, England, 1874.

Wood--Insects at Home by Rev. J.G. Wood. New York, N.Y., 1872.

C Bibliographies

Edwards, W.H.---Bibliography of the Described Transformations of North American Lepidoptera. Bulletin of the United States National Museum No. 35, 1889, Washington, D.C.

Index to the Literature of American Economic Entomology. Published by the American Association of Economic Entomologists.

Vol. I Covers the literature from Jan. 1, 1905 to Dec. 31, 1914. Compiled by Nathan Banks, Melrose Highlands, Mass., 1917.

Vol. II From Jan. 1, 1915 to Dec. 31, 1919. Compiled by M. Colcord and E.P. Felt, Melrose Highlands, Mass., 1921.

Vol. III From Jan. 1, 1920 to Dec. 31, 1924. Compiled by M. Colcord and E.P. Felt, Melrose Highlands, Mass., 1925.

Vol. IV From Jan. 1, 1925 to Dec. 31, 1929. Compiled by M. Colcord and E.P. Felt, Melrose Highlands, Mass., 1930.

Vol. V From Jan. 1, 1930 to Dec. 31, 1934. Compiled by M. Colcord and E. P. Felt, College Park, Md., 1938.

Vol. VI From Jan. 1, 1935 to Dec. 31, 1939. Compiled by M. Colcord, E. P. Felt and Stanley Bromley, College Park, Md., 1942.

Vol. VII From Jan. 1, 1940 to Dec. 31, 1944. Compiled by Ina L. Hawes, Wm. H. Mitchell, Ethel L. Coon, and Stanley W. Bromley, College Park, Md., 1948.

PART I–INSECTS

SECTION B---INSECTS--COMMON NAMES

ENTOMOLOGICAL

Abbot's Pine Sphinx
Lapara coniferarum (Abbot & Smith)

Abbot's Sphinx
Sphecodina abbottii (Swainson

Abutilon Moth
Anomis erosa Hubner

Acadian Hairstreak
Strymon acadica (Edwards)

*Achemon Sphinx
Pholous achemon (Drury)

Acmon Blue
Plebeius acmon (Doubleday & Hewitson)

Afranius Dusky-wing
Erynnis afranius (Lintner)

Ailanthus Silk Moth
Philosamia walkeri (Felder & Felder)

Alderman Butterfly
Vanessa atalanta (Linnaeus)

*Alfalfa Caterpillar
Colias eurytheme Boisduval

*Alfalfa Looper
Autographa californica (Speyer)

Alfalfa Semi-looper
Autographa californica (Speyer)

American Copper
Lycaena hypophlaeas (Boisduval)

American Copper Hindwing
Amphipyra pyramidoides Guenee

*American Dagger Moth
Acronicta americana Harris

American Lackey Moth
Malacosoma americana (Fabricius)

American Lappet Moth
Epicnaptera americana (Harris)

American Maple Moth
Epicnaptera americana (Harris)

American Painted Lady
Vanessa virginiensis (Drury)

American Swordgrass Moth
Xylena nupera (Lintner)

American Tiger Moth
Arctia caja americana (Harris)

Ampelos Ringlet
Coenonympha ampelos Edwards

Amputating Brocade Moth
Septis arctica (Boisduval)

Anise Swallowtail
Papilio zelicaon Lucas

Antique Tussock Moth
Notolophus antiqua (Linnaeus)

Aphrodite
Argynnis (Speyeria) aphrodite (Fabricius)

Apple Handmaid Moth
Datana ministra (Drury)

Apple Sphinx
Sphinx gordius Cramer

Apple Tree Tent Caterpillar
Malacosoma americana (Fabricius)

Arctic Blue
Plebeius aquilo (Boisduval)

INSECTS---COMMON NAMES

ARCTIC SKIPPER
CARTEROCEPHALUS PALAEMON (PALLAS)

ARCTIC SULPHUR
COLIAS NASTES BOISDUVAL

*ARMY CUTWORM
CHORIZAGROTIS AUXILLIARIS (GROTE)

*ARMY WORM
LEUCANIS UNIPUNCTA (HAWORTH)

AROTA COPPER
THARSALEA AROTA (BOISDUVAL)

ARROW-HEAD BLUE
PHAEDROTES PIASUS (BOISDUVAL)

ASH-GRAY PINION
LITHOPHANE ANTENNATA (WALKER)

ASH LOOPER
ELLOPIA SOMNIARIA HULST

ASH SPHINX
CHALAENOGRAMMA JASMINEARUM (GUERIN)
SPHINX KALMIAE ABBOT & SMITH

ASHEN PINION
LITHOPHANE ANTENNATA (WALKER)

ASTEROID, THE
CUCULLIA ASTEROIDES GUENEE

ATLANTIS FRITILLARY
ARGYNNIS (SPEYERIA) ATLANTIS EDWARDS

AZALEA SPHINX
DARAPSA PHOLUS (CRAMER)

B

BAD WING, THE
DYSPTERIS ABORTIVARIA HERRICH-SCHAEFFER

BAIRD'S SWALLOWTAIL
PAPILIO BAIRDII EDWARDS

BALTIMORE, THE
EUPHYDRYAS PHAETON (DRURY)

BANDED ELFIN
INCISALIA NIPHON (HUBNER)

BANDED HAIRSTREAK
STRYMON CALANUS (HUBNER)
SYRYMON FALACER (GODART)

BANDED PURPLE
BASILARCHIA ARTHEMIS (DRURY)

BANDED TIGER MOTH
APANTESIS VITTATA (FABRICIUS)

BANDED TUSSOCK MOTH
HALISIDOTA TESSELLARIS (ABBOT & SMITH)

*BANDED WOOLYBEAR
ISIA ISABELLA (ABBOT & SMITH)

BARBERRY GEOMETER
CORYPHISTA MEADI (PACKARD)

BARNE'S CLIMBING CUTWORM
ABAGROTIS BARNESI (BENJAMIN)

BARON'S CHECKER SPOT
EUPHYDRYAS BARONI (EDWARDS)

BARRED SULPHUR
EUREMA DELIA (CRAMER)

BATE'S CHECKER SPOT
PHYCIODES BATESII (REAKIRT)

BEAN CUTWORM
OGDOCONTA CINEREOLA (GUENEE)

BEAN LEAF-ROLLER
URBANUS PROTEUS (LINNAEUS)

BEAN LYCAENID
STRYMON MELINUS (HUBNER)

INSECTS--COMMON NAMES

BEAR SPHINX
ARCTONOTUS LUCIDUS BOISDUVAL

BEAUTIFUL UTETHEISA
UTETHEISA BELLA (LINNAEUS)

BEAUTIFUL WOOD NYMPH
EUTHISANOTIA GRATA (FABRICIUS)

BEAUTIFUL WOODLING
LYCANADES PULCHELLA (SMITH)

BECKER'S WHITE
PIERIS BECKERII EDWARDS

BEECH SPANWORM
HYPERETIS AMICARIA (HERRICH-SCHEAFFER)

*BEET ARMYWORM
LAPHYGMA EXIGUA (HUBNER)

BEGGAR, THE
EUDULE MENDICA (WALKER)

BEHR'S BLUE
PHILOTES BATTOIDES (BEHR)

BEHR'S SULPHUR
EURYMUS (COLIAS) BEHRII (EDWARDS)

*BELLA MOTH
UTETHEISA BELLA (LINNAEUS)

BELOVED, THE
CATOCALA ILIA OSCULATA HULST

*BERTHA ARMYWORM
MAMESTRA CONFIGURATA WALKER

BETROTHED, THE
CATOCALA INNUBENS GUENEE

BIG CRANBERRY SPANWORM
ABBOTTANA CLEMATARIA (ABBOT & SMITH)

BIG POPLAR SPHINX
PACHYSPHINX MODESTA (HARRIS)

BLACK ALDER SPHINX
DOLBA HYLAEUS (DRURY)

*BLACK ARMY CUTWORM
ACTEBIA FENNICA (TAUSCHER)

BLACK C DART
GRAPHIPHORA C-NIGRUM (LINNAEUS)

BLACK C OWLET
GRAPHIPHORA C-NIGRUM (LINNAEUS)

*BLACK CUTWORM
AGROTIS YPSILON (ROTTENBURG)

BLACK DASH
ATRYTONE CONSPICUA (EDWARDS)

BLACK-FACED CLIMBING CUTWORM
EUPSILIA SIDUS (GUENEE)

BLACK-LINED CUTWORM
ACTEBIA FENNICA (TAUSCHER)

BLACK RUSTIC
GRAPHIPHORA C-NIGRUM (LINNAEUS)

BLACK SWALLOWTAIL
PAPILIO ASTERIUS CRAMER

BLACK-TIPPED BEAR
ISIA ISABELLA (ABBOT & SMITH)

BLACK WALNUT CATERPILLAR
DATANA INTEGERRIMA GROTE & ROBINSON

BLACK WITCH
EREBUS ODORA (LINNAEUS)

BLACKBERRY LOOPER
CHLOROCHLAMYS CHLOROLEUCARIA (GUENEE)

BLIND-EYED SATYR
MINOIS ALOPE NEPHELE KIRBY

INSECTS---COMMON NAMES

BLIND-EYED SPHINX
PAONIAS EXCAECATA (ABBOT & SMITH)

BLUE-EYED SATYR
MINOIS ALOPE (FABRICIUS)

BLUE-SPANGLED CATERPILLAR
HAPLOA LECONTEI (GUERIN)

BLUE-SPANGLED PEACH TREE CATERPILLAR
HAPLOA LECONTEI (GUERIN)

BLUE SWALLOWTAIL
PAPILIO PHILENOR LINNAEUS

BOG COPPER
LYCAENA EPIXANTHE (BOISDUVAL & LECONTE)

BOISDUVAL'S BLUE
PLEBEIUS ICARIOIDES (BOISDUVAL)

BOISDUVAL'S HAIRSTREAK
HABRODAIS GRUNUS (BOISDUVAL)

BOISDUVAL'S MARBLE
ANTHOCHARIS LANCEOLATA BOISDUVAL

BOISDUVAL'S SULPHUR
EURYMUS (COLIAS) EURYTHEME BOISDUVAL

*BOLLWORM
HELIOTHIS OBSOLETA (FABRICIUS)
HELIOTHIS ZEAE BODDIE

BOMBYX SPHINX
LAPARA BOMBYCOIDES WALKER

BORDERED PATCH
CHLOSYNE LACINIA (GEYER)

BRASSY CUTWORM
ORTHODES RUFULA (GROTE)

BRAZILIAN SKIPPER
CALPODES ETHLIUS (CRAMER)

BRIDE, THE
CATOCALA NEOGAMA (ABBOT & SMITH)

*BRISTLY CUTWORM
LACINIPOLIA RENIGERA (STEPHENS)

BROAD ASHEN PINION
LITHOPHANE LATICINEREA GROTE

BROAD-WINGED SKIPPER
POANES VIATOR (EDWARDS)

BROKEN DASH
CATIA OTHO (ABBOT & SMITH)

BROME GRASS CUTWORM
SEPTIS FINITIMA (GUENEE)

BRONZE COPPER
LYCAENA THOE (GUERIN)

*BRONZED CUTWORM
NEPHELODES EMMEDONIA (CRAMER)

BROTHER, THE
RAPHIA FRATER GROTE

BROWN ANISOTA
ANISOTA VIRGINIENSIS (DRURY)

BROWN ARMY WORM
LEUCANIA PSEUDARGYRIA GUENEE

BROWN-BORDERED CUCULLIA
CUCULLIA CONVEXIPENNIS GROTE & ROBINSON

BROWN CRANBERRY SPANWORM
EMATURGA AMITARIA (GUENEE)

BROWN CUTWORM
RYNCHAGROTIS CUPIDA GROTE

BROWN DAY MOTH
PSEUDOHAZIS EGLANTERINA (BOISDUVAL)

BROWN ELFIN
INCISALIA AUGUSTUS (KIRBY)

INSECTS---COMMON NAMES

BROWN-STREAKED RUSTIC
CAPNODES RUFINANS GUENEE

*BROWN-TAIL MOTH
NYGMIA PHAEORRHOEA (DONOVAN)

BROWN-WINGED CTENUCH
CTENUCHA BRUNNEA STRETCH

BRUCE'S ARCTIC
OENEIS BRUCEI (EDWARDS)

BRUCE'S CANKERWORM
OPEROPHTERA BRUCEATA (HULST)

BRUCE'S MEASURING WORM
OPEROPHTERA BRUCEATA (HULST)

*BRUCE'S SPANWORM
OPEROPHTERA BRUCEATA (HULST)

BUCK MOTH
HEMILEUCA MAIA (DRURY)

BUCKEYE
JUNONIA COENIA (HUBNER)

BUMBLE BEE MOTH
HEMARIS DIFFINIS (BOISDUVAL)

BURDOCK BORER
PAPAIPEMA CATAPHRACTA (GROTE)

BURGLAR DAGGER MOTH
ACRONICTA NOCTIVAGA GROTE

C

CABBAGE BUTTERFLY
PIERIS RAPAE (LINNAEUS)

*CABBAGE LOOPER
AUTOGRAPHA NI (HUBNER)

CALIFORNIA DOG FACE BUTTERFLY
ZERENE EURYDICA (BOISDUVAL)

CALIFORNIA FORESTER
ALYPIA MARIPOSA GROTE & ROBINSON

*CALIFORNIA OAK WORM
PHRYGANIDIA CALIFORNICA PACKARD

CALIFORNIA RINGLET
COENONYMPHA CALIFORNICA WESTWOOD & HEWITSON

CALIFORNIA TENT CATERPILLAR
MALACOSOMA CALIFORNICA (PACKARD)

CALIFORNIA TORTOISE SHELL
NYMPHALIS CALIFORNICA (BOISDUVAL)

CALIFORNIA TUSSOCK MOTH
HEMEROCAMPA VETUSTA (BOISDUVAL)

CALIFORNIA WHITE
PIERIS SISYMBRII BOISDUVAL

CANADIAN SKIPPER
HESPERIA COMMA MANITOBA SCUDDER

CANNA LEAF-ROLLER
CALPODES ETHLIUS (CRAMER)

CANNA SKIPPER
CALPODES ETHLIUS (CRAMER)

CAROLINA SATYR
MEGISTO HERMES (FABRICIUS)

CAROLINA SPHINX
PHLEGETHONTIUS SEXTA (JOHANSSEN)

CATALPA SPHINX
CERATOMIA CATALPAE (BOISDUVAL)

CATOCALINE ANARTA
ANARTA CORDIGERA (THUNBERG)

CEANOTHUS SILK MOTH
PLATYSAMIA EURYALUS (BOISDUVAL)

INSECTS---COMMON NAMES

CECROPIA MOTH
PLATYSAMIA CECROPIA (LINNAEUS)

CELERY LOOPER
AUTOGRAPHA FALCIFERA (KIRBY)

CELERY WORM
PAPILIO ASTERIAS CRAMER

CHAIN-DOTTED GEOMETER
CINGILIA CATENARIA (DRURY)

*CHAIN-SPOTTED GEOMETER
CINGILIA CATENARIA (DRURY)

CHALCEDON CHECKER-SPOT
EUPHYDRYAS CHALCEDONA (DOUBLEDAY & HEWITSON)

CHECKER-SPOT
EUPHYDRYAS CHALCEDONA (DOUBLEDAY & HEWITSON)

CHECKERED RUSTIC
EUXOA TESSELLATA (HARRIS)

CHECKERED SKIPPER
PYRGUS COMMUNIS (GROTE)

CHECKERED TUSSOCK MOTH
HALISIDOTA TESSELLARIS (ABBOT & SMITH)

CHECKERED WHITE
PIERIS PROTODICE (BOISDUVAL & LECONTE)

CHERRY SCALLOP-SHELL MOTH
CALOCALPE UNDULATA (LINNAEUS)

CHERRY TREE THECLA
STRYMON TITUS (FABRICIUS)

CHEVRON MOTH
LYGRIS TESTATA (LINNAEUS)

CHICKWEED GEOMETER
HAEMATOPIS GRATARIA (FABRICIUS)

CHICKWEED MOTH
HAEMATOPIS GRATARIA (FABRICIUS)

CHIEFTAIN DAGGER MOTH
ACRONICTA SUPERANS GUENEE

CHOCOLATE-STRIPED CUTWORM
GRAPHIPHORA BICARNEA (GUENEE)

CHRYXUS ARCTIC
OENEIS CHRYXUS (DOUBLEDAY & HEWITSON

CIVIC RUSTIC
PLATYPERIGEA EXTIMA (WALKER)

CLANDESTINE DART
SPAELOTIS CLANDESTINA (HARRIS)

CLANDESTINE OWLET MOTH
SPAELOTIS CLANDESTINA (HARRIS)

CLARK'S DAY SPHINX
PROSERPINUS CLARKIAE (BOISDUVAL)

CLAY-BACKED CUTWORM
AGROTIS GLADIARIA MORRISON

CLEFT-HEADED SPANWORM
AMPHIDASIS COGNATARIA GUENEE

CLEMEN'S SPHINX
SPHINX LUSCITIOSA CLEMENS

CLIMBING CUTWORM
EUXOA SCANDENS (RILEY)

CLIMBING RUSTIC
EUXOA SCANDENS (RILEY)

CLODIUS PARNASSIAN
PARNASSIUS CLODIUS MENETRIES

CLOSE-BANDED COLOCASIA
COLOCASIA PROPINQUILINIA (GROTE)

CLOUDED SKIPPER
LEREMA ACCIUS (ABBOT & SMITH)

INSECTS---COMMON NAMES

*CLOUDED SULPHUR
COLIAS PHILODICE (GODART)

CLOUDED UNDERWING
CATOCALA NEBULOSA EDWARDS

CLOUDLESS SULPHUR
PHOEBIS SENNAE (LINNAEUS)

*CLOVER CUTWORM
SCOTOGRAMMA TRIFOLII (ROTTENBURG)

CLOVER DRASTERIA
CAENURGINA ERECHTEA (CRAMER)

CLOVER LOOPER
CAENURGINA ERECHTEA (CRAMER)

COAST TENT CATERPILLAR
MALACOSOMA PLUVIALIS (DYAR)

COBWEB SKIPPER
HESPERIA METEA SCUDDER

COLUMBIAN SILK MOTH
PLATYSAMIA COLUMBIA (SMITH)

*COLUMBINE BORER
PAPAIPEMA PURPURIFASCIA (GROTE & ROBINSON)

COLUMBINE DUSKY-WING
ERYNNIS LUCILIUS (SCUDDER & BURGESS)

COMMA BUTTERFLY
POLYGONIA COMMA (HARRIS)

COMMON ALPINE
EREBIA EPIPSODEA BUTLER

COMMON BLUE
LYCAENOPSIS PSEUDARGIOLUS (BOISDUVAL & LECONTE)

COMMON CLEARWING
HERARIS THYSBE (FABRICIUS)

COMMON COPPER
LYCAENA HYPOPHLAEAS (BOISDUVAL)

COMMON GRASS-NYMPH
SATYRODES EURYDICE (JOHANSSEN)

COMMON HAIR-STREAK
STRYMON MELINUS HUBNER

COMMON HYPPA
HYPPA XYLINOIDES (GUENEE)

COMMON PUG MOTH
EUPITHECIA SOBRINATA (HUBNER)

COMMON SOOTY-WING
PHOLISORA CATULLUS (FABRICIUS)

COMMON STRIPED CUTWORM
EUXOA TESSELLATA (HARRIS)

COMMON SULPHUR
COLIAS PHILODICE (GODART)

COMMON WAINSCOT
LEUCANIA COMMOIDES GUENEE

COMMON WHITE
PIERIS PROTODICE (BOISDUVAL & LECONTE)

COMMON WOOD NYMPH
MINOIS ALOPE (FABRICIUS)

COMPTON TORTOISE-SHELL
NYMPHALIS J ALBUM (BOISDUVAL & LECONTE)

CONCUMBINE, THE
CATOCALA CONCUMBENS WALKER

CONNECTED DAGGER MOTH
ACRONICTA CONNECTA GROTE

CONSORT, THE
CATOCALA CONSORS (ABBOT & SMITH)

COPPER UNDERWING
AMPHIPYRA PYRAMIDOIDES GUENEE

CORAL HAIR-STREAK
STRYMON TITUS (FABRICIUS)

*Corn Earworm
Heliothis obsoleta (Fabricius)
Heliothis zeae Boddie

Cosmopolite, The
Vanessa cardui (Linnaeus)

Cotton Boll Cutworm
Prodenia ornithogalli Guenee

Cotton Bollworm
Heliothis obsoleta (Fabricius)
Heliothis zeae Boddie

Cotton Cutworm
Prodenia ornithogalli Guenee

*Cotton Leafworm
Alabama argillacea (Hubner)

Cotton Spanworm
Anavitrinella pampinaria (Guenee)

Cotton Square Borer
Strymon melinus (Hubner)

Cotton Worm
Alabama argillacea (Hubner)

*Cottonwood Dagger Moth
Acronicta lepusculina Guenee

Cranberry Blossom-worm
Epiglaea apiata (Grote)

Cranberry Looper
Xanthotype sospeta (Drury)

*Cranberry Spanworm
Anavitrinella pampinaria (Guenee)

Crape Moth
Hemileuca maia (Drury)

Crimson-bordered Lichen Moth
Lerena incarnata Walker

Crimson Patch
Chlosyne janais (Drury)

Crocus Geometer
Xanthotype sospeta (Drury)

Crossline Skipper
Polites manataaqua (Scudder)

Cuban Crescent
Anthanassa frisia (Poey)

*Currant Spanworm
Itame ribearia (Fitch)

*Cynthia Moth
Philosamia walkeri (Felder & Felder)

D

Dainty Sulphur
Nathalis iole Boisduval

*Dark-sided Cutworm
Euxoa messoria (Harris)

Dark Tussock Moth
Olene achatina (Abbot & Smith)

Dark-veined Morning Sphinx
Celerio galli (Rottenburg)

Dark-winged Wainscot
Leucania latiuscula Herrich-Schaeffer

Dark Wood Nymph
Minois oetus (Boisduval)

Darkish Dagger Wing
Acronicta morula Grote & Robinson

Darling Underwing
Catocala cara Guenee

Dash-marked Cutworm
Protolampra brunneicollis (Grote)

Daunus Swallowtail
Papilio daunus Boisduval

DEFINITE-MARKED TUSSOCK MOTH
HEMEROCAMPA PLAGIATA (WALKER)

DELAWARE SKIPPER
ATRYTONE LOGAN (EDWARDS)

DELIGHTFUL DAGGER MOTH
ACRONICTA VINNULA (GROTE)

DEVASTATING DART
CRYMODES DEVASTATOR (BRACE)

DIANA
ARGYNNIS (SPEYERIA) DIANA (CRAMER)

*DINGY CUTWORM
FELTIA SUBGOTHICA (HAWORTH)

DINGY PURPLE-WING
EUNICA MONIMA (CRAMER)

DIVERSE-LINED MOTH
LYGRIS DIVERSILINEATA (HUBNER)

DOG FACE
ZERENE CAESONIA (STOLL)

DOT AND DASH SWORDGRASS
XYLENA CURVIMACULA (MORRISON)

DOTTED ANGLE OF THE WOOD
PARAPHIA SUBATOMARIA WOOD

DOTTED BLUE
PHILOTES ENOPTES (BOISDUVAL)

*DOUGLAS FIR TUSSOCK MOTH
HEMEROCAMPA PSEUDOTSUGATA McDUNNOUGH

DRAB CUTWORM
EUXOA BOSTONIENSIS (GROTE)

DRURY'S HAIR-STREAK
STRYMON ACIS (DRURY)

DUN SKIPPER
ATRYTONE RURICOLA (BOISDUVAL)

DUSKY CUTWORM
AGROTIS VENERABILIS WALKER

DUSKY METAL MARK
CALEPHELIS NEMESIS (EDWARDS)

DUSTED SKIPPER
ATRYTONOPSIS HIANNA (SCUDDER)

DWARF YELLOW
NATHALIS IOLE BOISDUVAL

E

EARLY HAIR-STREAK
ERORA LAETA (EDWARDS)

EASTERN PANTHEA
PANTHEA FURCILLA (PACKARD)

EASTERN TAILED BLUE
EVERES COMYNTAS (GODART)

*EASTERN TENT CATERPILLAR
MALACOSOMA AMERICANA (FABRICIUS)

EDITHA CHECKER SPOT
EUPHYDRYAS EDITHA (BOISDUVAL)

EDWARD'S BLUE
HEMIARGUS GYAS (EDWARDS)

EDWARD'S GRASSY WING
HEMIHYALEA EDWARDSII (PACKARD)

EDWARD'S WASP MOTH
LYMIRE EDWARDSII (GROTE)

*EIGHT SPOTTED FORESTER
ALYPIA OCTOMACULATA (FABRICIUS)

*ELDER SHOOT BORER
ACHATODES ZEAE (HARRIS)

ELEGANT EUCHAETIAS
EUCHAETIAS ELEGANS (STRETCH)

ELEGANT SPHINX
SPHINX PERELEGANS HENRY EDWARDS

ELKO RINGLET
COENONYMPHA AMPELOS EDWARDS

*ELM SPANWORM
ENNOMOS SUBSIGNARIUS (HUBNER)

ELM SPHINX
CERATOMIA AMYNTOR (HUBNER)

EUFALA SKIPPER
LERODEA EUFALA (EDWARDS)

EUROPEAN CABBAGE BUTTERFLY
PIERIS RAPAE (LINNAEUS)

EVENING PRIMROSE MOTH
RHODOPHORA FLORIDA GUENEE

EVEN-LINED SALLOW
IPIMORPHA PLEONECTUSA GROTE

EVERGREEN NEPYTIA
NEPYTIA CANOSARIA (WALKER)

EVERGREEN SPANWORM
THERA CONTRACTATA (PACKARD)

EYED BLUE
GLAUCOPSYCHE XERCES (BOISDUVAL)

EYED BROWN
SATYRODES EURYDICE (JOHANSSEN)

EYED EMPEROR
ASTEROCAMPA CELTIS (BOISDUVAL)

EYED HAWK MOTH
SMERINTHUS CERISYI KIRBY

F

FALCATE ORANGE TIP
ANTHOCHARIS MIDEA (HUBNER)

*FALL ARMYWORM
LAPHYGMA FRUGIPERDA (ABBOT & SMITH)

*FALL CANKERWORM
ALSOPHILA POMETARIA (HARRIS)

*FALL WEBWORM
HYPHANTRIA CUNEA (DRURY)

FALSE ARMY WORM
XYLENA NUPERA (LINTNER)

FALSE BUDWORM OF TOBACCO
HELIOTHIS OBSOLETA (FABRICIUS)
HELIOTHIS ZEAE BODDIE

FALSE WAINSCOT
LEUCANIA PSEUDARGYRIA GUENEE

FATAL METALMARK
CALEPHELIS NEMESIS (EDWARDS)

FAUN, THE
POLYGONIA FAUNUS (EDWARDS)

FAWNEY-EDGED SKIPPER
POLITES THEMISTOCLES (LATREILLE)

FELDER'S ORANGE TIP
ANTHOCHARIS CETHURA FELDER & FELDER

FERN MOTH
EUPLEXIA BENESIMILIS MCDUNNOUGH

FIERY SKIPPER
HYLEPHILA PHYLEUS (DRURY)

FIG SPHINX
PACHYLIA FICUS (LINNAEUS)

FIGURE EIGHT MOTH
LACINIPOLIA RENIGERA (STEPHENS)

FIGURED TIGER MOTH
APANTESIS FIGURATA (DRURY)

FILAMENT BEARER
NAMATOCAMPA LIMBATA (HAWORTH)

FILLET DART MOTH
EUXOA REDIMICULA (MORRISON)

FINGERED DAGGER MOTH
ACRONICTA DACTYLINA (GROTE)

INSECTS---COMMON NAMES

FINLAND DART
ACTEBIA FENNICA (TAUSCHER)

FIR NEEDLE INCH WORM
EUPITHECIA LUTEATA PACKARD

FIR TUSSOCK MOTH
HEMEROCAMPA PSEUDOTSUGATA McDUNNOUGH

FIRST BORN GEOMETER
BREPHOS INFANS MOESCHLER

FIVE-LINED GEOMETER
SCOPULA QUINQUELINEARIA (PACKARD)

FIVE-SPOTTED SPHINX
PHLEGETHONTIUS QUINQUEMACULATA (HAWORTH)

FLAME-SHOULDERED DART
OCHROPLEURA PLECTA (LINNAEUS)

* FLAX BOLLWORM
HELIOTHIS ONONIS (SCHIFFERMULLER)

FLORIDA BLUE
HEMIARGUS HANNO (STOLL)

* FLORIDA FERN CATERPILLAR
CALLOPISTRIA FLORIDENSIS (GUENEE)

FLORIDA WHITE
APPIAS ILAIRE (GODART)

FODDER WORM
EPIZEUXIS AEMULA (HUBNER)

* FORAGE LOOPER
CAENURGINA ERECHTEA (CRAMER)

* FOREST TENT CATERPILLAR
MALACOSOMA DISSTRIA HUBNER

FORKED DAGGER MOTH
ACRONICTA FURCIFERA GUENEE

FOUR-HORNED SPHINX
CERATOMIA AMYNTOR (HUBNER)

FOUR-LINED BORER
LUPERINA STIPATA (MORRISON)

FOUR-LINED SPHINX
CERATOMIA AMYNTOR (HUBNER)

FOUR-TOOTHED DART
EUXOA QUADRIDENTATA (GROTE & ROBINSON)

FRAGILE DAGGER MOTH
ACRONICTA FRAGILIS (GUENEE)

FROSTED ELFIN
INCISALIA IRUS (GODART)

FROSTED SKIPPER
ACHALARUS LYCIDAS (ABBOT & SMITH)

FROSTED DAGGER MOTH
ACRONICTA BRUMOSA GUENEE

FUNEREAL DUSKY WING
ERYNNIS FUNERALIS SCUDDER & BURGESS

G

GALIUM SPHINX
CELERIO GALLI INTERMEDIA (KIRBY)

GARDEN MAMESTRA
SCOTOGRAMMA TRIFOLII (ROTTENBURG)

GARTERED HALISIDOTA
HALISIDOTA CINCTIPES HENRY EDWARDS

GAUDY SPHINX
PHOLUS LABRUSCAE (LINNAEUS)

GEM, THE
PERCNOPTILOTA OBSTIPATA (FABRICIUS)

GEMMED SATYR
NEONYMPHA GEMMA HUBNER

GEORGIA SATYR
NEONYMPHA PHOCION (FABRICIUS)

INSECTS---COMMON NAMES

GIANT GRAY SPHINX
Pseudosphinx tetrio (Linnaeus)

GIANT SPHINX
Cocytius antaeus (Drury)

GIANT SWALLOWTAIL
Papilio cresphontes Cramer

*GLASSY CUTWORM
Crymodes devastator (Brace)

GLITTERING UNDERWING
Catocala innubens scintillans Grote

GLOOMY UNDERWING
Catocala andromedae (Guenee)

GLOVER'S PURSLANE MOTH
Copidryas gloveri (Grote & Robinson)

GOAT SALLOW
Homoglaea hircina Morrison

GOAT WEED BUTTERFLY
Anaea andria (Scudder)

GOLD AND BLACK SULPHUR
Colias occidentalis chrysomelas Henry Edwards

GOLDEN BANDED SKIPPER
Cecropterus cellus (Boisduval & Leconte)

GOLDENROD MOTH
Cisseps fulvicollis (Hubner)

GOOSEBERRY SPANWORM
Itame ribearia (Fitch)

GORDIAN SPHINX
Sphinx gordius Cramer

GORGONE CHECKERSPOT
Phyciodes gorgone (Hubner)

GOTHIC DART
Feltia subgothica (Haworth)

GRACEFUL ANGLE WING
Polygonia gracilis (Grote & Robinson)

GRACEFUL CLEARWING
Hemaris gracilis (Grote & Robinson)

GRACEFUL UNDERWING
Catocala gracilis Edwards

GRANITE MOTH
Semiothisa granitata (Guenee)

*GRANULATED CUTWORM
Feltia subterranea (Fabricius)

GRAPE VINE EPIMENIS
Psychomorpha epimenis (Drury)

GRAPE VINE HOG CATERPILLAR
Ampeloeca myron (Cramer)

*GRAPE VINE LOOPER
Lygris diversilineata (Hubner)

GRASS WORM
Laphygma frugiperda (Abbot & Smith)

GRATEFUL MIDGET
Elaphria grata (Hubner)

GRAY BLUE
Plebeius aquilo (Felder)

GRAY COMMA
Polygonia progne (Cramer)

GRAY DAGGER MOTH
Acronicta grisea Walker

GRAY HAIR-STREAK
Strymon adenostomatis (Henry Edwards) (west)

INSECTS---COMMON NAMES

<u>Strymon melinus</u> Hubner (east)

Gray Skipper
<u>Pyrgus communis</u> (Grote)

Gray-winged Pygarctia
<u>Pygarctia eglenensis</u> (Clemens)

Greasy Cutworm
<u>Agrotis ypsilon</u> (Rottenburg)

Great Arctic
<u>Oeneis nevadensis</u> (Felder & Felder)

Great Ash Sphinx
<u>Sphinx chersis</u> (Hubner)

Great Basin Tent Caterpillar
<u>Malacosoma fragilis</u> (Stretch)

Great Brown Dart
<u>Eurois astricta</u> Morrison

Great Copper
<u>Lycaena xanthoides</u> (Boisduval)

Great Gray Dart Moth
<u>Eurois occulta</u> (Linnaeus)

Great Leopard Moth
<u>Ecpantheria deflorata</u> (Fabricius)

Great Purple Hair-streak
<u>Atlides halesus</u> (Cramer)

Great Southern White
<u>Ascia monuste</u> (Linnaeus)

Great Spangled Fritillary
<u>Argynnis</u> (Speyeria) <u>cybele</u> (Fabricius)

Great Tiger Moth
<u>Arctia caja</u> (Linnaeus)

Greater Red Dart Moth
<u>Abagrotis placida minimalis</u> (Grote)

Green-clouded Swallowtail
<u>Papilio troilus</u> Linnaeus

*Green Cloverworm
<u>Plathypena scabra</u> (Fabricius)

Green Comma
<u>Polygonia faunus</u> (Edwards)

Green Cranberry Spanworm
<u>Itame sulphurea</u> (Packard)

Green Cutworm
<u>Agrotis gladiaria</u> Morrison

*Green Fruitworm
<u>Lithophane antennata</u> (Walker)

Green Grape Vine Sphinx
<u>Ampeloeca myron</u> (Cramer)

Green-gray Pinion
<u>Lithophane disposita</u> Morrison

Green Hair-streak
<u>Mitoura damon</u> (Cramer)

Green Marvel
<u>Agriopodes fallax</u> (Herrich-Scheaffer)

Green Oak Caterpillar
<u>Nadata gibbosa</u> (Abbot & Smith)

*Green-striped Maple Worm
<u>Anisota rubicunda</u> (Fabricius)

Green Swallowtail
<u>Papilio philenor</u> Linnaeus

Green-white Spotted Hair-streak
<u>Callophrys dumetorum</u> (Boisduval)

Greenish Blue
<u>Plebeius saepiolus</u> (Boisduval)

Grizzled Skipper
<u>Pyrgus centaureae</u> (Rambur)

GULF FRUITILLARY
DIONE VANILLAE (LINNAEUS)

*GYPSY MOTH
LIPARIS DISPAR (LINNAEUS)

HACKBERRY BUTTERFLY
ASTEROCAMPA CELTIS (BOISDUVAL & LECONTE)

HALF-WING GEOMETER
PHIGALIA TITEA (CRAMER)

HAMMOCK SKIPPER
POLYGONUS AMYNTAS (FABRICIUS)

HARDENED WOODLING
XYLOMYGES CURIALIS GROTE

HARFORD'S SULPHUR
COLIAS HARFORDII HENRY EDWARDS

HARLEQUIN MILKWEED CATERPILLAR
EUCHAETIAS EGLE (DRURY)

HARNESSED MOTH
APANTESIS VITTATA (FABRICIUS)

HARRIS' CHECKERSPOT
MELITAEA HARRISII SCUDDER

HARRIS' PINE HAWKMOTH
LAPARA BOMBYCOIDES WALKER

HARRIS' THREE SPOT
HARRISIMEMNA TRISIGNATA (WALKER)

HARVESTER, THE
FENISECA TARQUINIUS (FABRICIUS)

HEBREW, THE
POLYGRAMMATE HEBRAEICUM HUBNER

HECLA ORANGE
COLIAS HECLA LEFEBRE

HEDGEHOG CATERPILLAR
ISIA ISABELLA (ABBOT & SMITH)

HEDGE ROW HAIR-STREAK
STRYMON SAEPIUM (BOISDUVAL)

HELLEBORE GEOMETER
EUPITHECIA CRETACEATA (PACKARD)

*HEMLOCK LOOPER
ELLOPIA FISCELLARIA GUENEE

HEMLOCK SPANWORM
ELLOPIA FISCELLARIA GUENEE

HENRY'S ELFIN
INCISALIA HENRICI (GROTE & ROBINSON)

HENSHAW'S BROWN
NEONYMPHA HENSHAWI (EDWARDS)

HERALD, THE
SCOLIOPTERYX LIBATRIX (LINNAEUS)

HERBARIUM MOTH
STERRHA BONIFATA (HULST)

HERMIT-LIKE SPHINX
SPHINX EREMITOIDES STRECKER

HERMIT SPHINX
SPHINX EREMITUS (HUBNER)

HICKORY HAIR-STREAK
STRYMON CARYAEVORUS MCDUNNOUGH

*HICKORY HORNED DEVIL
CITHERONIA REGALIS (FABRICIUS)

HICKORY TIGER MOTH
HALISIDOTA CARYAE (HARRIS)

HOARY COMMA
POLYGONIA GRACILIS (GROTE & ROBINSON)

HOARY-EDGE SKIPPER
ACHALARUS LYCIDAS (ABBOT & SMITH)

HOARY ELFIN
INCISALIA IRUS (GODART)
INCISALIA POLIOS COOK & WATSON

Hog Caterpillar
Ampeloeca myron (Cramer)

Hog Sphinx
Ampeloeca myron (Cramer)

Honey Locust Moth
Adelocephala bicolor (Harris)

Hop Borer
Hydroecia immanis Guenee

Hop Butterfly
Polygonia satyrus (Edwards)

*Hop Looper
Hypena humuli (Harris)

Hop Merchant
Polygonia comma (Harris)

Hop-plant Borer
Hydroecia immanis Guenee

Hop Plusia
Plusia balluca (Geyer)

Hop Snout Moth
Hypene humuli (Harris)

Hop-vine Hydroecia
Hydroecia immanis Guenee

Horace's Dusky Wing
Erynnis horatius (Scudder & Burgess)

Horned Spanworm
Nematocampa limbata (Haworth)

Huckleberry Sphinx
Paonias astylus (Drury)

Hummingbird Clearwing
Hemaris thysbe (Fabricius)

Hunter's Butterfly
Vanessa virginiensis (Drury)

Hydrangea Sphinx
Ampeloeca versicolor (Harris)

I

*Imperial Moth
Eacles imperialis (Drury)

*Imported Cabbage Worm
Pieris rapae (Linnaeus)

Inconsolable Underwing
Catocala insolabilis Guenee

Indian River Blue
Hemiargus ammon (Lucas)

Indian Skipper
Hesperia sassacus Harris

Inscribed Thyatira
Habrosyne scripta (Gosse)

Interrupted Dagger Moth
Acronicta interrupta Guenee

*Io Moth
Automeris io (Fabricius)

*Iris Borer
Macronoctua onusta Grote

Isabella Tiger Moth
Isia isabella (Abbot & Smith

J

Jessemine Sphinx
Chalaenogramma jasminearum (Guerin)

Joker, The
Feralia jocosa (Guenee)

Juniper Twig Inch Worm
Syssaura puber (Grote & Robinson)

Jutta Arctic
Oeneis jutta (Hubner)

Juvenal's Dusky Wing
Erynnis juvenalis (Fabricius)

INSECTS---COMMON NAMES

Labrador Apantesis
<u>Apantesis quenselii</u> (Paykull)

Labrador Blue
<u>Plebeius aquilo</u> (Boisduval)

Lance Owlet
<u>Agrotis ypsilon</u> (Rottenburg)

Lance Rustic
<u>Agrotis ypsilon</u> (Rottenburg)

Langton's Forester
<u>Alypia langtoni</u> Couper

*Lappet Moth
<u>Epicnaptera americana</u> (Harris)

Larch Lappet
<u>Tolype laricis</u> (Fitch)

Large Blue-striped Looper
<u>Nacophora ypsilon</u> (Forbes)

Larger Copper
<u>Lycaena thoe</u> (Guerin)

Large Maple Spanworm
<u>Prochoerodes transversata</u> (Drury)

Large Marble
<u>Euchloe ausonides</u> (Boisduval)

Large Orange Sulphur
<u>Phoebis agarithe</u> (Boisduval)

Large Scallop-wing Geometer
<u>Stenotrachelys approximaria</u> (Hubner)

Large Tortoise Shell
<u>Nymphalis j album</u> (Boisduval & Leconte)

Large White Skipper
<u>Heliopetes ericetorum</u> (Boisduval)

Large Yellow Bear
<u>Estigmene acrea</u> (Drury)

*Large Canna-Leaf Roller
<u>Calpodes ethlius</u> (Stoll.)

Larger Cotton Cutworm
<u>Prodenia dolichos</u> (Fabricius)

Laugher, The
<u>Charadra deridens</u> (Guenee)

Laurel Sphinx
<u>Sphinx kalmiae</u> Abbot & Smith

Laenira Checker Spot
<u>Melitaea leanira</u> Felder & Felder

Least Copper
<u>Lycaena epixanthe</u> (Boisduval & Leconte)

Least Eubaphe
<u>Eubaphe aurantiaca</u> Hubner

Least Satyr
<u>Minois oetus</u> (Boisduval)

Least Skipper
<u>Ancyloxypha numitor</u> (Fabricus)

Leopard Spot
<u>Diaethria clymena</u> (Cramer)

Lesser Grape Vine Looper
<u>Lygris diversilineata</u> (Hubner)

Lesser Maple Spanworm
<u>Physostegania pustularia</u> (Guenee)

Lesser Red Dart
<u>Abagrotis placida minimalis</u> (Grote)

Lesser Vine Sphinx
<u>Pholus fasciatus</u> (Sulzer)

Lesser Wasp Moth
<u>Pseudocharis minima</u> (Grote)

Lettered Sphinx
<u>Deidamia inscriptum</u> (Harris)

Lime Tree Spanworm
Erannis tiliaria (Harris)

Lime Tree Winter Moth
Erannis tiliaria (Harris)

*Linden Looper
Erannis tiliaria (Harris)

Linden Moth
Erannis tiliaria (Harris)

Lined Corn Borer
Oligia fractilinea (Grote)

*Lined Stalk Borer
Oligia fractilinea (Grote)

Little Glassy-wing
Polites verna (Edwards)

Little Orange Sulphur
Eurema nicippe (Cramer)

Little Silver Spot
Brenthis myrina (Cramer)

Little Sister
Catocala micronmypha fractercula Grote & Robinson

Little Sulphur
Eurema lisa (Boisduval & Leconte)

Little Underwing
Catocala minuta Edwards

Little White Lichen Moth
Clemensia albata Packard

Little Wife
Catocala muliercula Guenee

Little Wood Satyr
Megisto eurytus (Fabricius)

Live Oak Thecla
Strymon favonius (Abbot & Smith)

Locust-leaf Folder
Epargyreus clarus (Cramer)

Long Dash
Polites mystic (Scudder)

Long-horned Prominent
Schizura ipomoeae Doubleday

Long-tailed Skipper
Urbanus proteus (Linnaeus)

Lorquin's Admiral
Basilarchia lorquini (Boisduval)

Lost Sallow
Eupsilia devia (Grote)

Lucilius Dusky Wing
Erynnis lucilius (Scudder & Burgess)

*Luna Moth
Actias luna (Linnaeus)

MacCulloch's Forester
Androloma maccullochi (Kirby)

Macoun's Arctic
Oeneis macounii (Edwards)

Magdalen Underwing
Catocala illecta Walker

Magpie Moth
Pseudohazis hera picta (Walker)

Mallow Caterpillar
Anomis erosa Hubner

Malva Butterfly
Vanessa carye (Hubner)

Mangrove Skipper
Phocides batabano Lucas

Many-banded Dagger Wing
Timetes chiron (Fabricius)

Many-dotted Apple Worm
Balsa malana (Fitch)

Many-lined Wainscot
Leucania multilinea Walker

Many-spotted Apple Worm
Balsa malana (Fitch)

Many-spotted Tiger Moth
Turuptiana permaculata (Packard)

Maple Prominent
Heterocampa guttivitta (Walker)

Maple Spanworm
Deuteronomos magnarius (Guenee)

Marbled Comma
Polygonia faunus (Edwards)

Marbled Green
Leuconycta lepidula (Grote)

Marbled Leuconycta
Leuconycta dipteroides (Guenee)

Marbled Skipper
Erynnis martialis (Scudder)

Marine Blue
Leptotes marina (Reakirt)

Marsh Fritillary
Brenthis bellona (Fabricius)

Martial Hair-streak
Strymon martialis (Herrich-Schaeffer)

Master Dart
Feltia herilis (Grote)

Mead's Satyr
Minois meadii (Edwards)

Meadow Crescent Spot
Phyciodes campestris (Behr)

Meadow Fritillary
Brenthis bellona (Fabricius)

Melissa Arctic
Oeneis melissa (Fabricius)

Melsheimer's Sack Bearer
Cincinnus melsheimeri (Harris)

Mexican Forester
Androloma disparata (Henry Edwards)

Mexican Fritillary
Euptoieta hegesia (Cramer)

Mexican Sulphur
Eurema mexicana (Boisduval)

Mexican Tiger Moth
Apantesis proxima (Guerin-Meneville)

Mexican Walnut Moth
Citheronia mexicana Grote & Robinson

Miami Blue
Hemiargus ammon (Lucas)

Milbert's Tortoise Shell
Nymphalis milberti (Godart)

Milkweed Butterfly
Danaus plexippus (Linnaeus)

Milkweed Moth
Euchaetias egle (Drury)

Milkweed Tussock Moth
Euchaetias egle (Drury)

Mimic, The
Hypolimnas misippus (Linnaeus)

Modest Sphinx
Pachysphinx modesta (Harris)

Mohave Sooty Wing
Pholisora libya Scudder

INSECTS---COMMON NAMES

* Monarch Butterfly
Danaus plexippus (Linnaeus)

Mormon Metal-mark
Apodemia mormo (Felder & Felder)

Mother Underwing
Catocala parta Guenee

Mottled Dusky Wing
Erynnis martialis (Scudder)

Mottled-gray Cutworm
Abagrotis alternata (Grote)

Mottled Plum Tree Dagger
Acronicta interrupta Guenee

Mottled Umber Moth
Erannis vancouverensis Hulst

Mountain Crescent
Phyciodes montana Behr

Mountain Silver Spot
Argynnis (Speyeria) atlantis Edwards

Mountain Vagabond
Argynnis (Speyeria) montivaga Behr

* Mourning Cloak
Nymphalis antiopa (Linnaeus)

Mourning Sphinx
Enyo lugubris (Linnaeus)

Mourning Underwing
Catocala flebilis Grote

Mouse-colored Amphipyra
Amphipyra tragopoginis (Linnaeus)

Mouse-colored Pygarctia
Pygarctia murina (Stretch)

Muddy Dart
Euxoa quinquelinea (Smith)

Mulberry Wing
Poanes massasoit (Scudder)

Mustard White
Pieris napi (Linnaeus)

N

Nelson's Hair-streak
Mitoura nelsoni (Boisduval)

Nettle Tortoise Shell
Nymphalis milberti (Godart)

Neumoegen's Checker Spot
Melitaea neumoegeni Skinner

Nevada Buck Moth
Hemileuca nevadensis Stretch

Nevada Copper
Tharsalea virginiensis (Edwards)

New Mexico Range Caterpillar
Hemileuca oliviae Cockerell

Nicippe Yellow
Eurema nicippe (Cramer)

Norman's Dart
Graphiphora normaniana (Grote)

Northern Checker Spot
Melitaea palla Boisduval

Northern Cloudy Wing
Thorybes pylades (Scudder)

Northern Dusky Wing
Thorybes pylades (Scudder)

Northern Grass Worm
Caenurgina erechtea (Cramer)

Northern Hair-streak
Strymon ontario (Edwards)

Northern Metal Mark
Calephelis borealis (Grote & Robinson

INSECTS---COMMON NAMES

Northern Smoky Skipper
Thorybes pylades (Scudder)

Northern Tobacco Worm
Phlegethontius quinquemaculata (Haworth)

Northern Tomato Worm
Phlegethontius quinquemaculata (Haworth)

Northwestern Climbing Cutworm
Hemigraphiphora plebeia (Smith)

Notch Wing
Deuteronomos magnarius (Guenee)

Notch-wing Geometer
Deuteronomos magnarius (Guenee)

Nova Scotian, The
Oeneis jutta (Hubner)

November Moth
Oporinia autumnata (Guenee)

Oak Beauty
Nacophora quernaria (Abbot & Smith)

Oak Tussock Caterpillar
Halisidota maculata (Harris)

Oblique Arzama
Arzama obliqua (Walker)

Okra Caterpillar
Anomis erosa Hubner

Old Maid
Catocala coelebs Grote

Old Man Dart
Agrotis vetusta (Walker)

Oldwife Underwing
Catocala palaeogama Guenee

Old World Swallowtail
Papilio machaon Linnaeus

Olive-green Cutworm
Dargida procincta (Grote)

Olive Hair-streak
Mitoura damon (Cramer)

Olympia
Euchloe olympia (Edwards)

Omniverous Looper
Sabulodes caberata Guenee

Once-married Underwing
Catocala unijuga Walker

Orange-barred Sulphur
Phoebis philea (Linnaeus)

*Orange Dog
Papilio cresphontes Cramer

Orange-margined Blue
Plebeius melissa (Edwards)

Orange Skipperling
Copaeodes aurantiaca (Hewitson)

Orange-skirted Calico
Ageronia fornax Hubner

*Orange-striped Oakworm
Anisota senatoria (Abbot & Smith)

Orange Sulphur
Colias eurytheme Boisduval

Orange Tip
Anthocharis midea (Hubner)

Orchard Tent Caterpillar
Malacosoma americana (Fabricius)

Orizaba Silk Moth
Rothschildia orizaba (Westwood)

P

Painted Arachnis
Arachnis picta Packard

INSECTS---COMMON NAMES

*PAINTED BEAUTY
VANESSA VIRGINIENSIS (DRURY)

PAINTED CRESCENT
PHYCIODES PICTA (EDWARDS)

PAINTED DESERT MOTH
ARACHNIS PICTA PACKARD

*PAINTED LADY
VANESSA CARDUI (LINNAEUS)

PAINTED LICHEN MOTH
HYPOPREPIA FOCOSA HUBNER

PAINTED MAMESTRA
CERAMICA PICTA (HARRIS)

PALE BRANDED MOTH
GRAPHIPHORA BADINODIS (GROTE)

PALE LICHEN MOTH
CRAMBIDIA PALLIDA PACKARD

*PALE-SIDED CUTWORM
AGROTIS MALEFIDA GUENEE

PALE SWALLOWTAIL
PAPILIO EURYMEDON LUCAS

*PALE TUSSOCK MOTH
HALISIDOTA TESSELLARIS (ABBOT & SMITH

*PALE WESTERN CUTWORM
AGROTIS ORTHOGONIA MORRISON

PALMER'S METAL MARK
APODEMIA PALMERI (EDWARDS)

PALSIED CATERPILLAR
HARRISIMEMNA TRISIGNATA (WALKER)

*PANDORA MOTH
COLORADIA PANDORA BLAKE

PANDORA SPHINX
PHOLUS SATELLITIA (LINNAEUS)

PAWPAW SPHINX
DOLBA HYLAEUS (DRURY)

*PARSLEY WORM
PAPILIO ASTERIUS CRAMER

PARSNIP SWALLOWTAIL
PAPILIO ASTERIAS CRAMER

PASSION VINE BUTTERFLY
DIONE VANILLAE (LINNAEUS)

PEARL CRESCENT
PHYCIODES THAROS (DRURY)

PEARLY-EYED GRAYLING
ENODIA PORTLANDIA (FABRICIUS)

PEARLY WOOD NYMPH
EUTHISANOTIA UNIO HUBNER

PENITENT UNDERWING
CATOCALA PIATRIX GROTE

*PEPPER AND SALT MOTH
AMPHIDASIS COGNATARIA (GUENEE)

PEPPER AND SALT CURRANT MOTH
AMPHIDASIS COGNATARIA (GUENEE)

PEPPER AND SALT SKIPPER
AMBLYSCIRTES HEGON (SCUDDER)

PERPLEXING HAIR-STREAK
CALLOPHRYS DUMETORUM PERPLEXA

PERSIUS DUSKY WING
ERYNNIS PERSIUS (SCUDDER)

PIGMY BLUE
BREPHIDIUM EXILIS (BOISDUVAL)

PIGWEED SKIPPER
PHOLISORA CATULLUS (FABRICIUS)

PIMA ORANGE TIP
ANTHOCHARIS PIMA EDWARDS

*PINE BUTTERFLY
NEOPHASIA MONAPIA (FELDER & FELDER)

PINE DEVIL
CITHERONIA SEPULCHRALIS GROTE & ROBINSON

PINE ELFIN
INCISALIA NIPHON *HUBNER)

PINE TREE SPHINX
LAPARA BOMBYCOIDES WALKER

PINE MEASURING WORM
PARAPHIA PINIATA PACKARD

PINE WHITE
NEOPHASIA MENAPIA (FELDER & FELDER)

PINK-BACKED CUTWORM
LACINIPOLIA MEDITATA (GROTE)

PINK-EDGED SULPHUR
COLIAS INTERIOR SCUDDER

PINK-SPOTTED DART
GRAPHIPHORA BICARNEA (GUENEE)

PINK-SPOTTED HAWK MOTH
HERSE CINGULATA (FABRICIUS)

PINK-STRIPED WILLOW SPANWORM
DEILINIA VARIOLARIA (GUENEE)

*PIPEVINE SWALLOWTAIL
PAPILIO PHILENOR LINNAEUS

PISTACHIO
NEMORIA PISTACIARIA (PACKARD)

PLACID DART
ABAGROTIS PLACIDA (GROTE)

PLAIN RINGLET
COENONYMPHA INORNATA EDWARDS

PLEATED UFEUS
UFEUS PLICATUS GROTE

PLEBEIAN SPHINX
ATREIDES PLEBEIA (FABRICIUS)

PLUM GEOMETER
NEMORIA BRUNNEARIA (PACKARD)

PLUM TREE CATOCALA
CATOCALA ULTRONIA (HUBNER)

PLUM TREE SPHINX
SPHINX DRUPIFERARUM ABBOT & SMITH

POLIXENES ARCTIC
OENEIS POLIXENES (FABRICIUS)

POLKA-DOT MOTH
SYNTOMEIDA EPILAIS (WALKER)

POLYDAMAS SWALLOWTAIL
PAPILIO POLYDAMAS LINNAEUS

POLYGAMIST, THE
CATOCALA GRYNEA (CRAMER)

*POLYPHEMUS MOTH
TELEA POLYPHEMUS (CRAMER)

POPLAR HAWK MOTH
PACHYSPHINX MODESTA (HARRIS)

*POPLAR TENT MAKER
ICHTHYURA INCLUSA HUBNER

*POTATO STEM BORER
HYDROECIA MICACEA (ESPER)

POT-HERB BUTTERFLY
PIERIS NAPI (LINNAEUS)

*PROMETHEA MOTH
CALLOSAMIA PROMETHEA (DRURY)

PROPERTIUS DUSKY WING
ERYNNIS PROPERTIUS (SCUDDER & BURGESS

PURBLIND SPHINX
PAONIAS MYOPS (ABBOT & SMITH)

PURPLE LESSER FRITILLARY
BRENTHIS MONTINUS (SCUDDER)

PURPLISH COPPER
LYCAENA HELLOIDES (BOISDUVAL)

PYGMY BLUE
BREPHIDIUM EXILIS (BOISDUVAL)

PYRAMID GRAPE VINE CATERPILLAR
AMPHIPYRA PYRAMIDOIDES GUENEE

Q

QUEEN, THE
DANAUS BERENICE (CRAMER)

QUESTION SIGN BUTTERFLY
POLYGONIA INTERROGATIONIS (FABRICIUS)

R

RANCHMAN'S TIGER MOTH
PLATYPREPIA GUTTATA VIRGINALIS (BOISDUVAL)

*RANGE CATERPILLAR
HEMILEUCA OLIVIAE COCKERELL

RANNOCH LOOPER
ITAME FULVARIA (VILLERS)

RASPBERRY GEOMETER
SYNCHLORA AERATA (FABRICIUS)

RASPBERRY SPANWORM
SYNCHLORA AERATA (FABRICIUS)

RATTLEBOX MOTH
UTETHEISA BELLA (LINNAEUS)

REAKIRT'S COPPER
LYCAENA MARIPOSA (REAKIRT)

REAPER DART
EUXOA MESSORIA (HARRIS)

REAPING OWLET
EUXOA MESSORIA (HARRIS)

REAPING RUSTIC
EUXOA MESSORIA (HARRIS)

*RED ADMIRAL
VANESSA ATALANTA (LINNAEUS)

*RED-BACKED CUTWORM
EUXOA OCHROGASTER (GUENEE)

RED-BACKED HAIR-STREAK
STRYMON CECROPS (FABRICIUS)

RED-BARRED SULPHUR
PHOEBIS PHILEA (LINNAEUS)

RED-BREASTED DART
PROTOLAMPRA RUFIPECTUS (MORRISON)

RED CUTWORM
ABAGROTIS PLACIDA (GROTE)

RED DISKED ALPINE
EREBIA DISCOIDALIS KIRBY

RED GROUNDLING
PERIGEA XANTHIOIDES GUENEE

RED-HEADED INCH WORM
SEMIOTHISA BISIGNATA (WALKER)

RED-HUMPED APPLE TREE CATERPILLAR
SCHIZURA COCINNA (ABBOT & SMITH)

*RED-HUMPED CATERPILLAR
SCHIZURA CONCINNA (ABBOT & SMITH)

RED-HUMPED MAPLE WORM
SYMMERISTA ALBIFRONS (ABBOT & SMITH)

RED-HUMPED OAK WORM
SYMMERISTA ALBIFRONS (ABBOT & SMITH)

RED-LEGGED DIACRISIA
DIACRISIA LATIPENNIS (STRETCH)

RED SPOTTED PURPLE
BASILARCHIA ASTYANAX (FABRICIUS)

RED-STREAKED ALPINE
EREBIA DISCOIDALIS KIRBY

RED-WINGED SALLOW
XYSTOPEPLUS RUFAGO (HUBNER)

REDDISH-WHITE DAGGER MOTH
ACRONICTA ALBARUFA (GROTE)

REGAL FRITILLARY
ARGYNNIS (SPEYERIA) IDALIA (DRURY)

*REGAL MOTH
CITHERONIA REGALIS (FABRICIUS)

RIDING'S SATYR
EUMENIS RIDINGSII (EDWARDS)

RINGLESS RINGLET
COENONYMPHA AMPELOS EDWARDS

ROADSIDE SKIPPER
AMBLYSCIRTES VIALIS (EDWARDS)

ROSY DYROCAMPA
ANISOTA RUBICUNDA (FABRICIUS)

ROSY MAPLE MOTH
ANISOTA RUBICUNDA (FABRICIUS)

ROSY-STRIPED OAK WORM
ANISOTA VIRGINIENSIS (DRURY)

ROYAL WALNUT MOTH
CITHERONIA REGALIS (FABRICIUS)

RUBBED DART
EUXOA DETERSA (WALKER)

RUBY TIGER MOTH
PHRAGMATOBIA FULIGINOSA (LINNAEUS)

RUDDY DAGGER WING
ATHENA PETREUS (CRAMER)

RUDDY DIACRISIA
DIACRISIA PTERIDIS RUBRA NEUMOEGEN

RUSTIC BLUE
PLEBEIUS AQUILO RUSTICA EDWARDS

RUSTIC SPHINX
PHLEGETHONTIUS RUSTICA (FABRICIUS)

*RUSTY TUSSOCK MOTH
NOTOLOPHUS ANTIQUA (LINNAEUS)

S

SACHEM, THE
ATALOPEDES CAMPESTRIS (BOISDUVAL)

*SADDLED PROMINENT
HETEROCAMPA GUTTIVITTA (WALKER)

SAGA FRITILLARY
BRENTHIS FRIGGA SAGA STAUDINGER

SAINT LAWERENCE TIGER MOTH
PARASEMIA PARTHENOS (HARRIS)

*SALT MARSH CATERPILLAR
ESTIGMENE ACREA (DRURY)

SALT MARSH SKIPPER
PANOQUINA PANOQUIN (SCUDDER)

SAND HILL CUTWORM
EUXOA DETERSA (WALKER)

SAND HILL SKIPPER
POLITES SABULETI (BOISDUVAL)

SARA ORANGE-TIP
ANTHOCHARIS SARA BOISDUVAL

SATELLITE SPHINX
PHOLUS SATELLITIA (LINNAEUS)

*SATIN MOTH
STILPNOTIA SALICIS (LINNAEUS)

SATYR, THE
POLYGONIA SATYRUS (EDWARDS)

INSECTS---COMMON NAMES

Saw Wing
Euchlaena serrata (Drury)

Scallop Shell Moth
Calocalpe undulata (Linnaeus)

Scallop Sack Bearer
Lacosoma chiridota Grote

Scarlet-bodied Wasp Moth
Cosmosoma auge (Linnaeus)

Scarlet Underwing
Catocala coccinata Grote

Scarlet-winged Lichen Moth
Hypoprepia miniata (Kirby)

Scudder's Blue
Plebeius scudderii (Edwards)
(probably melissa var Edwards)

Scudder's Patched Butterfly
Chlosyne lacinia adjutrix Scudder

Semicolon, The
Polygonia interrogationis (Fabricius)

Semilooper Maple Worm
Zale lunata (Drury)

Semitropical Armyworm
Prodenia eridania (Cramer)

Senna Sulphur
Phoebis sennae (Linnaeus)

Sequoia Sphinx
Sphinx sequoia Boisduval

Shagreened Caterpillar
Agrotis malefida Guenee

Shield-marked Cutworm
Rusina bicolorago (Guenee)

Short-tailed Swallowtail
Papilio brevicauda Saunders

Short-winged Dart
Euxoa brevipennis (Smith)

Sierra Checker Spot
Euphydryas sierra (Wright)

Sigmoid Dart
Eueretagrotis sigmoides (Guenee)

Silky Climbing Cutworm
Polia detracta (Walker)

Silky Pyrophila
Amphipyra tragopoginis (Linnaeus)

Silky Sallow
Harpaglaea sericea (Morrison)

Silver-bordered Fritillary
Brethis myrina (Cramer)

Silver Crescent
Phyciodes nycteis (Doubleday & Hewitson)

Silver-spotted Halisidota
Halisidota argentata Packard

*Silver-spotted Skipper
Epargyreus clarus (Cramer)

Silvery Blue
Glaucopsyche lygdamus (Doubleday)

Silvery Checker Spot
Phyciodes nycteis (Doubleday)

Simple Woodling
Xylomyges simplex (Walker)

Six-spotted Sphinx
Phlegethontius rustica (Fabricius)

Sleppy Dusky Wing
Erynnis brizo (Boisduval)

Sleppy Orange
Eurema nicippe (Cramer)

INSECTS---COMMON NAMES

Sleppy Underwing
Catocala concumbens Walker

Slippery Dart
Euagrotis lubricans (Guenee)

Small Blue
Philotes speciosa (Henry Edwards)

Small Brown Quaker
Orthodes vecors Guenee

Small Checkered Skipper
Pyrgus scriptura (Boisduval)

Small Copper
Lycaena epixanthe (Boisduval & Leconte)

Small-eyed Sphinx
Paonias myops (Abbot & Smith)

Small Striped Cutworm
Euxoa niveilinea (Grote)

Small Tiger Moth
Nemeophila plantaginis (Linnaeus)

Small Tortoise Shell
Nymphalis milberti (Godart)

Small Wood Nymph
Minois silvestris (Edwards)

*Smeared Dagger Moth
Acronicta oblinita (Abbot & Smith)

Smudged Sallow
Homoglaea carbonaria (Harvey)

Snout Butterfly
Libythea bachmanii Kirtland

Snowberry Clearwing
Hemaris diffinis (Boisduval)

Snow-white Linden Moth
Ennomos subsignarius (Hubner)

Snow's Copper
Lycaena snowi (Edwards)

Snowy Eupseudosoma
Eupseudosoma involutum (Sepp)

Sonora Blue
Philotes sonorensis (Felder & Felder)

Soothsayer Dart
Pseudospaelotis haruspica (Grote)

Sooty Gossamer-wing
Satyrium fuliginosa (Edwards)

Sooty Wing
Pholisora catullus (Fabricius)

*Southern Armyworm
Prodenia eridania (Cramer)

*Southern Cabbage Worm
Pieris protodice (Boisduval & Leconte)

Southern Cloudy Wing
Thorybes bathyllus (Abbot & Smith)

Southern Hair-streak
Strymon favonius (Abbot & Smith)

Southern Hornworm
Phlegethontius sexta (Johanssen)

Southern Tobacco Worm
Phlegethontius sexta (Johanssen)

Southern Tomato Worm
Phlegethontius sexta (Johanssen)

Southwestern Armyworm
Prodenia praefica Grote

Sovereign, The
Basilarchia astyanax (Fabricius)

Spear-marked Black Moth
Eulype hastata (Linnaeus)

INSECTS---COMMON NAMES

Speckled Cutworm
<u>Polia subjuncta</u> (Grote & Robinson)

Spice Bush Silk Moth
<u>Callosamia promethea</u> (Drury)

*Spice Bush Swallowtail
<u>Papilio troilus</u> Linnaeus

Spider Owlet Moth
<u>Prodenia dolichos</u> (Fabricius)

Spindle Worm
<u>Achatodes zeae</u> (Harris)

Spinous Currant Caterpillar
<u>Polygonia progne</u> (Cramer)

Spiny Elm Caterpillar
<u>Nymphalis antiopa</u> (Linnaeus)

Spiny Maple Worm
<u>Anisota rubicunda</u> (Fabricius)

*Spiny Oak Worm
<u>Anisota stigma</u> (Fabricius)

Spotless Fall Webworm
<u>Hyphantria textor</u> Harris

*Spotted Cutworm
<u>Graphiphora c-nigrum</u> (Linnaeus)

Spotted Halisidota
<u>Halisidota maculata</u> (Harris)

Spotted-legged Cutworm
<u>Agrotis vetusta</u> (Walker)

Spotted-sided Cutworm
<u>Graphiphora badinodis</u> (Grote)

*Spotted Tussock Moth
<u>Halisidota maculata</u> (Harris)

Spring Azure
<u>Lycaenopsis pseudargiolus</u> (Boisduval & Leconte)

*Spring Cankerworm
<u>Paleacrita vernata</u> (Peck)

*Stalk Borer
<u>Papaipema nebris</u> (Guenee)

Stout Spanworm
<u>Lycia ursaria</u> (Walker)

Strawberry Tiger Moth
<u>Haploa colona</u> (Hubner)

Streaked Calidota
<u>Calidota strigosa</u> (Walker)

Streaked Dagger Moth
<u>Acronicta lithospila</u> Grote

Strecker's Day Sphinx
<u>Proserpinus juanita</u> (Strecker)

Striped Beet Caterpillar
<u>Scotogramma trifolii</u> (Rottenburg)

*Striped Cutworm
<u>Euxoa tessellata</u> (Harris)

*Striped Garden Caterpillar
<u>Polia legitima</u> (Grote)

Striped Grass Looper
<u>Mocis latipes</u> (Guenee)

Striped Green Bean Caterpillar
<u>Ogdoconta cinereola</u> (Guenee)

Striped Hair-streak
<u>Strymon liparops</u> (Boisduval & Leconte)

Striped Hickory Caterpillar
<u>Datana angusi</u> Grote & Robinson

Striped Morning Sphinx
<u>Celerio lineata</u> (Fabricius)

Subject Lichen Moth
<u>Cisthene subjecta</u> Walker

INSECTS---COMMON NAMES

SUMAC DATANA
Datana perspicua Grote & Robinson

SWALLOWTAIL SKIPPER
Urbanus proteus (Linnaeus)

SWAMP METAL MARK
Calephelis muticum McAlpine

SWARTHY SKIPPER
Lerodea l'herminieri (Godart)

SWEET FERN GEOMETER
Cosymbia pendulinaria (Guenee)

SWEETHEART, THE
Catocala amatrix (Hubner)

SWEET POTATO HORNWORM
Herse cingulata (Fabricius)

SWORDSMAN'S DART
Agrotis gladiaria Morrison

*SYCAMORE TUSSOCK MOTH
Halisidota harrisii Walsh

TAILED BLUE
Everes comyntas (Godart)

TAWNY CRESCENT
Phyciodes batesii (Reakirt)

TAWNY EDGE
Polites themistocles (Latreille)

TAWNY EMPEROR
Asterocampa clyton (Boisduval & Leconte)

TAWNY EUBAPHE
Eubaphe opella (Grote)

TEARFUL UNDERWING
Catocala lacrymosa Guenee

TEN-LINED INCH WORM
Erannis tiliaria (Harris)

TESSELLATE DART
Euxoa tessellata (Harris)

TEXAN CRESCENT
Anthanassa texana (Edwards)

TEXAN HAIR-STREAK
Strymon autolycus (Edwards)

THICKET HAIR STREAK
Mitoura spinetorum (Hewitson)

THISTLE BUTTERFLY
Vanessa cardui (Linnaeus)

THREAD-BEARING SPANWORM
Nematocampa limbata (Haworth)

THREE-SPOTTED FILLIP
Heterophleps triguttaria Herrich-Schaeffer

THYSBE CLEARWING
Hemaris thysbe (Fabricius)

*TIGER SWALLOWTAIL
Papilio glaucus Linnaeus

TOBACCO BUDWORM
Heliothis vierscens (Fabricius)

*TOBACCO HORNWORM
Phlegethontius sexta (Johanssen)

TOMATO FRUITWORM
Heliothis obsoleta (Fabricius)
Heliothis zeae Boddie

*TOMATO HORNWORM
Phlegethontius quinquemaculata (Haworth)

TOMATO SPHINX
Phlegethontius quinquemaculata (Haworth)
Phlegethontius sexta (Johanssen)

TRIM PROMINENT
Schizura concinna (Abbot & Smith)

TROPICAL CHECKERED SKIPPER
<u>PYRGUS SYRICHTUS</u> (FABRICIUS)

TROPICAL LEAF WING
<u>ANAEA AIDEA</u> (GUERIN)

TULIP TREE BEAUTY
<u>EPIMECIS VIRGINARIA</u> (CRAMER)

TULIP TREE SILK MOTH
<u>CALLOSAMIA ANGULIFERA</u> (WALKER)

TWILIGHT MOTH
<u>POECILOPSIS RACHELAE</u> (HUBNER)

TWIN-SPOTTED SKIPPER
<u>OLIGORIA MACULATA</u> (EDWARDS)

TWIN SPOT SPHINX
<u>SMERINTHUS JAMAICENSIS</u> (DRURY)

TWO-BANDED SKIPPER
<u>PYRGUS RURALIS</u> (BOISDUVAL)

TWO SPOT DART MOTH
<u>EUERETAGROTIS PERATTENTA</u> (GROTE)

TWO-SPOTTED SKIPPER
<u>ATRYTONE BIMACULA</u> (GROTE & ROBINSON)

U

UHLER'S ARCTIC
<u>OENEIS UHLERI</u> (REAKIRT)

UMBER MOTH
<u>POANES MELANE</u> (EDWARDS)

UNARMED OWLET
<u>PERIDROMA MARGARITOSA</u> (HAWORTH)

UNARMED RUSTIC
<u>SPAELOTIS CLANDESTINA</u> (HARRIS)
<u>PERIDROMA MARGARITOSA</u> (HAWORTH)

UNCIVIL DART
<u>ANICLA INFECTA</u> (OCHSENHEIMER)

*UNICORN CATERPILLAR
<u>SCHIZURA UNICORNIS</u> (ABBOT & SMITH)

UNMARKED CUTWORM
<u>HEPTAGROTIS PHYLLOPHORA</u> (GROTE)

UNMARKED DAGGER MOTH
<u>ACRONICTA INNOTATA</u> GUENEE

UNSATED SALLOW
<u>METAXAGLAEA INULTA</u> (GROTE)

V

VELLEDA LAPPET
<u>TOLYPE VELLEDA</u> (STOLL)

VANCOUVER DART
<u>AGROTIS VANCOUVERENSIS</u> GROTE

VARIABLE DART
<u>ANOMOGYNA ELIMATA</u> (GUENEE)

*VARIABLE OAK LEAF CATERPILLAR
<u>HETEROCAMPA MANTEO</u> (DOUBLEDAY)

VARIED BLUE
<u>LYCAENA HETERONEA</u> BOISDUVAL

*VARIEGATED CUTWORM
<u>PERIDROMA MARGARITOSA</u> (HAWORTH)

VARIEGATED FRITILLARY
<u>EUPTOIETA CLAUDIA</u> (CRAMER)

VEILED APAMEA
<u>APAMEA VELATA</u> WALKER

*VELVET BEAN CATERPILLAR
<u>ANTICARSIA GEMMATILIS</u> HUBNER

VENERABLE DART
<u>AGROTIS VENERABILIS</u> WALKER

VESTA CRESCENT
<u>PHYCIODES VESTA</u> EDWARDS

VESTAL TIGER MOTH
<u>MAENAS VESTALIS</u> (PACKARD)

*VICEROY, THE
BASILARCHIA ARCHIPPUS (CRAMER)

VINE SPHINX
PHOLOS VITIS (LINNAEUS)

VIOLET TIP
POLYGONIA INTERROGATIONIS (FABRICIUS)

VIRGIN TIGER MOTH
APANTESIS VIRGO (LINNAEUS)

*VIRGINIA CREEPER SPHINX
AMPELOECA MYRON (CRAMER)

VIRGINIA ERMINE MOTH
DIACRISIA VIRGINICA (FABRICIUS)

VIRGINIA LADY
VANESSA VIRGINIENSIS (DRURY)

VIRGINIA TIGER MOTH
DIACRISIA VIRGINICA (FABRICIUS)

VOLCANIC SKIPPER
CATIA OTHO (ABBOT & SMITH)

VOLUBLE DART
AGROTIS VOLUBILIS HARVEY

W

WALKER'S SALLOW
EUPSILIA SIDUS (GUENEE)

*WALNUT CATERPILLAR
DATANA INTEGERRIMA GROTE & ROBINSON

WALNUT SPANWORM
CONIOIDES PLUMOGERARIA (HULST)

WALNUT SPHINX
CRESSONIA JUGLANDIS (ABBOT & SMITH)

WANDERER, THE
FENISECA TARQUINIUS (FABRICIUS)

WANDERING DIACRISIA
DIACRISIA VAGANS (BOISDUVAL)

WANDERING SKIPPER
PANOQUINA ERRANS (SKINNER)

WAVED SPHINX
CERATOMIA UNDULOSA (WALKER)

WAVY ASH SPHINX
CERATOMIA UNDULOSA (WALKER)

WAYWARD NYMPH
CATOCALA ANTINYMPHA (HUBNER)

WEIDEMEYER'S ADMIRAL
BASILARCHIA WEIDEMEYERII (EDWARDS)

WELL MARKED CUTWORM
SPAELOTIS CLANDESTINA (HARRIS)

WEST COAST LADY
VANESSA CARYE (HUBNER)

WEST INDIAN BLUE
LEPTOTES THEONUS (LUCAS)

WEST VIRGINIA WHITE
PIERIS VIRGINIENSIS EDWARDS

WESTERN ARMY CUTWORM
CHORIZAGROTIS AUXILIARIS (GROTE)

WESTERN CHECKERED SKIPPER
PYRGUS COMMUNIS (GROTE)

WESTERN ELDER MOTH
ZOTHECA TRANQUILLA GROTE

WESTERN PANTHEA
PANTHEA PORTLANDIA GROTE

WESTERN PARSLEY CATERPILLAR
PAPILIO ZELICAON

WESTERN PIGMY BLUE
BREPHIDIUM EXILIS (BOISDUVAL)

WESTERN STRIPED CUTWORM
FELTIA SUBGOTHICA (HAWORTH)

WESTERN SQALLOWTAIL
PAPILIO RUTULUS LUCAS

INSECTS---COMMON NAMES

Western Tailed Blue
Everes amyntula (Boisduval)

* Western Tent Caterpillar
Malacosoma pluvialis (Dyar)

Western Tiger Swallowtail
Papilio rutulus Lucas

* Western Tussock Moth
Hemerocampa vetusta (Boisduval)

Western Yellow-striped Armyworm
Prodenia praefica Grote

Western White
Pieris occidentalis Reakirt

* Wheat Head Armyworm
Protoleucania albilinea (Hubner)

Wheeler's Checker Spot
Euphydryas wheeleri (Henry Edwards)

Whirlabout, The
Polites brettus (Boisduval & Leconte)

White Admiral
Basilarchia arthemis (Drury)

White-bodied Estigmene
Estigmene congrua (Walker)

White-collared Scape Moth
Scepsis wrightii Stretch

* White Cutworm
Euxoa scandens (Riley)

White-dotted Apple Worm
Balsa malana (Fitch)

White-lined Morning Sphinx
Celerio lineata (Fabricius)

* White-lined Sphinx
Celerio lineata (Fabricius)

White-lined Wainscot
Protoleucania albilinea (Hubner)

White M Hair-streak
Strymon m-album (Boisduval & Leconte)

* White-marked Tussock Moth
Hemerocampa leucostigma (Abbot & Smith)

White Mountain Butterfly
Oeneis melissa (Fabricius)

White Mountain Fritillary
Brenthis montinus (Scudder)

White Peacock
Anartia jatrophae (Johanssen)

White Pine Tufted Caterpillar
Panthea furcilla (Packard)

White-spotted Cutworm
Homohadena badistriga (Grote)

White Striped Black
Neodezia albovittata (Guenee)

White Tipped Moth
Symmerista albifrons (Abbot & Smith)

White Underwing
Catocala relicta Walker

White-veined Dagger
Simyra henrici (Grote)

Widow Underwing
Catocala vidua (Abbot & Smith)

Wife, The
Catocala ilia uxor Guenee

Wild Cherry Sphinx
Sphinx drupiferarum Abbot & Smith

Willow Dagger
Acronicta dactylina (Grote)

Winter Moth
Erannis tiliaria (Harris)

Witch Hazel Dagger
Acronicta hamamelis Guenee

*W-marked Cutworm
Spaelotis clandestina (Harris)

Woodland Skipper
Ochlodes sylvanoides (Boisduval)

Wood Nymph
Minois alope (Fabricius)

X

Xerces Blue
Glaucopsyche xerces (Boisduval)

Y

Yellow Armyworm
Leucania phragmatidicola Guenee

Yellow-banded Day Sphinx
Proserpinus flavofasciata (Walker)

Yellow-banded Underwing
Catocala cerogama Guenee

Yellow-banded Wasp Moth
Syntomeida ipomoeae Harris

Yellow-barred Heliconian
Heliconius charitonia (Linnaeus)

Yellow Bear
Diacrisia virginica (Fabricius)

Yellow-bellied Dart
Euxoa ochrogaster (Guenee)

Yellow-blotched Lichen Moth
Cisthene nexa (Boisduval)

Yellow-collared Scape Moth
Cisseps fulvicollis (Hubner)

Yellow-gray Underwing
Catocala retecta Grote

Yellow-haired Dagger Moth
Acronicta impleta Walker

*Yellow-headed Cutworm
Septis arctica (Boisduval)

Yellow-necked Apple Tree Caterpillar
Datana ministra (Drury)

Yellow-necked Apple Worm
Datana ministra (Drury)

*Yellow-necked Caterpillar
Datana ministra (Drury)

Yellow Poplar Caterpillar
Acronicta lepusculina Guenee

*Yellow-striped Armyworm
Prodenia ornithogalli Guenee

Yellow-striped Oak Caterpillar
Anisota senatoria (Abbot & Smith)

Yellow-winged Cycnia
Cycnia insulata (Walker)

*Yellow Wooly Bear
Diacrisia virginica (Fabricius)

Youthful Underwing
Catocalla subnata Grote

Ypsilon Dart
Agrotis ypsilon (Rottenburg)

Z

Zarucco Dusky Wing
Erynnis zarucco (Lucas)

Zebra, The
Heliconius charitonia (Linnaeus)

ZEBRA CATERPILLAR
CARAMICA PICTA (HARRIS)

ZEBRA SWALLOWTAIL
PAPILIO MARCELLUS CRAMER

ZEPHYR, THE
POLYGONIA ZEPHYRUS (EDWARDS)

PART I - INSECTS

SECTION C---MACROLEPIDOPTERA

(INCLUDES LIFE HISTORY REFERENCES AND FOOD PLANTS)

AARONI (SKINNER), POANES–HESPERIIDAE
A HOWARDI SKINNER
LIFE HISTORY
ENT. NEWS 19:415, 1908
LAURENT EGG, EARLY STAGES
FOOD PLANTS
GRASSES

ABACTA HULST
SYN. SEE ALBIFASCIATA PACKARD

ABALIENALIS (WALKER), BOMOLOCHA–NOCTUIDAE
LIFE HISTORY
NEW YORK STATE MUS., MEM. 8:724, 1905
FELT BRIEF LARVAL DESCR.
CAN. ENT. 23:157, 1891
DYAR LARVA
FOOD PLANTS
ULMUS RUBRA MUHL. (SLIPPERY ELM)

ABAR STRECKER
SSP. SEE DIVERGENS WALKER EUXOA

ABBOTII EDWARDS
SSP. SEE MARCELLUS CRAMER

ABBOTTII (SWAINSON), SPHECODINA
SPHINGIDAE
LIFE HISTORY
AMER. ENT. 1:252, 1869
BRIEF LARVAL DESCR.
LOC. CIT. 2:123, 1870
LARVA* PUPA
AMER. JOURN. SCI. 36:307, 1839
HARRIS LARVA, PUPA
AMER. MUS. NAT. HIST., BULL. 7:283, 1895 BEUTENMULLER LARVA
CAN. ENT. 10:130, 1878
SAUNDERS LARVA* PUPA
LOC. CIT. 13:2, 1881
SAUNDERS LARVA*
ENT. NEWS 9:150, 1898
SOULE DURATION EGG STAGE
ENT. SOC. AMER., ANNALS 11:436, 1918 MOSHER PUPAL STRUCT.
ILLINOIS DEPT. AGRI., TRANS. 15:164, 1877 FRENCH LARVA
LOC. CIT. 18:APPEND:100, 1880
MARTIN LARVA*
MICHIGAN STATE COLLEGE, SPEC. BULL. 293:48, 1933
MCDANIEL LARVA
MINNESOTA AGRI. EXP. STA., BULL. 61:86, 1899 LUGGER LARVA*
PL. 5, F. 32; PUPA
MISSOURI INS. 2ND REPT. P. 78, 1870 RILEY LIFE HISTORY WITH FF.
NEW YORK STATE AGRI. EXP. STA. (CORNELL), MEM. 274:198, 1948
FORBES LARVA
NEW YORK STATE CABIN. NAT. HIST. ANN. REPT. NO. 24:114, 1871
LINTNER LARVA, PUPA
SEXUAL DIMORPHISM IN LARVA?
NEW YORK STATE MUS., MEM. 8: 739, 1905 FELT BRIEF LARVAL DESCR.
NOVITATES ZOOLOGICAE (TRING) 9 SUPPLEMENT P. 602, 1903
ROTHSCHILD & JORDAN
LARVA, PUPA
ONTARIO ENT. SOC., ANN. REPT. P. 13, 1872 SAUNDERS LARVA* PUPA
LOC. CIT. P. 42, 1880
SAUNDERS LARVA*
LOC. CIT. P. 53, 1881
REED LARVA*
LOC. CIT. P. 24, 1888
FYLES LARVA*
PAPILIO 2:147, 1882
SAUNDERS LIFE HISTORY
PHILADELPHIA ACAD. NAT. SCI., JOURN. 5:135, 1859
CLEMENS LARVA, PUPA
PHILADELPHIA ACAD. NAT. SCI., PROC. 9:148, 1857
KIRTLAND LARVA
PSYCHE 6:149, 1891
SOULE ABNORMAL LARVA
SMITHSONIAN MISC. COLLECTIONS 4:ART. 2:156, 1862
MORRIS LARVA, PUPA
UNITED STATES DEPT. AGRI., MISC. PUB. 657:359,376, 1950
CRAIGHEAD LARVA* FF. 71, A, B

Vermont Agri. Soc. Rept., No. 5: 281, 1878 Perkins larva

Boisduval-Spec. Gen. Lep., Sphing. p. 331, 1874 larva, pupa

Chenu-Demarests--Encyc. Hist. Nat., Papillons 1:257, 1856 larva*

Eliot-Soule--Caterpillars and their Moths p. 84, 1902 egg, larva*, pupa

Fernald-Sphing, New England p. 23, 1886 larva

Harris-Ent. Corresp. p. 284, 1869 larva* Pl. 3, f. 1

Holland-Moth Book p. 70, 1905 larva* f. 33

Kellogg-Amer. Ins. p. 437, 1908 larva* f. 629

Morris-Synop. Lep. NorthAmerica p. 156, 1862 larva, pupa

O'Kane-Inj. Ins. p. 273, 1912 larva* f. 380

Saunders-Ins. Inj. Fruits p. 253, 1883 & 1889 larva* pupa*

Swainson-Zool. Illustr. part 2, 1821 larva* Pl. 60

\- - -

Index Amer. Econ. Ent., Vol. 1:288; Vol. 4:444

FOOD PLANTS

Crataegus sp. (thorns)

Parthenocissus quinquefolia Linn. (Virginia creeper)

Vitis sp. (grape)

ABBREVIATELLA Grote, CATOCALA-Noctuidae

LIFE HISTORY

Amer. Mus. Nat. Hist., Bull. 38:169, 1918 Barnes & McDunnough egg

FOOD PLANTS

ABJECTA Guenee

syn. see DEVASTATOR Brace

ABJECTARIUS Hulst

syn. see SCOLOPACINARIUS Guenee

ABNORMA Wright

ssp. see HOFFMANNI Behr

ABNORMALIS Hulst

ssp. see TIGRINARIA Guenee

ABORTIVARIA Herrich-Schaeffer-DYSPTERIS-Geometridae

LIFE HISTORY

Illinois Dept. Agri., Trans. 15: 238, 1877 French larva

Psyche 19:14, 1912 Barnes & McDunnough egg, larva

FOOD PLANTS

Parthenocissus quinquefolia Linn. (Virginia creeper)

Prunus serotina Ehrh. (do not thrive)

Vitis sp. (grape)

ABRASA Guenee

syn. see SCRIPTA Gosse

ABRASARIA (Herrich-Schaeffer), ZANTHORHOE-Geometridae

a CONGREGATA Walker

syn. NIGROFASCIATA Packard

b TRILINEATA Warren

syn. SALVATA Pearsall

c AQUILONARIA Cassino & Swett

LIFE HISTORY

United States Nat. Mus., Proc. 27:900, 1904 Dyar egg, larva

FOOD PLANTS

ABRASOIDES Barnes & Benjamin

ssp. see SCRIPTA Gosse

ABRAXARIA Walker

ssp. see CREPUSCULARIA Schiffermuller

ABROSTOLELLA Walker

syn. see PYGMAEA Hubner (Paectes)

ABSCONDENS Oberthur

syn. see MUZINA Oberthur (Ecpantheria)

ABSYNTHIATA GUENEE
SYN. SEE COAGULATA GUENEE

ACADICA EDWARDS
SSP. SEE NAPI LINNEAUS (PIERIS)

ACADICA(EDWARDS), STRYMON LYCAENIDAE
SYN. SOUHEGAN WHITNEY
AB MUSKOKA WATSON & COMSTOCK
AB SWETTI WATSON & COMSTOCK
A COOLINENSIS WATSON & COMSTOCK
B MONTANENSIS WATSON & COMSTOCK
LIFE HISTORY
BUFFALO SOC. NAT. SCI., BULL. 19: 28, 1939 WILD LARVA
CAN. ENT. 1:95, 1869 SAUNDERS LARVA, CHRYSALIS
ILLINOIS DEPT. AGRI., TRANS. 18: APPEND:93, 1880 MIDDLETON LARVA
MONTANA UNIV. BULL. 30:128, 1906 ELROD LARVA
NEW YORK STATE MUS., MEM. 8: 742, 1905 FELT BRIEF LARVAL DESCR.
COMSTOCK-HOW TO KNOW BUTTER. P. 224, 1904 LARVA
W. H. EDWARDS-BUTTER. NORTH AMER., 1869, 1ST SERIES PL. THECLA NO. 1 LARVA* COLOR
FRENCH-BUTTER. EAST. UNITED STATES P. 261, 1886 LARVA, CHRYSALIS
HOLLAND-BUTTER. BOOK 1905 CHRYSALIS* PL. 5, F. 35
MAYNARD-BUTTER. NEW ENGLAND P. 34, 1886 LARVA
SCUDDER-BUTTER. EAST. UNITED STATES & CANADA PP. 900-902, 1889 LIFE HISTORY* PL. 65, F. 1; PL. 75, FF. 16-18; PL. 79, F. 25; PL. 84, F. 35
SCUDDER-GUIDE TO COMMON BUTTER. P. 122, 1893 LARVA, CHRYSALIS
FOOD PLANTS
SALIX DISCOLOR MUHL.
IN ANTS NESTS DURING THE DAY

ACANOOTUS SCUDDER
SYN. SEE BIMACULA GROTE & ROBINSON

ACAUDA OBERTHUR
SSP. SEE PHILENOR LINNEAUS

ACCEPTA HENRY EDWARDS
SYN. SEE APICELLA GROTE

ACCESSARIA HUBNER
SYN. SEE COLORARIA FABRICIUS

ACCIUS(ABBOT & SMITH), LEREMA-HESPERIIDAE
SYN. PATTENII SCUDDER
SYN. PARUMPUNCTATA HERRICH-SCHAEFFER
SYN. MONOCO SCUDDER
SYN. PUNCTELLA GROTE & ROBINSON
SYN. NORTONII EDWARDS
LIFE HISTORY
AMER. MUS. NAT. HIST., BULL. 5: 295, 1893 BEUTEN. LARVA
CAN. ENT. 9:191, 1879 EDWARDS LARVA, CHRYSALIS
FLORIDA ENT. 6:2, 11, 1922 AINSLIE EGG, LARVA, CHRYSALIS
ABBOT & SMITH-LEP. INS. GEORGIA I:45, 1797 LARVA* PUPA* PL. 23 COLOR
HOLLAND-BUTTERFLY BOOK 1905 CHRYSALIS* PL. 6, F. 46
SCUDDER-BUTTER. EAST. UNITED STATES & CANADA PP. 1769-1771, 1889 LIFE HISTORY* PL. 66, F. 36; PL. 77, F. 36; PL. 85, F. 46
SCUDDER-BUTTER. THEIR STRUCTURE, ETC. P. 42, 1881 CHRYSALIS*
- - -
INDEX AMER. ECON. ENT., VOL. 3:235
FOOD PLANTS
ERIANTHUS ALOPECUROIDES(LINN.)
ORYZOPSIS SP.
PANICUM CAPILLARE LINN.
SORGHUM SP.
SORGHUM HALEPENSE (LINN.)
WISTERIA FRUTESCENS (LINN.)
ZEA MAYS LINN. (CORN)

ACCURATA(HENRY EDWARDS), ORTHODES-NOCTUIDAE
LIFE HISTORY
SOUTHERN CALIFORNIA ACAD. SCI., BULL. 46:124, 1947 EDWARDS LARVA* PL. 28, A, B; PUPA* PL. 28, C, D, E,
FOOD PLANTS
BRICKELLIA RUSBYI GRAY

ACERICOLA GUENEE
SYN. SEE AMERICANA HARRIS (ACRONICTA)

ACESIAS FELDER
SYN. SEE PHLOXIPHAGA GROTE & ROBINSON

ACHAIA GROTE & ROBINSON
SSP. SEE ORNATA PACKARD

ACHATINA (ABBOT & SMITH), OLENE-LIPARIDAE
LIFE HISTORY
AMER. MUS. NAT. HIST., BULL. 10: 384, 1898 BEUTEN. LARVA
MUS. COMP. ZOOL., BULL. 87: NO. 6, 1941 DEITHER LARVAL ANTENNA* F. 106
NEW YORK STATE MUS., MEM. 8: 524, 1905 FELT LARVA
PSYCHE 7:136, 1894 DYAR LARVA
LOC. CIT. 18:157, 1911 BARNES & MCDUNNOUGH LARVA
ABBOT & SMITH-LEP. INS. GEORGIA 2:153, 1797 LARVA* PUPA* PL. 77 COLOR
FOOD PLANTS
CARYA SP. (HICKORIES)
CASTANEA DENTATA (MARSH.) (CHESTNUT)
PRUNUS CERASUS LINN. (CHERRY)
PRUNUS DOMESTICA LINN. (PLUM)
PRUNUS SEROTINA EHRH. (WILD CHERRY)
PYRUS COMMUNIS LINN. (PEAR)
QUERCUS SP. (OAKS)
MANY OTHER TREES AND SHRUBS

ACHATINALIS ZELLER
SYN. SEE MADEFACTALIS GUENEE

ACHATINATA HUBNER
SYN. SEE TESTATA LINNEAUS

ACHEMON DRURY, (PHOLOS)-SPHINGIDAE
SYN. CRANTOR CRAMER
LIFE HISTORY
AMER. ENT. 2:54, 1869 LARVA* PUPA*
AMER. JOURN. SCI., 36:300, 1839 HARRIS LARVA, PUPA
AMER. MUS. NAT. HIST., BULL. 7: 290, 1895 BEUTEN. LARVA, PUPA
CAN. ENT. 10:101, 1878 SAUNDERS LARVA* PUPA*
ENT. SOC. AMERICA, ANNALS 11:433 1918 MOSHER PUPAL STRUCTURE
ILLINOIS DEPT. AGRI., TRANS. 15: 165, 1877 FRENCH LARVA
LOC. CIT. 18: APPEND: 101, 182, 1880 COQUILLETT, MARTIN LARVA*
MINNESOTA AGRI. EXP. STA., BULL. 61:93, 1899 LUGGER LARVA* PL. 7, FF. 38, 39; PUPA* F. 38
MISSOURI INS., REPT. 2:74-76, 1870 RILEY LARVA* F. 49
NEW YORK STATE AGRI. EXP. STA. (CORNELL), MEM. 274:201, 1948 FORBES LARVA
NEW YORK STATE CABIN. NAT. HIST. ANN. REPT. NO. 24:117, 1871 LINTNER LARVA, PUPA
NEW YORK STATE MUS., MEM. 8: 739, 1905 FELT BRIEF LARVAL DESCR.
NOVITATES ZOOLOGICAE (TRING) 9 SUPPLEMENT P. 490, 1903 ROTHSCHILD & JORDAN LARVA
ONTARIO ENT. SOC., ANN. REPT. P. 11, 1872 SAUNDERS LARVA* PUPA*
LOC. CIT. P. 73, 1879 SAUNDERS LARVA* PUPA*
LOC. CIT. P. 58, 1881 REED LARVA* PUPA*
LOC. CIT. P. 27, 1888 FYLES LARVA*

LOC.CIT.P.25, 1901
YOUNG LARVA*
PAPILIO 2:147, 1882
SAUNDERS LIFE HISTORY
PHILADELPHIA ACAD. NAT. SCI.,
JOURN.5:155, 1859
CLEMENS LARVA
PHILADELPHIA ENT. SOC., PROC.3:
660, 1864 LINTNER LARVA, PUPA
SMITHSONIAN MISC. COLLECTIONS
4:ART.2:178, 1862
MORRIS LARVA
UNITED STATES DEPT. AGRI.,
FARMER'S BULL.70:17, 1898
MARLATT LARVA* F.10 C,D;
PUPA* F.10 E
LOC.CIT.,BULL.1220:34, 1921
QUAINTANCE LARVA* F.35
LOC.CIT.,BULL.1942:21, 1893
DEMAREE-RUNNER LARVA* F.20
UNITED STATES DEPT. AGRI. MISC.
PUB.657:359, 1950
CRAIGHEAD LARVA
UNITED STATES DEPT. AGRI. YEAR-
BOOK P.399, 1895 MARLATT
LARVA* F.103
VERMONT BOARD AGRI.,REPT.5:
279, 1878 PERKINS LARVA
ABBOT & SMITH-LEP. INS. GEORGIA
I:81, 1797 LARVA* PL.41
COLOR
BOISDUVAL-SPEC. GEN. LEP. SPHING.
P.200, 1874 LARVA
CRAGEN-OUR INS. FRIENDS & FOES
P.142, 1899 LARVA* F.77
ELIOT & SOULE-CATERPILLARS &
THEIR MOTHS P.102, 1902
EGG, LARVA*, PUPA*
ESSIG-INJ. & BENEF. INS.,P.384
1915 LARVA, PUPA* F.383
ESSIG-INS. WEST. NORTH AMERICA
P.665, 1926 EGG* F.536;
LARVA* F.537; PUPA* F.538
FERNALD-SPHINGIDAE NEW ENGLAND
P.61, 1886 LARVA* PUPA*
GLOVER-ILLUS. NORTH AMER. ENT.,
1878 LARVA* PL.2,F.20;
PL.5,F.5
HARRIS-ENT. CORRESP. P.283, 1869
LARVA* PL.111,F.11
HARRIS-INS. INJ. VEG. 1ST EDIT:
228, 1841 LARVA, PUPA
LOC.CIT. FLINT EDIT:325, 1862
LARVA* PUPA* F.150
HOLLAND-MOTH BOOK P.66, 1905
LARVA* F.28
KELLOGG-AMER. INS.,P.434, 1908
LARVA* FF.619,623
MORRIS-SYNOP. LEP. NORTH AMERICA
P.178, 1862 LARVA
O'KANE-INJ. INS. P.273, 1912
LARVA* F.387
PEAIRS-PESTS FARM, GARDEN,
ORCHARD P.328, 1941
EGG* LARVA* PUPA* F.391
QUAYLE-CITRUS & SUBTROPICAL
FRUITS P.370, 1938
EGG, LARVA* F.263
SAUNDERS-INS. INJ. FRUITS P.251,
1883 LARVA* PUPA*
LOC.CIT.P.251, 1889
LARVA* F.260; PUPA* F.261

- - -

INDEX AMER. ECON. ENT.,
VOL.1:241; VOL.2:275;
VOL.3:309; VOL.4:358;
VOL.6:582; VOL.7:769

FOOD PLANTS

PARTHENOCISSUS QUINQUEFOLIA
(LINN.) (VIRGINIA CREEPER)
VITIS SP. (GRAPE)
VITIS VINIFERA LINN.

ACIS (DRURY), STRYMON-LYCAENIDAE
SYN. MARS FABRICIUS
A BARTRAMI COMSTOCK & HUNTINGTON

LIFE HISTORY

CAN. ENT.79:144, 1947
CHERMOCK & CHERMOCK EGG,
LARVA* F.E; CHRYSALIS*
FF.C,D
COMSTOCK, J.A.-BUTTER. CALIF.
P.188, 1927 EGG* (A,62);
LARVA

FOOD PLANTS

CROTON LINEARIS MICHX.

<u>ACMON</u>(DOUBLEDAY & HEWITSON),
<u>PLEBEIUS</u>-LYCAENIDAE
SYN. <u>ANTAEGON</u> BOISDUVAL
AB <u>KELSEYI</u> WRIGHT
A <u>COTTLEI</u> GRINNELL
AB <u>LABECULA</u> WATSON & COMSTOCK
LIFE HISTORY
CALIFORNIA ACAD. SCI., PROC. 7: 20, 1876 HY. EDW. LARVA (<u>ANTAEGON</u>)

- - -

INDEX AMER. ECON. ENT. VOL. 5:501

FOOD PLANTS
ASTRAGALUS SP.
ASTRAGALUS ALLOCHROUS GRAY
ERIOGONUM SP.
ERIOGONUM FASCICULATUM BENTH.
LOTUS SP.
LOTUS GRANDIFLORUS (BENTH.)
LOTUS MICRANTHUS BENTH.

<u>ACONTIOIDES</u>(GUENEE), <u>MELIPOTIS</u>-NOCTUIDAE
SYN. <u>SINUALIS</u> HARVEY
LIFE HISTORY

FOOD PLANTS
DELONIX REGIA (BOJ.)

<u>ACREA</u>(DRURY), <u>ESTIGMENE</u>-ARCTIIDAE
SYN. SEE MCDUNNOUGH'S LIST
A <u>DUBIA</u> WALKER
B <u>ARIZONENSIS</u> ROTHSCHILD
LIFE HISTORY
AMER. MUS. NAT. HIST., BULL. 10; 373, 1898 BEUTEN. LARVA
CALIFORNIA ACAD. SCI., PROC. 5: 368, 1873 HY. EDW. LARVA
CALIFORNIA AGRI. EXP. STA., BULL. 660:36, 1942 SMITH LARVA* F. 26, A, B
CAN. JOURN. INDUSTRY & ART, 8: 363, 1863 SAUNDERS LARVA
ENT. NEWS 6:186, 1895 RICKER, LARVA (<u>RICKERSECKERI</u>)
LOC. CIT. 9:203, 1898 TITUS EGG
ILLINOIS AGRI. EXP. STA., BULL. 60:505, 1900 FORBES LARVAL HEAD MARKINGS* F. 84, A, B
ILLINOIS DEPT. AGRI., TRANS. 15: 79, 183, 1877 FRENCH-THOMAS LARVA
LOC. CIT. 18:APPEND:115, 170, 1880 COQUILLETT-MARTIN LARVA
JOURN. ECON. ENT. 24:838, 1931 STRACHNER EGG* PL. 33, FF. 2, 3; LARVA* PL. 33, F. 4
MASSACHUSETTS AGRI. REPOSITORY P. 322, 1823 HARRIS LARVA* PUPA*
MUS. COMP. ZOOL., BULL. 87: NO. 6, 1941 DETHIER LARVAL ANTENNA* FF. 88, 92
NATURALISTE CANADIEN P. 19, 1929 HUARD LARVA
NEW YORK ACAD. SCI., ANNALS 8:74 1893 PACKARD EGGS, LARVA
NEW YORK ENT. SOC., JOURN. 38:45, 1930 STEWART EGG
NOVA SCOTIA ENT. SOC., 4:24, 1918 PAYNE LIFE HISTORY* PL. 1 (ALL STAGES)
ONTARIO ENT. SOC., ANN. REPT. P. 100 1896 FYLES LARVA* PUPA*
LOC. CIT. P. 113, 1905 GIBSON LARVA*
OTTAWA NATURALIST 19:160, 1905 GIBSON EGG* LARVA*
SMITHSONIAN MISC. COLLECTIONS 4:ART. 2:342, 1862 MORRIS LARVA
UNITED STATES DEPT. AGRI. BUR. ENT., BULL. NEW SERIES 43:43 1903 CHITTENDEN EGG* F. 41, E; LARVA* F. 41, B, C, D
LOC. CIT. 44:80, 1904 HINDS EGG* F. 19E; LARVA* F. 19 B, C, D; PUPA
LOC. CIT. 57:34, 1906 SANDERSON EGG* F. 18E; LARVA* F. 18, B, C, D
UNITED STATES DEPT. AGRI. FARMER'S BULL. 223:14, 1905 SANDERSON EGG* F. 12 E; LARVA* F. 12 B, C, D

Loc.cit.890:15, 1917
PIERCE EGG* F.16 E;
LARVA* F.16 B,C,D
Loc.cit.1688:20, 1932
FOLSOM EGG* F.26 E;
LARVA* F.26 B,C,D
ABBOT & SMITH-LEP. INS. GEORGIA
1797 PL.67 LARVA* COLOR
BALLARD-INS. LIVES P.57, 1879
LARVA* PUPA*
CRAGEN-OUR INS. FRIENDS & FOES
P.151, 1899 LARVA
CROSBY & LEONARD-MANUAL VEG.
INS.P.359, 1918
EGG, LARVA, PUPA
DUNCAN-NATURALIST'S LIBRARY 22:
172, 1858 LARVA* PL.20,F.2;
PUPA* PL.20,F.3 COLOR
ELIOT & SOULE-CATERPILLARS &
THEIR MOTHS P.291, 1902 LARVA*
EMMONS-NATURAL HISTORY NEW
YORK 5:226, 1854 LARVA
ESSIG-INS. WEST. NORTH AMERICA
P.677, 1926 EGG& F.548;
LARVA* F.549
FERNALD-APPLIED ENTOM.P.280
1926 LARVA
FERNALD & SHEPARD-APPLIED
ENTOM.P.277, 1942 LARVA
GLOVER-ILLUS. NORTH AMER. ENT.,
1878 LARVA* PL.8,FF.4,13,14
HARRIS-INS. INJ. VEGET. 1ST EDIT.,
P.249, 1841 LARVA* PUPA
Loc.cit.2ND.EDIT.P.269, 1852
LARVA, PUPA
Loc.cit.FLINT EDIT.P.354, 1862
LARVA
HART-INS. COLLECTIONS ILLINOIS
HIGH SCHOOLS P.21, 1903
LARVA* PUPA
MORRIS-SYNOP. LEP. NORTH AMERICA
P.342, 1862 LARVA
O'KANE-INJ. INS.P.182, 1912
LARVA* F.217
OSBORN-MEADOW AND PASTURE INS.
P.219, 1939 LARVA
STRETCH-ILLUST. ZYGAEN. &BOMBY.
NORTH AMERICA P.100, 1873
LARVA* PL.10,F.6 COLOR
TENNY-MANUAL NAT. HIST.
P.398, 1867 LARVA* PUPA*
PL. 294 (BOTH)

\- - -

INDEX AMER. ECON. ENT.,
VOL.1:123; VOL.2:138;
VOL.3:153; VOL.4:176;
VOL.5:235; VOL.6:276;
VOL.7:364

FOOD PLANTS

ALLIUM CEPA LINN. (ONION)
AMARANTHUS SPINOSUS LINN.
ARTEMISIA STELLERIANA BES.
ASPARAGUS OFFICINALIS LINN. (ASPARAGUS)
BETA VULGARIS LINN. (BEET)
BRASSICA OLERACEA LINN. (CABBAGE)
BRASSICA OLERACEA V. BOTYRTIS (CAULIFLOWER)
CANNA SP.
CASSIA TORA LINN.
DAHLIA VARIABILIS DESF. (DAHLIA)
EUPATORIUM CAPILLIFOLIUM(LAM.)
GLYCINE MAX (LINN.)
GOSSYPIUM HERBACEUM LINN. (COTTON)
HELIANTHUS SP.
HYDRANGEA SP.
LACTUCA SATIVA LINN. (LETTUCE)
MALUS PUMILA MILL. (APPLE)
MELILOTUS SP.
MELILOTUS ALBA DESR.
NASTURTIUM SP.
NICOTIANA TABACUM LINN. (TOBACCO)
PARTHENIUM ARGENTATUM GRAY
PHASEOLUS VULGARIS LINN. (BEAN)
PHYSALIS ANGULATA V. LINKIANA NEES
PISUM SATIVUM LINN. (PEAS)
PLANTAGO SP.
POLYGONUM AMPHIBIUM LINN.
PRUNUS DOMESTICA LINN. (PLUM)
RUMEX SP.
SENECIO SP.
SOLANUM TUBEROSUM LINN. (POTATO)
TRIFOLIUM SP. (CLOVERS)
VACCINIUM SP. (HUCKLEBERRY)
VERNONIA NOVEBORACENSIS(LINN.)

ZEA MAYS LINN. (CORN)
ALSO MANY LOW PLANTS

ACRONYCTOIDES(WALKER), PANTHEA-NOCTUIDAE
SYN. LEUCOMELANA MORRISON
A ALBOSUFFUSA MCDUNNOUGH
LIFE HISTORY
CAN. ENT. 74:8, 1942
BROWN-MCGUFFIN LARVA*
INCLUDING STRUCTURES
- - -
INDEX AMER. ECON. ENT., VOL. 7:738
FOOD PLANTS
ABIES BALSAMEA(LINN.)(FIR)
PICEA SP. (SPRUCES)

ACUTALIS WALKER
SEE CALLITRICHOIDES GROTE

ACUTARIA HERRICH-SCHAEFFER
SYN. SEE BISTRIALIS GEYER

ACUTIFRONS SMITH
SSP. SEE TESSELLATA HARRIS

ADAPA STRAND
SSP. SEE BIMACULATA STEPHENS

ADENOSTOMATIS(HENRY EDWARDS), STRYMON-LYCAENIDAE
LIFE HISTORY
SOUTHERN CALIFORNIA ACAD. SCI., BULL. 34:211, 1935 COMSTOCK & DAMMERS LARVA, PUPA* PL. 52
LOC. CIT. 36:19, 1937 COMSTOCK EGG* PL. 6
FOOD PLANTS
ADENOSTOMA SP.
CERCOCARPUS BETULOIDES NUTT. EX. T. &G.

ADIANTE BOISDUVAL
SYN. SEE ADIASTE BEHR

ADIASTE BEHR, ARGYNNIS-NYMPHALIDAE
SYN. ADIANTE BOISDUVAL
A CLEMENCEI COMSTOCK
B ATOSSA EDWARDS
FORM TEJONICA COMSTOCK
C SEMIRAMIS EDWARDS
AB BERNARDENSIS GUNDER
LIFE HISTORY
SOUTHERN CALIFORNIA ACAD. SCI., BULL. 30:42,44, 1931 COMSTOCK & DAMMERS LARVA ATOSSA P. 42; LARVA SEMIRAMIS P. 44
FOOD PLANTS
VIOLA SP. (VIOLETS)

ADJUNCTA(BOISDUVAL), POLIA-NOCTUIDAE
FORM BENJAMINI LEMMER
LIFE HISTORY
AMER. MUS. NAT. HIST., BULL. 14: 296, 1901 BEUTEN. LARVA
CAN. ENT. 19:61, 1877 GODELL LARVA, PUPA
ILLINOIS DEPT. AGRI., TRANS. 18: APPEND. 136, 1880 MARTIN LARVA
NATURALISTE CANADIEN P. 7, 1936 DAVIAULT PUPA
NEW YORK STATE CABIN. NAT. HIST. NO. 26:161, 1874 LINTNER LARVA
ROYAL SOC. OF CANADA, TRANS. I SERIES 3:59, 1907 FLETCHER-GIBSON EGG, LARVA, PUPA
UNITED STATES NAT. MUS., PROC. 27:852, 1904 DYAR EGG
FOOD PLANTS
APIUM YRAVEOLANS VAR, RAPACEUM (MILL.)
ASPARAGUS OFFICINALIS LINN. (ASPARAGUS)
EUPATORIUM RUGOSUM HOUTT.
FRAGARIA VIRGINIANA DUSCH.
PHLOX SP.
PRUNUS EMARGINATA (DOUGL.)
PTERIDIUM AQUILINUM V. LATIUSCULUM (DESV.)
SOLIDAGO ALTISSIMA LINN.
WEIGELA FLORIDA (S & Z.)

ADJUTRIX SCUDDER
SSP. SEE LACINIA GEYER

ADMIRALIS RETZIUS
SYN. SEE ATALANTA LINN.

ADNATA BOISDUVAL
SYN. SEE CALIFORNIAE WALKER, LEPTARCTIA

ADNIXA (GROTE), ASEPTIS-NOCTUIDAE
LIFE HISTORY
ENT. SOC. WASHINGTON, PROC. 4: 321, 1898 DYAR LARVA
FOOD PLANTS
PRUNUS SEROTINA EHRH.

ADONEA GROTE
SYN. SEE INSUETA GUENEE

ADOPTIVA GROTE
SYN. SEE DELILAH STRECKER

ADORNATA PROUT
SSP. SEE ENUCLEATA GUENEE

ADRIANA HENRY EDWARDS
SSP. SEE ULTRONIA HUBNER

ADUMBRATA (BEHR), SYNEDOIDA-NOCTUIDAE
A SAXEA HENRY EDWARDS
B ALENI GROTE
LIFE HISTORY
JOURN. ECON. ENT. 22:137, 1929 PHIPPS EGG, LARVA, PUPA
FOOD PLANTS
VACINIUM SP. (BLUEBERRIES)
VACCINIUM ANGUSTIFOLIUM AIT.
"OTHER PLANTS"

ADUSTA BARNES & MCDUNNOUGH
SSP. SEE HENRIETTA SMITH

ADUSTARIA WALKER
SYN. SEE DUARIA GUENEE

ADVENA ELLSWORTH
SSP. SEE ARCHIPPUS CRAMER

ADVENA PACKARD
SSP. SEE WALKERI FELDER & FELDER

AEGROTATA GUENEE
SSP. SEE CABERATA GUENEE

AEHAJA BEHR
SSP. SEE SAEPIOLUS BOISDUVAL

AELIARIA WALKER
SYN. SEE QUERCIVORARIA GUENEE

AEMILEA SKINNER
SYN. SEE NEVADA SCUDDER (THORYBES)

AEMULA (HUBNER), EPIZEUXIS-NOCTUIDAE
SYN. UNDULALIS STEPHENS
SYN. MOLLIFERA WALKER
SYN. HERMINOIDES WALKER
SYN. CONCISA WALKER
LIFE HISTORY
AMER. NAT. 4:229, 1871 LARVA* PL. 2, F. 10A; PUPA* PL. 2, F. 10B
ENTOMOLOGICA AMERICANA 14:159, 1934 CRUMB LARVA* PL. 24
INS. LIFE 4:111, 1891 RILEY LARVA*
NEW YORK STATE MUS., MEM. 8: 754, 1905 FELT LARVA (BRIEF)
UNITED STATES DEPT. AGRI. MISC. PUB. 657:395 CRAIGHEAD LARVA
UNITED STATES DEPT. AGRI. REPT. OF ENTOMOLOGIST P. 326, 1886 PACK LARVA, PUPA
HART-INS. COLLECTION ILLINOIS HIGH SCHOOLS P. 44, 1903 PUPA
HOLLAND-MOTH BOOK P. 280, 1905 LARVA* F. 169
- - -
INDEX AMER. ECON. ENT., VOL. 1:120; VOL. 4:170; VOL. 7:359
FOOD PLANTS
CORN FODDER
DEAD LEAVES, DEAD SPRUCE NEEDLES
PHLOX SPECIOSA PURSH
PICEA PUNGENS ENGELM.
PSEUDOTSUGA MENZIESII (MIRB)

AEMULATARIA(WALKER),PHILOBIA-GEOMETRIDAE
SYN. SECTOMACULATA MORRISON
SYN. ENOTATA GUENEE
LIFE HISTORY
GUMPPENBERG-SYSTEMA GEOMETRARUM 7:435, 1895 LARVA PUPA
(ENOTATA)
PACKARD-MONOGRAPH GEOMETRIDAE p.564, 1876, LARVA* PL.13, F.24; PUPA* PL.13,F.24A
FOOD PLANTS
LACTUCA GRAMINIFOLIA MICHX.
ROBINIA PSEUDO-ACACIA LINN.

AENEIPENNIS GROTE,AGROTIS-NOCTUIDAE
SYN. CLODIANA GROTE
FORM DENTILINEA SMITH
LIFE HISTORY
UNITED STATES NAT.MUS.,PROC. 27:832, 1904 DYAR EGG, LARVA
FOOD PLANTS

AEQUALIARIA GROTE,CARIPETA-GEOMETRIDAE
SYN. INTERALBICANS WARREN
LIFE HISTORY

FOOD PLANTS
PSEUDOTSUGA MENZIESII(MIRB.)

AEQUALIARIA WALKER
SYN.SEE ATHASARIA WALKER

AEQUALIS HARVEY
SSP.SEE WILSONI GROTE,(EUXOA)

AEQUILINEA SMITH
SYN.SEE PLENOECTUSA GROTE

AERATA(FABRICIUS),SYNCHLORA-GEOMETRIDAE
FOR MANY SYN. OF THIS SP. SEE McDUNNOUGH'S LIST
LIFE HISTORY
AMER.ENT.2:205, 1870 LARVA* PUPA
ILLINOIS DEPT.AGRI.,TRANS.15: 238, 1877 FRENCH LARVA
MINNESOTA AGRI.EXP.STA., BULL.61:235, 1899 LUGGER LARVA* F.183
NEW YORK STATE AGRI.EXP.STA., (CORNELL),MEM.274:115, 1948 FORBES LARVA
NEW YORK STATE CABIN.NAT.HIST. ANN.REPT.NO.45:131, 1891 LINTNER LARVA*
OHIO AGRI.EXP.STA.,BULL.45: 178, 1893 WEBSTER LARVA
ONTARIO ENT.SOC.,ANN.REPT. P.15, 1873 SAUNDERS LARVA*
LOC.CIT.P.75, 1903 LOCKHEAD LARVA*
PSYCHE 9:93, 1900 DYAR EGG, LARVA (SYN. GLAUCARIA)
UNITED STATES DEPT.AGRI., MISC.PUB.657:361,424, 1950 CRAIGHEAD LARVA
CRAGEN-OUR INS.FRIENDS & FOES P.157, 1899 LARVA
GUMPPENBERG-SYSTEMA GEOMETRARUM 7:500, 1895 LARVA, PUPA
O'KANE-INJ.INS.P.327, 1912 LARVA* F.508
PACKARD-MONOGRAPH GEOMETRIDAE P.382, 1876 LARVA
RILEY-REPT.NOX.INS.MISSOURI 1:139, 1869 LARVA
SAUNDERS-INS.INJ.FRUITS P.316, 1883 LARVA* PUPA*
LOC.CIT.P.316, 1889 PUPA* F.330 1; RUBIVORA RILEY (SYN.)
- - -
INDEX AMER.ECON.ENT., VOL.1:298 AERATA & RUBIVORA
FOOD PLANTS
ASTER SP. (FLOWERS)
COMPOSITAE SP.
MALUS PUMILA MILL. (APPLE)
RUBUS ALLEGHENIENSIS PORTER
RUBUS IDAEUS V. STRIGOSUS (MICHX.) (BLACKBERRY)
SALIX SP. (WILLOWS)
DECAYING FRUITS

AERATA LYMAN
SYN.SEE LIMPIDA GUENEE

AEREOFUSCUS GUNDER
SSP. SEE CELLUS BOISDUVAL & LECONTE

AEREOIDES GROTE, PLUSIA-NOCTUIDAE
LIFE HISTORY
NEW YORK STATE CABIN. NAT. HIST. ANN. REPT. NO. 26:164, 1874 LINTNER LARVA BRIEF DESCR.
PSYCHE 1:188, 1876 THAXTER LARVA BRIEF DESCR.
FOOD PLANTS
SPIRAEA SALICIFOLIA LINN.
LOW PLANTS

AERUGINOSA (GUENEE), ZALE-NOCTUIDAE
SYN. PLENIPENNIS WALKER
SYN. NIGRIOR STRAND
LIFE HISTORY
CAN. JOURN. INDUSTRY, SCI. AND ART 10:260, 1865 BETHUNE LARVA, PUPA
GUENEE-SPEC. GEN. LEP. NOCT. 3:17, 1852 LARVA, PUPA
FOOD PLANTS
CORNUS ALTNERIFOLIA LINN.

AESCULAPIUS FABRICIUS
SYN. SEE PECKIUS KIRBY

AESIONARIA WALKER
SYN. SEE AMICARIA HERRICH-SCHAEFFER

AESTIVA EDWARDS
SYN. SEE PHAON EDWARDS

AESTIVA EDWARDS
SYN. SEE VESTA EDWARDS

AESTIVA EDWARDS
SYN. SEE NAPI LINNEAUS

AESTIVUS VERITY
SYN. SEE RAPE LINNEAUS

AETHRA STRECKER
SSP. SEE DIFFINIS BOISDUVAL

AETNA SCUDDER
SYN. SEE OTHO ABBOT & SMITH

AFFINIS GOEZE
SYN. SEE CINGULATA FABRICIUS

AFFINIS GROTE
SYN. SEE INSULATA WALKER

AFFINIS WESTWOOD
SYN. SEE ANTINYMPHA HUBNER

AFFIRMARIA WALKER
SSP. SEE HAESITATA GUENEE

AFFLICTA GROTE, ACRONICTA-NOCTUIDAE
SYN. DOLENS DRUCE
FORM SCHMALZRIEDI LEMMER
LIFE HISTORY
AMER. MUS. NAT. HIST., BULL. 14: 253, 1901 BEUTEN. LARVA
NEW YORK STATE MUS., MEM. 8: 704, 1905 FELT BRIEF LARVAL DESCR.
PAPILIO 3:17, 1883 THAXTER LARVA
UNITED STATES NAT. MUS., PROC. 21:128, 1898 SMITH & DYAR LARVA* PL. 5, FF. 1,2 COLOR & UNCOLORED; PUPA; COCOON
FOOD PLANTS
JUGLANS NIGRA LINN. (WALNUT)
QUERCUS SP. (OAKS)

AFFURATA HAMPSON
SYN. SEE COMMUNIS DYAR

AFRA EDWARDS
SYN. SEE LYGDAMUS DOUBLEDAY

AFRANIUS (LINTNER), ERYNNIS, HESPERIIDAE
LIFE HISTORY
DENISON UNIV. SCIENTIFIC LABORATORIES, JOURN. 22:111, 1927 LINDSEY LARVAL HEAD CAPSULE* F. 2B
SOUTHERN CALIFORNIA ACAD. SCI., BULL. 31:94, 1932 COMSTOCK & DAMMERS EGG* PL. 29; LARVA* AND PUPA* PL. 30
COMSTOCK, J. A.-BUTTER. CALIFORNIA P. 212, 1927 EGG, LARVA, PUPA

FOOD PLANTS
CEANOTHUS DIVARICATUS NUTT.
LOTUS AMERICANUS (NUTT.)
POPULUS SP. (POPLARS)
SALIX SP. (WILLOWS)
TRIFOLIUM SP. (CLOVERS)

AGARISTA CRAMER
SYN. SEE ODORA LINNEAUS

AGARITHE (BOISDUVAL), PHOEBIS-PIERIDAE
A MAXIMA NEUMOEGEN
‡FLORIDENSIS ROEBER
FORM ALBARITHE BROWN
LIFE HISTORY
ENT. NEWS 11:618, 1900
DYAR LIFE HISTORY
PENNSYLVANIA ACAD. SCI., PROC. 19:40, 1945 CHERMOCK EGG, LARVA, CHRYSALIS* F.1 D
BUTLER-LEPIDOPTERA EXOTICA 1874 CHRYSALIS* PL.45, F.4A
HOLLAND-BUTTERFLY BOOK P.287, 1905 LARVA, CHRYSALIS (BRIEF)
- - -
INDEX AMER. ECON. ENT., VOL.5:116 (CATOPSILIA)
FOOD PLANTS
PITHECELLOBIUM UNGUISCATI (LINN.)

AGASSIZII PACKARD
SSP. SEE MACULATA HARRIS

AGATHA BEUTENMULLER
SSP. SEE UNIJUGA WALKER

AGILIS GROTE
SYN. SEE VANCOUVERENSIS GROTE

AGREASARIA WALKER
SYN. SEE DUARIA GUENEE

AGRESTIS GROTE
SSP. SEE AUXILIARIS GROTE

AGRICOLA (BOISDUVAL), OCHLODES-HESPERIIDAE
SYN. YREKA EDWARDS
SYN. VERUS EDWARDS
SYN. MILO EDWARDS
LIFE HISTORY
SOUTHERN CALIFORNIA ACAD. SCI., BULL. 26:70, 1927 COMSTOCK EGG* LARVA*
FOOD PLANTS

AGRICOLA GROTE & ROBINSON
SYN. SEE ERECHTEA CRAMER

AGRIPPINA STRECKER, CATOCALA-NOCTUIDAE
LIFE HISTORY

FOOD PLANTS
CARYA SP. (HICKORY)

AGYLLARIA WALKER
SSP. SEE MACULARIA HARRIS

AHATON HARRIS
SYN. SEE THEMISTOCLES LATREILLE

AHOLAH STRECKER
SSP. SEE SIMILIS EDWARDS

AHOLIBAH STRECKER, CATOCALA-NOCTUIDAE
AB ELLENENSIS REIFF
FORM COLORADENSIS BEUTENMULLER
LIFE HISTORY
AMER. MUS. NAT. HIST., MEM. NEW SERIES 3, PART 1, 1918
BARNES & MCDUNNOUGH LARVA* PL.14, F.6; PL.15, FF.9,38; PL.16, F.13 ALL COLORED
ENT. NEWS 25:164, 1914
ROWLEY-BERRY EARLY STAGES
PSYCHE 20:191, 1913 BARNES & MCDUNNOUGH EGG, LARVA
ESSIG-INSECTS WESTERN NORTH AMERICA P.687, 1926 LARVA
FOOD PLANTS
QUERCUS SP. (OAKS)
QUERCUS MACROCARPUS MICHX. (BURR AND CATKINS)

AHWASHTEE FOX
SSP. SEE VIRGINIENSIS DRURY (VANESSA)

AIDEA (GUERIN-MENEVILLE), ANAEA-NYMPHALIDAE

SYN. MORRISONI HOLLAND
A FLORIDALIS JOHNSON & COMSTOCK
LIFE HISTORY
PENNSYLVANIA ACAD. SCI., PROC. 19:42, 1945 CHERMOCK EGG, LARVA, CHRYSALIS* F.1 G,H.
FOOD PLANTS
CROTON LINEARIS MICHX.

AJAX LINNEAUS SEE GLAUCUS LINNEAUS

ALABAMAE FREEMAN
SSP. SEE YUCCAE BOISDUVAL & LECONTE

ALABAMAE GROTE
SYN. SEE INFECTA OCHSENHEIMER

ALABAMAE LINDSEY
SSP. SEE DION EDWARDS

ALABASTARIA HULST
SEE ENUCLEATA GUENEE

ALAMEDA (SMITH), ARCHANARA-NOCTUIDAE
LIFE HISTORY

FOOD PLANTS
TYPHA SP.

ALASCENSIS PACKARD SSP. SEE AMERICANA HARRIS (EPICNAPTERA)

ALASCENSIS STRETCH
SYN. SEE PLANTAGINIS LINNEAUS

ALASKAE BANG-HAAS
SSP. SEE NASTES BOISDUVAL

ALASKAE CASSINO & SWETT
SSP. SEE FERRUGATA LINNEAUS

ALASKENSIS BARNES & BENJAMIN
SSP. SEE ARCUATA WALKER (DREPANA)

ALASKENSIS CHERMOCK
SSP. SEE SCUDDERII EDWARDS

ALASKENSIS GUNDER
SSP. SEE SARA BOISDUVAL

ALASKENSIS HOLLAND
SSP. SEE JUTTA HUBNER

ALASKENSIS LEHMAN
SEE FRIGGA THUNBERG

ALB STRECKER
SSP. SEE CELTIS BOISDUVAL & LECONTE

ALBA CHERMOCK
SYN. SEE PHILODICE GODART

ALBA EDWARDS
SYN. SEE ERICETORUM BOISDUVAL

ALBA GROTE
SSP. SEE RUBICUNDA FABRICIUS

ALBA STRECKER
SSP. SEE ALEXANDRA EDWARDS (COLIAS)

ALBA STRECKER
SSP. SEE EURYTHEME BOISDUVAL

ALBA STRECKER
SSP. SEE PHILODICE GODART

ALBA STRECKER
SSP. SEE LISA BOISDUVAL & LECONTE

ALBACLAUDIA FIELD
SSP. SEE CLAUDIA CRAMER

ALBANUS FELDER & FELDER
SSP. SEE EURYMEDON LUCAS

ALBARITHE BROWN
SSP. SEE AGARITHE BOISDUVAL

ALBARUFA (GROTE), ACRONICTA-NOCTUIDAE
SYN. WALKERI ANDREWS
LIFE HISTORY
AMER. MUS. NAT. HIST., BULL. 14:254, 1901 BEUTEN. LARVA
FOOD PLANTS

ALBATA GUENEE
SYN. SEE HERMINIATA GUENEE

ALBATA PACKARD, CLEMENSIA--ARCTIIDAE
SYN. ALBIDA WALKER
SYN. CANA WALKER
SYN. IRRORATA HENRY EDWARDS
FORM UMBRATA PACKARD
LIFE HISTORY
NATURALISTE CANADIEN P.275, 1929 HUARD LARVA
NEW YORK ENT. SOC., JOURN. 3:176, 1895 PACKARD LARVA (FIRST STAGE)
UNITED STATES NAT. MUS., PROC. 27:793, 1904 DYAR EGG, LARVA
FOOD PLANTS
LICHENS

ALBEQUINA HOLLAND
SYN. SEE MYRINA CRAMER

ALBERTA BOWMAN
SSP. SEE EURYTHEME BOISDUVAL

ALBERTA (EDWARDS), BRENTHIS-NYMPHALIDAE
AB BANFFENSIS GUNDER
LIFE HISTORY
EDWARDS-BUTTER. NORTH AMERICA 3RD SERIES, 1897 ARGYNNIS EGG* LARVA* PL.7, FF.A, B.
FOOD PLANTS

ALBERTA ELWES & ELWES
SSP. SEE DAURA STRECKER

ALBERTA TAYLOR
SSP. SEE INNOTATA HUFNAGEL

ALBERTAE STRAND
SSP. SEE SOLIDAGINIS HUBNER

ALBESCENS BARNES & MCDUNNOUGH
SSP. FRACTILINEA GROTE

ALBESCENS PLOETZ
SSP. SEE COMMUNIS GROTE

ALBESCENS TEPPER
SSP. SEE VASHTI STRECKER

ALBICOLLIS OBERTHUR
SYN. SEE MUZINA OBERTHUR

ALBICOMA STRECKER
SSP. SEE SCOLOPENDRINA BOISDUVAL

ALBICOSTA HUBNER
SSP. SEE ALBIFRONS ABBOT & SMITH

ALBICOSTA (SMITH), LOXAGROTIS-NOCTUIDAE
LIFE HISTORY
JOURN. ECON. ENT. 41:631, 1948 HOERNER EGG* LARVA* PUPA* (ALL F. 1)
FOOD PLANTS
PHASEOLUS VULGARIS LINN. (BEAN)

ALBICOSTA (WALKER), EUCHAETIAS-ARCTIIDAE
SYN. FUMIDUS HY. EDW.
LIFE HISTORY
AMER. MUS. NAT. HIST., BULL. 5:92, 1893 BEUTEN. COCOON
WASHINGTON (D.C.) ENT. SOC., PROC. 13:229, 1911 DYAR LARVA
FOOD PLANTS

ALBIDA BEUTENMULLER
SYN. SEE BRISEIS EDWARDS

ALBIDA CHERMOCK
SSP. SEE PHILODICA GODART

ALBIDA WALKER
SYN. SEE ALBATA PACKARD

ALBIDIOR STRAND
SSP. SEE POPULI STRECKER

ALBIFASCIA FRENCH
SYN. SEE CALIFORNIAE WALKER

ALBIFASCIATA (PACKARD), HYDRIOMENA GEOMETRIDAE
FORM RESECTA SWETT
FORM PUNCTICAUDATA BARNES & MCDUNNOUGH
A VICTORIA BARNES & MCDUNNOUGH
B REFLATA GROTE
SYN. ABACTA HULST

LIFE HISTORY
GUMPPENBERG-SYSTEMA GEOMETRARUM 3:453, 1890 LARVA PUPA
FOOD PLANTS
ACER SP. (MAPLES)

ALBIFERA(WALKER), HYDRELIA-GEOMETRIDAE
SYN. ALBOGILVARIA MORRISON
SYN. TRISERIATA PACKARD
LIFE HISTORY

FOOD PLANTS
ULMUS SP. (ELMS)

ALBIFRONS (ABBOT & SMITH), SYMMERISTA-NOTODONTIDAE
A ALBICOSTA HUBNER
B SUAVIS BARNES
LIFE HISTORY
AMER. ENT. 1:39, 1868 LARVA (BRIEF DESCR.)
AMER. MUS. NAT. HIST., BULL. 10: 418, 1898 BEUTEN. LARVA
AMER. NATURALIST 19:715, 1885 PACKARD PUPA (BRIEF)
BOSTON SOC. NAT. HIST., PROC. 24: 526, 1890 PACKARD LARVA
CAN. ENT. 78:100, 1946 FRANCELMONT LARVA
ENTOMOLOGICA AMERICANA 6:75, 1890 BEUTEN. EGG, LARVA, COCOON
ILLINOIS DEPT. AGRI., TRANS. 15; 191, 1877 FRENCH LARVA
LOC. CIT. 18:APPEND:120, 1880 MARTIN LARVA
MAINE AGRI. EXP. STA., BULL. 259: 68, 1917 MOSHER LARVA; PUPA* F. 6 A, B
LOC. CIT. 162:354, 1908 BATCH LARVA
MICHIGAN AGRI. EXP. STA., BULL. 243:23, 1933 MCDANIEL LARVA* F. 19 (GOOD)
MICHIGAN STATE COLLEGE, SPEC. BULL. 243:22, 1933 MCDANIEL LIFE HISTORY, LARVA*
MINNESOTA AGRI. EXP. STA., BULL. 61:187, 1899 LUGGER LARVA* F. 134, NO. 4 NO DESCR.
MUS. COMP. ZOOL., BULL. 87: NO. 6, 1941 DETHIER LARVA* F. 108 (ANTENNA ONLY)
NAT. ACAD. SCI., MEM. 7:180, 1895 PACKARD EGG; LARVA* PL. 22, FF. 1-4 COLOR; PUPA*
LOC. CIT. 9:179, 1905 PACKARD LARVA* PL. 44, F. 9 (HEAD)
NEW HAMPSHIRE AGRI. EXP. STA., BULL. 59:206, 1898 WEED LARVA* PUPA*
NEW YORK AGRI. EXP. STA. (CORNELL), MEM. 274:224, 1948 FORBES LARVA
NEW YORK STATE MUS., MEM. 8:519, 1905 FELT EGG, LARVA
OHIO AGRI. EXP. STA., BULL. 332: 253, 1918 HOUSER LARVA* PL. 38, F. 3
ONTARIO ENT. SOC., ANN. REPT. P. 137, 1905 GIBSON LARVA
PSYCHE 6:147, 1891 DYAR LARVAL HEAD SIZES (ALBICOSTA)
UNITED STATES DEPT. AGRI., MISC. PUB. 657:402, 403, 1950 CRAIGHEAD LARVA* F. 81A (ALBICOSTA)
ABBOT & SMITH-LEP. INS. GEORGIA P. 159, 1797 LARVA* PUPA* PL. 80 COLOR
CRAGEN-OUR INS. FRIENDS & FOES P. 161, 1899 LARVA
DOANE-FOREST INS., P. 295, 1936 LARVA
EMMONS-NAT. HIST. NEW YORK 5:242, 1854 LARVA* PUPA* PL. 37
GLOVER-ILLUS. NORTH AMER. ENT., 1878 LARVA* PL. 6, F. 10
HARRIS-ENTOMOLOGICAL CORRESP. P. 304, 1869 LARVA
HERRICK-INSECT ENEMIES SHADE TREES P. 226, 1935 LARVA
SMITH, J. B.-INSECTS NEW JERSEY P. 453, 1900 LARVA* F. 207

- - -

INDEX AMER. ECON. ENT.,
VOL. 1:297; VOL. 2:355; VOL. 4:472;
VOL. 5:630; VOL. 6:747; VOL. 7:982.

FOOD PLANTS

ACER SP. (MAPLES)
BETULA PAPYRIFERA MARSH.
FAGUS SP. (BEECH)
QUERCUS SP. (OAK)
QUERCUS ALBA LINN.
QUERCUS COCCINEA MUENCH.
QUERCUS MACROCARPA MICHX.
TILIA SP. (LINDENS)
ULMUS SP. (ELMS)

ALBIFUSA WALKER
SSP. SEE TRIFOLII ROTTENBERG

ALBIGENARIA WALKER
SYN. SEE DEFECTARIA GUENEE

ALBILINEA (HUBNER), PROTOLEUCANIA
NOCTUIDAE

SYN. MODERATA WALKER
SYN. HARVEYI GROTE
A DIFFUSA WALKER
B OBSCURIOR SMITH
C NEPTIS SMITH
D LIMITATA SMITH

LIFE HISTORY

AMER. MUS. NAT. HIST., BULL. 16: 445, 1902 BEUTEN. LARVA
BROOKLYN ENT. SOC., BULL. 22:47, 1927 CRUMB MATURE LARVA* PL. 4, E; PL. 5, MP.
ILLINOIS DEPT. AGRI., TRANS. 15: 223, 1877 FRENCH LARVA
LOC. CIT. 18: APPEND. 186, 1880 COQUILLET LARVA
IOWA AGRI. EXP. STA., BULL. 23: 882, 1893 OSBORN-SURRINE EGG* LARVA* (BOTH F. 1)
LOC. CIT. BULL. 122:337, 1911 WEBSTER EGG* F. 8; LARVA* FF. 9, 13, 14; PUPA* F. 4
MINNESOTA AGRI. EXP. STA., BULL. 123, 1911 WASHBURN LARVA* FF. 1, 2 COLOR DIFFUSA
MISSOURI, REPT. NOX. INS. 9:50, 1877 RILEY LIFE HIST. FF.
NEBRASKA AGRI. EXP. STA., RESEARCH BULL. 81:24, 1935 WHELAN LARVA TECH. DESCR.
ONTARIO ENT. SOC., ANN. REPT. P. 101, 1896 FYLES EGG* LARVA*
HART-INSECT COLLECTIONS FOR ILLINOIS HIGH SCHOOLS P. 36, 1903 EGG* LARVA* F. 50
HOLLAND-MOTH BOOK P. 202, 1905 EGG* LARVA* F. 195
O'KANE-INJURIOUS INSECTS P. 214, 1912 LARVA* F. 265
PEAIRS-PESTS FARM, GARDEN, ORCHARD P. 106, 1941 LARVA
SMITH-INSECTS NEW JERSEY P. 418, 1900 EGG* LARVA* F. 195

\- - -

INDEX AMER. ECON. ENT.
VOL. 1:149 HELIOPHILA DIFFUSA AND LIMITADA
VOL. 1:205 MELIANA ALBILINEA
VOL. 2:231 MELIANA ALBILINEA AND DIFFUSA
VOL. 2:249 NELEUCANIA ALBILINREA, LIMITADA
VOL. 3:256 MELIANA ALBILINEA
VOL. 5:444 NELEUCANIA ALBILINEA
VOL. 7:817

FOOD PLANTS

AGROPYRON TRACHYCAULUM VAR. GLAUCUM P. & M.
ANDROPOGON GERARDI VITMAN
AVENA SATIVA LINN. (OATS)
CALAMAGROSTIS CANADENSIS (MICHX.)
CONVOLVULUS SP.
DACTYLIS GLOMERATA LINN.
ECHINOCHLOA CRUS-GALLI (LINN.)
ELYMUS CANADENSIS LINN.
GRAMINEAE SP.
HORDEUM JUBATUM LINN.
HORDEUM VULGARE LINN. (BARLEY)
ORYZA SATIVA LINN. (RICE)
PHLEUM PRACTENSE LINN. (TIMOTHY)
POLYGONUM CONVOLVULUS LINN.
SECALE CEREALE LINN. (RYE)
SETARIA GLAUCA (LINN.)
SETARIA VIRIDIS (LINN.)
SORGHASTRUM NUTANS (LINN.)

SORGHUM BICOLOR (LINN.)(SORGHUM)
TRIFOLIUM SP. (CLOVERS)
TRITICUM AESTIVUM LINN. (WHEAT)
TRITICUM SPELTA LINN.
ZEA MAYS LINN. (CORN)
- - -
CEREALS
GRASSES

ALBIMONTANA AVINOFF
SSP. SEE HARRISII SCUDDER

ALBINA MCDUNNOUGH
SSP. SEE CONNECTA GROTE

ALBINA STRECKER
SYN. SEE DEDUCTA MORRISON

ALBINIC SKINNER
SYN. SEE PHILODICE GODART

ALBIPLAGA WALKER
SYN. SEE GUTTIVITTA WALKER

ALBIPUNCTA BARNES & BENJAMIN
SSP. SEE RANUNCULI HENRY EDWARDS

ALBIPUNCTA SMITH
SSP. SEE VIDENS GUENEE

ALBIPUNCTATA(HAWORTH), EUPITHECIA-GEOMETRIDAE
SYN. TRIPUNCTARIA HERRICH-SCHAEFFER
LIFE HISTORY
NEW YORK AGRI. EXP. STA. (CORNELL), MEM. 274:165, 1948 FORBES LARVA (TRIPUNCTARIA)
FOOD PLANTS
UMBELLIFEREA
VIBURNUM SP.

ALBIRADIATA GUNDER
SSP. SEE NUBIGENA BEHR.

ALBISTRIGATA BARNES & MCDUNNOUGH
SSP. MURINA STRETCH

ALBOCENTRATA BRYK & EISNER
SYN. SEE CLODIUS MENETRIES

ALBOFASCIA HENRY EDWARDS
SSP. SEE SEPTENTRIONALIS WALKER GLUPHISIA

ALBOFASCIATA BETHUNE
SSP. SEE MINEREA GUENEE

ALBOFASCIATA NEWCOMB
SSP. SEE ARTHEMIS DRURY

ALBOFASCIATA SAGEDER
SSP. SEE CECROPIA LINNEAUS

ALBOGILVARIA MORRISON
SYN. SEE ALBIFERA WALKER

ALBOGUTTATA GROTE
SSP. SEE ILLAUDABILIS GROTE

ALBOLINEATA PACKARD
SSP. SEE SILACEATA SCHIFFERMULLER

ALBOLINEATA PACKARD
SYN. SEE AERATA FABRICIUS

ALBOMACULA BUTLER
SYN. SEE ILIA CRAMER

ALBOMACULATA STOLL
SSP. SEE OCTOMACULATA FABRICIUS

ALBOMARGINATA CASSINO
SSP. SEE LACRYMOSA GUENEE

ALBOPUNCTATA MORRISON
SYN. SEE DIVISATA WALKER

ALBOPUNCTATA TEPPER
SSP. SEE HUMULI HARRIS(HYPENA)

ALBOSIGMA(FITCH), ICHTHYURA-NOTODONTIDAE
A SPECIFICA DYAR
LIFE HISTORY
AMER. MUS. NAT. HIST., BULL. 10: 410, 1898 BEUTEN. LARVA
NAT. ACAD. SCI., MEM. 7:139, 280, 1895 PACKARD EGG; LARVA* PL. 16, FF. 3, 4, 5 COLOR; PUPA; COCOON

New York Agri.Exp.Sta. (Cornell),Mem.274:211, 1948 Forbes larva
New York Ent.Soc., Journ.1:27, 1893 Packard larva
New York State Mus.,Mem.8:747, 1905 Felt larva (brief)
New York State,Rept.Nox.Ins. 2:274, 1855 Fitch egg, larva, cocoon
Loc.cit.5:814,64,sec.350, 1858 Fitch larva
Packard–Ins.Inj.Forest Trees p.122, 1881 larva

FOOD PLANTS
Populus tremuloides Michx.
Salix sp.

ALBOSIGNATA Packard
syn.see STELLATA Guenee

ALBOSQUAMULATA Strand
syn.see UNDULARIS Drury

ALBOSTIGMA Cockerell
ssp.see TEXANA Riley

ALBOSUFFUSA Freeman
ssp.see PYLADES Scudder

ALBOSUFFUSA McDunnough
ssp.see ACRONYCTOIDES Walker

ALBOVITTATA Grote
syn.see COMMUNIS Grote(Pyrgus)

ALBOVITTATA(Guenee),NEODEZIA–Geometridae
syn. PROPRIARIA Walker
syn. RECIPROCATA Walker
a TENUIFASCIATA Barnes & McDunnough

LIFE HISTORY

FOOD PLANTS
Impatiens sp.

ALBRIGHTI Clench
ssp.see COMYNTAS Godart

ALCESTA Smith
syn.see PALLIPENNIS Smith

ALCESTIS Edwards
ssp.see APHRODITE Fabricius

ALCIDAMUS Cramer
syn.see GLAUCUS Linneaus

ALCINA Skinner
syn.see MANATAAQUA Scudder

ALCIPHEARIA(Walker),SELENIA–Geometridae
syn. PERANGULATA Hulst
a ORNATA Barnes & McDunnough

LIFE HISTORY
New York Agri.Exp.Sta. (Cornell),Mem.274:91, 1948 Forbes larva
New York Ent.Soc., Journ.53: 198, 1945 Forbes pupa
New York State Mus.,Mem.8:728 1905 Felt larva (brief)
Psyche 8:395, 1899 Dyar egg, larva

FOOD PLANTS
Acer sp. (maples)
Alnus sp. (alders)
Crataegus sp.
Crataegus douglasii Lindl.
Salix sp. (willows)

ALCOOLARIA(Guenee),PLAGODIS–Geometridae

LIFE HISTORY
Can.Ent.82:206, 1950 McGuffin larva

FOOD PLANTS
Betula papyrifera Marsh.
Castanea dentata (Marsh.)
Fagus sp. (beech)
Quercus sp. (oaks)
Quercus rubra Linn.

ALETHE Neumoegen & Dyar
syn.see BRUCEI Henry Edwards (Ichthyura)

ALEXANDRA(Edwards),COLIAS–Pieridae
ab ALBA Strecker
syn. PALLIDA Cockerell

A EDWARDSII EDWARDS
FORM HATUI BARNES & BENJAMIN
B EMILIA EDWARDS
LIFE HISTORY
CAN. ENT. 19:226, 1887 EDWARDS EGG, LARVA, CHRYSALIS
MONTANA UNIVERSITY, BULL. 30:40, 1906 ELROD LARVA
COMSTOCK, J. A.-BUTTER. CALIF. P.54, 1927 LARVA, CHRYSALIS (EDWARDSII)
W.H. EDWARDS-BUTTER. NORTH AMERICA SUPPLEMENTARY PART 1872 EGG
HOLLAND-BUTTER. BOOK P.292, 1905 LARVA; 3RD EDIT. PP.294, 295, 1931 LARVA, CHRYSALIS
MEAD-WHELLER'S SURVEY WEST OF THE 100TH MERIDIAN, REPT. ON LEP. P.749, 1875 LARVA
FOOD PLANTS
ASTRAGALUS AMERICANUS (HOOK.)
LUPINUS PERENNIS LINN.
THERMOPSIS MONTANA NUTT.
THERMOPSIS RHOMBIFOLIA (NUTT.)
TRIFOLIUM SP. (CLOVERS)
TRIFOLIUM REPENS LINN.

ALFARATA STRECKER, CUCULLIA-NOCTUIDAE
SYN. PHILA SMITH
LIFE HISTORY
INSECUTOR INSCITIAE MENSTRUUS 11:5, 1923 DYAR LARVA
NEW YORK ACAD. SCI., ANNALS, 18: 118, 1908 SMITH LARVA (PHILA)
FOOD PLANTS
ASTER NOVAE-ANGLIAE LINN.

ALFARATTA HOLLAND
SSP. SEE HOBOMOK HARRIS

ALIA AUCT. SEE HIBISCI GUENEE
SYN. SUFFUSCA MORRISON
FORM RORULENTA SMITH
LIFE HISTORY
CAN. ENT. 23:156, 1891 DYAR LARVA, PUPA
LOC. CIT. 30:294, 1898 FYLES EGG, LARVA, PUPA
NEW YORK STATE AGRI. EXP. STA. (GENEVA), BULL. 423:385, 1916 PARROT-HODGKISS EGG* F.1; LARVA* PL.8, AND F.2
NEW YORK STATE MUS., MEM. 8: 704, 1905 FELT LARVA (BRIEF)

\- - -

INDEX AMER. ECON. ENT., VOL. 1: 300
FOOD PLANTS
AMELANCHIER CANADENSIS (LINN.)
BETULA SP. (BIRCHES)
CARYA SP. (HICKORY SP.)
MALUS PUMILA MILL. (APPLE)
QUERCUS SP. (OAKS)
SALIX SP. (WILLOWS)
SASSAFRAS ALBIDUM V. MOLLE (RAF.)

ALIAS OTTOLENGUI, AUTOGRAPHA-NOCTUIDAE
A INTERALIA OTTOLENGUI
LIFE HISTORY

FOOD PLANTS
PICEA GLAUCA (MUENCH)
PICEA MARIANA (MILL.)

ALIASKA SCUDDER
SSP. SEE MACHAON LINNEAUS

ALICIA EDWARDS
SSP. SEE CELTIS BOISDUVAL & LECONTE

ALIENARIA HERRICH-SCHAEFFER
SSP. SEE AMICARIA HERRICH-SCHAEFFER

ALLENI GROTE
SSP. SEE ADUMBRATA BEHR

ALNI HENRY EDWARDS
SSP. SEE MACULATA HARRIS

ALNIARIA PACKARD
SYN. SEE MAGNARIUS GUENEE

ALOPE (DRURY), ERINNYIS-SPHINGIDAE
SYN. FLAVICANS GOEZE
SYN. FASCIATA SWAINSON
SYN. EDWARDSII BUTLER

LIFE HISTORY
CAN. ENT. 24:19, 1892 COCKERELL LARVA
ENTOMOLOGICA AMERICANA 3:165, 1887 HY. EDW. LARVA, PUPA
LOC. CIT. 6:141, 1890 DYAR LARVA, PUPA
ENT. SOC. AMERICA, ANNALS 11: 430, 1918 MOSHER PUPAL STRUCTURE
NOVITATES ZOOLOGICAS (TRING) 9:SUPPLEMENT P. 363, 1903 ROTHSCHILD-JORDAN LARVA, PUPA
GUENEE-SPECIES GEN. LEP. SPHING. P. 121, 1874 LARVA
HOLLAND-MOTH BOOK P. 58, 1905 LARVA (BRIEF)

\- - -

INDEX AMER. ECON. ENT., VOL. 7: 360

FOOD PLANTS
CARICA PAPAYA LINN.

ALOPE (FABRICIUS), MINOIS-SATYRIDAE
THE VALID NAME IS PROBABLY PEGALA (FABRICIUS)
A MARITIMA EDWARDS
B TEXANA EDWARDS
C NEPHELE KIRBY
SYN. BOREALIS CHERMOCK
D OLYMPUS EDWARDS
SYN. INO HALL
E BOOPIS BEHR
SYN. ARIANE BOISDUVAL
FORM BARONI EDWARDS
FORM INCANA EDWARDS

LIFE HISTORY
AMER. MUS. NAT. HIST., BULL. 5: 276, 1893 BEUTEN. LARVA, CHRYSALIS
BUFFALO SOC. NAT. SCI., BULL. 19: 21-22, 1939 WILD LARVA
CAN. ENT. 9:142, 1877 EDWARDS EGG, LARVA, CHRYSALIS
LOC. CIT. 12:21, 1880 EDWARDS LARVA OF NEPHELE LIKE THAT OF ALOPE
ILLINOIS DEPT. AGRI., TRANS. 15: 156, 1877 FRENCH LARVA
LOC. CIT. 18:APPEND. PP. 92, 180, 1880 COQUILLETT & MIDDLETON EGG, LARVA, CHRYSALIS
MONTANA UNIVERSITY, BULL. 30; 105, 1906 ELROD EGG, LARVA, CHRYSALIS
ONTARIO ENT. SOC., ANN. REPT. P. 98, 1896 FYLES BRIEF LARVAL DESCR. NEPHELE
BOISDUVAL-SPEC. GEN. LEP. SPHING. P. 121, 1874 LARVA, PUPA
BOISDUVAL & LECONTE-HIST. LEP. NORTH AMERICA P. 228, 1833 LARVA* CHRYSALIS* PL. 59
COMSTOCK-HOW TO KNOW THE BUTTER. P. 186, 1904 LARVA
EDWARDS-BUTTER. NORTH AMERICA 2ND SERIES, 1882 LIFE HIST.* PL. SATYRUS, FF. 2,3 COLOR
EMMONS-NAT. HIST. NEW YORK ENT. 5:213, 1854 LARVA* CHRYSALIS* PL. 33 COLOR
FERNALD-BUTTER. MAINE P. 74, 1884 EGG, LARVA, CHRYSALIS, FF
FRENCH-BUTTER. EAST. UNITED STATES P. 246, 1886 LIFE HISTORY CHRYSALIS*
HOLLAND-BUTTERFLY BOOK 1905 CHRYSALIS* PL. 4, FF. 7,8
MAYNARD-BUTTER. NEW ENGLAND P. 4, 1886 LARVA, CHRYSALIS
SCUDDER-BUTTER. EAST. UNITED STATES & CANADA PP. 166-169, 1889 LIFE HISTORY* PL. 67, F. 1; PL. 70, F. 6; PL. 74, F. 18; PL. 78, FF. 12,13; PL. 86, F. 39
LOC. CIT. PP. 173,174 LIFE HIST.* PL. 64, F. 3; PL. 83, FF. 7,8 HEPHELE
SCUDDER-EVERYDAY BUTTER. PP. 269,275 EGG, LARVA, CHRYSALIS
SCUDDER-GUIDE TO COMMON BUTTER PP. 110,111, 1893 LARVA, CHRYSALIS
SCUDDER-BUTTER. THEIR STRUCTURE P. 43, 1881 CHRYSALIS
VANCE-LIFE AMONG THE BUTTER. P. 44, 1925 EGG, LARVA, CHRYSALIS

FOOD PLANTS
AVENA FATUA LINN.
STIPA SPARTEA TRIN. (OLYMPUS)
GRASSES

ALPHEUS (EDWARDS), PHOLISORA-HESPERIIDAE
SYN. ORICUS EDWARDS
SYN. ARIZONENSIS MABILLE & BOULLETT
LIFE HISTORY
SOUTHERN CALIFORNIA ACAD. SCI., BULL. 28:25, 1929 COMSTOCK LIFE HISTORY, EGG*, LARVA*, CHRYSALIS*
FOOD PLANTS
ATRIPLEX ARDENTEA V. EXPANSA (WATS.)

ALPINA CLARK
SSP. SEE JUGLANDIS ABBOT & SMITH

ALPINATA PACKARD
SSP. SEE PALUDATA THUNBERG

ALTAURUS DYAR
SSP. SEE CLODIUS MENETRIES

ALTERA (SMITH), PONOMETIA-NOCTUIDAE
SYN. CONOCHARODES HAMPSON
LIFE HISTORY
SOUTHERN CALIFORNIA ACAD. SCI., BULL. 32:83, 1933 COMSTOCK-DAMMERS LARVA* PL. 26; PUPA* PL. 27
FOOD PLANTS
HAPLOPAPPUS SP.

ALTERNA STRECKER
SYN. SEE AMPLA WALKER

ALTERNATA (GROTE), ABAGROTIS-NOCTUIDAE
AB ALTERNATELLA STRAND
AB UNIFORMIS STRAND
LIFE HISTORY
BROOKLYN ENT. SOC., BULL. 27: 90, 1932 CRUMB LARVA
MICHIGAN AGRI. EXP. STA., SPEC. BULL. 183:9, 1931 PETTIT LARVA* F. 3
LOC. CIT. 214:79, 1931 MCDANIEL LARVA*
MICHIGAN AGRI. EXP. STA., CIRC. 137:12, 1931 PETTIT-HUTSON LARVA* F. 9
MICHIGAN STATE COLLEGE., SPEC. BULL. 243:37, 1933 MCDANIEL LARVA*
NEW YORK STATE MUS., MEM. 8: 704 1905 FELT LARVA (BRIEF)
UNITED STATES DEPT. AGRI. ENT., BULL. OLD SERIES 13:60, 1887 MURTFELDT LARVA (BRIEF)
WASHINGTON ENT. SOC., PROC. 4: 316, 1898 DYAR LARVA

- - -

INDEX AMER. ECON. ENT.,
VOL. 4:260 LAMPRA
VOL. 5:372 LAMPRA
VOL. 7:2
FOOD PLANTS
CAMPSIS RADICANS (LINN.)
CARYA SP. (HICKORY)
FRAXINUS SP. (ASH)
JUGLANS NIGRA LINN. (WALNUT)
MALUS PUMILA MILL. (APPLE)
PRUNUS CERASUS LINN. (CHERRY)
PRUNUS DOMESTICUS LINN. (PLUM)
PRUNUS PERSICA (LINN.) (PEACH)
QUERCUS SP. (OAK)
QUERCUS ALBA LINN.
OTHER TREES AND HERBACEOUS SHRUBS

ALTERNATA (MULLER), EPIRRHOE-GEOMETRIDAE
SYN. SOCIATA BORKHAUSEN
LIFE HISTORY
NEW YORK AGRI. EXP. STA. (CORNELL), MEM. 274:152, 1948 FORBES LARVA
FOOD PLANTS
GALIUM SP.

ALTERNATA WALKER
SYN. SEE OBSTIPATA FABRICIUS

ALTERNATELLA STRAND
SSP. SEE ALTERNATA GROTE

ALTICOLA SMITH
SSP.SEE INTRITA MORRISON

ALTICOLA WALKER
SSP.SEE DEVERGENS HUBNER

ALTRUARIA PEARSALL
SYN.SEE KEUTZINGARIA PACKARD

ALTUA SMITH
SSP.SEE OLIVACEA MORRISON

ALUNATA SKINNER & AARON
SSP.SEE ASTERIUS CRAMER (AJAX LINN.)

ALUTALIS GROTE
SYN.SEE FACTIOSALIS WALKER

AMASIA ABBOT & SMITH
SEE CONNUBIALIS GUENEE

AMATRIX(HUBNER),CATOCALA-NOCTUIDAE
SYN. NURUS WALKER
FORM SELECTA WALKER
A EDITHA EDWARDS
FORM PALLIDA BARNES & MCDUNNOUGH
LIFE HISTORY
AMER.MUS.NAT.HIST.,BULL.16: 386, 1902 BEUTENMULLER LARVAL HEAD* F.8
AMER.MUS.NAT.HIST.,MEM.NEW SERIES 3:PART 1, 1918 BARNES & MCD. LARVA* PL.10, F.36;PL.11,F.3 COLOR
BROOKLYN ENT.SOC.,BULL.7:48, 1884 HULST LARVA
BUFFALO SOC.NAT.SCI.,BULL.4: 62, 1882 FISHER LARVA
ENT.NEWS 20:13 AND P.135 FOR CORRECTION, 1909 ROWLEY EGG
NEW YORK STATE MUS.,MEM.8: 747, 1905 FELT LARVA (BRIEF)
PAPILIO 1:142, 1881 KELLICOTT LARVA
LOC.CIT.4:8, 1884 FRENCH EGG, LARVA, PUPA
GUENEE-SPECIES GEN.LEP.NOCT. 3:86, 1852 LARVA
FOOD PLANTS
JUGLANS SP.
POPULUS SP. (POPLARS)
POPULUS DELTOIDES MARSH
POPULUS FREMONTI S. WATS.
POPULUS GRANDIDENTATA MICHX.
POPULUS NIGRA V.ITALICA MUENCH.
POPULUS TREMULOIDES MICHX.
SALIX SP.

AMATURARIA WALKER,TIMANDRA-GEOMETRIDAE
AB EFFUSARIA PROUT
LIFE HISTORY
CAN.ENT.24:242, 1892 KELLICOTT LARVA, PUPA
INSECT LIFE 4:382, 1892 PACKARD EGG, LARVA*, PUPA
NEW YORK AGRI.EXP.STA. (CORNELL),MEM.274:119, 1948 FORBES LARVA
FOOD PLANTS
FAGOPYRUM SAGITTATUM GILIB.
POLYGONUM CRISTATUM ENGEL.
RUMEX SP.

AMBIQUALIS(WALKER),LASCORIA-NOCTUIDAE
LIFE HISTORY
ENTOMOLOGICA AMERICANA 14:173, 1934 CRUMB LARVA
- - -
INDEX AMER.ECON.ENT., VOL.3:172 GABERASA
FOOD PLANTS
CHRYSANTHEMUM SP.

AMBISIMILIS DYAR,MALACOSOMA-LASIOCAMPIDAE
LIFE HISTORY
CAN.ENT.25:40, 1893 DYAR LARVA
FOOD PLANTS
FRUIT TREES

AMELAINA DYAR
SSP.SEE VESTALIS PACKARD

AMELLA(GUENEE),SAFIA-NOCTUIDAE
SYN. INTEGERRIMA WALKER

SYN. STYLOBATA HARVEY
FORM BLATCHLEYI HAIMBACH
LIFE HISTORY
WASHINGTON ENT. SOC., PROC. 4: 328, 1898 DYAR LARVA
FOOD PLANTS
QUERCUS VIRGINIANA MILL.

AMERICALIS (GUENEE), EPIZEUXIS-NOCTUIDAE
SYN. SCRIPTIPENNIS WALKER
LIFE HISTORY
ENTOMOLOGICA AMERICANA 14:159, 1934 CRUMB LARVA* PL. 23E; PL. 24 N-C
INSECT LIFE 4:111, 1891 RILEY LARVA*, NESTS*
UNITED STATES DEPT. AGRI., MISC. PUB. 657:396, 1950 CRAIGHEAD LARVA
GUENEE-SPEC. GEN. LEP. DELTOID & PYRALID. 8:78, 1854 LARVA, PUPA
HOLLAND-MOTH BOOK P. 380, 1905 LARVA* F. 168
- - -
INDEX AMER. ECON. ENT., VOL. 7: 359
FOOD PLANTS
HEDYSARUM SP.
LEGUMINOSAE ?
IN NESTS OF ANTS FORMICA RUFA (F. SMITH ENT. NEWS 52:109, 1941)
DEAD LEAVES

AMERICANA D'URBAN
SYN. SEE HYPOPHLAEAS BOISDUVAL

AMERICANA (FABRICIUS), MALACOSOMA-LASIOCAMPIDAE
SYN. DECIPIENS WALKER
SYN. FRUTETORUM BOISDUVAL
SYN. PENSYLVANICA GUERIN-MENEVILLE
LIFE HISTORY
AMER. ENT. 1:208, 1869 LARVA* COCOON* EGGS* F. 145 (ALL STAGES)
LOC. CIT. 2:143, 1870 EGG* LARVA* PUPA*
AMER. MUS. NAT. HIST., BULL. 10: 442, 1898 BEUTEN. LARVA
AMER. PHILOSOPHICAL SOC., PROC. 31:182, 1893 PACKARD LARVA* PL. 11, F. 21
ARKANSAS AGRI. EXP. STA., BULL. 317:22, 1935 BAERG EGG* FF. 14, 15; LARVA* F. 16
CAN. DEPT. AGRI. REPT. ENT. P. 24, 1887 FLETCHER EGG* LARVA* PUPA*
CAN. DEPT. AGRI. REPT. BOTANIST & ENT. P. 24, 1888 FLETCHER EGG* LARVA* COCOON*
CAN. ENT. 4:134, 1872 SAUNDERS EGG* LARVA* COCOON*
LOC. CIT. 5:142, 1873 SAUNDERS EGG* LARVA* COCOON*
LOC. CIT. 10:22, 1878 SAUNDERS LARVA IN EGG OVER WINTER
LOC. CIT. 21:74, 1889 FLETCHER EGG* LARVA* COCOON*
LOC. CIT. 25:39, 1893 DYAR LARVA
LOC. CIT. 75:204, 1943 ATWOOD LARVA*
ENT. NEWS 15:2, 1904 GIRAULT NUMBER OF EGGS
LOC. CIT. 50:45, 69, 1939 WILLIAMS EGG LAYING
ILLINOIS DEPT. AGRI., TRANS. 15: 197, 1877 FRENCH EGG, LARVA
LOC. CIT. 18: APPEND. PP. 122, 155, 1880 COQUILLETT-MARTIN EGG* LARVA* COCOON*
MAINE BOARD AGRI., REPT. 7:170, 1862 PACKARD LARVA (BRIEF)
MASSACHUSETTS AGRI. SOC., PROC. NO PAGE, 1796
MICHIGAN AGRI. EXP. STA., CIRC. 137:15, 1935 PETTIT LARVA* F. 11
MICHIGAN STATE BOARD AGRI., REPT. 13:131, 1874 COOK EGG* LARVA* COCOON* FF.
MICHIGAN STATE COLLEGE., SPEC. BULL. 243:26, 1933 MCDANIEL LIFE HISTORY EGG* LARVA*
LOC. CIT. 24:16, 1904 PETTIT EGG* LARVA*
MINNESOTA AGRI. EXP. STA., BULL.

61:192, 1899 Lugger egg* f.136; larva* f.163 & Pl.17, f.137; cocoon* f.136

Missouri,Rept.Nox.Ins.3:118, 1871 Riley egg* larva* cocoon*

Loc.cit.5:56, 1873 Riley egg*

Mus.Comp.Zool.,Bull.87,No.6, 1941 Dethier larval antenna* f.95

Naturaliste Canadien p.26,1921 Neault egg* larva* cocoon*

New England Farmer,4:354, 1826 Harris larva

New Hampshire Agri.Exp.Sta., Bull.38:49, 1896 Weed egg* larva* cocoon*

Loc.cit.59:201, 1898 Weed egg* larva*

New Hampshire Board Agri.,8th Rept.p.177, 1879 Cutting egg* larva* cocoon*

New York State Agri.Soc., Trans.3:152, 1843 Gaylord larva

New York State Mus.,Bull.27: 47, 1899 Felt egg* larva*

Loc.cit.37:7, 1900 Felt egg*

New York State Mus.,Mem.8: 109, 550, 1905 Felt egg* Pl.7,f.12; larva* Pl.7,ff.13, 16; pupa* Pl.7,f.15; cocoon* Pl.7,f.14 ALL IN COLOR

New York Rept.Nox.Ins.2:185, 1856 Fitch egg* larva, pupa*

Ontario Ent.Soc.,Ann.Rept.p. 15, 1871 Bethune egg, larva*, cocoon

Loc.cit.p.29, 1875 Saunders egg* larva* cocoon*

Loc.cit.p.41, 1877 Gott egg* larva, cocoon

Loc.cit.p.71, 1888 Bethune egg* larva* cocoon*

Loc.cit.p.33, 1889 Harrington egg* larva*

Ontario Ent.Soc.,Ann.Rept. p.54, 1895 Forbush egg* larva* cocoon*

Loc.cit.p.57, 1896 Bethune egg* larva* cocoon*

Loc.cit.p.19, 1897 Dearners egg* larva* cocoon*

Loc.cit.p.30, 1897 Harrington egg* larva*

Loc.cit.p.84, 1898 Fletcher egg* larva* cocoon*

Loc.cit.p.95, 1898 Hutt egg, larva

Loc.cit.p.67, 1899 Lockhead egg, larva, cocoon

Loc.cit.p.106, 1902 Lockhead egg, larva, cocoon

Loc.cit.p.55, 1905 Stevenson egg* larva* cocoon*

Loc.cit.p.13, 1911 Gibson egg* larva* cocoon*

Loc.cit.p.13, 1912 Gibson egg* larva* cocoon*

United States Dept.Agri.,Entom. Bull.34:30, 1902 new series Banks egg* f.19-c; larva* f.19 a-b; cocoon* f.19 d

United States Dept.Agri.Ent., Circular 98:2, 1908 Quaintance egg* f.1 a; larva* f.1 b & 2; pupa* f.1-c; cocoon* f.1 d

United States Dept Agri.Farmers Bull.662:4, 1915 Quaintance egg* f.3; larva* f.4; pupa* and cocoon* f.5

Loc.cit.1270:37, 1922 Quaintance & Siegler egg* f.69; larva* ff.70,71; pupa* f.72

United States Dept.Agri.Misc. Pub.657:353,416, 1950 Craighead all states* f.85

United States Dept.Agri.,Rept. p.83, 1870 Glover egg, larva

Utah Agri.Exp.Sta.,Bull.65:80, 1900 Close egg* larva* cocoon*

Vermont Agri.Exp.Sta.,Bul.76: 123, 1900 Perkins larval segment markings

Vermont Board Agri.,Rept.2: 586, 1876 Perkins egg* larva* cocoon*

ABBOT & SMITH LEP. INS. GEORGIA, II:119, 1797 LARVA* PL. 60 IN COLOR AS CASTRENIS
CRAGEN-OUR INSECT FRIENDS & FOES P. 173, 1899 LARVA* F. 121
CUVIER-ANIMAL KINGDOM XV, 1832 LARVA PL. 84
DICKERSON-MOTHS & BUTTERFLIES P. 201, 1901 EGG* F. 176; LARVA* F. 178; NEST* F. 177; COCOON* F. 179
DOANE, ETC-FOREST INSECTS, P. 281, 1936 LARVA* F. 162
EMMONS-NAT. HIST. NEW YORK 5:235, 1854 EGG* LARVA
ESSIG-INSECTS WESTERN NORTH AMERICA, P. 696, 1926 LARVA
FERNALD-APPLIED ENTOM. P. 260, 1926 EGG* F. 258; LARVA* F. 260; NEST* F. 259
FERNALD & SHEPARD-APPLIED ENT. P. 257, 1942 EGG* F. 255; LARVA* F. 257; NEST* F. 256
GLOVER-ILLUS. NORTH AMERICAN ENT. 1878 LARVA* PL. 12, F. 8
HARRIS-INSECT. INJURIOUS TO VEGET. 1ST EDIT., P. 268, 1841 LARVA, PUPA
LOC. CIT. 2ND EDIT., P. 286, 1852 LARVA
LOC. CIT. FLINT EDIT., P. 371, 1862 EGG* LARVA* COCOON*
HART-INSECT COLLECTIONS FOR ILLINOIS HIGH SCHOOLS P. 47, 1903 EGG* LARVA* COCOON* FF. 58, 59
HERRICK-INSECT ENEMIES SHADE TREES P. 316, 1935 EGG* LARVA* PUPA* ALL F. 268
HOLLAND-MOTH BOOK P. 313, 1905 EGG* LARVA* COCOON* F. 188
JAEGER-LIFE NORTH AMER. INSECTS P. 171, 1864 LARVA
KELLOGG-AMER. INSECTS P. 416, 1908 LARVA* PL. 599-E
LUTZ-A LOT OF INSECTS P. 160, 1941 EGG* LARVA* PUPA* COCOON*
METCALF & FLINT-DESTRUCTIVE & USEFUL INSECTS P. 563, 1939 EGG* F. 347; LARVA* FF. 348, 349; PUPA
O'KANE-INJURIOUS INSECTS P. 267, 1912 EGG* F. 375; LARVA* F. 373
PACKARD-HALF HOURS WITH INSECTS P. 171, 1877
PACKARD-INJURIOUS INSECTS IN THE WEST, HAYDEN*S REPT. P. 793, 1877 EGG* LARVA* PUPA*
PEAIRS-PESTS FARM, GARDEN, ORCHARD P. 377, 1941 EGG* F. 451; LARVA* F. 450; COCOON* F. 453
SAUNDERS-INSECTS, INJ. FRUITS P. 48, 1883 EGG* LARVA* COCOON*
SAUNDERS-LOC. CIT. P. 48, 1889 EGG* FF. 40-41; LARVA* F. 41; PUPA* AND COCOON* F. 41
SLINGERLAND & CROSBY MANUAL FRUIT INSECTS P. 114, 1914 EGG* F. 122; LARVA* F. 123; PUPA* F. 125; COCOON* F. 124
SMITH-INSECTS NEW JERSEY P. 393 1900 EGG* LARVA* COCOON* F. 178
TENNEY-NATURAL HISTORY P. 404, 1867 COCOON* PL. 299

– – –

INDEX AMER. ECON. ENT., VOL. 1:196; VOL. 2:223; VOL. 3:248; VOL. 4:287; VOL. 5:403; VOL. 6:480; VOL. 7:634

FOOD PLANTS

ACER SP. (MAPLES)
ACER SACCHAROPHORUM KOCH
AMELANCHIER CANADENSIS (LINN.)
ARBUTUS MENZIESII PURSH
BERBERIS VULGARIS LINN.
BETULA SP. (BIRCHES)
BETULA ALBA LINN.
BETULA NIGRA LINN.
BETULA PAPYRIFERA MARSH.
BETULA POPULIFOLIA MARSH.
CARPINUS CAROLINIANA WALT.
CARYA SP. (HICKORIES)
CEANOTHUS SP.
CERCIS CANADENSIS LINN.

Cornus florida Linn.
Corylus sp. (hazels)
Crataegus sp. (thorns)
Fagus sp. (beeches)
Fagus grandifolia Ehrh.
Fragaria chiloensis (Linn)
Fraxinus sp. (ashes)
Hamamelis virginiana Linn.
Juglans nigra Linn.
Liquidambar styraciflua Linn.
Malus pumila Mill. (apple)
Nyssa sylvatica Marsh.
Ostrya virginiana (Mill.)
Photina arbutifolia Lindl.
Populus sp. (poplars)
Populus fremonti S. Wats.
Prunus americana Marsh.
Prunus armeniaca Linn. (apricot)
Prunus cerasus Linn. (cherry)
Prunus domestica Linn. (plum)
Prunus domestica var. galatensis ex Hook (prune)
Prunus ilicifolia (Nutt. ex. H. & A.)
Prunus pensylvanica Linn.
Prunus persica (Linn.) (peach)
Prunus serotina Ehrh.
Prunus virginiana Linn.
Pyrus communis Linn. (pear)
Quercus sp. (oaks)
Quercus alba Linn.
Quercus stellata Wang.
Quercus velutina Lam.
Ribes sativum Reich. (currant)
Robinia pseudo-acacia Linn.
Rosa sp. (roses)
Salix sp. (willows)
Sorbus americana
Tilia sp. (lindens)
Ulmus sp. (elms)
Vaccinium sp. (blueberries)

<u>americana</u> Harris, <u>Acronicta</u>, Noctuidae
syn. <u>acericola</u> Guenee
a <u>obscura</u> Henry Edwards
b <u>eldora</u> Smith
LIFE HISTORY
Amer. Ent. 1:166, 1869 larva <u>acericola</u> (brief)
Loc. cit. 2:61, 1869 larva (brief)
Amer. Mus. Nat. Hist., Bull. 14: 238, 1901 Beuten. larva
Boston Soc. Nat. Hist., Proc. 27: 134, 1896 Dyar larval structure
Buffalo Soc. Nat. Sci., Bull. 2: 154, 1874 Grote larva
Illinois Dept. Agri., Trans. 15: 200, 1877 French larva
Loc. cit. 18: Append. p. 130, 1879 Martin larva
Mus. Comp. Zoology, Bull. 87: No. 6, 1941 Dethier larval antenna* f. 102
New York Agri. Exp. Sta. (Cornell), Mem. 56:1073, 1922 Wellhouse larva (brief)
New York State Mus., Mem. 8: 525, 1905 Felt larva
Ontario Ent. Soc., Ann. Rept. p. 41, 1872 Reed larva
Loc. cit. p. 49, 1889 Reed larva
Loc. cit. p. 75, 1902 Gibson larva*
Loc. cit. p. 57, 1903 Gibson larva*
Papilio 1:6, 1881 Coquillet larva
United States Dept. Agri., Misc. Pub. 657:356, 392, 1950 Craighead larva
United States Nat. Mus., Proc. 21:46, 1898 Smith & Dyar egg* larva* Pl. 6, f. 17 - in some editions this plate is colored, in others not; pupa, cocoon
Abbot & Smith - Lep. Ins. Georgia 2:185, 1797 larva* Pl. 93 color
Doane, etc - Forest Ins. p. 284, 1936 larva (brief)
Eliot & Soule - Caterpillars and their moths p. 297, 1902 larva*
Glover - Illus. North Amer. Ent., 1878 larva Pl. 10, f. 3
Guenee - Spec. Gen. Lep. Noct. 1: 48 1852 larva (<u>acericola</u>)
Harris - Ent. Corresp. p. 311, 1869 larva* Pl. 111

HARRIS-INS.INJ.VEGET. 1ST EDIT. P.317, 1841 LARVA
LOC.CIT.2ND EDIT.P.337, 1852 LARVA
LOC.CIT.FLINT EDIT.P.436, 1862 LARVA* PUPA*
HART-INS.COLLECTIONS FOR ILLINOIS HIGH SCHOOLS P.24, 1903 COCOON
KELLOGG-AMER.INS.P.404, 1908 LARVA
PACKARD-INS.INJ.FOREST TREES P.111, 1881 LARVA

- - -

INDEX AMER.ECON.ENT., VOL.3:3;VOL.4:4;VOL.7:8

FOOD PLANTS

ACER NEGUNDO LINN.
ACER PSEUDOPLATANUS LINN.
ACER RUBRUM LINN.
ACER SACCHARINUM LINN.
AESCULUS HIPPOCASTANUM LINN.
ALNUS SP. (ALDERS)
AMELANCHIER CANADENSIS (LINN.)
BETULA ALBA LINN.
BETULA POPULIFOLIA MARSH.
CARPINUS CAROLINIANA WALT.
CARYA CORDIFORMIS (WANG.)
CARYA GLABRA (MILL.)
CARYA TOMENTOSA NUTT.
CASTANEA DENTATA (MARSH.)
CASTANES SATIVA MILL.
CORNUS MAS LINN.
CRATAEGUS SP. (HAWS)
FAGUS GRANDIFOLIA EHRH.
FAGUS SYLVATICA LINN.
FRAXINUS SP. (ASHES)
JUGLANS SP. (WALNUTS)
MALUS PUMILA MILL. (APPLE)
PLATANUS OCCIDENTALIS LINN.
POPULUS SP. (POPLARS)
POPULUS NIGRA V. ITALICA MUENCH.
PRUNUS SEROTINA EHRH.
PRUNUS VIRGINIANA LINN.
QUERCUS ALBA LINN.
QUERCUS COCCINEA MUENCH
QUERCUS MACROCARPA MICHX.
QUERCUS PALUSTRIS MUENCH.
QUERCUS RUBRA LINN.
QUERCUS VELUTINA LAM.
ROBINIA PSEUDO-ACACIA LINN.
RUBUS ALLEGHENIENSIS PORTER
SALIX SP. (WILLOWS)
SORBUS AMERICANA
TILIA AMERICANA LINN.
TILIA EUROPAEA LINN.
TILIA HETEROPHYLLA VENT.
ULMUS AMERICANA LINN.
ULMUS PROCERA SALSIB.
ULMUS RUBRA MUHL.
VACCINIUM CORYMBOSUM LINN.

AMERICANA HARRIS
SSP.SEE CAJA LINN.

AMERICANA(HARRIS),EPICNAPTERA-LASIOCAMPIDAE
SYN. ILICIFOLIA AUCT.
SYN. OCCIDENTIS WALKER
FORM FERRUGINEA PACKARD
FORM MILDEI STRETCH
SYN. ROSEATA STRETCH
A CARPINIFOLIA BOISDUVAL
B ALASCENSIS PACKARD
C DYARI RIVERS.

LIFE HISTORY

AMER.ENT.2:32, 1869 LARVA(BRIEF)
AMER.MUS.NAT.HIST.,BULL.10: 444, 1898 BEUTEN. LARVA
AMER.PHILOSOPHICAL SOC.,PROC. 31:184, 1893 PACKARD LARVA*
CAN.ENT.6:158, 1874 LYMAN EGG
ILLINOIS DEPT.AGRI.,TRANS.15: 179, 1877 FRENCH LARVA
LOC.CIT.18:APPEND.P.166, 1880 COQUILLETT LARVA
MINNESOTA AGRI.EXP.STA.,BULL. 61:126, 1899 LUGGER LARVA* F.71, COCOON* F.71
NEW YORK ACAD.SCI.,ANNALS 8: 230, 1894 DYAR LARVAL SETAL PATTERN
NEW YORK AGRI.EXP.STA. (CORNELL),MEM.56:1075, 1922 WELLHOUSE LARVA (BRIEF)

New York State Cabin. Nat. Hist. Ann. Rept. No. 26:154, 1874 Lintner larva
New York State Mus., Mem. 8: 525, 1905 Felt larva
New York State, Rept. Nox. Ins. 3:337,380, 1856 Fitch larva
See also p.19, sec.30, & p.80, sec.121
Smithsonian Scientific Series 5:262, 1930 Snodgrass COLORED Pl. showing all stages
United States Dept. Agri., Misc. Pub. 657:355,419, 1950 Craighead larva
Abbot & Smith Lep. Ins. Georgia I:101, 1797 larva* Pl. 51 COLOR
Doane, etc-Forest Insects, p.282, 1936 larva
Harris-Entomological Corresp. p.292, 1869, larva
Harris-Ins. Inj. Veget. 1st edit. p.273, 1841 larva, pupa
Harris-Loc. cit. 2nd edit. p.293, 1852 larva
Harris-Loc. cit. Flint edit. p.377, 1862 larva
Kellogg-American Insects, p.416, 1908 larva
Saunders-Ins. Inj. Fruits, p.87, 1883 egg, larva, cocoon
Saunders-Loc. cit. p.87, 1889 egg, larva

FOOD PLANTS
Acer sp. (maples)
Betula lenta Linn.
Carya sp. (hickories)
Ceanothus sp.
Celtis occidentalis Linn.
Crataegus sp.
Fraxinus sp. (ashes)
Malus pumila Mill. (apple)
Parthenocissus quinquefolia (Linn.)
Populus sp. (poplars)
Prunus sp.
Prunus cerasus Linn. (cherry)
Prunus domestica Linn. (plum)
Prunus pensylvanica Linn.
Prunus serotina Ehrh.
Prunus virginiana Linn.
Pyrus sp.
Pyrus communis Linn. (pear)
Quercus sp. (oaks)
Quercus garryana Dougl.
Quercus ilicifolia Wang.
Rubus allegheniensis Porter
Salix hookeriana Barr.
Salis scouleriana Barr.
Vitis sp. (grapes)

americana D'Urban
syn. see hypophlaeas Boisduval

americana Harris
syn. see inclusa Hubner

americana Harvey, Lophopteryx-Notodontidae
syn. capusina Neumoegen & Dyar
LIFE HISTORY
Amer. Mus. Nat. Hist., Bull. 10: 415, 1898 Beuten. larva (camelina Linn.)
Nat. Acad. Sci., Mem. 7:157, 1895 Packard larva
FOOD PLANTS
Alnus sp. (alders)
Corylus sp. (hazels)
Populus sp. (poplars)
Quercus sp. (oaks)
Tilia sp. (lindens)

americana Herrich-Schaeffer
syn. see rosea Walker (Oreta)

americana (Speyer), Apamea-Noctuidae
syn. lusca Smith
syn. atlantica Smith
a pacifica Smith
LIFE HISTORY
Index Amer. Econ. Ent., Vol. 3:33 nictitans
Vol. 7:74

FOOD PLANTS
CAREX SP. (SEDGES) ROOTS
ZEA MAYS LINN. (CORN)
GRASSES, ROOTS

AMERICANUS GUNDER
SSP. SEE PLEXIPPUS LINNEAUS

AMERICUS KOLL
SSP. SEE ASTERIUS CRAMER (AJAX LINN.)

AMESTRIS STRECKER, CATOCALA-NOCTUIDAE
SYN. ANNA GROTE
FORM WESTCOTTI GROTE
LIFE HISTORY
AMER. MUS. NAT. HIST., BULL. 16: 393, 1902 BEUTEN. LARVAL HEAD* F.20
AMER. MUS. NAT. HIST., MEM. NEW SERIES 3: PART 1, 1918 BARNES & McDUNNOUGH LARVA* PL.10, F.33; PL.12, F.7 COLOR
WASHINGTON ENT. SOC., PROC. 4: 327, 1898 DYAR LARVA
FOOD PLANTS
AMORPHA FRUTICOSA LINN.

AMETHYSTARIA STRECKER
SYN. SEE HYPOCHRARIA HERRICH-SCHAEFFER

AMICA EDWARDS
SSP. SEE SAEPIOLUS BOISDUVAL

AMICA HARRIS
SYN. SEE ARCTICA BOISDUVAL

AMICA (HUBNER), CATOCALA-NOCTUIDAE
SYN. ANDROPHILA GUENEE
FORM NERISSA HENRY EDWARDS
FORM SUFFUSA BEUTENMULLER
A LINEELLA GROTE
SYN. NOVANGLIAE REIFF
AB AURANTIACA REIFF
FORM CURVIFASCIA BROWER
FORM MELANOTICA REIFF.
LIFE HISTORY
AMER. MUS. NAT. HIST., BULL. 16: 384, 1902 BEUTEN. LARVAL HEAD* F.3
AMER. MUS. NAT. HIST., MEM. NEW SERIES 3: PART 1, 1918 BARNES & McDUNNOUGH LARVA* PL.10, F.39; PL.12, F.12 COLOR
BROOKLYN ENT. SOC., BULL. 7:35, 1884 HULST LARVA
ENT. NEWS 21:448, 1910 ROWLEY LARVA
ILLINOIS DEPT. AGRI., TRANS. 18: APPEND. P. 182, 1880 COQUILLETT LARVA
NEW YORK STATE MUS., MEM. 8:705 1905 FELT LARVA
PAPILIO 1:7, 1881 COQUILLETT LARVA
GUENEE-SPEC. GEN. LEP. NOCT. 3: 106, 1852 LARVA
FOOD PLANTS
QUERCUS SP. (OAKS)
QUERCUS MACROCARPA MICHX.

AMICARIA (HERRICH-SCHAEFFER), HYPERETIS-GEOMETRIDAE
FOR SYNS. AND SSPS. SEE McDUNNOUGH'S LIST
LIFE HISTORY
AMER. MUS. NAT. HIST., BULL. 4:79, 1892 BEUTEN. PUPA
CAN. ENT. 3:209, 1871 SAUNDERS LARVA (ALIENARIA HERRICH-SCHAEFFER)
LOC. CIT. 82:209, 1950 McGUFFIN LARVA* F.1 C
ENT. NEWS 5:62, 1894 DYAR LARVA
ENT. NEWS 56:273, 1945 FORBES PUPAL STRUCTURE
NEW YORK ENT. SOC., JOURN. 53:205 1945 FORBES PUPA (ALIENARIA)
NEW YORK STATE AGRI. EXP. STA. (CORNELL), MEM. 274:97, 1948 FORBES LARVA OF AMICARIA, NYSSARIA GUENEE, AND ALIENARIA HERRICH-SCHAEFFER
NEW YORK STATE MUS., MEM. 8:722, 1905 FELT LARVA
PSYCHE 11:75, 1904 DYAR EGG, LARVA
UNITED STATES DEPT. AGRI. ENT., BULL. OLD SERIES 13:29, 1887 PACKARD LARVA, PUPA

United States Dept. Agri., Misc. Pub. 657:361, 433, 1950 Craighead larva
Gumppenberg-Systema Geometrarum 7:422, 1895 larva pupa (nyssaria)
Packard-Ins. Injur. Forest Trees p. 129, 1881 larva
Packard-Monograph Geometridae p. 464, 1876 larva* pupa* Pl. 13, f. 19, 19a, 25, 25a
Saunders-Ins. Inj. Fruits p. 129, 1881 larva

FOOD PLANTS
Acer sp. (maples)
Alnus sp. (alders)
Amelanchier florida Lindl.
Betula sp. (birches)
Betula alleghaniensis Britt.
Carpinus caroliniana Walt.
Castanea dentata (Marsh.)
Cornus florida Linn.
Cornus sericea Linn.
Fagus sp. (beeches)
Hypericum sp.
Hypericum myrtifolium Lam.
Kalmia sp. (laurels)
Prunus serotina Ehrh.
Quercus sp. (oaks)
Rosa carolina Linn.
Salix sp. (willows)
Spiraea latifolia (Ait.)
Spiraea salicifolia Linn.
Tilia sp. (lindens)
Ulmus sp. (elms)
Vaccinium sp.

amicetus Scudder
ssp. see epixanthe Boisduval

amicora Smith
syn. see lepusculina Guenee

amicula Henry Edwards
syn. see decia Grote

amitaria (Guenee), Ematurga-Geometridae
syn. faxoni Minot

LIFE HISTORY
Massachusetts Agri. Exp. Sta., Bull. 239:41, 1928 Franklin larva* Pl. 3, ff. 7a, 8a, 8b color
New York Agri. Exp. Sta. (Cornell), Mem. 274:50, 1948 Forbes larva
New York Ent. Soc., Journ. 53: 185, 1945 Forbes pupa (faxoni)

FOOD PLANTS
Arctostaphylos uva-ursi (Linn.)
Thuja occidentalis Linn.
Vaccinium macrocarpon Ait. (cranberry)
Shrubs

ammiralis Hemming
syn. see bellona Fabricius

ammon (Lucas), Hemiargus-Lycaenidae

LIFE HISTORY
Washington Ent. Soc., Proc. 4: 448, 1900 Dyar egg, larva

FOOD PLANTS
Caesalpinia bonduc (Linn.)
Pithecellobium unguiscati (Linn.)

ammoni Behrens
ssp. see rutulus Lucas

amorata Hulst
syn. see defensaria Guenee

amorata Packard, Callizzia-Epiplemidae

LIFE HISTORY
Washington Ent. Soc., Proc. 5: 131, 1902 Dyar egg, larva, pupa

FOOD PLANTS
Lonicera dioica Linn.

amorphae Henry Edwards
syn. see eurydice Boisduval

ampelophaga Walker
syn. see satellitia Linn.

AMPELOS EDWARDS, COENONYMPHA-SATYRIDAE
GEN. VERN. EUNOMIA FIELD
A ELKO EDWARDS

LIFE HISTORY

CAN. ENT. 19:41, 1887 W.H. EDWARDS EGG, LARVA, CHRYSALIS

COMSTOCK, J.A.-BUTTERFLIES CALIFORNIA P.68, 1927 EGG, LARVA, CHRYSALIS

FOOD PLANTS

GRASSES

ENT. SOC. AMER., ANNALS 11:416 1918 MOSHER PUPAL STRUCTURES

ILLINOIS DEPT. AGRI., TRANS. 15: 167, 1877 FRENCH LARVA

LOC. CIT. 18:APPEND. P.102, 1880 MARTIN LARVA

NAT. ACAD. SCI., MEM. 9: 1905 PACKARD LARVA* PL.34 COLOR; PL.42, FF.1,6 STRUCTURE

NATURALISTE CANADIEN P.154, 1929 HUARD LARVA

NEW YORK AGRI. EXP. STA. (CORNELL), MEM. 274:188, 1948 FORBES LARVA

NEW YORK ENT. SOC., JOURN. 9:88, 1901 BEUTEN. LARVA

NEW YORK STATE MUS., MEM. 8:546, 1905 FELT LARVA

NOVITATES ZOOLOGICAE (TRING) 9:SUPPLEMENT P.107, 1903 ROTHSCHILD & JORDAN LARVA

PHILADELPHIA ACAD. NAT. SCI., JOURN. 5:179, 1859 CLEMENS LARVA, PUPA (QUADRICORNIS)

PHILADELPHIA ENT. SOC., PROC. 1: 286 1862 LINTNER LIFE HISTORY (QUADRICORNIS)

LOC. CIT. 3:664, 1964 LINTNER PUPA

SMITHSONIAN MISC. COLLECTIONS 4: ART. 2:205 1862 MORRIS LARVA, PUPA

UNITED STATES DEPT. AGRI., MISC. PUB. 657:358, 372, 1950 CRAIGHEAD LARVA* F.71 H

BOISDUVAL SPEC. GEN. LEP. SPHING. P.54, 1874 LARVA* PUPA* PL.1 COLOR

ELIOT & SOULE-CATERPILLARS AND THEIR MOTHS P.164, 1902 EGG, LARVA* PUPA

FERNALD-SPHINGIDAE OF NEW ENGLAND P.25, 1886 EGG, LARVA

GLOVER-ILLUS. NORTH AMER. ENT. 1878 LARVA, PUPA PL.7, F.1

HARRIS-ENT. CORRESP. P.282, 1869 LARVA, PUPA

HARRIS-INS. INJ. VEGET. 1ST EDIT., P.227, 1841 LARVA, PUPA

HARRIS-LOC. CIT. 2ND EDIT. P.247, 1852 LARVA, PUPA

HARRIS-LOC. CIT. FLINT EDIT., P.324, 1862 LARVA*

JAEGER-LIFE NORTH AMER. INS. PL.192, 1864 LARVA

MORRIS-SYNOP. LEP. NORTH AMERICA P.205, 1862 LARVA, PUPA

PACKARD-SECOND REPT. NAT. HIST. & GEOLOGY MAINE P.167, 1862 LARVA

PACKARD-INS. INJ. FOREST TREES P.67, 1881 LARVA

- - -

INDEX AMER. ECON. ENT., VOL. 6: 137

FOOD PLANTS

BETULA SP. (BIRCHES)
BETULA ALBA LINN.
FAGUS SP. (BEECHES)
TILIA SP. (LINDENS)
TILIA AMERICANA LINN.
ULMUS AMERICANA LINN.

AMYNTULA (BOISDUVAL), EVERES-LYCAENIDAE
AB DODGEI GUNDER

LIFE HISTORY

ENT. NEWS 24:97, 1913 BAKER LIFE HISTORY AND HABITS

NEW YORK ENT. SOC., JOURN. 7:211, 1899 COQUILLETT LARVA

PAPILIO 4:126, 1884 WRIGHT HABITS OF LARVA IN SEED PODS

SOUTHERN CALIFORNIA ACAD. SCI., BULL. 34:218, 1935 COMSTOCK & DAMMERS EGG, LARVA* PL.57; PUPA* PL.58

Comstock, J.A.-Butterflies of California p.182, 1927 egg, larva, chrysalis

FOOD PLANTS

Aesculus sp.
Astragalus sp.
Astragalus crotalariae (Benth.)
Lathyrus ochroleucus Hook.
Lathyrus venosus Muhl.

amyrisaria Walker
syn. see duaria Guenee

anargyrea (Dyar), Papaipema-Noctuidae

LIFE HISTORY

FOOD PLANTS

Eupatorium sp.

anataria (Swett), Itame-Geometridae

LIFE HISTORY

Maine Agri. Exp. Sta., Bull. 259: 52, 1917 Mosher larva, pupa* f.6, G, H
New York Ent. Soc., Journ. 53: 190, 1945 Forbes pupa

FOOD PLANTS

Betula allegheniensis Britt.
Betula populifolia Marsh.

ancilla Barnes & McDunnough
ssp. see enoptes Boisduval

ancilla Smith
ssp. see georgii Grote

ancocisconensis Morrison
syn. see xylinoides Guenee (Hypa)

ancora Ploetz
syn. see phyleus Drury

andersoni Gunder
ssp. see anicia Doubleday & Hewitson

andersoni Swett
ssp. see occiduaria Packard

andrewsi Mart
ssp. see ausonides Boisduval

andria (Scudder), Anaea-Nymphalidae
syn. ops Drury
troglodyta Auct.

LIFE HISTORY

Can. Ent. 20:41, 1888 W.H. Edwards life history
Illinois Dept. Agri., Trans. 18: Append. p.89, 1880 Middleton larva* chrysalis*
Missouri, Rept. of Insects 2:125-128, 1870 Riley larva* chrysalis*
Missouri, Rept. of Insects 5:145-149, 1873 Riley life history ff.
Comstock-How to Know the Butterflies p.178, 1904 larva
Cragen-Our Insect Friends & Foes p.91, 1899 larva, chrysalis
Edwards-Butterflies of North America, 2nd series, 1871 larva* chrysalis* Pl. Paphia No.1 COLOR
French-Butterflies East. United States p.227, 1886 larva* f.69 chrysalis* f.69
Holland-Butterfly Book p.9, 1905 larva* chrysalis* f.21
Kellogg-Amer. Insects p.455, 1908 larva
Scudder-Butter. East United States & Canada pp.1795-1798, 1889 life history
Scudder-Butter. their structure p.111, 1881 larva, chrysalis, nest
Scudder-Guide to Common Butter. p.104, 1893 larva, chrysalis

- - -

Index Amer. Econ. Ent., Vol. 6:27; Vol. 7:35

FOOD PLANTS

Croton capitatus Michx.
Croton linearis Michx.
Croton monanthogynus Michx.

ANDROCARDIA HUBNER
SYN. SEE PORTLANDIA FABRICIUS

ANDROMACHA HUBNER
SYN. SEE PORTLANDIA FABRICIUS

ANDROMEDAE BOISDUVAL
SYN. SEE LUGENS WALKER

ANDROMEDAE(GUENEE), CATOCALA- NOCTUIDAE
A TRISTIS EDWARDS
LIFE HISTORY
BROOKLYN ENT. SOC., BULL. 17:114, 1922 MAYFIELD EGG, LARVA
FOOD PLANTS
CARYA SP. (HICKORIES)
QUERCUS SP. (OAKS)
VACCINIUM
VACCINIUM CORYMBOSUM LINN.

ANDROPHILA GUENEE
SYN. SEE AMICA HUBNER

ANFRACTATA HULST
SYN. SEE FORFICARIA GUENEE

ANGELICA(GROTE), APATELODES- ZANOLIDAE
SYN. HYALINOPUNCTA PACKARD
A INDISTINCTA HENRY EDWARDS
B SERAPHICA DYAR
LIFE HISTORY
AMER. MUS. NAT. HIST., BULL. 10: 403, 1898 BEUTEN. LARVA
BROOKLYN MUS. ARTS & SCI., BULL. 1:197, 1906 DYAR LARVA
MUS. COMP. ZOOL., BULL. 87: NO. 6, 1941 DETHIER
LARVAL ANTENNAE FF. 146, 148
NAT. ACAD. SCI., MEM. 7:104, 1895 PACKARD LARVA* PL. 49, F. 1
NEW YORK STATE MUS., MEM. 8: 727, 1905 FELT LARVA (BRIEF)
UNITED STATES DEPT. AGRI., MISC. PUB. 657:356, 421, 1950 CRAIGHEAD LARVA
FOOD PLANTS
FRAXINUS SP. (AHES)
SYRINGA SP. (LILACS)

ANGELICA(SMITH), PAPAIPEMA- NOCTUIDAE
LIFE HISTORY
CAN. ENT. 63:184, 1931 BIRD LIFE HISTORY, EGG, LARVA
NEW YORK ENT. SOC., JOURN. 7:209 1899 COQUILLETT LARVA, PUPA
FOOD PLANTS
PSORALEA MACROSTACHYA DC.

ANGELINA BOISDUVAL
SYN. SEE CETHURA FELDER & FELDER

ANGUINA(ABBOT & SMITH), DASYLOPHIA- NOTODONTIDAE
SYN. CUCULIFERA HERRICH-SCHAEFFER
SYN. PUNCTATA WALKER
SYN. CANA WALKER
SYN. SIGNATA WALKER
A PUNTAGORDA SLOSSON
B SATURATA BARNES
LIFE HISTORY
AMER. MUS. NAT. HIST., BULL. 10: 417, 1898 BEUTEN. LARVA
BOSTON SOC. NAT. HIST., PROC. 24: 529, 1890 PACKARD EGG, LARVA* PL. 3, F. E
ENTOMOLOGICA AMERICANA 5:55, 1899 DYAR EGG, LARVA, PUPA
NAT. ACAD. SCI., MEM. 7:174, 1895 PACKARD EGG, LARVA* PUPA* PL. 21, FF. 1-6 COLOR
NEW YORK AGRI. EXP. ST. (CORNELL), MEM. 274:225, 1948 FORBES LARVA
NEW YORK STATE MUS., MEM. 8:730 1905 FELT LARVA
ABBOT & SMITH-LEP. INS. GEORGIA 2:167, 1797 LARVA* PL. 84 COLOR
GLOVER-ILLUS. NORTH AMER. ENT., 1908 LARVA PL. 6, FF. 14, 15
HARRIS-ENT. CORRESP., P. 304, 1869 LARVA* PL. 1
FOOD PLANTS
AMORPHA FRUTICOSA LINN.
BAPTISIA TINCTORIA (LINN.)
LESPEDEZA CAPITATA MICHX.
ROBINIA PSEUDO-ACACIA LINN.
TRIFOLIUM SP. (CLOVERS)

ANGULALIS(HUBNER), PALTHIS-NOCTUIDAE
SYN. ARACINTHUSALIS WALKER
LIFE HISTORY
CAN. ENT. 74:53, 1942
McGUFFIN LARVA* PL.7, F.B
(ALSO LARVAL MANDIBLE)
UNITED STATES DEPT. AGRI. BUR. ENT., BULL. 22:79, 1890 OLD SERIES MURTFELDT LARVA
UNITED STATES NAT. MUS., PROC. 27:881, 1904 DYAR EGG, LARVA
- - -
INDEX AMER. ECON. ENT., VOL. 7: 737
FOOD PLANTS
ABIES BALSAMEA (LINN.)
ACER SACCHAROPHORUM KOCH
AMELANCHIER FLORIDA LINDL.
BETULA PAPYRIFERA MARSH.
BETULA POPULIFOLIA MARSH.
MALUS PUMILA MILL. (APPLE)
PICEA ENGELMANNI (PARRY)
PICEA GLAUCA (MOENCH)
PICEA MARIANA (MILL.)
PICEA SITCHENSIS (BONG.)
PRUNUS VIRGINIANA LINN.
RUBUS ALLEGHENIENSIS PORTER
SPIRAEA VANHOUTII ZABEL

ANGULARIA BARNES & McDUNNOUGH, METARRANTHIS-GEOMETRIDAE
LIFE HISTORY
NEW YORK AGRI. EXP. STA. (CORNELL), MEM. 274:93, 1948 FORBES LARVA
NEW YORK ENT. SOC., JOURN. 51:142, 1943 RUPERT EGG, LARVA
FOOD PLANTS
PURNUS SEROTINA EHRH.
ROSACEAE
SHRUBS

ANGULATA GROTE
SYN. SEE UMBRA HUFNAGEL

ANGULATA WARREN
SYN. SEE TAYLORATA HULST

ANGULIFERA(WALKER), CALLOSAMIA-SATURNIIDAE
A SECURIFERA MAASSEN
SYN. CAROLINA JONES
AB AURANTIACA IGEL
LIFE HISTORY
AMER. ACAD. ARTS & SCI., PROC. 28:70, 1893 PACKARD EGG, LARVA
AMER. MUS. NAT. HIST., BULL. 10: 436, 1898 BEUTEN. LARVA
ENTOMOLOGICA AMERICANA 5:200, 1889 BEUTEN. EGG, LARVA, PUPA
ENT. NEWS 1:20, 1890 SKINNER COCOON
LOC. CIT. 1:58, 1890 BEUTENMULLER NOTE ON COCOON
ENT. SOC. AMERICA, ANNALS 9:145; 1916 MOSHER PUPAL STRUCTURES
MARYLAND NAT. HIST. SOC., PUB. NO PAGE NUMBER, 1930 HAYDON EGG, LARVA, PUPA
NAT. ACAD. SCI., MEM. 12:232, 1914 PACKARD LIFE HISTORY EGG, LARVA* PL. 13, FF. 4-6; PL. 14, FF. 1-5 COLOR
NEW YORK STATE MUS., MEM. 8: 732, 1905 FELT LARVA (BRIEF)
PSYCHE 5:260, 1889 ELIOT & SOULE EGG, LARVA, PUPA, COCOON
ELIOT & SOULE-CATERPILLARS & THEIR MOTHS P. 241, 1902 EGG, LARVA* COCOON
KELLOGG-AMER. INSECTS P. 423, 1908 COCOON* F. 608
- - -
INDEX AMER. ECON. ENT., VOL. 2:64; VOL. 4:79; VOL. 7:154
FOOD PLANTS
LIRIODENDRON TULIPIFERA LINN.
PRUNUS SEROTINA EHRH.
SASSAFRAS ALBIDUM VAR. MOLLE(RAF.)

ANGULIFERA WALKER
SSP. SEE MACULATA HARRIS

ANGULIFERA WALKER
SYN. SEE SOCIALIS FEIST

ANGULIRENA SMITH
SSP. SEE BREVIPENNIS SMITH

ANGULOSA (ABBOT & SMITH,) LOPHODONTA-NOTODONTIDAE
LIFE HISTORY
AMER. MUS. NAT. HIST., BULL. 10: 412, 1898 BEUTEN. LARVA
AMER. NATURALIST 19:715, 1885 PACKARD PUPA (BRIEF)
BOSTON SOC. NAT. HIST., PROC. 26: 162, 1893 DYAR EGG, LARVA, PUPA
NATIONAL ACAD. SCI., MEM. 7:148, 1895 PACKARD EGG, LARVA* PL. 17, F. 5 COLOR; PUPA* COCOON
LOC. CIT. 9: 1905 PACKARD LARVA PL. 44, F. 8 HEAD
NEW YORK AGRI. EXP. STA. (CORNELL) MEM. 274:219, 1948 FORBES LARVA
NEW YORK ENT. SOC., JOURN. 1:58, 1893 PACKARD LARVA
NEW YORK STATE MUS., MEM. 8:705, 1905 FELT LARVA
UNITED STATES DEPT. AGRI., MICS. PUB. 657:364, 401, 1950 CRAIGHEAD LARVA
ABBOT & SMITH-LEP. INS. GEORGIA P. 165, 1797 LARVA* PL. 83 COLOR
FOOD PLANTS
CARYA ILLINOENSIS (WANG.)
QUERCUS SP. (OAKS)

ANGUSI GROTE, CATOCALA-NOCTUIDAE
SYN. FEMALE EDNA BEUTENMULLER
FORM LUCETTA HENRY EDWARDS
LIFE HISTORY
ENT. NEWS 20:135, 1909 ROWLEY LARVAL NOTE
FOOD PLANTS
CARYA SP. (HICKORIES)

ANGUSI GROTE & ROBINSON, DATANA-NOTODONTIDAE
LIFE HISTORY
AMER. MUS. NAT. HIST., BULL. 10: 407, 1898 BEUTEN. LARVA
CAN. ENT. 20:135, 1888 BEUTEN. LARVA AND KEY TO OTHER DATANA LARVA
ENTOMOLOGICA AMERICANA 6:219, 1890 BEUTEN. EGG, LARVA
INS. LIFE 2:149, 1889 BINKARD LARVA (BRIEF)
MAINE AGRI. EXP. STA., BULL. 259: 65, 1917 MOSHER PUPA* F. 5, D, E
MINNESOTA AGRI. EXP. STA., BULL. 61:163, 1899 LUGGER LARVA
NAT. ACAD. SCI., MEM. 7:111, 1895 PACKARD EGG, LARVA* PL. 10, F. 2 COLOR; PUPA*
NEW YORK AGRI. EXP. STA. (CORNELL), MEM. 274:215, 1948 FORBES LARVA
NEW YORK ENT. SOC., JOURN. 1:62, 1893 PACKARD LARVA
NEW YORK STATE MUS., MEM. 8: 535, 1905 FELT LARVA
PHILADELPHIA ENT. SOC., PROC. 6: 10, 1866 GROTE & ROBINSON MATURE LARVA
PSYCHE 6:573, 1893 DYAR KEY TO LARVA
UNITED STATES DEPT. AGRI., MISC. PUB. 657:355, 398, 1950 CRAIGHEAD LARVA
HERRICK-INS. ENEMIES SHADE TREES P. 127, 1935 EGG, LARVA
SMITH-INS. NEW JERSEY P. 454, 1900 EGG* LARVA* F. 208
- - -
INDEX AMER. ECON. ENT., VOL. 1:102; VOL. 2:109; VOL. 7:246
FOOD PLANTS
BETULA SP. (BIRCHES)
BETULA POPULIFOLIA MARSH.
CARYA SP. (HICKORIES)
CARYA ILLINOENSIS (WANG.)
FAGUS SP. (BEECHES)
HAMAMELIS VIRGINIANA LINN.
JUGLANS CINEREA LINN.
JUGLANS NIGRA LINN.

Quercus sp. (oaks)
Tilia sp. (lindens)
Tilia americana Linn.
Vaccinium sp.

ANGUSTIFASCIA Barnes & McDunnough
ssp. WEIDEMEYERII Edwards

ANGUSTIORATA Walker, CARIPETA-GEOMETRIDAE
syn. CRIMINOSA Swett
LIFE HISTORY
Amer. Naturalist 18:1045, 1884 Packard larva, pupa
New York Agri. Exp. Sta. (Cornell) Mem. 274:105, 1948 Forbes larva
New York State Mus., Mem. 8: 753, 1905 Felt larva (brief)
United States Dept. Agri. Bur. Ent., Bull. old series 3:25, 1883 Packard larva, pupa
FOOD PLANTS
Picea sp. (spruces)
Pinus strobus Linn.

ANICIA (Doubleday & Hewitson) EUPHYDRYAS-NYMPHALIDAE
For the many ssp. of this species see McDunnough's List
LIFE HISTORY
British Columbia Ent. Soc., Proc. 1 new series:10, 1911 Venables egg, young larva
Mead-Rept. Lep. Wheeler Exped. West 100th Meridian p. 758, 1875 larva (brief)
FOOD PLANTS
Erodium cicutarium (Linn.)
Penstemon sp.
Symphoricarpos albus (Linn.)
Trifolium sp. (clovers)
Viola sp. (violets)

ANILIS (Drury), ARGYROSTROTIS-GEOMETRIDAE
syn. SESQUISTRIARIS Hubner
LIFE HISTORY
Guenee-Spec. Gen. Lep. Noct. 3: 274, 1852 larva, pupa
FOOD PLANTS
Sabbatia sp.

ANITA Reiff
ssp. see DISSTRIA Hubner

ANIUSARIA Walker
syn. see RIBEARIA Fitch

ANNA (Edwards), PLEBEIUS-LYCAENIDAE
syn. CAJONA Reakirt
syn. PHILEMON Boisduval
syn. RICEI Cross
ab ARGYROTOXUS Behr
LIFE HISTORY
Ontario Ent. Soc., Ann. Rept. p. 59, 1899 Fletcher brief descr. of larva and of pupal habits
FOOD PLANTS
Lupinus sp. (lupines)

ANNA (Grote), APANTESIS-ARCTIIDAE
form PERSEPHONE Grote
LIFE HISTORY
Amer. Mus. Nat. Hist., Bull. 10: 378, 1898 Beuten. larva
New York Ent. Soc., Journ. 3: 178, 1895 Packard larva (PERSEPHONE)
Loc. cit. 8:40, 1900 Dyar larva (brief)
Psyche 8:53 1897 Dyar egg, larva
FOOD PLANTS
Low plants

ANNA Grote
syn. see AMESTRIS Strecker

ANNETTA Edwards
ssp. see SCUDDERI Edwards

ANNATEAE dosPassos
ssp. see IROIDES Boisduval

ANNETTEARIA Haimbach
ssp. see GRATARIA Fabricius

ANNEXA TREITSCHKE, FELTIA-NOCTUIDAE
SEE SUBTERRANEA FABRICIUS

ANNIDA FAGER
SSP. SEE PALAEOGAMA GUENEE

ANNISARIA WALKER
SYN. SEE RIBEARIA FITCH

ANNULIFASCIA WALKER
SYN. SEE CARYAE HARRIS

ANNULOSA WALKER, HALISIDOTA-ARCTIIDAE
SYN. NIVEIGUTTA WALKER
SYN. NUMBIFACTA DYAR
LIFE HISTORY
INSECUTOR INSCITIAE MENSTRUUS 5:130, 1917 DYAR LARVA
FOOD PLANTS

ANONA OTTOLENGUI
SSP. SEE GALBINA CLEMENS

ANONAE SHAW
SYN. SEE ANTAEUS DRURY

ANTAEGON BOISDUVAL
SYN. SEE ACMON DOUBLEDAY & HEWITSON

ANTAEUS (DRURY), COCYTIUS-SPHINGIDAE
SYN. JATROPHAE FABRICIUS
A HYDASPUS CRAMER
SYN. MEDOR STOLL
SYN. ANONAE SHAW
SYN. TAPAYUSA MOORE
LIFE HISTORY
AMER. ENT. SOC., TRANS. 15:154, 1888 SMITH LARVA (BRIEF)
ENTOMOLOGICA AMERICANA 3:164, 1887 HENRY EDWARDS PUPA
ENT. SOC. AMERICA, ANNALS 11:411, 1918 MOSHER PUPAL STRUCTURE
NOVITATES ZOOLOGICAE 9, SUPPLEMENT:58, 1903 ROTHSCHILD & JORDAN LARVA, PUPA
UNITED STATES NAT. MUS., PROC. 23: 256, 1901 DYAR EGG, LARVA
BOISDUVAL-SPEC. GEN. LEP. SPHING. P. 65, 1874 LARVA, PUPA
GUNDLACH-ENTOM. CUBANA P. 206, 1886, LARVA, PUPA
MERIAN-INS. SURINAM 1719 LARVA* PUPA* PL. 2 COLOR
SMITH-MONOG. SPHINGIDAE P. 154, 1888 LARVA (BRIEF)
FOOD PLANTS
ANNONA GLABRA LINN.
JATROPHA SP.

ANTAPICA SMITH
SSP. SEE PURPUREA GROTE

ANTAURATA WALKER
SYN. SEE DISTRIBUARIA HUBNER

ANTENNATA (WALKER), LITHOPHANE-NOCTUIDAE
SYN. CINEREA RILEY
LIFE HISTORY
ILLINOIS DEPT. AGRI., TRANS. 15: 227, 1877 FRENCH LARVA
JOURNAL ECON. ENT. 7:189, 1914 HERRICK EGG* F. 17, LARVA
MINNESOTA AGRI. EXP. STA., BULL. 61:225, 1899 LUGGER LARVA* F. 172
MISSOURI, REPT. NOX. INS. 3:135, 1871 RILEY LIFE HISTORY FF.
NEW YORK AGRI. EXP. STA. (CORNELL), BULL. 123:516, 1896 SLINGERLAND LARVA* PUPA* PL. 1
NEW YORK STATE MUS., BULL. 155: 49, 1912 FELT LARVA* PL. 17, F. 2
UNITED STATES DEPT. AGRI., MISC. PUB. 657:363, 394, 1950 CRAIGHEAD LARVA
HERRICK-INS. ENEMIES SHADE TREES P. 187, 1935 LARVA
HOLLAND-MOTH BOOK P. 206, 1905 LARVA* F. 117 (POOR)
SAUNDERS-INS. INJ. FRUITS P. 138, 1883 LARVA*

– – –

INDEX AMER. ECON. ENT.,
VOL. 1:32 XYLINA
VOL. 2:163; VOL. 3:180;
VOL. 3:437 XYLINA
VOL. 4:514; VOL. 5:281
VOL. 7:618 GRAPTOLITHA

FOOD PLANTS
Acer negundo Linn.
Acer saccharinum Linn.
Acer saccharophorum Koch
Carya sp. (hickories)
Fraxinus sp. (ashes)
Malus pumila Mill. (apple)
Populus sp. (poplars)
Rosa sp.
Salix sp. (willows)
Tilia sp. (lindens)
Fruit trees

ANTEPOSITA Guenee
syn. see SUBTERRANEA Fabricius

ANTHEDON Clark
ssp. see PORTLANDIA Fabricius

ANTHELLE Scudder
syn. see DORCAS Kirby

ANTHYALE Hubner
ssp. see PHILODICE Godart

ANTIACIS Boisduval
ssp. see XERCES Boisduval

ANTIBUBASTUS Hubner
syn. see HANNO Stoll

ANTICA Smith, TRICHOCLEA-Noctuidae
LIFE HISTORY
Southern California Acad. Sci., Bull. 28:58, 1929 Comstock larva
FOOD PLANTS
Atriplex canescens (Pursh)

ANTICA Walker
syn. see TENERA Hubner

ANTICARIA (Walker), AETHALURA-Geometridae
syn. SUBMURARIA Walker
a FUMATA Barnes & McDunnough
LIFE HISTORY

FOOD PLANTS
Alnus sp. (alders)
Aquilegia sp.
Betula sp. (birches)
Cytisus sp.
Genista sp.
Ligustrum sp.
Lonicera sp.
Populus sp. (poplars)
Prunus domestica Linn. (plum)
Quercus sp. (oaks)
Salix sp. (willows)
Sambucus sp.
Spartina sp.
Ulmus sp. (elms)

ANTICARIA (Walker), EUPITHECIA-Geometridae
syn. IMPLICATA Walker
syn. EXPLANATA Walker
LIFE HISTORY

FOOD PLANTS
Ribes sativum Syme. (currant)

ANTICARIA Walker
syn. see INDUCTATA Guenee

ANTICOSTIATA Strecker
syn. see MUNITATA Hubner

ANTICOSTIENSIS Grote
ssp. see PUDENS Guenee

ANTICOSTIENSIS Strecker
syn. see BREVICAUDA Saunders

ANTIGONE Strecker
syn. see CONGRUA Walker (Estigmene)

ANTILOCHUS Linneaus
syn. see GLAUCUS Linneaus

ANTINYMPHA (Hubner), CATOCALA-Noctuidae
syn. AFFINIS Westwood
syn. MELANYMPHA Guenee
form MULTOCONSPICUA Reiff
LIFE HISTORY
Amer. Mus. Nat. Hist., Bull. 16: 392, 1902 Beuten. larval head* f. 19

Amer. Mus. Nat. Hist., Mem. new series 3: part 1, 1918 Barnes & McD larva* Pl. 12, f. 14 color

FOOD PLANTS

Comptonia peregrina (Linn.)

<u>antiopa</u>(Linnaeus), <u>Nymphalis</u>-Nymphalidae

syn. <u>pompadour</u> Pollich

ab <u>grandis</u> Ehrmann

ab <u>obscura</u> Cockerell

ab <u>hippolyta</u> Lyman

ab <u>hygiaea</u> Heydenreich

syn. <u>lintnerii</u> Fitch

a <u>hyperborea</u> Seitz

LIFE HISTORY

Amer. Assoc. Advancement Sci., Proc. 28:458, 1879 Riley chrysalis

Amer. Ent. 2:307, 371, 1870 larva, pupa

Amer. Mus. Nat. Hist., Bull. 5: 267, 1893 Beuten. larva, chrysalis

Amer. Mus. Journ. p. 532, 1916 Lutz egg*

Buffalo Soc. Nat. Sci., Bull. 19: 26, 1939 Wild larva

California Acad. Sci., Proc. 5: 169, 1873 Hy. Edwards larva, chrysalis

Can. Ent. 1:75, 1869 Saunders larva

Can. Naturalist, p. 294, 1840 Gossee larva, chrysalis

Can. Record of Science 8:345, 1901 Norris life history

Ent. News 15:6, 1904 Field pupation notes

Entomologists Record 19:275-277 1907 Florsheim life history

Illinois Dept. Agri., Trans. 18: Append. pp. 85, 163, 1880 Coquillett & Midleton larva

Maine Naturalist 4:54, 1924 Kilburn egg* larva* pupa*

Michigan State College, Spec. Bull. 243:43, 1933 McDaniel larva* life history

Minnesota Agri. Exp. Sta., Bull. 6:84, 1903 Washburn larva* f. 8 color

Montana Univ., Bull. 30:96, 1906 Elrod egg, larva, chrysalis

Mus. Comp. Zoology, Bull. 87: No. 6, 1941 Dethier larval antenna* ff. 13, 16, 36

Naturaliste Canadien p. 169, 1921 Lissard life history

Loc. cit. p. 141, 1927 Huard larva

Nature Study Review p. 252, Sept., 1917 larva*

New Hampshire Agri. Exp. Sta., Bull. 67, 1899 Weed egg* larva* pupa*

New York Ent. Soc., Journ. 8:181, 1900 Weeks oviposition

New York State Mus., Bull. 37: 21, 1900 Felt larva*

New York State Mus., Mem. 8: 159, 1905 Felt egg* Pl. 10, ff. 7, 8; larva* Pl. 10, f. 9; chrysalis* Pl. 10, f. 10 color (all)

Ohio Agri. Exp. Sta., Bull. 332: 271, 1918 Houser larva, pupa

Ontario Ent. Soc., Ann. Rept. p. 136, 1905 Lockhead eggs, larva

Peabody Acad. Sci., Rept. p. 34, 1871 Scudder egg, larva, chrysalis

Philadelphia Ent. Soc., Proc. 3: 59, 1864 Lintner larva chrysalis

Psyche 26:13, 1919 Walker larval habits

United States Dept. Agri. Farmers Bull. 1169:48, 1921 Kotinsky egg* f. 29-7, 8; larva* f. 29-9; chrysalis* f. 29-10

United States Dept. Agri., Misc. Pub. 657:357, 369, 1950 Craighead larva f. 70 E

Vermont Board Agri., Fifth Rept. p. 265, 1878 Perkins larva, chrysalis

BALLARD-INS.LIVES P.40, 1879 LARVA, CHRYSALIS*

CHENU-LUCAS--ENCYC.NAT.HIST. PAPILLONS 1:103, 1856 LARVA, CHRYSALIS

COMSTOCK, J.A.-BUTTER.CALIF. P.126, 1927 LIFE HIST., CHRYSALIS* PL.A-41;PL.63,F.3 COLOR

COMSTOCK, J.H.-HOW TO KNOW BUTTER.P.148, 1904 EGG* LARVA* PL.25 COLOR

CRAGEN-OUR INS.FRIENDS & FOES P.87, 1899 EGG

DICKERSON-MOTHS & BUTTER.P.72, 1901 EGG* F.64; LARVA* FF. 65,66; CHRYSALIS* FF.67,68

DOANE ETC.-FOREST INS.P.332, 1936 LARVA* F.194; CHRYSALIS* F.194

DUNCAN-NATUR.LIBRARY 3:169, 1835 LARVA (BRIEF)

EDWARDS-KINGSLEY'S STAND.NAT. HIST.2:487, 1885 EGG, LARVA

EMMONS-NAT.HIST.NEW YORK 5: 207, 1854 LARVA* CHRYSALIS* PL.6 COLOR

ESSIG-INS.WEST.NORTH AMERICA P.643, 1926 EGG, LARVA* F.521; CHRYSALIS

ESSIG-INJ.&BENEF.INS.P.460, 1915 LARVA* F.472; CHRYSALIS

FERNALD-BUTTER.OF MAINE P.56, 1884 EGG, LARVA*, CHRYSALIS

FRENCH-BUTTER.EAST.UNITED STATES P.194, 1886 EGG*, LARVA, CHRYSALIS

GLOVER-ILLUS.NORTH AMER.ENT., 1878 LARVA, PUPA* PL.1,F.12, PL.21,F.2

GRAHAM-PRINCIPLES FOREST ENTOM. P.168, 1939 LARVA* F.57; CHRYSALIS* F.58

HARRIS-ENT.CORRESP.P.280, 1869 LARVA

HARRIS-INS.INJ.VEGET.1ST EDIT. P.219, 1841 LARVA, CHRYSALIS

HARRIS-LOC.CIT.2ND EDIT.P.238, 1852 LARVA, CHRYSALIS

HARRIS-LOC.CIT.FLINT EDIT.P.297 1862 LARVA, CHRYSALIS

HERRICK-INS.ENEMIES SHADE TREES P.88, 1935 EGG* F.57; LARVA* F.56

HOLLAND-BUTTER.BOOK P.5, 1905 EGG* F.11; LARVA* PL.3,F.28; CHRYSALIS* PL.4,FF.51,58,59

JAEGER-LIFE AMER.INS.P.202, 1864 LARVA, CHRYSALIS

KELLOGG-AMER.INS.P.455, 1908 LARVA

KIRBY-EUROPEAN BUTTER.& MOTHS P.12, 1882 LARVA* CHRYSALIS*

MAYNARD-BUTTER.NEW ENGLAND P.17, 1886 LARVA

PACKARD-INS.INJ. FOREST TREES P.64, 1881 LARVA (BRIEF)

SCUDDER-BUTTER.EAST.UNITED STATES & CANADA PP.400-407, 1889 PL.64,FF.26,33;PL.67, FF.3,6;PL.70,F.12;PL.74,F.28; PL.78,FF.50,51;PL.81,F.1;PL. 86,FF.83,84;PL.87,FF.13,21; PL.83,FF.51,58,59 LARVA, CHRYSALIS

SCUDDER-BUTTER.THEIR STRUCTURE PP.38,98, 1881 EGG* LARVA* CHRYSALIS*

SCUDDER-EVERYDAY BUTTER.P.4, 1889 EGG, LARVA, CHRYSALIS

SCUDDER-GUIDE TO COMMON BUTTER. P.91, 1893 LARVA, CHRYSALIS

STAINTON-BRITISH BUTTER.& MOTHS 1:39, 1857 LARVA, CHRYSALIS

STEPHENS-ILLUSTRATIONS BRITISH ENT.HAUSTELLATE 1:45, 1828 LARVA, CHRYSALIS

VANCE-LIFE AMONG THE BUTTER. P.42 1925 EGG, LARVA, CHRYSALIS

WESTWOOD-BRITISH BUTTER.P.50, 1855 LIFE HISTORY*

INDEX AMER.ECON.ENT.
VOL.1:132 EUVANESSA
VOL.2:9 AGLAIS
VOL.3:12 AGLAIS
VOL.4:12 AGLAIS

Vol.4:507 Vanessa
Vol.5:15 Aglais
Vol.5:678 Vanessa
Vol.6:536 Nymphalis
Vol.7:711 Nymphalis

FOOD PLANTS

Betula sp. (birches)
Betula papyrifera Marsh
Celtis occidentalis Linn.
Humulus lupulus Linn. (hops)
Populus sp. (poplars)
Populus gileadensis Roul.
Populus tremuloides Michx.
Pyrus communis Linn. (pear)
Rosa sp. (roses)
Salix sp. (willows)
Salix discolor Muhl.
Tilia sp. (lindens)
Ulmus sp. (elms)
Ulmus americana Linn.

ANTIPHOLA Walsh
syn. see TESSELLARIS Abbot & Smith

ANTIPHOLELLA Strand
syn. see TESSELLARIS Abbot & Smith

ANTIQUA (Linnaeus), NOTOLOPHUS-Liparidae
syn. NOVA Fitch
a BADIA Henry Edwards

LIFE HISTORY

Amer. Ent. 1: second series pp.77, 132, 1880 egg, cocoon
Amer. Mus. Nat. Hist., Bull. 10: 384 1898 Beutenmuller larva
Annals and Magazine of Nat. Hist. 2:460, 1829 Bree egg
British Columbia Ent. Soc., Proc. 36: new series p.14, 1940 Llewellyn Jones antiqua crossed with Hemerocampa pseudotsugata larva and pupa resulting
Loc. cit. 42:4, 1945 Hardy egg, larva, pupa, cocoon
California Acad. Sci., Proc. 5: 188, 1873 Henry Edwards larva, egg
Entomologica Americana 3:146, 1887 Henry Edwards larva (NOVA)
Insect Life 3:390, 1891 Dyar comparison of the eggs of ANTIQUA, DEFINITA, and LEUCOSTIGMA
Michigan State College, Spec. Bull. 243:18, 1933 McDaniel larva (brief)
New York Acad. Sci., Annals 8: 55, 1893 Packard egg, larva
New York Agri. Exp. Sta. (Cornell), Mem. 274:245, 1948 Forbes larva
New York State Mus., Mem. 8:524, 1905 Felt larva
New York State, Rept. on Insects 8:193, 1862 Fitch life hist.
Nova Scotia Ent. Soc., Proc. 3: 54, 1917 Payne egg, larva, pupa, cocoon
Ontario Ent. Soc., Ann. Rept. p. 55, 1903 Gibson larva
Oregon Biennial Crop Pest. Rept. 2:174, 1913 larva* pupa* Pl. 8
Papilio 3:38, 1883 Stretch egg, larva (BADIA)
Loc. cit. 3:189, 1883 Henry Edwards larva BADIA same as ANTIQUA
Psyche 6:419, 1893 Dyar egg, larva, pupa, cocoon (BADIA)
Loc. cit. 7:341, 1896 Dyar larva
Smithsonian Misc. Collections 4: Art. 2:250, 1862 Morris larva
United States Dept. Agri., Misc. Pub. 657:356, 407, 1950 Craighead larva
Washington Ent. Soc., Proc. 1:88, 1887 Riley larval glands
Doane-Forest Insects p.274, 1936 larva* f.156
Essig-Injur. & Benef. Ins. p.410, 1915 egg f.415
Essig-Ins. Western North America p.94, 1926 larva
Fernald-Applied Ent. p.265, 1926 larva

Glover-Illus. North Amer. Ent., 1878 larva* pupa* cocoon* Pl. 12, f. 18
Harris-Ins. Injur. Veget. 1st edit. p. 263, 1841 larva, pupa
Loc. cit. 2nd edit. p. 284, 1852 larva, cocoon
Loc. cit. Flint edit. p. 369, 1862 larva, cocoon
Herrick-Ins. Enemies Shade Trees p. 279, 1935 larva
Morris-Synop. Lep. North America p. 250, 1862 larva
O;Kane-Injur. Insects p. 295, 1912 egg* ff. 16, 440; larva* f. 16; pupa* f. 16
Slingerland & Crosby-Manual Fruit Ins. p. 105, 1914 larva* f. 115

\- - -

Index Amer. Econ. Ent.
Vol. 1:152 Hemerocampa
Vol. 2:253; Vol. 3:279; Vol. 4:325;
Vol. 5:450; Vol. 6:535; Vol. 7:709.

FOOD PLANTS
Acer sp. (maples)
Alnus sp. (alders)
Amelanchier florida Lindl.
Betula sp. (birches)
Betula populifolia Marsh.
Corylus sp. (hazels)
Crataegus sp.
Crataegus douglasii Lindl.
Cydonia oblonga Mill. (quince)
Fagus sp. (beeches)
Fraxinus sp. (ashes)
Humulus lupulus Linn. (hops)
Larix sp. (larches)
Malus pumila Mill. (apple)
Picea pungens Engelm.
Pinus strobus Linn.
Populus sp. (poplars)
Populus tremloides Michx.
Prunus americana Marsh.
Prunus armeniaca Linn. (apricot)
Prunus domestica Linn. (plum)
Prunus domestica v. gelatensis ex. Hook (prune)
Prunus serotina Ehrh.
Pyrus commumis Linn. (pear)
Quercus sp. (oaks)
Ribes sanguineum Pursh.
Rosa sp. (roses)
Rubus sp. (blackberries)
Rubus allegheniensis Porter
Rubus vitifolius C. & S.
Salix sp. (willows)
Salix hookeriana Barr.
Salix scouleriana Barr.
Sorbus americana Marsh.
Tilia sp. (lindens)
Ulmus sp. (elms)
Vaccinium sp. (blueberries)
Wisteria sinensis Sweet

<u>antonia</u> Edwards
ssp. see <u>celtis</u> Boisduval & Leconte

<u>apache</u> Poling
ssp. see <u>mubilis</u> Hubner

<u>apacheana</u> Skinner
syn. see <u>nokomis</u> Edwards

<u>apamiformis</u> (Guenee), <u>Septis</u>-Noctuidae
syn. <u>contenta</u> Walker
LIFE HISTORY
Washington Ent. Soc., Proc. 4:320, 1898 Dyar larva
FOOD PLANTS

<u>apera</u> Druce, <u>Nola</u>-Nolidae
syn. <u>involuta</u> Dyar
a <u>exposita</u> Dyar
LIFE HISTORY
New York Ent. Soc., Journ. 6:249, 1898 Coquillett larva incorrectly called <u>miniuscula</u> Zeller
Loc. cit. 7:210, 1899 for correction of above
Washington Ent. Soc., Proc. 4:465, 1900 Dyar larva
FOOD PLANTS
Laguncularia racemosa Gaertn.
Salix sp. (willows)

APHRODITE(FABRICIUS), ARGYNNIS-NYMPHALIDAE

‡ DAPHNIS MARTYN
A ALCESTIS EDWARDS
AB SUFFUSA WOLLCOTT
B CYPRIS EDWARDS
SYN. ETHNE HEMMING
AB MAYAE GUNDER
C COLUMBIA HENRY EDWARDS
D BYBLIS BARNES & BENJAMIN
E WINNI GUNDER

LIFE HISTORY

AMER. ENT. SOC., TRANS. 5:289, 1876 EDWARDS EGGS
AMER. MUS. NAT. HIST., BULL. 5: 255, 1893 BEUTENMULLER LARVA
BUFFALO SOC. NAT. SCI., BULL. 19: 22, 1929 WILD LARVA
CAN. ENT. 6:121, 1874 W.H. EDWARDS LARVA
LOC. CIT. 12:69, 1880 W.H. EDWARDS EGG, LARVA, CHRYSALIS
ILLINOIS DEPT. AGRI., TRANS. 18: APPEND. P. 81, 82, 1880 MIDDLETON LARVA
MONTANA UNIV., BULL. 30:57, 1906 ELROD EGG, LARVA, CHRYSALIS
NATURALISTE CANADIEN P. 116, 1927 HUARD LARVA
NEW YORK ENT. SOC., JOURN. 15: 56, 1907 FORBES LARVA
COMSTOCK-HOW TO KNOW BUTTER., P. 117, 1904 LARVA
EDWARDS, W.H.-BUTTER. NORTH AMERICA, 1ST SERIES, 1868 LARVA PL. ARGYNNIS NO. 3
EDWARDS, W.H.-LOC. CIT. 2ND SERIES, 1876 LIFE HISTORY* PL. ARGYNNIS NO. 7
EDWARDS, W.H.-LOC. CIT. 3RD SERIES, 1879 LIFE HISTORY* PL. ARGYNNIS NO. 5, FF. A, H
FRENCH-BUTTER. EASTERN UNITED STATES P. 159, 1886 LIFE HISTORY (ALCESTIS)
HOLLAND-BUTTER. BOOK P. 107 1905 CHRYSALIS* PL. 5, F. 5
HOLLAND-LOC. CIT. 3RD EDITION P. 88, 1931 LARVA, CHRYSALIS
MAYNARD-BUTTER. NEW ENGLAND P. 24, 1886 LARVA
SCUDDER-BUTTER. EASTERN UNITED STATES & CANADA PP. 566-569, 1889 LIFE HISTORY* PL. 64, F. 35; PL. 67, F. 11; PL. 72, F. 5; PL. 79, F. 3; PL. 85, F. 5
SCUDDER-EVERYDAY BUTTER. P. 263, 1899 EGG, LARVA, PUPA
SCUDDER-GUIDE TO COMMON BUTTER. PP. 77, 78, 1893 LARVA, CHRYSALIS

- - -

INDEX AMER. ECON. ENT., VOL. 7:105

FOOD PLANTS

PASSIFLORA INCARNATA LINN.
PODOPHYLLUM PELTATUM LINN.
PORTULACA OLERACEA LINN.
VIOLA SP. (VIOLETS)

APIATA(GROTE), EPIGLAEA-NOCTUIDAE

LIFE HISTORY

MASSACHUSETTS AGRI. EXP. STA., BULL. 239:27, 1928 FRANKLIN LARVA, CHRYSALIS FF. 2A, 2B, 4A, 4B COLOR
NEW YORK ENT. SOC., JOURN. 37:409 1929 BECKWITH LIFE HIST., LARVA*
UNITED STATES DEPT. AGRI. FARMERS BULL. 860:23, 1917 SCAMMILL LARVA* F. 21; PUPA

- - -

INDEX AMER. ECON. ENT., VOL. 2:132; VOL. 5:223

FOOD PLANTS

VACCINIUM SP.
VACCINIUM MACROCARPON AIT.

APICALIS(GROTE & ROBINSON), SCHIZURA-NOCTODONTIDAE

LIFE HISTORY

NEW YORK AGRI. EXP. STA. (CORNELL) MEM. 274:229, 1948 FORBES LARVA

FOOD PLANTS

MYRICA CERIFERA LINN.
POPULUS SP. (POPLARS)
SALIX SP. (WILLOWS)
VACCINIUM SP. (BLUEBERRIES)

APICALIS WALKER, ICHTHYURA-NOTODONTIDAE
SYN. VAU FITCH
SYN. INDENTATA PACKARD
A ORNATA GROTE & ROBINSON
SYN. INCARCERATA BOISDUVAL
B ASTORIAE HENRY EDWARDS
C BIFIRIA HENRY EDWARDS
LIFE HISTORY
AMER. MUS. NAT. HIST., BULL. 4:67, 1892 BEUTEN. LARVA
BOSTON SOC. NAT. HIST., PROC. 27: 140, 1896 DYAR LARVAL SETAE
CAN. ENT. 17:248, 1885 FRENCH EGG, LARVA, PUPA (ORNATA)
LOC. CIT. 25:159, 185, 1893 DYAR EGG, LARVA (VAU)
ENT. NEWS 3:5, 1892 DYAR EGG, LARVA, PUPA, COCOON (BIFERA)
NAT. ACAD. SCI., MEM. 7:127, 1895 PACKARD EGG, LARVA* PL. 15, FF. 1,2,3 COLOR
NEW YORK ENT. SOC., JOURN. 1:26, 1893 PACKARD LARVA
NEW YORK STATE AGRI. EXP. STA. (CORNELL), MEM. 274:211, 1948 FORBES LARVA
NEW YORK STATE MUS., MEM. 8:747, 1905 FELT LARVA (BRIEF)
NEW YORK STATE, REPT. NOX. INS. 5:65, SEC. 351, 1858 FITCH LARVA
HART-INSECT COLLECTIONS FOR ILLINOIS HIGH SCHOOLS P. 45, 1903 COCOON
FOOD PLANTS
MYRICA CERIFERA LINN.
POPULUS SP. (POPLARS)
POPULUS TREMULOIDES MICHX.
SALIX SP. (WILLOWS)
SALIX BABYLONICA LINN.

APICELLA (GROTE), HELIOCONTIA-NOCTUIDAE
SYN. TRUNCATULA ZELLER
SYN. ACCEPTA HENRY EDWARDS
LIFE HISTORY
NEW YORK ENT. SOC., JOURN. 8:33, 1900 SWAINSON LARVA, PUPA
FOOD PLANTS
GUTIERREZIA SAROTHRAE (PURSH) "YELLOW FLOWER"

APPALACHIA CHERMOCK
SSP. SEE EURYDICE JOHANSSON

APPASSIONATA (HARVEY), PAPAIPEMA-NOCTUIDAE
SYN. HORNI STRAND
LIFE HISTORY
CAN. ENT. 35:91, 1903 BIRD LARVA
ENT. NEWS 19:150, 1908 JONES LIFE HISTORY PL. SHOWS BURROWS
- - -
INDEX AMER. ECON. ENT., VOL. 3:294
FOOD PLANTS
SARRACENIA SP. (PITCHER PLANTS)

APPOSITA (GROTE), ABAGROTIS, NOCTUIDAE
LIFE HISTORY
UNITED STATES NAT. MUS., PROC. 27:820, 1904 DYAR EGG
FOOD PLANTS

APPROXIMARIA DYAR
SSP. SEE PHLOGOSARIA GUENEE

APPROXIMARIA (HUBNER), STENOTRACHELYS-GEOMETRIDAE
LIFE HISTORY
NEW YORK AGRI. EXP. STA. (CORNELL) MEM. 274:79, 1948 FORBES LARVA
GUENEE-SPEC. GEN. LEP. PHAL. I:290, 1857 LARVA
GUMPPENBERG-SYSTEMA GEOMETRARUM 6:370, 1893 LARVA
PACKARD-INS. INJUR. FOREST TREES P. 49, 1881 LARVA
PACKARD-MONOGRAPH GEOMETRIDAE, P. 450, 1876 LARVA
FOOD PLANTS
QUERCUS SP. (OAKS)
QUERCUS VIRGINIANA MILL.
RHEXIA CILIOSA MICHX.
SMILAX LAURIFOLIA LINN.
SMILAX ROTUNDIFOLIA LINN.

APPROXIMARIA PACKARD
SYN. SEE PYROLARIA GUENEE

APPROXIMARIA PACKARD
SYN. SEE MIMOSARIA GUENEE

APPROXIMATA STRECKER
SYN. SEE MINUTA EDWARDS (EUPHYDRYAS)

APPROXIMATA STRETCH
SSP. SEE PARTHENICE KIRBY

APPROXIMARIA (HUBNER), STENOTRACHELYS-GEOMETRIDAE
LIFE HISTORY
NEW YORK AGRI. EXP. STA. (CORNELL), MEM. 274:79, 1948 FORBES LARVA
GUENEE-SPEC. GEN. LEP. PHAL. I: 290, 1857 LARVA
GUMPPENBERG-SYSTEMA GEOMETRARUM 6:370, 1893 LARVA
PACKARD-INS. INJUR. FOREST TREES P.49, 1881 LARVA
PACKARD-MONOGRAPH GEOMETRIDAE, P.450, 1876 LARVA
FOOD PLANTS
QUERCUS SP. (OAKS)
QUERCUS VIRGINIANA MILL.
RHEXIA CILIOSA MICHX.
SMILAX LAURIFOLIA LINN.
SMILAX ROTUNDIFOLIA LINN.

APPROXIMARIA PACKARD
SYN. SEE PYROLARIA GUENEE

APPROXIMARIA PACKARD
SYN. SEE MIMOSARIA GUENEE

APPROXIMATA STRECKER
SYN. SEE MINUTA EDWARDS (EUPHYDRYAS)

APPROXIMATA STRETCH
SSP. SEE PARTHENICE KIRBY

APRICATUS STICHEL
SSP. SEE SMINTHEUS DOUBLEDAY & HEWITSON

APURPURA BARNES & MCDUNNOUGH
SSP. PURPURISSATA GROTE (POLIA)

AQUILO (BOISDUVAL), PLEBEIUS-LYCAENIDAE
SYN. FRANKLINII CURTIS
SYN. SUTTONI HOLLAND
AB KOHLSAATI GUNDER
A RUSTICA EDWARDS
B PODARCE FELDER & FELDER
SYN. TEHAMA REAKIRT
SYN. CILLA BEHR
SYN. NESTOS BOISDUVAL
C MEGALO MCDUNNOUGH
D BRYANTI LEUSSLER
LIFE HISTORY
BROOKLYN ENT. SOC., BULL. 30:59, 1935 LEUSSLER BRIEF LIFE HISTORY BRYANTI
FOOD PLANTS
DODECATHEON MEADIA LINN.

AQUILONARIA CASSINO & SWETT
SSP. SEE ABRASARIA HERRICH-SCHAEFFER

AQUILONARIS LINTNER
SSP. SEE SCOLOPENDRINA BOISDUVAL

AQUOSUS GROTE & ROBINSON
SSP. SEE PUBER GROTE & ROBINSON

ARACHNE EDWARDS
SSP. SEE POLA BOISDUVAL

ARACINTHUSALIS WALKER
SYN. SEE ANGULALIS HUBNER

ARALIAE BIRD, PAPAIPEMA-NOCTUIDAE
LIFE HISTORY
CAN. ENT. 53:138, 1921 BIRD-JONES LARVA* PL.6, FF.1.3
FOOD PLANTS
ARALIA SPINOSA LINN.
ROSA SP.

ARAXES (HEWITSON), APYRROTHRIX-HESPERIIDAE
SYN. CYRILLUS PLOETZ
A ARIZONAE GODMAN & SALVIN
LIFE HISTORY
BROOKLYN MUS. ARTS & SCI., BULL. 1:193, 1906 DYAR LARVA (ARIZONAE)

FOOD PLANTS
Loc.cit.11:224, 1879 W.H. Edwards larval habits
Loc.cit.12:248, 1880 W.H. Edwards larva, chrysalis
Illinois Dept.Agri.,Trans.15: 154, 1877 French larva
Minnesota Agri.Exp.Sta.,Bull. 61:73, 1899 Lugger egg* larva* chrysalis* all f.19
Missouri Rept.on Insects 3:153-157, 1871 Riley egg* larva* chrysalis*
Mus.Comp.Zoology,Bull.87: No.6, 1941 Dethier larval antennae* ff.14,25
Naturaliste Canadien p.163, 1928 Huard larva, chrysalis
Nature Study Review p.240, 1917 larva* pupa
New York State Mus.Handbook 13:19, 1932 Saunders larva ff.2,3
New York State Mus.,Mem.8:559 1905 Felt larva
Loc.cit.8:704, 1905 Felt larva (brief)
New York State,Rept.Ins.3:379, also p.61,sec.82, 1856 Fitch larva
Ohio Agri.Exp.Sta.,Bull.332: 253, 1918 Houser larva* pl.38,f.1
Ontario Ent.Soc.,Rept.pp.51: 54, 1872 Saunders egg* larva* chrysalis*
Loc.cit.p.45, 1889 Harrington egg* larva* chrysalis
Loc.cit.p.34, 1894 Bethune larva* chrysalis
Loc.cit.p.31, 1896 Hoyes-Panton larva* chrysalis
Loc.cit.p.77, 1901 Lockhead chrysalis*
Loc.cit.p.110, 1902 Lockhead larva* chrysalis
Papilio 3:26, 1883 Mundt larva of dissipus same as ursula

arcasaria(Walker),**Antepione**-Geometridae
syn. **depontanata** Grote
syn. **sulphuraria** Packard
form **furciferata** Packard
LIFE HISTORY
Can.Ent.36:103, 1904 Seifert egg, larva, pupa
New York Agri.Exp.Sta. (Cornell),Mem.274:108, 1948 Forbes egg, larva, pupa
New York State Mus.,Mem.8:750, 1905 Felt larva (brief)
FOOD PLANTS
Acer sp. (maples)
Alnus sp. (alder)
Diospyros virginiana Linn.
Rhus copallina Linn.
Rhus glabra Linn.
Rhus typhina Linn.

archesilea R. Felder
syn.see **frisia** Poey

archippus(Cramer),**Basilarchia**-Nymphalidae
For the many ssp. and syns. of this species see McDunnough's List
LIFE HISTORY
Amer.Ent.1:192, 1869 larva* pupa* f.134
Loc.cit.2:246, 1870 pupa*
Amer.Mus.Nat.Hist.,Bull.5: 271, 1893 Beutenmuller larva, chrysalis
Buffalo Soc.Nat.Sci.,Bull.29: 27, 1939 Wild larva
California Acad.Sci.,Proc.5: 329, 1873 Henry Edwards larva, chrysalis (**misippus**)
Can.Ent.1:94, 1869 Saunders larva
Loc.cit.3:52, 1871 Riley larva, chrysalis and differences between those of **ursula**
Loc.cit.6:46, 1874 Saunders egg* larva* chrysalis*

PHILADELPHIA ENT. SOC., PROC. 3: 63, 1864 LINTNER LARVA, CHRYSALIS
SMITHSONIAN MISC. COLLECTIONS 4:ART. 2:66, 1862 MORRIS LARVA, CHRYSALIS
SOUTHERN CALIFORNIA ACAD. SCI., BULL. 32:27, 1933 COMSTOCK-DAMMERS LIFE HIST. (OBSOLETA), EGG* LARVA* CHRYSALIS* PL. 5 (ALL)
UNITED STATES DEPT. AGRI., MISC. PUB. 657:370, 1950 CRAIGHEAD LARVA* F. 70 F
ABBOT & SMITH-LEP. INS. GEORGIA I:11, 1797 LARVA* PUPA* PL. 6 COLOR
BOISDUVAL & LECONTE-HIST. LEP. NORTH AMERICA P. 204, 1833 LARVA* CHRYSALIS* PL. 55 COLOR
COMSTOCK-HOW TO KNOW BUTTER. P. 170, 1904 LARVA
CRAGEN-OUR INS. FRIENDS & FOES P. 89, 1899 LARVA* F. 26
EDWARDS, W. H.-BUTTER. NORTH AMERICA 2ND SERIES 1882 LIFE HISTORY PL. LIMENITIS NO. 2 COLOR (EROS)
FERNALD-BUTTER. MAINE PP. 63, 64, 1884 EGG, LARVA*, CHRYSALIS*
FERNALD-APPLIED ENT. P. 294, 1926 CHRYSALIS
FRENCH-BUTTER. EAST. UNITED STATES P. 211, 1886 EGG* LARVA* CHRYSALIS*
FRENCH-LOC. CIT. P. 214 LIFE HISTORY EROS
GLOVER-ILLUS. NORTH AMER. ENT., 1878 LARVA, PUPA PL. 1, FF. 2, 11; PL. 22, F. 15
HARRIS-INS. INJUR. VEGET. FLINT EDIT. P. 282, 1862 LARVA, CHRYSALIS
HART-INS. COLLECTIONS FOR ILLINOIS HIGH SCHOOLS P. 11, 1903 LARVA* CHRYSALIS* HIBERNACULUM* F. 6
HOLLAND-BUTTER. BOOK 1905 LARVA* CHRYSALIS* PL. 3, FF. 19, 22, 24; PL. 4, FF. 18-20
MAYNARD-BUTTER. NEW ENGLAND P. 9, 1886 LARVA
MORRIS-SYNOP. LEP. NORTH AMERICA P. 66, 1862 LARVA, CHRYSALIS
SAUNDERS-INS. INJ. FRUITS PP. 169, 170, 1883 EGG* LARVA* CHRYSALIS*
SAUNDERS-LOC. CIT. P. 169, 1889 EGG* F. 179; LARVA* F. 180; CHRYSALIS* F. 180.
SCUDDER-BUTTER. THEIR STRUCTURE PP. 10, 22, 45, 112, 1881 EGG* LARVA* CHRYSALIS* HIBERNACULUM*
SCUDDER-EVERYDAY BUTTER. P. 301, 1899 EGG, LARVA, CHRYSALIS
SCUDDER-GUIDE TO COMMON BUTTER. P. 102, 1893 LARVA, CHRYSALIS
VANCE-LIFE AMONG THE BUTTER. P. 43, 1925 EGG, LARVA, CHRYSALIS

- - -

INDEX AMER. ECON. ENT.
VOL. 1:47; VOL. 5:72 DISSIPUS
VOL. 6:83 FLORIDENSIS
VOL. 7:122

FOOD PLANTS

CHRYSOBALANUS OBLONGIFOLIUS MICHX.
CRATAEGUS SP. (THORNS)
MALUS PUMILA MILL. (APPLE)
POPULUS SP. (POPLARS)
POPULUS FREMONTI S. WATS.
POPULUS GILEADENSIS ROUL.
POPULUS HETEROPHYLLA LINN.
POPULUS TREMULOIDES MICHX.
PRUNUS SP.
PRUNUS DOMESTICA LINN. (PLUM)
PRUNUS SEROTINA EHRH.
PYRUS COMMUNIS LINN. (PEAR)
QUERCUS SP. (OAKS)
SALIX SP. (WILLOWS)
SALIX INTERIOR ROWLEE
SALIX SERICEA MARSH.
VACCINIUM SP. (HUCKLEBERRIES)

ARCHIPPUS FABRICIUS
SYN. SEE PLEXIPPUS LINNEAUS

ARCIGERA(GUENEE), SCHINIS–NOCTUIDAE
SYN. FEMALE SPRAGUEI GROTE
A FERRICOSTA SMITH
LIFE HISTORY
ENT. NEWS 34:197, 1923
BIRD LARVA, PUPA
FOOD PLANTS
ASTER ERICOIDES LINN.
ASTER LAEVIS LINN.
ASTER PUNICEUS LINN.

ARCTICA DOSPASSOS
SSP. SEE THAROS DRURY

ARCTICA(BOISDUVAL), SEPTIS–NOCTUIDAE
SYN. AMPUTATRIX FITCH
SYN. AMICA HARRIS
SYN. PLUVIOSA WALKER
AB FORMOSUS ELLSWORTH
LIFE HISTORY
AMER. MUS. NAT. HIST., BULL. 16: 306, 1901 BEUTENMULLER LARVA
CAN. ENT. 2:28, 1869 MINOT EGG
LOC. CIT. 2:75, 1869 SAUNDERS LARVA
ILLINOIS AGRI. EXP. STA., BULL. 95: 351, 1904 FORBES LARVA
ILLINOIS DEPT. AGRI., TRANS. 15: 96, 1877 THOMAS LARVA
LOC. CIT. 15:217, 1877 FRENCH LARVA
LOC. CIT. 18: APPEND. P. 137, 1880 MARTIN LARVA
NEW YORK, REPT. NOX. INS. 3:425, 1856 FITCH LARVA ALSO, P. 107, SEC. 138
ONTARIO ENT. SOC., ANN. REPT. P. 39, 1879 BOWLES LARVA
LOC. CIT. P. 49, 1890 FYLES LARVA (AMICA)
HARRIS–ENT. CORRESP. P. 316, 1869 LARVA
HARRIS–INS. INJ. VEGET. 2ND EDIT. P. 350, 1852 LARVA
HARRIS–LOC. CIT. FLINT EDIT. P. 450, 1862 LARVA
– – –
INDEX AMER. ECON. ENT. VOL. 1:146; VOL. 4:212; VOL. 5:288; VOL. 7:875.
FOOD PLANTS
AVENA SATIVA LINN. (OATS)
BRASSICA OLERACEA LINN. (CABBAGE)
LACTUCA SATIVA LINN. (LETTUCE)
RIBES SATIVUM SYME (CURRANT)
ROSA SP. (ROSES)
SPINACIA OLERACEA LINN. (SPINACH)
TRITICUM AESTIVUM LINN. (WHEAT)
ZEA MAYS LINN. (CORN)
GRASSES
YOUNG TREES, FRUIT TREES

ARCTICA GIBSON
SYN. SEE MELISSA FABRICIUS

ARCTICA MOESCHLER
SYN. SEE INTERROGATIONIS LINNEAUS

ARCTICA VERITY
SYN. SEE NAPI LINNEAUS

ARCTICUS LEUSSLER
SSP. SEE FAUNUS EDWARDS

ARCTICUS SKINNER
SYN. SEE GLAUCUS LINNEAUS

ARCTIVORENS HAMPSON, PAPAIPEMA–NOCTUIDAE
LIFE HISTORY
CAN. ENT. 43:45, 1911
BIRD EGG, LARVA, PUPA
FOOD PLANTS
ARCTIUM LAPPA LINN.
CIRSIUM SP.
CIRSIUM ARVENSE (LINN.)
CIRSIUM VULAGRE (SAVI.)
DIPSACUS SYLVESTRIS HUDSON

ARCUATA WALKER, DREPANA–DREPANIDAE
SYN. FABULA GROTE
FORM GENICULA GROTE
A SICULIFERA PACKARD
B ALASKENSIS BARNES & BENJAMIN
FORM GROTEI BARNES & BENJAMIN
LIFE HISTORY

AMER. MUS. NAT. HIST., BULL. 10: 389, 1898 BEUTENMULLER LARVA
BOSTON SOC. NAT. HIST., PROC. 24: 483, 1890 PACKARD EGG, LARVA
ENTOMOLOGICA AMERICANA 5:38, 1889 BEUTENMULLER LARVA
MAINE AGRI. EXP. ST., BULL. 259:77, 1917 MOSHER LARVA, PUPA* F. 6E
NATURALISTE CANADIEN P.323, 1935 DAVICAULT EGG, LARVA, PUPA
NEW YORK ACAD. SCI., ANNALS 8: 227, 1894 DYAR LARVAL SETAE
NEW YORK ENT. SOC., JOURN. 3:67, 1895 DYAR LARVA
NEW YORK STATE MUS., MEM. 8:749 1905 FELT LARVA (BRIEF)

- - -

INDEX AMER. ECON. ENT., VOL. 6:22

FOOD PLANTS
ALNUS SP. (ALDERS)
BETULA SP. (BIRCHES)
BETULA ALBA LINN.

ARCUATA WALKER
SYN. SEE TURBULENTA HUBNER (PHOSPHILA)

ARDEA EDWARDS
SSP. SEE ICARIOIDES BOISDUVAL

ARELI STRECKER, ACONTIA-NOCTUIDAE
SYN. MOSTROSA STRAND
LIFE HISTORY
UNITED STATES NAT. MUS., PROC. 27:876, 1904 DYAR EGG, LARVA
FOOD PLANTS

AREOLATUS ABBOT & SMITH
SYN. SEE PHOCION FABRICIUS

ARETARIA (WALKER), CARIPETA-GEOMETRIDAE
SYN. SUBOCHREARIA GROTE
LIFE HISTORY

FOOD PLANTS
PUPA FOUND ON PINUS RESINOSA AIT.

ARETHUSA DOD
SSP. SEE HYPOPHLAEAS BOISDUVAL

ARGE (DRURY) APANTESIS-ARCTIIDAE
SYN. DIONE FABRICIUS
SYN. INCARNATORUBRA GOEZE
SYN. COELEBS MARTYN
AB NERVOSA NEUMOEGEN & DYAR
‡ STRIGOSA STRETCH
LIFE HISTORY
AMER. MUS. NAT. HIST., BULL. 10: 380, 1898 BEUTENMULLER LARVA
CAN. ENT. 6:98, 1874 PEABODY LARVA
LOC. CIT. 35:122, 1903 GIBSON LARVA
CAN. JOURN. INDUSTRY, SCI. & ART 8:356, 1863 SAUNDERS LARVA
ILLINOIS DEPT. AGRI., TRANS. 15: 182, 1877 FRENCH LARVA
LOC. CIT. 18: APPEND. PP. 115, 170, 1880 COQUILLETT & MARTIN LARVA
NEW YORK ENT. SOC., JOURN. 8:36, 1900 DYAR LARVA
LOC. CIT. 10:1, 1902 SEIFERT EGG, LARVA, PUPA
NEW YORK STATE CABIN. NAT. HIST. ANN. REPT. NO. 26:143, 1874 LINTNER LARVA
PSYCHE 2:79, 1877 ANDREWS LARVA
SMITHSONIAN MISC. COLLECTIONS 4: ART. 2:340, 1862 MORRIS LARVA (DIONE)
ABBOT & SMITH-LEP. INS. GEORGIA P. 125, 1797 LARVA* PL. 63 COLOR
DUNCAN-NATURALISTS LIBRARY 22: 175, 1858 LARVA* PL. 19, F. 4 COLOR; PUPA* PL. 19, F. 5 COLOR
EMMONS-NATURAL HIST. NEW YORK 5:226, 1854 LARVA* PUPA* PL. 41
GLOVER-ILLUS. NORTH AMER. ENT. 1878 LARVA* PL. 8, F. 1; P. 34, 1837 LARVA

HARRIS-ENT. CORRESP. P. 286, 1869 LARVA
HARRIS-INS. INJ. VEGET. 1ST EDIT. P. 244, 1841 LARVA
HARRIS-LOC. CIT. 2ND EDIT. P. 265, 1852 LARVA
HARRIS-LOC. CIT. FLINT EDIT. P. 346, 1862 LARVA
JAEGER-LIFE NORTH AMER. INS. P. 168, 1864 LARVA
MORRIS-SYNOP. LEP. NORTH AMERICA P. 340, 1862 LARVA
STRETCH-ZYGAEN. & BOMBYC. NORTH AMERICA P. 225, 1873 LARVA
WESTWOOD-ILLUS. EXOTIC ENT. 1:34, 1837 LARVA
INDEX AMER. ECON. ENT., VOL. 1:20; VOL. 2:29

FOOD PLANTS

CHENOPODIUM SP.
GOSSYPIUM HERBACEUM LINN. (COTTON)
OENOTHERA BIENNIS LINN.
PLANTAGO SP.
PLANTAGO LANCEOLATA LINN.
POLYGONUM SP.
RUMEX SP.
ZEA MAYS LINN. (CORN)
CACTUS; LOW PLANTS

ARGE STRECKER
SYN. SEE MORMONIA BOISDUVAL (ARGYNNIS)

ARGENTATA (DRURY), PHRYGIONIS-GEOMETRIDAE
SYN. ARGENTISTRIATA STRECKER
SYN. CERUSSARIA GROTE
SYN. OBRUSSATA GROTE

LIFE HISTORY

NEW YORK ENT. SOC., JOURN. 53: 194, 1945 FORBES LARVA, PUPA
WASHINGTON ENT. SOC., PROC. 4: 458, 1900 DYAR LARVA

FOOD PLANTS

ARDISIA ESCALLONIOIDES SCHIEDE & DEEPE

ARGENTATA FLETCHER
SSP. SEE PSEUDARGIOLUS BOISDUVAL & LECONTE

ARGENTATA PACKARD, HALISIDOTA-ARCTIIDAE
A SUBALPINA FRENCH
B SOBRINA STRETCH

LIFE HISTORY

CALIFORNIA ACAD. SCI., PROC. 5:369, 1873 HENRY EDWARDS LARVA, CHRYSALIS OF SOBRINA AND ARGENTATA
CAN. DEPT. AGRI., REPT. ENT. P. 34, 1887 FLETCHER LARVA (SOBRINA) BRIEF
SOUTHERN CALIFORNIA ACAD. SCI., BULL. 36:119, 1937 COMSTOCK LARVA* PL. 52 SUBALPINA
DOANE ETC-FOREST INS. P. 287, 1936 COCOON* F. 167
ESSIG-INS. WEST. NORTH AMERICA P. 673, 1926 LARVA

- - -

INDEX AMER. ECON. ENT., VOL. 4:213; VOL. 5:290 SOBRINA VOL. 6:340; VOL. 7:448

FOOD PLANTS

ABIES BALSAMEA (LINN.)
ABIES CONCOLOR (GORD. & GLAND.)
JUNIPERUS SP.
PICEA ABIES (LINN.)
PINUS CONTORTA DOUGL.
PINUS LAMBERTIANA DOUGL.
PINUS PONDEROSA LAWS
PINUS RADIATA DON
PLANTAGO SP.
PRUNUS CERASUS LINN. (CHERRY)
PSEUDOTSUGA MENZIESII (MIRB.)
RIBES SANGUINEUM PURSH.
RUBUS PARVIFLORUS NUTT.
SALIX SP. (WILLOW)
SALIX HOOKERIANA BARR.
TSUGA HETEROPHYLLA (RAF.)

ARGENTILINEA DRUCE, DIDUGUA-NOTODONTIDAE

LIFE HISTORY

BROOKLYN MUS. ARTS, SCI., BULL. 1:196, 1906 DYAR LARVA

FOOD PLANTS

ARGENTISTRIATA STRECKER
SYN. SEE ARGENTATA DRURY

ARGILLACEA(HUBNER), ALABAMA-NOCTUIDAE
SYN. XYLINA SAY
SYN. GRANDIPUNCTA GUENEE
SYN. BIPUNCTINA GUENEE
LIFE HISTORY
AMER. ASSOC. ADVANCEMENT SCI. BIOL. SEC., PROC. B:13, 1874 GROTE NOTE ON LIFE HISTORY
AMER. ENT. 1:210, 1869 LIFE HIST. ALL STAGES* F. 148
LOC. CIT. 3:6, 1880 RILEY PUPA* LARVA* F.40 NO DESCR.
BOSTON SOC. NAT. HIST., PROC. 4: 318, 1853 BURNETT EGG, LARVA
ILLINOIS DEPT. AGRI., TRANS. 15: 228, 1877 FRENCH EGG, LARVA
INSECUTOR INSCITIAE MENSTRUUS 1: 11, 1913 DYAR LARVA (BRIEF)
JOURN. ECON. ENT. 23:714, 1930 HINDS LARVA* PL. 25, F. 5; PL. 26, F. 8; PUPA* PL. 25, F. 6
LOC. CIT. 29:95, 1936 RAINWATER DIFFERENCES BETWEEN LARVA OF ARGILLACEA AND FIMBRIAGO STEPHENS
MISSOURI, REPT. NOX. INX. 2:38, 1870 RILEY LIFE HISTORY*
NEW YORK STATE AGRI. SOC., TRANS. P. 71, 1855 GLOVER EGG, LARVA, PUPA
OKLAHOMA AGRI. EXP. STA., CIRC. 96:19, 1942 FENTON-CHESTER LARVA* F. 10
UNITED STATES DEPT. AGRI. DIV. ENT., CIRC. 153:4, 1912 HUNTER EGG; LARVA* F. 1; PUPA* F. 1
UNITED STATES DEPT. AGRI. FARMERS BULL. 47:3, 1897 HOWARD EGG* F. 1; LARVA* F. 2; PUPA
LOC. CIT. 890:10, 1917 PIERCE EGG; LARVA* F. 8; PUPA* F. 8.
LOC. CIT. 1688:5, 1932 FOLSOM LARVA* F. 4; PUPA* F. 4
UNITED STATES ENT. COMM. REPT. 4: 1-344, 1855 RILEY LIFE HIST. ALL STAGES* COLOR
COMSTOCK-UNITED STATES DEPT. AGRI., REPT. ON COTTON INS. P. 75, 1879 EGG* LARVA* PUPA* COLOR
CRAGEN-OUR INS. FRIENDS & FOES P. 129, 1899 EGG, LARVA
FERNALD-IN KINGSLEY'S STANDARD NAT. HIST. 2:450, 1885 EGG, LARVA
FERNALD-APPLIED ENT. P. 272, 1926 LARVA
FERNALD & SHEPARD-APPLIED ENT. P. 269, 1942 LARVA
GLOVER-ILLUS. NORTH AMER. ENT., 1878 LARVA* PUPA* PL. 18, F. 4
GROTE-GEOLOGICAL SURV. ALABAMA PP. 199-204, 1876 LIFE HIST. NO FF.
GROTE-REPT. GEOL. ALABAMA P. 199, 1875 LARVA, PUPA
HARRIS-INS. INJ. VEGET. 2ND. EDIT. P. 356, 1852 LARVA
HARRIS-LOC. CIT. FLINT EDIT. P. 457, 1862 LARVA
HOLLAND-MOTH BOOK P. 243, 1905 EGG* LARVA* PUPA* ALL F. 151
METCALF & FLINT-DESTRUCTIVE & USEFUL INS. P. 444, 1939 LARVA* F. 278; PUPA* F. 278
O'KANE-INJ. INS. P. 175, 1912 LARVA* F. 201
PACKARD-GUIDE TO STUDY OF INS. P. 313, 1869 LARVA
PACKARD-HALF-HOURS WITH INS. P. 221, 1877 EGG, LARVA*
PACKARD-INS. INJ. IN WEST. (HAYDEN'S REPT.) P. 775, 1877 EGG* LARVA*

- - -

INDEX AMER. ECON. ENT., VOL. 1:4; VOL. 2:11; VOL. 3:14; VOL. 4:15; VOL. 5:18; VOL. 6:18; VOL. 7:27

FOOD PLANTS
GOSSYPIUM HERBACEUM LINN. (COTTON)
MALVA SP.
SOLANIUM LYCOPERSICUM LINN. (TOMATO)

ARGILLACEA ROTHSCHILD
SYN. SEE EDWARDSII PACKARD

ARGILLACEARIA(PACKARD), ITAME-GEOMETRIDAE
SYN. MODESTARIA HULST
SYN. OLIVALIS HULST
LIFE HISTORY
INDEX AMER. ECON. ENT., VOL. 6:435
FOOD PLANTS
VACCINIUM SP.?

ARGILLOCOSTA FRANCLEMONT
SSP. SEE DISPOSITA MORRISON

ARGUS NEUMOEGEN & DYAR
SSP. SEE IO FABRICIUS

ARGUS SMITH
SSP. SEE CURIALIS GROTE

ARGYOTOXUS BEHR
SSP. SEE ANNA EDWARDS

ARIADNE BARNES & MCDUNNOUGH
SSP. SEE DIFFINIS BOISDUVAL

ARIADNE EDWARDS
SYN. SEE EURYTHEME BOISDUVAL

ARIANE BOISDUVAL
SYN. SEE ALOPE FABRICIUS

ARICYE CRAMER
SYN. SEE PHILEA JOHANSSON

ARIDA COCKERELL
SSP. SEE VENERABILIS WALKER

ARIDA SKINNER
SYN. SEE MYLITTA EDWARDS

ARIDENSIS BENJAMIN
SSP. SEE CINNAMOMEA GROTE & ROBINSON

ARISTION FRUHST
SYN. SEE SMINTHEUS DOUBLEDAY & HEWITSON

ARIZONAE FRENCH
SYN. SEE LATIPENNIS BOISDUVAL

ARIZONAE GODMAN & SALVIN
SSP. SEE ARAXES HEWITSON

ARIZONAE GROTE, CATOCALA-NOCTUIDAE
ASPASIA AUCT (NEC STRECKER)
FORM ROSEATA CASSINO
FORM SARA FRENCH
FORM HUACHUCA BEUTENMULLER
SYN. ARIZONENSIS STRAND
? SYN. JUNCTURANA STRAND
? SYN. JUNCTURELLA STRAND
A AUGUSTA HENRY EDWARDS
LIFE HISTORY
AMER. MUS. NAT. HIST., MEM. NEW SERIES 3: PART 1, 1918 BARNES & MCD. LARVA* PL. 14, F. 7; PL. 15, FF. 12, 37; PL. 16, F. 9 COLOR ASPASIA
FOOD PLANTS

ARIZONENSIS EDWARDS
SSP. SEE ASTYANAX FABRICIUS

ARIZONENSIS EDWARDS
SSP. SEE PSEUDARGIOLUS BOISDUVAL & LECONTE

ARIZONENSIS EDWARDS
SSP. SEE RUTULUS LUCAS

ARIZONENSIS GUNDER
SSP. SEE COMYNTAS GODART

ARIZONENSIS MABILLE & BOULLETT
SYN. ALPHEUS EDWARDS

ARIZONENSIS MCDUNNOUGH
SSP. SEE LYGDAMUS DOUBLEDAY

ARIZONENSIS PACKARD, GLOVERIA-LASIOCAMPIDAE
SYN. DENTATA HENRY EDWARDS
FORM DOLORES NEUMOEGEN & DYAR
LIFE HISTORY

SOUTHERN CALIFORNIA ACAD. SCI., BULL. 29:26, 1930 COMSTOCK & COOLIDGE EGG, LARVA*, PUPA*

FOOD PLANTS

CUPRESSUS GUADELUPENSIS S. WATS.
JUNIPERUS CALIFORNICA CARR.

ARIZONENSIS ROTHSCHILD
SSP. SEE ACREA DRURY

ARIZONENSIS SKINNER
SSP. SEE AMYNTAS FABRICIUS

ARIZONENSIS STRECKER
SSP. SEE EGLANTERINA BOISDUVAL

ARIZONENSIS STRAND
SYN. SEE ARIZONAE GROTE

ARIZONENSIS STRETCH
SYN. SEE PROXIMA GUERIN-MENEVILLE

ARMATARIA (HERRICH-SCHAEFFER), PRIOCYCLA-GEOMETRIDAE

LIFE HISTORY

CAN. ENT. 3:130, 1871 SAUNDERS LARVA
MINNESOTA AGRI. EXP. STA., BULL. 61:231, 1899 LUGGER LARVA
NEW YORK ENT. SOC., JOURN. 53: 200, 1945 FORBES LARVA, PUPA
NEW YORK AGRI. EXP. STA. (CORNELL) MEM. 274:95, 1948 FORBES EGG, LARVA
NEW YORK STATE AGRI. SOC., TRANS. P. 83, 1872 GLOVER LARVA
NEW YORK STATE MUS., MEM. 8: 749, 1905 FELT LARVA (BRIEF)
ONTARIO ENT. SOC., ANN. REPT. P. 378, 1871 LARVA, PUPA
PSYCHE 4:272, 1885 DIMMOCK EGG
SAUNDERS-INS. INJ. FRUITS P. 354, 1883 LARVA, PUPA
SAUNDERS-INS. INJ. FRUITS P. 354, 1889 LARVA, PUPA

- - -

FOOD PLANTS

ACER SP. (MAPLES)
BETULA SP. (BIRCHES)
MALUS PUMILS MILL. (APPLE)
RIBES LACUSTRE (PERS.) (GOOSEBERRY)
RIBES NIGRUM LINN.
RIBES RUBRUM LINN.

ARMIGER HUBNER
SYN. SEE OBSOLETA FABRICIUS

ARNA GUENEE
SYN. SEE CHALCEDONIA HUBNER

AROTA (BOISDUVAL), THARSALEA-LYCAENIDAE
A NUBILA COMSTOCK

LIFE HISTORY

CAN. ENT. 23:204, 1891 DYAR LARVA, CHRYSALIS
LOC. CIT. 26:45, 1894 LEMBERT EGG
SOUTHERN CALIFORNIA ACAD. SCI., BULL. 26:67, 1927 COMSTOCK EGG*
LOC. CIT. 27:64, 1928 COMSTOCK LARVA* CHRYSALIS*
COMSTOCK, J. A.-BUTTER. CALIF. PP. 170, 171, 1927 EGG* (A 55) LARVA, CHRYSALIS

FOOD PLANTS

RIBES CEREUM DOUGL.
RIBES CYNOSBATI LINN.
RIBES MISSOURIENSE NUTT.
VACCINIUM SP.

ARPA (BOISDUVAL & LECONTE) ATRYTONE-HESPERIIDAE

LIFE HISTORY

CAN. ENT. 11:191, 1879 W. H. EDWARDS LARVA, CHRYSALIS
BOISDUVAL & LECONTE-HIST. LEP. NORTH AMERICA 1833 LARVA* CHRYSALIS* PL. 68 COLOR
FRENCH-BUTTER. EAST UNITED STATES P. 339, 1886 LARVA, CHRYSALIS

FOOD PLANTS
CLADIUM MARISCOIDES (MUHL.)

ARSACE BOISDUVAL & LECONTE
SYN. SEE IRUS GODART

ARSESARIA WALKER
SYN. SEE CABERATA GUENEE

ARTEMIS PACKARD
SYN. SEE NEVADENSIS STRETCH

ARTHECHIPPUS SCUDDER HYBRID
SEE ARCHIPPUS CRAMER

ARTHEMIS(DRURY), BASILARCHIA-NYMPHALIDAE
SYN. LAMINA FABRICIUS
SYN. RUFESCENS COCKERELL
FORM VIRITHEMIS FIELD
FORM ALBOFASCIATA NEWCOMB
FORM PROSPERPINA EDWARDS
SYN. CERULEA EHRMAN
SYN. BENJAMINI NAKAHARA
A RUBROFASCIATA BARNES & MCDUNNOUGH

LIFE HISTORY

BUFFALO SOC. NAT. SCI., BULL. 19: 26, 1939 WILD LARVA

BUTTERFLY PARK NATURE CLUB 1: NO PAGE NUMBER, 1929 LARVAL HABITS

CAN. ENT. 11:224, 1879 W.H. EDWARDS LARVAL HABITS

LOC. CIT. 13:237, 1881 W.H. EDWARDS LIFE HISTORY NOTES

CAN. NATURALIST PP. 220-221, 1840 GOSSE LARVA* CHRYSALIS*

MAINE NATURALIST 2:67, 1922 KILBURN LARVA* CHRYSALIS* WINTER QUARTERS*

MICHIGAN AGRI. EXP. STA., SPEC. BULL. 17:24, 1903 PETTIT LARVA* HIBERNACULUM* F. 24

MICHIGAN AGRI. EXP. STA., SPEC. BULL. 24:24, 1904 PETTIT LARVA* F. 22; HIBERNACULUM* F. 24

NATURALISTE CANADIEN P. 162, 1928 HUARD LARVA

NEW YORK STATE MUS., MEM. 8: 742, 1905 FELT LARVA (BRIEF)

PHILADELPHIA ENT. SOC., PROC. 3: 62, 1864 LINTNER LARVA, CHRYSALIS

COMSTOCK-HOW TO KNOW BUTTER., P. 167, 1904 LARVA

EDWARDS, W.H.-BUTTER. NORTH AMERICA, 2ND. SERIES, 1879 LIFE HISTORY* PL. LIMENITIS NO. 1 COLOR

FERNALD-BUTTER. MAINE P. 62, 1884 EGG, LARVA, CHRYSALIS

FRENCH-BUTTER. EAST. UNITED STATES P. 208, 1886 LIFE HISTORY

HOLLAND-BUTTERFLY BOOK, 1ST EDIT. P. 184, 1905 LARVA* PL. 3, F. 26 COLOR; PUPA* PL. 4, FF. 14, 23 COLOR

HOLLAND-LOC. CIT. 3RD EDIT. P. 164, 1931 LIFE HISTORY LARVA* PL. 3; CHRYSALIS* PL. 4

MAYNARD-BUTTER. NEW ENGLAND P. 10, 1886 LARVA

SCUDDER-BUTTER. EAST. UNITED STATES AND CANADA PP. 297-303, 1889 LIFE HISTORY* PL. 54, F. 15; PL. 67, F. 5; PL. 74, F. 26; PL. 78, FF. 21-24; PL. 81, F. 5; PL. 83, FF. 14, 23

SCUDDER-BUTTER. THEIR STRUCTURE ETC. P. 113, 1881 NEST*

SCUDDER-EVERYDAY BUTTER. P. 226, 1889 EGG, LARVA, CHRYSALIS

SCUDDER-GUIDE TO COMMON BUTTER P. 99, 1893 LARVA, CHRYSALIS

- - -

INDEX AMER. ECON. ENT. VOL. 4:59; VOL. 7:122

FOOD PLANTS
AMELANCHIER SP.
BETULA ALLEGHANIENSIS BRITT.
BETULA LENTA LINN.
CRATAEGUS SP. (THORNS)
MALUS PUMILA MILL. (APPLE)
POPULUS SP. (POPLARS)
POPULUS BALSAMIFERA LINN.

PRUNUS PENSYLVANICA LINN.
PRUNUS SEROTINA EHRH.
PYRUS COMMUNIS LINN. (PEAR)
SALIX SP. (WILLOWS)
TILIA AMERICANA LINN.
ULMUS SP. (ELMS)

ARTONIS EDWARDS
SSP. SEE EURYNOME EDWARDS

ARVALIS HENRY EDWARDS, ANNAPHILA-NOCTUIDAE
SYN. SALICIS HENRY EDWARDS
LIFE HISTORY
PSYCHE 6:325, 1892
DYAR EGG (SALICIS)
FOOD PLANTS
ACER MACROPHYLLUM PURSH
JUGLANS SP. (WALNUTS)
MALUS PUMILA MILL. (APPLE)
SALIX SP. (WILLOWS)

ARYXNA DYAR
SYN. SEE NEUMOEGENI EDWARDS

ASELLUS ROTHSCHILD & JORDAN
SSP. SEE PERELEGANS HENRY EDWARDS

ASHTAROTH FISCHER
SSP. SEE IDALIA DRURY

ASHTAROTH STRECKER
SYN. SEE RIDINGSII EDWARDS

ASIATICUS BUTLER
SYN. SEE PINASTRI LINNEAUS

ASOPIALIS(GUENEE), PALTHIS-NOCTUIDAE
SYN. INSIGNALIS WALKER
LIFE HISTORY

FOOD PLANTS
QUERCUS SP. (OAKS)
SYMPHORICARPOS ORBICULATUS MOENCH

ASPASIA AUCT.
SEE ARIZONAE GROTE

ASPASIA STRECKER
SYN. SEE ELECTILIS WALKER

ASPILATATA GUENEE
SSP. SEE CROCALLATA GUENEE

ASSIMILANS WALKER, PHRAGMATOBIA-ARCTIIDAE
FORM FRANCONIA SLOSSON
LIFE HISTORY
CAN. ENT. P. 125, 1911 GIBSON
EGG, LARVA, PUPA (FRANCONIA)
FOOD PLANTS
MERTENSIA SP.
PLANTAGO SP.
TARAXACUM OFFICINALE WEBER

ASSIMILIS BOISDUVAL
SYN. SEE GRATA FABRICIUS

ASSIMILIS BUTLER
SSP. SEE MELISSA FABRICIUS

ASSIMILIS(MORRISON), POLIA-NOCTUIDAE
LIFE HISTORY
AMER. MUS. NAT. HIST., BULL. 14: 295, 1901 BEUTENMULLER LARVA
PAPILIO 1:15, 1881
GOODELL LARVA
UNITED STATES NAT. MUS., PROC. 27:850, 1904 DYAR EGG, LARVA
- - -
INDEX AMER. ECON. ENT., VOL. 7:797
FOOD PLANTS
HYPERICUM PERFORATUM LINN.
SALIX SP. (WILLOWS)
SOLIDAGO SP.

ASSINIBOIA LYMAN
SSP. SEE COMMA LINNEAUS

ASSOCIANS WALKER
SYN. SEE LUBRICANS GUENEE

ASSOCIATA WALKER
SYN. SEE BILINEATA WALKER (HETEROCAMPA)

ASTARTE DOUBLEDAY, HETEROCAMPA
NOTODONTIDAE
SYN. MENAS HARRIS
SYN. FEMALE CHAPMANI GROTE
SYN. PICTA FELDER
FORM PEROLIVATA PACKARD
LIFE HISTORY
ENTOMOLOGIST 4:57, 1841 DOUBLEDAY LARVA* PUPA* PLATE FACING P.60, FF.1,2
NAT. ACAD. SCI., MEM. 7:248, 1895 PACKARD LARVA* PL.33, F.9 COLOR
NEW YORK AGRI. EXP. STA. (CORNELL) MEM. 274:234, 1948 FORBES LARVA
HARRIS-ENT. CORRESPOND. P.132, 1869 LARVA
FOOD PLANTS
QUERCUS SP. (OAKS)

ASTARTE FISHER SEE IDALIA DRURY

ASTARTE STRECKER
SSP. SEE CERISVI KIRBY

ASTER EDWARDS
SSP. SEE SCUDDERI EDWARDS

ASTERIOIDES REAKIRT
SYN. SEE ASTERIUS CRAMER

ASTERIUS CRAMER, PAPILIO-PAPILIONIDAE
(AJAX PARTIM OF LINNEAUS)
SYN. VIRIDIS COCKERELL
SYN. SEMIALBA EHRMANN
FORM CURVIFASCIA SKINNER
FORM AMPLIATA MENETRIES
SYN. ASTERIOIDES REAKIRT
AB ALUNATA SKINNER & AARON
AB EHRMANNI EHRMANN
AB CALVERLEYI GROTE
AB STRECKERI HOLLAND
AB FORSYTHEA WOOD
A AMERICUS KOLL
B STABILIS ROTHSCHILD & JORDAN
LIFE HISTORY
AMER. ENT. 1:59, 1868 LARVA (BRIEF)
LOC. CIT. 2:30, 1869 LARVA (BRIEF)
LOC. CIT. 2:341, 1870 LARVA (BRIEF)
AMER. MUS. NAT. HIST., BULL. 5: 243, 1893 BEUTENMULLER LARVA, CHRYSALIS
AMER. NAT. 22:1033, 1888 COMSTOCK NOTE ON PUPAL COLORATION
ARCHIVES NATURGESCH. (WEIGEMANN) 44:4, 1878 DEWITZ LARVA* CHRYSALIS* PL.1
BUFFALO SOC. NAT. SCI., BULL. 19: 17, 1939 WILDE LARVA
CALIFORNIA ACAD. SCI., PROC. 5: 164, 1873 HENRY EDWARDS LARVA, CHRYSALIS
CAN. ENT. 71:153, 1939 MCDUNNOUGH COMPARES LARVAE OF ASTERIUS* AND BREVICAUDA
CAN. NAT. & GEOL. 2:222, 1857 ANON. LARVA, CHRYSALIS
ENT. NEWS 47:269, 1936 MEINERS HABITS AND FOOD PLANTS OF ASTERIUS AND AMERICUS
ILLINOIS DEPT. AGRI., TRANS. 15: 137, 1877 FRENCH LARVA
LOC. CIT. 18:APPEND. P.74, 1880 MIDDLETON LARVA
LOC. CIT. 18:APPEND. P.173, 1880 COQUILLETT LARVA
MARYLAND ACAD. SCI., BULL. 3:(2) P.7, 1923 MOORE EGG, LARVA, CHRYSALIS
MUS. COMP. ZOOLOGY., BULL. 87:NO. 6, 1941 DETHIER LARVAL ANTENNAE* FF.31,37
NATURALISTE CANADIEN P.44, 1927 HUARD LARVA, CHRYSALIS
NATURE STUDY REVIEW P.219, SEPT. 1917 EGG, LARVA, CHRYSALIS
NEW YORK AGRI. EXP. STA. (GENEVA), BULL. 212:24, 1902 LOWE LARVA*
ONTARIO ENT. SOC., ANN. REPT. P.38, 1876 SAUNDERS LARVA

PAPILIO 4:84 1884
GRUBER LARVA*
PHILADELPHIA ENT. SOC., PROC. 3:
51, 1864 HENRY EDWARDS
LARVA, CHRYSALIS
PSYCHE 8:433, 1898
SCUDDER LARVA
LOC. CIT. P. 435, 1899 SOULE
COLOR VAR. IN LARVA
SMITHSONIAN MISC. COLLECTIONS
4:ART. 2:6, 1862 MORRIS
LARVA, CHRYSALIS
SOUTHERN CALIFORNIA ACAD. SCI.,
BULL. 38:200, 1939 BROWN
LARVA OF AMERICUS
UNITED STATES DEPT. AGRI. BUR.
ENT. BULL. OLD SERIES 18:69,
1888 LARVA* PUPA* PL. 1 COLOR
UNITED STATES DEPT. AGRI. BUR.
ENT. BULL. 82 NEW SERIES PART
2:20, 1909 CHITTENDEN EGG* F. 5D;
LARVA* F. 5A, 5B; CHRYSALIS* F. 5F
UNITED STATES DEPT. AGRI.,
FARMERS BULL. 856:41, 1917
CHITTENDEN EGG* F. 46D; LARVA*
F. 46A, 46B; CHRYSALIS* F. 46F
UNITED STATES NAT. MUS., BULL.
157:196, 1932 AUSTIN CLARK
LARVA, PUPA
ZEITSCHRIFT GESAMENT. NATUR.
1878 DEWITZ LARVA, CHRYSALIS
ABBOT & SMITH-LEP. INS. GEORGIA,
1797 LARVA* PL. 1 COLOR
BALLARD-INSECT LIVES P. 32, 1879
LARVA* CHRYSALIS*
BOISDUVAL-SPEC. GEN. DIURNAE 1:
333, 1836 LARVA, CHRYSALIS
BOISDUVAL & LECONTE-HIST. LEP.
NORTH AMERICA P. 16, 1833
LARVA* CHRYSALIS* PL. 4 COLOR
CHITTENDEN-INS. INJ. VEGET. P. 179,
1907 LARVA
COMSTOCK-HOW TO KNOW BUTTER.
P. 63, 1904 LARVA* CHRYSALIS*
PL. 11 COLOR
CRAGEN-OUR INS. FRIENDS & FOES
P. 99, 1899 EGG, LARVA* F. 37,
CHRYSALIS
CROSBY & LEONARD-MANUAL VEGET.
GARDEN INS. P. 187, 1918 EGG*
F. 110; LARVA* FF. 111, 112;
CHRYSALIS* F. 113
DICKERSON-MOTHS & BUTTER. P. 39,
1901 LARVA* FF. 29-32, 35-38;
CHRYSALIS* FF. 39-42
EMMONS-NAT. HIST. NEW YORK
5:200, 1854 LARVA, PUPA
FERNALD-BUTTER. MAINE P. 22,
1884 EGG, LARVA, CHRYSALIS
FERNALD-APPLIED ENTOM. P. 302,
1926 LARVA* CHRYSALIS* F. 317
FERNALD-SHEPARD--APPLIED
ENTOM. P. 294, 1942 LARVA*
CHRYSALIS* F. 311
FRENCH-BUTTER. EAST. UNITED
STATES P. 92, 1886 LARVA*
CHRYSALIS*
GLOVER-ILLUS. NORTH AMER. ENT.,
1878 LARVA, PUPA PL. 1, F. 1
HARRIS-ENT. CORRESP. P. 270,
1869 LARVA
HARRIS-INS. INJ. VEGET., 1ST EDIT.
P. 212, 1841 LIFE HIST. (BRIEF)
HARRIS-LOC. CIT. 2ND EDIT. P. 231,
1852 LIFE HISTORY (BRIEF)
HARRIS-LOC. CIT. FLINT EDIT. P. 263,
1862 LIFE HIST* PL. 4 COLOR
HOLLAND-BUTTERFLY BOOK 1905
LARVA* PL. 2, FF. 17, 24, 27
COLOR; CHRYSALIS* PL. 6, FF. 13,
18, 19 COLOR
JAEGER-LIFE NORTH AMER. INS. P.
208, 1864 LARVA*
KELLOGG-AMER. INS. P. 450, 1908
LARVA
MAYNARD-BUTTER. NEW ENGLAND
P. 51, 1886 LARVA, CHRYSALIS
METCALF & FLINT-DESTRUCTIVE &
USEFUL INS. PP. 218, 545, 1939
EGG* LARVA* CHRYSALIS* F. 148
MORRIS-SYNOP. NORTH AMER. LEP.
P. 6, 1862 LARVA
O'KANE-INJ. INS. P. 170, 1912
LARVA* F. 192
PACKARD-INS. INJ. WEST, (HAYDEN'S
REPT.) P. 765, 1877 LARVA

Scudder-Butter. East. United States & Canada pp. 1357-1362, 1889 life history* Pl. 66, f. 2; Pl. 72, f. 11; Pl. 76, ff. 17, 24, 27; Pl. 79, ff. 56-60; Pl. 85, ff. 13, 18

Scudder-Butter. Their Structure etc. p. 24, 1881 larva*

Scudder-Everyday Butter. p. 133, 1899 egg, larva, chrysalis (polyxenes)

Scudder-Guide to Common Butter. p. 153, 1893 larva, chrysalis

Strecker-Rhop. et Heter. 6:49, 1873 larva* Pl. 8

Tenney-Manual Zoology p. 383, 1867 larva* chrysalis* ff. 281, 282

- - -

Index Amer. Econ. Ent.
Vol. 1:230 polyxenes
Vol. 2:264 polyxenes
Vol. 3:294 calverleyi
Vol. 3:295 polyxenes
Vol. 4:342 asterius
Vol. 4:343 polyxenes
Vol. 5:469 polyxenes
Vol. 6:559 polyxenes
Vol. 7:740

FOOD PLANTS

Anethum graveolens Linn. (dill)
Angelica sp.
Angelica atropurpurea Linn.
Apium graveolens Linn. (celery)
Apium petroselinum Linn. (parsley)
Arracacia xanthorrhiza Bancr.
Carum carvi Linn. (caraway)
Citcuta bulbifera Linn.
Cicuta maculata Linn.
Conium maculatum Linn.
Cosmos bipinnatus Cav.
Daucus carota Linn.
Daucus carota v. sativas Hoff. (carrot)
Dictamnus albus Linn.
Echinophora spinosa Linn.
Hydrocotyle umbellata Linn.
Osmorhiza claytoni (Michx.) (anise)
Osmorhiza longistylis (Torr.)
Oxypolis filiformis (Walt.)
Pastinaca sativa Linn. (parsnip)
Ptilimnium capillaceum (Michx.)
Ruta graveolens Linn.
Sium suave Walt.
Spermolepis divaricata (Walt.)
Tanacetum vulgare Linn.

asteroides Guenee, Cucullia-Noctuidae

LIFE HISTORY

Can. Ent. 76:161, 1944 Deither larva* pupa

New York State Cabin. Nat. Hist., Ann. Rept. No. 26, p. 139, 1874 Lintner larva

- - -

Index Amer. Econ. Ent., Vol. 7:228

FOOD PLANTS

Compositae (flower heads)
Solidago sp.

asthenaria Walker
syn. see insulsaria Guenee

astinous Druce
syn. see philenor Linneaus

astoriae Henry Edwards
ssp. see apicalis Walker

astraea Edwards
ssp. see christina Edwards

astragala Wright
syn. see gyas Edwards

astriata Reiff
ssp. see disstria Hubner

astricta Morrison, Eurois-Noctuidae
a subjugata Dyar
b elenae Barnes and Benjamin

LIFE HISTORY

Maine Agri. Exp. Sta., Bull. 356: 164, 1930 Phipps larva* f. 24b; pupa

United States Nat. Mus., Proc. 27:823, 1904 Dyar egg, larva

- - -

Index Amer. Econ. Ent., Vol. 5:393

FOOD PLANTS

Acer sp. (maples)
Alnus rugosa (DuRoi)
Alnus serrulata (Ait.)
Betula populifolia Marsh.
Fragaria sp. (strawberries)
Helianthus sp.
Prunus pensylvanica Linn.
Spiraea latifolia (Ait.)
Vaccinium angustifolium Ait.
Vaccinium myrtilloides Michx.
Viburnum sp.
Viburnum cassinoides Linn.

<u>Astriotes</u> Fruhst

syn. see <u>Smintheus</u> Doubleday & Hewitson

<u>Astuta</u> Bird, <u>Papaipema</u>-Noctuidae

LIFE HISTORY

Can. Ent. 39:272, 1907 Bird larva, pupa

FOOD PLANTS

Collinsonia canadensis Linn.

<u>Astyanax</u>(Fabricius), <u>Basilarchia</u>-Nymphalidae

syn. <u>ephestion</u> Stoll
syn. <u>ursula</u> Fabricius
syn. <u>ephestiaena</u> Hubner
form <u>viridis</u> Strecker
form <u>purpuratus</u> Gunder
form <u>atlantis</u> Nakahara
form <u>inornata</u> Nakahara
form <u>tildeni</u> Field
form <u>arizonensis</u> Edwards
form <u>duodoroffi</u> Gunder

LIFE HISTORY

Amer. Mus. Nat. Hist., Bull. 5: 272, 1893 Beuten. larva, chrysalis

Buffalo Soc. Nat. Sci., Bull. 19: 27, 1939 Wild larva

Can. Ent. 3:52, 1871 Riley larva, chrysalis (difference between this sp. and <u>archippus</u> Cramer)

Loc. cit. 13:242, 1881 Edwards larval habits

Illinois Dept. Agri., Trans. 15: 154, 1877 French larva

Loc. cit. 18:Append. pp. 87, 162, 1880 Coquillett-Middleton larva

Minnesota Agri. Exp. Sta., Bull. 61:72, 1899 Lugger larva

New York State Mus., Mem. 8: 704, 1905 Felt larva (brief)

Papilio 3:26, 1883 Mund <u>ursula</u> and <u>dissipus</u> larvae are identical

Psyche 8:210, 1898 Scudder larva

Smithsonian Misc. Collections 4:art. 2:65, 1862 Morris larva, chrysalis

Abbot & Smith-Lep. Ins. Georgia 1:19, 1797 larva, pupa Pl. 10

Boisduval & Leconte-Hist. Lep. North America p. 199, 1833 larva* chrysalis* Pl. 53 COLOR

Comstock-How to Know Butter. p. 166, 1904 larva

Dickerson-Moths & Butter. p. 97 1901 larva* ff. 84, 85, 86

French-Butter. East. United States p. 207, 1886 larva

Gosse-Letters from Alabama p. 123, 1859 larva

Harris-Ent. Corresp. p. 276, 1869 larva* chrysalis* Pl. 4

Harris-Ins. Inj. Veget. Flint edit. p. 283, 1862 larva, chrysalis

Holland-Butter. Book p. 184, 1905 life history larva* Pl. 3, ff. 17, 21, 25; chrysalis* Pl. 4, ff. 12, 13

Holland-loc. cit. 3rd edit. p. 164, 1931 life history

MAYNARD-BUTTER. NEW ENGLAND P. 10, 1886 LARVA
MORRIS-SYNOP. LEP. NORTH AMERICA P. 65, 1862 LARVA, CHRYSALIS
SAUNDERS-INS. INJ. FRUITS P. 218, 1889 LARVA, CHRYSALIS
SCUDDER-BUTTER. EAST UNITED STATES & CANADA PP. 283-287, 1889 LIFE HIST* PL. 64, F. 12; PL. 74, FF. 17, 21, 25; PL. 78, F. 30; PL. 86, F. 60; PL. 83, FF. 12, 13
SCUDDER-EVERYDAY BUTTER. P. 218, 1899 EGG, LARVA, CHRYSALIS
SCUDDER-GUIDE TO COMMON BUTTER. P. 101, 1893 LARVA, CHRYSALIS

- - -

INDEX AMER. ECON. ENT. VOL. 1:47; VOL. 6:83; VOL. :122

FOOD PLANTS

CARPINUS CAROLINIANA WALT.
CRATAEGUS SP. (THORNS)
CYDONIA SP.
CYDONIA OBLONGA MILL. (QUINCE)
MALUS PUMILA MILL. (APPLE)
POPULUS BALSAMIFERA LINN.
POPULUS DELTOIDES MARSH.
POPULUS HETEROPHYLLA LINN.
POPULUS NIGRA V. ITALICA MUENCH.
POPULUS TREMULOIDES MICHX.
PRUNUS SP.
PRUNUS CERASUS LINN. (CHERRY)
PRUNUS DOMESTICA LINN (PLUM)
PRUNUS SEROTINA EHRH.
PYRUS COMMUNIS LINN. (PEAR)
QUERCUS ILICIFOLIA WANG.
RIBES SP.
RIBES LACUSTRE(PERS.)(GOOSEBERRY)
RIBES SATIVUM SYME (CURRANT)
SALIX SP.
TILIA AMERICANA LINN.
VACCINIUM SP.

ASTYLUS(DRURY), PAONIAS-SPHINGIDAE

SYN. IO GRAY
SYN. INTEGERRIMA HARRIS

LIFE HISTORY

AMER. ENT. SOC., TRANS. 15:229, 1888 SMITH LARVA (BRIEF)
AMER. MUS. NAT. HIST., BULL. 7: 316, 1895 BEUTEN. LARVA, PUPA
CAN. ENT. 8:239, 1876 PECK LARVAL NOTE
ENT. NEWS 1:21, 1890 JONES LIFE HISTORY NOTE
LOC. CIT. 9:150, 1898 SOULE DURATION EGG STAGE
ENT. SOC. AMERICA ANNALS 11:426, 1918 MOSHER PUPAL STRUCTURES
NEW YORK AGRI. EXP. STA. (CORNELL) MEM. 274:194, 1948 FORBES LARVA
NEW YORK STATE MUS., MEM. 8: 738, 1905 FELT LARVA (BRIEF)
NOVITATES ZOOLOGICAE (TRING) 9: SUPPLEMENT P. 331, 1903 ROTHSCHILD & JORDAN LARVA, PUPA
PSYCHE 5:265, 1889 ELIOT & SOULE EGG, LARVA, PUPA
LOC. CIT. 6:31, 1891 ELIOT & SOULE LARVAL VARIATIONS
BOISDUVAL-SPEC. GEN. LEP., SPHING. P. 41, 1874 LARVA
ELIOT & SOULE-CATERPILLARS & THEIR MOTHS P. 195, 1902 EGG, LARVA* PUPA
FERNALD-SPHINGIDAE NEW ENGLAND P. 77, 1886 LARVA
SMITH, J. B.-MONOGRAPH OF SPHING. P. 229, 1888 LARVA
STRECKER-LEP. RHOP. ET HETER. 8: 56, 1873 LARVA

FOOD PLANTS

GAYLUSSACIA FRONDOSA (LINN.)
PRUNUS CERASUS LINN. (CHERRY)
PRUNUS SEROTINA EHRH.
ROSACEAE SP.
SALIX SP. (WILLOWS)
VACCINIUM CORYMBOSUM LINN.
VACCINIUM ULIGINOSUM LINN.
XOLISMA LIGUSTRINA (LINN.)

ASTYLUSARIA(WALKER), EUCHLAENA-GEOMETRIDAE

SYN. MADUSARIA WALKER
SYN. VINOSARIA GROTE & ROBINSON

A OPONEARIA WALKER
SYN. TIVIARIA WALKER
B DEPLANARIA WALKER
LIFE HISTORY
NEW YORK AGRI. EXP. STA. (CORNELL)
MEM. 274:76, 1948
FORBES LARVA
PSYCHE 12:43, 1905
DYAR EGG, LARVA
FOOD PLANTS
GENERAL FEEDER ON TREES

ASTYMONE OLIVER
SYN. SEE VIRGINIENSIS DRURY (ANISOTA)

ATA STRAND
SSP. SEE CINCTIPES HENRY EDWARDS

ATALA CASSINO
SSP. SEE SEMIRELICTA GROTE

ATALA POEY, EUMAEUS-LYCAENIDAE
LIFE HISTORY
BOSTON SOC. NAT. HIST., MEM. 2: 416, 1875 SCUDDER LIFE HISTORY LARVA* CHRYSALIS* PL. 14
ENT. NEWS 21:179,327, 1910 HEALY LARVA
INSECT LIFE 1:38, 1888 SCHWARZ EGG* LARVA* CHRYSALIS*
- - -
INDEX AMER. ECON. ENT., VOL. 2:140
FOOD PLANTS
MANIHOT SP.
MINIHOT DULCIS V. AIPI (POHL.)
MANIHOT ESCULENTA CRANTZ
MANIHOT PALMATA MUELL.
PRIMULA SP.
ZAMIA INTEGRIFOLIA AIT.
ZORNIA SP.

ATLANTA (LINNEAUS), VANESSA-NYMPHALIDAE
SYN. ADMIRALIS RETZIUS
AB EDWARDSI GRINNELL
LIFE HISTORY
AMER. JOURN. SCI. 33:323, 1887 POULTON PUPAL COLORATION
AMER. MUS. NAT. HIST., BULL. 5: 268, 1893 BEUTEN. LARVA, CHRYSALIS
BUFFALO SOC. NAT. SCI., BULL. 19:26, 1939 WILD LARVA
CALIFORNIA ACAD. SCI., PROC. 5: 170, 1873 HENRY EDWARDS LARVA, CHRYSALIS
CAN. ENT. 14:229, 1882 W.H. EDWARDS EGG, LARVA, CHRYSALIS
LOC. CIT. 15:14, 1883 W.H. EDWARDS LARVAL HABITS
ENTOMOLOGIST'S MONTHLY MAG. 13:209, 1877 BIRCH LARVA
ILLINOIS DEPT. AGRI., TRANS. 18: APPEND. P. 86, 1880 MIDDLETON LARVA
MAINE NATURALIST 2:123, 1922 KILBURN LARVA* CHRYSALIS*
MONTANA UNIV., BULL. 30:99, 1906 ELROD EGG, LARVA, CHRYSALIS
MUS. COMP. ZOOLOGY, BULL. 87:NO. 6, 1941 DETHIER LARVAL ANTENNA* F. 42
NATURALISTE CANADIEN P. 158, 1928 HUARD LARVA, CHRYSALIS
SMITHSONIAN MISC. COLLECTIONS 4:ART. 2:59, 1862 MORRISON LARVA, CHRYSALIS
BOISDUVAL & LECONTE-HIST. LEP. NORTH AMERICA P. 176, 1833 LARVA, PUPA
CHENUS-LUCAS--ENCYC. NAT. HIST. PAPILLONS 1:103, 1856 LARVA, CHRYSALIS
COMSTOCK, J.A.-BUTTERFLIES CALIFORNIA P. 131, 1927 EGG, LARVA, CHRYSALIS* PL. 63, F. 5 COLOR
COMSTOCK, J.H.-HOW TO KNOW BUTTER. P. 154, 1904 LARVA
DUNCAN-NATUR. LIBRARY ENT. 3: 171, 1835 LARVA (BRIEF)
ESSIG-INS. WEST. NORTH AMERICA P. 644, 1926 LARVA, CHRYSALIS
FERNALD-BUTTER. MAINE P. 58, 1884 EGG* LARVA

French-Butter. East. United States p. 197, 1886 life hist. egg*
Glover-Illus. North Amer. Ent., 1878 larva Pl. 22, f. 1
Harris-Ent. Corresp. p. 279, 1869 larva
Harris-Ins. Inj. Veget., Flint edit. p. 294, 1862 larva, chrysalis
Holland-Butter. Book 1905 larva* Pl. 3, f. 35; chrysalis* Pl. 4, ff. 52, 53, 55
Jaeger-Life North Amer. Ins. p. 203, 1864 larva
Kirby-European Butter. & Moths p. 11, 1882 larva, chrysalis
Maynard-Butter. New England p. 18, 1886 larva
Morris-Synop. Lep. North America p. 59, 1862 larva, chrysalis
Saunders-Butter. Allegheny State Park (New York State Mus. Handbook 13) p. 20, 1932
Scudder-Butter. East. United States & Canada pp. 445-452, 1889 life history* Pl. 64, f. 24; Pl. 70, f. 10; Pl. 74, f. 35; Pl. 78, ff. 58, 59; Pl. 81, ff. 2, 6; Pl. 86, ff. 61-65; Pl. 83, ff. 52, 53, 55
Scudder-Butter. Their Structure pp. 8, 109, 1881 egg* nest*
Scudder-Everyday Butter. p. 80, 1899 egg, larva
Scudder-Guide to Common Butter. p. 87, 1893 larva, chrysalis
Stainton-Manual British Butter. 1:38, 1857 larva
Stephens-Illus. British Ent. Haust. 1:46, 1828 larva, chrysalis
Westwood-British Butter. p. 53, 1855 larva* chrysalis* Pl. 8 color

- - -

Index Amer. Econ. Ent. Vol. 1:318; Vol. 5:678; Vol. 6:800; Vol. 7:1042

FOOD PLANTS

Ambrosia artemisiifolia Linn.
Ambrosia trifida Linn.
Boelmeria cylindrica (Linn.)
Humulus lupulus Linn. (hops)
Parietaria floridana Nutt.
Parietaria pensylvanica Muhl.
Ulmus sp. (elms)
Urtica sp. (nettles)

<u>atarah</u> Strecker
ssp. see <u>micronympha</u> Guenee

<u>ate</u> Strecker
ssp. see <u>cardui</u> Linneaus

<u>athasaria</u> Walker, <u>Ellopia</u>-Geometridae
syn. <u>siccaria</u> Walker
syn. <u>seminudata</u> Walker
syn. <u>aequaliaria</u> Walker
syn. <u>bibularia</u> Grote & Robinson

LIFE HISTORY

Journ. Econ. Entom. 20:300, 1927 Houser larva
New York Agri. Exp. Sta. (Cornell), Mem. 274:101, 1948 Forbes larva
New York State Mus., Mem. 8: 706, 1905 Felt larva
Psyche 19:10, 1900 Dyar egg, larva
United States Dept. Agri., Misc. Pub. 657:362, 436, 1950 Craighead larva* f. 91J
United States Dept. Agri., Rept. p. 329, 1886 Packard pupa

FOOD PLANTS

Acer sp. (maples)
Acer saccharinum Linn.
Fagus sp. (beeches)
Ostrya virginiana (Mill.)
Pinus sp. (pines)
Quercus sp. (oaks)
Quercus alba Linn.
Quercus rubra Linn.
Tsuga canadensis (Linn.)

ATHENA STRECKER
SSP. SEE CONGRUA WALKER(ESTIGMENE)

ATHERO HARRIS
SYN. SEE UMBRATA WALKER

ATLANTICA BARNES & MCDUNNOUGH
SSP. SEE GRAPHICA HUBNER

ATLANTICA(GROTE), POLIA--NOCTUIDAE
SYN. DISCOLOR SPEYER
LIFE HISTORY
CAN. ENT. 34:279-284, 1902 FLETCHER-GIBSON EGG, LARVA, PUPA
NEW YORK STATE MUS., MEM. 8:704, 1905 FELT LARVA (BRIEF)
- - -
INDEX AMER. ECON. ENT. VOL. 2:225 MAMESTRA VOL. 6:602; VOL. 7:797
FOOD PLANTS
LONICERA CAPRIFOLIUM LINN.
MEDICAGO SATIVA LINN. (ALFALFA)
QUERCUS SP. (OAKS)
QUERCUS ALBA LINN.
TRIFOLIUM SP. (CLOVERS)

ATLANTICA SMITH
SYN. SEE AMERICANA SPEYER

ATLANTIS EDWARDS, ARGYNNIS-NYMPHALIDAE
AB CHEMO SCUDDER
A CANADENSIS DOS PASSOS
B NIKIAS EHRMANN
C BEANI BARNES & BENJAMIN
AB HUTCHINSI GUNDER
LIFE HISTORY
BUFFALO SOC. NAT. SCI., BULL. 19:22, 1939 WILD LARVA
CAN. ENT. 20:1, 1888 W. H. EDWARDS EGG, LARVA, CHRYSALIS
ILLINOIS DEPT. AGRI., TRANS 18: APPEND. P. 82, 1880 MIDDLETON LARVA
MONTANA UNIV., BULL. 30:64, 1906 ELROD EGG, LARVA, CHRYSALIS
NATURALISTE CANADIEN P. 115, 1927 HUARD LARVA, CHRYSALIS
COMSTOCK-HOW TO KNOW BUTTER. P. 118, 1904 LARVA
EDWARDS, W. H.-BUTTER. NORTH AMERICA, 1ST SERIES, 1868 LARVA PL. ARGYNNIS NO. 5
FRENCH-BUTTER. EAST. UNITED STATES P. 161, 1886 CHRYSALIS*
HOLLAND-BUTTER. BOOK, P. 109, 1905 CHRYSALIS* PL. 5, F. 6
MAYNARD-BUTTER. NEW ENGLAND P. 25, 1886 LARVA
PACKARD-GUIDE STUDY INS. P. 252, 1869 LARVA, CHRYSALIS
SCUDDER-BUTTER. EAST. UNITED STATES & CANADA PP. 474-577, 1889 LIFE HISTORY PL. 64, F. 22; PL. 67, F. 13; PL. 84, F. 6
SCUDDER-BUTTER. THEIR STRUCTURE ETC. P. 150, 1881 CHRYSALIS*
SCUDDER-EVERYDAY BUTTER. P. 214, 1899 EGG, LARVA, PUPA
SCUDDER-GUIDE TO COMMON BUTTER. P. 76, 1893 LARVA, CHRYSALIS
- - -
INDEX AMER. ECON. ENT., VOL. 7: 105
FOOD PLANTS
VIOLA SP. (VIOLETS)

ATLANTIS NAKAHARA
SSP. SEE ASTYANAX FABRICIUS

ATOMA SMITH
SSP. SEE DILUCIDA MORRISON

ATOMARIA WALKER
SYN. SEE CINEREOLA GUENEE

ATOMARIA(WALKER), OLENE-LIPARIDAE
SYN. OBLIQUATA GROTE & ROBINSON
FORM PARALLELA GROTE & ROBINSON
A LEMMERI BARNES & BENJAMIN
LIFE HISTORY
ENT. AMER. 3:93, 1887 SEIFERT EGG, LARVA, PUPA, COCOON

ILLINOIS DEPT. AGRI., TRANS. 18: APPEND. P. 166, 1880 COQUILLETT LARVA

MINNESOTA AGRI. EXP. STA., BULL. 61:146, 1899 LUGGER LARVA (PARALLELA)

NEW YORK AGRI. EXP. STA. (CORNELL) MEM. 274:244, 1948 FORBES LARVA

NEW YORK STATE CABIN. NAT. HIST., ANN. REPT. NO. 26, P. 129, 1874 LINTNER LARVA

PSYCHE 7:136, 1894 DYAR LARVA

WASHINGTON ENT. SOC., PROC. 13: 16, 1911 DYAR LARVA (OBLIQUATA) NOTE

GLOVER–ILLUS. NORTH AMER. ENT., 1878 LARVA PL. 9, F. 12

PACKARD–INS. INJ. FOREST TREES P. 204, 1881 LARVA

SAUNDERS–INS. INJ. FRUITS P. 179, 1883 LARVA

SAUNDERS–LOC. CIT. P. 179, 1889 LARVA

FOOD PLANTS

AESCULUS SP. (HORSECHESTNUTS)
BETULA SP.
CARYA SP. (HICKORY)
MALUS PUMILA MILL. (APPLE)
PRUNUS DOMESTICA LINN. (PLUM)
PRUNUS SEROTINA EHRH.
PYRUS SP.
QUERCUS SP. (OAKS)
SALIX SP. (WILLOWS)
ULMUS SP. (ELMS)
DECIDIOUS TREES AND SHRUBS

ATOMARIS (SMITH), EUXOA–NOCTUIDAE

A DETESTA SMITH

LIFE HISTORY

CALIFORNIA STATE COMMISSION OF HORTICULTURE, MONTHLY BULL. 2, (1–2):163, 1913 ESSIG LIFE HISTORY LARVA* (BRIEF DESCR. AND F.)

CROSBY & LEONARD–MANUAL VEGET. & GARDEN INS. P. 298, 1918 LARVA

ESSIG–INJ. & BENEF. INS. P. 399, 1915 LARVA* F. 399

FOOD PLANTS

PRUNUS DOMESTICUS V. GALATENSIS EX HOOK (PRUNE)
VITIS SP. (GRAPES)
GENERAL FEEDER MOST ANYTHING

ATTOSA EDWARDS
SSP. SEE ADIASTE BEHR

ATRA GROTE
SSP. SEE RENIFORMIS GROTE

ATRATA HENRY EDWARDS
SSP. SEE CROTCHI HENRY EDWARDS (PSEUDALYPIA)

ATRESCENS HULST
SSP. SEE QUERNARIA ABBOT & SMITH

ATRICILIATA GROTE
SYN. SEE VIDENS GUENEE

ATRIFERA GROTE
SSP. SEE MESSORIA HARRIS

ATRIOR FRANCLEMONT
SSP. SEE FRAGILIS GUENEE

ATRITES GROTE
SYN. SEE BREVIS GROTE

ATROCOLORATA (HULST), STENASPILATES–GEOMETRIDAE

LIFE HISTORY

ENT. NEWS 11:583, 1900 CAUDELL LARVA (PEPLARIA)

FOOD PLANTS

PRUNUS SEROTINA EHRH.

ATROCOSTALIS HUARD
SSP. SEE MYRINA CRAMER

ATROFASCIATA MCDUNNOUGH
SSP. SEE MELINUS HUBNER

ATROPUNCTATA(PACKARD), APAECASIA GEOMETRIDAE
SYN. FERNALDI GROTE
FORM DARLINGTONI LEMMER
LIFE HISTORY

FOOD PLANTS
BETULA SP. (BIRCHES)

ATROPURPUREA GROTE
SSP. SEE TESSELLATA HARRIS

ATRPSOGMATA WALKER
SYN. SEE BITACTATA WALKER

AUBURNIANA HARRIS
SYN. SEE DAMON CRAMER

AUGE(LINNEAUS), COSMOSOMA-AMATIDAE
SYN. OMPHALE HUBNER
SYN. MELITLA MOESCHLER
LIFE HISTORY
ENTOMOLOGICA AMERICANA 1:186, 1886 SMITH PUPA, COCOON, (OMPHALA)

NEW YORK ENT. SOC., JOURN. 5:132, 1897 DYAR LARVAL SETAE ONLY* PL. 8, F. 6
PSYCHE 7:414, 1896 DYAR EGG, LARVA, PUPA, COCOON
GUNDLACH-ENTOM. CUBANA P. 239, 1886 LARVA, COCOON
FOOD PLANTS
MIKANIA SCANDENS (LINN.)

AUGUSTA EDWARDS
SSP. SEE EDITHA BOISDUVAL

AUGUSTA HENRY EDWARDS
SSP. SEE ARIZONAE GROTE

AUGUSTINA WRIGHT
SSP. SEE EDITHA BOISDUVAL

AUGUSTINUS WESTWOOD
SYN. SEE AUGUSTUS KIRBY

AUGUSTUS(KIRBY), INCISALIA-LYCAENIDAE
SYN. AUGUSTINUS WESTWOOD
SYN. CROESIOIDES SCUDDER
LIFE HISTORY
CAN. ENT. 36:136, 1904 COOK LARVA
LOC. CIT. 38:214, 1906 COOK LIFE HISTORY
LOC. CIT. 39:145, 1907 COOK CHRYSALIS*
ENT. NEWS 15:107, 1904 GRUNDEL LARVA LIKE THAT OF GORGON
MONTANA UNIV., BULL. 30:130, 1906 ELROD LARVA, CHRYSALIS
NATURALISTE CANADIEN P. 205, 1928 HUARD LARVA, CHRYSALIS
COMSTOCK-HOW TO KNOW BUTTER. P. 232, 1904 LARVA
SCUDDER-BUTTER. EAST. UNITED STATES & CANADA PP. 844-845, 1889 LARVA, CHRYSALIS
SCUDDER-GUIDE TO COMMON BUTTER. P. 117, 1893 LARVA, CHRYSALIS
FOOD PLANTS
ERIODYCTION CALIFORNICUM(H&A)
KALMIA AUGUSTIFOLIA LINN.
VACCINIUM AUGUSTIFOLIUM AIT.
VACCINIUM CORYMBOSUM LINN.
VACCINIUM VACILLANS TORR.

AULAEA(GEYER), ARACHNIS-ARCTIIDAE
SYN. INCARNATA WALKER
LIFE HISTORY
CAN. ENT. 26:307, 1894 DYAR EGG, LARVA
ENT. NEWS 4:312, 1893 DYAR-DOLL LARVA
PAPILIO 3:188, 1883 SCHAUS LARVA, PUPA
FOOD PLANTS
GENERAL FEEDER ON LOW PLANTS

AURANTIACA HENRY EDWARDS, INCITA-NOCTUIDAE
SYN. CALIFORNICA HAMPSON
A TENUIMARGO BARNES & MCDUNNOUGH

LIFE HISTORY
PAN PACIFIC ENT. 26:21, 1950
EVANS LARVA, LIFE HISTORY
FOOD PLANTS
GILIA VIRGATA V. DASYANTHA(BRAND)

AURANTIACA(HEWITSON), COPAEODES-HESPERIIDAE
SYN. WACO EDWARDS
SYN. PROCRIS EDWARDS
SYN. MACRA PLOETZ
SYN. CANDIDA WRIGHT
LIFE HISTORY
SOUTHERN CALIFORNIA ACAD. SCI., BULL. 28:28, 1929 COMSTOCK LIFE HISTORY EGG* LARVA* PUPA*
FOOD PLANTS
CYNODON DACTYLON (LINN.)
DIGITARIA SANGUINALIS (LINN.)

AURANTIACA HUBNER, EUBAPHE-ARCTIIDAE
FORM QUINARIA GROTE
SYN. CHROIONA REAKIRT
SYN. BIMACULATA SAUNDERS
A RUBICUNDARIA HUBNER
SYN. ROSA FRENCH
SYN. DIMINUTIVA GRAEF
B BREVICORNIS WALKER
SYN. BELFRAGEI STRETCH
LIFE HISTORY
CAN. ENT. 2:4, 1869 SAUNDERS LARVA (BIMACULATA)
NATURALISTE CANADIEN P. 11, 1929 HUARD EGG, LARVA, COCOON
UNITED STATES NAT. MUS., PROC. 25:373, 1903 DYAR EGG, LARVA
\- - -
INDEX AMER. ECON. ENT. VOL. 1:123 ROSA
FOOD PLANTS
CHENOPODIUM ALBUM LINN.
TARAXACUM OFFICINALE WEBER
ZEA MAYS LINN. (CORN)
GENERAL FEEDER ON LOW PLANTS

AURANTIACA IGEL
SSP. SEE ANGULIFERA WALKER(CALLOSOMIA)

AURANTIACA REIFF
SSP. SEE AMICA HUBNER

AURANTIACARIA PACKARD
SYN. SEE CERVINARIA PACKARD (SYNAXIS)

AURANTIAGO(GUENEE), RHODOECIA-NOCTUIDAE
SYN. ILLITERATA GROTE
SYN. DIFFERTA MORRISON
SYN. ILLINOISENSIS FRENCH
LIFE HISTORY
BROOKLYN ENT. SOC., BULL. 33:93, 1938 WYATT LARVA (BRIEF)
FOOD PLANTS
ANTIRRHINUM SP. IN CAPTIVITY
AUREOLARIA GRANDIFLORA(BENTH.)
LINARIA VULGARIS HILL

AURANTICOLORATA STRECKER
SYN. SEE SUBCROCEATA WALKER

AURATA BUTLER
SYN. SEE INSULATA WALKER

AURELIA DRURY
SYN. SEE POLYPHEMUS CRAMER

AURELLA FISCHER
SYN. SEE CEROGAMA GUENEE

AUREOMARGINATA COCKERELL
SSP. SEE J-ALBUM BOISDUVAL & LECONTE

AURETORUM(BOISDUVAL), STRYMON-LYCAENIDAE
SYN. TETRA EDWARDS
SYN. TACITA HENRY EDWARDS
A SPADIX HENRY EDWARDS
LIFE HISTORY
SOUTHERN CALIFORNIA ACAD. SCI., BULL. 33:79, 1934 COMSTOCK-DAMMERS EGG* PL. 23; LARVA* PLS. 24, 25; PUPA* PL. 25
FOOD PLANTS
QUERCUS SP. (OAKS)

AURICINCTARIA GROTE, MELANOMMA-NOCTUIDAE

LIFE HISTORY
CAN. ENT. 30:257, 1898 DYAR LARVA
FOOD PLANTS
CEPHALANTHUS OCCIDENTALIS LINN.
VACCINIUM SP. (BLUEBERRIES)

AURILACUS GUNDER
SSP. SEE NUBIGENA BEHR.

AURIPENNIS GROTE
SYN. SEE REPANDA WALKER

AURORA (ABBOT & SMITH), HYPARPAX-NOTODONTIDAE
SYN. ROSEA WALKER
SYN. VENUSTA WALKER
LIFE HISTORY
AMER. MUS. NAT. HIST., BULL. 10; 418, 1898 BEUTEN. LARVA
ENTOMOLOGICA AMERICANA 3:169, 1887 HENRY EDWARDS LARVA
NAT. ACAD. SCI., MEM. 7:185, 281, 1895 PACKARD LARVA* PL. 24, FF. 1-6 COLOR
NEW YORK AGRI. EXP. STA. (CORNELL) MEM. 274:226, 1948 FORBES LARVA
NEW YORK ENT. SOC., JOURN. 1:73, 1893 PACKARD LARVA
NEW YORK STATE MUS., MEM. 8:250, 1905 FELT LARVA
PSYCHE 6:179 1891 DYAR LARVA, PUPA, COCOON (ROSEA)
ABBOT & SMITH-LEP. INS. GEORGIA 2:173, 1797 LARVA* PL. 87 COLOR
DOANE ETC-FOREST INSECTS P. 295, 1936 LARVA
GLOVER-ILLUST. NORTH AMER. ENT. 1878 LARVA PL. 4, F. 16
FOOD PLANTS
QUERCUS SP. (OAKS)
VIBURNUM ACERIFOLIUM LINN.
VIBURNUM DENTATUM LINN.
VIBURNUM LENTAGO LINN.

AUROSEA NEUMOEGEN
SSP. SEE PAMINA NEUMOEGEN

AUROSTRIATA GRAEF
SSP. SEE PEROPHOROIDES STRECKER

AURULENTA (SMITH), EUXOA-NOCTUIDAE
FORM AURULENTOIDES STRAND
LIFE HISTORY

FOOD PLANTS
SALIX INTERIOR ROWLEE (NOT SURE OF THE IDENTITY OF LARVA)

AURULENTOIDES STRAND
SSP. SEE AURULENTA SMITH

AUSONIDES (BOISDUVAL), EUCHLOE-PIERIDAE
AB SEMIFLAVA COMSTOCK
AB FLAVIDALIS COMSTOCK
AB BOHARTI DOUDOROFF
A ANDREWSI MARTEN
B COLORADENSIS HENRY EDWARDS
SYN. MONTANA VERITY
AB HEMIFLAVA FIELD
LIFE HISTORY
AMER. MUS. NAT. HIST., BULL. 10: 240, 1898 BEUTENMULLER LARVA, CHRYSALIS
CALIFORNIA ACAD. SCI., PROC. 5:326, 1873 HY. EDWARDS LARVA, CHRYSALIS
ENT. NEWS 19:204, 1908 COOLIDGE EGG, LARVA, CHRYSALIS
MONTANA UNIV., BULL. 30:33 1906 ELROD LARVA
COMSTOCK, J. A.-BUTTERFLIES OF CALIFORNIA P. 38, 1927 EGG, LARVA* CHRYSALIS* PL. A13
EDWARDS, W. H.-BUTTER. NORTH AMERICA, 2ND SERIES, 1874 EGG* LARVA* CHRYSALIS* PL. 2 ANTHOCHARIS COLOR
EDWARDS, W. H.-LOC. CIT., 2ND SERIES 1885 LARVA* CHRYSALIS* PL. 13
EDWARDS, W. H.-LOC. CIT., 3RD SERIES 1897 SUPPLEMENTARY NOTES ON BREEDING EXPERIMENTS

HOLLAND-BUTTER. BOOK P. 283, 1905
HOLLAND-BUTTER. BOOK P. 284, 1931
MEAD-REPT. WHEELER EXPED. WEST 100 MERIDIAN P. 747, 1875 LARVA, CHRYSALIS

FOOD PLANTS

ARABIS SP. (CRESS)
CARDAMINE PAUCISECTA BENTH.
CRUCIFERAE SP. (CABBAGE, ETC.)

AUSONIUS LINTNER
SSP. SEE MARTIALIS SCUDDER

AUSTRALIS EDWARDS
SYN. SEE NEMESIS EDWARDS

AUSTRALIS GRINNELL
SSP. SEE LANCEOLATA BOISDUVAL

AUSTRALIS GRINNELL
SSP. SEE LYGDAMUS DOUBLEDAY

AUSTRALIS MAYNARD
SSP. SEE GLAUCUS LINNEAUS

AUTHOLEA BOISDUVAL
SSP. SEE PROXIMA GUERIN-MENEVILLE

AUTOLYCUS (EDWARDS), STRYMON-LYCAENIDAE
A ILAVIA BEUTENMULLER
SYN. MIRABELLE BARNES

LIFE HISTORY

FOOD PLANTS

QUERCUS STELLATA WANG

AUTUMNALIS COCKERELL
SSP. SEE EURYTHEME BOISDUVAL

AUTUMNALIS RILEY
SYN. SEE FRUGIPERDA ABBOT & SMITH

AUTUMNATA (GUENEE), OPORINIA-GEOMETRIDAE
SYN. ? DILUTATA DENNIS & SCHIFFERMULLER
A HENSHAWI SWETT
B OMISSA HARRIS
C OBLITA HARRIS
FORM DAYI HARRIS
FORM COARCTATA HARRIS

LIFE HISTORY

BRITISH COLUMBIA ENT. SOC., PROC. 5: NEW SERIES P. 95, 1915 DAY EGG, LARVA, PUPA (DILUTATA)
NEW YORK STATE MUS., MEM. 8: 706, 1905 FELT LARVA (BRIEF)
NEW YORK STATE, REPT. NOX. INS. 5:842, AND P. 62, SEC. 344, 1858 FITCH LARVA
GUMPPENBERG-SYSTEMA GEOMETRARUM 3:395, 1890 LARVA, PUPA
NEWMAN-BRITISH MOTHS P. 109, 1869 LARVA
PACKARD-INS. INJ. FOREST TREES P. 63, 1881 LARVA
PACKARD MONOGRAPH GEOMETRIDAE P. 88, 1876 LARVA

FOOD PLANTS

ALNUS SP. (ALDERS) HENSHAWI
CARPINUS CAROLINIANA WALT.
CRATAEGUS SP.
CRATAEGUS DOUGLASII LINDL.
CRATAEGUS OXYCANTHA LINN.
PINUS SP. (PINES)
PRUNUS SP.
PRUNUS AMERICANA MARSH.
PRUNUS ILICIFOLIA (NUTT. EX H. & A.)
QUERCUS SP. (OAKS)
ULMUS SP. (ELMS)
VACCINIUM SP.
VIBURNUM LENTAGO LINN.
CONIFERS, GENERAL FEEDER ON FOREST TREES

AUTUMNATA PACKARD
SYN. SEE POMETARIA HARRIS

AUTUMNATA PACKARD
SYN. SEE VERNATA PECK.

AUXILIARIS(GROTE), CHORIZAGROTIS-NOCTUIDAE

SYN. INTROFERENS GROTE
SYN. SOROR SMITH
FORM AGRESTIS GROTE
SYN. MERCENARIA GROTE
FORM TEGULARIS STRAND
FORM MONTANUS COOK

LIFE HISTORY

BROOKLYN ENT. SOC., BULL. 22:43, 1927 CRUMB LARVA PL. 4, C
LOC. CIT. 27:82, 1932 CRUMB LARVA
JOURN. AGRI. RESEARCH 6:877, 1916 COOLEY LIFE HISTORY
LOC. CIT. 6:808, 1916 DAVIS & SATTERWAIT LARVA* PL. 107-D; PUPA* FF. 2A, 2B; PUPAL CELLS* PL. 107-E, F
NEBRASKA AGRI. EXP. STA., RESEARCH BULL. 81:15, 1935 WHELAN TECH. DESCR. LARVA* F. 1A (LARVAL HEAD)
OKLAHOMA AGRI. EXP. STA., BULL. 275:19, 1944 FENTON WHITEHEAD LARVA
UNITED STATES DEPT. AGRI. BUR. ENT., BULL. 43:32, 1903 NEW SERIES CHITTENDEN LARVA* F. 28
UNITED STATES DEPT. AGRI. TECH. BULL. 88:97, 1929 CRUMB TECH. DESCR. EGG, LARVA* PL. 7, A, I; PUPA
CROSBY & LEONARD-MANUAL VEGET. INS. P. 287, 1918 EGG, LARVA PUPA
ESSIG-INJ. & BENEF. INS. P. 398, 1915 LARVA* F. 298
ESSIG-INS. WEST. NORTH AMERICA P. 682, 1926 LARVA (AGRESTIS AND AUXILIARIS)
METCALF & FLINT-DESTRUCT. & USEFUL INS. P. 349, 1939 LARVA

- - -

INDEX AMER. ECON. ENT.
VOL. 7:78 AUXILIARIS, AGRESTIS AND INTROFERENS
VOL. 2:83 AUXILIARIS, AGRESTIS AND INTROFERENS
VOL. 3:85 AGRESTIS, AUXILIARIS
VOL. 4:103; VOL. 5:129;
VOL. 6:148; VOL. 7:184

FOOD PLANTS

ACER SP. (MAPLES)
AGROPYRON SMITHII RYDB.
AGROSTIS SP.
ALLIUM CEPA LINN. (ONION)
APIUM GRAVEOLENS LINN. (CELERY)
ARMORACIA LAPATHIFOLIA GILB.
AVENA SATIVA LINN. (OATS)
BALSAMORRHIZA SAGITTATA (PURSH)
BETA VULGARIS LINN. (BEET)
BRASSICA HIRTA MOENCH (MUSTARD)
BRASSICA NIGRA (LINN.)(MUSTARD)
BRASSICA OLERACEA LINN. (CABBAGE)
BRASSICA RAPA LINN. (TURNIP)
CAPSELLA BURSA-PASTORIS(LINN.)
CIRSIUM SP.
DELPHINIUM SP.
DESCURAINEA RICHARDSONII SSP. INCISA (ENGELM)
FRAGARIA CHILOENSIS (LINN.)
GEUM TRIFLORUM PURSH
GOSSYPIUM HERBACIUM LINN. (COTTON)
HORDEUM VULGARE LINN. (BARLEY)
LINUM SULCATUM RIDD. (FLAX)
LUPINUS SP.
MALUS PUMILA MILL. (APPLE)
MEDICAGO SATIVA LINN. (ALFALFA)
ONOBRYCHIS VICIIFOLIA (SCOP.)
OPUNTIA SP. (CACTI)
PHLEUM PRATENSE LINN. (TIMOTHY)
PISUM SATIVUM LINN. (PEAS)
POA PRATENSIS LINN.
PRUNUS AMERICANA MARSH
PRUNUS ARMENIACA LINN. (APRICOT)
PRUNUS CERASUS LINN. (CHERRY)
PRUNUS DOMESTICA LINN. (PLUM)
P. DOMESTICA V. GALATENSIS EX HOOK (PRUNE)
PRUNUS PERSICA LINN. (PEACH)
RAPHANUS SATIVUS LINN. (RADISH)
RHEUM RHAPONTICUM LINN. (RHUBARB)
RIBES LACUSTRE PERS.
RIBES SATIVUM SYME (CURRANT)
RUBUS ALLEGHENIENSIS PORTER
RUBUS IDAEUS V. STRIGOSUS(MICHX.)

Secale cereale Linn. (rye)
Solanum lycopersicum Linn. (tomato)
Solanum tuberosum Linn. (potato)
Taraxacum officinale Weber
Thlaspi arvense Linn.
Trifolium sp. (clovers)
Triticum aestivum Linn. (wheat)
Zea mays Linn. (corn)
Berries, Cacti, Field Crops, Forage Crops, Forest Trees, Fruit Trees
General feeder
Grains
Grasses
Herbaceous plants
Vegetables
Weeds

AVALONA (Wright), STRYMON-Lycaenidae
LIFE HISTORY
Brooklyn Ent. Soc., Bull. 18: 160, 1923 Coolidge egg
Southern California Acad. Sci., Bull. 32:107, 1933 Comstock-Dammers life history egg* Pl. 49; larva* & pupa* Pl. 50
- - -
Index Amer. Econ. Ent., Vol. 6:744
FOOD PLANTS
Eriogonum giganteum S. Wats.
Lotus argophyllus v. ornithopus (Guenee)

AVIMACULA Hudson
ssp. see LINTNERI Grote (Gluphisia)

AVINOFFI Chermock
ssp. see MACHAON Linneaus

AVIRIDA Smith
ssp. see LEPIDULA Grote

AUXILLARIS Grote & Robinson
ssp. see DIFFINIS Boisduval

AZALEAE Abbot & Smith
syn. see PHOLUS Cramer

AZIF Strecker
ssp. see DETERSA Walker

-B-

BAAL Strecker
ssp. see CYBELE Fabricius

BABAYAGA Strecker, CATOCALA-Noctuidae
LIFE HISTORY

FOOD PLANTS
Salix sp. (willows)

BACCHUS Scudder
syn. see HYPOPHLAEAS Boisduval

BACHMANNI Kirtland, LIBYTHEA Libytheidae
syn. MOTYA Boisduval & Leconte
LIFE HISTORY
Amer. Mus. Nat. Hist., Bull. 5: 277, 1893 Beutenmuller larva
Can. Ent. 13:226, 1881 W.H. Edwards egg, larva, chrysalis
New York State Mus., Mem. 8:726, 1905 Felt larva
Boisduval & Leconte-Hist. Lep. North America 1833 larva* chrysalis* Pl. 64 COLOR
Comstock, J.A.-Butter. Calif. p. 147, 1927 egg* Pl. A 45; larva, chrysalis
Comstock, J.H.-How to Know Butter., p. 211, 1904 larva
Edwards, W.H.-Butter. North America, 2nd series, 1874 life history* Pl. Libythea No. 1 COLOR
French-Butter. East. United States p. 251, 1886 life history
Holland-Butterfly Book, 1905 chrysalis* Pl. 5, ff. 23, 24

MAYNARD-BUTTER. NEW ENGLAND P. 31, 1886 LARVA, CHRYSALIS
SCUDDER-BUTTER. EAST. UNITED STATES & CANADA PP. 762-766, 1889 LIFE HISTORY* PL. 64, F. 42; PL. 75, F. 9; PL. 84, FF. 23, 24
VANCE-LIFE AMONG BUTTER. P. 45, 1925 EGG, LARVA, CHRYSALIS

- - -

INDEX AMER. ECON. ENTOM., VOL. 6: 458

FOOD PLANTS

CELTIS OCCIDENTALIS LINN.
SYMPHORICARPOS OCCIDENTALIS HOOK

BADIA GROTE & ROBINSON, CATOCALA-NOCTUIDAE

LIFE HISTROY

AMER. MUS. NAT. HIST., BULL. 16: 391, 1902 BEUTENMULLER LARVAL HEAD* F. 17
AMER. MUS. NAT. HIST., MEM., NEW SERIES 3: PART 1, 1918 BARNES & MCDUNNOUGH LARVA* PL. 10, F. 32; PL. 12, FF. 2-6 COLOR (QUESTIONABLE SP.)

FOOD PLANTS

AZALEA NUDIFLOR LINN.
LONICERA DIOICA LINN.
LONICERA FLAVA SIMS
LONICERA PERICLYMENUM LINN.
MALUS PUMILA MILL. (APPLE)
PRUNUS CERASUS LINN. (CHERRY)
PYRUS COMMUNIS LINN. (PEAR)
QUERCUS SP. (OAKS)
QUERCUS ALBA LINN.
ULMUS SP. (ELMS)

BAEIS SCUDDER
SYN. SEE RURICOLA BOISDUVAL

BAILEYANA GROTE
SYN. SEE PERQUIRITATA MORRISON

BAIRDII EDWARDS, PAPILIO-PAPILIONIDAE
SYN. UTAHENSIS STRECKER
A OREGONIA EDWARDS
B HOLLANDII EDWARDS
FORM BRUCEI EDWARDS
C RUDKINI COMSTOCK
FORM CLARKI CHERMOCK
FORM COMSTOCKI CHERMOCK

LIFE HISTORY

CAN. ENT. 27:230, 1895 W. H. EDWARDS LARVA (BRIEF)
PAPILIO 2:119, 1882 STRETCH LARVA, CHRYSALIS (OREGONIA)
SOUTHERN CALIFORNIA ACAD. SCI., BULL. 39:116, 1930 INGHAM LARVA* PUPA*
LOC. CIT. 36:14, 1937 COMSTOCK-DAMMERS LIFE HISTORY EGG, LARVA* PLS. 2, 3, 4; PUPA* PL. 15
COMSTOCK, J. A.-BUTTER. CALIF. P. 19, 1927 EGG, LARVA
EDWARDS, W. H.-BUTTER. NORTH AMERICA, 3RD SERIES, 1897 PAPILIO NO. 4 (RUDKINI)

FOOD PLANTS

ANETHUM GRAVEOLENS LINN. (DILL)
APIUM PETROSELINUM LINN. (PARSLEY)
PASTINACA SATIVA LINN. (PARSNIPS)
THAMNOSOMA MONTANA T. & FREM.

BAJA AUCT
SEE SMITHI SNELL

BAJOIDES BARNES & BENJAMIN
SSP. SEE PLEBEIA SMITH

BAKERI CLARK
SSP. SEE APHRODITE FABRICIUS

BAKERI CLENCH
SSP. SEE PSEUDARGIOLUS BOISDUVAL & LECONTE

BALDER GUERIN-MENEVILLE
SYN. SEE JUTTA HUBNER

BALDUR EDWARDS
SSP. SEE CLODIUS MENETRIES

BALDUS EHRMANN
SSP. SEE CLODIUS MENETRIES

BALDYENSIS GUNDER
SSP. SEE BATTOIDES BEHR.

BALLUCA (GEYER), PLUSIA-NOCTUIDAE
LIFE HISTORY
CAN. ENT. 5:10, 1873
SAUNDERS LIFE HISTORY
ILLINOIS DEPT. AGRI., TRANS. 15: 228, 1877 FRENCH LARVA
NEW YORK STATE CABIN. NAT. HIST., ANN. REPT. NO. 26, P. 163, 1874
LINTNER COCOON & PUPA CASE
ONTARIO ENT. SOC., ANN. REPT. P. 32, 1872 BETHUNE LARVA
PHILADELPHIA ENT. SOC., PROC. 2:29, 1863 SAUNDERS LARVA
FOOD PLANTS
HUMULUS LUPULUS LINN. (HOPS)

BALTEARIA HULST
SSP. SEE SULPHURARIA PACKARD (HESPERUMIA)

BALTEATA SCUDDER
SSP. SEE IRUS GODART

BALTIMORALIS (GUENEE), BOMOLOCHA-NOCTUIDAE
SYN. BENIGNALIS WALKER
SYN. LACINIOSA ZELLER
LIFE HISTORY
NEW YORK STATE MUS., MEM. 8:727, 1905 FELT LARVA (BRIEF)
PAPILIO 2:181, 1882
PACKARD LARVA, PUPA
FOOD PLANTS
ACER SP. (MAPLES)
ACER RUBRUM LINN.

BANAVAHRATA STRECKER
SYN. SEE NUBILOFASCIATA PACKARD

BANFFENSIS GUNDER
SSP. SEE ALBERTA EDWARDS

BANKSARIA SPERRY, CHLOROSEA-GEOMETRIDAE
LIFE HISTORY
PAN PACIFIC ENT. 25:24, 1949
RINDGE LARVA, LIFE HISTORY
FOOD PLANTS
CEANOTHUS THYRSIFLORUS ESCHS.

BANKSI WATSON & COMSTOCK
SSP. SEE HYPOPHLAEAS BOISDUVAL

BAPTISIAE BIRD, PAPAIPEMA-NOCTUIDAE
LIFE HISTORY
CAN. ENT. 34:111, 1902
BIRD LARVA, PUPA
FOOD PLANTS
BAPTISIA TINCTORIA (LINN.)

BARBARA HENRY EDWARDS
SSP. SEE OCCIDENTALIS SCUDDER

BARDA HENRY EDWARDS
SYN. SEE ORNATA PACKARD

BARNESI (BENJAMIN), ABAGROTIS-NOCTUIDAE
SYN. NEVADENSIS BENJAMIN
LIFE HISTORY
BROOKLYN ENT. SOC., BULL. 27:90, 1932 CRUMB LARVA* PL. 11, H
- - -
INDEX AMER. ECON. ENT., VOL. 6:1
FOOD PLANTS
PRUNUS PERSICA (LINN.)(PEACH)
MALUS PUMILA MILL. (APPLE BUDS)

BARNESI CASSINO & SWETT
SYN. SEE MARMORATUS GROSSBECK

BARNESI FRENCH
SYN. SEE AGRIPPINA STRECKER

BARNESI GUNDER
SSP. SEE TAYLORI EDWARDS

BARNESI GUNDER
SSP. SEE XERCES BOISDUVAL

BARNESI SKINNER
SSP. SEE MYLITTA EDWARDS

BAROMETRICUS GOOSSENS
SYN. SEE PLICATUS GROTE

BARONI EDWARDS
SSP. SEE LILIANA HENRY EDWARDS

BARONI(EDWARDS), EUPHYDRYAS-NYMPHALIDAE
AB MIRABILIS WRIGHT
AB DUNNI GUNDER
AB STERNITZKYI GUNDER
LIFE HISTORY
CAN. ENT. 11:130, 1879
W. H. EDWARDS EGG, LARVA, CHRYSALIS
ILLINOIS DEPT. AGRI., TRANS. 18: APPEND. P. 163, 1880
COQUILLETT LARVA
COMSTOCK, J. A.-BUTTER. CALIF. P. 101, 1927 EGG* LARVA* CHRYSALIS* PL. A. 36
EDWARDS-BUTTER. NORTH AMERICA, 3RD SERIES, 1897 LIFE HIST. (MELITEA NO. 1, FF. A-F
FOOD PLANTS
COLLINSIA TORREYI GRAY
SCROPHULARIACEAE SP.

BARONI EDWARDS
SSP. SEE ALOPE FABRICIUS

BARTRAMI COMSTOCK & HUNTINGTON
SSP. SEE ACIS DRURY

BARTSCHI REIFF
SSP. SEE CYBELE FABRICIUS

BARYSPUS DYAR
SYN. SEE VARIA WALKER(HETEROCAMPA)

BASILIS GROTE
SYN. SEE HABILIS GROTE

BASALIS WALKER, MARATHYSSA-NOCTUIDAE
SYN. VENTILATOR GROTE
LIFE HISTORY

FOOD PLANTS
RHUS SP. (SUMACS)
RHUS TOXICODENDRON LINN. (POISON IVY)
TSUGA CANADENSIS (LINN.)
ULMUS SP. (ELMS)

BASIARIA(WALKER), GUENERIA-GEOMETRIDAE
SYN. INCOLORARIA WALKER
SYN. CONGRUA WALKER
LIFE HISTORY
WASHINGTON ENT. SOC., PROC. 51: 142, 1939 RUPERT LARVA
FOOD PLANTS
DRYOPTERIS FILIX-MAS (LINN.)
THELYPTERIS NOVEBORACENSIS (LINN.)

BASIFLAVA(PACKARD), OLENE-LIPARIDAE
SYN. CLINTONI GROTE & ROBINSON
A MERIDIONALIS BARNES & MCDUNNOUGH
LIFE HISTORY
CAN. ENT. 12:45, 1880
COQUILLETT LARVA (CLINTONI)
ILLINOIS DEPT. AGRI., TRANS. 18: APPEND P. 165, 1880
COQUILLETT LARVA
NEW YORK STATE AGRI. EXP. STA., (CORNELL), MEM. 274:242, 1948
FORBES LARVA
\- - -
INDEX AMER. ECON. ENT., VOL. 7:721
FOOD PLANTS
BETULA SP. (BIRCHES)
CARYA SP. (HICKORIES)
MALUS PUMILA MILL. (APPLE)
QUERCUS SP. (OAKS)
TILIA SP. (LINDENS)

BASIPUNCTA BARNES & MCDUNNOUGH, COPICUCULLIA-NOCTUIDAE
LIFE HISTORY
SOUTHERN CALIFORNIA ACAD. SCI., BULL. 42:45, 1943
COMSTOCK-HENNE LARVA* PL. 4
FOOD PLANTS
STEPHANOMERIA PAUCIFLORA(TORREY)

BASIRUFA STRAND
SSP. SEE CUNEATA GROTE

B-ATA GOEZE
SYN. SEE PHYLLIRA DRURY

BATABANO LUCAS, PHOCIDES-HESPERIIDAE
SYN. MACINUS HERRICH-SCHAEFFER
SYN. OKEECHOBEE WORTHINGTON
LIFE HISTORY
CAN. ENT. 22:211, 1890
DYAR EGG, LARVA, CHRYSALIS
NEW YORK ENT. SOC., JOURN. 13:142, 1905 DYAR CHANGE IN LARVAL COLOR
OHIO JOURN. SCI. 18:294, 1938
STROHECKER LARVA
FOOD PLANTS
RHIZOPHORA MANGLE LINN.

BATESII (REAKIRT), PHYCIODES-NYMPHALIDAE
AB HARPERI GUNDER
LIFE HISTORY
CAN. ENT. 52:56, 1920
MCDUNN EGG, LARVA, CHRYSALIS* PL. 1, FF. 8, 9, 10
FOOD PLANTS
ASTER UNDULATUS LINN.

BATHYLLUS (ABBOT & SMITH), THORYBES-HESPERIIDAE
SYN. DAUNUS CRAMER
LIFE HISTORY
AMER. ENT. SOC., TRANS. 37:179, 1911 SKINNER LARVA, CHRYSALIS
AMER. MUS. NAT. HIST., BULL. 5:303, 1893 BEUTENMULLER LARVA, CHRYSALIS
ILLINOIS DEPT. AGRI., TRANS. 18: APPEND. P. 98, 1880
MIDDLETON LARVA
ABBOT & SMITH-LEP. INS. GEORGIA 1:43, 1797 LARVA* CHRYSALIS* PL. 22 COLOR
BOISDUVAL & LECONTE-HIST. LEP. NORTH AMERICA P. 74, 1833 LARVA* CHRYSALIS* COLOR
HOLLAND BUTTERFLY BOOK, 1905 LARVA* PL. 2, F. 32; CHRYSALIS* PL. 6, F. 24 COLOR
SCUDDER-BUTTER. EAST. UNITED STATES & CANADA PP. 1434-1436, 1889 LIFE HISTORY* PL. 66, F. 9; PL. 76, F. 32; PL. 80, F. 6; PL. 85, F. 24
FOOD PLANTS
APIOS AMERICANA MEDIC
CENTROSEMA VIRGINIANUM (LINN.)
DESMODIUM SP.
LEGUMINOSAE SP.
LESPEDEZA CAPITATA MICHX.
LESPEDEZA HIRTA (LINN.)
LESPEDEZA VIRGINICA (LINN.)
RHYNOSIA TOMENTOSA (LINN.)
TEPHROSIA AMBIGUA CURTIS
TRIFOLIUM PRATENSE LINN.

BATTOIDES (BEHR), PHILOTES-LYCAENIDAE
A OREGONENSIS BARNES & MCDUNNOUGH
B BERNARDINO BARNES & MCDUNNOUGH
AB BALDYENSIS GUNDER
LIFE HISTORY
SOUTHERN CALIFORNIA ACAD. SCI., BULL. 33:27, 1934 COMSTOCK-DAMMERS EGG, LARVA* PL. 6; PUPA* PL. 7 (BERNARDINO)
FOOD PLANTS
ASTRAGALUS BOLANDERI GRAY
ERIOGONUM FASCICULATUM BENTH
ERIGONUM MARIFOLIUM V. INCANUM (T. & G.)

BAUERARIA SPERRY, DREPANULATRIX-GEOMETRIDAE
LIFE HISTORY
AMER. MUS. NAT. HIST., BULL. 94, 5:284, 1939 RINDGE EGG, LARVA, PUPA
FOOD PLANTS
CEANOTHUS THYRSIFLORUS ESCHS.

BAUTISTA PLOETZ
SYN. SEE ICELUS LINTNER

BAXTERI HOLLAND
SSP. SEE MYRINA CRAMER

BAYENSIS STERNITZKY
SSP. SEE EDITHA BOISDUVAL

BEANI BARNES & BENJAMIN
SSP. SEE ATLANTIS EDWARDS

BEANII (NEUMOEGEN), NEOARCTIA-ARCTIIDAE
FORM FUSCOSA NEUMOEGEN
LIFE HISTORY
CAN. ENT. 23:124, 1891
NEUMOEGEN LARVA
LOC. CIT. 41:400, 1909
GIBSON LARVA, PUPA
FOOD PLANTS
SALIX SP. (WILLOWS)
TARAXACUM OFFICINALE WEBER

BEANIANA GROTE
SYN. SEE MESKEI GROTE

BECKERII EDWARDS, PIERIS-PIERIDAE
AB GUNDERI INGHAM
A PSEUDOCHLORIDICE MCDUNNOUGH
LIFE HISTORY
ENT. NEWS 34:225, 1923
COOLIDGE LIFE HISTORY
MONTANA UNIV., BULL. 30:32
1906 ELROD MATURE LARVA
PSYCHE 2:183, 1878
MEAD LARVA
SOUTHERN CALIFORNIA ACAD. SCI.,
BULL. 25:86, 1926 COMSTOCK
LARVA* CHRYSALIS*
COMSTOCK-BUTTER. CALIFORNIA
P. 30, 1927 LARVA*
CHRYSALIS* PL. A8
EDWARDS-BUTTER. NORTH AMERICA,
2ND SERIES, 1883 LIFE HIST.*
PL. PIERIS 1 COLOR
FOOD PLANTS
CRUCIFERAE
ISOMERIS ARBOREA NUTT.

BEDDEKI HAMPSON
SSP. SEE PERQUIRITATA MORRISON

BEERIANA BIRD, PAPAIPEMA-NOCTUIDAE
FORM LACINARIAE BIRD
LIFE HISTORY
CAN. ENT. 51:108, 1923
BIRD LARVA, PUPA
FOOD PLANTS
LIATRIS PYCNOSTACHYA MICHX.

BEHRENSARIA HULST
SYN. SEE UNICALCARARIA GUENEE

BEHRENSI HENRY EDWARDS
SSP. SEE GIBBOSA ABBOT & SMITH

BEHRII (EDWARDS), CALLIPSYCHE-LYCAENIDAE
SYN. KALI STRECKER
AB NIGROINITA GUNDER
LIFE HISTORY
ENT. NEWS 19:478, 1908
WILLIAMS LARVA, CHRYSALIS
SOUTHERN CALIFORNIA ACAD. SCI.,
BULL. 26:68, 1927 COMSTOCK
EGG*
LOC. CIT. 27:63, 1928
COMSTOCK LARVA* CHRYSALIS*
FOOD PLANTS
ASTRAGALUS SP.
LOTUS GRANDIFLORUS (BENTH)
LUPINUS SP.
PURSHIA GLANDULOSA CURRAN

BEHRII (EDWARDS), COLIAS-PIERIDAE
SYN. CANESCENS COMSTOCK
LIFE HISTORY
CAN. ENT. 25:158, 1893
DYAR EGG, LARVA, CHRYSALIS
SIERRA CLUB BULL. 9:NO. 2:99-102,
1913 GRINNELL, JR. LIFE HIST.,
EGG, LARVA, CHRYSALIS
COMSTOCK, J. A.-BUTTER. CALIF.
P. 55, 1927 EGG, LARVA,
CHRYSALIS
FOOD PLANTS
GENTIANA NEWBERRYI GRAY
VACCINIUM CAESPITOSUM MICHX.
GROUND HUCKLEBERRY

BEHRII EDWARDS
SSP. SEE LYGDAMUS SCUDDER

BEHRII Edwards
ssp. see SMINTHEUS Doubleday & Hewitson

BEHRII Stretch
syn. see NEVADENSIS Grote & Robinson

BELFRAGEI Grote
syn. see FLAVIPUNCTALIS Geyer

BELFRAGEI Stretch
syn. see AURANTIACA Hubner

BELIOIDES Verity
syn. see creusa DOUBLEDAY & Hewitson

BELLA Grote
syn. see ECTYPA Morrison

BELLA(Linneaus), UTETHEISA-Arctiidae
form HYBRIDA Butler
form INTERMEDIA Butler
form TERMINALIS Neumoegen & Dyar
form NOVA Smith
syn. GROSSBECKI Strand
a VENUSTA Dalman
syn. SPECIOSA Walker

LIFE HISTORY
Amer. Mus. Nat. Hist., Bull. 10: 367, 1898 Beutenmuller larva
Illinois Dept. Agri., Trans. 18: Append. p. 113, 1880 Martin larva
Insecutor Inscitiae Menstruus 2:130, 1914 Dyar larva (brief)
New York Acad. Sci., Annals 8:82, 1893 Packard egg, larva
New York State Mus., Mem. 8:724, 1905 Felt larva (brief)
Papilio 3:128, 1883 Edwards-Elliot larva
United States Dept. Agri., Rept. p. 80, 1870 Glover larva* pupa*
Gundlach-Entom. Cubana p. 258, 1886 larva, pupa
Stretch-Illus. Zygaenidae & Bomby., North America p. 236, 1873 larva

- - -

Index Amer. Econ. Ent., Vol. 5:678; Vol. 6:800; Vol. 7:1042

FOOD PLANTS
Crotalaria sagittalis Linn.
Epilobium angustifolium (Linn.)
Lespedeza sp.
Lupinus sp.
Myrica gale Linn.
Prunus sp.
Ulmus sp. (elms)

BELLI Gunder
ssp. see ANICIA Doubleday

BELLIOIDES Verity
syn. see CREUSA Doubleday & Hewitson

BELLONA(Fabricius, BRENTHIS-Nymphalidae
MYRINA Martyn
syn. AMMIRALIS Hemming
ab FASCIATA Cockerell
syn. KLEENEI Watson
ab PARDOPSIS Holland
a TODDI Holland

LIFE HISTORY
Amer. Mus. Nat. Hist., Bull. 5:257, 1893 Beutenmuller larva, chrysalis
Amer. Naturalist 6:513, 1872 Scudder life hist. (brief)
Buffalo Soc. Nat. Sci., Bull. 19: 23, 1939 Wild larva
Butterfly Park Nature Club 2: no page, 1931 egg, larva
Can. Ent. 7:189, 1875 W. H. Edwards life hist. note
Entomologica Americana 3:162, 1887 Edwards chrysalis
Ent. News 37:57, 1926 McMurray oviposition
Illinois Dept. Agri., Trans. 18: Append. pp. 83, 183, 1880 Coquillett-Middleton larva
Montana Univ., Bull. 30:70 1906 Elrod larva
Papilio 1:134, 1881 Edwards life history notes
Peabody Acad. Sci., Rept. p. 46, 1871 Scudder egg

COMSTOCK-HOW TO KNOW BUTTER. P.120, 1904 LARVA
FERNALD-BUTTERFLIES MAINE P.41, 1884 LARVA
HOLLAND-BUTTERFLY BOOK 1905 P.134, CHRYSALIS* PL.5,FF.10, 11; 3RD EDITION P.112, 1931 LIFE HISTORY* PL.5,FF.10,11
MAYNARD-BUTTER. NEW ENGLAND P.27, 1886 LARVA
SCUDDER-BUTTER. EAST UNITED STATES & CANADA PP.611-617, 1889 LIFE HISTORY* PL.64, F.27;PL.67,F.17;PL.75,FF.1, 5;PL.79,FF.10,11;PL.84,FF. 10,11
SCUDDER-EVERYDAY BUTTER. P.315, 1899 EGG, LARVA, CHRYSALIS
SCUDDER-GUIDE TO COMMON BUTTER. P.73, 1893 LARVA, CHRYSALIS

FOOD PLANTS

VIOLA SP. (VIOLETS)
GRASS

BELLUS (EDWARDS), AMBLYSCIRTES-HESPERIIDAE

LIFE HISTORY

CAN. ENT. 43:7,8, 1911 COOLIDGE EGG, LARVA

FOOD PLANTS

GRASSES

BELMARIA EHRMANN

SSP. SEE OPELLA GROTE

BENESIGNATA(HARVEY), ZALE-NOCTUIDAE

A LARGERA SMITH
B SUBMEDIANA STRAND

LIFE HISTORY

FOOD PLANTS

PINUS BANKSIANA LAMB.

BENESIMILIS MCDUNNOUGH, EUPLEXIA-NOCTUIDAE

LUCIPARA AUCT.

LIFE HISTORY

AMER. MUS. NAT. HIST., BULL. 16: 425, 1902 BEUTEN. LARVA
CAN. ENT. 21:137, 1889 DYAR EGG, LARVA (LUCIPARA)
NEW YORK STATE MUS., MEM. 8:749, 1905 FELT LARVA (LUCIPARA)
PAPILIO 3:133, 1883 EDWARDS-ELLIOT LARVA
ESSIG-INS. WEST. NORTH AMERICA P.685, 1926 LARVA

THE FOLLOWING FOREIGN REFERENCES DEAL WITH THE TRUE EUROPEAN LUCIPARA OF LINNEAUS WHICH MCDUNNOUGH CONSIDERS DISTINCT FROM OUR SPECIES. OURS HE CALLS BENESIMILIS.

CURTIS-BRITISH ENTOM. 13:619, 1862 LARVA* PL.619 COLOR
GUENEE-SPEC. GENER. LEP. NOCT. 2:68, 1852 LARVA
HUMPHREYS-GENERA BRITISH MOTHS 1:49, 1859 LARVA* PL.20 COLOR
KIRBY-EUROPEAN BUTTER. & MOTHS P.237, 1882 LARVA
STAINTON-BRITISH BUTTER. & MOTHS 1:270, 1857 LARVA
STEPHENS-ILLUS. BRITISH ENT. HAUSTELLATE 3:4, 1829 LARVA

- - -

INDEX AMER. ECON. ENT. (ALL THE AMER. FORM) VOL. 5:240 (LUCIPARA)

FOOD PLANTS

BETULA SP. (BIRCHES)
PTERIDIUM AQUILINUM V. LATIUSCULUM (DESV.)
VIRBURNUM SP.
FERNS

BENIGNALIS WALKER

SYN. SEE BALTIMORALIS GUENEE

BENJAMINI CLARK

SSP. SEE LUCITIOSA CLEMENS

BENJAMINI GUNDER

SSP. SEE MORMONIA BOISDUVAL(ARGYNNIS)

BENJAMINI GUNDER
SSP. SEE THEONA MENETRIES

BENJAMINI LEMMER
SSP. SEE ADJUNCTA BOISDUVAL

BENJAMINI McDUNNOUGH
SSP. SEE INORNATA EDWARDS (COENONYMPHA)

BENJAMINI NAKAHARA
SYN. SEE ARTHEMIS DRURY

BERENICE(CRAMER), DANAUS-DANAIDAE
A STRIGOSA BATES
AB KERRI COMSTOCK
LIFE HISTORY
AMER. ENT. SOC., TRANS. 51:27, 1926 COOLIDGE LARVA, PUPA
CALIFORNIA ACAD. SCI., PROC. 5:327, 1873 EDWARDS, HY. LARVA, CHRYSALIS
MUS. COMP. ZOOL., BULL. 87:NO. 6, 1941 DETHIER
LARVAL ANTENNA* F. 26
SMITHSONIAN MISC. COLLECTIONS 4:37, 1862 MORRIS LARVA, CHRYSALIS
SOUTHERN CALIFORNIA ACAD. SCI., BULL. 31:16, 1932 COMSTOCK LARVA* PL. 8; CHRYSALIS* PL. 9
ABBOT & SMITH-LEP. INS. GEORGIA 1:13, 1797 PUPA* PL. 7 COLOR
COMSTOCK-BUTTER. CALIFORNIA P. 60, 1927 LARVA (STRIGOSA)
COMSTOCK-J. H.-HOW TO KNOW BUTTER. P. 208, 1904 LARVA
BOISDUVAL & LECONTE-HIST. LEP. NORTH AMERICA P. 134, 1833 LARVA* CHRYSALIS* PL. 39
FRENCH-BUTTER. EAST. UNITED STATES P. 146, 1886 LARVA, CHRYSALIS
GUNDLACH-ENTOM. CUBANA P. 25, 1881 LARVA, CHRYSALIS
MORRIS-SYNOPSIS NORTH AMER. LEP. P. 37, 1862 LARVA, CHRYSALIS
- - -
INDEX AMER. ECON. ENT., VOL. 7:245
FOOD PLANTS
ASCLEPIAS AMPLEXICAULIS SM.
ASCLEPIAS MEXICANA CAV.
CYNANCHUM PALUSTRE (GRAY)
GONOLOBUS CAROLINENSIS (JACQ.)
NERIUM SP.
SARASTEMMA CRISPUM BENTH.
SARCOSTEMMA CYNANCHOIDES V. HARTWEGII (VAIL)

BERNADENSIS GUNDER
SSP. SEE ADIASTE BEHR.

BERNADETTA LEUSSLER
SSP. SEE ANICIA DOUBLEDAY & HEWITSON

BERNARDINO BARNES & McDUNNOUGH
SSP. SEE BATTOIDES BEHR

BERNARDINO EDWARDS
SSP. SEE EURYDICE BOISDUVAL

BESKEI(HUBNER), CRINODES-NOTODONTIDAE
LIFE HISTORY
BROOKLYN MUS. ARTS & SCI., BULL. 1:199, 1906 DYAR LARVA
FOOD PLANTS

BETHUNEI(GROTE & ROBINSON), LITHOPHANE-
LIFE HISTORY
CAN. ENT. 33:1, 1901 LYMAN EGG, LARVA
NOVA SCOTIA ENT. SOC., PROC. 1: 87 1915 SANDERS LIFE HISTORY
ONTARIO ENT. SOC., ANN. REPT. P. 54, 1903 GIBSON LARVA
- - -
INDEX AMER. ECON. ENT., VOL. 2:164; VOL. 4:208
FOOD PLANTS
ACER SACCHARINUM LINN.
BETULA PAPYRIFERA MARSH.
CRATAEGUS SP. (THORNS)
FRAXINUM SP. (ASHES)
GUTIERREZIA SAROTHRAE (PURSH.)
MALUS PUMILA MILL. (APPLE)
PLANTAGO SP.

PRUNUS CERASUS LINN.
QUERCUS RUBRA LINN.
SALIX SP. (WILLOWS)
TILIA SP. (LINDENS)

BETULAE RILEY, ACRONICTA NOCTUIDAE
LIFE HISTORY
AMER. MUS. NAT. HIST., BULL. 14:241, 1901 BEUTEN. LARVA
BOSTON SOC. NAT. HIST., PROC. 27:132, 1896 DYAR LARVAL SETAE
BROOKLYN ENT. SOC., BULL. 7:2 1884 RILEY LARVA* PUPA
NEW YORK STATE MUS., MEM. 8: 749, 1905 FELT LARVA
UNITED STATES NAT. MUS., PROC. 21:76, 1898 SMITH-DYAR LARVA* PL. 5, FF. 4,5 COLOR IN SOME EDITIONS; PUPA, COCOON
FOOD PLANTS
BETULA NIGRA LINN.

BEUTENMULLERI BARNES & MCDUNNOUGH
SSP. SEE VERRILLIANA GROTE

BIANCA HENRY EDWARDS
SYN. SEE RELICTA WALKER

BIBULARIA GROTE & ROBINSON
SYN. ATHASARIA WALKER

BICARNEA (GUENEE), GRAPHIPHORA-NOCTUIDAE
SYN. PLAGIATA WALKER
LIFE HISTORY
ENT. SOC. AMERICA, ANNALS 23:396, 1930 HAWKINS LARVAL CLAWS
MARINE AGRI. EXP. STA., BULL., 356:152, 1930 PHIPPS LARVA* F. 23A; PUPA
WASHINGTON ENT. SOC., PROC. 4: 317, 1898 DYAR LARVA
HART-INSECT COLLECTIONS FOR ILLINOIS HIGH SCHOOLS P. 29, 1903 LARVA* F. 38
INDEX AMER. ECON. ENT. VOL. 5:17 AGROTIS
FOOD PLANTS
ACER SP. (MAPLES)
BETULA POPULIFOLIA MARSH.
SPIREA LATIFOLIA (AIT.)
SPIREA SALICIFOLIA LINN.
TARAXACUM OFFICINALE WEBER
VACCINIUM ANGUSTIFOLIUM AIT.
GRASS
LOW PLANTS

BICOLOR COCKERELL
SYN. LACINIA GEYER

BICOLOR (HARRIS), ADELOCEPHALA-CITHERONIIDAE
SYN. DISTIGMA WALSH
FORM SUPREMA NEUMOEGEN
FORM IMMACULATA JEWETT
LIFE HISTORY
AMER. PHILOSOPHICAL SOC., PROC. 31:147, 1893 PACKARD EGG, LARVA* PL. 5, F. 2
BOSTON SOC. NAT. HIST., PROC. 9:292, 1863 WALSH LARVA, PUPA (DISTIGMA)
ENT. SOC. AMERICA, ANNALS 7:283, 1914 MOSHER PUPAL STRUCTURE
LOC. CIT. 9:1916 PUPA* PL. 5, F. 3
NAT. ACAD. SCI., MEM. 9:67, 1905 PACKARD EGG, LARVA* PLS. 1,2 COLOR; 45,46; LARVAL STRUCTURE; PUPA* PL. 56, FF. 3,4
NEW YORK STATE MUS., MEM. 8:729, 1905 FELT LARVA (BRIEF)
PAPILIO 2:38, 1882 JEWETT EGG, LARVA, PUPA
PHILADELPHIA ENT. SOC., PROC. 3:425, 1864 WALSH LARVA
GLOVER-ILLUS. NORTH AMER. ENT. 1878 LARVA, PUPA PL. 68
- - -
INDEX AMER. ECON. ENT., VOL. 7:9
FOOD PLANTS
GLEDITSIA TRIACANTHOS LINN.
GYMNOCALADUS DIOICA (LINN.)
QUERCUS SP. (OAKS)

BICOLOR ROTHSCHILD
SYN. SEE INVOLUTUM SEPP

BICOLOR WALKER
SYN. SEE VAGANS BOISDUVAL

BICOLORAGO(GUENEE), RUSINA-NOCTUIDAE
SYN. SPURCATA WALKER
FORM FERRUGINEOIDES GUENEE
LIFE HISTORY
UNITED STATES DEPT. AGRI., TECH. BULL. 88:129, 1929 CRUMB EGG, LARVA* PL. 7, E, F; PUPA* F. 18, E, F.
WASHINGTON ENT. SOC., PROC. 4:324, 1898 DYAR LARVA
- - -
INDEX AMER. ECON. ENT. VOL. 4:345 PARASTICHTIS
FOOD PLANTS
ACER SP. (MAPLES)
ACER SACCHARINUM LINN.
BRASSICA OLERACEA LINN. (CABBAGE)
NICOTIANA TABACUM LINN. (TOBACCO)
RUMEX CRISPUS LINN.

BICOLORARIA MINOT
SYN. SEE NOTATARIA WALKER

BICOLORATA(FABRICIUS), SEMIOTHISA-GEOMETRIDAE
SYN. PRAEATOMATA HAWORTH
SYN. CONSEPTA WALKER
SYN. CONSIMILATA ZELLER
SYN. GRASSATA HULST
LIFE HISTORY
NEW YORK STATE MUS., MEM. 8:738, 1905 FELT LARVA (BRIEF)
GUENEE-SPEC. GEN. LEP. PHAL. 2:76, 1857 LARVA
GUMPPENBERG-SYSTEMA GEOMETRARUM 7:440, 1895 LARVA PUPA (PRAEATOMATA)
PACKARD-MONOGRAPH GEOMETRIDAE P. 293, 1876 LARVA
FOOD PLANTS
PINUS BANKSIANA LAMB.
PINUS STROBUS LINN.
VACCINIUM SP.
VACCINIUM ULIGINOSUM LINN.

BIDENTATA WALKER, NERICE-NOTODONTIDAE
LIFE HISTORY
AMER. MUS. NAT. HIST., BULL. 10: 416, 1898 BEUTEN. LARVA
INSECT LIFE 3:62, 1890 DYAR LARVA
KANSAS ACAD. SCI., TRANS. 11:110, 1888 MARLATT EGG* LARVA* PUPA*
NAT. ACAD. SCI., MEM. 7:171, 280, 1895 PACKARD EGG, LARVA* PL. 19, F. 4; PL. 23, F. 1 COLOR, PUPA* COCOON*
NATURALISTE CANADIEN 69:154, 1942 DAVIAULT EGG, LARVA* F. 2; PUPA
NEW YORK STATE AGRI. EXP. STA. (CORNELL), MEM. 274:222, 1948 FORBES LARVA (BRIEF)
NEW YORK STATE MUS., MEM. 8: 725, 1905 FELT LARVA (BRIEF)
PSYCHE 6:276, 1892 SOULE EGG, LARVA, PUPA
UNITED STATES DEPT. AGRI., MISC. PUB. 657:360, 401, 1950 CRAIGHEAD LARVA
ELIOT & SOULE-CATERPILLARS AND THEIR MOTHS P. 225, 1902 EGG, LARVA* PUPA
HOLLAND-MOTH BOOK, 1905 LARVA* PL. I, F. 15 COLOR
FOOD PLANTS
ULMUS SP. (ELMS)

BIEDERMANI BARNES & MCDUNNOUGH, SCHIZURA, NOTODONTIDAE
SYN. CLAMMENHOA DYAR
LIFE HISTORY
NEW YORK ENT. SOC., JOURN. 19: 82, 1911 BARNES & MCD LARVA
FOOD PLANTS

BIEDERMANNI EHRMANN
SSP. SEE MEXICANA BOISDUVAL

BIFASCIA HUBNER, SCHINIA-NOCTUIDAE
SYN. DIVERGENS WALKER
LIFE HISTORY

FOOD PLANTS
EUPATORIUM PURPUREUM LINN.

BIFASCIATA DYAR
SSP. SEE LUTEATA PACKARD

BIFILATA WALKER
SYN. SEE BISTRIARIA HUBNER

BIFILATA(HULST), DREPANULATRIX-GEOMETRIDAE
SYN. MALE PERPALLIDARIA GROTE
SYN. CORNEATA WARREN
LIFE HISTORY

FOOD PLANTS
CERCOCARPUS SP.

BIFIRIA HENRY EDWARDS
SSP. SEE APICALIS WALKER

BIGUTTATUS PACKARD
SYN. SEE IPOMOEAE DOUBLEDAY

BILINEARIA PACKARD
SYN. SEE JOHNSONARIA FITCH.

BILINEATA(PACKARD), DREPANA-DREPANIDAE
FORM LEVIS HUDSON
LIFE HISTORY
AMER. MUS. NAT. HIST., BULL. 10: 389, 1898 BEUTEN. LARVA
BOSTON SOC. NAT. HIST., PROC. 6: 402, 1894 DYAR LARVA
MAINE AGRI. EXP. STA., BULL. 259:75, 1917 MOSHER LARVA, PUPA* F.6, F, H
NATURALISTE CANADIEN P. 324, 1935 DAVIAULT LARVA, PUPA
NEW YORK ENT. SOC., JOURN. 2: 108, 1894 DYAR LARVA
NEW YORK STATE MUS., MEM. 8: 749, 1905 FELT LARVA (BRIEF)
PSYCHE 4:278, 1885 DIMMOCK LARVA
- - -
INDEX AMER. ECON. ENT., VOL. 6:284 FALCARIA
FOOD PLANTS
ALNUS SP. (ALDERS)
BETULA ALBA LINN.
BETULA PAPYRIFERA MARSH.
ULMUS SP. (ELMS)
ULMUS AMERICANA LINN.

BILINEATA(PACKARD), HETEROCAMPA-NOTODONTIDAE
SYN. TURBIDA WALKER
SYN. ASSOCIATA WALKER
SYN. ULMI HARRIS
A EXSANGUIS DYAR
LIFE HISTORY
AMER. MUS. NAT. HIST., BULL. 10: 424, 1898 BEUTEN. LARVA
CAN. ENT. 18:49, 1886 FRENCH LARVA
ENT. NEWS 4:262, 1893 DYAR LARVA COMPARED WITH MANTEO
MAINE AGRI. EXP. STA., BULL. 259: 73, 1917 MOSHER PUPA
MUS. COMP. ZOOL., BULL. 87:NO. 6, 1941 DETHIER LARVAL ANTENNA* F. 99
NAT. ACAD. SCI., MEM. 7:218, 1895 PACKARD LARVA* PL. 29, F. 1; PL. 35, FF. 1-3
NATURALISTE CANADIEN P. 146, 1942 DAVIAULT EGG, LARVA* F. 1 C; PUPA* F. 1 D
NEW YORK STATE MUS., MEM. 8: 725, 1905 FELT LARVA
GLOVER-ILLUS. NORTH AMER. ENT. 1878 LARVA* PL. 13, FF. 15, 16; PL. 19, FF. 1, 4
HARRIS-ENT. CORRESP. PP. 201, 302, 1869 LARVA* PLS. 1, 11 (ALSO AS ULMI)
- - -
INDEX AMER. ECON. ENT., VOL. 5: 298; VOL. 7:461

FOOD PLANTS
Betula sp. (birches)
Fagus sp. (beeches)
Quercus sp. (oaks)
Tilia sp. (lindens)
Tilia americana Linn.
Ulmus sp. (elms)

biloba(Stephens),Autographa-Noctuidae
LIFE HISTORY
Papilio 2:113, 1882
French life history
- - -
Index Amer. Econ. Ent.,
Vol. 1:45;Vol. 2:46;Vol. 3:51
FOOD PLANTS
Centaurea solstitialis Linn.
Cirsium horridulum Michx.
Helianthus sp. (sunflowers)
Lactuca sativa Linn. (lettuce)

bimacula(Grote & Robinson,Atrytone-Hesperiidae
syn. acanootus Scudder
syn. illinois Dodge
ab contradicta Leussler
LIFE HISTORY
Ent. News 19:415, 1908
Laurent egg
FOOD PLANTS
Grasses

bimaculata Gmelin
syn. see octomaculata Fabricius

bimaculata Saunders
syn. see aurantiaca Hubner

bimaculata(Stephens),Autographa-Noctuidae
syn. u-brevis Guenee
ab adapa Strand
LIFE HISTORY
Ent. News 4:225, 1893
Fyles egg, larva, pupa
FOOD PLANTS
Taraxacum officinale Weber

binigrimaculella Gunder
syn. see clodius Menetries

binominalis Smith
syn. see exsertistigma Morrison

binotata(Walker),Aseptis-Noctuidae
syn. extersa Walker
syn. rubiginosa Walker
a curvata Grote
LIFE HISTORY
Southern California Acad. Sci.,
Bull. 41:91, 1942 Comstock-Dammers larva* Pl. 21;
pupa* Pl. 21
United States Nat. Mus., Proc.
27:810, 1904 Dyar egg, larva (curvata)
FOOD PLANTS
Fragaria chiloensis (Linn.) (strawberry)
Quercus sp. (oaks)
Ribes sanguineum Pursh.

biplaga Guenee
ssp. see aprica Hubner

biplagiata Fruhst
ssp. see stelenes Linneaus

biplaga Walker
syn. see carneola Guenee

bipuncta(Morrison),Gabara-Noctuidae
LIFE HISTORY

FOOD PLANTS
? Arundo donax Linn.
Phragmites communis Trin.
Marsh grasses ?
Marsh reeds ?

bipunctina Guenee
syn. see argillacea Hubner

birdi Dyar
syn. see marginidens Guenee

BIRNATA SMITH
SSP. SEE PASSER GUENEE

BISCHOFFII EDWARDS
SSP. SEE EURYNOME EDWARDS

BISECTA(LINTNER), ADELOCEPHALA-CITHERONIIDAE
SYN. NEBULOSA NEUMOEGEN
LIFE HISTORY
ENT. NEWS 25:156, 1914 EYER-MENKE LIFE HISTORY LARVA
ENT. SOC. AMERICA, ANNALS 7:284, 1914 MOSHER PUPAL STRUCTURE
NAT. ACAD. SCI., MEM. 19:75, 1905 PACKARD LARVA
WASHINGTON ENT. SOC., PROC. 4:428, 1900 DYAR LARVA
FOOD PLANTS
GLEDITSIA TRIACANTHOS LINN.

BISERIATA HERRICH-SCHAEFFER
SYN. SEE MENDICA WALKER

BISERIATA PACKARD
SSP. SEE HERMINIATA GUENEE

BISIGNATA(WALKER), SEMIOTHISA-GEOMETRIDAE
SYN. GALBINEATA ZELLER
LIFE HISTORY
NEW YORK STATE AGRI. EXP. STA. (CORNELL), MEM. 274:46, 1948 FORBES LARVA
NEW YORK STATE MUS., MEM. 8:753, 1905 FELT LARVA
FOOD PLANTS
ABIES SP. (FIRS)
BETULA SP. (BIRCHES)
PINUS SP. (PINES)

BISTRIALIS (GEYER), DORYODES-NOCTUIDAE
SYN. ACUTARIA HERRICH-SCHAEFFER
LIFE HISTORY
AMER. MUS. NAT. HIST., BULL. 4:78, 1892 BEUTEN. LARVA, PUPA
WASHINGTON ENT. SOC., PROC. 4:454, 1900 DYAR EGG, LARVA
FOOD PLANTS
SPARTINA PATENS (AIT.)
SALT MARSH GRASSES

BISTRIARIA HUBNER, NUMORIA-GEOMETRIDAE
SYN. BIFILATA WALKER
LIFE HISTORY
NEW YORK STATE MUS., MEM. 8:720, 1905 FELT LARVA (BRIEF)
FOOD PLANTS
JUGLANS NIGRA LINN.
QUERCUS SP. (OAKS)

BISTRIARIA(PACKARD), CHETEOSCELIS-GEOMETRIDAE
SYN. UNDINARIA STRECKER
LIFE HISTORY
ENT. AMERICANA 3:193, 1888 HULST LARVA
FOOD PLANTS
SOLIDAGO FLOWERS

BISTRIARIS HUBNER, PARALLELIA-NOCTUIDAE
SYN. AMPLISSIMA WALKER
LIFE HISTORY
CAN. ENT. 2:130, 1870 SAUNDERS LARVA
ONTARIO ENT. SOC., ANN. REPT. P. 41, 1872 REED LARVA
PAPILIO 3:136, 1883 EDWARDS-ELLIOTT LARVA
GLOVER-ILLUS. NORTH AMER. ENT. 1878 LARVA PL. 19, F. 10
GUENEE-SPEC. GEN. LEP. NOCT. 3:268, 1852 LARVA
HART-INSECT COLLECTIONS FOR ILLINOIS HIGH SCHOOLS P. 43, 1903 PUPATION HABITS
PACKARD-INS. INJ. FOREST TREES P. 113, 1881 LARVA
FOOD PLANTS
ACER SP. (MAPLES)
ACER RUBRUM LINN.
ACER SACCHARINUM LINN.
JUGLANS NIGRA LINN.

BISTRIGALIS STEPHENS
SYN. SEE LUBRICALIS GEYER

BITACTATA(WALKER),ITAME-GEOMETRIDAE
SYN. ATROSIGNATA WALKER
A PACKARDARIA MOESCHLER
B EPIGENATA BARNES & MCDUNNOUGH
? C WAVARIA LINNEAUS
LIFE HISTORY
NEW YORK AGRI. EXP. STA. (CORNELL), MEM. 274:41, 1948 FORBES LARVA
FOOD PLANTS
RIBES CYNOSBATI LINN.
RIBES SATIVUM SYME (CURRANT)

BIUNDATA WALKER, HETEROCAMPA-NOTODONTIDAE
SYN. SEMIPLAGA WALKER
SYN. OLIVATA PACKARD
SYN. MOLLIS WALKER
LIFE HISTORY
AMER. MUS. NAT. HIST., BULL. 10: 426, 1898 BEUTEN. LARVA
MUS. COMP. ZOOLOGY, BULL. 87:NO. 6, 1941 DETHIER LARVAL ANTENNA* F. 100
NAT. ACAD. SCI., MEM. 7:236, 1895 PACKARD LARVA* PL. 32, FF. 1-4 COLOR; PUPA*
LOC. CIT. 9: 1905 PACKARD LARVAL HEAD* PL. 44, F. 10
NEW YORK STATE AGRI. EXP. STA. (CORNELL), MEM. 274:232, 1948 FORBES LARVA
NEW YORK STATE MUS., MEM. 8: 727, 1905 FELT LARVA (BRIEF)
FOOD PLANTS
ACER SP. (MAPLES)
ACER RUBRUM LINN.
ACER SACCHAROPHORUM KOCH
BETULA SP. (BIRCHES)
BETULA ALLEGHANIENSIS BRITT.
BETULA PAPYRIFERA MARSH.
CARYA SP. (HICKORIES)
CORNUS FLORIDA LINN.
FAGUS SP. (BEECHES)
HAMAMELIS VIRGINIANA LINN.
PRUNUS CERASUS LINN.
PRUNUS PENSYLVANICA LINN.
PRUNUS SEROTINA EHRH.
SALIX SP. (WILLOWS)

BIVITTATA (HULST), EUPITHECIA-GEOMETRIDAE
LIFE HISTORY

FOOD PLANTS
CEANOTHUS SP.

BLACKMOREI BARNES & MCDUNNOUGH
SSP. SEE ICARIOIDES BOISDUVAL

BLACKMOREI CASSINO & SWETT
SSP. SEE OCCIDENTALIS HULST(PERO)

BLACKMOREI GUNDER
SSP. SEE PALLA BOISDUVAL

BLAKEI(GROTE), APANTESIS-ARCTIIDAE
LIFE HISTORY
NEW YORK ENT. SOC., JOURN. 8:46, 1900 DYAR LARVA
PSYCHE 6:380, 1893 DYAR DESCRIBES EGG, LARVA WHICH HE INCORRECTLY DETERMINES AS BLAKEI, BUT IN SAME JOURNAL P. 556, SAYS IT IS ORNATA
FOOD PLANTS
TRITICUM AESTIVUM LINN. (WHEAT)

BLANDA GROTE
SYN. SEE OLIVATA HARVEY

BLANDULA HULST, CATOCALA-NOCTUIDAE
A MANITOBENSIS CASSINO
LIFE HISTORY
AMER. MUS. NAT. HIST., BULL. 38: 173, 1918 BARNES & MCDUNNOUGH EGG, LARVA
AMER. MUS. NAT. HIST., MEM. NEW WERIES 3:PART 1, 1918 BARNES & MCDUNNOUGH LARVA* PL. 14, F. 12; PL. 15, FF. 23, 39; PL. 16, F. 3 ALL IN COLOR
BROOKLYN ENT. SOC., BULL. 7:38, 1884 HULST LARVA

FOOD PLANTS
CRATAEGUS SP. (THORNS)

BLATCHLEYI HAIMBACH
SSP. SEE AMELLA GUENEE

BOCHUS MORRISON
SYN. SEE BROCHA MORRISON

BOHARTI DOUDOROFF
SSP. SEE AUSONIDES BOISDUVAL

BOHARTI GUNDER
SSP. SEE NUBIGENA BEHR

BOHARTI GUNDER
SSP. SEE MONTIVAGA BEHR

BOHARTI GUNDER
SSP. SEE NEUMOEGENI SKINNER

BOHARTI GUNDER
SSP. SEE SAEPIOLUS BOISDUVAL

BOISDUVALI OBERTHUR
SYN. SEE EGLANTERINA BOISDUVAL

BOISDUVALII BUTLER
SYN. SEE CALIFORNIAE WALKER

BOISDUVALII HARRIS
SYN. SEE EURYDICE JOHANSSON

BOLANDERI (STRETCH), APANTESIS-ARCTIIDAE
A CONFLUENTIS STRAND
LIFE HISTORY
BOSTON SOC. NAT. HIST., PROC. 26:153, 1893 DYAR EGG, LARVA, PUPA
FOOD PLANTS
LOW HERBS

BOLLII EDWARDS
SSP. SEE THEONA MENETRIES

BOMBAX BARNES & BENJAMIN
SSP. SEE LUSCITIOSA CLEMENS

BOMBYCIFORMIS SMITH
SYN. SEE DEPILIS GROTE

BOMBYCOIDES WALKER, LAPARA-SPHINGIDAE
SYN. HARRISII CLEMENS
A PINEUM LINTNER
LIFE HISTORY
AMER. MUS. NAT. HIST., BULL. 7: 310, 1895 BEUTEN. LARVA, PUPA
AMER. NAT. 16:745, 1882 RILEY PUPA (BRIEF)
ENT. NEWS 9:150, 1898 SOULE DURATION OF EGG
ENT. SOC. AMERICA, ANNAL 11:423, 1918 MOSHER PUPA
ILLINOIS DEPT. AGRI., TRANS. 18: APPEND. P. 105, 1880 MARTIN LARVAE (BOMBYCOIDES & PINEUM)
MAINE AGRI. EXP. STA., BULL. 162, 1908 PATCH LARVA F. 52
NEW YORK STATE AGRI. EXP. STA. (CORNELL), MEM. 274:192, 1948 FORBES LARVA
NEW YORK STATE CABIN. NAT. HIST., ANN. REPT. NO. 23:170, 171, 1873 LINTNER LARVA* PL. 8, FF. 8, 9 (PINEUM)
NEW YORK STATE MUS., MEM. 8:679 1905 FELT LARVA
NOVITATES ZOOLOGICAE (TRING) 9: SUPPLEMENT P. 152, 1903 ROTHSCHILD & JORDAN LARVA, PUPA
ONTARIO ENT. SOC., ANN. REPT. P. 70, 1881 REED LARVA, PUPA
PHILADELPHIA ACAD. NAT. SCI., JOURN. 5:188, 1859 CLEMENS LARVA, PUPA
PHILADELPHIA ENT. SOC., PROC. 3:669, 1864 LINTNER LARVA, PUPA
SMITHSONIAN MISC. COLLECTIONS 4:ART. 2:216, 1862 MORRIS LARVA (BRIEF)
UNITED STATES DEPT. AGRI. MISC. PUB. 657:364, 375, 1950 CRAIGHEAD LARVA F. 17 J

Boisduval-Spec. Gen., Lep. Sphing. pp. 106, 107, 1874 larva, pupa (pineum also)
Fernald-Sphingidae New England p. 84, 1886 larva
Glover-Illus. North Amer. Ent. 1878 larva Pl. 4, f. 6
Morris-Synop. Lep. North America p. 216, 1862 larva (brief)
Packard-Guide Study Insects p. 272, 1883 larva
Packard-Ins. Inj. Forest Trees p. 202, 1881 larva (pineum also)
Strecker-Lep. Rhop. et. Heter. 13:117, 1876 larva

- - -

Index Amer. Econ. Ent., Vol. 1:175

FOOD PLANTS

Pinus banksiana Lamb.
Pinus rigida Mill.
Pinus strobus Linn.

bonifata (Hulst), Sterrha-Geometridae

syn. ptelearia Riley
syn. delicata Hulst

LIFE HISTORY

Botanical Gazette 16:334, 1891 Riley egg* larva* pupa*
Ent. News 9:15, 1898 Kellogg duration of stages
Insect Life 4:108, 1891 Riley larva* pupa* (ptelearia)
New York State Agri. Exp. Sta. (Cornell), Mem. 274:127, 1948 Forbes larva, pupa
Holland-Moth Book p. 334, 1905 egg* larva* pupa* ff. 199, 200

- - -

Index Amer. Econ. Ent., Vol. 3:144 Eois ptelearia

FOOD PLANTS

Herbarium specimens

boopis Behr

ssp. see alope Fabricius

borealis (Boisduval), Cerura-Notodontidae

LIFE HISTORY

Amer. Mus. Nat. Hist., Bull. 10: 429, 1898 Beuten. larva
Amer. Nat. 8:691, 1874 Packard larva* f. 106
Boston Soc. Nat. Hist., Proc. 24: 553, 1890 Packard larva
Can. Ent. 13:145, 1881 French larva
Loc. cit. 23:83, 1891 Dyar egg, larva, pupa
Maine Agri. Exp. Sta., Bull. 259:63, 1917 Mosher larva, pupa* f. 5 J
Minnesota Agri. Exp. Sta., Bull. 61:172, 1899 Lugger larva* f. 120
Nat. Acad. Sci., Mem. 7:266, 1895 Packard egg, larva* Pl. 36, ff. 1-3 COLOR; Pl. 49, f. 2; pupa; cocoon
New York State Agri. Exp. Sta. (Cornell), Mem. 274:237, 1948 Forbes larva
New York State Cabin. Nat. Hist., Ann. Rept. No. 26:151, 1874 Lintner larva*
New York State Mus., Mem. 8:743 1905 Felt larva (brief)
Abbot & Smith-Lep. Ins. Georgia p. 141, 1797 larva* Pl. 71 COLOR
Cuvier-Animal Kingdom 15, 1832 larva Pl. 82
Glover-Illus. North Amer. Ent., 1878 larva Pl. 2, f. 4; Pl. 20, f. 35
Harris-Ins. Inj. Veget., 1st edit. p. 305, 1841 larva

FOOD PLANTS

Populus sp. (poplars)
Prunus sp. (cherries)
Prunus serotina Ehrh.
Salix sp. (willows)

borealis Cary

ssp. see propertius Scudder & Burgess

BOREALIS CHERMOCK
SYN. SEE ALOPE FABRICIUS

BOREALIS CLARK
SSP. SEE UNDULOSA WALKER

BOREALIS CLARK
SSP. SEE PORTLANDIA FABRICIUS

BOREALIS CLARK
SSP. SEE MODESTA HARRIS

BOREALIS CLARK
SSP. SEE EXCAECATA ABBOT & SMITH

BOREALIS CLARK
SSP. SEE CERISVI KIRBY

BOREALIS CLARK
SSP. SEE GORDIUS CRAMER

BOREALIS CLARK
SEE LUSCITIOSA CLEMENS

BOREALIS FITCH
SYN. SEE LEUCOSTIGMA ABBOT & SMITH

BOREALIS GROTE
SYN. SEE NAPI LINNEAUS

BOREALIS(GROTE & ROBINSON)CALEPHELIS-RIODINIDAE
SYN. GEDA SCUDDER
LIFE HISTORY
CAN. ENT. 68:167, 1936
DOS PASSOS LIFE HISTORY
EGG* LARVA* PUPA*
FOOD PLANTS
PRUNUS ILICIFOLIA (NUTT. EX H & A)
SENECIO OBOVATUS MUHL.

BOREALIS MOESCHLER
SYN. SEE PARTHENOS HARRIS

BOREALIS SMITH
SYN. SEE ROTUNDALIS WALKER

BOREALIS SMITH
SSP. SEE STYRACIS GUENEE

BOREALIS STAUDINGER
SYN. SEE FULIGINOSA LINNEAUS

BOREATA(HUBNER), OPOROPHTERA-GEOMETRIDAE
LIFE HISTORY
ENTOMOLOGICA AMERICANA 3:49, 1887 BRUCE LARVA
FOOD PLANTS
ACER SP. (MAPLES)
BETULA SP. (BIRCHES)
MALUS PUMILA MILL. (APPLE)
PYRUS COMMUNIS LINN. (PEAR)
ULMUS SP. (ELMS)

BOREATA PACKARD
SYN. SEE PALUDATA THUNBERG

BORUS BOISDUVAL
SYN. SEE CALIFORNICA EDWARDS(STRYMON)

BOSTONIENSIS(GROTE), EUXOA-NOCTUIDAE
LIFE HISTORY
UNITED STATES DEPT. AGRI., TECH. BULL. 88:89, 1929 CRUMB
EGG; LARVA* F. 16B; PUPA
FOOD PLANTS
NICOTIANA TABACUM LINN. (TOBACCO)

BOUCARDI GODMAN & SALVIN
SSP. SEE VESTA EDWARDS

BOWERI CHERMOCK
SSP. SEE EURYDICE JOHANSSON

BOWMANI CASSINO & SWETT
SSP. SEE DESTINATA MOESCHLER

BRADORATA MCDUNNOUGH, EUPITHECIA GEOMETRIDAE
LIFE HISTORY

FOOD PLANTS
PICEA GLAUCA (MUENCH.)

BRASSICAE RILEY
SEE NI HUBNER

BREDOWII(HUBNER), HETEROCHROA-NYMPHALIDAE
SYN. EULALIA DOUBLEDAY & HEWITSON
A CALIFORNICA BUTLER
LIFE HISTORY
CALIFORNIA ACAD. SCI., PROC. 7:20, 1876 HY. EDWARDS LARVA, CHRYSALIS (EULALIA)
CAN. ENT. 23:278, 1891 DYAR LIFE HISTORY EGG, LARVA, CHRYSALIS
SOUTHERN CALIFORNIA ACAD. SCI., BULL. 30:83, 1931 COMSTOCK-DRUMMERS LIFE HISTORY, EGG* LARVA*
COMSTOCK, J. A. -BUTTER CALIF. P. 146, 1927 LARVA, CHRYSALIS* F. A 44 (CALIFORNICA)
ESSIG-INS. WEST. NORTH AMERICA P. 648, 1926 EGG, LARVA, CHRYSALIS
FOOD PLANTS
QUERCUS CHRYSOLEPIS LIEBM.
QUERCUS DOUGLASII H. & A.
QUERCUS KELLOGGII NEWB.
QUERCUS VIRGINIANA MILL.

BREHMEI BARNES & BENJAMIN
SSP. SEE DAMON CRAMER

BREPHOIDES(WALKER), LEUCOBREPHOS-GEOMETRIDAE
SYN. RESOLUTA ZELLER
SYN. HOYI GROTE
LIFE HISTORY
CAN. ENT. 48:133, 1916 CRIDDLE EGG, LARVA, PUPA
NEW YORK ENT. SOC., JOURN. 53: 209, 1945 FORBES PUPA
NEW YORK STATE AGRI. EXP. STA. (CORNELL), MEM. 274:15, 1948 FORBES EGG, LARVA
\- - -
INDEX AMER. ECON. ENT., VOL. 1:184
FOOD PLANTS
POPULUS TREMULOIDES MICHX.

BRETONENSIS MCDUNNOUGH
SSP. SEE BREVICAUDA SAUNDERS

BRETTOIDES EDWARDS
SSP. SEE BRETTUS BOISDUVAL & LECONTE

BRETTOIDES WRIGHT
SYN. SEE PHYLEUS DRURY

BRETTUS(BOISDUVAL & LECONTE), POLITES-HESPERIIDAE
SYN. WINGINA SCUDDER
SYN. UNNA PLOETZ
A BRETTOIDES EDWARDS
SYN. CLARA PLOETZ
LIFE HISTORY
CAN. ENT. 11:190, 1879 EDWARDS, W. H. EGG, LARVA, CHRYSALIS
FRENCH-BUTTER. EAST. UNITED STATES P. 315, 1886 EGG, LARVA, CHRYSALIS
GUNDLACH-ENTOM. CUBANA P. 164, 1881 LARVA, CHRYSALIS
SCUDDER-BUTTER. EAST. UNITED STATES & CANADA PP. 1703-1704, 1889 LIFE HISTORY
FOOD PLANTS
PASPALUM CILIATIFOLIUM MICHX.
PASPALUM SETACEUM MICHX.

BRETTUS HOLLAND
SYN. SEE PHYLEUS DRURY

BRETTUS WRIGHT
SYN. SEE PHYLEUS DRURY

BREVICAUDA SAUNDERS, PAPILIO-PAPILIONIDAE
SYN. ANTICOSTIENSIS STRECKER
A GASPEENSIS MCDUNNOUGH
B BRETONENSIS MCDUNNOUGH
LIFE HISTORY
BUTTERFLY PARK NATURE CLUB 2:NO PAGE NUMBERS, 1931 EGG, LARVA
CAN. ENT. 15:45, 1883 GOSSE LARVA, CHRYSALIS

AMER. MUS. NAT. HIST., BULL. 38: 162, 1918 BARNES & MCD. EGG, LARVA

AMER. MUS. NAT. HIST., MEM. NEW SERIES 3: PART 1, 1918 BARNES & MCD. LARVA* PL. 13, F. 7; PL. 15, F. 17; PL. 16, F. 17; PL. 17, F. 4 COLOR

NEW YORK STATE MUS., MEM. 8:743 1905 FELT LARVA (BRIEF)

FOOD PLANTS

POPULUS SP. (POPLARS)

SALIX SP. (WILLOW)

BRIZO (BOISDUVAL), ERYNNIS-HESPERIIDAE

A SOMNUS LINTNER

LIFE HISTORY

AMER. MUS. NAT. HIST., BULL. 5: 299, 1893 BEUTENMULLER LARVA, CHRYSALIS

BOISDUVAL & LECONTE-HIST. LEP. NORTH AMERICA 1833 LARVA* CHRYSALIS* PL. 66 COLOR

HOLLAND-BUTTER. BOOK 1905 CHRYSALIS* PL. 6, F. 38

SCUDDER-BUTTER. EAST. UNITED STATES & CANADA PP. 1503-1506, 1889 LIFE HISTORY* PL. 66, F. 12; PL. 69, F. 4; PL. 77, FF. 12, 18; PL. 85, F. 38

SCUDDER-EVERYDAY BUTTER. P. 50, 1899 EGG, LARVA

SCUDDER-GUIDE TO COMMON BUTTER. P. 162, 1893 LARVA, CHRYSALIS

FOOD PLANTS

BAPTISIA SP.

CYNOGLOSSUM SP.

GALACTIA REGULARIS (LINN.)

QUERCUS ILICIFOLIA WANG.

BROCHA (MORRISON), EUXOA-NOCTUIDAE

SYN. BOCHUS MORRISON

LIFE HISTORY

CAN. ENT. 27:332, 1895 FRENCH EGG, LARVA

FOOD PLANTS

RAPHANUS SATIVUS LINN. (RADISH) (IN CAPTIVITY)

TRIFOLIUM SP. (CLOVER) (IN CAPTIVITY)

BRODIEI CLARK

SSP. SEE PHOLUS CRAMER

BRONTES (DRURY), PHLEGETHONTIUS-SPHINGIDAE

SYN. PAMPHILIUS STOLL

SYN. COLARIS WALKER

A CUBENSIS GROTE

LIFE HISTORY

GUENEE-SPEC. GEN. LEP. SPHING. P. 116, 1874 LARVA, PUPA

FOOD PLANTS

BRONTES DENIS & SCHIFFERMULLER

SYN. SEE PALAEMON PALLAS

BROWERI GUNDER

SSP. SEE SARA BOISDUVAL

BROWERI GUNDER

SSP. SEE MARCELLUS CRAMER

BROWERI RUPERT, METARRANTHIS-GEOMETRIDAE

LIFE HISTORY

ENT. NEWS 5:61, 1894 DYAR EARLY STAGES AS HYPOCHRARIA

PSYCHE 8:415, 1899 EARLY STAGES AS HYPOCHARIA BY DYAR. SEE NOTE BY RUPERT IN JOURN. NEW YORK ENT. SOC., JOURN. 51: 150, 1943

FOOD PLANTS

DIOSPYROS VIRGINIANA LINN. (PERSIMMON)

MALUS PUMILA MILL. (APPLE)

PRUNUS SEROTINA EHRH.

SASSAFRAS ALBIDUM V. MOLLE (RAF.)

BROWNI FIELD

SSP. SEE SENNAE LINNEAUS

BROWNINGI SKINNER

SSP. SEE SARA BOISDUVAL

Loc. cit. 71:153, 1939
McDunnough larva*
compared with asterias
Ontario Ent. Soc., Ann. Rept.
p. 36, 1898 Winn egg, chrysalis
Papilio 4:85, 1884
Gruber larva
Comstock-How to Know Butter.
p. 61, 1904 larva
Edwards, W. H.-Butter. North
America 2nd series 1875 egg*
Pl. 13; larva* Pl. 8 Papilio
COLOR
Loc. cit. 2nd. series 1880
Part 9 life history* Pl. 8
Papilio COLOR
Loc. cit. 1885
egg* larva* Pl. 13
Scudder-Butter. East. United
States & Canada pp. 1852-1854,
1889 life history
Strecker-Rhop. et Heter.
pp. 49-69, 1873 egg* young
larva* Pl. 8

FOOD PLANTS

Angelica sp.
Apium petroselinum Linn. (parsley)
Ligusticum scothicum Linn.
Pastinaca sativa Linn. (parsnip)
Umbelliferae sp.

brevicornis Walker
ssp. see aurantiaca Hubner

brevipennis (Smith), Euxoa-Noctuidae
ab angulirena Smith
a brevistriga Smith

LIFE HISTORY

Can. Ent. 62:261, 1930
Cook larva (brief)

FOOD PLANTS

Conringia orientalis (Linn.)
Melilotus alba Desr.
Salsola tenuifolia
v. tragus Tausch
Sisymbrium altissimum Linn.

brevipennis (Stretch), Euthisanotia-Noctuidae

LIFE HISTORY

Southern California Acad. Sci.,
Bull. 37:118, 1938 Comstock-
Dammers egg* Pl. 41; larva*
Pls. 42, 43; pupa* Pls. 43, 44

FOOD PLANTS

Epilobium californicum Haussk.
Epilobium watsonii v.
franciscanum (Barley)
Oenothera biennis Linn.
Eggs laid on dry Typha stems.
Do larvae then migrate to
Epilobium sp.?

brevirostralis Grote
syn. see salusalis Walker

brevis (Grote), Schinia-Noctuidae
syn. female atrites Grote

LIFE HISTORY

Ent. News 34:199, 1923
Bird larva, pupa

FOOD PLANTS

Aster laevis Linn.
Aster novae-angliae Linn.
Aster puniceus Linn.

brevistriga Smith
ssp. see brevipennis Smith

bridgei Comstock
syn. see hoffmanni Behr

brillians Ottolengui
syn. see thoracica Putnam-Cramer

briseana Strand
ssp. see briseis Edwards

briseis Edwards, Catocala-Noctuidae
form briseana Strand
form clarissima Beutenmuller
form albida Beutenmuller
a minerva Cassino

LIFE HISTORY

BRUCEATA(HULST), OPOROPHTERA-
GEOMETRIDAE
LIFE HISTORY
NEW YORK STATE AGRI. EXP. STA.
(CORNELL), MEM. 274:158, 1948
FORBES EGG, LARVA
UNITED STATES DEPT. AGRI.,
MISC. PUB. 657:361, 426, 1950
CRAIGHEAD LARVA
ESSIG-INS. WEST. NORTH AMERICA
P. 699, 1926 LARVA
SLINGERLAND & CROSBY-MANUAL
FRUIT INS. P. 94, 1914 EGG*
FF. 102, 103; LARVA* F. 104;
PUPA* F. 104A; COCOON* F. 105
- - -
INDEX AMER. ECON. ENT.,
VOL. 1:265 RACHELA
VOL. 3:352 RACHELA
VOL. 4:417 RACHELA
VOL. 5:540 RACHELA
FOOD PLANTS
ACER SACCHAROPHORUM KOCH
TAGUS SP. (BEECH)
MALUS PUMILA MILL. (APPLE)
POPULUS TREMULOIDES MICHX.
SALIX SP. (WILLOWS)

BRUCEI EDWARDS(EUPHYDRYAS)
SYN. SEE ANICIA DOUBLEDAY & HEWITSON

BRUCEI(EDWARDS), OENEIS-SATYRIDAE
A YUKONENSIS GIBSON
LIFE HISTORY
EDWARDS-BUTTER. NORTH AMERICA
3RD SERIES 1897 LIFE HIST*
(CHIONOBAS 6 FF. A-G)
FOOD PLANTS

BRUCEI EDWARDS(PAPILIO)
SSP. SEE BAIRDII EDWARDS

BRUCEI ELWES
SSP. SEE EPIPSODEA BUTLER

BRUCEI GUNDER
SSP. SEE EURYNOME EDWARDS

BRUCEI HENRY EDWARDS, ICHTHYURA-
NOTODONTIDAE
A MULTNOMA DYAR
SYN. ALETHE NEUMOEGEN & DYAR
LIFE HISTORY
NAT. ACAD. SCI., MEM. 7:137, 1895
PACKARD LARVA, PUPA, COCOON
NEW YORK STATE AGRI. EXP. STA.
(CORNELL), MEM. 274:212, 1948
FORBES LARVA
NEW YORK STATE MUS., MEM. 8:743
1905 FELT LARVA (BRIEF)
PSYCHE 6:403, 1893 DYAR
LARVA, PUPA, COCOON (MULTNOMA)
FOOD PLANTS
POPULUS SP. (POPLARS)
SALIX SP. (WILLOWS)

BRUCEI HENRY EDWARDS, NEOARCTIA-
ARCTIIDAE
LIFE HISTORY
ENTOMOLOGICA AMERICANA 3:219,
1888 BRUCE EGG, LARVA
FOOD PLANTS
PLANTAGO SP.
POLYGONUM SP.

BRUCEI SMITH
SSP. SEE HIBISCI GUENEE

BRUMOSA GUENEE, ACRONICTA-NOCTUIDAE
A PERSUASA HARVEY
B LITURATA SMITH
LIFE HISTORY
AMER. MUS. NAT. HIST., BULL. 14:
250, 1901 BEUTEN. LARVA
NEW YORK STATE MUS., MEM. 8:
704, 1905 FELT LARVA
OHIO AGRI. EXP. STA., BULL. 45:
174, 1893 WEBER LARVA
PAPILIO 1:56, 1881
COQUILLETT LARVA
UNITED STATES NAT. MUS., PROC.
21:118, 1898 SMITH & DYAR
LARVA, PUPA, COCOON
GUENEE-SPEC. GEN. LEP. NOCT. 1:
52, 1852 LARVA* PL. 2 COLOR

SAUNDERS–INS. INJ. FRUITS P. 313, 1883 LARVA, PUPA
SAUNDERS–LOC. CIT. P. 313, 1889 LARVA

- - -

INDEX AMER. ECON. ENT., VOL. 7:8

FOOD PLANTS

BETULA SP. (BIRCHES)
CARYA SP. (HICKORIES)
CORYLUS SP. (HAZELS)
HAMAMELIS VIRGINIANA LINN.
PRUNUS DOMESTICA LINN. (PLUM)
QUERCUS SP. (OAKS)
RUBUS ALLEGHENIENSIS PORTER
RUBUS IDAEUS V. STRIGOSUS (MICHX.)
SALIX SP. (WILLOWS)

BRUNNEA GROTE & ROBINSON
SYN. SEE OBLIQUA PACKARD

BRUNNEA SAGED
SSP. SEE POLYPHEMUS CRAMER

BRUNNEA STRECKER
SSP. SEE CUNEA DRURY

BRUNNEA STRETCH, CTENUCHA–AMATIDAE

LIFE HISTORY

JOURN. ENT. & ZOOLOGY 7: (POMONA COLLEGE) P. 241, 1915 ESSIG EGG*, LARVA*, PUPA
SOUTHERN CALIFORNIA ACAD. SCI., BULL. 40:11, 1941 COMSTOCK-DAMMERS EGG, LARVA* PL. 3; PUPA* PL. 3

- - -

INDEX AMER. ECON. ENT., VOL. 2:103

FOOD PLANTS

CAREX SP.
ELYMUS CONDENSATUS PRESL.
SORGHUM HALEPENSE (LINN.)

BRUNNEA TUTT
SSP. SEE PSEUDARGIOLUS BOISDUVAL & LECONTE

BRUNNEARIA (PACKARD), NEMORIA–GEOMETRIDAE

LIFE HISTORY

CAN. ENT. 10:66, 1878 GOODELL LARVA, PUPA
LOC. CIT. 17:32, 1885 KELLICOTT LARVA, PUPA
MINNESOTA AGRI. EXP. STA., BULL. 61:237, 1899 LUGGER LARVA, PUPA
NEW YORK STATE AGRI. EXP. STA. (CORNELL), MEM. 274:114, 1948 FORBES LARVAL COLOR
GUMPPENBERG–SYSTEMA GEOMATRARUM 7:498, 1895 LARVA, PUPA
PACKARD–MONOGRAPH GEOMETRIDAE P. 388, 1876 LARVA

FOOD PLANTS

JUGLANS NIGRA LINN.
MALUS PUMILA MILL. (APPLE)
PRUNUS DOMESTICA LINN. (PLUM)
PRUNUS PENSYLVANICA LINN.
QUERCUS SP. (OAKS)

BRUNNEATA CASSINO & SWETT
SSP. SEE VANCOUVERENSIS HULST

BRUNNEATA PACKARD
SYN. SEE LATIFERRUGATA WALKER

BRUNNEATA THUNBERG
SYN. SEE FULVARIA VILLERS

BRUNNEICILIATA PACKARD
SYN. SEE GRATULATA WALKER

BRUNNEICOLLIS (GROTE), PROTOLAMPRA–NOCTUIDAE

LIFE HISTORY

BROOKLYN ENT. SOC., BULL. 27:87, 1932 CRUMB LARVA
ENT. SOC. AMERICA, ANNALS 23:396, 1930 HAWKINS LARVAL CLAWS
MAINE AGRI. EXP. STA., BULL. 356:168, 1930 PHIPPS LARVA* F. 23F; PUPA

- - -

INDEX AMER. ECON. ENT., VOL. 5:372 LAMPRA

FOOD PLANTS
COMPTONIA PEREGRINA(LINN.)
NICOTIANA TABACUM LINN. (TOBACCO)
TARAXACUM OFFICINALE WEBER
VACCINIUM ANGUSTIFOLIUM AIT.
VACCINIUM MYRTILLOIDES MICHX.
LOW HERBS

BRUNNEIGERA(GROTE), EUXOA-NOCTUIDAE
A LATEBRA BENJAMIN
LIFE HISTORY
WASHINGTON ENT. SOC., PROC.
4:318, 1898 DYAR LARVA
FOOD PLANTS

BRUNNEIPENNIS GROTE
SSP. SEE CUPIDA GROTE

BRUNNEIPENNIS HULST
SYN. SEE OBSTIPATA FABRICIUS

BRUNNEOSUFFUSA STRAND
SYN. SEE COMMUNIS DYAR

BRYANTI LEUSSLER
SSP. SEE AQUILO BOISDUVAL

BUCEPHALUS STEPHENS
SYN. SEE PHYLEUS DRURY

BUCHHOLZARIA LEMMER, PSEUDOBOARMIA-GEOMETRIDAE
LIFE HISTORY

FOOD PLANTS
COMTONIA PEREGRINA(LINN.)

BUDEA HUBNER
SSP. SEE CUNEA DRURY

BUFFALOENSIS(GROTE), PARAPEMEA-NOCTUIDAE
SYN. LATIA STRECKER
FORM SIMPLICISSIMA BIRD
LIFE HISTORY
NEW YORK ENT. SOC., JOURN.
24:89, 1916
BIRD LARVA, PUPA
FOOD PLANTS
SAURURUS CERNUUS LINN.

BUFFALOENSIS GROTE & ROBINSON
SYN. SEE THYSBE FABRICIUS

BUNKERI GROTE
SSP. SEE CEROGAMA GUENEE

BURDICKI HENNE
SSP. SEE MARINA REAKIRT

BURNSI WATSON, HEMILEUCA-SATURNIIDAE
AB ILMAE WATSON
AB NIGROVENOSA WATSON
AB CONJUNCTA WATSON
AB PARADOXA WATSON
LIFE HISTORY
ENT. SOC. AMERICA, ANNALS
7:294, 1914
MOSHER PUPAL STRUCTURE
NAT. ACAD. SCI., MEM. 12:124,
1914 PACKARD LARVA*
PL. 24, F. 2 COLOR
SOUTHERN CALIFORNIA ACAD. SCI.,
BULL. 36:69, 1937 COMSTOCK-DAMMERS LIFE HISTORY; EGG*
PL. 29; LARVA* PLS. 30, 31;
PUPA* PL. 32
FOOD PLANTS
DALEA CALIFORNICA S. WATS.
ERIOGONUM SP.
TETRADYMIA SPINOSA H. & A.

BURRISONII MAYNARD
SSP. SEE LORQUINI BOISDUVAL
(BASILARCHIA)

BURSERAE(DYAR), PAECTES-NOCTUIDAE
LIFE HISTORY
UNITED STATES NAT. MUS., PROC.
23:271, 1901 DYAR LARVA
WASHINGTON ENT. SOC., PROC.
4:455, 1900 DYAR LARVA(BRIEF)
FOOD PLANTS
BURSERA SIMARUBA (LINN.)

BUTLERI DYAR
SYN. SEE ORNATRIX LINNEAUS

BUXEA (GROTE), ATETHMIA-NOCTUIDAE
SYN. GUSTATA DYAR
FORM DELICIA DYAR
LIFE HISTORY
SOUTHERN CALIFORNIA ACAD. SCI., BULL. 34:138, 1935 COMSTOCK-DAMMERS LARVA* PLS. 32, 33; PUPA* PL. 34
FOOD PLANTS
SPHAERALCEA SP.

BYBLIS BARNES & BENJAMIN
SSP. SEE APHRODITE FABRICIUS

C

CABERATA GUENEE, SABULODES-GEOMETRIDAE
SYN. ARSESARIA WALKER
FORM AEGROTATA GUENEE
FORM COTTLEI BARNES & BENJAMIN
LIFE HISTORY
CALIFORNIA ACAD. SCI., PROC. 5:372, 1873 HY. EDWARDS LARVA, PUPA (AEGROTATA)
NEW YORK ENT. SOC., JOURN. 53: 193, 1945 FORBES PUPA
SOUTHERN CALIFORNIA ACAD. SCI., BULL. 28:56, 1929 COMSTOCK-LARVA* PUPA*
ESSIG-INJ. BENEF. INS. P. 418, 1915 EGG* F. 424; LARVA* F. 425; PUPA* F. 427
ESSIG-INS. WEST. NORTH AMERICA P. 701, 1926 EGG* F. 578; LARVA* F. 579; PUPA* AND COCOON* F. 580
QUAYLE-CITRUS & SUBTROPICAL FRUITS P. 347, 1938 EGG, LARVA* F. 240; PUPA* F. 240
- - -
INDEX AMER. ECON. ENT., VOL. 5:556; VOL. 6:651; VOL. 7:860
FOOD PLANTS
ACACIA SP.
ACER SP. (MAPLES)
ACER NEGUNDO LINN.
AESCULUS CALIFORNICA (SPACH)
ALNUS SP. (ALDERS)
ARALIA SP.
CAMPSIS RADICANS (LINN.)
CARYA ILLINOENSIS (WANG)
CASTANEA DENTATA (MARSH.) (CHESTNUT)
CHRYSANTHEMUM MAXIMUS RAY.
CITRUS LIMON (LINN.)
CITRUS SINENSIS (LINN.)
CLEMATIS SP.
EUCALYPTUS SP.
GINKGO BILOBA LINN.
GREVILLEA ROBUSTA CUNN.
HEDERA HELIX LINN.
JUGLANS SP. (WALNUTS)
JUGLANS NIGRA LINN.
KALMIA SP. (LAURELS)
LIGUSTRUM SP. (PRIVETS)
LONICERA SP. (HONEYSUCKLES)
MAGNOLIA SP.
NEMOPANTHUS MUCRONATA (LINN.)
OLEA EUROPEA LINN. (OLIVES)
PASSIFLORA INCARNATA LINN.
PELARGONIUM SP.
PERSEA AMERICANA MILL. (AVOCADO)
PHOTINA ARBUTIFOLIA LINDL.
PLATANUS OCCIDENTALIS LINN.
PRUNUS SEROTINA EHRH.
RHUS SP. (SUMACS)
ROSA SP. (ROSES)
SALIX SP. (WILLOWS)
SCHINUS MOLLE LINN.
SENECIO VULGARIS LINN.
ULMUS SP. (ELMS)
UMBELLULARIA CALIFORNICA (HOOK. ARN.)
VERBENA SP.
VIOLA SP. (VIOLETS)
CALIFORNIA LAUREL
FERNS

CABLEI REIZENSTEIN
SYN. SEE MODESTA HARRIS

CACHEXIATA GUENEE
SYN. SEE LORATA GROTE (TETRACIS)

CACUMINARIA(MORRISON), SCOPULA-GEOMETRIDAE
LIFE HISTORY
NEW YORK STATE AGRI. EXP. STA. (CORNELL), MEM. 274:122, 1948
FORBES EGG, LARVA
FOOD PLANTS
LACTUCA SATIVA LINN. (LETTUCE)
GENERAL FEEDER, VEGETABLES

CADAVEROSA STRECKER
SYN. SEE INSULATA WALKER

CADUCA(GROTE), NEOERASTRIA-NOCTUIDAE
SYN. RETIS GROTE
LIFE HISTORY
AMER. NAT. 24:971, 1890
KELLICOTT EGG, COCOON
INS. LIFE 3:321, 1891
KELLICOTT EGG, LARVA, PUPA
WASHINGTON ENT. SOC., PROC. 11: 200, 1909 DYAR EGG, LARVA
FOOD PLANTS
NUPHAR ADVENA (AIT.)
SAGITTARIA SP.

CADUCALIS WALKER
SYN. SEE MADEFACTALIS GUENEE

CAECA COCKERELL
SSP. SEE PROMETHEA DRURY

CAECA IGEL
SSP. SEE IO FABRICIUS

CAECA REIFF
SYN. SEE HYPOPHLAEAS BOISDUVAL

CAECA STRECKER
SYN. SEE PERMACULATA PACKARD

CAELARIA HULST
SYN. SEE SOSPETA DRURY

CAERULESCENS HOLLAND
SSP. SEE NITOCRIS EDWARDS

CAESIA SMITH
SSP. SEE DIVERGENS WALKER

CAESIARIA HULST
SYN. SEE GNOPHOSARIA GUENEE

CAESIATA(BORKHAUSEN), SPARGANIA-GEOMETRIDAE
LIFE HISTORY
NEWMAN-BRITISH MOTHS P. 110, 1869 LIFE HISTORY
PACKARD-MONOGRAPH OF GEOMETRIDAE P. 69, 1876 LARVA
FOOD PLANTS
VACCINIUM MYRTILLUS LINN.
VACCINIUM VITIS-IDAEA LINN.

CAESONIA(STOLL), ZERENE-PIERIDAE
FORM FEMALE IMMACULSECUNDA GUNDER
FORM FEMALE STAINKEAE FIELD
FORM ROSEA ROEBER
FORM ROSA MCNEILL
SYN. CAROLINIANA PETIVER
LIFE HISTORY
BUTTERFLY PARK NATURE CLUB 2:NO PAGE NUMBERS, 1931
EGG, LARVA
CALIFORNIA ACAD. SCI., PROC. 5: 327, 1873 HY. EDWARDS LARVA
CAN. ENT. 20:21, 1888 EDWARDS EGG, LARVA, CHRYSALIS
ENT. NEWS 2:134, 1891
ROWLEY EGG, LARVA
ILLINOIS DEPT. AGRI., TRANS. 18:78, APPEND., 1880
MIDDLETON LARVA
SMITHSONIAN MISC. COLLECTIONS 4:ART. 2:27, 1862
MORRIS LARVA
BOISDUVAL-SPEC. GEN. LEP. DIURNAE 1:636, 1863 LARVA
BOISDUVAL & LECONTE-HIST. LEP. NORTH AMERICA P. 69, 1833
LARVA* PL. 22

COMSTOCK, J. H.-HOW TO KNOW BUTTER. P. 90, 1904 LARVA
FRENCH-BUTTER. EAST. UNITED STATES P. 128, 1886 LARVA
GUNDLACH-ENTOM. CUBANA P. 106, 1881 LARVA, CHRYSALIS
MORRIS-SYNOP. LEP. NORTH AMERICA P. 27, 1862 LARVA
SCUDDER-BUTTER. EAST. UNITED STATES & CANADA PP. 1837-1839, 1889 LIFE HISTORY
SCUDDER-GUIDE TO COMMON BUTTER. P. 133, 1893 LARVA, CHRYSALIS

- - -

INDEX AMER. ECON. ENT., VOL. 7: 1061

FOOD PLANTS

AMORPHA CALIFORNICA NUTT. EX. T. & G.
AMORPHA FRUTICOSA LINN.
DYSSODIA PAPPOSA (VENT.)
GLYCINE SP.
TRIFOLIUM SP. (CLOVERS)

CAESPITALIS BOISDUVAL
SYN. SEE RURALIS BOISDUVAL

CAESPITIS GROTE & ROBINSON
SYN. SEE PLANTAGINIS LINNEAUS

CAJA (LINNEAUS), ARCTIA-ARCTIIDAE
SYN. ERINACEA RETZIUS
A OPULENTA HENRY EDWARDS
B VIRGINIVIR DYAR
C WAROI BARNES & BENJAMIN
D AMERICANA HARRIS
E UTAHENSIS HENRY EDWARDS
AB TRANSMONTANA NEUMOEGEN & DYAR
F PARVA ROTHSCHILD

LIFE HISTORY

AMER. MUS. NAT. HIST., BULL. 10: 377, 1898 BEUTENMULLER LARVA (AMERICANA)
BOSTON SOC. NAT. HIST., PROC. 11: 34, 1866 PACKARD LARVA
CAN. ENT. 7:149, 1875 BUNKER EGG, LARVA (AMERICANA)
LOC. CIT. 8:20, 1876 LYMAN LARVAL HABITS
LOC. CIT. 32:321, 1900 GIBSON EGG, LARVA, PUPA (AMERICANA)
CAN. JOURN. INDUSTRY, SCI., ARTS 8:352, 1863 SAUNDERS LARVA
NEW YORK STATE CABIN. NAT. HIST., REPT. NO. 24:134, 1871 LINTNER EGG, LARVA (AMERICANA)
NEW YORK STATE, REPT. INS. 9:234, 1863 FITCH LARVA
OTTAWA NAT. 14:16, 1900 GIBSON LARVA (BRIEF)
PHILADELPHIA ENT. SOC., PROC. 2:28, 1863 SAUNDERS LARVA
HARRIS-ENT. CORRESP. P. 287, 1869 LARVA
STRETCH-ZYGAEN. & BOMBY. NORTH AMERICA P. 96, 1873 LARVA

FOOD PLANTS

CHENOPODIUM ALBUM LINN.
JUGLANS SP. (WALNUTS)
LACTUCA SATIVA LINN. (LETTUCE)
MALUS PUMILA MILL. (APPLE)
PLANTAGO SP.
POPULUS SP.
POPULUS BALSAMIFERA LINN.
POPULUS TREMULOIDES MICHX.
POPULUS TRICHOCARPA T. & G.
PTERIDUM AQUILINUM V. LATIUSCULUM (DESV.)
TARAXACUM OFFICINALE WEBER
URTICA DIOICA LINN.
URTICA LYALLII S. WATS.

CAJONA REAKIRT
SYN. SEE ANNA EDWARDS

CALAIS SCUDDER
SSP. SEE CHRYXUS DOUBLEDAY & HEWITSON

CALAMI HARVEY
SSP. SEE CANESCENS BEHR.

CALANUS (HUBNER), STRYMON-LYCAENIDAE
SYN. WITTFELDII EDWARDS

LIFE HISTORY
Amer. Mus. Nat. Hist., Bull. 5:278, 1893 Beuten. larva, chrysalis
Buffalo Soc. Nat. Sci., Bull. 19:28, 1939 Wild larva
Can. Ent. 2:61, 1870 Saunders larva (inorata)
Illinois Dept. Agri., Trans. 18:Append.:93, 1880 Middleton larva
Montana Univ., Bull. 30:128 1906 Elrod larva
New York State Mus., Mem. 8: 704, 1905 Felt larva
Boisduval & Leconte-Hist. Lep. North America p. 92, 1833 larva* chrysalis* Pl. 29 as falacer color
Comstock-How to Know Butter. p. 226, 1904 larva
French-Butter. East. United States p. 264, 1886 larva, chrysalis
Holland-Butterfly Book 1905 chrysalis* Pl. 5, ff. 25, 27
Maynard-Butter. New England p. 34, 1886 larva
Morris-Synop. Lep. North America p. 95, 1862 larva chrysalis
Packard-Ins. Inj. Forest Trees p. 83, 1881 larva
Scudder-Butter, East. United States & Canada pp. 887-890, 1889 life history* Pl. 65, f. 3; Pl. 68, f. 1; Pl. 75, ff. 20, 26; Pl. 84, ff. 25, 27
Scudder-Everyday Butterflies p. 259, 1889 egg, larva, chrysalis
Scudder-Guide to Common Butter. p. 120, 1893 larva, chrysalis

- - -

Index Amer. Econ. Ent., Vol. 6:744

FOOD PLANTS
Carya sp. (hickories)
Castanea dentata (Marsh.)
Crataegus sp. (thorns)
Juglans cinerea Linn.
Juglans nigra Linn.
Malus pumila Mill. (apple)
Quercus sp. (oaks)

calgary (Smith), Diarsia-Noctuidae
LIFE HISTORY

FOOD PLANTS
Salix sp. (willows)

caliente Wright
ssp. see cethura Felder & Felder

california Westwood & Hewitson, Coenonympha-Satyridae
syn. ceres Butler
form galactinus Boisduval
ab pulla Henry Edwards
a eryngii Henry Edwards
gen. vern. siskiyouensis Comstock
LIFE HISTORY
Can. Ent. 18:201, 1886 Edwards egg, larva, chrysalis
Comstock, J. A.-Butterflies California p. 66, 1927 egg, larva, chrysalis f. A-27
Edwards-Butter. North America 3rd series, 1897 life history* Pl. 12, ff. a-g (Coenonympha)
FOOD PLANTS
Melica bulbosa Geyer
Poa annua Linn.
Poa pratensis Linn.
Other grasses

californiae (Walker), Leptarctia-Arctiidae
For many ssp and syns of this species see McDunnough's List
LIFE HISTORY
Can. Ent. 21:210, 221, 1889 French egg, larva, pupa
Southern California Acad. Sci., Bull. 42:67, 1943 Comstock-Dammers life history; larva* Pl. 9; pupa* Pls. 9, 10

United States Nat. Mus., Proc.
25:273, 1903 Dyar egg, larva
Loc. cit. 27:794, 1904
Dyar larva
FOOD PLANTS
Castilleja sp.
Croton californicus Mull. Arg.
Lotus grandiflorus (Benth.)
Penstemon sp.
Plantago sp.
Quercus sp. (oaks)
Ribes aureum Pursh.
Rubus sp. (blackberries)
Salix sp. (willows)
Stellaria media (Linn.)
Taraxacum officinale Weber
Low plants

californiana Menetries
syn. see eurytheme Boisduval

californiaria Packard
syn. see pulveraria Linneaus

californiaria Packard
ssp. see juturnaria Guenee

californiata Packard
syn. see defensaria Guenee

californica Behr, Hypena-Noctuidae
LIFE HISTORY
Entomologica Americana
14:146, 1934 Crumb
larva Pl. 22, B; Pl. 25, H
FOOD PLANTS
Rubus allegheniensis Porter
Urtica lyallii S. Wats.

californica (Boisduval), Nymphalis-Nymphalidae
A herri Field
LIFE HISTORY
California Acad. Sci., Proc. 3:
123, 1864 Behr larval note
Loc. cit. 6:146, 1875
Hy. Edw. larva, chrysalis
Southern California Acad. Sci.,
Bull. 31:17, 1932 Comstock
larva* Pl. 10; pupa* Pl. 11
Comstock, J. A.-Butterflies
California p. 125, 1927
larva, chrysalis
Doane etc.-Forest Ins. p. 332,
1936 chrysalis* f. 195
Essig-Ins. West. North America
p. 642, 1926 larva* chrysalis*
f. 520

- - -

Index Amer. Econ. Ent.,
Vol. 1:318 Vanessa
Vol. 5:15 Aglais
Vol. 6:537; Vol. 7:711
FOOD PLANTS
Arctostaphylos glauca Lindl.
Ceanothus cordulatus Kell.
Ceanothus integerrimus H. & A.
Ceanothus thyrsiflorus Eschs
Medicago sativa Linn. (alfalfa)

californica Boisduval
syn. see nevadensis Felder & Felder

californica Butler
ssp. see bredowii Geyer

californica Cassino
ssp. see ornata Packard

californica Dyar
ssp. see ministra Drury

californica Edwards, Catocala-Noctuidae
syn. mariana Strecker
syn. eldoradensis Beuten.
A edwardsi Kusnezov
B elizabeth Cassino
LIFE HISTORY
Amer. Mus. Nat. Hist., Mem. New
Series 3: part 1, 1918 Barnes
& McDunnough egg, larva
Essig-Ins. West. North America
p. 687, 1926 larva
FOOD PLANTS
Juglans nigra Linn.
Salix sp. (willows)

CALIFORNICA(EDWARDS), STRYMON-
LYCAENIDAE
SYN. BORUS BOISDUVAL
SYN. CYGNUS EDWARDS
LIFE HISTORY
ENT. NEWS 5:329, 1894
DYAR LIFE HISTORY
SOUTHERN CALIFORNIA ACAD. SCI., BULL. 32:114, 1933
COMSTOCK LARVA* PL. 54; CHRYSALIS* PL. 55
FOOD PLANTS
QUERCUS SP. (OAKS)
SALIX SP. (WILLOWS)

CALIFORNICA GROTE
SYN. SEE EURYALUS BDV.

CALIFORNICA HAMPSON
SYN. SEE AURANTIACA HENRY EDWARDS

CALIFORNICA MABILLE
SYN. SEE RURICOLA BOISDUVAL

CALIFORNICA MENETRIES
SYN. SEE ZELICAON LUCAS

CALIFORNICA PACKARD
SSP. SEE AMERICANA HARRIS(EPICNAPTERA)

CALIFORNICA PACKARD
SYN. SEE ACREA DRURY

CALIFORNICA PACKARD
SYN. SEE ISABELLA ABBOT & SMITH

CALIFORNICA PACKARD, PHRYGANIDIA-
DIOPTIDAE
LIFE HISTORY
CALIFORNIA ACAD. SCI., PROC. 5:367, 1873 HY. EDW. EGG, LARVA
LOC. CIT. 2ND SERIES, 5:562, 1895 KELLOGG-JACK EGG* PL. 41A; LARVA* PL. 46, B-C; PUPA* PL. 41-D
NEW YORK ACAD. SCI., ANNALS 8: 89, 1893 PACKARD LARVA
NEW YORK ENT. SOC., JOURN. 11:125, 1903 LETCHER LARVA, PUPA
UNITED STATES DEPT. AGRI. FARMERS BULL. 1076:5, 1920 BURKE-HERBERT EGG* F. 2; LARVA* F. 3; PUPA* F. 4
WASHINGTON ENT. SOC., PROC. 21: 124, 1919 BURKE LIFE HISTORY NOTE
LOC. CIT. 22:193, 1920 HERBERT LARVAL INSTARS
DOANE ETC-FOREST INS. P. 291, 1936 LARVA
ESSIG-INJ. BENEF. INS. P. 403, 1915 EGG* F. 406; LARVA* F. 405; PUPA* F. 407
ESSIG-INS. WEST. NORTH AMERICA P. 688, 1926 EGG* F. 563; LARVA* & PUPA* F. 564
FERNALD-APPLIED ENTOM. P. 271, 1926 LARVA* F. 275
FERNALD-SHEPARD--APPLIED ENTOM. P. 267, 1942 LARVA* F. 270
HERRICK-INS. ENEMIES SHADE TREES P. 223, 1935 LARVA* F. 185
KELLOGG-AMER. INS. P. 407, 1908 EGG* LARVA* PUPA* F. 589
PACKARD-INS. INJ. WEST P. 808, 1877 EGG, LARVA
STRETCH-ZYGAEN. ET BOMBYC. NORTH AMERICA P. 91, 1873 LARVA* PUPA* PL. 10, FF. 4, 5 COLOR
- - -
INDEX AMER. ECON. ENT., VOL. 1:243; VOL. 2:278; VOL. 3:311; VOL. 4:359; VOL. 5:489; VOL. 6:583
FOOD PLANTS
CASTANEA DENTATA (MARSH.)
EUCALYPTUS GLOBULUS LABILL.
QUERCUS CHRYSOLEPIS LIEBM.
QUERCUS DOUGLASII H. & A.
QUERCUS DUMOSA NUTT.
QUERCUS KELLOGGII NEWB.
QUERCUS LOBATA NEE.
QUERCUS SUBER LINN.
QUERCUS VELUTINA LAM.
QUERCUS VIRGINIANA MILL.

CALIFORNICA(PACKARD), MALACOSOMA–
LASIOCAMPIDAE
SYN. CALIFORNICA WALKER
SYN. PSEUDONEUSTRIA BOISDUVAL
LIFE HISTORY
AMER. PHILOSOPHICAL SOC.,
PROC. 31:179, 1893
PACKARD LARVA
CAN. ENT. 25:41, 1893
DYAR LARVA
OHIO AGRI. EXP. STA., BULL. 45:
173, 1893 WEBSTER LARVA
PAPILIO 1:64, 1881
STRETCH LARVA, COCOON
UNITED STATES DEPT. AGRI.
BUR. ENT., BULL. 32:28, 1894
OLD SERIES COQUILLETT
LIFE HISTORY
DOANE ETC–FOREST INS. P. 281,
1936 LARVA* F. 164
ESSIG–INJ. & BENEF. INS.
P. 411, 1915 LARVA
ESSIG–INS. WEST. NORTH AMERICA
P. 697, 1926 LARVA
HERRICK–INS. ENEMIES SHADE
TREES P. 33, 1935 LARVA
PACKARD–INS. INJ. FOREST TREES
P. 41, 1881 LARVA
– – –
INDEX AMER. ECON. ENT.,
VOL. 1:197; VOL. 5:403; VOL. 6:480
FOOD PLANTS
AESCULUS SP. (BUCKEYES)
ARBUTUS MENZIESII PURSH.
CERCIS CANADENSIS LINN.
CORYLUS SP. (HAZELS)
FRAXINUS SP. (ASHES)
MALUS PUMILA MILL. (APPLE)
PHOTINA ARBUTIFOLIA LINDL.
POPULUS FREMONTI S. WATS.
PRUNUS CERASUS LINN. (CHERRY)
PYRUS COMMUNIS LINN. (PEAR)
QUERCUS CHRYSOLEPIS LIEBM.
QUERCUS LOBATA NEE.
QUERCUS VIRGINIANA MILL.
RHAMNUS PURSHIANA DC.
RIBES GROSSULARIA LINN.
(GOOSEBERRY)
RUBUS ALLEGHENIENSIS PORTER
RUBUS IDAEUS V. STRIGOSUS (MICHX.)
SALIX SP. (WILLOWS)

CALIFORNICA PROUT
SYN. SEE PUNCTULARIA BARNES &
MCDUNNOUGH

CALIFORNICA SMITH
SYN. SEE CLARUS CRAMER

CALIFORNICA(SPEYER), AUTOGRAPHA–NOCTUIDAE
FORM RUSSEA HENRY EDWARDS
LIFE HISTORY
AGRICULTURAL GAZETTE CANADA
5:NO. 2:132, 1918 GIBSON EGG,
LARVA*, PUPA
BRITISH COLUMBIA ENT. SOC.,
PROC. NO. 32 SUPPLEMATIC
SERIES P. 26, 1936 SPENCER
EGG, LARVA, PUPA
CALIFORNIA AGRI. EXP. STA.,
BULL. 660;32, 1942
SMITH LARVA* F. 25B
LOC. CIT. BULL. 707:10, 1948
MICHELBACHER LARVA F. 6B
ENT. AMER. 6:14, 1890
DYAR EGG, LARVA, PUPA
OREGON AGRI. EXP. STA., BIENNIAL
CROP PEST REPT. HORT. REPT.
2:185, 1913–1914 WILSON EGG;
LARVA* FF. 41, 42; PUPA* F. 42
SOUTHERN CALIFORNIA ACAD. SCI.,
BULL. 29:25, 1930
COMSTOCK–COOLIDGE PUPA*
UNITED STATES DEPT. AGRI.,
BULL. 95: PART 7:112, 1911
HYSLOP EGG; LARVA F. 45 A, B;
PUPA* PL. 11, F. 4
ESSIG–INJ. & BENEF. INS. P. 394,
1915 EGG, LARVA* F. 394
ESSIG–INS. WEST. NORTH AMERICA
P. 687, 1926 EGG, LARVA
METCALF–FLINT––DESTR. & USEFUL
INS. P. 149, 1939 LARVA F. 88P
– – –
INDEX AMER. ECON. ENT.,

Vol. 1:45; Vol. 2:47; Vol. 3:51;
Vol. 5:68; Vol. 6:79; Vol. 7:117

FOOD PLANTS

Allium cepa Linn. (onion)
Beta vulgaris Linn. (beet)
Brassica oleracea Linn. (cabbage)
Brassica rapa Linn. (turnip)
Ceanothus sp.
Chenopodium album Linn.
Citrullus vulgaris Schrad. (watermelon)
Citrus sp. (oranges, lemons, etc)
Cucumis melo Linn. (melons)
Cucumis sativus Linn. (cucumber)
Cucurbita maxima Dusch. (squash)
Dahlia variabilis Desf.
Daucus carota v. sativus Hoff. (carrot)
Doronicum sp.
Gossypium herbacium Linn. (cotton)
Helianthus sp. (sunflowers)
Hordeum vulgare Linn. (barley)
Ilex opaca Ait.
Kalmia sp. (laurels)
Lactuca canadensis Linn.
Lactuca sativa Linn. (lettuce)
Linum sulcatum Ridd. (flax)
Lupinus sp. (lupines)
Malus pumila Mill. (apple)
Malva sp.
Malva rotundifolia Linn.
Medicago sativa Linn. (alfalfa)
Melilotus alba Desr.
Mentha sp.
Mentha spicata Linn. (mint)
Phaseolus vulgaris Linn. (bean)
Pisum sativum Linn. (peas)
Plantago sp.
Rhododendron sp.
Ribes grossularia Linn. (gooseberry)
Ribes sativum Syme (currant)
Rosa sp. (roses)
Rubus idaeus v. strigosus (Michx.) (raspberry)
Rumex sp.
Sambucus canadensis Linn.
Solanum lycopersicum Linn. (tomato)
Solanum tuberosum Linn. (potato)
Trifolium sp. (clovers)
Trifolium hybridum Linn.
Trifolium monanthum Gray
Trifolium pratense Linn.
Trifolium repens Linn.
Vicia sativa Linn. (vetch)
Viguiera deltoidea A. Gray
Vitis sp. (grapes)
Zea mays Linn. (corn)
Market Garden Crops

californica Strecker
syn. see hiemalis Grote

californica Stretch
syn. see dimidiata Herrich-Schaeffer

californica Wright, Chlosyne-Nymphalidae
ab chinoi Gunder

LIFE HISTORY

Southern California Acad. Sci., Bull. 28:53, 1929
Comstock larva* pupa*

FOOD PLANTS

Ceanothus sp.
Viguiera deltoidea Gray

californica Wright
ssp. see nevadensis Stretch

californica Walker
syn. see maculata Harris

californica Walker
syn. see californica Packard

californicus Boisduval
syn. see claiforniae Walker

caliginosa Comstock
ssp. see hydaspe Boisduval

CALIPUSARIA WALKER
SYN. SEE MACULARIA HARRIS

CALLAE FIELD
SYN. SEE NICIPPE CRAMER

CALLETA(WESTWOOD), EUPACKARDIA-SATURNIIDAE
SYN. POLYOMMATA TEPPER
LIFE HISTORY
AMER. ACAD. ARTS. SCI., PROC. 39:547, 1904 PACKARD EGG, LARVA
AMER. MUS. NAT. HIST., BULL. 5:92, 1893 BEUTEN. COCOON
ENT. SOC. AMERICA, ANNALS 9:146, 1916 MOSHER PUPAL STRUCTURE
NAT. ACAD. SCI., MEM. 12:236, 1914 PACKARD LIFE HISTORY, EGG, LARVA
- - -
INDEX AMER. ECON. ENT. VOL. 3:68 CALLOSAMIA
FOOD PLANTS
PRUNUS SEROTINA EHRH.

CALLIDA GROTE
SSP. SEE PSEUDARGYRIA GUENEE

CALLIPPE BOISDUVAL, ARGYNNIS-NYMPHALIDAE
A COMSTOCKI GUNDER
AB CREELMANI GUNDER
LIFE HISTORY
SOUTHERN CALIFORNIA ACAD. SCI., BULL. 30:42, 1931 COMSTOCK-DAMMERS EGG, LARVA*, CHRYSALIS*
LOC. CIT. 39:75, 1940 COMSTOCK EGG* PL. 8-C
FOOD PLANTS
VIOLA PEDUNCULATA T. & G.

CALLITRICHOIDES GROTE, PHYPROSOPUS-NOCTUIDAE
SYN. NASUTARIA ZELLER
‡ ACUTALIS WALKER
LIFE HISTORY
ENT. AMER. 14:152, 1934 CRUMB LARVA
NEW YORK ENT. SOC., JOURN. 11:105, 1903 DASCHE LARVA
NEW YORK STATE MUS., MEM. 8:740, 1905 FELT LARVA
FOOD PLANTS
SMILAX ROTUNDIFOLIA LINN.

CALPHURNIA HENRY EDWARDS
SSP. SEE DELILAH STRECKER

CALVERLEYI GROTE
SSP. SEE ASTERIUS LINN. (AJAX PARTIM)

CALYCANTHATA(ABBOT & SMITH), ZALE-NOCTUIDAE
FORM DEALBATA STRAND
LIFE HISTORY
CAN. JOURN. INDUSTRY, SCIENCE & ART 10:251, 1865 BETHUNE LARVA
UNITED STATES NAT. MUS., PROC. 27:879, 1904 DYAR EGG, LARVA
ABBOT & SMITH-LEP. INS. GEORGIA P. 207, 1797 LARVA* PUPA* PL. 207 COLOR
GUENEE-SPEC. GEN. LEP. NOCT. 3:15, 1852 LARVA, PUPA
FOOD PLANTS
BETULA SP. (BIRCHES)
CALYCANTHUS FLORIDUS LINN.
QUERCUS SP. (OAKS)

CALYCE EDWARDS
SSP. SEE OCCIDENTALIS REAKIRT

CALYDON MEAD
SSP. SEE PALLA BOISDUVAL

CAMBRICA CURTIS, VENUSIA-GEOMETRIDAE
SYN. ERUTARIA DUPONCHEL
SYN. NEBULOSARIA FREYER
SYN. SCITULARIA WALKER
LIFE HISTORY
NEW YORK STATE AGRI. EXP. STA. (CORNELL), MEM. 274:157, 1948 FORBES LARVA
NEWMAN-BRITISH MOTHS P. 76, 1869 LARVA

Packard–Monograph Geometridae p.86, 1876 larva

FOOD PLANTS

Sorbus americana Marsh.
Sorbus acuparia Linn.

camillus Edwards
ssp. see campestris Behr.

campestris (Boisduval), Atalopedes– Hesperiidae

syn. sylvanoides Boisduval (female)
syn. huron Edwards

LIFE HISTORY

Amer. Mus. Nat. Hist., Bull. 5:290, 1893 Beutenmuller larva and chrysalis (huron)
Southern California Acad. Sci., Bull. 28:30,78, 1929 Comstock life history, egg*, larva* chrysalis*
Loc. cit. 29:136, 1930 Comstock egg* larva* pupa*
Holland Butterfly Book 1905 chrysalis* Pl. 6, ff. 43,47 (huron)
Scudder–Butter. East. United States & Canada pp. 1664–1666, 1889 life history* Pl. 66, f. 31; Pl. 77, ff. 28,30,31; Pl. 80, ff. 68–71; Pl. 85, ff. 43,47

- - -

Index Amer. Econ. Ent., Vol. 7:113

FOOD PLANTS

Cynodon dactylon (Linn.)
Digitaria sanguinalis (Linn.)
Distichlis spicata (Linn.)

campestris (Behr), Phyciodes–Nymphalidae

syn. pratensis Behr. (male)
ab macdunnoughi Gunder
a camillus Edwards
syn. emissa Edwards
ab tristis Cockerell
ab rohweri Cockerell

LIFE HISTORY

Papilio 4:128, 1884 Edwards egg, larva, chrysalis
Southern California Acad. Sci., Bull. 28:53, 1929 Comstock egg*
Loc. cit. 29:136, 1930 Comstock egg* larva* chrysalis*

FOOD PLANTS

Artemisia vulgaris Linn. (pratensis)
Aster praeltus Poir. (pratensis)
Cirsium sp. (thistles)

campestris McDunnough
ssp. see gordius Cramer

cana Henry Edwards
ssp. see vetusta Boisduval

cana Martyn
syn. see coniferarum Abbot & Smith

cana Walker
ssp. see albata Packard

cana Walker
syn. see anguina Abbot & Smith

canace Edwards
syn. see picta Edwards

canadaria (Guenee), Melanolophia– Geometridae

LIFE HISTORY

Can. Ent. 65:123, 1933 McD. egg, larva, pupa
Loc. cit. 76:125, 1944 McGuffin larva ff. mandible
Psyche 12:58, 1905 Dyar egg, larva
United States Dept. Agri., Misc. Pub. 657:362,429, 1950 Craighead larva

FOOD PLANTS

Abies balsamea (Linn.)
Acer rubrum Linn.
Betula allegheniensis Britt.
Betula papyifera Marsh.
Epilobium sp.
Larix laricina (DuRoi)

MYRICA SP.
MYRICA GALE LINN.
PICEA SP. (SPRUCES)
PINUS SP. (PINES)
POLYGONUM SP.
PRUNUS SEROTINA EHRH.
QUERCUS SP. (OAKS)
RHUS TYPHINA LINN.
TILIA SP. (LINDENS)
TSUGA CANADENSIS (LINN.)
ULMUS THOMASII SARG.

CANADENSIS BETHUNE, CALPE-NOCTUIDAE
SYN. PURPURASCENS WALKER
SYN. SOBRIA WALKER
LIFE HISTORY
CAN. ENT. 12:44, 1880
COQUILLETT LARVA
ILLINOIS DEPT. AGRI., TRANS. 18: APPEND:137,172, 1880
COQUILLETT-MARTIN LARVA
PSYCHE 2:123, 1878
THAXTER LARVA
FOOD PLANTS
THALICTRUM POLYGAMUM MUHL.

CANADENSIS BOISDUVAL, SPHINX-SPHINGIDAE
SYN. PLOTA STRECKER
LIFE HISTORY
ENT. NEWS 49:69, 1938
CLAYTON EARLY STAGES
FOOD PLANTS
FRAXINUS SP. (ASHES)
RUBUS CHAMAEMORUS LINN. (QUESTIONABLE)
VACCINIUM AUGUSTIFOLIUM AIT.

CANADENSIS DOS PASSOS
SSP. SEE ATLANTIS EDWARDS

CANADENSIS GUENEE
SYN. SEE GALLI ROTTENBURG

CANADENSIS MCDUNNOUGH
SSP. SEE SCITISCRIPTA WALKER

CANADENSIS ROTHSCHILD & JORDAN
SSP. SEE GLAUCUS LINNEAUS

CANADENSIS SMITH
SSP. SEE LEPUSCULINA GUENEE

CANADENSIS SMITH
SSP. SEE NEVADAE GROTE

CANDEFACTA (HUBNER), TARACHIDIA-NOCTUIDAE
SYN. DEBILIS WALKER
A NEOMEXICANA SMITH
SYN. CANDEFACTELLA STRAND
LIFE HISTORY
CAN. ENT. 4:198, 1872
DODGE LARVA
PAPILIO 3:84, 1883
COQUILLETT COMPARES LARVA OF CANDEFACTA AND ERASTEROIDES
GLOVER-ILLUS. NORTH AMER. ENT. 1878 LARVA PL. 16, F. 3
GUENEE-SPEC. GEN. LEP. NOCT. 2: 216, 1852 LARVA* PL. 2 COLOR
PACKARD-GUIDE TO STUDY INS. P. 258, 1869 LARVA (AS MELITEA HARRISII SCUDDER)
FOOD PLANTS
AMBROSIA ARTEMISIIFOLIA LINN.
ARCTIUM LAPPA LINN.
ASTER DUMOSUS LINN.

CANDEFACTELLA STRAND
SYN. SEE CANDEFACTA HUBNER

CANDIDA LINTNER, CERURA-NOTODONTIDAE
LIFE HISTORY
SOUTHERN CALIFORNIA ACAD. SCI., BULL. 38:184, 1939 COMSTOCK-DAMMERS LARVA* PL. 48
FOOD PLANTS
SALIX SP. (WILLOWS)

CANDIDA WALKER
SYN. SEE TEXTOR HARRIS

CANDIDA WRIGHT
SYN. SEE AURANTIACA HEWITSON

CANDIDARIA PACKARD
SYN. SEE ORDINATA WALKER (SCOPULA)

CANESCENS(BEHR), COSMIA-NOCTUIDAE
FORM CALAMI HARVEY
FORM ORINELLA STRAND
FORM ORINULA STRAND
LIFE HISTORY
AMER. MUS. NAT. HIST., BULL. 16: 457, 1902 BEUTENMULLER LARVA (ORINA GN.)
CALIFORNIA ACAD. SCI., BULL. 3:61, 1885 BEHR LARVA
NEW YORK STATE MUS., MEM. 8: 704, 1905 FELT LARVA (ORINA)
- - -
INDEX AMER. ECON. ENT., VOL. 7:216
FOOD PLANTS
QUERCUS KELLOGGII NEWB.

CANESCENS COMSTOCK
SYN. SEE BEHRII EDWARDS (COLIAS)

CANICOSTA FRANCLEMONT, SYMMERISTA-NOTODONTIDAE
LIFE HISTORY
CAN. ENT. 78:100, 1946 FRANCELEMONT LARVA
NEW YORK STATE AGRI. EXP. STA. (CORNELL), MEM. 274:224, 1948 FORBES LARVA
FOOD PLANTS
ACER SP. (MAPLES)
BETULA SP. (BIRCHES)
FAGUS SP. (BEECHES)
QUERCUS SP. (OAKS)

CANIPLAGA(WALKER), ELLIDA-NOTODONTIDAE
SYN. TRANSVERSATA WALKER
SYN. GELIDA GROTE
LIFE HISTORY
NEW YORK ENT. SOC., JOURN. 10: 143, 1902 DYAR EGG, LARVA, PUPA
NEW YORK STATE AGRI. EXP. STA. (CORNELL), MEM. 274:218, 1948 FORBES LARVA
NEW YORK STATE MUS., MEM. 8:731, 1905 FELT LARVA
FOOD PLANTS
TILIA AMERICANA LINN.

CANOSARIA(WALKER), NEPYTIA-GEOMETRIDAE
SYN. PULCHRARIA MINOT
FORM PINIARIA PACKARD
FORM FUSCARIA BARNES & BENJAMIN
LIFE HISTORY
CAN. ENT. 74:227, 1942 DETHIER EGG, LARVA* PUPA* PL. 14, FF. 3, 6, 8; PL. 15, F. 9; PL. 16, F. 1
LOC. CIT. 75:188, 1943 MCGUFFIN LIFE HISTORY LARVAL MANDIBLE* PL. 15, F. B
NEW YORK ENT. SOC., JOURN. 53: 195, 1945 FORBES PUPA
NEW YORK STATE AGRI. EXP. STA., (CORNELL), MEM. 274:103, 1948 FORBES EGG, LARVA, PUPA
UNITED STATES DEPT. AGRI., MISC. PUB. 657:361, 434, 1950 CRAIGHEAD LARVA
UNITED STATES DEPT. AGRI., REPT. OF ENT. P. 328, 1886 PACKARD LARVA, PUPA
GUMPPENBERG-SYSTEMA GEOMETRARUM 5:316, 1892 LARVA, PUPA (PULCHRARIA)
PACKARD-INJ. INS. NEW OR LITTLE KNOWN P. 14, 1870 LARVA, PUPA
PACKARD-INS. INJ. FOREST TREES P. 205, 1881 LARVA, PUPA
PACKARD-MONOGRAPH GEOMETRIDAE P. 453, 1876 PUPA* PL. 13, F. 13
- - -
INDEX AMER. ECON. ENT., VOL. 7:706
FOOD PLANTS
ABIES SP. (FIRS)
ABIES BALSAMEA (LINN.)
BETULA PAPYRIFERA MARSH.
LARIX LARICINA (DUROI)
PICEA ENGELMANNI (PARRY)
PICEA GLAUCA (MOENCH.)
PICEA MARIANA (MILL.)
PICEA RUBENS SARG.
PINUS BANKSIANA LAMB.
PINUS STROBUS LINN.
PSEUDOTSUGA MENZIESII (MIRB.)
THUJA OCCIDENTALIS LINN.
TSUGA CANADENSIS (LINN.)
FOREST TREES

CANTHE HUBNER
SYN. SEE HERMES FABRICIUS

CANTHEUS GODART
SYN. SEE EURYDICE JOHANSSON

CANTHUS LINNEAUS
SYN. SEE EURYDICE JOHANSSON

CAPELLA BARNES
SSP. SEE ANICIA DOUBLEDAY

CAPPSARIA RUPERT
SSP. SEE WARNERI HARVEY

CAPROTINA DRURY
SYN. SEE ACREA DRURY

CAPSELLA GROTE
SYN. SEE OVIDUCA GUENEE

CAPTICOLA WALKER
SYN. SEE GRAPHICA HUBNER

CAPUSINA NEUMOEGEN & DYAR
SYN. SEE AMERICANA HARVEY

CARA GUENEE, CATOCALA-NOCTUIDAE
A CARISSIMA HULST
SYN. SYLVIA HENRY EDWARDS
LIFE HISTORY
AMER. MUS. NAT. HIST., BULL. 16: 386, 1902 BEUTENMULLER LARVAL HEAD* F.7
AMER. MUS. NAT. HIST., MEM. NEW SERIES 3:PART 1, 1918 BARNES & MCDUNNOUGH LARVA* P.,10, F.38;PL.11,FF.1,2 COLOR
BROOKLYN ENT. SOC., BULL. 4:22, 1881 KOEBLE LARVA
LOC. CIT. 7:47, 1884 HULST LARVA
ENT. NEWS 19:116, 1908 ROWLEY LARVA, PUPA
LOC. CIT. 20:1909 ROWLEY EGG AND TEXT CORRECTION P.135
NEW YORK STATE MUS., MEM. 8: 743, 1905 FELT LARVA (BRIEF)
FOOD PLANTS
POPULUS SP. (POPLARS)
SALIX SP. (WILLOWS)

CARADRINALIS GUENEE, BLEPTINA-NOCTUIDAE
SYN. CLONIASALIS WALKER
LIFE HISTORY
ENTOMOLOGICAL AMERICANA 14:171, 1934 CRUMB LARVA* PL. 21E
UNITED STATES NAT. MUS., PROC. 27:881, 1904 DYAR EGG, LARVA
FOOD PLANTS
DEAD LEAVES

CARBONARIA FRANCLEMONT
SSP. SEE PYRAMIDOIDES GUENEE

CARBONARIA HAIMBACH
SSP. SEE VIRGINARIA CRAMER

CARBONARIA(HARVEY), HOMOGLAEA-NOCTUIDAE
LIFE HISTORY
WASHINGTON ENT. SOC., PROC. 4: 325, 1898 DYAR LARVA
FOOD PLANTS
SALIX COMMUTATA BEBB.

CARBONATA CASSINO & SWETT
SSP. SEE SIGNATARIA WALKER

CARDUELIS SEBA
SYN. SEE CARDUI LINNEAUS

CARDUI(LINNEAUS), VANESSA-NYMPHALIDAE
SYN. CARDUELIS SEBA
AB MINOR COCKERELL
AB ELYMI RAMBUR
AB ATE STRECKER
LIFE HISTORY
AMER. MUS. NAT. HIST., BULL. 5:269, 1893 BEUTEN. LARVA
AMER. NAT. 10:607, 1876 SCUDDER LIFE HISTORY
BUFFALO SOC. NAT. SCI., BULL. 19:26, 1939 WILD LARVA
CALIFORNIA ACAD. SCI., PROC. 5:170, 1873 HY. EDW. LARVA, CHRYSALIS
CAN. ENT. 1:93, 1869 SAUNDERS LARVA

Loc. cit. 24:6, 1892 Dyar
egg, larva, chrysalis
Illinois Dept. Agri., Trans.
18:Append:87,153, 1880
Coquillett & Middleton larva
Montana Univ., Bull. 30:100,
1906 Elrod larva
Naturaliste Canadien p. 160,
1928 Huard larva, chrysalis
Pan Pacific Ent. 18:176, 1942
Woodberry life cycle note
Smithsonian Misc. Collections
4:Art. 2:60, 1862 Morris
larva, chrysalis
Southern California Acad. Sci.,
Bull. 27:68, 1928 Schrader
life history note
United States Dept. Agri.,
Farmer's Bull. 1495:12, 1926
Weigel-Middleton larva* f. 11
Chenu-Lucas--Encyc. Nat. Hist.
Papillons 1:104, 1856, larva,
chrysalis
Comstock, J. A.-Butter. Calif.
p. 133, 1927 egg, larva,
chrysalis* A 42; Pl. 63, f. 6
COLOR
Comstock, J. H.-How to Know
Butter., p. 158, 1904 larva
Duncan-Natur. Library Entom. 3:
126, 1835 larva, chrysalis
Emmons-Nat. Hist. New York
5:210, 1854 larva* chrysalis*
Pl. 46 COLOR
Essig-Inj. & Benef. Ins. p. 461,
1915 larva* f. 473; chrysalis
Essig-Ins. West. North America
p. 645, 1926 egg, larva* f. 523;
chrysalis
Fernald-Butter. Maine p. 59,
1884 larva
French-Butter. East. United
States p. 200, 1886 larva
Glover-Illus. North Amer. Ent.
1878 larva, pupa Pl. 21, f. 3
Gundlach-Entom. Cubana p. 40,
1881 larva (brief)
Harris-Ent. Corresp. p. 277,
1869 larva* Pl. 1
Harris-Ins. Inj. Veget. Flint
edit. p. 292, 1862 larva,
chrysalis
Holland-Butterfly Book 1905
larva* Pl. 3, f. 37; chrysalis*
Pl. 4, ff. 60-62
Jaeger-Life North Amer. Ins.
p. 204, 1864 larva
Kirby-European Butter. & Moths
p. 13, 1882 chrysalis
Maynard-Butter. New England
p. 20, 1886 larva
Morris-Synop. Lep. North
America p. 60, 1862 larva,
chrysalis
Scudder-Butter. East. United
States & Canada pp. 475-477,
481-485 life history* Pl. 64,
f. 30; Pl. 67, f. 7; Pl. 74, f. 37;
Pl. 78, ff. 61-63; Pl. 81, f. 10;
Pl. 86; ff. 69-73; Pl. 83, ff. 60-
62
Scudder-Everyday Butter. p. 112,
1899 larva, chrysalis
Scudder-Guide to Common Butter.
p. 84, 1893 larva, chrysalis
Stainton-Manual British Butter.
1:37, 1857 larva
Stephens-Illus. British Entom.
Haustell. 1:47, 1828 larva,
chrysalis

- - -

Index Amer. Econ. Ent.,
Vol. 1:318; Vol. 2:381; Vol. 3:431,
Vol. 4:403; Vol. 5:678; Vol. 6:800,
Vol. 7:1042

FOOD PLANTS

Althaea officinalis Linn.
Althaea rosea Cav.
Amsinckia sp.
Anaphalis margaritacea (Linn.)
Arctium lappa Linn.
Arctium minus (Hill.)
Artemisia sp.
Artemisia stelleriana Bess.

BORAGO OFFICINALIS LINN.
CALENDULA OFFICINALIS LINN.
CARDUUS NUTANS LINN.
CHENOPODIUM ALBUM LINN.
CIRSIUM ARVENSE (LINN.)
CIRSIUM VULGARE (SAVI.)
CNICUS BENEDICTUS LINN.
ECHIUM VULGARE LINN.
ERYNGIUM SP.
GNAPHALIUM SP.
HELIANTHUS SP.
HELICHRYSUM SP.
LUPINUS SP.
MALVA SYLVESTRIS LINN.
ONOPORDUM ACANTHIUM LINN.
PARIETARIA SP.
PARTHENIUM ARGENTATUM GRAY
PHASEOLUS VULGARIS LINN. (BEAN)
SERRATULA SP.
SILYBUM MARIANUM (LINN.)
STACHYS SIEBOLDII MIQ.
URTICA SP.
URTICA LYALLII S. WATS.

C-ARGENTEUM KIRBY
SYN. SEE PROGNE CRAMER

CARIN HUBNER
SYN. SEE PHYLEUS DRURY

CARINATOR DYAR
SYN. SEE CINCTIPES HENRY EDWARDS

CARIOSA SMITH
SYN. SEE TESSELLATA HARRIS

CARISSIMA HARVEY
SSP. SEE EXSERTISTIGMA MORRISON

CARISSIMA HULST
SSP. SEE CARA GUENEE

CARLOTA BEUTENMULLER
SSP. SEE FAUSTINA STRECKER

CARLOTA REAKIRT
SYN. SEE GORGONE HUBNER

CARLOTTA HULST
SSP. SEE YPSILON FORBES

CARMELITA MORRISON
SYN. SEE MINIMA GUENEE

CARNARIA PACKARD
SYN. SEE DETERMINATA WALKER

CARNATA PACKARD
SSP. SEE CUSTODIATA GUENEE

CARNEARIA (HULST), DREPANULATRIX-GEOMETRIDAE

LIFE HISTORY
AMER. MUS. NAT. HIST., BULL. 94:5:276, 1949
RINDGE EGG, LARVA, PUPA
PSYCHE 9:419, 1902
DYAR EGG, LARVA

FOOD PLANTS
CEANOTHUS SP.
CEANOTHUS CORDULATUS KELL.

CARNEATA WARREN
SYN. SEE BIFILATA HULST-DREPANULATRIX

CARNEICOSTA GUENEE, PANOPODA-NOCTUIDAE
SYN. SCISSA WALKER
SYN. COMBINATA WALKER

LIFE HISTORY
AMER. MUS. NAT. HIST., BULL. 4: 68, 1892 BEUTEN. LARVA

FOOD PLANTS
CARYA SP. (HICKORIES)
QUERCUS SP. (OAKS)
SALIX SP. (WILLOWS)

CARNEOLA (GUENEE), ERASTRIA-NOCTUIDAE
SYN. BIPLAGA WALKER

LIFE HISTORY
PAPILIO 1:7, 1881
COQUILLETT LARVA
LOC. CIT. 2:57, 1882
COQUILLETT EGG, LARVA, PUPA

- - -

INDEX AMER. ECON. ENT., VOL. 7:359

FOOD PLANTS
- POLYGONUM PENSYLVANICUM LINN.
- RUMEX SP.
- RUMEX CRISPUS LINN.
- SOLIDAGO SP.

CAROLINA HARRIS
SYN. SEE COLONA HUBNER

CAROLINA(HENRY EDWARDS), EUCEREON-AMATIDAE
SYN. CUBENSIS SCHAUS
SYN. CONFUSUM ROTHSCHILD
LIFE HISTORY
- UNITED STATES NAT. MUS., PROC. 23:262, 1901 DYAR EGG, LARVA

FOOD PLANTS
- CYNANCHUM PALUSTRE (GRAY)
- SARCUSTEMMA VIMINALE (LINN.)

CAROLINA HOLLAND
SYN. SEE FLEBILIS GROTE

CAROLINA JONES
SYN. SEE ANGULIFERA WALKER(CALLOSAMIA)

CAROLINA LINNEAUS
SYN. SEE SEXTA JOHANSSON

CAROLINIANA PETIVER
SYN. SEE CAESONIA STOLL

CAROLINIANUS HOLLAND
SYN. SEE MARCELLUS CRAMER

CAROLYNAE GUNDER
SSP. SEE LEANIRA FELDER & FELDER

CARPENTERII EDWARDS
SSP. SEE CYBELE FABRICIUS

CARPINIFOLIA BOISDUVAL
SSP. SEE AMERICANA HARRIS (EPICNAPTERA)

CARYAE(HARRIS), HALISIDOTA-ARCTIIDAE
SYN. ANNULIFASCIA WALKER
SYN. PORPHYREA HERRICH-SCHAEFFER
LIFE HISTORY
- AGRICULTURAL GAZETTE OF CANADA 4:1043, 1917 SWAINE EGG, LARVA* PUPA* COCOON*
- AMER. MUS. NAT. HIST., BULL. 10: 371, 1898 BEUTEN. LARVA
- BOSTON SOC. NAT. HIST., PROC. 9:289, 1863 WALSH LARVA
- CAN. JOURN. INDUSTRY, SCI.&ART 8:368, 1863 SAUNDERS LARVA
- ILLINOIS DEPT. AGRI., TRANS. 18:APPEND:168, 1880 COQUILLETT LARVA
- MICHIGAN STATE COLLEGE, SPEC. BULL. 243:39, 1933 McDANIEL LIFE HISTORY EGG*
- MINNESOTA AGRI. EXP. STA., BULL. 61:140, 1889 LUGGER LARVA
- NATURALISTE CANADIEN P. 18, 1930 HUARD LARVA, COCOON
- NEW YORK ACAD. SCI., ANNALS 8:65, 1893 PACKARD EGG, LARVA
- NEW YORK ENT. SOC., JOURN. 5:132, 1897 DYAR LARVAL STRUCTURE* PL. 8, F. 5
- NEW YORK STATE AGRI. EXP. STA. (CORNELL), MEM. 56:1073, 1922 WELLHOUSE LARVA (BRIEF)
- NEW YORK STATE MUS., MEM. 8: 315, 1905 FELT LARVA* PL. 15, F. 5 COLOR
- NEW YORK STATE, REPT. INS. 1:160, 1854 FITCH LARVA* PUPA*
- LOC. CIT. 3:458, 140 SEC. 183, 1856 LARVA
- OHIO AGRI. EXP. STA., BULL. 332: 247, 1918 HOUSER LARVA* PL. 36, F. 3
- ONTARIO ENT. SOC., ANN. REPT. P. 56, 1903 GIBSON LARVA
- LOC. CIT. P. 84, 1907 GIBSON LARVA
- LOC. CIT. P. 42, 1917 CAESAR LARVA*

Ottawa Naturalist 19:51
1905 Gibson larva
Psyche 6:158, 1891
Soule egg, larva, pupa
Smithsonian Misc. Collections
4:Art. 2:348, 1862
Morris larva
United States Dept. Agri., Bull.
598:13, 1918 Isley egg*
Pl. II, f. 4; larva* Pl. I,
Pl. II, ff. 1,2,3; pupa*
Pl. III, f. 1
United States Dept. Agri., Bur.
Ent. Bull. Old series 4:42,
1884 Smith larva, pupa
United States Dept. Agri.,
Farmer's Bull. 1270:48, 1922
Siegler egg; larva*
ff. 94,96; cocoon* f. 97
United States Dept. Agri., Misc.
Pub. 657:356,385, 1950
Craighead larva f. 75 A
Cragen–Our Ins. Friends & Foes
p. 149, 1899 larva
Dickerson–Moths & Butter.
p. 195, 1901 larva* f. 172;
cocoon* f. 173
Doane etc–Forest Ins. p. 285,
1936 larva* f. 165
Eliot & Soule–Caterpillars &
Moths p. 206, 1902 egg,
larva* cocoon*
Fernald–Applied Entom. p. 281,
1926 larva* f. 287
Fernald–Shepard––Applied
Entom. p. 278, 1942 larva*
f. 285
Harris–Ent. Corresp. p. 289,
1869 larva
Harris–Ins. Inj. Veget. 1st edit.
p. 258, 1841 larva, pupa
Harris–Loc. cit. 2nd edit.
p. 279, 1852 larva, pupa
Harris–Loc. cit. Flint Edit.
p. 361, 1862 larva* cocoon*
Pl. 6 COLOR
Herrick–Ins. Enemies Shade Trees
p. 123, 1935 larva* f. 31
Morris–Synop. Lep. North America
p. 348, 1862 larva
Packard–Ins. Inj. Forest Trees
p. 76, 1881 larva
Stretch–Illus. Zygaen. & Bombyc.
North America p. 141, 1873
larva

- - -

Index Amer. Econ. Ent.,
Vol. 1:147; Vol. 2:169; Vol. 3:186;
Vol. 4:213; Vol. 5:290; Vol. 6:340;
Vol. 7:448

FOOD PLANTS

Acer negundo Linn.
Acer pseudoplatanus Linn.
v. purpurascens Pax
Acer rubrum Linn.
Acer saccharinum Linn.
Acer saccharophorum Koch.
Aesculus glabra Willd.
Alnus sp. (alders)
Alnus serrulata (Ait.)
Betula alba Linn.
Betula papyrifera Marsh.
Betula populifolia Ait.
Carpinus caroliniana Walt.
Carya glabra (Mill.)
Carya tomentosa Nutt.
Castanea dentata (Marsh.)
Celtis occidentalis Linn.
Crataegus sp. (thorns)
Cydonia japonica (Thunb.)
Fagus sp. (beeches)
Fagus grandifolia Ehrh.
Fraxinus americana Linn.
Hamamelis virginiana Linn.
Humulus lupulus Linn.
Juglans cinerea Linn.
Juglans cordifolia
v. ailanthifolia (Carr.)
Juglans nigra Linn.
Juglans regia Linn.
Larix decidua Mill.
Larix laricina (DuRoi.)
Malus pumila Mill. (apple)
Ostrya virginiana (Mill.)
Parthenocissus quinquefolia
(Linn.)
Parthenocissus tricuspidata
(S & Z)
Platanus occidentalis Linn.
Platanus orientalis Linn.

Populus sp. (poplars)
Prunus cerasus Linn.
Prunus serotina Ehrh.
Prunus virginiana Linn.
Pyrus communis Linn. (pear)
Quercus sp. (oaks)
Quercus alba Linn.
Quercus palustris Muench.
Quercus rubra Linn.
Rhus sp. (sumacs)
Robinia pseudo-acacia Linn.
Rosa sp. (roses)
Rubus idaes v. strigosus (Michx.)(raspberry)
Salix sp. (willows)
Salix alba Linn.
Tilia americana Linn.
Tilia europea Linn.
Ulmus americana Linn.
Ulmus procera Salisb.
Ulmus rubra Muhl.

caryaevorus McDunnough, Strymon-Lycaenidae

LIFE HISTORY

FOOD PLANTS

Carya sp. (hickories)

carye (Hubner), Vanessa-Nymphalidae

syn. charie Blanchard
ab intermedia Grinnell
ab muelleri Letcher
ab letcheri Grinnell
ab schraderi Gunder
ab nivosa Gunder

LIFE HISTORY

Argentina, Ann. Soc. Cient. 78: 172, 1914 Giacomelli larva, chrysalis
Brooklyn Ent. Soc., Bull. 20: 147, 1925 Coolidge egg
California Acad. Sci., Proc. 3:125, 1863 Behr larva (brief)
Loc. cit. 5:329, 1873 Hy. Edw. larva, chrysalis
Can. Ent. 21:237, 1889 Dyar egg, larva, chrysalis
Ent. News 32:216, 1921 Huguenin life history
Comstock, J. A.-Butter. Calif. p. 135, 1927 larva, chrysalis* Pl. A. 42; Pl. 63, f. 9 COLOR
Essig-Ins. West. North America p. 647, 1926 egg, larva, chrysalis

- - -

Index Amer. Econ. Ent., Vol. 3:431; Vol. 7:1042

FOOD PLANTS

Althaea rosea Cav.
Fabaceae sp.
Lavandula spica Linn.
Lavatera assurgentiflora Kell.
Ligustrum ovalifolium Hassk.
Ligustrum vulgare Linn.
Lupinus arboreus Sims.
Malva rotundifolia Linn.
Malvaceae sp.
Sidalcea malvaeflora (DC.)
Spharalacea ambigua Gray
Urtica holosericea Nutt.

caryi Dyar

ssp. see chryxus Doubleday & Hewitson

cassandra Henry Edwards

syn. see electilis Walker

cassinoi Barnes & Benjamin

ssp. see leporina Linneaus

cassinoi Beutenmuller

ssp. see unijuga Walker

casta Kirby

syn. see napi Linneaus

castalis Edwards

ssp. see damon Cramer

castanea (Grote), Septis-Noctuidae

syn. cymosa Grote
form cymosana Strand

LIFE HISTORY

FOOD PLANTS
- ACER SP. (MAPLES)
- ALNUS SP. (ALDERS)
- PRUNUS PENSYLVANICA LINN.
- PRUNUS SEROTINA EHRH.
- RIBES SANGUINEUM PURSH.
- SALIX SP. (WILLOWS)
- SYMPHORICARPOS ALBUS (LINN.)

CASTANEA STRAND
SYN. SEE MORRISONI GROTE

CASTIGATA(HUBNER), EUPITHECIA-GEOMETRIDAE
SYN. LATIPENNIS HULST
LIFE HISTORY
- NEW YORK STATE AGRI. EXP. STA. (CORNELL), MEM. 274:167, 1948 FORBES LARVA

FOOD PLANTS
- COMPOSITAE FLOWERS
- GENERAL FEEDER EXCEPT CONIFERS, SHRUBS

CASTORIA REAKIRT
SSP. SEE NAPI LINNEAUS

CASTRENSIS ABBOT & SMITH
SYN. SEE AMERICANA FABRICIUS

CASTRO REAKIRT
SYN. SEE HELLOIDES BOISDUVAL

CASTRUS BRYK
SYN. SEE CLODIUS MENETRIES

CATALINA DYAR
SYN. SEE DIFFUSA BARNES

CATALINA MEADOWS
SSP. SEE CETHURA FELDER & FELDER

CATALINA REAKIRT
SSP. SEE PIASUS BOISDUVAL

CATALPAE(BOISDUVAL), CERATOMIA-SPHINGIDAE

LIFE HISTORY
- AMER. NAT. 16:745, 1882 RILEY PUPA (BRIEF)
- LOC. CIT. 17:400, 1883 ALL STAGES* PL. 7 (NO DESCR.)
- ARKANSAS AGRI. EXP. STA., BULL. 317:16, 1935 BAERG EGG* F. 9; LARVA* FF. 10, 11, 12 (GOOD)
- BROOKLYN ENT. SOC., BULL. 4:20 1881 KOEBELE EGG, LARVA
- ENT. SOC. AMERICA, ANNALS 11: 417, 1918 MOSHER PUPAL STRUCTURE* PL. 36, F. 4
- ILLINOIS AGRI. EXP. STA., BULL. 151:464, 1911 FORBES LARVA* PUPA*
- ILLINOIS DEPT. AGRI., TRANS. 18:APPEND:104, 1880 MARTIN LARVA
- MICHIGAN STATE AGRI. COLLEGE SPEC. BULL. 243:47, 1933 MCDANIEL LARVA
- MUS. COMP. ZOOLOGY, BULL. 87: NO. 6, 1941 DETHIER LARVAL ANTENNAE* FF. 65, 71
- NEW YORK STATE AGRI. EXP. STA. (CORNELL), MEM. 274:189, 1948 FORBES LARVA
- NEW YORK STATE MUS., MEM. 8: 735, 1905 FELT LARVA (BRIEF)
- NOVITATES ZOOLOGICA (TRING) 9:SUPPLEMENT:109, 1903 ROTHSCHILD & JORDAN LARVA, PUPA
- OHIO AGRI. EXP. STA., BULL. 332: 238, 1918 HOUSER LARVA, PUPA
- UNITED STATES DEPT. AGRI. ENT. CIRC. 96:3, 1907 HOWARD EGG* F. 1 A, 1; LARVA* F. 1 B-H; PUPA* F. 1 J
- UNITED STATES DEPT. AGRI. FARMER'S BULL. 705:2, 1916 HOWARD-CHITTENDEN EGG* F. 1 A; LARVA* F. 1 B-I; PUPA* F. 1 J
- LOC. CIT. 1169:34, 1921 KOTINSKY EGG* F. 18 A-1; LARVA* F. 18 B-1; PUPA* F. 18 J

United States Dept. Agri., Misc. Pub. 657:359,373, 1950
Craighead larva ff. 71 C,D
United States Dept. Agri., Rept. pp. 192,193, 1882 Riley
life hisotry* Pl. 13
Boisduval-Spec. Gen. Lep. Sphing. p. 103, 1874 larva* pupa*
Pl. 2 color
Doane etc.-Forest Insects p. 275, 1936 larva
Harris-Ent. Corresp. p. 127, 1869 larva (undertermined)
Herrick-Ins. Enemies Shade Trees p. 60, 1935 larva
Metcalf-Flint--Destr. & Useful Ins. p. 695, 1939 egg* larva* pupa* f. 458

\- - -

Index Amer. Econ. Ent., Vol. 1:70; Vol. 2:74; Vol. 3:76; Vol. 4:94; Vol. 5:122; Vol. 6:138; Vol. 7:175

FOOD PLANTS

Catalpa bignonioides Walt.
Catalpa speciosa Warder

cataphracta (Grote), Papaipema-Noctuidae

ab fluxa Bird
a sulphurata Bird

LIFE HISTORY

Amer. Mus. Nat. Hist., Bull. 5:94, 1893 Beuten. larva
Can. Ent. 23:157, 1891 Dyar larva, pupa
Loc. cit. 30:128, 1898 Bird larva, pupa
New York State Cabin. Nat. Hist., Ann. Rept. No. 48:374, 1894 Lintner larva
Crosby-Leonard--Manual Veget. Garden Ins. p. 161, 1918 larva

\- - -

Index Amer. Econ. Ent., Vol. 1:230; Vol. 2:263; Vol. 3:294; Vol. 4:342; Vol. 5:468; Vol. 6:558; Vol. 7:740

FOOD PLANTS

Acer negundo Linn.
Althaea rosea Cav.
Ambrosia sp.
Arctium lappa Linn.
Aster sp.
Cannabis sativa Linn. (hemp)
Cirsium sp.
Cirsium iowense (Pammel)
Cosmos bipinnatus Cav.
Dahlia variabilis Desf.
Decodon verticillatus (Linn.)
Delphinium sp.
Elymus canadensis Linn.
Helianthus sp.
Iris sp.
Lactuca canadensis Linn.
Lilium sp.
Lilium superbum Linn.
Lilium tigrinum Andr.
Paeonia officinalis Linn.
Phalaris arundinacea Linn.
Rheum rhaponticum Linn.
Ribes lacustre (Pers.)
Ribes grossularia Linn.
Rubus allegheniensis Porter
Rubus daeus v. strigosus (Michx.)
Rudbeckia sp.
Sambucus canadensis Linn.
Solanum lycopersicum Linn. (tomato)
Solanum tuberosum Linn. (potato)
Zea mays Linn. (corn)
Giant Bur Elder

catenaria (Drury), Cingilia-Geometridae

syn. devinctaria Guenee
syn. humeralis Walker
ab immacularia Swett
a rubiferaria Swett

LIFE HISTORY

Connecticut Agri. Exp. Sta., Bull. 256:313, 1923 Britton larva* Pl. 24, f. e; pupa* and cocoon* Pl. 24, f. b
Maine Agri. Exp. Sta., Bull.

259:43, 1917 Mosher
larva, pupa* f.4,c,f
Loc. cit. Bull. 356:144, 1930
Phipps egg, larva* f.21 c,d;
pupa* f.21 e,f
Minnesota Agri. Exp. Sta., Bull.
61:241, 1899 Lugger larva
Mus. Comp. Zool., Bull. 87:No.6,
1942 Dethier larval antennae*
f.96
Naturaliste Canadien p.327,
1935 Daviault larva, pupa
New York Ent. Soc., Journ. 53:
195, 1945 Forbes larva, pupa
New York State Agri. Exp. Sta.
(Cornell), Mem. 274:103, 1948
Forbes larva, pupa
New York State Mus., Mem. 8:706,
1905 Felt larva (brief)
Ohio Agri. Exp. Sta., Bull. 45:
179, 1893 Webster larva
Papilio 1:56, 1881
Coquillett larva
Psyche 6:124, 1891
Scudder egg, larva, pupa
Psyche 9:250, 1901
Dyar egg, larva
United States Dept. Agri. Bur.
Ent., Bull. 4:31, 1884 Old
Series Smith larva
United States Dept. Agri.,
Misc. Pub. 657:362,426, 1950
Craighead larva* f.91 h;
pupa; cocoon p.22; larva* f.29
Wisconsin Agri. Exp. Sta., Bull.
159:17, 1908 Hardenberg
larva* pl.2, ff.19a, 19b; pupa*
pl.2, f.21 color
Gumppenberg–Systema
Geometrarum
1:370, 1887 larva, pupa
Harris–Ent. Corresp. p.321,
1869 larva, pupa
Harris–Ins. Inj. Veget. 2nd edit.
p.358, 1852 cocoon
Harris–Loc. cit. Flint edit.
p.460, 1862 larva* pupa*
cocoon*
O'Kane–Inj. Ins. p.285, 1912
larva* f.415
Packard–Monograph Geometridae
p.218, 1876 larva* pl.13,
f.12
Peairs–Pests Farm, Garden,
Orchard p.22, 1941 larva* f.29

- - -

Index Amer. Econ. Ent.,
Vol. 1:82; Vol. 3:91; Vol. 4:108;
Vol. 5:138; Vol. 6:157; Vol. 7:193

FOOD PLANTS

Abies balsamea (Linn.)
Acer rubrum Linn.
Acer saccharinum Linn.
Acer saccharophorum Koch
Aesculus hippocastanum Linn.
Alnus sp. (alders)
Anaphalis margaritacea (Linn.)
Baptisia tinctoria (Linn.)
Betula papyrifera Marsh.
Betula populifolia Marsh.
Carex sp.
Carex pensylvanica Lam.
Castanea dentata (Marsh.)
Comptonia peregrina (Linn.)
Corylus sp. (hazels)
Corylus americana Walt.
Crataegus sp. (thorns)
Dennstaedtia punctilobula (Michx.)
Fagus sp. (beeches)
Fraxinus sp. (ashes)
Gaylussacia baccata (Wang.)
Genista sp.
Hoya carnosa R. Br.
Juniperus sp. (junipers)
Kalmia angustifolia Linn.
Lactuca canadensis Linn.
Larix laricina (DuRoi.)
Malus pumila Mill. (apple)
Myrica sp.
Myrica cerifera Linn.
Myrica gale Linn.
Myrica pensylvanica Loisel
Picea rubens Sarg.
Pinus sp. (pines)
Pinus strobus Linn.

Platanus occidentalis Linn.
Populus sp. (poplars)
Prenanthes alba Linn.
Prunus domestica Linn. (plum)
Prunus serotina Ehrh.
Pyrus communis Linn.
Quercus sp. (oaks)
Quercus ilicifolia Wang.
Rhododendron sp.
Rhodora canadense (Linn.)
Rhus sp. (sumacs)
Rhus toxicodendron Linn.
Rosa sp. (roses)
Rubus sp.
Rubus allegheniensis Porter
Rubus idaeus v. strigosus (Michx.)
Salix sp. (willows)
Salix canida Flugge
Salsola kali v. tenuifolia Tausch
Solidago sp.
Spiraea latifolia (Ait.)
Spiraea salicifolia Linn.
Spiraea tomentosa Linn.
Thuja occidentalis Linn.
Ulmus sp. (elms)
Vaccinium sp.
Vaccinium angustifolium Ait.
Vaccinium macrocarpon Ait.
Vaccinium myrtilloides (Michx.)
Vaccinium uliginosum Linn.
Xolisma ligustrina (Linn.)
Grass

catenula, (Grote), Euxoa-Noctuidae
syn. contagionis Smith
LIFE HISTORY

FOOD PLANTS
Conringia orientalis (Linn.)
Lupinus sp.
Melilotus alba Desr.
Salsola kali v. tenuifolia Tausch
Viola pedatifida Don.

catenulata Grote
ssp. see forficaria Guenee

catenuloides Smith
ssp. see vetusta Walker

catskillata Pearsall
syn. see luteata Packard

catullius Fruhst
syn. see smintheus Doubleday & Hewitson

catullus (Fabricius), Pholisora-Hesperiidae
LIFE HISTORY
Amer. Mus. Nat. Hist., Bull. 5: 298, 1893 Beuten. larva, chrysalis
Can. Ent. 17:245, 1885 Edwards egg, larva, chrysalis
Illinois Dept. Agri., Trans. 18: Append:98, 1880 Middleton larva
Naturaliste Canadien p. 44, 1928 Huard larva
Southern California Acad. Sci., Bull. 33:140, 1934 Comstock-Dammers egg* Pl. 45
United States Nat. Mus., Bull. 157:208, 1932 Austin Clark larva
Abbot & Smith-Lep. Ins. Georgia 1:47, 1797 larva, chrysalis Pl. 24 COLOR
Comstock, J. A.-Butter. Calif. p. 211, 1927 egg, larva, chrysalis
French-Butter. East. United States p. 367, 1886 egg*
Holland-Butterfly Book 1905 chrysalis* Pl. 6, ff. 29, 36, 41
Scudder-Butter. East. United States & Canada pp. 1521-1522, 1889 life history* Pl. 66, f. 21; Pl. 69, f. 2; Pl. 77, ff. 16, 21, 23; Pl. 80, ff. 43, 44; Pl. 85, ff. 29, 36, 41
Scudder-Butter. Their Structure etc. p. 9, 1881 egg

SCUDDER-GUIDE TO COMMON BUTTER. P. 164, 1893 LARVA, CHRYSALIS

FOOD PLANTS

AMARANTHUS SP.
AMBROSIA SP.
CHENOPODIUM ALBUM LINN.
MONARDA PUNCTATA LINN.
ORIGANUM VULGARE LINN.
SYMPHORICARPOS ALBUS (LINN.)

CAVERNA GUNDER
SSP. SEE NUBIGENA BEHR

CAYUGA NAKAHARA
SYN. SEE ARCHIPPUS CRAMER

CEANOTHI BEHR
SYN. SEE EURYALUS BOISDUVAL

CECROPIA (LINNAEUS) PLATYSAMIA-SATURNIIDAE
AB MACULA REIFF
A ALBOFASIATA SAGEDER

LIFE HISTORY

AMER. ACAD. ARTS, SCI., PROC. 28: 58, 1893 PACKARD EGG, LARVA
AMER. ENT. 2:26, 1869 LARVA (BRIEF)
LOC. CIT. 2:96, 1870 LARVA* PUPA* COCOON*
AMER. MUS. NAT. HIST., BULL. 10: 435, 1898 BEUTEN. LARVA
CAN. ENT. 2:82, 1870 SPRAGUE EGG
LOC. CIT. 3:149, 1871 SAUNDERS LIFE HISTORY LARVA*
LOC. CIT. 9:41, 1877 GENTRY LIFE HISTORY
DAVENPORT ACAD. SCI., PROC. 2:277, 1878 STRECKER COMPARES THE LARVA OF GLOVERI AND CECROPIA
ENT. NEWS. 1:19, 1898 SKINNER SEX INDICATED BY COCOON
LOC. CIT. 9:150, 1898 SOULE DURATION OF EGG STAGE
LOC. CIT. 17:368, 1906 DAVIS NUMBER OF EGGS
ENT. SOC. AMERICA, ANNALS 9:152, 1916 MOSHER PUPAL STRUCT* PL. 6, FF. 8, 9, 10
ILLINOIS DEPT. AGRI., TRANS. 15: 193, 1877 FRENCH LARVA
LOC. CIT. 18: APPEND: 126, 177, 1880 COQUILLETT-MARTIN LARVA*
MAINE NAT. 2:16, 1922 DECOSTER LARVA*
MARYLAND NAT. HIST. SOC., BULL. 8:50, 1938 MASTERS COCOON
LOC. CIT. 9:41, 1939 REED COCOON* PUPA*
MARYLAND NAT. HIST. SOC., PUBLICATION 1930 NO PAGE HAYDON EGG, LARVA, PUPA
MICHIGAN STATE COLLEGE, SPEC. BULL. 243:51, 1933 MCDANIEL LIFE HISTORY
MINNESOTA AGRI. EXP. STA., BULL. 61:174, 1899 LUGGER LARVA* PL. 14, FF. 122, 124; COCOON* PL. 13, F. 123
LOC. CIT. BULL. 84(5):70, 1903 WASHBURN LARVA* F. 1 COLOR
MISSOURI, REPT. INS. 4:103, 1872 RILEY LIFE HISTORY*
MUS. COMP. ZOOL., BULL. 87: NO. 6, 1941 DETHIER LARVAL ANTENNA* FF. 48, 52, 54, 56, 58
NAT. ACAD. SCI., MEM. 12:214, 1914 PACKARD LIFE HISTORY EGG, LARVA* PL. 5; FF. 4-6; PL. 7; PL. 8, F. 1 COLOR
LOC. CIT. P. 267 PACKARD HYBRID LARVA WITH GLOVERI
LOC. CIT. P. 267 PACKARD HYBRID LARVA* WITH COLUMBIA PL. 5, F. 3 COLOR
NATURALISTE CANADIEN P. 211, 1929 HUARD LARVA* F. 28; COCOON* F. 29
NEW YORK STATE CABIN. NAT. HIST., ANN. REPT. NO. 26:125, 1874 LINTNER EGG, LARVA

New York State Mus., Mem. 8: 539, 1905 Felt larva
New York State, Rept. Ins. 3:338, 1856 Fitch larva
Ohio Agri. Exp. Sta., Bull. 45:170, 1893 Webster larva, cocoon
Ontario Ent. Soc., Ann. Rept. p. 23, 1871 Bethune larva* cocoon*
Loc. cit. p. 40, 1872 Reed larva
Loc. cit. p. 22, 1874 Saunders larva* cocoon*
Loc. cit. p. 26, 1886 Harrington larva
Loc. cit. p. 47, 1889 Harrington larva
Loc. cit. p. 56, 1898 Lockhead larva
Ottawa Nat. 20:149, 1906 Gibson larva
Loc. cit. 27:106, 1913 White larva
Outlook 64, No. 14, Apr. 17:828, 1900 Soule cocoon*
Practical Ent. 2:55, 1866 larva (brief)
Psyche 7:131, 1894 Huston life history
Loc. cit. 19:69, 1912 Rau duration egg state
Loc. cit. 20:161, 1913 Rau duration pupa stage
Smithsonian Misc. Collections 4: Art. 2:224, 1862 Morris larva, cocoon
United States Dept. Agri., Misc. Pub. 657:357, 377, 1950 Craighead larva f. 72 B
Abbot & Smith-Lep. Ins. Georgia 1:89, 1797 larva* Pl. 45 COLOR
Ballard-Ins. Lives p. 75, 1879 larva* pupa* cocoon*
Cragen-Our Ins. Friends & Foes p. 109, 1899 larva* f. 44
Dickerson-Moths & Butterflies p. 157, 1901 egg* ff. 133, 134; larva* ff. 136-139; pupa* ff. 141-143; cocoon* ff. 132, 140, 144, 146
Duncan-Naturalist's Library 22:133, 1858 larva
Eliot & Soule-Caterpillars & Moths p. 251, 1902 egg, larva*, cocoon*
Emmons-Nat. Hist. New York 5: 237, 1854 larva* Pl. 44 COLOR
Fernald-Kingslesy's Standard Nat. Hist. 2:456, 1885 egg, larva
Fernald-Applied Ent. 286, 1926 larva* f. 21
Fernald & Shepard-Applied Ent. p. 283, 1942 larva* f. 22
Glover-Illus. North Amer. Ent. 1878 larva, cocoon* Pl. 7, f. 2
Harris-Ent. Corresp. p. 294, 1869 larva
Harris-Ins. Inj. Veget. 1st. edit. p. 279, 1841 larva, pupa
Harris-Loc. cit. 2nd edit. p. 299, 1852 larva
Harris-Loc. cit. Flint's edit. p. 385, 1862 larva* pupa* cocoon*
Hart-Ins. Collection Illinois High Schools p. 17, 1903 larva* f. 12
Herrick-Ins. Enemies Shade Trees p. 155, 1935 larva* f. 128
Holland-Moth Book 1905 larva* Pl. 1, f. 8 COLOR; egg* p. 5, f. 3; cocoon* p. 83, f. 37
Jaeger-Life North Amer. Ins. p. 155, 1864 larva* cocoon*
Kellog-Amer. Ins. p. 418, 1908 larva* f. 604
Metcalf & Flint-Destr. & Useful Ins. p. 149, 1939 larva* f. 88F
Morris-Synop. Lep. North America p. 224, 1862 larva

Packard–Ins. Inj. Forest Trees p.113, 1881 larva*
Saunders–Ins. Inj. Fruits p.75, 1883 egg, larva* cocoon*
Saunders–Loc. cit. p.75, 1889 egg, larva* f.72; cocoon* f.71
Tenney–Nat. Hist. p.400, 1867 larva* pupa*
Weed–Life History Amer. Ins. p.99, 1897 larva f.38

- - -

Index Amer. Econ. Ent., Vol. 1:271; Vol. 2:321; Vol. 3:361; Vol. 4:429; Vol. 5:557; Vol. 6:652; Vol. 7:862

FOOD PLANTS

Acer negundo Linn.
Acer rubrum Linn.
Acer saccharinum Linn.
Acer spicatum Lam.
Alnus serrulata (Ait.)
Amelanchier canadensis (Linn.)
Aronia arbutifolia (Linn.)
Berberis vulgaris Linn.
Betula alba Linn.
Betula alleghaniensis Britt.
Betula lenta Linn.
Betula papyrifera Marsh.
Carya sp. (hickories)
Cephalanthus occidentalis Linn.
Corylus americana Walt.
Corylus cornuta Marsh.
Crataegus calpodendron (Ehrh.)
Crataegus crus-galli Linn.
Crataegus oxycantha Linn.
Crataegus pedicellata Sarg.
Decodon verticillatus (Linn.)
Fagus sp. (beeches)
Fraxinus sp. (ash)
Gaylussacia frondosa (Linn.)
Gleditsia triacanthos Linn.
Humulus lupulus Linn. (hops)
Malus coronaria (Linn.)
Malus pumila Mill. (apple)
Ostrya virginiana (Mill.)
Paeonia officinalis Linn.
Parthenocissus quinquefolia (Linn.)
Philadelphus inodorus Linn.
Physocarpus opulifolius (Linn.)
Picea sp. (spruces)
Populus sp. (poplars)
Populus balsamifera Linn.
Populus tremuloides Michx.
Prunus cerasus Linn. (cherry)
Prunus domestica Linn. (plum)
Prunus ilicifolia (Nutt. ex H&A.)
Prunus maritima Marsh.
Prunus pensylvanica Linn.
Prunus serotina Ehrh.
Prunus virginiana Linn.
Pyrus communis Linn. (pear)
Quercus alba Linn.
Ribes americanum Mill.
Ribes grossularia Linn.
Ribes nigrum Linn.
Ribes rubrum Linn.
Ribes sativum Syme (currant)
Rosa blanda Ait.
Rubus allegheniensis Porter
Rubus idaeus v. strigosus (Michx.)
Rubus occidentalis Linn.
Salix alba Linn.
Salix humilis Marsh.
Salix lucida Muhl.
Salix viminalis Linn.
Sambucus canadensis Linn.
Sambucus racemosa v. pubens (Michx.)
Sassafras albidum var. molle (Raf.)
Sorbus americana Marsh.
Spiraea betulifolia v. corymbosa (Raf.)
Spiraea salicifolia Linn.
Spiraea tomentosa Linn.
Symphoricarpos albus (Linn.)
Syringa vulgaris Linn.
Tilia americana Linn.
Tilia europaea Linn.
Ulmus americana Linn.
Ulmus rubra Muhl.
Ulmus thomasii Sarg.
Vaccinium sp.
Wisteria sinensis Sweet.

CECROPS(FABRICIUS),STRYMON-LYCAENIDAE
SYN. POEAS HUBNER
LIFE HISTORY
AMER. ENT. VOL. 1, 2ND SERIES, P. 201, 1880 RILEY LARVA
ILLINOIS DEPT. AGRI., TRANS. 18: APPEND:93, 1880 MIDDLETON LARVA
- - -
INDEX AMER. ECON. ENT., VOL. 7:155
FOOD PLANTS
GOSSYPIUM HERBACEUM LINN. (COTTON)

CELATARIA HULST
SYN. SEE MONICARIA GUENEE

CELEUS HUBNER
SYN. SEE QUINQUEMACULATA HAWORTH

CELIA HENRY EDWARDS
SSP. SEE ULTRONIA HUBNER

CELIA(SAUNDERS),APPANTESIS-ARCTIIDAE
FORM FRANCONIA HENRY EDWARDS
LIFE HISTORY
CAN. ENT. 2:74, 1870 SAUNDERS LARVA
CAN. ENT. 35:150, 1903 GIBSON LARVA
NEW YORK ENT. SOC., JOURN. 8:42, 1900 DYAR LARVA (BRIEF)
FOOD PLANTS

CELLUS(BOISDUVAL & LECONTE), CECROPTERUS-HESPERIIDAE
SYN. FESTUS GEYER
AB AEREOFUSCUS GUNDER
AB LEILAE CLARK
LIFE HISTORY
AMER. ENT. SOC., TRANS. 37:190, 1911 SKINNER LARVA, CHRYSALIS
AMER. MUS. NAT. HIST., BULL. 5: 304, 1893 BEUTEN. LARVA, CHRYSALIS
SCIENCE 80:163,164, 1934 CLARK LIFE HISTORY
SMITHSONIAN MISC. COLLECTIONS VOL. 95, ART. 7, 1936 AUSTIN CLARKE EGG* PL. 4, F. 17; PL. 6, FF. 28, 29; TEXT F. A-D P. 36; LARVA* PL. 6, FF. 30, 31, 33-36; PL. 8, FF. 43, 44, 48, 49; TEXT FF. E-H P. 29; CHRYSALIS* PL. 5, FF. 19-25 ALL IN COLOR
BOISDUVAL & LECONTE-HIST. LEP. NORTH AMERICA 1833 LARVA* CHRYSALIS* PL. 73 COLOR
SCUDDER-BUTTER. EAST. UNITED STATES & CANADA PP. 1855, 1856, 1889 LARVA, CHRYSALIS
FOOD PLANTS
AMPHICARPA BRACTEATA V. COMOSA (LINN.)
BREWERIA MICHAUXII FERN.
CONVOLCULACEAE

CELTIPHAGA HARVEY
SSP. SEE SUBROTATA HARVEY

CELTIS(BOISDUVAL & LECONTE),ASTEROCAMPA-NYMPHALIDAE
AB ALB STRECKER
AB INORNATA WOLLCOT
A ALICIA EDWARDS
B ANTONIA EDWARDS
C MONTIS EDWARDS
LIFE HISTORY
AMER. ENT. 3:206, 1880 LARVA (BRIEF)
BROOKLYN ENT. SOC., BULL. 6:113, 1884 SMITH LARVA, CHRYSALIS
ILLINOIS DEPT. AGRI., TRANS. 15: 155, 1877 FRENCH LARVA
LOC. CIT. 18:APPEND:88, 1880 MIDDLETON LARVA
MISSOURI, REPT. INS. 6:137-140, 1874 RILEY EGG* LARVA* CHRYSALIS*
NEW YORK STATE MUS., MEM. 8: 725, 1905 FELT LARVA (BRIEF)
PSYCHE 3:123, 1880 W.H. EDWARDS EGG, LARVA, CHRYSALIS
SAINT LOUIS ACAD. SCI., TRANS.

P. 195, 1875 RILEY EGG*
LARVA* CHRYSALIS* (LYCAON)
SMITHSONIAN MISC. COLLECTIONS
4:ART. 2:69, 1862 MORRIS
LARVA, CHRYSALIS
UNITED STATES DEPT. AGRI.,
FARMER'S BULL. 1169:49, 1921
KOTINSKY EGG* F. 30A; LARVA*
F. 30B; CHRYSALIS* F. 30 C,D
UNITED STATES DEPT. AGRI., MISC.
PUB. 657:370 1950 CRAIGHEAD
LARVA
BOISDUVAL & LECONTE-HIST. LEP.
NORTH AMERICA P. 210, 1833
LARVA* CHRYSALIS* PL. 57 COLOR
COMSTOCK-HOW TO KNOW BUTTER.
P. 174, 1904 LARVA
EDWARDS, W. H.-BUTTER. NORTH
AMERICA, 2ND SERIES, 1875
LIFE HISTORY* PL. APATURA NO. 1
COLOR
FRENCH-BUTTER. EAST. UNITED
STATES P. 216, 1886 LIFE HISTORY
HERRICK-INS. ENEMIES SHADE
TREES P. 107, 1935 LARVA
MORRIS-SYNOP. LEP. NORTH
AMERICA P. 69, 1862 LARVA,
CHRYSALIS
SCUDDER-BUTTER. EAST. UNITED
STATES & CANADA PP. 1789-1792
1889 LIFE HISTORY
SCUDDER-BUTTER. THEIR STRUCT.
P. 263, 1881 EGG, LARVA
SCUDDER-GUIDE TO COMMON
BUTTER. P. 106, 1893 LARVA,
CHRYSALIS
VANCE-LIFE AMONG BUTTER. P. 43,
1925 EGG, LARVA, CHRYSALIS

- - -

INDEX AMER. ECON. ENT.,
VOL. 3:84 (CHLORIPPE)
VOL. 7:112

FOOD PLANTS
CELTIS LAEVIGATA WILLD.
CELTIS OCCIDENTALIS LINN.

CENTAUREAE (RAMBUR), PYRGUS-HESPERIIDAE
SYN. WYANDOT EDWARDS
A FRIJA WARREN.
LIFE HISTORY

FOOD PLANTS
MENTHA SP.?

CENTRALIS GROTE
SYN. SEE FACTIOSALES WALKER

CENTRALIS MCDUNNOUGH
SSP. SEE FURCILLA PACKARD

CENTRIFERRUGINEA STRAND
SSP. SEE CLARESCENS GUENEE

CENTROSTRIGARIA (WOLLASTRON), EUPHYIA-GEOMETRIDAE
SYN. LATIRUPTA WALKER
SYN. LUSCINATA ZELLER
LIFE HISTORY

FOOD PLANTS
POLYGONUM AVICULARE LINN.
POLYGONUM HYDROPIPER LINN.

CEPHISARIA GROTE
SYN. SEE INCONSTANS GEYER

CERAMICA HUBNER
SYN. SEE FIGURATA DRURY

CERASI BOISDUVAL
SYN. SEE MYOPS ABBOT & SMITH

CERES BUTLER
SYN. SEE CALIFORNIA WESTWOOD & HEWITSON

CERINA (GROTE), PAPAIPEMA-NOCTUIDAE
LIFE HISTORY
CAN. ENT. 48:14, 1916
BIRD LARVA, PUPA
FOOD PLANTS
LILIUM CANADENSE LINN.
LILIUM SUPERBUM LINN.
PODOPHYLLUM PELTATUM LINN.
GRASS IN EARLY STAGES

CERINTHA(TREITSCHKE), CHAMYRIS-NOCTUIDAE
AB OBSCURA DYAR
LIFE HISTORY
AMER. MUS. NAT. HIST., BULL. 4:78, 1892 BEUTEN. LARVA, PUPA
NEW YORK STATE MUS., MEM. 8: 733, 1905 FELT LARVA (BRIEF)
PAPILIO 1:56, 1881 COQUILLETT LARVA
PSYCHE 8:349, 1899 DYAR LARVAL HEAD SETAE
- - -
INDEX AMER. ECON. ENT., VOL. 2:78
FOOD PLANTS
CRAETAGUS SP. (THORNS)
LIGUSTRUM SP. (PRIVETS)
MALUS PUMILA MILL. (APPLE)
PRUNUS DOMESTICA LINN. (PLUM)
PRUNUS SEROTINA EHRH.
PYRUS SP.

CERISYI KIRBY, SMERINTHUS-SPHINGIDAE
A ASTARTE STRECKER
B OPTHALMICUS BOISDUVAL
SYN. VANCOUVERENSIS BUTLER
C SALICETI BOISDUVAL
FORM PALLIDULUS HENRY EDWARDS
AB NIGRESCENS CLARK
D BOREALIS CLARK
LIFE HISTORY
CALIFORNIA ACAD. SCI., PROC. 7:3, 1876 HENRY EDWARDS EGG, LARVA (OPTHALMICUS)
CAN. ENT. 23:143, 1891 FRENCH LARVA, PUPA
LOC. CIT. 23:200, 1891 DYAR EGG, LARVA
ENT. NEWS 24:201, 1913 ROWLEY LIFE HISTORY
ENT. SOC. AMERICA, ANNALS 11:425, 1918 MOSHER PUPAL STRUCTURES
NEW YORK STATE AGRI. EXP. STA. (CORNELL), MEM. 274:193, 1948 FORBES LARVA
NOVITATES ZOOLOGICAE (TRING) 9:SUPPLEMENT P. 322, 1903 ROTHS. & JORD. LARVA, PUPA
PSYCHE 7:331, 1896 HARVEY EGG, LARVA
SOUTHERN CALIFORNIA ACAD. SCI., BULL. 42:42, 1943 COMSTOCK-DAMMERS LARVA* PUPA* PL. 3
BOISDUVAL-SPEC. GEN. LEP. SPHING. P. 34, 1874 LARVA* PL. 1 COLOR; PUPA
EDWARDS, HY-PACIFIC COAST LEP. 16:3, 1876 LARVA
ESSIG-INS. WEST. NORTH AMERICA P. 664, 1926 EGG, LARVA
FOOD PLANTS
POPULUS SP. (POPLARS)
POPULUS NIGRA V. ITALICA MUENCH.
POPULUS TREMULOIDES MICHX.
PRUNUS DOMESTICA LINN. (PLUM)
PYRUS COMMUNIS LINN. (PEAR)
SALIX SP. (WILLOWS)
SALIX SERICEA MARSH.
SYMPHORICARPOS ALBUS (LINN.)

CERIVANA SMITH
SSP. SEE FINITIMA GUENEE

CERNES BOISDUVAL & LECONTE
SYN. SEE THEMISTOCLES LATREILLE

CERNES HARRIS
SYN. SEE MANATAAQUA SCUDDER

CEROGAMA GUENEE, CATOCALA-NOCTUIDAE
SYN. AURELLA FISCHER
FORM BUNKERI GROTE
FORM ELIZA FISCHER
FORM RUPERTI FRANCLEMONT
LIFE HISTORY
CAN. ENT. 26:21, 1894 DYAR LARVA
ENT. NEWS 20:17, 1909 ROWLEY EGG
LOC. CIT. 21:105, 1910 ROWLEY-BERRY LARVA
NEW YORK STATE MUS., MEM. 8: 731, 1905 FELT LARVA (BRIEF)
ONTARIO ENT. SOC., ANN. REPT. P. 54, 1903 GIBSON LARVA

Ottawa Nat. 14:18, 1900
Gibson larva (brief)
FOOD PLANTS
Acer rubrum Linn.
Tilia americana Linn.

cerrita Wright
ssp. see leanira Boisduval

cerulea Ehrmann
syn. see arthemis Drury

cerussaria Grote
syn. see argentata Drury

cerussata (Grote), Papaipema-Noctuidae
LIFE HISTORY
Can. Ent. 32:231, 1900
Bird larva, pupa
- - -
Index Amer. Econ. Ent., Vol. 7:740
FOOD PLANTS
Arctium lappa Linn.
Cirsium sp.
Thalictrum sp.
Vernonia noveboracensis (Linn.)

cervantes Boisduval
syn. see persius Scudder

cervina Smith
syn. see lustralis Grote

cervinaria (Packard), Synaxis-Geometridae
syn. female aurantiacaria Packard
LIFE HISTORY
Southern California Acad. Sci., Bull. 33:36, 1934 Comstock larva* Pl. 18; pupa* Pl. 19
FOOD PLANTS
Quercus sp.

cervinicolor Hulst
ssp. see unicalcararia Guenee

cervinifascia Walker
ssp. see hersiliata Guenee

cethura Felder & Felder
syn. cooperii Behr
syn. angelina Boisduval
a morrisoni Edwards
b caliente Wright
c deserti Wright
d catalina Meadows
LIFE HISTORY
Can. Ent. 69:117, 1937
Doudoroff larva (cooperii)
Southern California Acad. Sci., Bull. 31:32, 1932
Comstock-Dammers egg; larva* Pl. 13; pupa* Pl. 14
Comstock, J. A.-Butter. California p. 41, 1927 egg
- - -
Index Amer. Econ. Ent., Vol. 6:42
FOOD PLANTS
Brassica kaber v. pinnatifida (Stokes)
Caulanthus inflatus Wats.
Descurainea pinnata (Walt.)
Streptanthella longirostris (Wats.)

chagnoni (Swett), Eupithecia-Geometridae
syn. gibsonata Taylor
LIFE HISTORY
Can. Ent. 74:152, 203 (McD), 1942
McGuffin larva* Pl. 12, f. b; also larval mandible
Loc. cit. 77:55, 1945
McGuffin larva (gibsonata)
FOOD PLANTS
Thuja occidentalis Linn.

chagnoni Swett
ssp. see limbata Haworth

chalcas Fabricius
syn. see palamedes Drury

chalcedona (Doubleday & Hewitson), Euphydryas-Nymphalidae

FOR THE MANY SSP. AND SYNS. OF THIS SP. SEE MCDUNNOUGH'S LIST

LIFE HISTORY

CALIFORNIA ACAD. SCI., PROC. 3:89, 1863 BEHR LARVA, CHRYSALIS
LOC. CIT. 5:167, 1873 HY. EDWARDS LARVA, CHRYSALIS
MONTANA UNIV., BULL. 30:72, 1906 ELROD EGG, LARVA, CHRYSALIS
LOC. CIT. P. 76, ELROD LARVA (BRIEF) (COOPERI)
PAPILIO 3:26, 1883 RIVERS LARVAL HABITS
LOC. CIT. 3:123, 1883 WRIGHT LARVAL NOTE
LOC. CIT. 4:63, 1884 EDWARDS LARVA, CHRYSALIS
PSYCHE 13:136, 1906 RAMSAY CHRYSALIS* PL. 8
SOUTHERN CALIFORNIA ACAD. SCI., BULL. 26:67, 1927 COMSTOCK EGG*
COMSTOCK, J. A. - BUTTER. CALIF. P. 97, 1927 EGG, LARVA, CHRYSALIS* PL. A 35
EDWARDS, W. H. - BUTTER. NORTH AMERICA, 2ND SERIES, 1871 LARVA, CHRYSALIS PL. MELITEA NO. 1 COLOR
EDWARDS, HENRY - KINGSLEY'S NAT. HIST. 2:486, 1885 CHRYSALIS
ESSIG - INS. WEST. NORTH AMERICA P. 640, 1926 LARVA* F. 517; CHRYSALIS* F. 518
HOLLAND - BUTTERFLY BOOK P. 139, 1905 LIFE HISTORY
HOLLAND - BUTTERFLY BOOK 3RD EDIT. P. 116, 1931 LIFE HISTORY

FOOD PLANTS

ACTINOMERIS ALTERNIFOLIA (LINN.)
ASTER SP.
BUDDLEIA SP. (IN CAPTIVITY)
CAPRIFOLIACAEA SP.
CASTELLEJA SP.
CASTELLEJA AFFINIS H. & A.
CASTELLEJA PARVIFLORA BONGARD
CHRYSANTHEMUM SP.
CHRYSANTHEMUM MAXIMUM RAMOND.
COLLINSIA HETEROPHYLLA GRAHAM
CYMBALARIA MURALIS GAERTN
GEUM TRIFLORUM PURSH
LONICERA CAPRIFOLIUM LINN.
MIMULUS AURANTIACUS CURT.
MIMULUS LUTEUS LINN.
MIMULUS RINGENS LINN.
PENSTEMON ANTIRRHINOIDES BENTH.
PENSTEMON BARBATUS (CAV.)
PENSTEMON CORDIFOLIUS BENTH.
PENSTEMON TERNATUS TURR. EX GRAY
RIBES CYNOSBATI LINN.
ROSA SP.
ROSA BLANDA AIT.
ROSA CAROLINA LINN.
ROSA MINUTIFOLIA ENGELM.
SCROPHULARIA CALIFORNICA CHAM.
SCROPHULARIA MARYLANDICA LINN.
STACHYS PALUSTRIS LINN.
VERONICA SP.

CHALCEDONIA (HUBNER), ELAPHRIA - NOCTUIDAE

SYN. ARNA GUENEE
SYN. EXPUNCTA WALKER
SYN. VINCTA WALKER
SYN. IRRESOLUTA WALKER
SYN. TRACTA WALKER

LIFE HISTORY

FOOD PLANTS

MIMULUS SP.
PENSTEMON ANTIRRHINOIDES BENTH.
PENSTEMON CORDIFOLIUS BENTH.
SCROPHULARIA CALIFORNICA CHAM.

CHAMAECHRYSARIA GROTE

SYN. SEE XANTHOMETATA WALKER

CHAMAENERII HARRIS

SYN. SEE GALLII ROTTENBERG

CHAMIS SCUDDER

SYN. SEE ACCIUS ABBOT & SMITH

CHAPMANI GROTE
SYN. SEE ASTARTE DOUBLEDAY

CHARA EDWARDS
SSP. SEE DYMAS EDWARDS

CHARIE BLANCHARD
SYN. SEE CARYE HUBNER

CHARITONIA(LINNAEUS), HELICONIUS-NYMPHALIDAE
LIFE HISTORY
CAN. ENT. 13:158, 1881 EDWARDS EGG, LARVA, CHRYSALIS
PAPILIO 1:210, 1881 EDWARDS LARVA, CHRYSALIS
COMSTOCK-HOW TO KNOW BUTTER. P. 201, 1904 LARVA
DEWITZ-ZEITSCH. GESAMENT. NATUR. 1879 LARVA, CHRYSALIS
EDWARDS, W. H. -BUTTER. NORTH AMERICA, 2ND SERIES, PART 10, 1882 LIFE HISTORY
FRENCH-BUTTER. EAST. UNITED STATES P. 142, 1886 LIFE HISTORY
GOSSE-DOUBLEDAY'S GEN. DIURN. LEP., VOL. 1:97, 1950 LARVA
GUNDLACH-ENTOM. CUBANA P. 21, 1881 LARVA, CHRYSALIS
HOLLAND-BUTTERFLY BOOK P. 92, 1905
HOLLAND-BUTTERFLY BOOK, 3RD EDIT. P. 75, 1931 CHRYSALIS
VANCE-LIFE AMONG BUTTERFLIES P. 40, 1925 EGG, LARVA, CHRYSALIS
FOOD PLANTS
PASSIFLORA CAERULEA LINN.

CHARLOTTENSIS HOLLAND
SSP. SEE MARIPOSA REAKIRT

CHARLOTTII BARNES
SSP. SEE LETO BEHR

CHARON EDWARDS
SYN. SEE OETUS EDWARDS

CHEMNIS FABRICIUS
SYN. SEE ETHLIUS CRAMER

CHEMO SCUDDER
SSP. SEE ATLANTIS EDWARDS

CHEPHISE(CRAMER), MELANCHROIA-GEOMETRIDAE
LIFE HISTORY
NEW YORK ENT. SOC., JOURN. 53: 190, 1945 FORBES PUPA
FOOD PLANTS

CHERSIS(HUBNER), SPHINX-SPHINGIDAE
SYN. CINEREA HARRIS
A PALLESCENS ROTHSCHILD & JORDAN
B OREODAPHNE HENRY EDWARDS
LIFE HISTORY
AMER. MUS. NAT. HIST., BULL. 7:303, 1895 BEUTEN. LARVA PUPA
ENT. SOC. AMERICA, ANNALS 11:420 1918 MOSHER PUPAL STRUCTURE
ILLINOIS DEPT. AGRI., TRANS. 18: APPEND. P. 105, 1880 MARTIN LARVA
MICHIGAN STATE COLLEGE, SPEC. BULL. 243:49, 1933 MCDANIEL LARVA
NATURALISTE CANADIEN P. 138, 1928 HUARD LARVA
NEW YORK STATE AGRI. EXP. STA. (CORNELL), MEM. 274:191, 1948 FORBES LARVA
NEW YORK STATE MUS., MEM. 8: 727, 1905 FELT LARVA
NOVITATES ZOOLOGICAE (TRING) 9: SUPPLEMENT P. 128, 1903 ROTHSCHILD & JORDAN LARVA, PUPA
ONTARIO ENT. SOC., ANN. REPT. P. 65, 1881 REED LARVA, PUPA
PAPILIO 3:127, 1883 EDWARDS-ELLIOT LARVA
PHILADELPHIA ENT. SOC., PROC. 3: 655, 1864 LINTNER LARVA, PUPA (CINEREA)

Psyche 2:76, 1877 Scudder
larva
Loc. cit. 6:435, 1893
Scudder pupa
United States Dept. Agri.,
Misc. Pub. 657:358, 374, 1950
Craighead larva f. 71 F
Dickerson-Moths & Butter.
p. 327, 1901 larva f. 231
(no descr.)
Eliot & Soule-Caterpillars &
Moths p. 150, 1902 egg, larva*
pupa
Essig-Ins. West. North America
p. 663, 1926 larva
Fernald-Sphing. New England
p. 42, 1886 larva, pupa
Harris-Ent. Corresp. p. 282,
1869 larva
Holland-Moth Book 1905
larva* Pl. 1, f. 1 COLOR
Kellogg-Amer. Ins. p. 606,
1908 larva f. 792 E

- - -

Index Amer. Econ. Ent.,
Vol. 4:444; Vol. 5:580;
Vol. 6:678

FOOD PLANTS
Fraxinus sp. (ash)
Ligustrum sp.
Ligustrum vulgare Linn.
Prunus serotina Ehrh.
Syringa persica Linn.
Syringa vulgaris Linn.

CHINATIENSIS Tinkham, PSEUDOHAZIS-Saturniidae
LIFE HISTORY
Can. Ent. 75:162, 1943
Tinkham egg, larva
FOOD PLANTS
Forestiera angustifolia Torr.

CHINOI Gunder
ssp. see CALIFORNICA Wright (Chlosyne)

CHIONANTHI (Abbot & Smith), ADITA-Noctuidae
LIFE HISTORY
Abbot & Smith-Lep. Ins. Georgia
2:195, 1797 larva* Pl. 98
COLOR
FOOD PLANTS
Chionanthus virginicus Linn.
Triosteum perfoliatum Linn.

CHIONANTHI Abbot & Smith
syn. see RUSTICA Fabricius

CHIONOCHROA Hampson
syn. see LEPUSCULINA Guenee

CHIQUITA Bartsch
ssp. see COCCINATA Grote

CHIRIDOTA Grote, LACOSOMA-Lacosomidae
LIFE HISTORY
Amer. Mus. Nat. Hist., Bull. 10:
433, 1898 Beuten. larva
New York Acad. Sci., Annals 8:
47, 1893 Packard larva
New York Ent. Soc., Journ. 8:
178, 1900 Dyar egg* Pl. 6,
ff. 1, 2; larva* Pl. 6, ff. 6, 7
New York State Mus., Mem. 8:
706, 1905 Felt larva
FOOD PLANTS
Quercus alba Linn.
Quercus coccinea Muench.
Quercus ilicifolia Wang.
Quercus palustris Muench.
Quercus stellata Wang.
Quercus velutina Lam.

CHIRON (Fabricius), TIMETES-Nymphalidae
syn. MARIUS Cramer
syn. CHIRONIAS Hubner
LIFE HISTORY

FOOD PLANTS
Chlorophora tinctoria (Linn.)
Morus sp.

CHIRONIAS Hubner
syn. see CHIRON Fabricius

CHISPA WRIGHT
SYN. SEE SABULETI BOISDUVAL

CHLORINDA MARTYN
SYN. SEE PHOLUS CRAMER

CHLORIS FIELD
SSP. SEE GRUNNUS BOISDUVAL

CHLOROLEUCARIA(GUENEE), CHLOROCHLAMYS-GEOMETRIDAE
SYN. RECTILINEA ZELLER
LIFE HISTORY
BROOKLYN ENT. SOC., BULL. 2:78, 1879 HULST LARVA, PUPA
CAN. ENT. 12:235, 1880 GOODELL LARVA, PUPA
LOC. CIT. 74:231, 1942 DETHIER LARVA* PL. 15, FF. 1, 2, 3; PL. 17, F. 4; PUPA
NEW YORK STATE AGRI. EXP. STA. (CORNESS), MEM. 274:116, 1948 FORBES LARVA
SOUTHERN CALIFORNIA ACAD., SCI., BULL. 33:29, 1934 COMSTOCK-DAMMERS LARVA* PL. 8; PUPA* PL. 9
GUMPPENBERG-SYSTEMA GEOMETRARUM 7:488, 1895 LARVA, PUPA
- - -
INDEX AMER. ECON. ENT., VOL. 3:84
FOOD PLANTS
ACHILLEA MILLEFOLIUM LINN.
APOCYNUM ANDROSAEMIFOLIUM(LINN.)
ASTER SP.
CHRYSANTHEMUM LEUCANTHEMUM (LINN.)
ERIOGONUM FASCICULATUM BENTH.
EUPATORIUM PERFOLIATUM LINN.
FRAGARIA CHILOENSIS (LINN.)
HELENIUM AUTUMNALE LINN.
HELIANTHUS SP.
RUBUS ALLEGHENIENSIS PORTER
RUBUS OCCIDENTALIS LINN.
RUBUS IDAEUS STRIGOSUS(MICHX.)
SOLIDAGO ALTISSIMA LINN.
ZINNIA ELEGANS JACQ.

CHLOROPHORA WATSON & COMSTOCK
SSP. SEE SAEPIUM BOISDUVAL

CHOERILUS CRAMER
SYN. SEE PHOLUS CRAMER

CHORIONA REAKIRT
SYN. SEE AURANTIACA HUBNER

CHRISTINA EDWARDS, COLIAS-PIERIDAE
AB FEMALE PALLIDA COCKERELL
FORM ASTRAEA EDWARDS
A KRAUTHII KLOTS
LIFE HISTORY
CAN. ENT. 24:111, 1892 W. H. EDWARDS LARVA, CHRYSALIS
EDWARDS-BUTTER. NORTH AMERICA 3RD SERIES, SUPPLEMENTARY NOTES, 1897 LARVA
FOOD PLANTS
LUPINUS PERENNIS LINN.
SALIX BRACHYCARPA NUTT. (CHRISTINA)
TRIFOLIUM SP.

CHRYSEIS OLIVIER
SYN. SEE DEFLORATA FABRICIUS

CHRYSOCARENA HARRIS
SYN. SEE HERA HARRIS

CHRYSOMELAS HENRY EDWARDS
SSP. SEE OCCIDENTALIS SCUDDER

CHRYSOPASA WALKER
SYN. SEE PEPITA GUENEE

CHRYSOPTERA WRIGHT
SSP. SEE SATYRUS EDWARDS

CHRYSORRHOEA AUCT.
SEE PHAEORRHOEA DONOVAN

CHRYSOTHEMOIDES VERITY
SYN. SEE HECLA LEFEBRE

CHRYXUS(DOUBLEDAY & HEWITSON), OENEIS-SATYRIDAE

A IVALLDA MEAD
B STANISLAUS HOVANITZ
C CALAIS SCUDDER
D CARYI DYAR
E STRIGULOSA McDUNNOUGH
LIFE HISTORY
CAN. ENT. 11:142, 1879
EDWARDS EGG, LARVA
COMSTOCK, J. A.-BUTTER.
CALIFORNIA P. 259, 1927
EGG (A 97) (IVALLDA)
EDWARDS-W. H.-BUTTER. NORTH
AMERICA, 2ND SERIES, 1879
EGG* LARVA* PL. CHIONOBAS
NO. 3 COLOR (IVALLDA)
EDWARDS-LOC. CIT. 3RD SERIES,
1897 LIFE HISTORY* CHIONOBAS
1, FF. A-H
EDWARDS-LOC. CIT. 3RD. SERIES,
SUPPLEMENTARY NOTES, 1897
EGG, LARVA (INVALLDA)
MEAD-REPT. LEP. WHEELER EXPED.
WEST 100 MERIDIAN 5:777, 1875
EGG
FOOD PLANTS
GRASSES

CHUSCA EDWARDS
SYN. SEE SABULETI BOISDUVAL

CICHORII GROTE & ROBINSON
SYN. SEE PLANTAGINIS LINNAEUS

CILLA BEHR
SYN. SEE AQUILO BOISDUVAL

CIMBICIFORMIS STEPHENS
SSP. SEE THYSBE FABRICIUS

CIMICIFUGATA PEARSALL, EUPITHECIA-GEOMETRIDAE
LIFE HISTORY

FOOD PLANTS
ACTAEA RUBRA (AIT.)
CIMICIFUGA RACEMOSA (LINN.)

CINCTARIA FRENCH
SYN. SEE TITEA CRAMER

CINCTIPES HENRY EDWARDS, HALISIDOTA-ARCTIIDAE
SYN. CARINATOR DYAR
AB ATA STRAND
AB META STRAND
LIFE HISTORY
PSYCHE 7:450, 1896
DYAR LARVA, COCOON
GUNDLACH-ENTOM. CUBANA
P. 269, 1886 LARVA
FOOD PLANTS
COCCOLOBA DIVERSIFOLIA JACQ.
COCCOLOBA UVIFERA LINN.
HIBISCUS SP.
TREMA MICRANTHA (LINN.)

CINCTUS TEPPER
SSP. SEE JORULLA WESTWOOD

CINDERELLA SMITH
SYN. SEE LEPUSCULINA GUENEE

CINDERELLA (SMITH), RAPHIA-NOCTUIDAE
LIFE HISTORY
SOUTHERN CALIFORNIA ACAD. SCI.,
BULL. 40:111, 1941 COMSTOCK
EGG, LARVA PL. 10; COCOON
FOOD PLANTS
POPULUS SP. (POPLARS)

CINERARIA WALKER
SYN. SEE CREPUSCULARIA SCHIFFERMULLER

CINERASCENS WALKER
SYN. SEE MANTEO DOUBLEDAY

CINEREA EDWARDS
SSP. SEE PSEUDARGIOLUS BOISDUVAL & LECONTE

CINEREA HARRIS
SYN. SEE CHERSIS HUBNER

CINEREA MAYFIELD
SSP. SEE GRACILIS EDWARDS

CINEREA MORRISON
SYN. SEE CORACIAS GUENEE

CINEREA RILEY
SYN. SEE ANTENNATA WALKER

CINEREA WALKER, CERURA-NOTODONTIDAE
A CINEREOIDES DYAR
B PARADOXA BEHR
C PLACIDA DYAR
D WILEYI DYAR
LIFE HISTORY
AMER. MUS. NAT. HIST., BULL. 10: 431 1898 BEUTEN. LARVA
BOSTON SOC. NAT. HIST., PROC. 24: 556, 1890 PACKARD LARVA
NAT. ACAD. SCI., MEM. 7:274, 1895 PACKARD EGG, LARVA* PL. 36, F. 4; PL. 37 COLOR; PUPA, COCOON
NEW YORK STATE AGRI. EXP. STA. (CORNELL), MEM. 274:236, 1948 FORBES LARVA
NEW YORK STATE MUS., MEM. 8: 559, 1905 FELT EGG* PL. 16, F. 11; LARVA* PL. 16, F. 10; COCOON* PL. 16, F. 12 COLOR
PAPILIO 3:130, 1883 EDWARD-ELLIOTT LARVA
PSYCHE 6:80, 1891 DYAR EGG, LARVA, COCOON
PSYCHE 6:82, 1891 DYAR EGG, LARVA, PUPA, COCOON (CINEREOIDES)
HART-INS. COLLECTIONS ILLINOIS HIGH SCHOOLS P. 46, 1903 LARVA, COCOON
FOOD PLANTS
POPULUS SP. (POPLARS)
SALIX SP. (WILLOWS)

CINEREANA NEUMOEGEN & DYAR
SSP. SEE REVAYANA SCOPOLI

CINEREATA BATES
SYN. SEE NEPTARIA GUENEE

CINEREOFRONS PACKARD
SSP. SEE IPOMOEAE DOUBLEDAY

CINEREOIDES DYAR
SSP. SEE CINEREA WALKER

CINEREOLA (GUENEE), OGDOCONTA-NOCTUIDAE
SYN. ATOMARIA WALKER
LIFE HISTORY
BROOKLYN ENT. SOC., BULL. 18: 155, 1923 CHITTENDEN LARVA
ILLINOIS DEPT. AGRI., TRANS. 18: APPEND:180, 1880 COQUILLETT LARVA
MAINE AGRI. EXP. STA., BULL. 162: 358, 1908 PATCH LARVA (BRIEF)
NORTH AMER. ENT. 1:52, 1880 COQUILLETT LARVA
UNITED STATES DEPT. AGRI. BUR. ENT., BULL. 14:22, 1887 OLD SERIES ASHMEAD LARVA, PUPA
LOC. CIT. BULL. 33:98, 1902 NEW SERIES CHITTENDEN LARVA* FF. 33, B, C; PUPA* F. 23 D
CHITTENDEN-INS. INJ. VEGET. P. 113, 1907 LARVA* F. 68; PUPA* F. 68
CROSBY & LEONARD-MANUAL VEGET. GARDEN INS. P. 83, 1918 LARVA
- - -
INDEX AMER. ECON. ENT., VOL. 1:224; VOL. 7:720
FOOD PLANTS
AMBROSIA SP.
AMBROSIA TRIFIDA LINN.
BETA VULGARIS LINN. (BEET)
HELIANTHUS SP.
HELIANTHUS ANNUUS LINN.
HELIANTHUS TUBEROSUS LINN. (ARTICHOKE)
TRITICUM AESTIVUM LINN. (WHEAT)
XANTHIUM STRUMARIUM LINN.

CINEREOMACULA MORRISON
SYN. SEE OCHROGASTER GUENEE

CINEREOVIRIDIS STRAND
SSP. SEE QUADRILINEATA GROTE

CINEREUS PACKARD
SYN. SEE GUTTIVITTA WALKER

CINERITIA(GROTE), XYLENA-NOCTUIDAE
A MERTENA SMITH
LIFE HISTORY
AMER. NAT. 20:168, 1886
CLARK LIFE HISTORY, LARVA, PUPA
BRITISH COLUMBIA ENT. SOC., PROC. 5:86, 1915 NEW SERIES
G. O. DAY EGG, LARVA
MAINE AGRI. EXP. STA., BULL. 356: 174, 1930 PHIPPS LARVA, PUPA
- - -
INDEX AMER. ECON. ENT., VOL. 2:386
FOOD PLANTS
ALNUS RUBRA BONG. (MERTENA)
AMELANCHIER FLORIDA LINDL. (MERTENA)
HOLODISCUS DISCOLOR(PURSH.) (MERTENA)
PRUNUS EMARGINATA(DOUGL.) (MERTENA)
ROSA GYMNOCARPA NUTT. EX T. &G. (MERTENA)
ROSA NUTKANA PRESL (MERTENA)
SALIX HOOKERIANA BARR(MERTENA)
SALIX SCOULERIANA BARR. (MERTENA)
VACCINIUM ANGUSTIFOLIUM AIT.
VACCINIUM MYRTILLOIDES MICHX.

CINEROSA GROTE & ROBINSON
SYN. SEE OBSCURA FABRICIUS

CINEROSA GROTE, LITHOPHANE-NOCTUIDAE
SYN. GROTEI RILEY
LIFE HISTORY
NEW YORK STATE AGRI. EXP. STA. (CORNELL), BULL. 123, 1896
SLINGERLAND LARVA* PL. 3
KELLOGG-AMER. INS. P. 402, 1908
LARVA* F. 582
- - -
INDEX AMER. ECON. ENT.,
VOL. 1:321 XYLINA
VOL. 2:164 GROTEI
VOL. 6:326
FOOD PLANTS
ACER NEGUNDO LINN.
ACER SACCHARINUM LINN.
MALUS PUMILA MILL. (APPLE)
PRUNUS CERASUS LINN. (CHERRY)
PRUNUS SEROTINA EHRH.
GENERAL FEEDER FRUIT AND FOREST TREES

CINGULATA(FABRICIUS), HERSE-SPHINGIDAE
SYN. AFFINIS GOEZE
SYN. PUNGENS ESCHSHOLTZ
SYN. DRURAEI DONOVAN
AB DECOLORA HENRY EDWARDS
LIFE HISTORY
AMER. ENT. 2:241, 1870
LARVA, PUPA
AMER. JOURN. SCI. 36:294, 1839
HARRIS LARVA, PUPA
AMER. MUS. NAT. HIST., BULL. 7: 299, 1895 BEUTEN LARVA, PUPA
ENT. NEWS 3:217, 1892 LARVA* PUPA* ADULT* (PLATE)
LOC. CIT. 6:95, 1895
DYAR LIFE HISTORY
LOC. CIT. 9:150, 1898 SOULE
DURATION OF EGG STAGE
ENT. SOC. AMERICA, ANNALS 11: 411, 1918 MOSHER PUPAL STRUCTURE* PL. 36, F. 2
FLORIDA ENT. 27:58, 1944
WATSON LARVA
ILLINOIS DEPT. AGRI., TRANS. 18:APPEND:104, 158, 1880
COQUILLETT-MARTEN LARVA
NEW YORK STATE AGRI. EXP. STA., (CORNELL), MEM. 274:186, 1948
FORBES LARVA
NOVITATES ZOOLOGICAE (TRING) 9:SUPPLEMENT P. 11, 1903
ROTHSCHILD & JORDAN
LARVA, PUPA
PHILADELPHIA ACAD. SCI., JOURN. 5:164, 1859 CLEMENS LARVA, PUPA
PHILADELPHIA ENT. SOC., PROC. 3:650, 1864 LINTNER LARVAL VARIATION

Smithsonian Misc. Collections 4:Art. 2:189, 1862 Morris larva, pupa

Abbot & Smith–Lep. Ins. Georgia p. 63, 1797 larva* pupa* Pl. 32 COLOR

Boisduval–Spec. Gen. Lep. Sphing. p. 96, 1874 larva, pupa

Fernald–Sphing. New England p. 36, 1886 larva, pupa

Glover–Illus. North Amer. Ent., 1878 larva, pupa* Pl. 2, f. 14; Pl. 22, f. 3

Gundlach–Ent. Cubana p. 212, 1886 larva, pupa

Harris–Ent. Corresp. p. 282, 1869 larva

Morris–Synop. Lep. North America p. 189, 1862 larva, pupa

Westwood–In Drury's Illust. Exotic Ent. 1:50, 1837 larva, pupa

- - -

Index Amer. Econ. Ent., Vol. 2:174 convolvuli & cingulata Vol. 3:193; Vol. 5:298

FOOD PLANTS

Asimina triloba (Linn.)
Datura stramonium Linn.
Ipomoea batatus (Linn.)

cinis Guenee
syn. see jucunda Hubner

cinna Ploetz
syn. see otho Abbot & Smith

cinnabarina Grote
ssp. see stricta Walker

cinnamomea (Grote & Robinson), Olene–Liparidae

a aridensis Benjamin

LIFE HISTORY

Amer. Mus. Nat. Hist., Bull. 10: 386, 1898 Beuten. cocoon

Brooklyn Ent. Soc., Bull. 22: 226, 1927 Barnes & Benjamin larva

New York State Agri. Exp. Sta. (Cornell), Mem. 274:243, 1948 Forbes larva (aridensis)

Psyche 7:137, 1894 Dyar larva probably this species

FOOD PLANTS

Castanea dentata (Marsh.)
Juglans nigra Linn.
Malus pumila Mill. (apple)
Ostrya virginiana (Mill.)
Prunus domestica Linn. (plum)
Prunus serotina Ehrh.
Quercus sp. (oaks)
Ulmus sp. (elms)
General feeder on forest trees, fruit trees and shrubs

circa Stretch
syn. see parthenice Kirby

circae Henry Edwards
syn. see gaurae Abbot & Smith (proserpinus)

circe Strecker
ssp. see coccinata Grote

circumlucens (Smith), Papaipema–Noctuidae

a vaha Benjamin

LIFE HISTORY

Can. Ent. 48:17, 1916 Bird larva

FOOD PLANTS

Apocynum androsaemifolium Linn. preferred
Cacalia tuberosa Nutt.
Vernonia fasciculata Michx.

cirphidia Hampson
syn. see colorada Smith (Benjaminiola)

cita Grote
syn. see laetifica Lintner

CITIMA GROTE
SSP. SEE LARENTIOIDES GROTE

CITRA NEUMOEGEN & DYAR
SSP. SEE PICTA PACKARD

CITRATA(LINNAEUS), DYSSTROMA-GEOMETRIDAE
A IMMANATA HAWORTH
B MULLEOLATA HULST
C HEWLETTARIA WRIGHT
LIFE HISTORY
NEW YORK STATE AGRI. EXP. STA. (CORNESS), MEM. 274:136, 1948 FORBES LARVA
FOOD PLANTS
RUBUS ALLEGHENIENSIS PORTER
RUBUS VITIFOLIUS C. & S.
GENERAL FEEDER

CITRINA HUBNER
SYN. SEE SOSPETA DRURY

CITRINARIA NEUMOEGEN & DYAR
SSP. SEE VIRGO LINNAEUS

CITRINELLUS(GROTE & ROBINSON), OXYLOS-NOCTUIDAE
LIFE HISTORY
SOUTHERN CALIFORNIA ACAD. SCI., BULL. 30:18, 1931 COMSTOCK LARVA* PUPA*
FOOD PLANTS
CROTON SP.

CITRIOR FRANCLEMONT
SSP. SEE FRAGILIS GUENEE

CIVICA GROTE
SYN. SEE EXTIMA WALKER

CLAMMENHOA DYAR
SYN. SEE BIEDERMANI BARNES & MCDUNNOUGH

CLANDESTINA(HARRIS), SPAELOTIS-NOCTUIDAE
SYN. UNICOLOR WALKER
SYN. NIGRICEPS WALKER
LIFE HISTORY
AMER. MUS. NAT. HIST., BULL. 14: 277, 1901 BEUTEN. LARVA
BROOKLYN ENT. SOC. BULL. 27:86 1932 CRUMB LARVA* PL. 11, A
CAN. ENT. 3:36, 1871 SAUNDERS LARVA
ENT. SOC. AMERICA, ANNALS 23:396, 1930 HAWKINS LARVAL CLAWS* F. 2
ILLINOIS AGRI. EXP. STA., BULL. 95:361, 1904 FORBES LARVA
ILLINOIS DEPT. AGRI., TRANS. 15: 95, 213, 1877 FRENCH-THOMAS LARVA
LOC. CIT. 18:APPEND:135, 1880 MARTEN LARVA
JOURN. ECON. ENT. 8:394, 1915 CRUMB KEY TO LARVA
MAINE AGRI. EXP. STA., BULL. 340:205, 1927 PHIPPS LARVA* F. 30 F
LOC. CIT. 356:163, 1930 PHIPPS LARVA* F. 23 E; PUPA UNICOLOR
MINNESOTA AGRI. EXP. STA., BULL. 61:215, 1899 LUGGER LARVA* F. 159
MISSOURI, REPT. NOX. INS., 1:79, 1869 RILEY LARVA
NEBRASKA AGRI. EXP. STA., RESEARCH BULL. 8:20, 1935 WHELAN LARVA TECHNICAL
NEW YORK STATE AGRI. EXP. STA. (CORNELL), BULL. 104:572, 1895 SLINGERLAND LARVA* PL. VI
NEW YORK STATE AGRI. SOC., REPT. 44:57, 1885 LINTNER LARVA*
NEW YORK STATE MUS., BULL. 6:4, 1888 LINTNER LARVA*
ONTARIO ENT. SOC., ANN. REPT. P. 44, 1879 BOWLES LARVA*
LOC. CIT. P. 50, 1890 FYLES LARVA
UNITED STATES DEPT. AGRI., TECH. BULL. 88:142, 1929 CRUMB LARVA* PL. 9, FF. F, I; PUPA* F. 17 B

United States Dept. Agri. Rept. Entom. p. 293, 1884 Riley larva* Pl. 2, f. 4
Vermont Board Agri., Rept. 2:597, 1876 Perkins larva
Chittenden--Ins. Inj. Veg. p. 51, 1907 larva* f. 25
Crosby & Leonard-Manual Veg. Garden Ins. p. 264, 1918 larva* f. 155
Harris-Ins. Inj. Veg. 1st edit. p. 325, 1841 larva
Hart-Ins. Collections Illinois High Schools p. 30, 1903 larva* f. 40
O'Kane-Inj. Ins. p. 140, 1912 larva f. 132
Packard-Half Hours with Ins. p. 23, 1877 larva*
Packard-Ins. Inj. West p. 717, 1877 larva
Saunders-Ins. Inj. Fruits p. 108, 1883 larva*
Saunders-Loc. cit. p. 108, 1889 larva* f. 106

- - -

Index Amer. Econ. Ent.,
Vol. 1:220 Noctua
Vol. 2:11 Agrotis unicolor
Vol. 4:14 Agrotis unicolor
Vol. 5:18 Agrotis unicolor
Vol. 6:676; Vol. 7:24

FOOD PLANTS

Acer sp. (maples)
Acer negundo Linn.
Acer saccharinum Linn.
Althea rosea Cav.
Apium graveolens Linn. (celery)
Artemisia sp.
Asparagus officinalis Linn. (asparagus)
Brassica oleracea Linn. (cabbage)
Brassica oleracea v. botyrtis DC. (cauliflower)
Cichorium endiva Linn.
Cichorium intybus Linn.
Cucurbita pepo Linn. (pumpkin)
Fagopyrum sagittatum Gilib (buckwheat)
Fragaria chiloensis (Linn.) (strawberry)
Lactuca sativa Linn. (lettuce)
Malus pumila Mill. (apple)
Nicotiana tabacum Linn. (tobacco)
Phaseolus vulgaris Linn. (bean)
Pinus sp. (pines)
Plantago sp.
Prunus persica (Linn.) (peach)
Ribes lacustre (Pers.) (gooseberry)
Ribes sativum Syme (currant)
Rubus idaeus v. strigosus (Michx.) (raspberry)
Trifolium sp. (clovers)
Triticum aestivum Linn. (wheat)
Vaccinium sp.
Vitis sp. (grapes)
Zea mays Linn. (corn)
Forest and fruit trees, garden crops, grass, low plants, most anything

clandestina Walker
syn. see septentrionalis Walker (Gluphisia)

clappiana Holland, Gnophaela-Pericopidae
LIFE HISTORY
Southern California Acad. Sci., Bull. 36:120, 1937 Comstock larva* Pl. 53; pupa* Pl. 54
FOOD PLANTS

clappii Maynard
ssp. see lisa Boisduval & Leconte

clara Beutenmuller
ssp. see relicta Walker

clara Harvey
ssp. see nucicolora Guenee

clara Henry Edwards
ssp. see heteronea Boisduval

CLARA PLOETZ
SYN. SEE BRETTUS BOISDUVAL & LECONTE

CLARESCENS GUENEE, ACRONICTA-NOCTUIDAE
AB CENTRIFERRUGINEA STRAND
LIFE HISTORY
NEW YORK STATE MUS., MEM. 8: 733, 1905 FELT LARVA (BRIEF)
HARRIS-ENT. CORRESP. P. 313, 1869 LARVA* PL. 4 AS PRUNI
FOOD PLANTS
BETULA ALLEGHANIENSIS BRITT.
MALUS PUMILA MILL. (APPLE)
PRUNUS SEROTINA EHRH.
SORBUS AMERICANA MARSH.

CLARISSIMA BEUTENMULLER
SSP. SEE BRISEIS EDWARDS

CLARKEI BENJAMIN
SSP. SEE PERQUIRITATA MORRISON

CLARKI BARNES & BENJAMIN
SSP. SEE LUCIDUS BOISDUVAL

CLARKI CHERMOCK
SSP. SEE BAIRDII EDWARDS

CLARKIAE (BOISDUVAL), PROSERPINUS-SPHINGIDAE
SYN. VICTORIA GROTE
LIFE HISTORY

FOOD PLANTS
CLARKIA RHOMBOIDEA DOUGL. EX HOOK
GAYOPHYTUM DIFFUSUM T. & G.

CLARKII FRANCK
SSP. SEE JAMAICENSIS DRURY

CLARUS (CRAMER), EPARGYREUS-HESPERIIDAE
TITYRUS FABRICIUS ?
AB OBLITERATUS SCUDDER
AB SMYTHI WILLIAMS
LIFE HISTORY
AMER. ENT. SOC., TRANS. 37:192, 1911 SKINNER LARVA, CHRYSALIS
AMER. MUS. NAT. HIST., BULL. 5: 302, 1893 BEUTEN. LARVA, CHRYSALIS
BUFFALO SOC. NAT. SCI., BULL. 19:30, 1939 WILD LARVA
ILLINOIS DEPT. AGRI., TRANS. 18: APPEND:98,153, 1880 COQUILLETT-MIDDLETON EGG, LARVA
LOC. CIT. 15:163, 1877 FRENCH LARVA
MICHIGAN STATE COLLEGE, SPEC. BULL. 243:45, 1933 MCDANIEL LIFE HISTORY LARVA*
MONTANA UNIV., BULL. 30:144, 1906 ELROD EGG, LARVA
NATURALISTE CANADIEN P. 45, 1928 HUARD LARVA
NEW YORK STATE MUS., BULL. 198: 71, 1918 FELT LARVA
NEW YORK STATE MUS., MEM. 8: 556, 1905 FELT LARVA
NEW YORK STATE, REPT. INS. 5:330 AND P. 52, SECTION 330, 1858 FITCH LARVA
SMITHSONIAN MISC. COLLECTIONS 4: ART. 2:112, 1862 MORRIS LARVA
UNITED STATES NAT. MUS., BULL. 157:203, 1932 A. CLARK LARVA PUPA
UNITED STATES DEPT. AGRI., MISC. PUB. 657:351,371, 1950 CRAIGHEAD LARVA* F. 70 G
ABBOT & SMITH-LEP. INS. GEORGIA 1:37, 1797 LARVA*, CHRYSALIS* PL. 19 COLOR
BOISDUVAL & LECONTE-HIST. LEP. NORTH AMERICA 1833 LARVA* CHRYSALIS* PL. 72 COLOR
COMSTOCK, J. A.-BUTTER. CALIF. P. 203, 1927 EGG, LARVA* (A. 65); CHRYSALIS
COMSTOCK-HOW TO KNOW BUTTER. P. 293, 1904 LARVA* CHRYSALIS* PL. 43, FF. 43,48 COLOR
CRAGEN-OUR INS. FRIENDS & FOES P. 104, 1899 LARVA* F. 40

Edwards-Kingsley's Stand. Nat. Hist. 2:475, 1885 larva
Emmons-Nat. Hist. New York 5: 215, 1854 larva* chrysalis* Pl. 832 COLOR
French-Butter. East. United States p. 376, 1886 life history, chrysalis*
Glover-Illus. North Amer. Ent. 1878 larva, pupa Pl. 1, f. 15
Gosse-Letters from Alabama p. 85, 1859 larva, chrysalis, nest
Harris-Ins. Inj. Veg. 1st edit: 224, 1841 larva, chrysalis
Harris-Loc. cit. 2nd edit: 244, 1852 larva
Harris-Loc. cit. Flint edit:311, 1862 larva* chrysalis*
Holland-Butterfly Book 1905 larva* chrysalis* Pl. 2, ff. 30,31,33; Pl. 6, ff. 22, 25,26 COLOR
Kellogg-Amer. Ins. p. 442, 1908 larva
Maynard-Butter. New England p. 52, 1886 larva, chrysalis
Morris Synop. Lep. North America p. 112, 1862 larva
Packard-Ins. Inj. Forest Trees p. 100, 1881 larva
Scudder-Butter. East. United States & Canada pp. 1402-1410, 1889 life history* Pl. 73, f. 8, Pl. 76, ff. 30,31,33,35; Pl. 80, ff. 19-22; Pl. 85, ff. 22, 25,26; Pl. 86, f. 26; Pl. 87, ff. 1, 11,12; Pl. 82, ff. 9-11
Scudder-Butter. Their Structure pp. 107,256, 1881 larva* chrysalis* cocoon* nest*
Scudder-Everyday Butter. p. 191, 1899 egg, larva, chrysalis
Scudder-Guide to Common Butter. p. 155, 1893 larva, chrysalis
Vance-Life Among Butter. p. 52, 1925 larva
Index Amer. Econ. Ent., (as tityrus) Vol. 1:117; Vol. 2: 130,139 (Eudamus) Vol. 5:220; Vol. 7:817

FOOD PLANTS

Amorpha fruticosa Linn.
Amphicarpa bracteata (Linn.)
Apios americana Medic.
Astragalus sp.
Desmodium canadensis (Linn.)
Desmodium marilandicum (Linn.)
Desmodium nudiflorum (Linn.)
Gleditsia triacanthos Linn.
Lathyrus palustris Linn.
Leguminosae
Lespedeza capitata Michx.
Lotus grandiflorus (Benth.)
Purerafia lobata (Willd.)
Robinia hispida Linn.
Robinia neomexicana Gray
Robinia pseudo-acacia Linn.
Robinia viscosa Linn.
Wisteria frutescens (Linn.)

- - -

claudia (Cramer), Euptoieta-Nymphalidae

syn. daunius Herbst
syn. marianne Scudder
syn. fumosa Field
ab albaclaudia Field
ab dodgei Gunder

LIFE HISTORY

Amer. Mus. Nat. Hist., Bull. 5: 258, 1893 Beuten. larva chrysalis
Argentine Soc. Cient., Ann. Vol. 78:171, 1914 Giocomoelli larva
Can. Ent. 2:163, 1870 W. H. Edwards larva
Loc. cit. 12:231, 1880 W. H. Edwards egg, larva, chrysalis
Illinois Dept. Agri., Trans. 18: Append:80, 1880 Middleton larva
Iowa Acad. Sci., Biological Survey 1936 Jaques How to Know Ins. No. 1:91 larva*

New York State Agri. Soc., Trans. p. 105, 1855 Glover larva, chrysalis
Smithsonian Misc. Collections 4:Art. 2:44, 1862 Morris larva, chrysalis
United States Dept. Agri., Bur. Ent. Bull. 27:81, 1901 new series Chittenden larva* f. 22b; chrysalis* f. 22 c
Boisduval & Leconte–Hist. Lep. North America p. 153, 1833 larva* chrysalis* Pl. 44
Comstock, J. A.–Butter. Calif. p. 80, 1927 egg, larva, chrysalis
Fernald–Butter. Maine p. 43, 1884 larva, chrysalis
French–Butter. East. United States p. 166, 1886 life history
Holland–Butterfly Book 1905 pp. 99, 100 3rd edit. pp. 80, 81, 1931 life history
Maynard Butter. New England p. 21, 1886 larva
Mead–Rept. Lep. Wheeler Exped. 5:751, 1875 egg, larva
Morris–Synop. Lep. North America p. 44, 1862 larva, chrysalis
Scudder–Butter. East. United States & Canada pp. 523–527, 1889 life history* Pl. 64, f. 23; Pl. 67, f. 8; Pl. 75, ff. 6, 7; Pl. 79, f. 1; Pl. 84, f. 89
Scudder–Everyday Butter. p. 357, 1899 egg, larva, chrysalis
Scudder–Guide to Common Butter. p. 81, 1893 larva, chrysalis

– – –

Index Amer. Econ. Ent., Vol. 5: 240; Vol. 6:282; Vol. 7:367

FOOD PLANTS

Cynoglossum sp.
Desmodium paniculatum (Linn.)
Linum sulcatum Riddell
Passiflora caerulea Linn.
Passiflora incarnata Linn.
Podophyllum peltatum Linn.
Portulaca sp.
Portulaca olercaea Linn.
Sedum sp.
Sedum purpureum (Linn.)
Turnera ulmifolia Linn.
Viola sp.
Viola tricolor Linn.

<u>claudianus</u> Stichel
ssp. see <u>clodius</u> Menetries

<u>clavipes</u> Rothschild & Jordan
ssp. see <u>tantalus</u> Linnaeus

<u>claviplene</u> Grote
syn. see <u>detracta</u> Walker

<u>clemataria</u> (Abbot & Smith), <u>Abbottana</u>–Geometridae
syn. <u>transducens</u> Walker
syn. <u>transferens</u> Walker
syn. <u>transfingens</u> Walker

LIFE HISTORY

Amer. Mus. Nat. Hist., Bull. 4: 79, 1892 Beuten. larva
Maine Agri. Exp. Sta., Bull. 259: 48, 1917 Mosher larva, pupa
Massachusetts Agri. Exp. Sta., Bull. 239:44, 1928 Franklin larva* Pl. 3, f. 9 color; pupa* Pl. 9, f. 10 color
Naturaliste Canadien p. 327, 1935 Daviault larva, pupa
New York State Agri. Exp. Sta. (Cornell), Mem. 274:109, 1948 Forbes egg, larva
New York Ent. Soc., Journ. 53: 201, 1945 Forbes pupa
New York State Mus., Mem. 8: 721, 1905 Felt larva (brief)
Nova Scotia Ent. Soc., Proc. 4: 82, 1918 Payne life hist* all stages Pl. 6

PSYCHE 9:142, 1900 DYAR
EGG, LARVA
UNITED STATES DEPT. AGRI.,
MISC. PUB. 657:363,437, 1950
CRAIGHEAD LARVA
ABBOT & SMITH-LEP. INS. GEORGIA
2:201, 1797 LARVA* PL. 101
COLOR
GUMPPENBERG-SYSTEMA
GEOMETRARUM 7:415, 1895
LARVA, PUPA
PACKARD-MONOGRAPH GEOMETRIDAE
P. 562, 1876 LARVA
- - -
INDEX AMER. ECON. ENT., VOL. 6:1

FOOD PLANTS
ACER SP.
CARYA SP. (HICKORIES)
CASTANEA DENTATA (MARSH.)
CELASTRUS SCANDENS LINN.
CLEMATIS SP.
CLEMATIS RETICULATA WALT.
CORNUS FLORIDA LINN.
MALUS PUMILA MILL. (APPLE)
PRUNUS SEROTINA EHRH.
PYRRHOPAPPUS CAROLINIANUS (WALT.) (IN CAPTIVITY)
PYRUS COMMUNIS LINN. (PEAR)
QUERCUS SP. (OAKS)
QUERCUS ALBA LINN.
QUERCUS VELUTINA LAM.
QUERCUS VIRGINIANA MILL.
SALIX SP. (WILLOWS)
SASSAFRAS ALBIDUM V. MOLLE (RAF.)
ULMUS SP. (ELMS)
VACCINIUM MACROCARPON AIT.

CLEMENCEI COMSTOCK
SSP. SEE ADIASTE BEHR

CLEMENTINA GUMPPENBERG
SSP. SEE EFFECTA WALKER

CLEOMES BOISDUVAL & LECONTE
SYN. SEE MONUSTE LINNAEUS

CLETHRAE (DYAR), CELAMA-NOLIDAE
LIFE HISTORY
CAN. ENT. 33:221, 1901 DODGE
EGG, LARVA, PUPA
WASHINGTON ENT. SOC., PROC.
1:88, 1887 RILEY LARVAL
SCENT GLANDS
FOOD PLANTS
GLEDITSIA TRIACANTHOS LINN.
MALUS PUMILA MILL. (APPLE)
PRUNUS ILICIVOLIA (NUTT. EX H&A)
ULMUS SP. (ELMS)

CLINTONI GROTE & ROBINSON
SYN. SEE BASIFLAVA PACKARD

CLIO BARNES & MCDUNNOUGH
SSP. SEE ELECTRA WRIGHT

CLIO EDWARDS
SSP. SEE EURYNOME EDWARDS

CLIO (PACKARD), ECTYPIA-ARCTIIDAE
AB THONA STRECKER
A JESSICA BARNES
LIFE HISTORY
PAPILIO 2:187, 1882 BEHR "LARVA LIKE THAT OF HALISIDOTA EDWARDSI"
SOUTHERN CALIFORNIA ACAD. SCI.,
BULL. 23:136, 1935 COMSTOCK-DAMMERS EGG; LARVA* PL. 30;
PUPA* PL. 31
FOOD PLANTS
APOCYNUM ANDROSAEMIFOLIUM LINN.
ASCLEPIAS MEXICANA CAV.
SARCOSTEMMA HIRTELLUM (GRAY)

CLITOSALIS WALKER
SYN. SEE FACTIOSALIS WALKER

CLITUS (EDWARDS), ERYNNIS-HESPERIIDAE
SYN. MAESTUS GODMAN & SALVIN
LIFE HISTORY
FOOD PLANTS
LOTUS SP.

CLODIANA GROTE
SYN. SEE AENEIPENNIS GROTE

CLODIUS MENETRIES, PARNASSIUS-PAPILIONIDAE

FOR THE MANY SYNS. AND SSPS. OF THIS SP. SEE MCDUNNOUGH'S LIST

LIFE HISTORY
CAN. ENT. 11:142, 1879
EDWARDS EGG, LARVA
LOC. CIT. 17:64, 1885
EDWARDS EGG, LARVA
LONDON ZOOL. SOC., PROC. P. 51, JAN., 1886
ELWES EGG, LARVA
COMSTOCK-BUTTER, CALIFORNIA P. 25, 1927 EGG, AND EGG OF BALDUR

FOOD PLANTS
RUBUS SP.
SAXIFRAGA SP.
SEDUM SP.
VACCINIUM SP.?
VITIS CALIFORNICA BENTH.
EGGS LAID ON MOST ANYTHING

CLONIASALIS WALKER
SYN. SEE CARADRINALIS GUENEE

CLOTHILDE EDWARDS
SYN. SEE LAETA EDWARDS

CLOTHO FABRICIUS
SEE LABRUSCAE LINNAEUS

CLYMENA(CRAMER), DIAETHRIA-NYMPHALIDAE

LIFE HISTORY
AMER. ACAD. SCI., PROC. 27:241, 1892 SCUDDER LIFE HISTORY

FOOD PLANTS
TREMA MICRANTHA(LINN.)

CLYMENE(BROWN), HAPLOA-ARCTIIDAE
SYN. INTERRUPTOMARGINATA BEAUVOIS
SYN. COMMA WALKER

LIFE HISTORY
AMER. MUS. NAT. HIST., BULL. 10:366, 1898 BEUTEN. LARVA
CAN. ENT. 10:84, 1878
SIEWERS LARVA
DAVENPORT ACAD. SCI., PROC. 2:276, 2878 STRECKER LARVA (INTERRUPTOMARGINATA BEAUVOIS)
MISSOURI, REPT. ON INS. 3:134 1871 RILEY LARVA(BRIEF)
PSYCHE 13:32, 1906 FOSTER
TABLE OF LARVAL DIFFERENCES IN GENUS HAPLOA

FOOD PLANTS
PRUNUS PERSICA(LINN.)(PEACH)
EUPATORIUM SP.
EUPATORIUM RUGOSUM HOUTT(LINN)
QUERCUS SP. (OAKS)
SALIX SP. (WILLOWS)
LOW PLANTS

CLYTON(BOISDUVAL & LECONTE), ASTEROCAMPA-NYMPHALIDAE
SYN. OCELLATA EDWARDS
AB NIG STRECKER
FORM PROSPERINA SCUDDER
A FLORA EDWARDS
B TEXANA SKINNER
C SUBPALLIDA BARNES & MCDUNNOUGH

LIFE HISTORY
BROOKLYN ENT. SOC., BULL 6:115, 1884 SMITH LARVA, CHRYSALIS
CAN. ENT. 13:81, 1881
W. H. EDWARDS EGG, LARVA CHRYSALIS
ILLINOIS DEPT. AGRI., TRANS. 15: 155, 1877 FRENCH LARVA
LOC. CIT. 18:APPEND:89, 1880
MIDDLETON EGG* CHRYSALIS*
MISSOURI, REPT. ON INS. 6:140-150, 1874 RILEY EGG* LARVA*
NEW YORK STATE MUS., MEM. 8: 725, 1905 FELT LARVA(BRIEF)
NEW YORK STATE, REPT. ON INS. 3:

380,1856 Fitch larva
also p.62, sec.83
Saint Louis Acad.Sci.,Trans.
p.199,1873 Riley egg*
larva* chrysalis*
Smithsonian Misc.collections
4:art.2:68,1962 Morris
larva, chrysalis
United States Dept. Agri.,
Misc.Pub.657:370,1950
Craighead larva
Boisduval & Leconte-Hist.Lep.
North America p.208,1833
larva* chrysalis* Pl.56
COLOR
Comstock-How to Know Butter.
p.176,1904 larva
Edwards-Butter,North America
2nd series 1876 life history
Pl.Apatura No.2 COLOR
Loc.cit.3rd series 1897 life
history* Pl.Apatura 1,
ff.A-H of <u>flora</u>
French-Butter.East.United
States pp.219,222,1886
egg* larva* chrysalis*
Gundlach-Ent.Cubana p.63,
1881 larva, chrysalis
(<u>idyja</u> Hubner)
Hart-Ins.Collections Illinois
High Schools p.11,1903
egg* larva* chrysalis*
Herrick-Inx.Enemies Shade
Trees p.108,1935 egg*
larva* chrysalis* f.76
Holland-Butterfly Book 1905
larva* chrysalis* Pl.3,
f.20;Pl.4,ff.15-17 COLOR
Morris-Synop.Lep.North America
p.68, 1862 larva, chrysalis
Scudder-Butter,East United
States & Canada pp.244-247,
1889 life history* Pl.64,
ff.6,7;Pl.74,f.20;Pl.87,
f.7; Pl.83, ff.15-17
Scudder-Butter,Their Structure,
etc.p.262,1881 egg* larvae*
chrysalis*
Scudder-Guide to common Butter.
p.105,1893 larva,chrysalis
Index Amer. Econ. Ent.
Vol.3:84 Chloripe
Vol.7:112

FOOD PLANTS
Aristolochia sp.(?)
Celtin laevigata Willd.
Celtis occidentalis Linn.
Prunus americana Marsh.
Prunus ilicifolia(Nutt.ex
H.&A.)

<u>c-nigrum</u>(Linnaeus),<u>Graphiphora</u>-
Noctuidae

LIFE H
LIFE HISTORY
Amer.Mus.Nat.Hist.,Bull.14:
274,1901 Beuten. larva
Brooklyn Ent.Soc.,Bull.22:48,
1927 Crumb mature larva*
Pl.4,G;Pl.5,C,G,R
Loc.cit.27:87, 1932
Crumb larva
Ent.Soc.America,Annals
23:396,1930
Hawkins larval claws* f.2
Illinois Agri.Exp.Sta.,Bull.
95:355, 1904 Forbes larva*
Illinois Dept.Agri.,Trans.15:
89,202,1877 French-Thomas
larva
Loc.cit.18:Append:132,184,1880
Coquillett-Marten larva
Journ.Econ.Ent.8:393,1915
Crumb key to larva
Maine Agri.Exp.Sta.,Bull.356:
153,1930 Phipps egg;
larva* f.23B; pupa
Massachusetts Agri.Exp.Sta.,
Bull.239:28,1928 Franklin
larva* Pl.2,ff.2A,2B
COLOR
Minnesota Agri.Exp.Sta.,Bull.
84,Art.5,1903 Washburn
larva*f.6 COLOR
Nebraska Agri.Exp.Sta.,Re-
search Bull.81:20,1935
Whelan larva(Tech.des.)

PAPILIO 3:133,1883 EDWARDS
LARVA
UNITED STATES DEPT. AGRI., BUR. ENT. BULL. NEW SERIES 27:55, 1901 EGG, LARVA* PL. 18B; PUPA
UNITED STATES DEPT. AGRI. BUR. ENT., BULL. NEW SERIES 43:31, 1903 CHITTENDEN LARVA* F. 27B
UNITED STATES DEPT. AGRI., CIRC. 270:33, 1933 NEWCOMER LARVA* F. 36 B, C NO DESCR.
UNITED STATES DEPT. AGRI., FARMER'S BULL. 1494:7, 1926 CRUMB LARVA* F. 8, B, C
UNITED STATES DEPT. AGRI., TECH. BULL 88:137, 1929 CRUMB EGG, LARVA* PL. 9 A, B, C, G; PUPA* F. 17A
CHITTENDEN-INS. INJ. VEG. P. 211, 1907 LARVA* F. 134
CROSBY & LEONARD-MANUAL VEG. GARDEN INS. P. 262, 1918 EGG, LARVA* F. 153; PUPA
HART-INS. COLLECTIONS ILLINOIS HIGH SCHOOLS P. 29, 1903 LARVA* F. 39
KIRBY-EUROPEAN BUTTER. & MOTHS P. 192, 1882 LARVA
METCALF & FLINT-DESTR. & USEFUL INS. P. 349, 1939 LARVA* F. 203
STAINTON-BRITISH BUTTER. & MOTHS 1:234, 1857 LARVA
STEPHENS-ILLUS. BRITISH ENT. HAUST. 2:136, 1829 LARVA, PUPA
INDES AMER. ECON. ENT.
VOL. 1:221 NOCTUA;
VOL 2:11; VOL 3:14; VOL. 4:14;
VOL. 5:277 GRAPHOLITHA;
VOL 6:23 AMATHES;
VOL. 7:24 AMATHES

FOOD PLANTS

ACER SP. (MAPLES)
ALIUM CEPA LINN. (ONION)
APIUM GRAVEOLENS LINN. (CELERY)
ARONIA MELANOCARPA (MICHX.)
AVENA SATIVA LINN. (OATS)
BETA VULGARIS LINN. (BEET)
BRASSICA OLERACEA LINN. (CABBAGE)
B. OLERACEA V. BOTYRTIS D.C. (CAULIFLOWER)
BRASSICA RAPA LINN. (TURNIP)
CICHORIUM INTYBUS LINN.
CIRSIUM ARVENSE (LINN.)
COMPTONIA PEREGRINA (LINN.)
DAUCUS CAROTA V. SATIVUS HOFF. (CARROT)
HELIANTHUS SP.
HORDEUM VULGARA LINN. (BARLEY)
LACTUCA SATIVA LINN. (LETTUCE)
LAMIUM SP.
LINUM SULCATUM RIDD. (FLAX)
LOBELIA SP.
MALUS PUMILA MILL. (APPLE)
MYOSOTIS SP.
NICOTIANA TABACUM LINN. (TOBACCO)
PISUM SATIVUM LINN. (PEAS)
PLANTAGO SP.
PRIMULA SP.
PYRUS COMMUNIS LINN. (PEAR)
RHEUM RHAPONTICUM LINN. (RHUBARB)
RIBES LACUSTRE (PERS.) (GOOSEBERRY)
RIBES SATIVUM SYME. (CURRANT)
RUMEX SP.
SENECIO VULGARIS LINN.
SOLANUM LYCOPERSICUM LINN.
SOLANUM TUBEROSUM LINN. (POTATO)
SOLIDAGO SP.
SPIRAEA LATIFOLIA (AIT.)
STELLARIS MEDIA (LINN.)
TARAXACUM OFFICINALE WEBER
THALICTRUM SP.
TRIFOLIUM SP.
TRIFOLIUM PRATENSE LINN.
TRITICUM AESTIVUM LINN. (WHEAT)
VACCINIUM SP.
VACCINIUM ANGUSTIFOLIUM AIT.
VACCINIUM MACROCARPON AIT. (CRANBERRY)
VACCINIUM MYRTILLOIDES MICHX.
VERBASCUM SP.
VIBURNUM SP.
VIOLA SP.
VIOLA TRICOLOR LINN.

ZEA MAYS LINN. (CORN) FERNS, FIELD CROPS, FRUITS & FRUIT TREES, LOW PLANTS, MOST ANYTHING, SHRUBS, VEGETABLES, AND GRASSES

CNOTUS HUBNER
SSP. SEE MYRON CRAMER

COACHELLA HILL, LITOPROSOPUS-NOCTUIDAE

LIFE HISTORY
SOUTHERN CALIFORNIA ACAD. SCI., BULL. 32:117, 1933 COMSTOCK EGG*PL. 56; LARVA* PL. 57; PUPA* PL. 58; COCOON*PL. 59

FOOD PLANTS
WASHINGTONIA FILIFERA (LINDL.)

COAGULATA GUENEE, EUPITHECIA-GEOMETRIDAE
SYN. FUMOSA HULST
SYN. ABSYNTHIATA GUENEE

LIFE HISTORY
EAN. ENT. 9:63, 1877 GOODELL LARVA, PUPA (ABSYNTHIATA)
CREW-NEWMAN'S BUTTER. & MOTHS P. 136, 1869 LARVA
KIRBY-EUROPEAN BUTTER. MOTHS P. 403, 1882 LARVA
PACKARD-MONOGRAPH GEOMETRIDAE P. 50, 1876 LARVA, PUPA
STAINTON-BRITISH BUTTER. & MOTHS II:69, 1859 LARVA

FOOD PLANTS
ARTEMISIA SP.
COMPOSITAE SP. (FLOWERS)
SENECIO SP.
SOLIDAGO SP.

COARCTATA HARRIS
SSP. SEE AUTUMNATA GUENEE

COBETA SMITH
SSP. SEE LUBRICALIS GEYER

COCANIDICIDES VERITY
SYN. SEE NASTES BOISDUVAL

COCCINATA GROTE, CATOCALA-NOCTUIDAE
AB CHIQUITA BARTSCH
FORM CIRCE STRECKER
A SINUOSA GROTE

LIFE HISTORY
AMER. MUS. NAT. HIST., BULL. 38, 170, 1918 BARNES & McDUNNOUGH EGG, LARVA
AMER. MUS. NAT. HIST., MEM. NEW SERIES 3, PART 1, 1918 BARNES & McDUNNOUGH LARVA* PL. 14, F. 1; PL. 15, FF. 20, 33; PL. 17, F. 11 COLOR
BROOKLYN ENT. SOC., BULL. 7: 42, 1884, HULST LARVA
CAN. ENT. 47:336, 1915 ROWLEY LARVA
NEW YORK STATE MUS., MEM. 8:705, 1905 FELT LARVA (BRIEF)
PAPILIO 1:56, 1881 COQUILLETT LARVA

FOOD PLANTS
QUERCUS SP.
QUERCUS MACROCARPA MICHX.

COSHISE DYAR
SSP. SEE INTEGERRIMA GROTE & ROBINSON

COCHLES SCUDDER
SYN. SEE PANOQUIN SCUDDER

COCHRANI RILEY
SNY. SEE MESSORIA HARRIS

COCKLEI GUNDER
SSP. SEE EURYMEDON LUCAS

COCYTA CRAMER
SNY. SEE THAROS DRURY

COELEBS GROTE, CATOCALA-NOCTUIDAE
FORM PHOEBE HENRY EDWARDS

FOOD PLANTS
MYRICA GALE LINN.

COELEBS MARTYN
SYN. SEE ARGE DRURY

COENIA(HUBNER), JUNONIA-NYMPHALIDAE
AB SCHRADERI GUNDER
AB WEIDENHAMMERI POLACEK
AB WILHELMI GUNDER
FORM RUBROSUFFUSA FIELD
A NIGROSUFFUSA BARNES & MCDUNNOUGH
LAVINIA CRAMER ACCORDING TO FORBES IS THE TRUE SPECIFIC NAME AND COENIA HUBNER IS A SUBSPECIES(JOURN. N. Y. ENT. SOC. 36:305, 1928)

LIFE HISTORY
AMER. MUS. NAT. HIST., BULL. 5: 270, 1893 BEUTEN. LARVA
CALIFORNIA ACAD. SCI., PROC. 5: 171, 1873 HY. EDWARDS LARVA, CHRYSALIS
CAN. ENT. 23:187, 1891 DYAR EGG, LARVA, CHRYSALIS
ILLINOIS DEPT. AGRI., TRANS. 18: APPEND:87, 1880 MIDDLETON LARVA
NATURALISTE CANADIEN P. 185, 1928 HUARD LARVA, CHRYSALIS
ONTARIO ENT. SOC., ANNUAL REPT. P. 40, 1880 SAUNDERS LARVA (BRIEF)
SMITHSONIAN MISC, COLLECTIONS R, ART. 2:62, 1862 MORRIS LARVA, CHRYSALIS
ABBOT & SMITH-LEP. INS. GEORGIA 1:15, 1797 LARVA* PL. 8 COLOR
BOISDUVAL & LECONTE-HIST. LEP. NORTH AMERICA P. 182, 1833 LARVA* CHRYSALIS* PL. 49 COLOR
COMSTOCK, J. A.-BUTTER, CALIFORNIA P. 136, 1927 EGG, LARVA, CHRYSALIS
COMSTOCK, J. H.-HOW KNOW BUTTER. P. 160, 1904 LARVA
ESSIG-INS. WEST. NORTH AMERICA P. 647, 1926 EGG, LARVA CHRYSALIS
FERNALD-BUTTER. MAINE P. 60, 1884 LARVA
FRENCH-BUTTER. EAST. UNITED SATES P. 201, 1886 LARVA, CHRYSALIS
GLOVER-ILLUS. NORTH AMER. ENT. 1878 LARVA, PUPA* PL. 2, F. 5
HARRIS-INS. INJ. VEG. FLINT EDIT. P. 293, 1862 LARVA, CHRYSALIS
HOLLAND-BUTTERFLY BOOK 1905 P. 173 LARVA* CHRYSALIS* PL. 3, FF. 29, 30; PL. 4, FF. 56, 1931 LIFE HISTORY; LARVA* PL. 3; CHRYSALIS* PL. 4
KELLOGG-AMER. INS. P. 455, 1908 LARVA
MAYNARD-BUTTER, NEW ENGLAND P. 20, 1886 LARVA
SCUDDER-BUTTER. EAST. UNITED STATES & CANADA PP. 497-498, 500, 1889 LIFE HISTORY* PL. 64, F. 14; PL. 74, FF. 29, 30; PL. 78, FF. 64, 66; PL. 83, FF. 56, 57, 65-67
SCUDDER-EVERYDAY BUTTER. P. 377, 1899 EGG, LARVA
SCUDDER-GUIDE TO COMMON BUTTER. P. 82, 1893 LARVA CHRYSALIS
VANCE-LIFE AMONG BUTTER. P. 42, 1925 EGG, LARVA, CHRYSALIS

INDES AMER. ECON. ENT. VOL. 6:436; VOL. 7:584

FOOD PLANTS
AGALINIS PURPUREA(LINN.)
ANTIRRHINUM SP.
ASTER SP.
DIGITALIS SP.
HEMIZONIA SP.
LANTANA INVOLUCRATA V. ODORATA (LINN.)
LINARIA CANADENSIS(LINN.)

LINARIA VULGARIS HILL
LUDWIGIA ALTERNIFOLIA LINN.
MIMULUS SP.
PLANTAGO LANCEOLATA LINN.
PLANTAGO VIRGINICA LINN.
SEDUM SP.

COFAQUI (STRECKER), MEGATHYMUS-MEGATHYMIDAE

LIFE HISTORY
ENT. NEWS 27:372, 1916
BONIWELL LOCATION OF PUPA
LOC. CIT. 28:232, 1917 SKINNER
PUPA COMPARED WITH THAT OF YUCCA

FOOD PLANTS

COGITATA SMITH
SSP. SEE DUBITANS WALKER

COGNATARIA GUENEE, AMPHIDASIS-GEOMETRIDAE
AB SWETTARIA BARNES & MCDUNNOUGH
A FORTITARIA BARNES & MCDUNNOUGH
B CONTRASTA BARNES & MCDUNNOUGH

LIFE HISTORY

BROOKLYN ENT. SOX., BULL. 6:48
1883 CRAMER EGG
CAN. ENT. 3:11, 1871
BOWLES LARVA, PUPA
LOC. CIT. 10:67, 1878
GOODELL LARVA, PUPA
MINNESOTA AGRI. EXP. STA., BULL,
61:247, 1899 LUGGER LARVA
NEW YORK ENT. SOC., JOURN. 53:
183, 1945 FORBES PUPA
NEW YORK STATE AGRI. EXP. STA.
(CORNNELL), MEM. 56:1076,
1922 WELLHOUSE LARVA
(BRIEF)
NEW YORK STATE AGRI. EXP. STA.
(CORNELL), MEM. 274:65, 1948
FORBES EGG, LARVA, PUPA
NEW YORK STATE CABIN. NAT. HIST.,
ANNUAL REPT. NO. 26:166, 1874
LINTNER LARVA
NEW YORK STATE MUS. MEM. 8:
728, 1905 FELT LARVA (BRIEF)
NEW YORK STATE, REPT. ENT. 2:99,
1885 LINTNER LARVA*
ONTARIO ENT. SOC., ANNUAL REPT.
P. 39, 1871 SAUNDERS LARVA
PUPA
PAPILIO 3:136, 1883
EDWARDS-ELLIOTT LARVA
PSYCHE 4:269, 1885
DIMMOCK LARVA (BRIEF)
PSYCHE 10:78, 1903
DYAR EGG, LARVA
UNITED STATES DEPT. AGRI., BUR.
ENT. BULL. OLD SERIES 13:30
1887 PACKARD LARVA, PUPA
UNITED STATES DEPT. AGRI.,
MISC. PUB. 657:363, 432,
1950 CRAIGHEAD LARVA*
F. 91A
GUENEE-SPEC. GEN. LEP.
PHAL. 1:208, 1857 LARVA
HOLLAND-MOTH BOOK 1905
LARVA* PL. 1, F. 17 COLOR
PACKARD-MONOGRAPH GEOMETRIDAE
P. 414, 1876 LARVA
SAUNDERS-INS. INJ. FRUITS
P. 349, 1889 LARVA, PUPA
SLINGERLAND & CROSBY-MANUAL
FRUIT INS. P. 346, 1914
EGG, LARVA* F. 306
INDEX AMER. ECON. ENT.
VOL. 1:191 LYCIA
VO. 2:15 AMPHIDASIS
VOL. 2:217 LYCIA
VOL. 3:19

FOOD PLANTS
ALCER SP. (MAPLES)
ACER MACROPHYLLUM PURSH
ALNUS SP. (ALDERS)
AMELANCHIER FLORIDA LINDL.
ASTER ERICOIDES LINN.
BETULA SP. (BIRCHES)
BETULA ALBA LINN.
BETULA LENTA LINN.
BETULA PAPYRIFERA MARSH.
CASTANEA DENTATA (MARSH.)

CASTANEA SATIVA MILL.
COMPTONIA PEREGRINA(LINN.)
CORYLUS SP.
CORYLUS EORNUTA MARSH.
CRATAEGUS SP. (THORNS)
GLADIOLUS SP.
GLEDITSIA TRIACANTHOS LINN.
HOLODISCUS DISCOLOR(PURSH)
LARIX LARICINA(DUROI)
MALUS PUMILA MILL. (APPLE(
MELILOTUS ALBA DESR.
POPULUS SP.
PRUNUS DOMESTICA LINN. (PLUM)
PRUNUS SEROTINA EHRH.
PYRUS COMMUNIS LINN. (PEAR)
QUERCUS ROBUR LINN.
RHUS TYPHINA LINN.
RIBES AUREUM PURSH.
RIBES GROSSULARIA LINN.
RIBES LACUSTRE(PERS.)
RIBES NIGRUM LINN.
RIBES SANGUINEUM PURSH
RIBES SATIVUM SYME. (CURRANT)
ROBINIA PSEUDO-ACACIA LINN.
ROSA SP. (ROSES)
RUBUS IDAEUS V. STRIGOSUS(MICHX)
SALIX SP. (WILLOWS)
SORBIA SORBIFOLIA(LINN.)
SPIRAEA TOMENTOSA LINN.
ULMUS SP. (ELMS)
VACCINIUM SP. (BLUEBERRIES)

COGNATARIA(HUBNER), GLENA-GEOMETRIDAE
SYN. INFIXARIA WALKER
SYN. LURIDULA HULST
SYN. MURICOLOR HULST
SYN. UMATILLARIA STRECKER
SYN. INSARIA DYAR
LIFE HISTORY
FOOD PLANTS
ROSA SP. (ROSES)
SPIRAEA SP.
TAMARIX CHINENSIS LOUR.
VACCINIUM ANGUSTIFOLIUM AIT.
VACCINIUM MYRTILLOIDES MICHX.

COLATA(GROTE), ITAME-GEOMETRIDAE
SYN. SERICEATA HULST
SYN. CORRELLATA HULST
LIFE HISTORY
SOUTHERN CALIFORNIA ACAD. SCI. BULL. 34:153, 1935 COMSTOCK LARVA* PL. 45; PUPA* PL. 46
FOOD PLANTS
PURSHIA GLANDULOSA CURRAN

COLLARIS FITCH
SYN. SEE TENERA HUBNER

COLLARIS WALKER
SYN. SEE BRONTES DRURY

COLLECTA WALKER
SYN. SEE PAMPINARIA GUENEE

COLLINA BEHR
SYN. SEE MYLITTA EDWARDS

COLLINSI GUNDER
SSP. SEE MYLITTA EDWARDS

COLON FABRICIUS
SYN. SEE PHYLEUS DRURY

COLONA(HUBNER), HAPLOA-ARCTIIDAE
SYN. CAROLINA HARRIS
FORM CONSCITA WALKER
SYN. LACTATA SMITH
FORM FULVICOSTA CLEMENS
SYN. DUPLICATA NEUMOEGEN & DYAR
FORM REVERSA STRETCH
SYN. SUFFUSA SMITH
LIFE HISTORY
ARKANSAS AGRI. EXP. STA., BULL. 183:9, 1923 BAERG EGG* F. 2; LARVA* F. 3; PUPA* F. 4 (REVERSA VERY GOOD); STRUCTURE OF LARVA* PL. 5
CAN. ENT. 29:98, 1897 DYAR EGG, LARVA
ILLINOIS DEPT. AGRI., TRANS. 15: 180, 1877 FRENCH LARVA*
LOC. CIT. 18:APPEND:113, 1880 MARTEN LARVA*
KANSAS ACAD. SCI., TRANS. 11:113,

1888 MARLATT EGG* LARVA* PUPA* SUFFUSA
MISSOURI, REPT. INS. 3:132, 1871 RILEY LARVA* PUPA*
PSYCHE 13:32, 1906 FOSTER TABLE OF LARVAL DIFFERENCES IN GENUS HAPLOA
STRETCH-ZYGAEN. & BOMBY. NORTH AMERICA P. 64, 1873 LARVA, PUPA
INDEX AMER. ECON. ENT.
VOL. 3:187 REVERSA
VOL. 7:448 REVERSA

FOOD PLANTS
FRAGARIA CHILOENSIS (LINN.) (STRAWBERRY)
FRAXINUS SP. (ASH)
MALUS PUMILA MILL. (APPLE)
PRUNUS SP.
PRUNUS PERSICA (LINN.)
PYRUS SP.
RUBUS ALLEGHENIENSIS PORTER

COLORADO (SMITH), BENJAMINIOLA-NOCTUIDAE
SYN. CIRPHIDIA HAMPSON
SYN. LEUCANIDIA HAMPSON
LIFE HISTORY
ESSIG-INS. WEST. NORTH AMERICA P. 686, 1926 LARVA* F. 561 NO DESCR.
FOOD PLANTS
ELYMUS CANADENSIS LINN.
ELYMUS CONDENSATUS PRESL.

COLORADARIA (GROTE & ROBINSON) EPIPLATYMETRA-GEOMETRIDAE
LIFE HISTORY
UNITED STATES NAT. MUS., PROC. 25:393, 1903 CYAR EGG, YOUNG LARVA
FOOD PLANTS

COLORADENSIS BEUTENMULLER
SSP. SEE AHOLIBAH STRECKER

COLORADENSIS COCKERELL
SSP. SEE IO FABRICIUS

COLORADENSIS GUNDER
SSP. SEE HETERONEA BOISDUVAL

COLORADENSIS HENRY EDWARDS
SSP. SEE AUSONIDES BOISDUVAL

COLORADENSIS PUTNAM-CRAMER, RAPHIA-NOCTUIDAE
LIFE HISTORY
CAN. ENT. 26:17, 1894 DYAR EGG, PUPA, COCOON
FOOD PLANTS
POPULUS SP. (POPLARS)

COLORADENSIS RILEY
SSP. SEE YCCAE BOISDUVAL & LECONTE

COLARADICOLA STRAND
SSP. SEE PLUSIAEFORMIS HENRY EDWARDS

COLORADO SCUDDER
SSP. SEE COMMA LINNAEUS

COLORADO SMITH
SSP. SEE SIDUS GUENEE

COLORARIA (FABRICIUS), CATOPYRRHA-GEOMETRIDAE
SYN. ACCESSARIA HUBNER
SYN. CRUENTARIA HUBNER
FORM AEST. DISSIMILARIA HUBNER
SYN. OLENUSARIA WALKER
LIFE HISTORY
AMER. MUS. NOVITATES 1469:17, 1950 RINDGE-RUPERT LARVA
NEW YORK STATE AGRI. EXP. STA. (CORNELL), MEM. 274:70, 1948 FORBES LARVA
GUENEE-SPEC. GEN. LEP. PHAL. 2:183, 1857 LARVA, PUPA
GUMPPENBERG-SYSTEMA GEOMETRARUM 5:279, 1892 LARVA
PACKARD-MONOGRAPH GEOMETRIDAE P. 207, 1876 LARVA* PL. 13, F. 35; PUPA* PL. 35A DISSIMILARIA
FOOD PLANTS
CEANOTHUS SP.
CEANOTHUS AMERICANUS LINN.
RUBUS SP.
RUBUS ALLEGHENIENSIS PORTER

RUBUS IDAEUS V.STRIGOSUS(MICHX.)
TRIFOLIUM SP.(CLOVERS)

COLORATA WALKER
SYN. SEE RADIANS WALKER

COLORO WRIGHT
SYN.SEE ZELICAON LUCAS

COLUMBA McDUNNOUGH
SSP.SEE BEHRII EDWARDS(CALLIPSYCHE)

COLUMBIA HENRY EDWARDS
SSP.SEE APHRODITE FABRICIUS

COLUMBIA SKINNER
SSP.SEE LYGDAMUS DOUBLEDAY

COLUMBIA SMITH
SSP.SEE MEDITATA GROTE

COLUMBIA(SMITH),PLATYSAMIA-SATURNIIDAE
A NOKOMIS BRODIE
B WINONAH BRODIE
LIFE HISTORY
AMER.NAT.15:752,1881
NOTE ON LARVA
LOC.CIT.19"809,1885
CHRISTY BRIEF LARVAL NOTE
BOSTON SOC.NAT.HIST.,PROC.9:
344,1863 S.I.SMITH COCOON
BUFFALO SOC.NAT.SCI.,BULL.2:
201,1879 HAGEN COCOON
CAN.ENT.3:201,1871
BOWLES LARVA,COCOON
LOC.CIT.10:41,1878 CAULFIELD
NOTES ON LARVA* COLOR
(GOOD)
LOC.CIT.10"43,1878 FERNALD
EGG,LARVA*(FRONTISPIECE)
COLOR; COCOON
LOC.CIT.26"281,1894 MOFFAT
COMPARES COCOON WITH
CECROPIA
LOC.CIT.40"350,1908 FREEDLEY
EGG,LARVA, COCOON
DAVENPORT ACAD.SCI.,PROC.2:
277,1878 STRECKER COMPARES
LARVAE OF GLOVERI AND COLUMBIA
ENTOMOLOGICA AMERICANA 2:18,
1886 CHRISTY COCOON
ENT.SOC.AMERICA,ANNALS 9:153,
1916 MOSHER PUPAL STRUCTURES
ILLINOIS DEPT.AGRI.,TRANS.
18:APPEND:177,1880
COQUILLETT LARVA
NAT.ACAD.SCI.,MEM.12:222,1914
PACKARD EGG,LARVA* PL.9,
FF.3-6,PL.10,F.1 COLOR
LOC.CIT.P.267 PACKARD
HYBRID LARVA WITH CECROPIA*
PL.5,F.3 COLOR
NEW YORK STATE MUS.,MEM.8:
755,
1905 FELT LARVA (BRIEF)
INDEX AMER.ECON.ENT.
VO. 2:231 SAMIA
VOL. 7:862
FOOD PLANTS
ABIES BALSAMEA(LINN.)
ACER SP.(MAPLES)
KALMIA ANGUSTIFOLIA LINN.
LARIX LARICINA(DUROI)
MYRICA PENSYLVANICA LOISEL
NEMOPANTHUS MUCRONATA(LINN.)
PICEA GLAUCA(MUENCH)
PICEA MARIANA(MILL.)
PRUNUS CERASUS LINN.(CHERRY)
PRUNUS PENSYLVANICA LINN.
PRUNUS VIRGINIANA LINN.
RHODORA CANADENSIS(LINN.)
SALIX ALBA LINN.
SHEPHERDIA ARGENTEA NUTT.
SPIRAEA TOMENTOSA LINN.
TSUGA CANADENSIS(LINN.)
ULMUS SP.(ELMS)

COLUMBIANA GUNDER
SSP.SEE EURYMEDON LUCAS

COLUMBIANA HENRY EDWARDS
SSP.SEE REVAYANA SCOPOLI

COLUMBIANA McDUNNOUGH
SSP.SEE INORNATA EDWARDS

columbiata(Dyar), Eupithecia-Geometridae
A erpata Pearsall
LIFE HISTORY
FOOD PLANTS
Rubus parviflorus Nutt.

columbiata McDunnough
ssp. see paludata Thunberg

columbina Godart
syn. see claudia Cramer

columboides Francelmont
ssp. see morula Grote & Robinson

columella(Fabricius), Strymon-Lycaenidae
syn. eurytulus Hubner
syn. istapa Reakirt
syn. salona Hewitson
syn. modesta Maynard
syn. ocellifera Grote
LIFE HISTORY
Southern California Aca. Sci. Bull. 34"120, 1935 Comstock-Dammers egg* Pl. 16; larva* Pl. 17; chrysalis* Pl. 18
FOOD PLANTS
Sida hederacea(Doug.)

combinata Walker
syn. see carneicosta Guenee

comis(Grote), Lacinipolia-Noctuidae
A obnigra Smith
B rectilinea Smith
LIFE HISTORY
United States Nat. Mus., Proc. 27:856, 1904 Dyar
egg, larva
FOOD PLANTS

comma(Harris), Polygonia-Nymphalidae
syn. harrisii Edwards
form dryas Edwards
LIFE HISTORY
Amer. Mus. Nat. Histl, Bull 5: 264, 1893 Beuten. larva, chrysalis
Buffalo Soc. Nat. Sci., Bull. 19: 24, 1939 Wild larva
Can. Ent. 14:189, 1882 Edwards egg, larva, chrysalis
Can. Nat. pp. 221, 226, 280, 1840 Gosse larva, chrysalis
Illinois Dept. Agri., Trans. 15: 152, 1877 French larva
Loc. cit. 18:Append:84, 163, 1880 Coquillett-Middleton larva
Naturaliste Canadien p. 136, 1927 Huard larva, chrysalis
New York State Mus., Mem. 8: 545, 1905 Felt larva
New York State, Rept. Ins. 3:432, 1856 Fitch larva
also p. 114, sec. 143
Philadelphia Ent. Soc, Proc. 1: 182, 1862 Edwards larva
Loc. cit. 3:55, 1864
Lintner larva, cheyaslis
Smithsonian Misc. Collections 4:art. 2:54, 1862 Morris larva, chrysalis
United States Dept. Agri., Bur. Ent. Bull. 4 old series:40, 1884 Smith larva
Loc. Cit. Bull7 new series:50, 1897 Howard egg* f. 38a; larva* f. 38b; pupa* f. 38c
United States Dept. Agri., Misc. Pub. 657, 356, 369, 1950 Craighead larva
Comstock-How to Know Butter. p. 140, 1904 larva
Edwards-Butter. North America 2nd series, 1871 life hist* Pls. Grapta No. 2 & 3; larva; chrysalis COLOR
Loc. cit. 3rd series 1897 life history* Pl. grapta 1
Fernald-Butter. maine p. 51, 1884p egg, larva, chrysalis
French-Butter, East United States p. 186, 1886 life history

HARRIS-INS.INJ.VEG.1ST EDIT.
P.221,1841 CHRYSALIS
LOC.CIT.2ND EDIT.P.241,1852
LARVA, CHRYSALIS
LOC.CIT.FLINT EDIT.P.301,1862
CHRYSALIS* PL.4
HOLLAND-BUTTERFLY BOOK 1905
LARVA* CHRYSALIS* P1.3,F.
38; P1.4,FF.27,29,30,46-48
MAYNARD-BUTTER,NEW ENGLAND
P.12,1886 LARVA
MORRIS-SYNOP.LEP.NORTH AMERICA
P.54,1862 CHRYSALIS
PACKARD-INS.INJ.WEST
P.774,1877 LARVA
PACKARD-INS.INJ.FOREST TREES
P.66,1881 LARVA,CHRYSALIS
SCUDDER-BUTTER.EAST.UNITED
STATES & CANADA PP.336-341,
1889 LIFE HISTORY* P1.64
FF.18-20;P1.74,F.38;P1.78,
FF.36;P1.81,F.9;P1.86,F.67;
P1.83,FF.27,29,29.29,46-48
SCUDDER-BUTTER.THEIR STRUCTURE
ETC.P.99, 1881 EGG*
SCUDDER-EVERYDAY BUTTER.P.370,
1899 EGG,LARVA,CHRYSALIS
SCUDDER-GUIDE TO COMMON BUTTER
P.96,1893 LARVA,CHRYSALIS
INDEX AMER.ECON.ENT.
VOL.1:251; VOL.2:293;
VOL.4:379; VOL.6:604;
VOL.7:799

FOOD PLANTS

ALTHEA ROSEA CAV.
AMBORIS ARTEMISIIFOLIA LINN.
AMBROSIA TRIFIDA LINN.
BOEHMERIA CYLINDRICA(LINN.)
CELTIS OCCIDENTALIS LINN.
HUMULUS LUPULUS LINN.(HOPS)
RIBES SP. (CURRANTS)
RUBUS ALLEGHENIENSIS PORTER
RUBUS IDAEUS V.STRIGOSUS(MICHX.)
(RASPBERRY)
TILIA SP.(LINDENS)
ULMUS SP.(ELMS)
URTICA SP.(NETTLES)

COMMA(LINNAEUS),HESPERIA-HESPERIIDAE
A COLORADO SCUDDER
B MANITOBA SCUDDER
SYN. LAURENTINA LYMAN
C IDAHO EDWARDS
D ASSINIBOIA LYMAN
E LEUSSLERI LINDS.

LIFE HISTORY

CAN.ENT.27"336,1895
FYLES EGG,LARVA,
CHRYSALIS(MANITOBA)

FOOD PLANTS

PHELEUM PRATENSE LINN.
POA PRATENSIS LINN.
PROBABLY MOST GRASSES

COMMA WALKER
SYN. SEE CLYMENE BROWN

COMMELINAE ABBOT & SMITH
SYN. SEE DOLICHOS FABRICIUS

COMMODA(BUTLER),ANOMIS-NOCTUIDAE

FOOD PLANTS

HIBISCUS SYRIACUS LINN.

COMMOIDES GUENEE,LEUCANIA-NOCTUIDAE

LIFE HISTORY

INDEX AMER.ECON.ENT.
VOL.7:607

FOOD PLANTS

GRASSES

COMMUNIS(DYAR),ORTHODES-NOCTUIDAE
SYN,AFFURATA HAMPSON
SYN. BRUNNEOSUFFUSA STRAND

LIFE HISTORY

UNITED STATES NAT.MUS.,
PROC.27:868,1905
DYAR EGG,LARVA

FOOD PLANTS

COMMUNIS GROTE
SYN.SEE NEOGAMA ABBOT & SMITH

COMMUNIS(GROTE),PYRGUS-HESPERIIDAE
SYN TESSELLATA SCUDDER
SYN ALBOVITTATA GROTE
A ALBESCENS PLOETS

SYN. OCCIDENTALIS SKINNER
AB SKINNERI GUNDER
LIFE HISTORY
BUFFALO SOC. NAT. SCI., BULL. 19:32, 1939 WILD LARVA
CAN. ENT. 29:283, 1897 FRENCH EGG, LARVA, CHRYSALIS
SOUTHERN CALIFORNIA ACAD, SCI. BULL. 43:30, 1944 DETHIER EGG; LARVA* P1.11; PUPA
COMSTOCK, J. A.-BUTTER CALIFORNIA P. 207, 1927 EGG* A67; LARVA* A68; CHRYSALIS A68; TESSELLATA
ESSIG-ONS. WEST. NORTH AMERICA P. 651, 1926 EGG
KELLOGG-AMER. INS. P. 442, 1908 LARVA
SCUDDER-GUIDE TO COMMON BUTTER. P. 165, 1893 LARVA, CHRYSALIS
INDEX AMER. ECON. ENT. VOL. 7:839; VOL. 7:1041 OCCIDENTALIS
FOOD PLANTS
ABUTILON THEOPHRASTI MEDIC
ALTHAEA ROSEA CAV.
HIBISCUS TRIONUM LINN.
MALVA SP.
MALVA ROTUNDIFOLIA LINN.
SIDA SPINOSA LINN.
SIDALCEA SP.

COMPACTA TAYLOR
SYNN SEE GELIDATA MOESCHLER

COMPLICATA STRECKER
SYN. SEE LATIUSCULA HERRICH-SCHAEFFER

COMPLICATA WALKER
SSP. SEE ORNATA PACKARD

COMPRESSIPALPUA GUENEE, PLUSIODONTA NOCTUIDAE
SYN. INSIGNIS WALKER
AB SUFFUSA HILL.
LIFE HISTORY
AMER. NAT. 17:1171, 1883 LARVA
CAN. ENT. 15:172, 1883 HOY LARVA (BRIEF)
ONTARIO ENT. SOC., ANNUAL REPT. 19, 1883 HOY LARVA
WASHINGTON ENT. SOC., PROC. 4: 326, 1898 DYAR LARVA
LOC. CIT. 5:292, 1903 DYAR EGG, LARVA
FOOD PLANTS
MENISPERMUM CANADENSE LINN.

COMPTA WALKER
SYN. SEE IPOMOEAE DOUBLEDAY

COMPTARIA (WALKER), VENUSIA-GEOMETRIDAE
SYN. PERLINEATA PACKARD
SYN. SALIENTA PEARSALL
A PALUMBES FRANCELEMONT
LIFE HISTORY
INDEX AMER. ECON. ENT. VOL. 1:12
FOOD PLANTS
ALNUS SP. (ALDERS)
BETULA SP. (BIRCHES)
FAGUS SP. (BEECHES)

COMSTOCKI CHERMOCK
SSP. SEE BAIRDII EDWARDS

COMSTOCKI GUNDER
SSP. SEE CALLIPPE BOISDUVAL

COMSTOCKI GUNDER
SSP. SEE LORQUINI BOISDUVAL (BASILARCHIA)

COMSTOCKI GUNDER
SSP. SEE VANILLAE LINNAEUS

COMSTOCKI GUNDER
SSP. SEE SONORENSIS FELDER & FELDER (PHILOTES)

COMSTOCKI GUNDER
SYN. SEE SABULETI BOISDUVAL

COMSTOCKI HENNE, CALLOPHRYS-LYCAENIDAE
LIFE HISTORY

AMER. MUS. NAT. HIST., BULL. 4:76
1892 BEUTEN. LARVA
LOC. CIT. 5:284, 1893
BEUTEN. LARVA, CHRYSALIS
BUFFALO SOC. NAT. SCI. BULL. 19:
30, 1939 WILDE LARVA
CAN. ENT. 8:202, 1876 EDWARDS
EGG, LARVA, CHRYSALIS
ENT. NEWS 24:97, 1913 BAKER
LIFE HISTORY AND HABITS
ILLINOIS DEPT. AGRI. TRANS. 18:
APPEND:95 1880
MIDDLETON LARVA
MUS. COMP. ZOOL. BULL. 87: NO. 6,
1941 DETHIER
LARVAL ANTENNA* F. 28
NATURALISTE CANADIEN P. 213,
1928 HUARD LARVA
SMITHSONIAN MISC. COLLECTIONS
4, ART, 2:83, 1862 MORRIS
LARVA, CHRYSALIS
BOISDUVAL & LECONTE-HIST. LEP.
NORTH AMERICA P. 120, 1833
LARVA* CHRYSALIS* P1.36
FF. 8, 9 COLOR
COMSTOCK-HOW TO KNOW BUTTER.
P. 254, 1904 LARVA
FERNALD-BUTTER. MAINE P. 92,
1884 EGG, LARVA
FRENCH-BUTTER. EAST. UNITED
STATES P. 292, 1886 LIFE HIST.
GLOVER-ILLUS. NORTH AMER. ENT.,
1878 LARVA* PUPA* P12, F. 3;
P1.9, F. 11
HARRIS-ENT. CORRESP. P. 275
1869 LARVA, CHRYSALIS
HARRIS-INS. INJ. VEG. FLINT EDIT.
P. 275, 1862 LARVA
HOLLAND-BUTTERFLY BOOK 1905
CHRYSALIS* P1.5, FF. 42, 47, 48
MORRIS-SYNOP. LEP. NORTH AMERICA
P. 83, 1862 LARVA, CHRYSALIS
SCUDDER-BUTTER. EAST. UNITED
STATES & CANADA PP. 913, 917
1889 LIFE HISTORY* P1.65,
F. 20; P1 68, FF. 5, 12; P1.71
F. 5; P175. FF. 37, 44; P1.79,
FF. 36-38, P1.84, FF. 42, 47, 48
SCUDDER-EVERYDAY BUTTER. P. 70,
1899 EGG, LARVA, CHRYSALIS
SCUDDER-GUIDE TO COMMON BUTTER
P. 124, 1893 LARVA,
CHRYSALIS
INDEX AMER. ECON. ENT.
VOL. 5:245; VOL. 7:783

FOOD PLANTS

AMBROSIA TRIFIDA LINN.
DESMODIUM MARILANDICA (LINN.)
GALACTIA SP.
LEGUMINOSAE SP.
LESPEDEZA CAPITATA MICHX.
PHASEOLUS PLLYSTACHIOS (LINN.)
TRIFOLIUM SP. (CLOVERS)

CONCHYLIATUS COMSTOCK
SSP. SEE ZERENE BOISDUVAL

CONCILIARIA SWETT
SSP. SEE DEFENSARIA GUENEE

CONCINNA (ABBOT & SMITH), SCHIZURA-
NOTODONTIDAE
SYN. NITIDA PACKARD
A SALICIS EDWARDS
SYN. RIVERSII BEHR.

LIFE HISTORY

AMER. ENT. 2:27, 1869
LARVA PUPA
AMER. MUS. NAT. HIST. BULL. 10:
424, 1898 BEUTEN. LARVA
BOSTON SOC. NAT. HIST. PROC. 24:
532, 1890 PACKARD LARVA
CALIFORNIA ACAD. SCI. PROC. 7:
121, 1876 EDWARDS LARVA,
PUPA (SALICIS)
LOC. CIT. 2: NEW SERIES: 95, 1889
BEHR LARVA (DRYOCAMPA
RIVERSII)
CAN. DEPT. AGRI. REPT. ENT. P. 28,
1888 FLETCHER LARVA*'
CAN. ENT. 13:138, 1881
SAUNDERS LARVA* PUPA*
ENT. NEWS 13:293, 1902
FOSTER LARVAL NOTE
ILLINOIS DEPT. AGRI. TRANS. 15:
190, 1877 FRENCH LARVA

Loc.cit.18:Append:12o,1880
Marten larva
Maine Agri.Exp.Sta.Bull.259:
71,1917 Mosher larva,
pupa* f.5A
Michigan State College,Spec.
Bull.243:23, 1933 McDaniel
life history
Michigan Agri.Exp.Sta.Circ.
137:19,1935 Pettit larva
Loc. cit. 84:65,1903
Washburn larva* f.47
Minnesota Agri.Exp.Sta.Bull.
61:165,1899 Lugger larva*
f.109; Pl.14, f.110
Mus.Comp.Zool.Bull.87:No.6,
1941 Dethier larval
antenna* ff.80,85
Nat.Acad.Sci.Mem.7:213,1895
Packard egg,larva*Pl.26,
f.5 COLOR; pupa;cocoon
New York Ent.Soc.Journ.1:68
1893 Packard egg, larva
New York Agri.Exp.Sta.
(Cornell),Mem.56:1074,1922
Wellhouse larva(brief)
Loc.cit.274:228 1948
Forbes larva
New York State Mus.Bull.97:
403,1905 Felt life hist.
Loc.cit 155:99, 1912
Felt larva
New York State Mus.Mem.8:743
1905 Felt larva (brief)
New York State,Rept.Ins.3:
342,1856 Fitch larva
also, p.24, sec.37
Ohio Agri.Exp.Sta.Bull.45:
167,1893 Webster larva
Loc.cit.332:255,1918
Houser larva* Pl.38, f.3
Ontario Ent.Soc.Ann.Rept.p.20,
1871 Bethune egg, larva*
Loc.cit.p.20,1881
Saunders larva* pupa*
Loc.cit.p.49,1889
Harrington larva*
Loc.cit.p.40,1893
Fyles larva*
Loc.cit.p.106,1902
Lockhead larva*
Loc.cit.p.16,1906
Young larva
Loc.cit.p.47, 1917
Corcoran larva
Papilio 3:130, 1883
Edwards-Williot larva
Pennsylvania Agri.Rept.p.70,
1883 Edge larva* pupa*
Psyche 6:177, 1891
Dyar larva (salicis)
Smithsonian Misc.Collections
4,art.2:242, 1862
Morris larva
Smithsonian Scienfitic Series
5:260,1930 Snodgrass
larva* COLOR
United States Dept.Agri.,
Farmer's Bull.1270:39,1922
Quaintance egg* f.79 c;
larva* f.79 d,e; pupa*
and cocoon* f.79 f
United States Dept. Agri.,
Misc.Pub.657:357,404,1950
Craighead larva* f.81 C
Abbot & Smith-Lep.Ins.Georgia
2:169,1797 larva* Pl.85
C COLOR
Cragen-Our Ins.Friends & Foes
p.161,1899 larva* f.105
Dickerson-Moths & Butter
P.219, 1901 larva* f.188 B;
cocoon* f.188 A
Doane-Forest Ins.p.293,1936
larva* f.171
Essig-Ins.West.North America
p.692,1926 larva
Fernald-Applied Ent.p.270,1926
larva* f.273
Harris-Ent.Corresp.p.303,1869
larva* Pl.1
Harris-Ins.Inj.Veg.1st edit.
p.309,1841 larva
Harris-loc.cit.Flint edit.
p.426,1862 egg* larva*
Herrick-Ins.Enemies Shade

TREES P.279,1935 LARVA*
F.236
KELLOGG-AMER.INS.P.393,1908
LARVA* F.562
METCALF & FLINT-DESTR.& USE-
FUL INS.P.564,1939 LARVA
MORRIS-SYNOP.LEP.NORTH AMERICA
P.242,1862 LARVA
O'KANE-INJ.INS.P.271,1912
EGG* F.383; LARVA* F.382
PEAIRS-PESTS FARM.GARD.ORCHARD
P.379,1941 LARVA* F.457
QUALE-INS.CITRUS.SUBTROPICAL
FRUITS P.387,1938 LARVA
SAUNDERS-INS.INJ.FRUITS
P.63,1883 LARVA* PUPA
SAUNDERS-LOC.CIT.P.63,1889
LARVA* F.57; PUPA* F.58
SLINGERLAND & CROSBY-MANUAL
FRUIT INS.P.126,1914
EGG* F.139; LARVA* F.140
INDEX AMER.ECON.ENT.
VOL. 1:279; VOL2:326;
VOL. 3:366; VOL.4:433;
VOL. 5:564; VOL.6:661;
VOL. 7:870

FOOD PLANTS

ACER SP.(MAPLES)
BETULA ALBA LINN.
BETULA PAPYRIFERA MARSH.
CARYA SP.(HICKORIES)
CARYA GLABRA(MILL.)
CARYA ILLINOENSIS(WANG.)
CARYA OVALIS(WANG.)
CARYS TOMENTOSA NUTT.
CORNUS FLORIDA LINN.
CRATAEGUS SP.
CRATAEGUS OXYACANTHA LINN.
DIOSPYROS KAKI LINN.
DIOSPYROS VIRGINIANA LINN.
HALESIA CAROLINA LINN.
JUGLANS CINEREA LINN.
JUGLANS NIGRA LINN.
JUGLANS REGIA LINN.
LIQUIDAMBAR STRYACIFLUA LINN.
LUPINUS SP.
MALUS PUMILA MILL.(APPLE)
MYRICA PENSYLVANICA LOISEL
PERSEA AMERICANA MILL.(AVOCADO)
POPULUS SP.(POPLARS)
POPULUS FREMONTI S.WATS
POPULUS TREMULOIDES MICHX.
PRUNUS AMERICANA MARSH.
PRUNUS ARMENIACA LINN.(APRICOT)
PRUNUS CERASUS LINN.(CHERRY)
PRUNUS DOMESTICA LINN.(PLUM)
PRUNUS DOMESTICA V.GELATENSIS
EX HOOK(PRUNE)
PRUNUS SEROTINA EHRH.
PYRUS COMMUNIS LINN.(PEAR)
QUERCUS SP.(OAKS)
ROBINIA PSEUDO-ACACIA LINN.
ROSA SP.(ROSE)
ROSA CANINA LINN.
RUBUS ALLEGHENIENSIS PORTER
RUBUS IDAEUS V. STRIGOSUS
(MICHX.)(RASPBERRY)
SALIX SP.(WILLOWS)
SALIX ALBA LINN.
SALIX BABYLONICA LINN.
SALIX LUCIDA MUHL.
ULMUS SP.(ELMS)
VACCINIUM SP.(BLUEBERRIES)
VIBRUNUM SP.
VIBURNUM LENTAGO LINN.
WISTERIA SINESIS SWEET

CONCISA WALKER
SYN.SEE AEMULA HUBNER

CONCISARIA WALKER
SYN. SEE SERRATA DRURY

ÇONCOLORATA MCDUNNOUGH
SSP.SEE MESKEI GROTE

CONCORDATA WALKER
SYN.SEE LUCTUATA SCHIFFERMULLER

CUNCUMBENS WALKER,ÇATOCALA-NOCTUIDA
AB HILLI GROTE
AB DIANA HENRY EDWARDS

LIFE HISTORY

AMER.MUS.NAT.HIST.,BULL.
38:168,1918 BARNES &
MCDUNNOUGH EGG,LARVA

AMER. MUS. NAT. HIST. MEM.
NEW SERIES 3, PART 1, 1918
BARNES & MCDUNNOUGH
LARVA* P1.11, F.5 COLOR
BROOKLYN ENT. SOC. BULL. 7:
47, 1884 HULST LARVA
ENT. NEWS 20:17, 1909
ROWLEY EGG
NEW YORK STATE MUS. MEM. 8:743,
1905 FELT LARVA(BRIEF)
PHILADELPHIA ENT. SOC. PROC.
2:29 1863 SAUNDERS LARVA
GLOVER-ILLUST. NORTH AMER. ENT.
1878 LARVA*PUPA* P1.15,
F.11

FOOD PLANTS

POPULUS SP. (POPLARS)
SALIX SP. (WILLOWS)

CONDITA(GUENEE), APLECTOIDES-NOCTUIDAE
SYN. TRABALIS GROTE

LIFE HISTORY

SOUTHERN CALIFORNIA ACAD. SCI.
BULL. 34:140, 1935 COMSTOCK-
DAMMERS LARVA* PLS. 35, 36;
PUPA* PLS. 35, 37

FOOD PLANTS

ERIOGONUM FASCICULATUM BENTH
PINUS STROBUS LINN.
COCOON FOUND UNDER PINE BARK

CONDUCTA WALKER
SYN. SEE EXACTA HUBNER

CONFIGURATA WALKER, MAMESTRA-NOCTUIDAE
SYN. OCCIDENTA GROTE

LIFE HISTORY

JOURN. ECON. ENT. 21:282, 1928
KING LARVA* PL. 6, FF. 1, 2, 4;
PUPA* FF. 19, 20
OTTOWA NAT. 14:17, 1900
GIBSON LARVA(BRIEF)
INDEX AMER. ECON. ENT.
VO VOL. 4:59 BARATHRA
VOL. 5:71 BARATHRA
VOL. 6:82 BARATHRA

FOOD PLANTS

BETA VULGARIS LINN. (BEET)
BRASSICA OLERACEA LINN. (CABBAGE)
BRASSICA RAPA LINN. (TURNIP)
CHENOPODIUM ALBUM LINN.
DELPHINIUM SP.
LINUM SULCATUM RIDD. (FLAX)
MEDICAGO SATIVA LINN.
MELILOTUS ALBA DESR.
PHASEOLUS VULGARIS LINN. (BEAN)
PISUM SATIVUM LINN. (PEAS)
SOLANUM LYCOPERSICUM LINN.
(TOMATO)
TRIFOLIUM SP. (CLOVERS)
TRITICUM AESTIVUM LINN. (WHEAT)
ZEA MAYS LINN. (CORN)
FLOWERS, VEGETABLES

CONFINE(HERRICH-SCHAEFFER),
EUCEREON-AMATIDAE

LIFE HISTORY

FOOD PLANTS

CYNANCHUM PALUSTRE(GRAY)
HELLER
SARCOSTEMMA VIMINALE(LINN.)
R. BR.

CONFINIS WALKER
SYN. SEE LECONTEI GUERIN-MENEVILLE

CONFLUENS MORRISON
SYN. SEE HIBISCI GUENEE

CONFLUENS OBERTHUR
SSP. SEE DEFLORATA FABRICIUS

CONFLUENTIS STRAND
SSP. SEE BOLANDERI STRETCH

CONFORMATA PEARSALL
SYN. SEE MISTURATA HUBNER

CONFRACTA MORRISON
SYN. SEE MESSORIA HARRIS

CONFUSA(HUBNER), MORRISONIA-NOCTUIDAE
SYN INFRUCTUOSA WALKER

LIFE HISTORY

INS. LIFE3:63, 1890 DYAR LARVA
NEW YORK STATE MUS. MEM. 8:

SYN. CANA MARTYN
LIFE HISTORY
AMER. JOURN. SCI. 36:297, 1839
HARRIS LARVA
AMER. MUS. NAT. HIST. BULL7:310
1895 BEUTEN. LARVA, PUPA
BROOKLYN ENT. SOC. BULL. 4:20
1881 KOEBELE EGG, LARVA
ENTOMOLOGICA AMERICANA 3:167,
1887 HY. EDWARDS PUPA
ENG. SOC. AMERICA, ANNUALS11:423
1918 MOSHER PUPAL STRUCTURE
ILLINOIS DEPT. AGRI.
TRANS. 18:APPEND:105, 1880
MARTEN LARVA
NEW YORK STATE AGRI. EXP. STA.
(CORNELL), MEM. 274:192, 1948
FORBES LARVA
NEW YORK STATE MUS. MEM. 8:753
1905 FELT LARVA (BRIEF)
NEW YORK STATE, REPT. INS. 4:472
56, SEC. 270, 1858 FITCH
LARVA (BRIEF)
NOVITATES ZOOLOGICAE(TRING) 9:
SUPPLEMENT P. 151, 1903
ROTHSCHILD & JORDAN LARVA
PUPA
PHILADELPHIA ACAD. NAT. SCI.,
JOURN. 5:174, 1859
CLEMENS LARVA
SMITHSONIAN MISC. COLLECTIONS
4:ART, 2:199, 1862
CLEMENS LARVA
ABBOT & SMITH-LEP. INS. GEORGIA
1:83, 1797 LARVA* PL. 42
COLOR
BOISDUVAL-SPEC. GEN. LEP. SPHING.
P. 105, 1874 LARVA, PUPA
FERNALD-SPHING. NEW ENGLAND
P. 85, 1886 EGG, LARVA
HARRIS-INS. INJ. VEG. FLINT
EDIT. P. 328, 1862 LARVA
PACKARD-INS. INJ. FOREST TREES
P. 201, 1881 LARVA (BRIEF)
STRECKER-LEP. RHOP. ET HETER.
13:116, 1876 LARVA
WESTOOD-INTROD. MODERN CLASSIF.
INS. 2:367, 1840 LARVA (BRIEF)

FOOD PLANTS
PINUS AUSTRALIS MICHX.
PINUS RESINOSA AIT.

CONJUNCTA HENRY EDWARDS
SSP. SEE MACCULLOCHI KIRBY

CONJUNCTA WATSON
SSP. SEE BURNSI WATSON

CONJUNGENS WALKER
SYN. SEE TALIDIFORMIS GUENEE

CONNECTA GROTE, ACRONICTA-NOCTUIDAE
A ALBINA MCDUNNOUGH
LIFE HISTORY
AMER. MUS. NAT. HIST., BULL. 14:
250, 1901 BEUTEN. LARVA
NEW YORK ENT. SOC., JOURN.
7:253, 1899 DYAR LARVA
NEW YORK STATE MUS. MEM. 8:
742, 1905 FELT LARVA
(BRIEF)
UNITED STATES NAT. MUS. PROC.
21:116, 1898 SMITH & DYAR
LARVA, PUPA, COCOON
FOOD PLANTS
SALIX SP. (WILLOWS)

CONNUBIALIS GUENEE, CATOCALA-NOCTUIDAE
SYN. SANCTA HULST
SYN. VIRENS FRENCH
FORM CORDELIA HENRY EDWARDS
~~AMA~~SIA ABBOT & SMITH
LIFE HISTORY
ABBOT & SMITH-LEP. INS. GEORGIA
2:179, 1797 LARVA* PL. 90
COLOR
DUNCAN-NATURAL. LIBRARY 22:205
1858 LARVA (AMASIA)
GUENEE-SPEC. GEN. LEP. NOCT.
3:104, 1852 LARVA
FOOD PLANTS
CEPHALANTHUS OCCIDENTALIS LINN.
MELIA AZEDARACH LINN.
QUERCUS SP. (OAKS)

CONOCHARODES HAMPSON
SYN SEE ALTERA SMITH

CONSCITA WALKER
SSP. SEE COLONA HUBNER

CONSECUTARIA WALKER
SYN. SEE INDUCTATA GUENEE

CONSEPTA WALKER
SYN. SEE BICOLORATA FABRICIUS

CONSEQUARIA HENRY EDWARDS
SYN. SEE IRIDARIA GUENEE

CONSIMILATA ZELLER
SYN. SEE BICOLORATA FABRICIUS

CONSIMILIS HEINRICH, GALENARA-GEOMETRIDAE
LIFE HISTORY
FOOD PLANTS
ABIES SP.
CARYA SP. (HICKORIES)

CONSORS(ABBOT & SMITH), CATOCALA-NOCTUIDAE
SYN. PENSACOLA REIFF.
A SORSCONI BARNES & BENJAMIN
LIFE HISTORY
AMER. MUS. NAT. HIST. BILL. 16: 391, 1902 BEUTEN. LARVA* F. 18
AMER. MUS. NAT HIST. MEM. NEW SERIES 3: PART 1, 1918 BARNES & MCDUNNOUGH EGG, LARVA* PL. 10, F. 31; PL. 12. F. 16 COLOR
BROOKLYN ENT. SOC. BULL. 7:40 1884 HULST LARVA
ABBOT & SMITH-LEP. INS. GEORGIA 2:177, 1797 LARVA* PL. 89 COLOR
GUENEE-SPEC. GEN. LEP. NOCTUIDAE 3:99, 1852 LARVA
FOOD PLANTS
?AMORPHA FRUTICOSA LINN.
CARYA SP. (HICKORIES)
?LEGUMINOSAE
?LYSIMACHIA NUMMULARIA LINN.

CONSPECTA HENRY EDWARDS
SSP. SEE UNICORNIS ABBOT & SMITH

CONSPERSA RILEY
SSP. SEE PYRAMIDOIBES GUENEE

CONSPICUA(EDWARDS), ATRYTONE-HESPERIIDAE
SYN. PONTIAC EDWARDS
SYN. ORONO SCUDDER
LIFE HISTORY
ENT. NEWS 19:415, 1908 LAURENT EGG, EARLY STAGES
FOOD PLANTS
GRASSES

CONSPICUA MORRISON
SSP. SEE PASSER GUENEE

CONSPICUA NEUMOEGEN
SYN. SEE SPRAGUEI GROTE(PYGARCTIA)

CONSPICUA STRETCH
SSP. SEE OITHONA STRECKER

CONSPICUA WORTHINGTON
SSP. SEE ILIA CRAMER

CONSTANS HULST
SSP. SEE GRYNEA CRAMER

CONSTRICTA HENRY EDWARDS
SSP. SEE MARGINATA HAWORTH

CONSTRICTA(STRETCH), MALACOSOMA-LASIOCAMPIDAE
A STRIGOSA STRETCH
LIFE HISTORY
AMER. PHILOSOPHICAL SOC. PROC. 31:177, 1893 PACKARD LARVA
CALIFORNIA ACAD. SCI. PROC. 5: 368, 1873 EDWARDS LARVA, PUPA, COCOON
CAN. ENT. 25:39 DYAR LARVA
PAPILIO 1:65, 1881 STRETCH LARVA
UNITED STATES DEPT. AGRI. BUR. ENT. BULL. NEW SERIES 37:

108,1902 COCKERELL LARVA
ESSIG-INJ.&BENEF.INS.
P.411,1915 LARVA
ESSIG-INS.WEST NORTH AMERICA
P.697,1926 LARVA
HERRICH-INS.ENEMIES SHADE
TREES P.34,1935 LARVA
PACKARD-INS.INJ.FOREST TREES
P.41,1881 LARVA
INDES AMER.ECON.ENT.
VOL.1:197
FOOD PLANTS
POPULUS FREMONTI S.WATS
WUERCUS EMORYII TORR
QUERCUS KELLOGGII NEWB.
QUERCUS LOBATA NEE
QUERCUS VIRGINIANA MILL.
FRUIT TREES

CONSTRICTINA NEUMOEGEN & DYAR
SSP.SEE FRAGILIS STRETCH

CONSUETA WALKER
SYN.SEE MALEFIDA GUENEE

CONSULARIS DYAR,ANISOTA-CITHERONIIDAE
LIFE HISTORY
ENT.SOC.AMERICA,ANNALS 7:
219,1914 MOSHER PUPAL
STRUCTURES
NAT.ACAD.SCI.MEM.9:107,1905
PACKARD LARVA* PL.5, F.8
COLOR
NEW YORK ENT.SOC.JOURN 4:166,
1896 DYAR LARVA
FOOD PLANTS
QUERCUS VIRGINIANA MILL.

CONTAGIONIS SMITH
SYN.SEE CATENULA GROTE

CONTENTA GROTE,LITHOPHANE-NOCTUIDAE
SYN.POMONA SMITH
LIFE HISTORY
FOOD PLANTS
QUERCUS SP.

CONTENTA WALKER
SYN.SEE DEVASTATOR BRACE

CONTENTA WALKER
SYN.SEE APAMIFORMIS GUENEE

CONTEXTA(GROTE),AUTOGRAPHA-NOCTUIDAE
LIFE HISTORY
PAPILIO 3:18,1883
THAXTER EGG,LARVA
INDES AMER.ECON.ENT.,
VOL.7:117
FOOD PLANTS
GRASS

CONTIGUA(WALKER),HAPLOA-ARCTIIDAE
FORM LUMBONIGERA DYAR
LIFE HISTORY
PSYCHE 13:32,1906 FOSTER
TABLE LARVAL DIFFERENCES
IN GENUS HAPLOA
FOOD PLANTS

CONTINGENS WALKER
SYN.SEE TRANSVERSATA DRURY

CONTINUA HENRY EDWARDS
SYN.SEE LATIPENNIS BOISDUVAL

CONTINUARIA WALKER
SYN.SEE ENUCLEATA GUENEE

CONTINUATA(WALKER),SEMIOTHISA-GEOMETRIDAE
A STRIGULARIA WALKER
LIFE HISTORY
FOOD PLANTS
?CELTIS SP.
?EUPATORIUM PERFOLIATUM LINN.
JUNIPERUS SP.
JUNIPERUS VIRGINIANA LINN.
MIKANIA SCANDENS(LINN.)
?THUJA OCCIDENTALIS LINN.

CONTRACTA WALKER,DATANA-NOTODONTIDAE
LIFE HISTORY
AMER.MUS.NAT.HIST.BULL.10:
408,1898 BEUTEN.LARVA
CAN.ENT.20:135,1888 BEUTEN.
KEY. TO DATANA LARVAE

ENT. NEWS 1:144, 1890
BEUTEN. EARLY STAGES
MINNESOTA AGRI. EXP. STA. BULL. 61:164, 1899 LUGGER LARVA
NAT. ACAD. SCI. MEM. 7:122, 1895 PACKARD LARVA* PL. 14, F. 7 COLOR
NEW YORK ENT. SOC. JOURN. 1: 60, 1893 PACKARD LARVA
NEW YORK STATE AGRI. EXP. STA. (CORNELL), MEM. 274:216, 1948 FORBES LARVA
NEW YORK STATE MUS. MEM. 8:725 1905 FELT LARVA (BRIEF)
PHILADELPHIA ENT. SOC. PROC. 6: 14, 1866 GROTE-ROBINSON MATURE LARVA
PSYCHE 6:573, 1893
DYAR KEY TO DATANA LARVAE
UNITED STATES DEPT. AGRI. MISC. PUB. 657:355, 400, 1950 CRAIGHEAD LARVA
GLOVER-ILLUS. NORTH AMER. ENT. 1878 LARVA* PL. 10

FOOD PLANTS
CARYA SP.
CASTANEA DENTATA (MARSH.)
FRAXINUS SP. (ASHES)
HAMAMELIS VIRGINIANA LINN.
JUGLANS NIGRA LINN.
PLATANUS OCCIDENTALIS LINN.
QUERCUS SP. (OAKS)
ROSA SP. (ROSES)
ULMUS SP. (ELMS)

CONTRACTA WALKER
SYN. SEE QUADRISIGNATA WALKER (METALECTRA)

CONTRACTA WALKER
SYN. SEE MARGINATA HAWORTH

CONTRACTATA (PACKARD), THERA-GEOMETRIDAE
LIFE HISTORY
NEW YORK STATE AGRI. EXP. STA. (CORNELL), MEM. 274:142, 1948 FORBES LARVA
NEW YORK STATE MUS. MEM. 8: 754, 1905 FELT LARVA (BRIEF)
UNITED STATES DEPT. AGRI. BUR. ENT. BULL., OLD SERIES 12: 22, 1886 PACKARD LARVA, PUPA

FOOD PLANTS
JUNIPERUS SP. (JUNIPERS)
LARIX SP. (LARCHES)
LARIX LARICINA (DUROI)
PICEA SP. (SPRUCES)
PINUS SP. (PINES)
PINUS STROBUS LINN.

CONTRADICTA LEUSSLER
SSP. SEE BIMACULA GROTE & ROBINSON

CONTRARIA WALKER
SYN. SEE PICTA HARRIS

CONTRARIA WALKER
SYN. SEE XYLINOIDES GUENEE (HYPPA)

CONTRASTA BARNES & BENJAMIN
SSP SEE CONNATARIA GUENEE

CONTRASTA BARNES & MCDUNNOUGH
SSP. SEE HYDROMELI HARVEY

CONTRASTA STRAND
SSP. SEE POPULI STRECKER

CONVALLARIA GUENEE
SSP. SEE MUNITATA HUBNER

CONVERGARIA WALKER
SSP. SEE NOTATARIA WALKER

CONVEXIPENNIS GROTE & ROBINSON, CUCULLIA-NOCTUIDAE
LIFE HISTORY
MUS. COMP. ZOOL. BULL. 87: NO. 6, 1941 DETHIER LARVAL ANTENNAE* F. 66
NEW YORK CABIN. NAT. HIST., ANNUAL REPT. NO. 23:215, 1873 LINTNER LARVA
LOC. CIT. NO. 26:138, 1874

LINTNER LARVA
ONTARIO ENT. SOC. ANNUAL REPT. P.119,1905 GIBSON LARVA
HOLLAND-MOTH BOOK 1905 LARVA* PL.1, F.3 COLOR
INDEX AMER. ECON. ENT. VOL.5:166, VOL.7:228
FOOD PLANTS
ASTER SP.
CALLISTEPHUS CHINENSIS(LINN.)
SOLIDAGE ALTISSIMA LINN. (FLOWERS)

COOLIDGEI GUNDER
SSP. SEE EXILIS BOISDUVAL

COOLINENSIS WATSON & COMSTOCK
SSP. SEE ACADICA EDWARDS (STRYMON)

COOPERI BEHR
SYN. SEE CHALCEDONA DOUBLEDAY & HEWITSON

COOPERII BEHR
SYN. SEE CETHURA FELDER & FELDER

COORTARIA(HULST), ITAME-GEOMETRIDAE
A ENIGMATA BARNES & MCDUNNOUGH
LIFE HISTORY
NEW YORK STATE AGRI. EXP. STA. (CORNELL), MEM. 274:41, 1948 FORBES LARVA
FOOD PLANTS
CRATAEGUS OXYACANTHA LINN.
MALUS PUMILA MILL. (APPLE)

CORACIAS(GUENEE), ZALE-NOCTUIDAE
SYN. CORNIX GUENEE
SYN. CINEREA MORRISON
LIFE HISTORY
AMER. MUS. NAT. HIST. BULL. 16: 397, 1902 BEUTEN. LARVA
GUENEE-SPEC. GEN. LEP. NOCT. 3:19, 1852 LARVA* PL.2 COLOR
FOOD PLANTS
QUERCUS SP. (OAKS)
QUERCUS ALBA LINN.
ROBINIA PSEUDO-ACACIA LINN.

CORAS AUCT.
SEE PECKIUS KIRBY

CORBIS GODMAN & SALVIN
SYN. SEE PHILENOR LINNAEUS

CORCORANI GUNDER
SSP. SEE SARA BOISDUVAL

CORCORANI GUNDER
SSP. SEE HALESUS CRAMER

CORDAY HUBNER
SYN. SEE PHILEA JOHANSSON

CORDELIA HENRY EDWARDS
SSP. SEE CONNUBIALIS GUENEE

CORDIGERA(THUNBERG), ANARTA-NOCTUIDAE
SYN LUTEOLA GROTE & ROBINSON
LIFE HISTORY
NEW YORK STATE MUS. MEM. 8: 738, 1905 FELT LARVA (BRIEF)
STAINTON-BRITISH BUTTER & MOTHS 1:293, 1857 LARVA
FOOD PLANTS
VACCINIUM SP.

CORNELIUS FABRICIUS
SYN. SEE GEMMA HUBNER

CORNIX GUENEE
SYN. SEE CORACIAS GUENEE

COROLLARIA PERRY
SYN. SEE IO FABRICIUS

CORRELLATA HULST
SYN. SEE COLATA GROTE

CORTICEA WALKER
SYN. SEE IPOMOEAE DOUBLEDAY

CORVINA BOISDUVAL
SYN. SEE REBROSCAPUS MENETRIES

COSMION DYAR, MEROPLEON-NOCTUIDAE

LIFE HISTORY
ENT. SOC. AMERICA, ANNALS 17: 467, 1924 BYNUM-HOLLOWAY LARVA*
FOOD PLANTS
ERIANTHUS GIGANTEUS (WALT.)
SACCHARUM OFFICINARIUM LINN. (SUGAR)
ZEA MAYS LINN. (CORN)
GRASSES

COSTALIS WESTWOOD & HEWITSON
SYN. SEE JUVENALIS FABRICIUS

COSTATA (GROTE), EUXOA-NOCTUIDAE
A IDAHOENSIS GROTE
SYN. FURTIVA SMITH
LIFE HISTORY
FOOD PLANTS
ASTRAGALUS BISULCATUS (HOOK.)
MELILOTUS ALBA DESR.
SALSOLA KALI
V. TENUIFOLIA TAUSCH.

COSTINOTATA WARREN
SSP. SEE TRIANGULIFERATA PACKARD

COSYRA (DRUCE), MISA-AGARISTIDAE
LIFE HISTORY
NEW YORK ENT. SOC. JOURN. 14: 133, 1906 KUNZE LARVA
FOOD PLANTS
OPUNTIA ARBUSCULA ENGELM.

COTTLEI BARNES & BENJAMIN
SSP. SEE CABERATA GUENEE

COTTLEI COMSTOCK
SYN. SEE IRENE BOISDUVAL

COTTLEI GRINNELL
SSP. SEE ACMON DOUBLEDAY & HEWITSON

COTTLEI GUNDER
SSP. SEE NAPI LINNAEUS

COUPERI GROTE
SSP. SEE LYGDAMUS DOUBLEDAY

COXEYI CABBURY
SSP. SEE GORDIUS CRAMER

CRAMBIS FREYER
SYN. SEE POLIXENS FABRICIUS

CARMERI HOLLAND
SSP. SEE MONUSTE LINNAEUS

CRAMERI MENETRIES
SYN. SEE FICUS LINNAEUS

CRAMERI SCUDDER
SYN. SEE INTERROGATIONIS FABRICIUS

CRANTOR CRAMER
SYN. SEE ACHEMON DRURY

CRASSIUSCULA (HAWORTH), CAENURGINA-NOCTUIDAE
SYN. ERICHTO GUENEE
SYN. SOBRIA WALKER
FORM OCHREA GROTE
LIFE HISTORY
CALIFORNIA ACAD. SCI. PROC. 7: 23, 1876 EDWARDS EGG (ERICHTO)
INSECT LIFE 5:87, 1892
SLINGERLAND LARVA COMPARED WITH ERECHTEA
FOOD PLANTS
ERODIUM SP.
LUPINUS SP.
TRIFOLIUM SP. (CLOVER)

CRATAEGI BOISDUVAL & LECONTE
SYN. SEE TARQUINIUS FABRICIUS

CRATAEGI SAUNDER, CATOCALA-NOCTUIDAE
FORM PRETIOSA LINTNER
LIFE HISTORY
AMER. MUS. NAT. MEM. NEW SERIES 3: PART 1, 1918 BARNES & McDUNNOUGH LARVA* PL. 10, F. 21; PL. 12, F. 18 COLOR
BROOKLYN ENT. SOC. BULL. 7:38, 1884 HULST LARVA
CAN. ENT. 8:72, 1876 SAUNDERS

LARVA
NEW YORK STATE MUS. MEM. 8: 734, 1905 FELT LARVA
FOOD PLANTS
CRATAEGUS SP.
MALUS PUMILA MILL. (APPLE)

CREELMANI GUNDER
SSP. SEE CALLIPPE BOISDUVAL

CRENULATA SMITH
SYN. SEE EXSERTISTIGMA MORRISON

CREPUSCULARIA (SCHIFFERMULLER), ECTROPIS–GEOMETRIDAE
SYN. OCCIDUARIA GUENEE
SYN. SIGNARIA WALKER
SYN. SPATIOSARIA WALKER
SYN. INTRATARIA WALKER
SYN. CINERARIA WALKER
SYN. CUNEARIA D'URBAN
FORM ABRAXARIA WALKER
SYN. DIVISARIA WALKER
SYN. FERNALDARIA HULST
AB FUMATARIA MINOT
LIFE HISTORY
CAN. ENT. 10:67, 1878 GOODELL LARVA, PUPA
MINNESOTA AGRI. EXP. STA. BULL 61:242, 1899 LUGGER LARVA
NEW YORK ENT. SOC. JOURN. 53: 188, 1945 FORBES PUPA
NEW YORK STATE MUS. MEM. 8: 743, 1905 FELT LARVA
PSYCHE 4:269, 1885 DIMMOCK LARVAL NOTE
PSYCHE 19:17, 1912 BARNES & McDUNNOUGH EGG, LARVA
UNITED STATES DEPT. AGRI, MISC. PUB. 657:361, 430, 1950 CRAIGHEAD LARVA
GUENEE–SPEC. GEN. LEP. PHAL. 1: 260, 1857 LARVA
GUMPPENBERG–SYSTEMA GEOMETRARUM 5:346, 1892, LARVA, PUPA; 7:438, 1895 LARVA, PUPA (SIGNARIA)
STAINTON–BRITISH BUTTER & MOTHS 2:28, 1859 LARVA
FOOD PLANTS
ACER SP. (MAPLES)
ALNUS SP. (ALDERS)
AQUILEGIA SP.
BETULA SP. (BIRCHES)
CYTISUS SP.
GENISTA SP.
JUGLANS SP. (WALNUTS)
LIGUSTRUM SP.
LONICERA SP.
MALUS PUMILA MILL. (APPLE)
POPULUS SP. (POPLARS)
PRUNUS DOMESTICA LINN. (PLUM)
PYRUS COMMUNIS LINN. (PEAR)
QUERCUS SP. (OAKS)
ROSA SP. (ROSES)
SALIX SP. (WILLOWS)
SAMBUCUS SP.
TRIFOLIUM SP. (CLOVERS)
TSUGA CANADENSIS (LINN.)
ULMUS SP. (ELMS)
GENERAL FEEDER MOST TREES
GENERAL FEEDER ORCHARD TREES

CRESPHONTES CRAMER. PAPILIO–PAPILIONIDAE
SNY. OXILUS HUBNER
AB NAXWELLI FRANCK
AB FORSYTHAE GUNDER
LIFE HISTORY
AMER. ENT. 2:308, 1870 LARVA (THOAS)
AMER. MUS. NAT. HIST. BULL. 4: 75, 1892 BEUTEN. EGG, LARVA
LOC. CIT. 5:245, 1893 BEUTEN LARVA, CHRYSALIS
BUFFALO SOC. NAT. SCI., BULL. 19:18, 1939 WILD LARVA
CAN. ENT. 10:154, 1878 BOLL LARVA
LOC. CIT. 10:48, 1878 SAUNDERS LARVA
LOC. CIT 13:177, 1881 FRENCH EGG, LARVA, CHRYSALIS
LOC. CIT. 15:234, 1883 SAUNDERS EGG, LARVA
ILLINOIS DEPT. AGRI. TRANS. 15:

140,1877 FRENCH LARVA
LOC.CIT.18:75,1880 MIDDLETON LARVA
ILLINOIS STATE NORMAL UNIV., ANNUAL REPT.6:43,1880 FRENCH LARVA
MARYLAND ACAD.SCI.BULL.3(2): 8,1923 MOORE EGG,LARVA, CHRYSALIS
MARYLANB NAT.HIST.SOC. PROC. P.7,1933 HAYDON EGG, LARVA, CHRYSALIS
NATURE STUDY REVIEW P.224, SEPT. 1917 LARVA*
NEW YORK STATE CABIN.NAT. HIST.,ANN.REPT.NO.46:337, 1892 LINTNER LARVA* F.ONLY
NEW YORK STATE MUST.BULL.53: 760,1901 FELT LARVA
NEW YORK STATE MUS.MEM.8: 554,729,1905 FELT LIFE HIST. LARVA* F.137
ONTARIO ENT.SOC.ANN.REPT. P. 61,1878 SAUNDERS LARVA, CHRYSALIS
LOC.CIT.P.41, 1880 SAUNDERS LARVA (BRIEF)
LOC.CIT.P.110,1896 BETHUNE LARVA* CHRYSALIS*
PSYCHE 8:210,1898 SCUDDER LARVA
SMITHSONIAN MISC.COLLECTIONS 4:ART.2:8,1862 MORRIS LARVA, CHRYSALIS
UNITED STATES NAT.MUSM.BULL. 157:178,1932 AUSTIN CLARK LARVA, CHRYSALIS
ASHMEAD-INS.INJ.ORANGE P.47, 1880 LARVA
BOISDUVAL-SPEC.GEN.LEP.DIURNAE 1:356,1836 LARVA,CHRYSALIS
BOISDUVAL & LECONTE-HIST.LEP. NORTH AMERICA P.34,1833 LARVA* PLS.12,13 (THOAS)
COMSTOCK-HOW TO KNOW BUTTER. P.57,1904 LARVA*CHRYSALIS* PL.8 COLOR
FRENCH-BUTTER.EAST UNITED STATES P.103,1886 LARVA*
GLOVER-ILLUST.NORTH AMER. ENT. 1878 LARVA* PUPA* PL.24
GOSSE-LETTERS FROM ALABAMA P.170,1859 LARVA (THOAS)
GUNDLACH-ENTOM.CUBANA P.132, 1881 LARVA, CHRYSALIS
HOLLAND-BUTTERFLY BOOK 1905 LARVA* CHRYSALIS* PL.2, F.16; PL.6,FF.8-10
HUBBARD-INS. AFFECT.ORANGE P.137,1885 LARVA*COLOR
KELLOGG-AMER.INS.P.450, 1908 LARVA
MAYNARD-BUTTER.NEW ENGLAND P.50,1886 LARVA,CHRYSALIS
MORRIS-SYNOP.LEP.NORTH AMERICA P.8,1862 LARVA
PEAIRS-POSTS FARM.GARD. ORCHARD 1941 LARVA* F.561
QUALE-INS.CITRUS SUBTROPICAL FRUITS P.204,1938 LARVA* F.136 CHRYSALIS
SAUNDERS-INS.INJ.FRUITS P.137, 1883 LIFE HISTORY*
SAUNDERS-LOC.CIT.P.317,1889 LARVA* F.390; CHRYSALIS* F.391
SCUDDER-BUTTER.EAST.UNITED STATES & CANADA PP.1337-1339 1341-1343,1889 LIFE HISTORY* PL.66,F.19;PL.75,F.1;PL. 76,F.16;PL.79,FF.62,66;PL 85,FF.8-10
SCUDDER-EVERYDAY BUTTER.P. 341,1899 EGG,LARVA, CHRYSALIS
SCUDDER-GUIDE TO COMMON BUTTER.P.152,1893 LARVA, CHRYSALIS
VANCE-LIFE AMONG BUTTER. P. 51,1925 LARVA
INDEX AMER.ECON.ENT.

CRESPHONTES AND THOAS ARE OFTEN CONFUSED IN THE LITERATURE SO REFERENCES TO BOTH SPECIES ARE GIVEN.

VOL. 1:230;VOL. 2:264;
VOL. 3:294;295;VOL4:342,343;
VOL. 5:469;VOL6:559;VOL. 7:
741

FOOD PLANTS

ANETHUM GRAVEOLENS LINN.
ATALANTIA CEYLONICA OLIVER
CITRUS SP.
CITRUS GRANDIS(LINN.)
CITRUS LIMON(LINN.)
CITRUS SINENSIS(LINN.)
DICTAMNUS ALBUS LINN.
FERONIA LIMONIA(LINN.)
FORTUNELLA CRASSIFOLIA SWINGLE
NYSSA SYLVATICA MARSH.
PERSEA BORBONIA(LINN.)
PIPER ADUNCUM LINN.
PIPER MOLLICOMUM KUNTH
PIPER PELTATUM LINN.
PIPER UMBELLATUM LINN.
POPULUS NIGER V. ITALICA MUENCH
?PRUNUS VIRGINIANA LINN.
PTELEA TRIFOLIATA LINN.
RUTA GRAVEOLENS LINN.
STAPHYLEA TRIFOLIA LINN.
THALICTRUM SP.
TRIPHASIA SP. (CUBA)
ZANTHOXYLUM CLAVA-HERCULES LINN.
ZANTHOXYLUM CORIACEUM A. RICH.
ZANTHOXYLUM FAGARA(LINN.)
ZANTHOXYLUM FRAXMEUM(WILLD.)

CRESSONANA GROTE, CTENUCHA-AMATIDAE

FORM SANGUINARIA STRECKER
FORM LUTEA GROTE

LIFE HISTORY

NEW YORK ACAD. SCI. ANNALS 8:
89,1893 PACKARD LARVA

FOOD PLANTS

CRESSONI GROTE

SYN. SEE RUFIMAGO HUBNER

CRETACEATA(PACKARD), EUPITHECIA-
GEOMETRIDAE

LIFE HISTORY

NEW YORK STATE AGRI. EXP. STA.
(CORNELL), MEM. 274:172,
1948 FORBES LARVA

FOOD PLANTS

VERATRUM SP.

CRETATA AUCT.

SEE LEPORINA LINNAEUS

CRETATA SMITH

SYN. SEE LEPUSCULINA GUENEE

CRETATOIDES BENJAMIN

SSP. SEE LEPORINA LINNAEUS

CREUSA(DOUBLEDAY & HEWITSON), EUCHLOE-
PIERIDAE

SYN. ELSA BEUTENMULLER
SYN. PUMILIO STRAND
A HYANTIS EDWARDS
SYN. PSEUDOAUSONIDES VERITY
B ORIENTALIDES VERITY
C. LOTTA BEUTENMULLER
SYN. BELLIOIDES VERITY

LIFE HISTORY

AMER. MUS. NAT. HIST. BULL. 10:
242,1898 BEUTEN. LARVA
ENT. NEWS 36:66,1925
COOLIDGE LIFE HISTORY
(HYANTIS)
PSYCHE 2:183,1878 MEADE LARVA
(HYANTIS)
SOUTHERN CALIFORNIA ACAD. SCI.
BULL. 31:35,1932 COMSTOCK
& DAMMERS LIFE HISTORY
(LOTTA) EGG, LARVA* PL. 15;
CHRYSALIS* PL. 16
COMSTOCK-BUTTER. CALIFORNIA
P. 37,1927 LARVA (HYANTIS)
INDEX AMER. ECON. ENT.
VOL. 6:276 (LOTTA)

FOOD PLANTS

ARABIS GLABRA LINN.
CAULANTHUS INFLATUS WATS.
CRUCIFERAE SP.
DESCURAINEA PINNATA(WALT),
BRITT.
SISYMBRIUM ALTISSIMUM LINN.
(CAPTIVITY)
STREPTANTHUS TORTUOSUS KELL.

727,1905 Felt larva
FOOD PLANTS
Acer sp. (maples)
Betula sp. (birches)
Carya sp. (hickories)
Malus pumila Mill. (apple)
Quercus sp. (oaks)
Quercus alba Linn.
Salix sp. (Willows)

confusa (Lyman), Haploa-Arctiidae
form lymani Dyar
form tringularis Smith
form suffusia Francelmont
LIFE HISTORY
Can. Ent. 19:185, 1887
Lyman egg, larva, pupa
Naturaliste Canadien p. 15,
New York Ent. Soc. Journ. 3:59
1895 Dyar egg, larva
Psyche 13:32, 1906 Poster
Table larval differences
in genus Haploa
FOOD PLANTS
Cynoglossum officinale Linn.

confusa Walker
syn. see ipomoeae Doubleday

confusa Worthington
syn. see ilia Cramer

confusaria (Hubner), Apicia-Geometridae
form metrocamparia Guenee
ab subochrearia Hulst
for many syns. see McDunnough's
List
LIFE HISTORY
New York Ent. Soc. Journ. 53:194
1945 Forbes larva, pupa
New York State Agri. Exp. Sta.
(Cornell), Mem. 274:106,
1948 Forbes egg, larva
Psyche 9:407, 1902 Dyar
egg, larva (subochrearia)
Psyche 14:92, 1907
Dyar egg, larva
Packard-Monograph Geometridae
p. 535, 1876 larva* Pl. 13,
f. 18 a; pupa*
FOOD PLANTS
Aster sp.
Polygonum sp.
Solidago sp.
Taraxacum officinale Weber
Trifolium sp. (clovers)
Many low plants

confusum Rothschild
syn. see carolina Henry Edwards

congregata Walker
ssp. see abrasaria Herrich-Schaeffer

congrua Walker
syn. see virginica Fabricius

congrua (Walker), Estigmene-Arctiidae
syn. antigone Strecker
a athena Strecker
LIFE HISTORY
Amer. Mus. Nat. Hist. Bull. 10;
375, 1898 Beuten. larva
New York Ent. Soc. Journ. 5:131,
1897 Dyar larval structure
only* Pl. 7, f. 2 antigone
FOOD PLANTS
Agaricus campestris Linn.
(mushrooms)
bores in the stems
Chenopodium album Linn.
Plantago sp.
Taraxacum officinale Weber
Low Plants

congrua Walker
syn see basiaria Walker

congruata Walker
syn. see herbaria Fabricius

coniferaria Packard
syn. see mimosaria Guenee

coniferarum (Abbot & Smith) Lepara-
Sphingidae

CRIBARIA(LJUNG), ARTACE-LASIOCAMPIDAE
SYN. PUNCTISTRIGA WALKER
SYN. RUBRIPALPIS FELDER
LIFE HISTORY
AMER. MUS. NAT. HIST. BULL. 10: 444, 1898 BEUTEN. LARVA
AMER. PHIL. SOC. PROC. 31:189, 1893 PACKARD EGG, LARVA* PL. 11, F. 22 RUBRIPALPIS
NEW YORK ACAD. SCI. ANNALS 8: 230, 1894 DYAR LARVAL SETAE
FOOD PLANTS
DIOSPYROS VIRGINIANA LINN.
PRUNUS PERSICA(LINN.) PEACH)
PRUNUS SEROTINA EHRH.
QUERCUS SP. (OAKS)

CRIBRATARIA(GUENEE), GLENA-GEOMETRIDAE
LIFE HISTORY
NEW YORK ENT. SOC., JOURN. 53:189, 1945 FORBES PUPA
NEW YORK STATE AGRI. EXP. STA. (CORNELL), MEM. 274-54, 1848 FORBES LARVA
GUMPPENBERG-SYSTEMA GEOMETRARUM 5:319, 1892 LARVA, PUPA
FOOD PLANTS
PICEA SP. (SPRUCES)
POPULUS SP. (POPLARS)
POPULUS NIGRA V. ITALICA MUENCH
POPULUS TREMULOIDES MICHX.
SALIX SP. (WILLOWS)

~~CRIMI~~NOSA SWETT
SYN. SEE ANGUSTIORATA WALKER

CRISPA HARVEY
SSP. SEE PURPUREA GROTE

CRISTIFERA(WALKER), POLIA-NOCTUIDAE
SYN. IMPOLITA MORRISON
LIFE HISTORY
FOOD PLANTS
BETULA SP. (BIRCHES)
RHUS SP. (SUMACS)
VACCINIUM SP. (HUCKLEBERRIES)

CROCALE EDWARDS
SSP. SEE LACINIA GEYER

CROCALLATA GUENEE, TETRACIS-GEOMETRIDAE
A ASPILATATA GUENEE
LIFE HISTORY
CAN. ENT. 12:193, 1879 GOODELL LARVA PUPA
NEW YORK AGRI. EXP. STA. (CORNELL), MEM. 274:107, 1948 FORBES LARVA
NEW YORK STATE MUS. MEM. 8: 721, 1905 FELT LARVA (BRIEF)
PACKARD-MONOGRAPH GEOMETRIDAE P. 548, 1876 LARVA PUPA
FOOD PLANTS
ALNUS SP. (ALDERS)
CASTANEA DENTATA(MARSH.)
LINDERA BENZOIN(LINN.)
RHUS SP. (SUMACS)
GENERAL FEEDER MOST TREES?

CROCATARIA FABRICIUS
SYN. SEE SOSPETA DRURY

CROCEARIA PACKARD
SSP. SEE MACULARIA HARRIS

CROESIOIDES SCUDDER
SYN. SEE AUGUSTUS KIRBY

CROESUS DALMAN
SYN. SEE PLUTO FABRICIUS

CROTCHI HENRY EDWARDS, PSEUDALYPIA-AGARISTIDAE
FORM ATRATA HENRY EDWARDS
LIFE HISTORY
SOUTHERN CALIFORNIA ACAD. SCI. BULL 42::59, 1943 COMSTOCK EGG* PL. 7; LARVA* PLS. 7, 8; PUPA* PL. 7
FOOD PLANTS
MALVASTRUM EXILE GRAY

CRUCIALIS HARVY, XYLOMYGES-NOCTUIDAE
A PERITALIS SMITH

LIFE HISTORY
FOOD PLANTS
QUERCUS GARRYANA DOUGL.

CRUCIFERARUM BOISDUVAL
SSP.SEE NAPI LINNAEUS

CRUENTARIA HUBNER
SYN.SEE COLORARIA FABRICIUS

CRUMBI(BENJAMIN),CYMATOPHORA-THYATIRIDAE
LIFE HISTORY
FOOD PLANTS
SORBUS OCCIDENTALIS (WATS.)

CRURALIS(GUENEE),ZANCLOGNATHA-NOCTUIDAE
LIFE HISTORY
ENT.AMER.14:181,1934 CRUMB LARVA* PL.23 A; PL.24, MC; PL.25 F
FOOD PLANTS
DEAD LEAVES

CRUZENSIS DYAR
SSP.SEE MACULARIA HARRIS

CRYDINA DYAR
SSP.SEE PURPURISSATA GROTE(POLIA)

CUBENSIS GROTE
SYN.SEE STRIGOSA WALKER

CUBENSIS GROTE
SSP.SEE BRONTES DRURY

CUBENSIS SCHAUS
SYN.SEE CAROLINA HENRY EDWARDS

CUBILIS GROTE
SSP.SEE STRACIS GUENEE

CUCULIFERA HERRICH-SCHAEFFER
SYN.SEE ANGUINA ABBOT & SMITH

CUCULLIFORMIS(GROTE),SEPTIS-NOCTUIDAE
LIFE HISTORY
WASHINGTON ENT.SOC.PROC. 4: 320,1898 DYAR LARVA
FOOD PLANTS
ELYMUS CONDENSATUS PRESL.

CULTA LINTNER
SYN.SEE OO CRAMER

CUMATILIS GROTE & ROBINSON
SYN.SEE MAGNOLIATA GUENEE

CUMULATA SWETT
SSP.SEE NUBILOFASCIATA PACKARD

CUNEA(DRURY),HYPHANTRIA-ARCTIIDAE
SYN.PUNCTATISSIMA ABBOT & SMITH
SYN. MUTANS WALKER
AB PUNCTATA FITCH
AB PALLIDA PACKARD
AB SUFFUSA STRUCKER
AB BRUNNEA STRECKER
FORM BUDEA HUBNER
SYN.PUNCTATISSIMA ABBOT & SMITH
LIFE HISTORY
AMER.MUS.NAT.HIST.BULL.10: 376,1898 BEUTEN.LARVA
CAN.ENT. 3:36,1871 SAUNDERS LARVA
LOC.CIT.70:83,1938 SWAINE NUMBER OF LARVAL MOULTS
ENT.NEWS 26:225,1915 GIRAULS LIFE HISTORY
ILLINOIS DEPT.AGRI.TRANS.15: 185,1877 FRENCH LARVA
LOC.CIT18:APPEND:116,154,1880 COQUILLETT-MARTEN LARVA* PUPA*
ILLINOIS,REPT.ON INS.1:18 1871 LEBARON LARVA,NEST
MICHIGAN AGRI.EXP.STA.CIRC. 137:18,1935 PETTIT LARVA
MICHIGAN STATE BOARD AGRI. REPT.13:152,1874 COOK LARVA* PUPA*
MICHIGAN STATE COLLEGE,SPEC. BULL.243:19,1933 MCDANIEL LIFE HISTORY LARVA* PUPA*

MINNESOTA ACAD. NAT. SCI., BULL 61:137, 1899 LUGGER LARVA* F. 82; PUPA* F. 82
MISSOURI, REPT. INS. 3:130, 1871 RILEY LARVA* PUPA*
NEW HAMPSHIRE BOARD AGRI., REPT. 8:182 1879 CUTTING LARVA* PUPA*
NEW YORK ENT. SOC. JOURN. 5:132, 1897 DYAR LARVAL STRUCTURE* PL. 7, F. 3
NEW YORK STATE MUS. BULL. 37:23, 1900 FELT LARVA* PUPA*
NEW YORK STATE, REPT. INS. P. 382, 1856 FITCH LARVA SEE ALSO PP. 64, 66 IN SEC 88, 91.
OHIO AGRI. EXP. STA. BULL. 45: 162, 1893 WEBSTER EGG, LARVA
LOC. CITP. 332, 213, 1918 HOUSER EGG* PL. 20, F. 1; LARVA
ONTARIO ENT. SOC. ANN. REPT. P. 27, 1893 HARRINGTON LARVA* PUPA*
LOC. CIT P. 58, 1901 LYMAN LARVA
OTTAWA NATURALIST 6:70, 1892 FLETCHER EGG, LARVA* PUPA*
UNITED STATES SEPT. AGRI. BUR. ENT. BULL. 10. OLD SERIES P. 37. 1888 RILEY EGG* F. 17B; LARVA* F. 18 A, B, C; PUPA* F. 18 D, 3
UNITED STATES DEPT. AGRI. BUR. ENT. BULL. 34 NEW SERIES P. 30, 1902 BANKS EGG, LARVA, PUPA, COCOON* F. 20
UNITED STATES DEPT. AGRI. DIV. ENT. BULL. 10-36-58, 1887 RILEY LIFE HISTORY*
UNITED STATES DEPT. AGRI. FARMER'S BULL. 99:20, 1899 HOWARD LARVA* F. 10 A, B, C; F. 11; PUPA* F. 10C; COCOON* F. 9
UNITED STATES DEPT. AGRI. MIXC. PUB. 657:353, 387, 1950 CRAIGHEAD LARVA* F. 76
UNITED STATES DEPT. AGRI. REPT. ENT. PP. 518-539, 1886 RILEY LIFE HISTORY* PLS. 10, 11
UNITED STATES DEPT. AGRI. YEARBOOK FOR 1895 P. 386 HOWARD LARVA* F. 92; PUPA* F. 92 NO DESCRIPTION
VERMONT STATE BOARD AGRI. REPT. 2:589, 1876 PERKINS LARVA
ABBOT & SMITH-LEP. INS. GEORGIA P. 139, 1797 LARVA* PL. 70 COLOR
ESSIG-INS. WEST. NORTH AMERICA P. 677, 1926 EGG, LARVA* COCOON
FERNALD & SHEPARD-APPLIED ENT. P. 285, 1942 LARVA
HARRIS-ENT. CORRESP P. 360, 1869 LARVA
HARRIS-INS. INJ. VEG. 1ST EDIT. P. 255 1841 LARVA, PUPA
HARRIS-LOC. CIT 2ND EDIT., P. 274, 1852 LARVA
HARRIS-LOC. CIT FLINT EDIT. P. 358, 1862 LARVA* PUPA* COCOON
HARRIS-NEW ENGLAND FARMER 7:33, 1828 LARVA
MORRIS-SYNOP. LEP. NORTH AMERICA P. 344, 1862 LARVA
PACKARD-INS. INJ. FORREST TREES P P. 67, 1881 LARVA* PUPA*
PACKARD-INS. INJ. IN WEST P. 794, 1877 LARVA* PUPA*
SAUNDERS-INS. INJ. FRUITS P. 71, 1883 LARVA*PUPA*COCOON*
SLINGERLAND & CROSBY-MANUAL FRUIT INS. P. 107, 1914 EGG; LARVA* F. 119; PUPA
SMITH-INS. NEW JERSEY P. 399, 1900 EGG*LARVA*PUPA* FF. 184, 186
STRETCH-ZYGAEN. & BOMBY. NORTH AMERICA P. 206, 1873 LARVA

Index Amer. Econ. Ent.
Vol. 1:60; Vol. 2:182;
Vol. 3:203; Vol. 4:229;
Vol. 5:313; Vol6:364;
Vol. 7:482

FOODPLANTS-n. p. not preferred

Acer sp. (maples)
Acer nagundo Linn.
Acer rubrum Linn. (N. P.)
Acer saccharinum Linn.
Acer saccharophorum Koch.
Aesculus glabra Willd. (n. p.)
Aesculus hippocastanum Linn.
Aesculus octandra Marsh. (n. p.)
Ailanthus altissima(Mill.)
Alnus sp. (alders)
Alnus maritima(Marsh.)
Althaea rosea Cav.
Amelanchier canadensis(Linn.)
Asimina triloba(Linn.)
Berberis canadensis Mill.
Betula sp. (b9rches)
Betula alba Linn.
Betula nigra Linn.
Betula papyrifera Marsh.
Buxus sp. (boxwoods)
Buxus semperivirens Linn.
Campsis radicans(Linn.)
Canna sp.
Carpinus carolinaiana Walt
Carya sp. (hickories)
Carya galbra(Mill.)
Carya illinoensis(Wand.)
Carys laciniosa(Michx.)
Castanea dentata(Marsh.)
Castanea pumila(Linn.)
Catalpa bignonioides Walt.
Catalpa speciosa Warder
Celtis occidentalis Linn.
Cephalanthus occidenalis Linn.
Cercis canadensis Linn.
Chamaecyprris thyoides(Linn.)
Chenopodium album Linn.
Chionanthus virginicus Linn. (n. p.)
Clematis sp.
Corus sp.
Cornus alternifolia Linn.
Cornus florida Linn.
Corylus americana Walt.
Cotinus coggygria Scop.
Crataegus sp. (thorns)
Cupressus sp. (cypress)
Cydonia oblonga Mill.
Diospyros kaki Linn.
Diospyros virginiana Linn.
Euonymus atropurpureus Jacq.
Fagus grandifolia Ehrh. (n. p.)
Ficus carica Linn.
Fraxinus sp. (ashes)
Fraxinus americana Linn.
Fraxinus excelsior Linn.
Gerranium sp.
Gleditsia triacanthos Linn. (n. p.)
Gossypium herbaceum Linn.
Gymocladus dioica(Linn.)(n. p.)
Hamamelis virginiana Linn.
Helianthus sp.
Hibiscus syriacus Linn.
Ilex opaca Ait.
Jasminium sp.
Juglans californica S. Wats.
Juglans nigra Linn.
Juglans regia Linn.
Juniperus virginiana Linn.
Kalmis sp. (laurels) n. p.
Larix decidua Mill.
Ligustrum vulgare Linn. (n. p.)
Liquidambar styraciflua Linn.
Liriodendron tulipifera Linn. n. p.
Lonicera sp.
Macluara pomifera(Raf.)
Magnolia sp. (n. p.)
Malus angustifolia(Ait.)
Malus coronaris(Linn.)
Malus diversifolia(Bong.)
Malus pumila Mill. (apple)
Morus sp.
Morus rubra Linn. n. p.
Nyssa sylvatica Marsh.
Ostrya virginiana (Mill.)
Oxydendrum arboreum(Linn.)
Parthenium argentatum Gray
Parthenocissus quinquefolia (Linn.)

PAULOWNIA TOMENTOSA(THUNB.)
PELARGONIUM SP.
PINUS SP. (PINES)N.P.
PLATANUS OCCIDENTALIS LINN.
POPULUS SP. (POPLARS)
POPULUS ALBA LINN.
POPULUS BALSAMIFERA LINN.
POPULUS DELTOIDES MARSH.
POPULUS FREMONTI S.WATS.
POPULUS NIGRA V.ITALICA MOENCH
POPULUS TREMULOIDES MICHX.
PORTULACA OLERACEA LINN.
PRUNUS AMERICANA MARSH.
PRUNUS AVIUM(LINN.)
PRUNUS CERASUS LINN. (CHERRY)
PRUNUS DOMESTICA LINN. (PLUM)
PRUNIS ILICIFOLIA(NUTT.EX. H.& A.)
PRUNUS PERSICA(LINN.)(PEACH)
PRUNUS SEROTINA EHRH.
PRUNUS VIRGINIANA LINN.
PYRUS COMMUNIS LINN. (PEAR)
QUERCUS SP. (OAKS)
QUERCUS ALBA LINN.
QUERCUS COCCINEA MOENCH
QUERCUS PHELLOS LINN.
QUERCUS PRINUS LINN.
QUERCUS RUBRA LINN.
RHAMNUS ALNIFOLIA L'HER
RHODODENDRON SP.N.P.
RIBES LACUSTRE(PERS.)
RIBES SATIVUM SYME(CURRANT)
RICINUS COMMUNIS LINN.
ROBINIA PSEUDO=ACACIA LINN.
ROSA SP. (ROSES)
RUBUS ALLEGHENIENSIS PORTER
RUBUS IDAEUS(MICHX.)V.STRIGOSUS (MICHX.)
SALIX SP. (WILLOWS)
SAMBUCUS CANADENSIS LINN.
SASSAFRAS ALBIDUM V.MOLLE (RAF.)
SPIREAA SP.
STAPHYLEA TRIFOLIA LINN.
SYMPHORICARPOS ALBUS(LINN.)
SYRINGA SP.
TAXUS SP.
TILIA SP.
TILIA AMERICANA LINN.
TILIA EUROPAEA LINN.
TRIFOLIUM SP. (CLOVERS)
ULMUS SP. (ELMS)
ULMUS AMERICANA LINN.
ULMUS RUBRA MUHL.
VIBURNUM SP.
VITIS VULPINA LINN.
WISTERIA FRUTESCENS(LINN.)
WISTERIA SINENSIS SWEET
ZANTHOXYLUM FRAXINCUM WILLD.
ZEA MAYS LINN. (CORN)

<u>CUNEARIA</u> D'URBAN
SYN. SEE <u>CREPUSCULARIA</u> SCHIFFER-MULLER

<u>CUNEATA</u> GOEZE
SYN. SEE <u>NAIS</u> DRURY

<u>CUNEATA</u>(GROTE), <u>LACINIPOLIA</u>-NOCTUIDAE
A <u>GERTANA</u> SMITH
AB <u>BASIRUFA</u> STRAND
AB <u>RUBICUNDA</u> STRAND
LIFE HISTORY
FOOD PLANTS
ALNUS SP. (ALDERS)
PRUNUS PENSYLVANICA LINN.
RIBES SANGUINEUM PURSH.
SALIX SP. (WILLOWS)

<u>CUNEGUNDA</u> CRAMER
SYN. SEE <u>DEFLORATA</u> FABRICIUS

<u>CUNNINGHAMI</u> OWEN
SSP. SEE <u>MORMONIA</u> BOISDUVAL (ARGYNNIS)

<u>CUPEDINARIA</u> GROTE, <u>RACHEOSPILA</u>-GEOMETRIDAE
SYN. <u>LOUISA</u> HULST
LIFE HISTORY
WASHINGTON ENT. SOC. PROC. 4:457, 1900 DYAR LARVE
FOOD PLANTS
LANTANA CAMARA LINN.

<u>CUPIDA</u>(GROTE), <u>RHYNCHARGTOTIS</u>-

NOCTUIDAE
SYN. VELATA WALKER
FORM BRUNNEIPENNIS GROTE
LIFE HISTORY
BROOKLYN ENT. SOC. BULL. 27:88 1932 CRUMB LARVA*PL. 11, F
ILLINOIS DEPT. AGRI. TRANS. 18: APPENDP. 136, 1880 MARTIN LARVA
FOOD PLANTS
MALUS PUMILA MILL, (APPLE)
PRUNUS PERSICA (LINN.) (PEACH)
VITIS SP.
VITIS VINIFERA LINN. (GRAPE)

CUPIDARIA GROTE
SYN. SEE QUERNARIA ABBOT & SMITH

CUPIDATA SWETT
SSP. SEE NUBILOFASCIATA PABKARD

CUPIDISSIMA GROTE
SSP. SEE EXSERTISTIGMA MORRISON

CUPRESSATA PEARSALL, EUPITHECIA-GEOMETDAE
LIFE HISTORY
FOOD PLANTS
CUPRESSUS SP.

CUPRESSI (BOISDUVAL), ISOPARCE-SPHINGIDAE
LIFE HISTORY
FLORIDA ENT. 12:20, 1928 BATES LARVA, PUPA
BOISDUVAL-SPEC. GEN. LEP. SPHING. P. 102, 1874 LARVA* PUPA* PL. 2 COLOR
HARRIS-ENT. CORRESP. P. 128, 1869 LARVA PROBABLY THIS SP.
INDEX AMER. ECON. ENT. VOL. 4:256
FOOD PLANTS
CHAMAECYPARIS THYOIDES (LINN.)
TAXODIUM DISTICHUM (LINN.)

CUPREUS (EDWARDS), LYCAENA-LYCAENIDAE
AB MACULINITA GUNDER
LIFE HISTORY
FOOD PLANTS
RUMEX PAUCIFOLIUS NUTT. EX WATS.

CURASSAVICAE FABRICIUS
SYN. SEE PLEXIPPUS LINNAUS

CURIALIS GROTE, XYLOMYGES-NOCTUIDAE
A INDURATA SMITH
B NICALIS SMITH
C TANTIVA SMITH
D ARGUS SMITH
LIFE HISTORY
JOURN. ECON. ENT. 28:1020, 1935 WOLGUM-LEWIS LARVA (BRIEF)
FOOD PLANTS
CITRUS SP.
STELLARIA MEDIA (LINN.)
COVER CROPS'

CURIALIS SMITH, MAMESTRA-NOCTUIDAE
LIFE HISTORY
CAN. ENT. 38:381, 1906 FLETCHER EGG, LARVA*PUPA
UNITED STATES NAT. MUS. PROC. 27:859, 1904 DYAR EGG
FOOD PLANTS
BRASSICA OLERACEA LINN. (CABBAGE)
DELPHINIUM SP.
DICENTRA SPECTABILIS (LINN.)
SPINACEA OLERACEA LINN.
TARAXACUM OFFICINALE WEBER
TRIFOLIUM SP. (CLOVERS)
GRASSES

CURTIUS SCUDDER
SEE ACCIUS ABBOT & SMITH

CURVATA FRENCH
SSP. SEE ROBINSONI GROTE

CURVATA GROTE
SSP. SEE BINOTATA WALKER

CURVIFASCIA BROWER

SSP. SEE AMICA HUBNER

CURVIFASCIA SKINNER
SSP. SEE ASTERIUS CRAMER

CURVIMACULA(MORRISON), XYLENA-NOCTUIDAE
LIFE HISTORY
NEW YORK STATE MUS. MEM. 8: 743, 1905 FELT LARVA(BRIEF)
PSYCHE 2:122, 1878 THAXTER LARVA
PSYCHE 8:336 1899 DYAR EGG, LARVA
INDEX AMER. ECON. ENT. VOL. 2:386
FOOD PLANTS
IRIS IN CAPTIVITY BUT NOT GOOD
SALIX SP. (WILLOWS)
"OTHER LEAVES"
GENERAL FEEDER FOREST TREES

CUSPIDEA(HUBNER), EUCLIDINA-NOCTUIDAE
LIFE HISTORY
CAN. ENT. 27:107, 1895 LEMBERT EGG, LARVA
WASHINGTON ENT. SOC. PROC. 4:326, 1898 DYAR LARVA
FOOD PLANTS
LUPINUS CHAMISSONIS ESCHESCH
TRIFOLIUM SP. (CLOVERS)
VITIS SP. (GRAPES)
VITIS VINIFERA LINN.
GRASSES

CUSTODIATA(GUENEE), PERIZOMA-GEOMETRIDAE
SYN. GUENEEATA PACKARD
FORM CARNATA PACKARD
FORM POLYGRAMMATA HULST
LIFE HISTORY
SOUTHERN CALIFORNIA ACAD. SCI. BULL. 38:32, 1939 COMSTOCK-DAMMERS LARVA* PL. 9; PUPA* PL. 9 POLYGRAMMATA
FOOD PLANTS
ATRIPLEX CANESCENS(PURSH)
ATRIPLEX SEMIBACCATA R. BR.
GRAYIA SPINOSA(HOOK.)
ALL THREE PLANTS LISTED AS FOOD FOR POLYGRAMMATA

CUYAMA COMSTOCK, SOMATOLOPHIA-GEOMETRIDAE
LIFE HISTORY
SOUTHERN CALIFORNIA ACAD. SCI. BULL. 38:172, 1939 COMSTOCK LARVA, PUPA* PL. 40
FOOD PLANTS
HYMENOCLEA SALSOLA GRAY

CUYAMACA WRIGHT
SYN. SEE SPINETORUM HEWITSON

CYANESCENS HAMPSON
SSP. SEE LEPUSCULINA GUENEE

CYBELE(FABRICIUS), AGRYNNIS-NYMPHALIDAE
SYN. DAPHNIS CRAMER
AB BAAL STRECKER
AB BARTSCHI REIFF
A NOVASCOTIAE MCDUNNOUGH
B KRAUTWORMI HOLLAND
C CARPENTERII EDWARDS
D PSEUDOCARPENTERI CHERMOCK
LIFE HISTORY
AMER. MUS. NAT. HIST. BULL. 5: 255, 1893 BEUTEN. LARVA, CHRYSALIS
BUFFALO SOC. NAT. SCI. BULL 19:22, 1939 WILD LARVA
BUTTERFLY PARK NATURE CLUB 2:NO PAGE NUMBER, 1931 EGG, LARVA
CAN. ENT. 4:121, 1872 SAUNDERS LARVA, CHRYSALIS
LOC. CIT. 6:121, 1874 EDWARDS LARVA
LOC. CIT. 12:141, 1880 EDWARDS EGG, LARVA, CHRYSALIS
ILLINOIS DEPT. AGRI. TRANS 18:APPEND. P. 81, 1880 MIDDLETON LARVA
MONTANA UNIV. BULL. 30:55, 1906 ELROD EGG, LARVA, CHRYSALIS
MUS. COMP. ZOOL BULL 38:NO. 6 1941 DETHIER LARVAL

ANTENNA* F.38
NATURALISTE CANADIEN P.115, 1927 HUARD LARVA, CHRYSALIS
COMSTOCK-HOW TO KNOW BUTTER. P.114, 1904 LARVA
EDWARDS, W.H.-BUTTER, NORTH AMERICA 2ND SERIES SUPPL. 1872 EGG
EDWARDS, W.H.-LOC.CIT. 2ND SERIES, 1876 LIFE HISTORY* PL. ARGYNNIS NO.7
EDWARDS, W.H.-LOC.CIT3RD SERIES, 1897 LIFE HISTORY* PL. ARGYNNIS NO.11 FF.A-I
EDWARDS, W.H.-LOC.CIT.3RD SERIES, 1897 LIFE HISTORY* PL. ARGYNNIS NO.11 <u>CARPENTERII</u>
FRENCH-BUTTER. EAST. UNITED STATES P.155, 1886 LIFE HISTORY
HOLLAND-BUTTERFLY BOOK 1905 P.106 CHRYSALIS* PL.5, FF. 1-3
HOLLAND-LOC.CIT.3RD EDIT. PP.86-87, 1931 LIFE HISTORY CHRYSALIS PL.4, FF.1-3
KELLOGG-AMER. INS. P.456, 1908 LARVA
MAYNARD-BUTTER. NEW ENGLAND P.23, 1886 LARVA, CHRYSALIS
SCUDDER-BUTTER. EAST. UNITED STATES & CANADA PP.558-562, 1889 LIFE HISTORY* PL. 64, F.39; PL.67, F.12; PL.75, F.4; PL.79, FF.4-6; PL.84, FF.1-3.
SCUDDER-EVERYDAY BUTTER. P. 205, 1899 EGG, LARVA, PUPA
SCUDDER-GUIDE TO COMMON BUTTER. P.79, 1893 LARVA, CHRYSALIS
INDEX. AMER. ECON. ENT. VOL.6:68; VOL.7:105

FOOD PLANTS
VIOLA SP. (VIOLETS)

<u>CYCLADATA</u> GROTE, <u>HELIOMATA</u>-GEOMETRIDAE

LIFE HISTORY
NEW YORK STATE AGRI. EXP. STA. (CORNELL), MEM. 274:35, 1948 FORBES EGG, LARVA, PUPA
NEW YORK ENT. SOC., JOURN. 53:183, 1945 FORBES PUPA
PSYCHE 9:177, 1901 DYAR EGG

FOOD PLANTS
ROBINIA PSEUDO-ACACIA LINN.

<u>CYCLICA</u> HENRY EDWARDS
SSP. SEE <u>EGLE</u> DRURY

<u>CYGNUS</u> EDWARDS
SYN. SEE <u>CALIFORNICA</u> EDWARDS (STRYMON)

<u>CYMATOPHOROIDES</u> (GUENEE), <u>PSEUDOTHYATIRA</u> THYATIRIDAE
FORM <u>EXPULTRIX</u> GROTE

LIFE HISTORY
AMER. MUS. NAT. HIST. BULL. 14: 231, 1901 BEUTENMULLER LARVA
NAT. ACAD. SCI. MEM. 7:168, 1895 PACKARD LARVA* PL. 19, F.5 COLOR
NEW YORK STATE MUS. MEM. 8: 706, 1905 FELT LARVA (BRIEF)
PAPILIO 3:10, 1883 THAXTER LARVA
PHILADELPHIA ENT. SOC. PROC. 2:134, 1863 GROTE LARVA INCLUDES <u>EXPULTRIX</u>
UNITED STATES NAT. MUS. PROC. 27:887, 1904 DYAR LARVA

FOOD PLANTS
ACER SP. (MAPLES)
ALNUS SP. (ALDERS)
AMELANCHIER FLORIDA LINDL.
BETULA SP. (BIRCHES)
BETULA ALLEGHANIENSIS BRITT.
CORYLUS CORNUTA MARSH.
CRATAEGUS SP.
CRATAEGUS OXYACANTHA LINN.
POPULUS SP. (POPLARS)

PRUNUS DOMESTICA LINN.
QUERCUS SP. (OAKS)
QUERCUS RUBRA LINN.
ROSA SP. (ROSES)
ROSA CANINA LINN.
RUBUS PARVIFLORUS NUTT.
SALIX SP. (WILLOWS)

CYMELA CRAMER
SYN. SEE EURYTUS FABRICIUS

CYMOSA GROTE
SYN. SEE CASTANEA GROTE

CYMOSANA STRAND
SSP. SEE CASTANEA GROTE

CYNOGLOSSUM HENRY EDWARDS
SSP. SEE DIFFINIS BOISDUVAL

CYNTHIS AUCT.
SEE WALKERI FELDER & FELDER

CYPRIS EDWARDS
SSP. SEE APHRODITE FABRICIUS

CYRILLUS PLOETZ
SNY. SEE ARAXES HEWITSON

CYTHERA EDWARDS
SYN. SEE MORMO FELDER & FELDER

D

DACOTAH EDWARDS
SSP. SEE MYSTIC SCUDDER

DACTYLINA (GROTE), ACRONICTA–NOCTUIDAE
LIFE HISTORY
AMER. MUS. NAT. HIST. BULL. 4: 77, 1892 BEUTEN. LARVA
LOC. CIT. 14:239, 1901 BEUTENMULLER LARVA
NEW YORK STATE AGRI. EXP. STA. (CORNELL), MEM. 56:1073, 1922 WELLHOUSE LARVE (BRIEF)
NEW YORK STATE MUS. MEM. 8: 742, 1905 FELT LARVA
UNITED STATES NAT. MUS. PROC. 21:152, 1898 SMITH & DYAR LARVA* PL. 7, FF. 18, 19 COLOR OR UNCLORED; PUPA COCOON
HART–INS. COLLECTIONS ILLINOIS HIGH SCHOOLS P. 24, 1903 COCOON
INDES AMER. ECON. ENT. VOL. 3:3
FOOD PLANTS
ALNUS SP. (ALDERS)
BETULA SP. (BIRCHES)
BETULA ALBA LINN.
BETULA PAPYRIFERA MARSH.
CRATAEGUS SP. (THORNS)
POPULUS SP. (POPLARS)
SALIX SP. (WILLOWS)
TRIFOLIUM PRATENSE LINN.

DAEDALUS BEHR
SSP. SEE ICARIOIDES BOISDUVAL

DAIRA GODART
SYN. SEE DELIA CRAMER

DAKOTAENSIS BRYK & EISNER
SSP. SEE SMINTHEUS DOUBLEDAY & HEWITSON

DAMARIS FELDER & FELDER
SYN. SEE MEXICANA BOISDUVAL

DAMASTUS GODART
SYN. SEE DAMON CRAMER

DAMMERSI COMSTOCK & HENNE
SSP SEE ENOPTES BOISDUVAL

DAMMERSI GUNDER
SSPSEE NICIPPE CRAMER

DAMMERSI MCDUNNOUGH, LATHOSEA–NOCTUIDAE
LIFE HISTORY
SOUTHERN CALIFORNIA ACAD. SCI. BULL. 34:22, 1935 COMSTOCK

DAMMERS LARVA* PLS.62,63; PUPA* PL.62; COCOON* PL.64

FOOD PLANTS

GUTIERREZIA SAROTHRAE(PURSH)
HAPLOPPAPUS LINEARIFOLIUS DC.

DAMNOSALIS WALKER
SYN. SEE MADEFACTALIS GUENEE

DAMON(CRAMER), MITOURA-LYCAENIDAE
SYN. GRYNEUS HUBNER
SYN. DAMASTUS GODART
SYN. AUBURNIANA HARRIS
FORM SMILACIS BOISDUVAL & LECONTE
SYN. PATERSONIA BREHME
A CASTALIS EDWARDS
SYN. DISCOIDALIS SKINNER
FORM BREHMEI BARNES & BENJAMIN

LIFE HISTORY

AMER. MUS. NAT. HIST. BULL. 5:279, 1893 BEUTEN. LARVA, CHRYSALIS
CAN. ENT. 16:52, 1884 SAUNDERS LARVA, CHRYSALIS (SMILACIS)
ILLINOIS DEPT. AGRI., TRANS. 18: APPEND. P.93, 1880 MIDDLETON LARVA, CHRYSALIS
NEW YORK STATE MUS. MEM. 8: 740, 1905 FELT LARVA (BRIEF)
ONTARIO ENT. SOC. ANN. REPT. P.29, 1884 SAUNDERS LARVA, CHRYSALIS
SMITHSONIAN MISC. COLLECTIONS 4:ART.2:99, 1862 MORRIS LARVA, CHRYSALIS (SMILACIS)
BOISDUVAL & LECONTE-HIST. LEP. NORTH AMERICA P.107, 1883 LARVA* CHRYSALIS* PL.33

COLOR

COMSTOCK-HOW TO KNOW BUTTER. P.230, 1904 LARVA
FRENCH-BUTTER. EAST. UNITED STATES P.268, 1886 LARVA, CHRYSALIS (SMILACIS)
HOLLAND-BUTTERFLY BOOK 1905 CHRYSALIS* PL.5, FF.30,31
MORRIS-SYNOP. LEP. NORTH AMERICA P.99, 1862 LARVA, CHRYSALIS
SCUDDER-BUTTER. EAST. UNITED STATES & CANADA PP.864-866, 1889 LIFE HISTORY* PL.65, F.4; PL.71, F.3; PL.75, FF.30,31; PL.79, F.27; PL.84, FF.30,31
SCUDDER-GUIDE TO COMMON BUTTER. P.118, 1893 LARVA CHRYSALIS

FOOD PLANTS

JUNIPERUS VIRGINIANA LINN.
SMILAX ROTUNDIFOLIA LINN.

DANA CASSINO
SSP. SEE MIRA GROTE

DANAE CRAMER
SYN. SEE ODIUS FABRICIUS

DANBYI HULST, PARAPTERA-GEOMETRIDAE

LIFE HISTORY

FOOD PLANTS

QUERCUS GARRYANA DOUGL.

DANBYI NEUMOEGEN
SSP SEE LINTNERI GROTE(GLUPHISIA)

DANBYI NEUMOEGEN(DIACRISIA)
SYN. SEE PTERIDIS HENRY EDWARDS

DAPHNIS CRAMER
SYN. SEE CYBELE FABRICIUS

DAPHNIS MARTYN
SEE APHRODITE FABRICIUS

DARGO(STRECKER), EUXOA-NOCTUIDAE
SYN. RUMATANA SMITH

LIFE HISTORY

FOOD PLANTS

ZEA MAYS LINN. (CORN)

DARLINGTONI LEMMER
SSP. SEE ATROPUNCTATA PACKARD

DARWINIATA(DYAR), NEMORIA-GEOMETRIDAE

LIFE HISTORY

PSYCHE 11:121,1904 DYAR LIFE HISTORY EGG,LARVA
FOOD PLANTS
SALIX SP.(WILLOWS)

DAUCUS CRAMER
SYN.SEE LINEATA FABRICIUS

DAUNIA EDWARDS
SSP.SEE PIASUS BOISDUVAL

DAUNIUS HERBST
SYN.SEE CLAUDIA CRAMER

DAUNUS BOISDUVAL,PAPILIO-PAPILIONIDAE
SYN.MULTICAUDATA KIRBY
AB RAGANI BARNES
LIFE HISTORY
CALIFORNIA ACAD.SCI.PROC.5: 325,1873 HENRY EDWARDS CHRYSALIS
PAPILIO 3:2,1883 EDWARDS COMPARES LARVA WITH TURNUS
LOC.CIT.4:100,1884 SCHAUS LARVA
COMSTOCK-BUTTER,CALIFORNIA P.23,1927 LARVA,CHRYSALIS
EDWARDS-BUTTER.NORTH AMERICA 3RD SERIES SUPPLNOTES. 1897 LARVA
PEALE-LEPIDOPTERA AMERICANA 1833 LARVA* PL.1 COLOR
INDEX AMER.ECON.ENT. VOL.6:559 (MULTICAUDATA)
FOOD PLANTS
FRAGARIS SP.(STRAWBERRIES)
FRAXINUS SP.(ASHES)
FRAXINUS LATIFOLIA BENT.
KALMIS SP.(LAURELS)
POPULUS SP.(POPLARS)
PRUNUS AMERICANA MARSH
PRUNUS CERASUS LINN.(CHERRY)
PRUNUS ILICIFOLIA(NUTT.EX. H & A.)DAVENA
PRUNUS SEROTINA EHRH.
PRUNUS VIRGINIANA LINN.
PTELEA TRIFOLIATA V.ANGUSTIFOLIA (BENTH.)
ROSACEAE
SALIX SP.(WILLOWS)
UMBELLULARIA CALIFORNICA (HOOK.&ARN.)

DAUNUS CRAMER
SYN.SEE BATHYLLUS ABBOT & SMITH

DAURA(STRECKER),OENEIS-SATYRIDAE
A ALBERTA ELWES & ELWES
AB DENNISI GUNDER
B OSLARI SKINNER
LIFE HISTORY
SOUTHERN CALIFORNIA ACAD.SCI. BULL.35:165,1936 COMSTOCK-SPERRY LIFE HISTORY,EGG* PL.37; LARVA* PL.38
EDWARDS-BUTTER.NORTH AMERICA 3RD SERIES,1897 LIFE HISTORY* PL.CHIONOBAS 14, FF.A-G ALBERTA
FOOD PLANTS
?PINUS SP.(PINES)
POPULUS SP.
?POPULUS TERMULOIDES MICHX.

DAVENA SMITH
SSP.SEE OLIVACEA MORRISON

DAVISI WATSON & COMSTOCK
SSP.SEE POLIOS WATSON & COMSTOCK

DAYI BLACKMORE,TOLYPE-LASIOCAMPIDAE
LIFE HISTORY
CAN.ENT.52:276,1920 BLACKMORE EGG, COCOON
FOOD PLANTS

DAYI HARRIS
SSP.SEE AUTUMNATA GUENEE

DEALBATA STRAND
SSP.SEE CALYCANTHATA ABBOT & SMITH

DEBILIS WALKER
SYN.SEE CANDEFACTA HUBNER

DECEPTA GROTE
SYN. SEE OLIVATA HARVEY

DECERNENS WALKER
SYN. SEE SUBTERRANEA FABRICIUS

DECIA BOISDUVAL
SSP. SEE CALIFORNIAE WALKER

DECIA GROTE, ANNAPHILA-NOCTUIDAE
SYN. AMICULA HENRY EDWARDS
LIFE HISTORY
FOOD PLANTS
MIMULUS DOUGLASII (BENTH.)

DECIPIENS WALKER
SYN. SEE AMERICANA FABRICIUS

DECISARIA WALKER
SYN. SEE OBTUSARIA HUBNER

DECLARATA (WALKER), EUXOA-NOCTUIDAE
SYN. DECOLOR MORRISON
LIFE HISTORY
CAN. ENT. 17:32, 1885 SAUNDERS LARVA
INDEX AMER. ECON. ENT. VOL. 2:144
FOOD PLANTS
ASTRAGALUS CROTALARIAE (BENTH.)
HELIANTHUS SP.
LUPINUS SP.
MELILOTUS ALBA DESR.
SOLANUM TUBEROSUM LINN. (POTATO)
ZEA MAYS LINN. (CORN)
GENERAL FEEDER ON FIELD CROPS AND VEGETABLES

DECOLOR MORRISON
SYN. SEE DECLARATA WALKER

DECOLOR (WALKER), ENARGIA-NOCTUIDAE
FORM INFUMATA GROTE
SYN. PUNCTIRENS SMITH
AB MIA STRAND
AB SIA STRAND
LIFE HISTORY
NEW YORK ENT. SOC. JOURN. 10: 169, 1902 COQUILLETT LARVA (PUNCTIRENA)
SOUTHERN CALIFORNIA ACAD. SCI. BULL. 36:21, 1937 COMSTOCK LARVA* PL. 7 PUPA* PL. 8
UNITED STATES NAT. MUS. PROC. 27, 873, 1904 DYAR LARVA (UNDER NAME OF COSMIA) PALLACEA ESP.)
FOOD PLANTS
BETULA SP. (PALLACEA)
POPULUS TREMULOIDES MICHX.
POPULUS TRICHOCARPA T. & G.

DECOLORA HENRY EDWARDS
SSP. SEE SINGULATA FABRICIUS

DECORALIS HUBNER, PANAGRAPTA-NOCTUIDAE
SYN. GEOMETROIDES GUENEE
SYN. EPIONOIDES GUENEE
SYN. ELEGANTALIS FITCH
SYN. RECUSANS WALKER
LIFE HISTORY
MAINE AGRI. EXP. STA. BULL 356:178, 1930 PHIPPS LARVA PUPA
FOOD PLANTS
VACCINIUM SP.
VACCINIUM ANGUSTIFOLIUM AIT.
VACCINIUM CORYMBOSUM LINN.

DECOLORARIA (HULST), PRIOCYCLA-GEOMETRIDAE
SYN. JUCUNDARIA STRECKER
LIFE HISTORY
NEW YORK ENT. SOC. JOURN. 53: 200, 1945 FORBES PUPA
NEW YORK STATE AGR, EX. PSTA. (CORNELL), MEM. 274:95, 1948 FORBES LARVA
FOOD PLANTS
RUBUS ALLEGHENIENSIS PORTER
RUBUS IDAEUS V. STRIGOSUS (MICHX)

DECORATA SAUNDERS
SYN. SEE VITTATA FABRICIUS

DECORATA SMITH, HYPENA-NOCTUIDAE
LIFE HISTORY

ENT.AMER.14:147,1934 CRUMB
LARVA* PL.21,F; PL.25,C
FOOD PLANTS
URT CA LYALLII S.WATS
RIBES SANGUINEUM PURSH
SALIX HOOKERIANA BARR.
STELLARIA MEDIA(LINN.)

DECORATA WORTHINGTON
SSP.SEE ILIA CRAMER

DEDUCTA(MORRISON),BULIA-NOCTUIDAE
SYN.PAVITENSIS MORRISON
SYN.INCANDESCENS GROTE
SYN.ALBINA STRECKER
FORM VULPINA HENRY EDWARDS
LIFE HISTORY
FOOD PLANTS
PROSOPIS GLANDULOSA TORR.

DEDUCTARIA WALKER
SYN.SEE PECTINARIA DENNIS & SHIFFERMULLER

DEE FREEMAN
SSP.SEE YUCCAE BOISDUVAL &LECONTE

DEFECTA(GROTE),HYPOCOENA-NOCTUIDAE
LIFE HISTORY
NEW YORK STATE AGRI.EXP.STA. (CORNELL),MEM.274:57,1948 FORBES LARVA?
FOOD PLANTS
POPULUS SP.(POPLARS)
SALIX SP.(WILLOWS)

DEFENSARIA(GUENEE),XANTHORHOE-GEOMETRIDAE
SYN.CALIFORNIATA PACKARD
SYN AMORATA HULST
SYN.THANATARIA SWETT
FORM MEPHISTARIA SWETT
SYN.SUPPURARIA SWETT
FORM CONCILLARIA SWETT
FORM GIGANTARIA SWETT
LIFE HISTORY
BRITISH COLUMBIA ENT.SOC.PROC. 45:17,1949 HARDY EGG,LARVA, PUPA
FOOD PLANTS
ACER MACROPHYLLUM PURSH
ALNUS RUBRA BONG.

DEFINITA PACKARD
SYN.SEE PLAGIATA WALKER(HEMEROCAMPA)

DEFLAVATA STAUDINGER
SSP.SEE SILACEATA SCHIFFERMULLER

DEFLORATA(FABRICIUS),ECPANTHERIA-ARCTIIDAE
SYN.SCRIBONIA STOLL
SYN.OCULATISSIMA ABBOT & SMITH
SYN.CHRYSEIS OLIVIER
SYN.OCULANA FABRICIUS
SYN.CUNEGUNDA CRAMER
FORM CONFLUENS OBERTHUR
A DENODATA SLOSSON
LIFE HISTORY
AMER.ENT.2:179,1870 LARVA (BRIEF)
AMER.MUS.NAT.HIST.BULL.10: 372,1898 BEUTEN.LARVA
CAN.ENT.14:113,1882 SAUNDERS LARVA* PUPA
LOC,CIT.23:106,1891 DYAR EGG, LARVA, PUPA
CAN.JOURN.INDUSTRY,SCI.&ART 8:370,1863 SAUNDERS LARVA (SCRIBONIA)
ENT.NEWS 19:497,1908 SCHROERS PUPAE ONLY IN LOGS INFESTED WITH ANTS
ILLINOIS DEPT.AGRI.TRANS.15: 184,1877 FRENCH LARVA
LOC.CIT.18:APPENT.P.116,1880 MARTEN LARVA
MISSOURI,REPT.INS.4:141,1872 RILEY LARVA* PUPA
NEW YORK ACAD.SCI.ANNALS 8:81,1893 PACKARD LARVA
NEW YORK STATE CABIN.NAT. HIST.,ANN.REPT.NO.50: 184,1896 LINTNER EGG, LARVA,PUPA LARVAL SPINES* PL.4,F.1

Ontario Ent. Soc. Ann. Rept. p. 14, 1882 Saunders larva* pupa
Loc. cit. p. 48, 1889 Harrington larva*
Loc. cit. p. 46, 1903 Bethune larva
Philadelphia Ent. Soc. Proc. 2:28, 1863 Saunders larva
Smithsonian Misc. Collections 4:art. 2:347, 1862 Morris larva (scribonia(
Abbot & Smith–Lep. Ins. Georgia 2:137, 1797 larva* Pl. 69
COLOR
Duncan–Naturalist Library 22:169, 1858 larva
Glover–Illus. North Amer. Lep/ 1878 larva* Pl. 8, f. 12
Harris–Ins. Inj. Veg. 1st edit. p. 247, 1841 larva
Holland–Moth Book p. 120, 1905 larva* f. 62 (poor)
Morris–Synop. Lep. Ins. North America p. 347, 1862 larva
Stretch–Illus. Zygaen. & Bomby. North America p. 176, 1873 larva
Index Amer. Econ. Ent. Vol. 5:207

FOOD PLANTS
Brassica oleracea Linn. (cabbage)
Euphorbia heterophylla Linn.
Helianthus decapetalus Linn.
Persea americana Mill (avacado)
Phytolacca americana Linn.
Plantago sp.
Ricinus communis Linn.
Robinia pseudo-acacia Linn.
Salix sp. (willows)
Viola sp. (violets)
General feeder most anything low plants

defloriana Martyn
syn. see nais Drury

defluata Walker
syn. see subaequaria Walker

defoliaria Clerck
syn. see vancouverensis Hulst

dejecta Strecker, Catocala–Noctuidae
LIFE HISTORY
FOOD PLANTS
Carya sp.
Quercus sp.

delaware Edwards
syn. see logan Edwards

delecta Walker, Acontia–Noctuidae
syn. metallica Grote
LIFE HISTORY
Amer. Mus. Nat. Hist. Bull. 4: 68, 1892 Beuten. larva
Amer. Nat. 4:229, 1871 Packard larva* Pl. 2, f. 4a; pupa* Pl. 2, f. 4b
Loc. cit 21:495, 1887 Weeks larva
Ent. Amer. 3:20, 1887 Weeks larva (brief)
Loc. cit 4:46, 1888 Weeks larva
New York Ent. Soc. Journ. 27: 57, 1919 Weiss–Dickerson egg, larva, pupa
Index Amer. Econ. Ent. Vol. 2:360, Trachae
FOOD PLANTS
Hibiscus moscheutos Linn.

deleta Herrich–Schaeffer
syn. see maculata Edwards

delia (Cramer), Eurema–Pieridae
syn. demoditas Hubner
syn. daira Godart
form delioides Haskin
LIFE HISTORY
California Acad. Sci. Proc. 5: 327, 1873 Henry Edwards larva, chrysalis (brief)
Ent. News 44:153, 1933 Haskin

EGG, LARVA, CHRYSALIS
SMITHSONIAN MISC. COLLECTIONS 4:ART, 2:35, 1862 MORRIS LARVA, CHRYSALIS
BOISDUVAL-SPEC. GEN. LEP. DIURNAE 1:664, 1836 LARVA, CHRYSALIS
BOISDUVAL & LECONTE-HIST. LEP. NORTH AMERICA P.51, 1883 LARVA* PL.18
DEWITZ-BESCHREIBUNGEN VON JUNGENSTADIEN EXOTISCHER LEPIDOPTEREN, HALLE P.248, 1882 LARVA* PL.8, FF.6, 6A
COLOR
FRENCH-BUTTER. EAST. UNITED STATES P.140, 1886 LARVA
MAYNARD-BUTTER. NEW ENGLAND P.46, 1886 LARVA, CHRYSALIS
MORRIS-SYNOP. LEP. NORTH AMERICA P.35, 1886 LARVA, CHRYSALIS

FOOD PLANTS
AESCHYNOMEME VISCIDULA MICHX.
CASSIA SP.
GLYCINE SP.
STYLOSANTHES BIFLORA (LINN.)
TRIFOLIUM SP. (CLOVERS)

DELICATA CASSINO
SSP. SEE INDUCTATA GUENEE

DELICATA DYAR
SSP. SEE LIGNICOLOR WALKER

DELICATA HULST
SYN. SEE BONIFATA HULST

DELICATARIA (DYAR), NEMORIA-GEOMETRIDAE
LIFE HISTORY
SOUTHERN CALIFORNIA ACAD. SCI. BULL. 44:20, 1945 COMSTOCK LARVA* PUPA*
FOOD PLANTS
PHOTINA ARBUTIFOLIA LINDL.

DELICIA DYAR
SSP. SEE BUXEA GROTE

DELILAH STRECKER, CATOCALA-NOCTUIDAE
SYN. ADOPTIVA GROTE
AB CALPHURNIA HENRY EDWARDS
FORM UMBELLA BARNES & BENJAMIN
A DESDEMONA HENRY EDWARDS
FORM UMBRA BARNES & BENJAMIN
B UTAHENSIS CASSINO
FORM SWETTI BARNES & BENJAMIN
LIFE HISTORY
AMER. MUS. NAT. HIST., MEM. NEW SERIES 3: PART 1, 1918 BARNES & MCDUNNOUGH LARVA* PL.14, F.2; PL.15, FF.18, 28; PL.16, F.2
COLOR
PSYCHE 20:195, 1913 BARNES & MCDUNNOUGH EGG, LARVA (DESDEMONA)
FOOD PLANTS
QUERCUS MACROCARPA MICHX. (BUR AND CATKINS)
SALIX SP. (WILLOWS)

DELINEATA (GUENEE), PAECTES-NOCTUIDAE
LIFE HISTORY
CAN. ENT. 31:27, 1899 DYAR EGG LARVA
NEW YORK STATE MUS. MEM. 8: 724, 1905 FELT LARVA (BRIEF)
GUENEE-SPECL. GEN. LEP. NOCT. 2:311, 1852 LARVA
GUENEE-LOC. CIT. 3:397, 1852 LARVA, PUPA
FOOD PLANTS
LIQUIDAMBAR STYRACIFLUA LINN.
TABERNAEMONTANA LAURIFOLIA LINN.

DELIOIDES HASKIN
SSP. SEE DELIA CRAMER

DELORATA SMITH
SSP. SEE ORTHOGONIA MORRISON

DEL-SUD WRIGHT
SYN. SEE HERMES EDWARDS

DELZURATA SWETT & CASSINO

SYN. SEE MISTURATA HULST

DEMACULATA STRAND
SSP. SEE RECTANGULA KIRBY

DEMODITAS HUBNER
SYN. SEE DELIA CRAMER

DENDRARIA GUENEE
SSP. SEE VIRGINIARIA CRAMER

DENNISI GUNDER
SSP. SEE DAURA STRECKER

DENSA WALKER, ARZAMA-NOCTUIDAE
SYN. OECOGENES DYAR
LIFE HISTORY
FOOD PLANTS
TYPHA LATIFOLIA LINN.

DENTATA HENRY EDWARDS
SYN. SEE ARIZONENSIS PACKARD

DENTICULARIA (WALKER), SYNCHLORA-GEOMETRIDAE
SYN. ESCURVARIA PACKARD
LIFE HISTORY
ENT. NEWS 5:62, 1894 DYAR
LARVA, PUPA, COCOON
(EXCURVARIA)
JOURN. AGRI. RESEARCH 33:801
1926 BOTTIMER LARVA
GUMPPENBERG-SYSTEMA GEOMETRARUM 7:500, 1895 LARVA, PUPA
PACKARD-MONOGRAPH GEOMETRIDAE
1876 LARVA* PL. 13, F. 23
INDEX AMER. ECON. ENT.
VOL. 4:474
FOOD PLANTS
HELIANTHUS SP.
RUDBECKIA MAXIMA NUTT.
RUDBECKIA NITIDA NUTT.
SPHAERALCEA INCANA TORR.
VERBESINA SP.

DENTICULATA GROTE, SEMIOTHISA-GEOMETRIDAE
A SEXPUNCTATA BATES
LIFE HISTORY
CAN. ENT. 62:112, 1930
MCDUNNOUGH LARVA (BRIEF)
FOOD PLANTS
LARIX LARICINA (DUROI)

DENTILINEA SMITH
SSP. SEE AENEIPENNIS GROTE

DENUDATA NEUMOEGEN
SSP. SEE EGLANTERINA BOISDUVAL

DENUDATA SLOSSON
SSP. SEE DEFLORATA FABRICIUS

DENUSSA EHRMAN
SSP. SEE PALAEOGAMA GUENEE

DECORUM DYAR
SYN. SEE OCCIDENTALIS LINTNER

DEPAUPERATA OBERTHUS
SYN. SEE MUZINA OBERTHUS

DEPILIS GROTE-EUTOLYPE-NOCTUIDAE
SYN. BOMBYCIFORMIS SMITH
LIFE HISTORY
CAN. ENT. 40:120, 1908 ENGEL
EGG, LARVA, PUPA
FOOD PLANTS
CARYA LACINOSA (MICHX.)
CARYA TOMENTOSA LINN.

DEPLANARIA GUENEE
SYN. SEE SUBATOMARIA WOOD

DEPLANARIA WALKER
SSP. SEE ASTYLUSARIA WALKER

DEPLORENS FRANCLEMONT
SSP. SEE TITEA CRAMER

DEPONTANATA GROTE
SYN. SEE ARCASARIA WALKER

DEPUISETI BOISDUVAL
SYN. SEE MEXICANA BOISDUVAL

DERIDENS (GUENEE), CHARADRA-NOCTUIDAE

SYN. NIGROSUFFUSANA STRAND
FORM FUMOSA DRAUDT
LIFE HISTORY
AMER. MUS. NAT. HIST. BUL. 14: 234, 1895 BEUTEN. LARVA
CAN. ENT. 2:145, 1870 SAUNDERS LARVA
NEW YORK STATE CABIN. NAT. HIST. ANN. REPT. NO. 26:157, 1874 LINTNER LARVA*
NEW YORK STATE MUS. MEM. 8: 704, 1905 FELT LARVA(BRIEF)
PAPILIO 3:11, 1883 THAXTER EGG LARVA
UNITED STATES DEPT. AGRI. MIXC. PUB. 657:355, 395, 1950 CRAIGHEAD LARVA
UNITED STATES NAT. MUS. PROC. 21:26, 1898 SMITH & DYAR EGG, LARVA
GLOVER-ILLUS. NORTH AMER. ENT. 1878 LARVA*PL. 11, FF. 1, 18
FOOD PLANTS
ACER SP.
BETULA SP. (BIRCHES)
QUERCUS SP. (OAKS)
ULMUS SP. (ELMS)

DERUFATA STRAND
SSP. SEE PSEUDARGYRIA GUENEE

DERUPTA MORRISON
SYN. SEE ERIDANIA CRAMER

DESCHEREI NEUMOEGEN
SYN. SEE PORTLANDIA HENRY EDWARDS

DESDEMONA HENRY EDWARDS
SSP. SEE DELILAH STRECKER

DESERTA BARNES
SSP. SEE UNICORNIS ABBOT & SMITH

DESERTA FELDER & FELDER
SYN. SEE NEXA BOISDUVAL

DESERTA HENRY EDWARDS
SSP. SEE SEXSIGNATA HARVEY

DESERTI BARNES & MCDUNNOUGH
SSP. SEE MORMO FELDER & FELDER

DESERTI WRIGHT
SSP. SEE CETHURA FELDER & FELDER

DESERTICOLA HILL
SSP. SEE EDWARDSI SMITH TRICHOCLEA

DESERTORUM GRINNELL
SSP. SEE SYLVINUS EDWARDS

DESIGNATA(HUFNAGEL), XANTHORHOE-GEOMETRIDAE
A EMENDATA PEARSALL
FORM GYNANDRATA PEARSALL
LIFE HISTORY
CAN. ENT. 12:235, 1880 GOODELL LARVA, PUPA
KIRBY-EUROPEAN BUTTER. & MOTHS P. 376, 1882 LARVA
NEWMAN-BRITISH MOTHS P. 168, 1869 LARVA
PACKARD-MONOGRAPH GEOMETRIDAE P. 146, 1876 LARVA
SEPP-DUTCH LEPIDOPTERA 1855 LARVA* COLOR
STAINTON-BRITISH BUTTER. & MOTHS 2, 1859 LARVA
FOOD PLANTS
ARMORACIA AQUATICA(EATON)
BRASSICA OLERACEA LINN. (CABBAGE)
LEPIDIUM VIRGINICUM LINN.
LOBULARIA MARITIMA(LINN.)
RAPHANUS SATIVUS LINN. (RADISH)
TSUGA CANADENSIS(LINN.)

DESIGNATA WALKER
SYN. SEE MARGINATA HAWORTH

DESPERATA GUENEE
SYN. SEE VIDUA ABBOT & SMITH

DESPERATA SMITH
SYN. SEE RADIX WALKER

DESTINATA MOESCHLER-LYGRIS-GEOMETRIDAE
SYN. REMOTATA WALKER

?SYN. PRUNATA LINN.
A LUGUBRATA MOESCHLER
B TRIANGULATA PACKARD
SYN. MONTANATA PACKARD
C SCHISTACEA WARREN
D BOWMANI CASSINO & SWETT
E GRISEATA CASSINO & SWETT
F SIMILIS WALKER
G HARVEYATA TAYLOR
LIFE HISTORY
GUMPPENBERG-SYSTEMA GEOMETRARUM 4:324, 1890 LARVA, PUPA
FOOD PLANTS
RIBES SATIVUM SYME(CURRANT)
GENERAL FEEDER FRUIT TREES

DETERMINATA NEUMOEGEN
SSP. SEE WILLIAMSII DODGE

DETERMINATA WALKER, METANEMA-GEOMETRIDAE
SYN. CARNARIA PACKARD
LIFE HISTORY
NEW YORK STATE AGRI. EXP. STA. (CORNELL), MEM. 274:92, 1948 FORBES LARVA
PSYCHE 19:18, 1912 BARNES & MCDUNNOUGH EGG, LARVA
FOOD PLANTS
SALIX SP.

DETERSA(WALKER), EUXOA-NOCTUIDAE
SYN. PITYCHROUS GROTE
A PERSONATA MORRISON
B AZIF STRECKER
LIFE HISTORY
AMER. MUS. NAT. HIST. BULL. 14: 283, 1901 BEUTEN. LARVA
ENT. AMER. 5:38, 1889 BEUTEN. LARVA (PITYCHROUS)
NEBRASKA AGRI. EXP. STA. RESEARCH BULL. 81:13, 1935 WHELAN LARVA TECHNICAL DESCR.
INDEX AMER. ECON. ENT.
VOL. 2:144 PERSONATA
VOL. 3:161 PERSONATA
VOL. 6:283; VOL. 7:372
FOOD PLANTS
CAKILE EDENTULA(BIGEL)
SALSOLA KALI LINN.
SPARTINA PATENS(AIT.)
VACCINIUM MACROCARPON AIT.
ZEA MAYS LINN. (CORN)
GRASSES MARATIME GRASSES

DETERSATA(GUENEE), APAECASIA-GEOMETRIDAE
LIFE HISTORY
NEW YORK STATE AGRI. EXP. STA. (CORNELL), MEM. 274:88, 1948 FORBES LARVA
WASHINGTON ENT. SOC. PROC. 51:149, 1949 RUPERT LARVA PUPA
FOOD PLANTS
ALNUS SP. (ALDERS)

DETESTA SMITH
SSP. SEE ATOMARIS SMITH

DETRACTA(WALKER), POLIA-NOCTUIDAE
SYN. CLAVIPLENA GROTE
A NEOTERICA SMITH
LIFE HISTORY
BROOKLYN ENT. SOC. BULL. 27:95 1932 CRUMB LARVA* PL. 11, G
MAINE AGRI. EXP. STA. BULL. 356:170, 1930 PHIPPS LARVA PUPA
NATURALISTE CANADIEN P. 6, 1936 DAVIAULT LARVA, PUPA
NEW YORK STATE MUS. MEM. 8: 716, 1905 FELT LARVA(BRIEF)
UNITED STATES NAT. MUS. PROC. 27:842, 1904 DYAR EGG, LARVA
WASHINGTON ENT. SOC. PROC. 4: 319, 1898 DYAR LARVA
FOOD PLANTS
AMELANCHIER FLORIDA LINDL.
BETULA POPULIFOLIA MARSH.
CARYA SP. (HICKORIES)D
QUERCUS SP. (OAKS)
QUERCUS ALBA LINN.
TARAXACUM OFFICINALE WEBER

VACCINIUM ANGUSTIFOLIUM AIT.
VACCINIUM MYRTILLOIDES MICHX.

DETRITA(GUERIN-MENEVILLA)HEMEROCAMPA-LIPARIDAE
?SYN. INORNATA BEUTENMULLER
LIFE HISTORY
AMER.MUS.NAT.HIST.BULL.12: 157,1900 BEUTEN.LARVA COCOON (INORNATA)
NEW YORK ENT.SOC.JOURN.8:207 1900 BEUTEN. EGG MASS BRIEF LARVAL DESCR. (INORNATA)
NEW YORK STATE AGRI.EXP.STA. (CORNELL),MEM.274:246, 1948 FORBES LARVA
PSYCHE 5:300,1890 BEUTEN. LARVA,COCOON (INORNATA)
PSYCHE 7:342,1896 DYAR KEY TO LARVA
INDEX AMER.ECON.ENT. VOL.1:153 INORNATA
FOOD PLANTS
CUPRESSUS SP.(CYPRESS)
QUERCUS SP.(OAKS)
QUERCUS VIRGINIANA MILL.

DEVASTATOR(BRACE),CRYMODES-NOCTUIDAE
SYN. ABJECTA GUENEE
SYN. ORDINARIA WALKER
SYN. CONTENTA WALKER
FORM SPECIOSA MORRISON
LIFE HISTORY
AMER.MUS.NAT.HIST.BULL.14: 306, 1901 BEUTEN.LARVA
ILLINOIS AGRI.EXP.STA.BULL. 95:349,1904 FORBES LARVA
ILLINOIS DEPT.AGRI.TRANS.15: 96,216,1877 THOMAS LARVA
LOC.CIT.18:APPEND;137,1880 MARTEN LARVA
MINNESOTA AGRI.EXP.STA.BULL 61:217,1899 LUGGER LARVA* F.162
LOC.CIT.BULL123:67,1911 WASHBURN LARVA* F.8;PUPA* F.15; COLOR
MISSOURI,REPT.INS.1:83,1869 RILEY LARVA*
NEBRASKA AGRI.EXP.STA.RESEARCH BULL 81:25,1935 WHELEN LARVA(TECHNICAL DESCR.)
NEW YORK ENT.SOC.JOURN.9: 90,1901 BEUTEN.LARVA
NEW YORK STATE AGRI.SOC. REPT.44:58,1885 LINTNER LARVA
NEW YORK STATE MUS.BULL. 6: 5,1888 LINTNER LARVA
ONTARIO ENT.SOC.ANN.REPT. P.38,1879 BOWLES LARVA*
LOC.CIT P.49,1890 FYLES LARVA*
LOC.CIT.P.101,1896 FYLES LARVA
LOC.CIT.P.75,1898 FLETCHER LARVA
UNITED STATES DEPT.AGRI.REPT. ENT.P.296,1884 RILEY LIFE HISTORY* PL.3,FF.3,4
UNITED STATES DEPT.AGRI.TECH. BULL.88:101,1929 CRUMB LARVA,PUPA
CROSBY & LEONARD-MANUAL VEG. GARDEN INS.P.280,1918 LARVA, PUPA
HARRIS-INS.INJ.VEG.,2ND EDIT. P.345,1852 LARVA
HARRIS-LOC.CIT.FLINT EDIT. P.445,1862 LARVA
HART-INS.COLLECTIONS ILLINOIS HIGH SCHOOLS P.26,1903 LARVA*
METCALF & FLINT-DESTR.& USEFUL INS.P.349,1939 LARVA* F.205
SAUNDERS-INS.INJ.FRUITS P.329 1889 LARVA* F.341
INDEX AMER.ECON.ENT.
VOL.1:146 HADENA
VOL.2:331 SIDEMA
VOL.3:370 SIDEMA
VOL.4:439 SIDEMA
VOL.5:572 SIDEMA
VOL.7:225

FOOD PLANTS
ALTHAEA ROSEA CAV.
AVENA SATIVA LINN. (OATS)
BETA VULGARIS LINN. (BEETS)
BRASSICA OLERACEA LINN. (CABBAGRE)
ELYMUS VIRGINICUS LINN.
FRAGARIA CHILOENSIS(LINN.) (STRAWBERRY)
GERANIUM SP.
HORDEUM JUBATUM LINN.
HORDEUM VULGARA LINN. (BARLEY)
LACTUCA SATIVA LINN. (LETTUCE)
MEDICAGO SATIVA LINN. (ALFALFA)
NICOTIANA TABACUM LINN. (TOBACCO)
PHASEOLUS VULGARIS LINN. (BEAN)
PHLEUM PRATENSE LINN.
POA ANNUA LINN.
POA PRATENSIS LINN.
PRUNUS PERSICA(LINN.)(PEACH)
RAPHANUS SATIVUS LINN.
TRITICUM AESTIVUS LINN. (WHEAT)
ZEA MAYS LINN. (CORN)
GRASSES

DEVERGENS(HUBNER), SYNGRAPHA-NOCTUIDAE
A ALTICOLA WALKER
LIFE HISTORY
KIRBY-EUROPEAN BUTTER. & MOTHS P. 266, 1882 LARVA
FOOD PLANTS

DEVIA(GROTE), EUPSILIA-NOCTUIDAE
LIFE HISTORY
CAN. ENT. 16:33, 1884 THAXTER EGG, LARVA
NEW YORK STATE MUS. MEM. 8: 705, 1905 FELT LARVA(BRIEF)
FOOD PLANTS
QUERCUS SP. (OAKS)

DEVINCTARIA GUENEE
SYN. SEE CATENARIA DRURY

DIA GROTE
SSP. SEE INSUETA GUENEE

DIANA(CRAMER), ARGYNNIS-NYMPHALIDAE
LIFE HISTORY
CAN. ENT. 6:121, 1874 W.H. EDWARDS LARVA
ILLINOIS DEPT. AGRI. TRANS. 18: APPEND:81, 1880 MIDDLETON LARVA
EDWARDS, W.H.-BUTTER, NORTH AMERICA, 2ND SERIES SUPPLEMENTARY PART, 1872 EGG
EDWARDS-BUTTER. NORTH AMERICA 2ND SERIES, 1876 LIFE HIST*' ARGYNNIS PL. 7 COLOR
FRENCH-BUTTER. EAST. UNITED STATES P. 153, 1886 LIFE HISTORY
HOLLAND-LOC. CIT. 3RD EDIT. PP. 84, 85, 1931 LIFE HISTORY
SCUDDER-BUTTER. EAST. UNITED STATES & CANADA P. 1800, 1889 LIFE HISTORY
INDEX AMER. ECON. ENT. VOL. 7:105
FOOD PLANTS
VERNONIA SP.
VIOLA SP. (VIOLETS)

DIANA HENRY EDWARDS
SSP. SEE CONCUMBENS WALKER

DIANTHA BEUTENMULLER
SSP. SEE HERMIA HENRY EDWARDS

DICTYNNA WALKER
SSP. SEE LUNA LINNAEUS

DIDONEA(SMITH), XYLOMOIA-NOCTUIDAE
A DRAUDTI BARNES & BENJAMIN
LIFE HISTORY
FOOD PLANTS
PHALARIS ARUNDINACEA LINN. (BORES IN THE ROOTS)

DIDYMA BEAUVOIS
SSPSEE IMPERIALIS DRURY

DIETZI GUNDER
SSP. SEE GLAUCUS LINNAEUS

DIFFERTA MORRISON
SYN. SEE AURANTIAGO GUENEE

DIFFINIS(BOISDUVAL), HEMARIS-SPHINGIDAE
FORM VERN. TENUIS GROTE
FORM AEST. AXILLARIS GROTE & ROBINSON
A AETHRA STRECKER
B ARIADANA BARNES & MCDUNNOUGH
C THETIS BOISDUVAL
FORM CYNOGLOSSUM HENRY EDWARDS
D MEDUNNOUGHI CLARK
E RUBENS HENRY EDWARDS
FOR SYNS. SEE MCDUNNOUGH'S LIST

LIFE HISTORY

AMER. JOURN. SCI. 36:308, 1839 HARRIS LARVA
AMER. MUS. NAT. HIST. BULL. 7: 279, 1895 BEUTEN. LARVA, PUPA (AXILLARIS ALSO)
BROOKLYN ENT. SOC. BULL. 4:17 1881 JEWETT EGG, LARVA, PUPA
CAN. ENT. 2:157, 1870 MEAD LARVA
LOC. CIT. 16:143, 1884 FISCHER LARVA
LOC. CIT. 18:101, 1886 HOLLAND LARVA, PUPA, COCOON
ENT. NEWS 5:178, 1894 ROWLEY LARVA
ENT. SOC. AMERICA, ANNALS 11: 431, 1918 MOSHER PUPA
ILLINOIS DEPT. AGRI. TRANS. 18: APPEND. PP. 88, 160, 1880 COQUILLETT-MARTEN LARVA (TENUIS ALSO)
MICHIGAN STATE COLLEGE SPEC. BULL. 243:50, 1833 MCDANIEL LARVA WITH THYSBE
NEW YORK STATE AGRI. EXP. STA. (CORNELL), MEM. 274:196, 1948 FORBES LARVA
NEW YORK STATE CABIN. NAT. HIST. ANN. REPT. NO. 24:109 1871 LINTNER LARVA
NEW YORK STATE MUS. MEM. 8: 741, 1905 FELT LARVA
NOVITATES ZOOLOGICA(TRING) 9: SUPPLEMENT P. 446, 1903 ROCHSCHILD & JORDAN LARVA
ONTARIO ENT. SOC. ANN. REPT. 12:50, 1881 REED LARVA
PHILADELPHIA ACAD. NAT. SCI. JOURN. 5:129, 1859 CLEMENS LARVA
PSYCHE 2:66, 1877 PICKMEN LARVA (TENUIS)
PSYCHE 6:142, 1891 ELIOT & SOULE LARVA, PUPA
SMITHSONIAN MISC. COLLECTIONS 4: ART, 2:149, 1862 MARRIS LARVA
UNITED STATES DEPT. AGRI. MISC PUB. 657:359, 1950 CRAIGHEAD LARVA
UNITED STATES NAT. MUS. PROC. 27:789, 1904 DYAR EGG, LARVA
ABBOT & SMITH-LEP. INS. GEORGIA P. 85, 1797 LARVA* PL. 43 COLOR
BOISDUVAL-SPEC. GEN. LEP. SPHING. P. 367, 1874 LARVA
ELIOT & SOULE-CATER. & THEIR MOTHS P. 71, 1902 LARVA* PUPA
FERNALD-SPHING. NEW ENGLAND P. 15, 1886 LARVA
HARRIS-CATALOGUE NORTH AMER. SPHING. P. 308, 1839 LARVA
KELLOGG-AMER. INS. P. 438, 1908 LARVA* F. 631
MORRIS-SYNOP. LEP. NORTH AMERICA P. 149, 1862 LARVA
INDEX AMER. ECON. ENT. VOL. 5:289; VOL. 7:458

FOOD PLANTS

APOCYNUM ANDROSAEMIFOLIUM LINN.
CLARKIA RHOMBOIDEA DOUGL. EX HOOK.
DIERVILLA LONICERA MILL.
GAYOPHYTUM DIFFUSUM T. & G.
LONICERA NITIDA WILSON
LONICERA SEMPERVIRENS LINN.
LONICERA TATARICA LINN.
SYMPHORICARPOS ALBUS(LINN.)
TRIOSTEUM PERFOLIATUM LINN.

VACCINIUM STAMINEUM LINN.

DIFFUSA(BARNES), THURBERIPHAGA-NOCTUIDAE
SYN. CATALINA DYAR
LIFE HISTORY
ARIZONA AGRI.EXP.STA.TECH. BULL.7:145,1926 VORHIES EGG* F.3;LARVA* F.2B; PUPA* F.2C
FOOD PLANTS
GOSSYPIUM THURBERI TOD.

DIFFUSA WALKER
SSP.SEE ALBILINEA HUBNER

DILUCIDA(MORRISON), ANOMOGYNA-NOCTUIDAE
SYN. JANUALIS GROTE
FORM YOUNGI SMITH
A ATOMA SMITH
LIFE HISTORY
ONTARIO ENT.SOC.ANN.REPT. P.79,1902 GIBSON LARVA
FOOD PLANTS
CHAMAEDAPHNE CALYCULATA(LINN)
LARIX LARICINA(DUROI)
PICEA MARIANA(MILL.)
VACCINIUM SP.

DILUCIDARIA ROTHKE
SYN.SEE PENDULINARIA GUENEE

DILUTATA DENNIS & SCHIFFERMULLER
SYN. AUTUMNATA GUENEE

DIMIDIATA(HERRICH-SCHAEFFER), PHEOSIA-NOTODONTIDAE
SYN. CALIFORNICA STRETCH
NOTE:THIS SPECIES MAY BE RIMOSA OF PACKARD
LIFE HISTORY
AMER.MUS.NAT.HIST.BULL.10: 415,1898 BEUTEN.LARVA (DIMIDIATA)
BOSTON SOC.NAT.HIST.PROC.24: 523,1890 PACKARD LARVA
DAN.ENT.14:73,1882 GOODHUE LARVA,PUPA
NAT.ACAD.SCI.MEM.7:159,1895 PACKARD EGG,LARVA*PL.20, FF.1-7 COLOR;PUPA, COCOON
NEW YORK ACAD.SCI.ANNALS 8: 227,1894 DYAR LARVAL SETAE
NEW YORKENT.SOC.JOURN.1:63, 1893 PACKARD EGG,LARVA
NEW YORK STATE AGRI.EXP.STA. (CORNELL),MEM.274:221, 1948 FORBES LARVA
NEW YORK STATE MUS.MEM.8: 747,1905 FELT LARVA(BRIEF)
ONTARIO ENT.SOC.ANN.REPT.P. 107,1906 FYLES LARVA*
PSYCHE 5:261,1889 ELIOT-SOULE EGG,LARVA,PUPA
PSYCHE 6:194,1891 DYAR,EGG, LARVA,PUPA,COCOON (DIMIDIATA)
UNITED STATES DEPT.AGRI.MIXC, PUB.657:358,400,1950 CRAIGHEAD LARVA(RIMOSA PACKARD)
ELIOT&SOULE-CATER.&THEIR MOTHS P.223,1902 EGG, LARVA* PUPA
STRETCH-ZYGAEN.&BOMBY. NORTH AMERICA P.117,1873 LARVA* PL.10,F.9 COLOR
FOOD PLANTS
BETULA SP.(BURCHES)
POPULUS SP.
POPULUS GILEADENSIS ROUL
POPULUS TREMULOIDES MICHX.
SALIX SP.(WILLOWS)
SYMPHORICARPOS ALBUS(LINN.)

DIMIDIATA STRETCH
SSP.SEE CALIFORNIAE WALKER

DIMINUTIVA GRAEF
SYN.SEE AURANTIACA HUBNER'

DIMMOCKI GROTE
SYN.SEE RADIX WALKER

DION(EDWARDS), ATRYTONE-HESPERIDAE

A ALABAMAE LINDSEY
LIFE HISTORY
ENT. NEWS 10:415, 1908 LAURENT
EGG, EARLY STAGES
FOOD PLANTS
GRASSES

DIONE FABRICIUS
SYN. SEE ARGE DRURY

DIONE (SCUDDER), LYCAENA-LYCAENIDAE
AB GIBBONI GUNDER
LIFE HISTORY
CAN. ENT. 25:22, 1893 SKINNER
LARVA, CHRYSALIS
FOOD PLANTS
RUMEX OCCIDENTALIS WATS.

DIONYSUS SCUDDER
SSP. SEE RIDINGSII EDWARDS

DIONYZA HENRY EDWARDS
SSP. SEE PIATRIX GUENEE

DIPHTEROIDES (GUENEE), LEUCONYCTA-NOCTUIDAE
FORM OBLITERATA GROTE
LIFE HISTORY
CAN. ENT. 30:16, 1898 DYAR LARVA
FOOD PLANTS
SOLIDAGO SP.

DIPLOCYMA HAMPSON
SYN. SEE STRAMENTOSA GUENEE

DISCALIS DYAR
SSP. SEE PERSPICUA GROTE & ROBINSON

DISCALIS (GROTE), POLIA-NOCTUIDAE
LIFE HISTORY
UNITED STATES NAT. MUS. PROC.
27:839, 1904 DYAR EGG, LARVA
FOOD PLANTS
SALIX SP. (WILLOWS)

DISCERPTA WALKER
SSP. SEE RUFINANS GUENEE

DISCOIDALIS KIRBY, EREBIA-SATYRIDAE
LIFE HISTORY
OTTAWA NATURALIST 9:92, 1895
FLETCHER LARVA
EDWARDS=BUTTER. NORTH AMERICA
3RD SERIES SUPPLEMENTARY
NOTES, 1897 EGG
FOOD PLANTS
CAREX SP.
POA PRATENSIS LINN.

DISCOIDALIS SKINNER
SYN. SEE DAMON CRAMER

DISCOLOR SPEYER
SYN. SEE ATLANTICA GROTE

DISCOLORALIS GUENEE, RENIA-NOCTUIDAE
SYN. FALLACIALIS WALKER
SYN. GENERALIS WALKER
SYN. THRAXALIS WALKER
LIFE HISTORY
ENT. AMER. 14:169, 1934 CRUMB
LARVA*PL. 20 E, F; PL. 23 B;
PL. 24 H; PL. 25 B; PL. 21 H
FOOD PLANTS
DEAD LEAVES

DISCOLORATA NEUMOEGEN
SSP. SEE FRAGAILIS STRETCH

DISCONVENTA WALKER
SNY. SEE FRITILLARIA GUENEE

DISCOSPILATA (WALKER), EUFIDONIA-GEOMETRIDAE
LIFE HISTORY
CAN. ENT. 77:199, 1945 MCGUFFIN
LARVA* F. 1 (MANDIBLE)
NEW YORK STATE AGRI. EXP. STA.
(CORNELL), MEM. 274:64,
1948 FORBES LARVA
FOOD PLANTS
BETULA PAPYRIFERA MARSH
KALMIA SP. (LAURELS)
SPIRAEA SP.
VACCINIUM SP. (BLUEBERRIES)

DISCRETA STRETCH
SSP. SEE LATIPENNIS BOISDUVAL

DISIPPE GODART
SYN. SEE ARCHIPPUS CRAMER

DISPAR(LINNAEUS), LIPARIS-LIPARIDAE
LIFE HISTORY
AMER. MUS. NAT. HIST. BULL. 10; 386, 1898 BEUTEN. LARVA
AMER. NATURALIST 27:750, 1893 LIFE HISTORY* COLOR FF. ALL STAGES (GOOD)
BOSTON SOC. NAT. HIST. PROC. 26:155, 1893 DYAR EGG, LARVA, PUPA
LOC. CIT. 27:142, 1895 DYAR LARVAL SETAE
ENT. SOC. AMERICA ANNALS 41: 202, 1948 LEE LARVAL GLAND BANDS* PL., F. 7
ILLINOIS AGRI. EXP. STA. BULL. 151:481, 1911 FORBES LARVA
INSECT LIFE 2:208, 1890 RILEY LARVA* PUPA*
MINNESOTA AGRI. EXP. STA. BULL. 61:147, 1899 LUGGER EGG* LARVA* PUPA* ALL F. 89
NATURALISTE CANADIEN P. 244, 1929 TESSIER LARVA
NEW YORK STATE AGRI. EXP. STA. (CORNELL), MEM. 274:240, 1948 FORBES LARVA
NEW YORK STATE MUS. BULL. 97: 382, 1905 FELT LIFE HIST.
LOC. CIT. 103:9, 1906 FELT LIFE HISTORY* ALL STAGES COLOR
LOC. CIT. 155:43, 1912 FELT EGG MAGNIFIED* PL. 9; LARVA; PUPA
NEW YORK STATE MUS. MEM. 8: 119, 1905 FELT EGG* PL. 9, F. 15; LARVA* PL. 9, F. 16; PUPA* PL. 9, F. 17 ALL FF COLOR
OHIO AGRI. EXP. STA. BULL. 332: 273, 1918 HOUSER EGG* LARVA* PL. 44
ONTARIO ENT. SOC. ANN. REPT. P. 69, 1894 FLETCHER EGG, LARVA* PUPA*
LOC. CIT. P. 56, 1903 GIBSON LARVA
LOC. CIT. P. 28, 1908 FELT LARVA
PARK MUS. PROVIDENCE, RHODE ISLAND, BULL. P. 2, 1913 EGG MASS* LARVA*
UNITED STATES DEPT. AGRI. BUR. ENT. BULL. 11, NEW SERIES P. 6, 1897 HOWARD EGG, LARVA* F. 3; PUPA* F. 4
LOC. CIT. 34:34, 1902 BANKS EGG, LARVA* F. 24; PUPA* F. 25
LOC. CIT. 87:12, 1910 ROGERS-BURGESS EGG* PL. 1, F. 5; LARVA* PL. 1, F. 6; PUPA* FF. 3, 4
UNITED STATES DEPT. AGRI. DIV. ENT. CIRC. 464:6, 1938 BURGESS-BAKER EGG* F. 2E; LARVA* F. 2C; PUPA* F. 2F
UNITED STATES DEPT. AGRI. FARMER'S BULL. 275:12, 1907 HOWARD EGG* F. 2; LARVA* F. 2; PUPA; COCOON* F. 4
LOC. CIT. 453:16, 1911 MARLATT EGG* F. 1; LARVA* F. 2; PUPA; COCOON* F. 3
LOC. CIT. 564:13, 1914 BURGESS EGG* LARVA* PUPA* ALL F. 2
LOC, CIT. 845:10, 1917 BURGESS EGG* LARVA* PUPA* ALL F. 2
LOC. CIT. 1270:43, 1922 QUAINTANCE EGG, LARVA* F. 82
LOC. CIT. 1335:6, 1923 BURGESS EGG* LARVA* PUPA* ALL F. 3
LOC. CIT. 1623:10, 1930 BURGESS EGG* F. 8E; LARVA* F. 8C12; PUPA* F. 8E
UNITED STATES DEPT. AGRI. MISC, PUB. 657:355, 408, 1950 CRAIGHEAD ALL STAGES* F. 82 B, C, D.
UNITED STATES DEPT. AGRI. YEARBOOK FOR 1905, P. 125 HOWARD LARVA* F. 2C, D

Loc.cit.for 1932 p.240 Burgess egg* larva*pupa* all f.89

Doane,etc.-Forest Insects p. 268,1936 larva*pupa*f.151

Fernald-Applied Ent.p.265, 1926 larva*pupa*f.268

Fernald & Shepard-Applied Dnt. p.263,1942 larva*pupa* f.265

Fernald & Forbush-Report on Gipsy Moth 1896 life history* COLOR

Graham-Principles Forest Ent. p.177,1939 egg*f.62; larva*ff.62;pupa*f.62

Herrich-Ins.Enemies Shade Trees p.81,1935 larva*f.39

Kellogg-Amer.Ins.p.406,1908 larva

Metcalf&Flint-Destr.and Useful Ins.p.702,1939 egg, larva*f.464;pupa

O'Kane-Inj.Ins.p.290,1912 egg*f.428;larva*f.427

Peairs-Pests Farm,Garden, Orchard p.365,1941 egg* larva* f.432

Slingerland & Crosby-Manual Fruit Ins.P.130,1914 egg larva* f.142;pupa

Index Amer.Econ.Ent. Vol.1:253;Vol.2:295; Vol.3:331;Vol.4:386; Vol.5:511,Vol.6:611; Vol.7:808

FOOD PLANTS

Key: 1 favored
2 not favorable for young larva
3 not favired but tolerated
4 not favored,unsuitable

Abies balsamea(Linn.)(4)
Acer negundo Linn.(1)
Acer pensylvanicum Linn.(4)
Acer platanoides Linn.(3)
Acer rubrum Linn.(3)
Acer saccharinum Linn.(3)
Acer saccharophorum Koch.(3)
Acer spictum Lam.(4)
Alnus serulata(Ait.)
Amelanchier florida Lindl.(1)
Aralia nudicaulis Linn.(4)
Aronis melanocarpa(Michx.)(3)
Berberis vulgaris Linn.(3)
Betula alleghaniensis Britt.(3)
Betula lenta Linn.(3)
Betula nigra Linn.(1)
Betula papyrifera Marsh.(1)
Betula populifolia Marsh
Carpinus caroliniana Walt.(3)
Carya cordiformia(Wang.)(2)
Carya glabra(Mill.)(3)
Carya ovata(Mill.)(3)
Carya tomentosa Nutt.(3)
Castanea dentata(Marsh.)(2)
Catalpa sp.(4)
Celtis occidentalis Linn.(4)
Chamaecypris thyodies(Linn.) (3)
Clethra alnifolia Linn.(4)
Comptonia peregrina(Linn.((3)
Cornus sp.(4)
Cornus florida Linn.(4)
Cornus sericea Linn.(4)
Corylus americana Walt.(1)
Corylus cornuta Marsh.(1)
Corylus cornuta Marsh.(1)
Crataegus sp.(1)
Diervilla lonicera Mill.(4)
Diospyros virginiana Linn.(4)
Fagus grandifolia Ehrh.(2)
Fraxinus americana Linn.(4)
Fraxinus nigra Marsh.(4)
Fraxinus profunda(Bush)(4)
Fraxinus quadrangulata Michx. (4)
Gaylussacia frondosa(Linn.)(4)
Gleditsia triacanthos Linn.(4)
Gymnocladus dioica(Linn.((4)
Hamamelis virginiana Linn.(1)
Ilex glabra(Linn.)(4)
Ilex laevigata(Pursh)(4)
Ilex opaca Ait.(4)
Iris versicolor Linn.(4)
Juglans cinerea Linn.(3)
Juglans nigra Linn.(4)

Juniperus virginiana Linn. (2)
Juniperus communis Linn. (4)
Kalmia angustifolia Linn. (4)
Kalmia latifolia Linn. (4)
Larix decidua Mill. (1)
Larix laricina(DuRoi)(1)
Linders Benzoin(Linn.)(4)
Leucothoe sp. (4)
Ligustrum sp. (4)
Liriodendron tulipfera Linn. (4)
Maclura pomifera(Raf.)(4)
Malus pumila Mill. (apple)(1)
Morus alba Linn. (4)
Morus rubra Linn. (4)
Myrica gale Linn. (3)
Myrica pensylvanica Loisel(3)
Nyssa sylvatica Marsh. (3)
Ostrya virginiana(Mill.)(3)
Picea abies(Linn. ((2)
Picea glauca(Moench)(2)
Picea mariana(Mill.)(2)
Picea pungens Engel. (1)
Picea rubens Sarg. (2)
Pinus banksiana Lamb(2)
Pinus flexilis James(2)
Pinus resinosa Ait. (2)
Pinus rigida Mill. (2)
Pinus strobus Linn. (2)
Pinus sylvestris Linn. (2)
Platanus occidentalis Linn. (3)
Populus alba Linn. (3)
Populus deltoides Marsh. (3)
Populus gileadensis Roul(1)
Populus grandidentata Michx. (1)
Populus nigra v. italica Muench(1)
Populus tremulides Michx. (1)
Prunus avium(linn.)(3)
Prunus maritima Marsh. (2(
Prunus pensylvanica Linn. (3)
Prunus persica Linn(peach)(3)
Prunus serotina Ehrh.
Prunus virginiana Linn. (3)
Pyrus communis Linn. (pear)(3)
Quercus alba Linn. (1)
Quercus bicolor Willd. (1)
Quercus coccinea Muench(1)
Quercus ilicifolia Wang. (1)
Quercus imbricaria Michx. (1)
Quercus macrocarpa Michx. (1)
Quercus palustris Muench(1)
Quercus prinoides Willd. (1)
Quercus prinus Linn. (1)
Quercus rubra Linn. (1)
Quercus stellata Wang. (1)
Quercus velutina Lam. (1)
Rhododendron calandulaceum (Michx.)(4)
Rhododendron viscosum(Linn.)
Rhus glabra Linn. (1)
Rhus toxicodendron Linn. (4)
Rhus typhina Linn. (1)
Ribes rubrum Linn(currnt)(4)
Ribes sativum Syme
Robinia pseudo-acacia Linn. (4)
Rosa carolina Linn. (1)
Rosa eglanteria Linn. (4)
Rubus nigrobaccus Bailey(4)
Rubus idaeus v. strigosus(Michx.) (raspberry)(4)
Rumex altissimus Wood(4)
Rumex crispus Linn. (4)
Salix alba Linn. (1)
Salix discolor Muhl. (1)
Salix interior Rowlee(1)
Sambucus canadensis Linn. (4)
Sassafras albidum v. molle (Raf.)(3)
Smilax rotundifolia Linn. (4)
Sorbus americana Marsh. (1)
Spiraea tomentosa Linn. (4)
Symplocarpus foetidus(Linn.) (4)
Taxodium distichum(linn.)(4)
Thuja occidentalis Linn. (4)
Tilia americana Linn. (1)
Tilia europaea Linn. (1)
Tsuga canadensis(Linn.)(2)
Ulmus americana Linn. (3)
Ulmus procera Salisb. (3)
Vaccinium angustifolium Ait. (1)
Vaccinium macrocarpon Ait. (3)
Viburnum acerifolium Linn.
Viburnum sassinoides Linn. (4)
Viburnum lentago Linn. (4)
Vitis vinifera Linn. (4)
Zea mays Linn. (corn)(3)

BAY-LEAVED WILLOW
GENERAL FEEDER ON FOREST AND
FRUIT TREES AND SHRUBS

DISPARATA (HENRY EDWARDS) ANDROLOMA-AGARASTIDAE
SYN. GRACILENTA GRAEF
LIFE HISTORY
BROOKLYN MUS. ARTS. &SCI. BULL 1:196, 1906 DYAR LARVA
FOOD PLANTS

DISPERSUS GUNDLACH
SYN. SEE EUFALA EDWARDS

DISPLICIENS WALKER
SYN. SEE MESSORIA HARRIS

DISPOSITA MORRISON, LITHOPHANE-NOCTUICAE
A ARGILLOCOSTA FRANCLEMONT
LIFE HISTORY
ROYAL SOC. CAN. TRANS. 1: SERIES 3:61, 1907 FLETCHER-GIBSON EGG, LARVA, PUPA
FOOD PLANTS

DISPUNCTA WALKER
SSP. SEE GRANITATA GUENEE

DISSECTA GROTE & ROBINSON
SYN. SEE RETARDATA WALKER

DISSERPTARIA WALKER
SYN. SEE VIRGINIARIA CRAMER

DISSIDENS FELDER, MAGUSA-NOCTUIDAE
LIFE HISTORY
WASHINGTON ENT. SOC. PROC. 4: 453, 1900 DYAR LARVA
FOOD PLANTS
CONDALIA FERREA GRISEB.

DISSIMILARIA HUBNER
SSP. SEE COLORARIA FABRICIUS

DISSTRIA HUBNER, MALACOSOMA-LASIOCAMPIDAE
SYN. DRUPACEARUM BOISDUVAL
SYN. NUBILIS GUERIN-MENEVILLE
FORM SYLVATICA HARRIS
FOR SYLVATICOIDES NEUMOEGEN & DYAR
AB ASTRIATA REIFF
AB ANITA REIFF
A EROSA STRETCH
FORM PERVERSA NEUMOEGEN & DYAR
LIFE HISTORY
AMER. ENT. 1:208, 1869 RILEY LARVA* FF. 146, 147
LOC. CIT. 2:261, 1870 RILEY EGG* LARVA*
AMER. MUS. NAT. HIST. BULL. 10: 443, 1898 BEUTEN. LARVA
AMER. PHILOSOPHICAL SOC. PROC. 31:175, 179, 1893 PACKARD LARVA (DISSTRIA & THORACICA)
BRITISH COLUMBIA ENT. SOC. PROC. 4, NEW SERIES P. 21, 1914 TREHERNE LARVA
CALIFORNIA ACAD. SCI. PROC. 1, NEW SERIES P. 105, 1888 EDWARDS LARVA
CAN. DEPT. AGRI. REPT. BOTANIST &ENTOMOLOGIST P. 24, 1889 FLETCHER EGG, LARVA, COCOON
CAN. DEPT. AGRI. REPT. ENTOMOLOGIST P. 24, 1887 FLETCHER EGG* LARVA*
CAN. ENT. 9:158, 1877 SAUNDERS LARVA* (SYLVATICA)
LOC. CIT. 10:22, 1878 SAUNDERS LARVA OVERWINTERS IN EGG (SYLVATICA)
LOC. CIT. 25:37, 1893 DYAR LARVA
LOC. CIT. 25:38, 1893 DYAR LARVA (EROSA)
LOC. CIT. 25:38, 1893 DYAR LARVA (THORACICA)
LOC. CIT. 75:204, 1943 ATWOOD LARVA*
ILLINOIS AGRI. EXP. STA., BULL. 151:489, 1911 FORBES LARVA*
ILLINOIS DEPT. AGRI. TRANS. 18: APPEND. P. 124, 1880 MARTEN LARVA*
ILLINOIS NORMAL UNIV. REPT. 6: 44, 1880 FRENCH LARVA

Michigan Agri. Exp. Sta. Bull. 175:349, 1899 Pettit egg* larva*

Michigan Agri. Exp. Ata. Circ. 137:16, 1935 Pettit larva

Michigan State College, Bull. 243:24, 1933 McDaniel life history, larva*

Minnesota Agri. Exp. Sta. Bull. 61:195, 1899 Lugger egg* larva* both f. 139; cocoon* Pl. 17, f. 140

Loc. cit. Bull. 84, 1903 Washburn larva* f. 11 COLOR

Missouri, Rept. Nox. Ins. 3:121, 1871 Riley egg* larva*

Mus. Comp. Zool., Bull. 87: No. 6, 1941 Dethier larval antenna* f. 116

New England Farmer 22:412, 1844 Harris larva

New Hampshire Board Agri. 8th Rept. p. 180, 1879 Cutting larva*

New York State Mus. Bull. 27: 47, 1899 Felt egg* larva*

Loc. cit. 37:18, 1900 Felt egg*

Loc. cit. 180L26, 1915 Felt egg* larva* pupa*

New York State, Rept. Nox. Inx. 2:198, 1856 Fitch larva

Loc. cit. 5:820, 1858 Fitch larval stages also, p. 40. sec. 321

Ohio Agri. Exp. Sta. Bull. 332: 241, 1918 Houser larva* Pl. 35, f. 2

Ontario Ent. Soc. Ann. Rept. p. 16, 1871 Bethune larva

Loc. cit. p. 30, 1875 Saunders larva*

Loc. cit. p. 42, 1877 Gott larva*

Loc. cit. p. 28, 1878 Saunders egg* larva

Loc. cit. p. 73, 1888 Bethune egg* larva*

Loc. cit. 33, 1889 Harrington egg* larva, pupa

Loc, cit. p. 88, 1898 Harrington larva*

Loc. cit. p. 95, 1898 Hutt egg, larva

Loc. cit. p. 93, 1904 Fyles larva*

Loc. cit. p. 13, 1911, Gibson egg* larva

Psyche 6:364, 1892 Dyar egg, larva, pupa, cocoon (erosa)

Smithsonian Misc. Collections 4: art. 2:236, 1862 Morris larva (sylvatica)

United States Dept. Agri. Bur. Ent. Bull. 32, old series p. 27, 1894 Coquillett egg, cocoon

United States Dept. Agri., Farmer's Bull., 1169:35, 1921 Kotinsky egg* f. 19, No. 12; larva* f. 19, No. 13; cocoon f. 19, No. 14

United States Dept. Agri. Misc. Pub. 657:355, 418, 1950 Craighead all stages* ff. 85, 86

Vermont Agri. Exp. Sta. Bull. 76:123, 1900 Perkins larva

Vermont Board Agri. Rept. 5: 258, 1878 Perkins egg* larva*

Abbot & Smith–Lep. Ins. Georgia 2:1717, 1797 larva* Pl. 59 COLOR

Doane, etc–Forest Ins. p. 279, 1936 egg* larva* f. 160

Emmons–Natural Hist. New York 5:240, 1854 larva* cocoon*

Essig–Inj. & Benef. Ins. p. 411, 1915 egg* f. 416; larva* ff. 417, 418

Essig–Ins. West. North. America p. 695, 1926 egg* f. 573

Fernald–Applied Ent. p. 262, 1926 egg* f. 262; larva* f. 263

Fernald & Shepard–Applied Ent. p. 259, 1942 egg* f. 259; larva* f. 260

GLOVER-ILLUS. NORTH. AMER. ENT. 1878 LARVA* PL. 12, F. 9
GRAHAM-PRINCIPLES FOREST ENT. P. 33, 1939 EGG* F. 12; LARVA* F. 75
HARRIS-ENT. CORRESP. P. 292, 1869 LARVA
HARRIS-INS. INJ. VEG. 1ST. EDIT. P. 271, 1841 LARVA, PUPA
HARRIS-LOC. CIT. 2ND EDIT. P. 291, 1852 LARVA, COCOON
HARRIS-LOC. CIT. FLINT EDIT. P. 375, 1862 LARVA*
HART-INSECT COLLECTIONS FOR ILLINOIS HIGH SCHOOLS P. 48, 1903 EGG, LARVA* F. 60
HERRICK-INS. ENEMIES SHADE TREES P. 180, 1935 LARVA* FF. 150, 151
HOLLAND-MOTH BOOK P. 313, 1905 EGG* LARVA* FF. 189, 190
JAEGER-LIFE NORTH AMER. INC. P 173, 1864 LARVA
KELLOGG-AMER. INS. P. 416, 1908 EGG* F. 599G; LARVA* F. 599
METCALF-& FLINT-DESTR. & USE FULL INS. P. 698, 1939 LARVA
MORRIS-SYNOP. NORTH AMER. LEP. P. 236, 1862 LARVA
O'KANE-INJ. INS. P. 269, 1912 EGG* F. 378; LARVA* F. 377
PACKARD-INJ. INS. WEST P. 794, 1877 EGG* LARVA*
PACKARD-INS. INJ. FOREST TREES P. 40, 1881 EGG* LARVA*
PEAIRS-PESTS FARM. GARDEN, ORCHARD P. 378, 1941 LARVA* F. 452
SAUNDERS-INS. INJ. FRUITS P. 53, 1883 EGG* LARVZ* COCOON*
SAUNDERS-LOC. CIT. P. 52, 1889 EGG* F. 43; LARVA* F. 44; PUPA; COCOON* F. 45
SLINGERLAND & CROSBY-MANUAL FRUIT INS. P. 119, 1914 EGG* F. 129: LARVA* FF. 130, 131; COCOON* F. 132
SMITH-INS. NEW JERSEY P. 393, 1900 EGG* LARVA* FF. 176, 177
WOOD-INS. ABROAD P. 680, 1874 LARVA*
INDEX AMER. ECON. ENT.
VOL. 1:197; VOL. 1:198 EROSA AND SYLVATICA
VOL. 2:224 DISSTRIA, EROSA
VOL. 3:249 DISSTRIA, EROSA
VOL. 4:287; VOL. 5:403;
VOL. 6:480; VOL. 7:634

FOOD PLANTS

ACER SP. (MAPLES)
ACER NEGUNDO LINN.
ACER SACCHAROPHORUM KOCH
ALNUS SP. (ALDERS)
ALNUS RUBRA BONG.
AMELANCHIER ALNIFOLIA NUTT.
BETULA PAPYRIFERA MARSH.
BETULA POPULIFOLIA MARSH.
CARPINUS CAROLINIANA WALT.
CARYA SP. (HICKORIES)
CRATAEGUS SP. (THORNS)
CYDONIA OBLONGA MILL. (QUINCE)
FAGUS SP. (BEECHES)
FRAXINUS SP. (ASHES)
HAMAMELIS VIRGINIANA LINN.
JUGLANS NIGRA LINN.
JUGLANS REGIA LINN.
LIQUIDAMBAR STYRACIFLUA LINN.
MALUS PUMILA MILL. (APPLE)
NYSSA SYLVATICA MARSH.
POPULUS SP. (POPLARS)
POPULUS TREMULOIDES MICHX.
PRUNUS CERASUS LINN. (CHERRY)
PRUNUS DOMESTICA LINN. (PLUM)
PRUNUS DOMESTICA V. GALATENSIS EX HOOK. (PRUNE)
PRUNUS PERSICA (LINN.) (PEACH)
PRUNUS SEROTINA EHRH.
PRUNUS VIRGINIANA LINN.
PYRUS COMMUNIS LINN. (PEAR)
QUERCUS SP. (OAKS)
QUERCUS STELLATA WANG.
QUERCUS VELUTINA LAM.
ROBINIA PSEUDO-ACACIA LINN.
ROSA BLANDA AIT.
ROSA CAROLINA LINN.

SALIX SP. (WILLOWS)
TILIA SP. (LINDENS)
ULMUS SP. (ELMS)
VACCINIUM ANGUSTIFOLIUM AIT.

DISTANS (GROTE), ACRONICTA–NOCTUIDAE
A DOLOROSA DYAR
B SCINTILLANS FRANCLEMONT
LIFE HISTORY
AMER. MUS. NAT. HIST. BULL. 14: 259, 1901 BEUTEN. LARVA
MAINE AGRI. EXP. STA. BULL. 356; 175, 1930 PHIPPS LARVA
NEW YORK STATE MUS. MEM. 8: 747, 1905 FELT LARVA (BRIEF)
UNITED STATES NAT. MUS. PROC. 21:162, 1898 SMITH & DYAR LARVA, PUPA, COCOON
LOC. CIT. 27:804, 1904 DYAR EGG LARVA (DOLOROSA)
FOOD PLANTS
ALNUS SP. (ALDERS)
ALNUS RUBRA BONG.
BETULA SP. (BIRCHES)
POPULUS SP. (POPLARS)
ROSA SP. (ROSES)
SALIX SP. (WILLOWS)
VACCINIUM ANGUSTIFOLIUM AIT.

DISTIGMA WALSH
SYN. SEE BICOLOR HARRIS

DISTINCTA (HUBNER), POLIA–NOCTUIDAE
SYN. VITIS FRENCH
LIFE HISTORY
AMER. MUS. NAT. HIST. BULL. 14: 291, 1901 BEUTEN. LARVA
CAN. ENT. 11:76, 1879 FRENCH LARVA, PUPA
ILLINOIS DEPT. AGRI. TRANS. 18: APPEND. P. 136, 1880 MARTEN LARVA, PUPA
ILLINOIS NORMAL UNIV. REPT. 6:45, 1880 FRENCH LARVA
FOOD PLANTS
VITIS SP. (GRAPES)
VITIS VINIFERA LINN. (GRAPE)

DISTINCTA SCHWARZ
SSP. SEE TITANIA DODGE

DISTRACTA SMITH
SYN. SEE E

DISTINCTA SCHWARZ
SSP. SEE TITANIA DODGE

DISTRACTA SMITH
SYN. SEE EXSERTISTIGMA MORRISON

DISTRIBUARIA (HUBNER), SEMIOTHISA
GEOMETRIDAE
SYN. OPPOSITARIA GUENEE
SYN. PROXANTHATA WALKER
SYN. ANTAURATA WALKER
A TRANSITARIA WALKER
LIFE HISTORY
FOOD PLANTS
PINUS VIRGINIANA MILL.

DIVERGENS (WALKER), EUXOA–NOCTUIDAE
SYN. VERSIPELLIS GROTE
AB FUSIMACULA SMITH
A ABAR STRECKER
SYN. FACTORIS SMITH
A CAESIA SMITH
LIFE HISTORY
UNITED STATES NAT. MUS. PROC. 27:838, 1904 DYAR EGG
WASHINGTON ENT. SOC. PROC. 4: 318, 1898 DYAR LARVA
FOOD PLANTS
MEDICAGO SATIVA LINN. (ALFALFA)

DIVERGENS WALKER
SYN. SEE BIFASCIA HUBNER

DIVERSICOLOR (MORRISON), OLIGIA–NOCTUIDAE
LIFE HISTORY
INSECUTOR INSCITIAE MENSTRUUS P. 123, 1913 BIRD LARVA
FOOD PLANTS
SCIRPUS CYPERINUS (LINN.)

DIVERSILINEATA (HUBNER), LYGRIS–GEOMETRIDAE
FORM GRACILINEATA GUENEE
LIFE HISTORY
CAN. DEPT. AGRI. REPT. BOTANIST &ENTOMOLOGIST P. 30, 1888 FLETCHER LARVA*
CAN. DEPT. AGRI. REPT. ENTOMOLOGIST P. 30, 1887 FLETCHER

LARVA*
CAN. ENT. 2:128, 1870 SAUNDERS
LARVA
LOC. CIT. 2:74, 1870 SAUNDERS
LARVA
ILLINOIS DEPT. AGRI. TRANS. 15: 137, 1877 FRENCH LARVA
MINNESOTA AGRI. EXP. STA. BULL. 61:259, 1899 LUGGER LARVA
NEW YORK STATE AGRI. EXP. STA. (CORNELL), MEM. 274:137, 1948 FORBES LARVA(& GRACILINEATA)
NEW YORK STATE MUS. MEM. 8: 706, 1905 FELT LARVA(BRIEF)
ONTARIO ENT. SOC. ANN. REPT. P. 39 1871 SAUNDERS LARVA, PUPA
UNITED STATES DEPT. AGRI. BULL 900:3, 1920 ISLEY EGG* PL. I A, B; LARVA* PL. II; PUPA* PL. IV A, B
UNITED STATES DEPT. AGRI. MISC. PUB. 657:362, 428, 1950 CRAIGHEAD LARVA
PACKARD–MONOGRAPH GEOMETRIDAE P. 138, 1876 EGG, LARVA, PUPA
WAUNDERS–INS. INJ. FRUITS P. 271, 1883 LARVA, PUPA
SAUNDERS–LOC. CIT. P. 270, 1889 LARVA* F. 280
INDEX AMER. ECON. ENT.
VOL. 2:81 CIDARIA
VOL. 1:130 EUSTROMA
VOL. 3:244

FOOD PLANTS

HEDERA HELIX LINN.
MALUS PUMILA MILL. (APPLE)
PARTHENOCISSUS QUINQUEFOLIA (LINN.)
?PRUNUS SP.
?PYRUS COMMUNIS LINN. (PEAR)
?QUERCUS SP. (OAKS)
ROSA SP. (ROSES)
ULMUS SP. (ELMS)
VACCINIUM SP.
VITIS SP. (GRAPES)
FRUIT TREES

DIVES SMITH, HOMOGLAEA–NOCTUIDAE

LIFE HISTORY

FOOD PLANTS

SALIX HOOKERIANA BARR.
SALIX SCOULERIANA BARR.

DIVIDUA(GROTE), NERASTRIA–NOCTUIDAE

FORM OPIPARA HENRY EDWARDS

LIFE HISTORY

SOUTHERN CALIFORNIA ACAD. SCI. BULL. 35:104, 1936 COMSTOCK–DAMMERS LARVA* PL. 31; PUPA* PL. 31

FOOD PLANTS

BELOPERONE CALIFORNICA BENTH

DIVINULA CASSINO & SWETT

SYN. SEE SATYRATA HUBNER

~~DIVI~~SARIA WALKER

SYN. SEE CREPUSCULARIA SCHIFFERMULLER

DIVISARIA(WALKER), HYDRIOMENA–GEOMETRIDAE

LIFE HISTORY

CAN. ENT. 74:153, 1942 MCGUFFIN LARVA* PL. 12, F. C (ALSO LARVAL MANDIBLE)
INDEX AMER. ECON. ENT. VOL. 7:475

FOOD PLANTS

ABIES BALSAMEA(LINN.)
LARIX LARICINA(DUROI)
PICEA GLAUCA(MOENCH)
PICEA MARIANA(MILL.)
PICEA RUBENS SARG.

DIVISATA WALKER, CARIPETA–GEOMETRIDAE

SYN. ALBOPUNCTATA MORRISON
AB NIGRARIA FORBES

LIFE HISTORY

CAN. ENT. 75:189, 1943 MCGUFFIN LIFE HISTORY LARVAL MANDIBLE* PL. 15, F. C
NEW YORK ENT. SOC. JOURN. 53: 203, 1945 FORBES LARVA, PUPA
NEW YORK STATE MUS. MEM. 8:

754,1905 Felt larva

FOOD PLANTS

Abies balsamea(Linn.)
Abies lasiocarpa(Hook)
Larix laricina(DuRoi)
Larix occidentalis Nutt.
Picea engelmanni Parry
Picea glauca(Moench)
Picea mariana(Mill.)
Picea rubens Sarg.
Picea sitchensis(Bong.)
Pinus strobus Linn.
Pseudotsuga Menziesii(Mirb.)
Tsuga heterophylla(Raf.)

docta Walker
syn. see proxima Guerin-Memeville

doctorium Dyar
syn. see impasta Guenee

dodata Taylor
ssp. see satyrata Hubner

dodgei Butler
syn. see phylliar Drury

dodgei Gunder
ssp. see irene Boisduval

dodgei Gunder
ssp. see claudia Cramer

dodgei Gunder
ssp. see amyntula Boisduval

dodgei Gunder
ssp. see clodius Menetries

dodi McDunnough
ssp. see machaon Linnaeus

dolens Druce
syn. see afflicta Grote

dolichos(Fabricius), Prodenia-Noctuidae
syn. commelinae Abbot & Smith

LIFE HISTORY

Amer. Entomologist 2:62,1869 Riley larva* ff. pattern
Loc. cit. 2:363,1870 Riley larva*
Amer. Mus. Nat. Hist. Bull. 16: 422,1902 Beuten. larva (Commelinae)
Illinois Dept. Agri. Trans. 15: 220,1877 French larva
Loc. cit. 18: Append. p. 138,1880 Marten larva
Journ. Econ. Ent. 8:393,1915 Crumb key to larvae (commelinae)
Missouri, Rept. Nox. Ins. ov 3: 113,1871 Riley larva*
United States Dept. Agri. Bur. Ent. Bull 27 new series p. 60, 1901 Chittenden egg* larva(Pl. 4, f. 1 b, c; pupa
United States Dept. Agri. Tech Bull. 88:146,1929 Crumb egg, larva* Pl. 8, E, F: pupa
Abbot & Smith-Lep. Ins. Georgia 2:189,1797 larva* Pl. 95 color
Glover-Illust. North Amer. Ent. 1878 larva* Pl. 14, f. 5; Pl. 16, f. 5; Pl. 19, f. 19; Pl. 21, f. 9
Guenee-Spec. Gen. Lep. Noct. 1:162,1852 larva
Index Amer. Econ. Ent.
Vol. 1:257 commelinae
Vol. 2:299; Vol. 3:334;
Vol. 4:390; Vol. 5:516

FOOD PLANTS

Asparagus officinalis Linn. (asparagus)
Brassica rapa Linn. (turnip)
Carya sp. (hickories)
Commelina communis Linn.
Datura stramonium Linn.
Gossypium herbacium Linn.
Gramineae(grains)
Ipomoea batatas(Linn.)(sweet potato)
Lupinus rivularis Dougl.

NICANDRA PHYSALODES(LINN.)
NICOTANA TABACUM LINN.
PHYSALIS SP.
RUBUS IDAEUS V.STRIGOSUS(MICHX.)
SOLANUM LYCOPERSICUM LINN. (TOMATO)
TRADESCANTIA VIRGINIANA LINN.
TRIFOLIUM SP.(CLOVERS)
VIOLA SP.(VIOLETS)
ZEA MAYS LINN.(CORN)
CEREALS GRASS

DOLICHOS HUBNER
SYN.SEE HALESUS CRAMER

DOLLI BEUTENMULLER
SSP.SEE OPHELIA HENRY EDWARDS

DOLLI GUNDER
SSP.SEE IDALIA DRURY

DOLLII DYAR
SSP.SEE FUCOSA HUBNER

DOLORES NEUMOEGEN & DYAR
SSP.SEE ARIZONENSIS PACKARD

DOLOROSA DYAR
SSP.SEE DISTANS GROTE

DORCAS KIRBY.LYCAENA-LYCAENIDAE
SYN.ANTHELLE SCUDDER
LIFE HISTORY
CAN.ENT.43:160,1911 NEWCOMB EGG,CHRYSALIS,LARVA
FOOD PLANTS
POTENTILLA FRUTICOSA LINN.

DORIS(BOISDUVAL),APANTESIS-ARCTIIDAE
FORM NEREA BOISDUVAL
SYN.MICHABO GROTE
A MINEA SLOSSON
LIFE HISTORY
CAN.ENT.35:116,1903 GIBSON LARVA (MICHABO)
LOC.CIT.47:271,1915,BARNES & MCDUNNOUGH EGG,LARVA (MICHABO)
NEW YORK ENT.SOC.JOURN.8:36, 1900 DYAR LARVA (MICHABO)
FOOD PLANTS
LACUTCA SATIVA LINN.(LETTUCE)
TARAXACUM OFFICINALE WEBER

DORMITANS(GUENEE),BAILEYA-NOCTUIDAE
SYN.LATEBRICOLA GROTE
LIFE HISTORY
FOOD PLANTS
CARYA SP.(HICKORIES)

DOROTHEA MOCCK
SSP.SEE ATLANTIS EDWARDS

DORSIPENNATA BARNES & MCDUNNOUGH, OLENE-LIPARIDAE
?SYN.TEPHRA HUBNER
LIFE HISTORY
NEW YORK STATE AGRI.EXP.STA. (CORNELL),MEM.274:243, 1948 FORBES LARVA(BRIEF)
FOOD PLANTS
CASTANEA DENTATA(MARSH.)
CORYLUS SP.(HAZELS)
QUERCUS SP.(OAKS)
GENERAL FEEDER DECIDIOUS TREES AND SHRUBS

DOSITHEATA GUENEE,SABULODES GEOMETRIDAE
LIFE HISTORY
ENT.NEWS 5:60,1894 DYAR EGG, LARVA,PUPA,COCOON
FOOD PLANTS
EUCALYPTUS SP.
RHAMNUS SP.
RICINUS SP.
SALIX SP.(WILLOWS) ETC

DOSPASSOSI MCDUNNOUGH
SSP.SEE DORCAS KIRBY

DOUBLEDAYI PACKARD
SSP.SEE GIBBOSA ABBOT & SMITH

DOUDOROFFI GUNDER
SSP.SEE ASTYANAX FABRICIUS

DOUDOROFFI GUNDER
SSP. SEE EURYDICE BOISDUVAL

DRAUDTI BARNES & BENJAMIN
SSP. SEE DIDONEA SMITH

DRAUDTI BENJAMIN
SSP. SEE FORBESI BENJAMIN (ROTHSCHILDIA)

DREXELI HENRY EDWARDS, DATANA- NOTODONTIDAE

LIFE HISTORY

AMER. MUS. NAT. HIST. BULL. 10: 406, 1898 BEUTEN. LARVA
CAN. ENT. 20:57, 1888 BEUTEN. EGG, LARVA, PUPA
LOC. CIT. 20:35, 1888 BEUTEN. KEY TO DATANA LARVA
ENT. NEWS 9:150, 1898 SOULE DURATION OF EGG STAGE
NAT. ACAD. SCI. MEM. 7:112, 1895 PACKARD EGG, LARVA*PL. 9, F. 23 COLOR:PUPA
NEW YORK STATE AGRI. EXP. STA. (CORNELL), MEM. 274:215, 1948 FORBES LARVA
NEW YORK STATE MUS. MEM. 8: 739, 1905 FELT LARVA(BRIEF)
OHIO AGRI. EXP. STA., BULL. 332; 230, 1918 HOUSER LARVA
PSYCHE 5:418, 1890 DYAR EGG, LARVA, PUPA
LOC. CIT. 6:575, 1893 DYAR KEY TO LARVA
UNITED STATES DEPT. AGRI. MISC. PUB. 657:355, 398, 1950 CRAIGHEAD LARVA
HERRICK-INS. ENEMIES SHADE TREES P. 137, 1935 LARVA
INBEX AMER. ECON. ENT. VOL. 1:109

FOOD PLANTS

BETULS SP. (BIRCHES)
GAYLUSSACIA FRONDOSA(LINN.)
HAMAMELIS VIRGINIANA LINN.
JUGLANS SP.
SASSAFRAS ALBIDUM V. MOLLE(RAF)
TILIA SP. (LINDENS)
VACCINIUM SP.
VACCINIUM CORYMBOSUM LINN.
VACCINIUM STAMINEUM LINN.
VACCINIUM ULIGINOSUM LINN.

DRUPACEARUM BOISDUVAL
SYN. SEE DISSTRIA HUBNER

DRUPIFERARUM ABBOT & SMITH, SPHINX SPHINGIDAE
A UTAHENSIS HENRY EDWARDS
B MARGINALIS CLARK

LIFE HISTORY

AMER. JOURN. SCI. 36:294, 1839 HARRIS LARVA, PUPA
AMER. MUS. NAT. HIST. BULL. 7: 301, 1895 BEUTEN. LARVA, PUPA
LOC, CIT. 8:291, 1896 BEUTEN. EGG, LARVA, PUPA
CAN. ENT. 3:5, 1871 REED EGG, LARVA*; PUPA*
ENT. NEWS 11:581, 1900 FERNALD LARVA
ILLINOIS DEPT. AGRI., TRANS. 18: APPEND. P. 104, 1880 MARTEN LARVA
MICHIGAN STATE COLLEGE, SPEC. BULL. 61:100, 1899 LUGGER LARVA*F. 49; PUPA*F. 50
MINNESOTA AGRI. EXP. STA., BULL. 61:100, 1899 LUGGER LARVA* F. 49; PUPA*F. 50
LOC. CIT. 84:86, 1903 WASHBURN LARVA*PUPA*BOTH F. 74
NEW YORK STATE AGRI. EXP. STA. (CORNELL), MEM. 274:191, 1948 FORBES LARVA
NEW YORK STATE MUS. MEM. 8: 726, 1905 FELD LARVA(BRIEF)
NEW YORK STATE, REPT. NOX. INX. 3:362, 1856 FITCH LARVA ALSO, P. 44, SEC. 65
NOVITATES ZOOLOGICAE 9 SUPPLEMENT P. 139, 1903 ROTHSCHILD & JORDAN LARVA, PUPA
ONTARIO ENT. SOC. ANN. REPT. P.

67,1881 REED LARVA*PUPA*
LOC.CIT.P.29,1888 FYLES LARVA*
PUPA
LOC.CIT.P.32,1899 LYMAN
LARVA*
PHILADELPHIA ACAD.NAT.SCI.,
JOURN.5:172,1859 CLEMENS
LARVA,PUPA
PHILADELPHIA ENT.SOC.PROC.3:
658,1864 LINTNER LARVA,
PUPA
SMITHSONIAN MISC.COLLECTIONS
4:ART.2:197,1862 MORRIS
LARVA,PUPA
UNITED STATES DEPT.AGRI.MISC.
PUB.657:358,374,1950
CRAIGHEAD LARVA
ABBOT & SMITH-LEP.INS.GEORGIA
1:71,1797 LARVA*PUPA* BOTH
PL.36 COLOR
BOISDUVAL-SPEC.GEN.LEP.SPHING.
P.98,1874 LARVA,PUPA
ELLIOT & SOULE-CATERPILLARS
&THEIR MOTHS P.139,1902
EGG,LARVA*PUPA
ESSIG-INS.WEST.NORTH AMERICA
P.663,1926 EGG,LARVA,PUPA
FERNALD-SPHING.NEW ENGLAND
P.40,1886 EGG,LARVA*PUPA
GLOVER-ILLUST.NORTH AMER.ENT.
1878 LARVA* PL.3,F.1
MORRIS-SYNOP.LEP.NORTH AMER-
ICA P.197,1862 LARVA,PUPA
O'KANE-INJ.INS.P.274,1912
LARVA
SAUNDERS-INS.INJ.FRUITS P.163,
1883 EGG,LARVA* PUPA
SAUNDERS-LOC.CIT.P.163,1889
LARVA*F.171;PUPA* F.172
INDEX AMER.ECON.ENT.
VOL.1:289; VOL.:333;
VOL.5:580; VOL.7:893

FOOD PLANTS
CELTIS OCCIDENTALIS LINN.
MALUS PUMILA MILL.(APPLE)
PRUNUS SP.
PRUNUS DOMESTICA LINN.(PLUM)
PRUNUS MARITIMA MARSH.
PRUNUS PSEUDOCERASUS LINDL.
PRUNUS PERSICA(LINN.)(PEACH)
PRUNUS SEROTINA EHRH.
SYRINGA SP.(LILACS)
VACCINIUM MACROCARPON AIT.

DRURAEI DONOVAN
SYN. SEE CINGULATA FABRICIUS

DRURY LATREILLE
SYN.SEE OTHO ABBOT & SMITH

DRUSIUS EDWARDS
SSP.SEE MYCTEIS DOUBLEDAY & HEWITSON

DRYA FABRICIUS
SSP.SEE STENNAE LINNAEUS

DRYAS EDWARDS
SSP.SEE COMMA HARRIA

DRYOPTERATA GROTE,CALLEDAPTERYX-EPIPLEMIDAE
SYN.EROSIATA PACKARD
LIFE HISTORY
CAN.ENT.30:157,1898 DYAR
LARVA
NEW YORK STATE MUS.MEM.8:
738,1905 FELT LARVA
WASHINGTON ENT.SOC.PROC.4:
414,1899 DYAR EGG,LARVA,
PUPA,COCOON
FOOD PLANTS
VIBURNUM NUDUM LINN.
VIBURNUM PRUNIFOLIUM LINN.

DUAE BARNES & BENJAMIN
SSP.SEE ORTHOGONIA MORRISON

DUARIA CASSINO * SWETT
SSP.SEE SULPHURARIA PACKARD (HESPERMIA)

DUARIA(GUENEE),METARRANTHIS-GEOMETRIDAE
SYN.PANISARIA WALKER
SYN. AMYEISARIS WALKER
SYN.ADUSTARIA WALKER

FORM HAMARIA GUENEE
A SEPTENTRIONARIA BARNES & McDUNNOUGH

LIFE HISTORY
- NEW YORK ENT. SOC. JOURN. 51: 140, 1943 RUPERT EGG, LARVA
- NEW YORK STATE AGRI. EXP. STA. (CORNELL), MEM. 274:93, 1948 FORBES LARVA
- PSYCHE 9:371, 1902 DYAR EGG, LARVA
- ROYAL SOC. CANADA, TRANS. 1: SERIES 3:68, 1907 FLETCHER GIBSON EGG, LARVA, PUPA

FOOD PLANTS
- ALNUS RUBRA BONG.
- PRUNUS SEROTINA EHRH.
- PRUNUS VIRGINIANA LINN.
- QUERCUS RUBRA LINN.
- ROSACEAE
- TILIA SP. (LINDENS)
- VACCINIUM SP.
  - GENERAL FEEDER ON SHRUBS

DUBIA WALKER
SSP. SEE ACREA DRURY

DUBITANS (WALKER), AGROPERINA-NOCTUIDAE
SYN. INSIGNATA WALKER
SYN. SPUTATOR GROTE
A COGITATA SMITH

LIFE HISTORY
- INDEX AMER. ECON. ENT.
  VOL. 3:185 HADENA
  VOL. 7:24

FOOD PLANTS
- CONRINGIA ORIENTALIS (LINN.)

DUBITATA PACKARD
SYN. SEE HAESITATA GUENEE

DUCENS WALKER, FELTIA-NOCTUIDAE
SYN. RADIATA HARRIS
NOTE: DUCENS WALKER AND SUBGOTHICA HAWORTH ARE OFTEN CONFUSED IN THE LITERATURE SO THAT SOME OF THE REFERENCES UNDER SUBGOTHICA MAY BELONG HERE

LIFE HISTORY
- BROOKLYN ENT. SOC. BULL. 22:44 1927 CRUMB MATURE LARVA
- LOC. CIT. 27:80, 1932 CRUMB LARVA
- ENT. SOC. AMERICA, ANNALS 23: 396, 1930 HAWKINS LARVAL CLAWS
- MAINE AGRI. EXP. STA. BULL. 356: 151, 1930 PHIPPS EGG, LARVA, PUPA
- NEBRASKA AGRI. EXP. STA. RESEARCH BULL. 81:18, 1935 WHELEN LARVA
- UNITED STATES DEPT. AGRI. FARMER'S BULL. 1494:5, 1926 CRUMB LARVA* F. 5 B, C
- UNITED STATES DEPT. AGRI. TECH BULL. 88:62, 1929 CRUMB EGG, LARVA* PL. 6 A, B; PUPA* PL. 19B
- CROSBY & LEONARD-MANUAL VEG. GARD. INS. P. 271, 1918 LARVA
- INDEX AMER. ECON. ENT.
  VOL. 1:134; VOL. 2:147;
  VOL. 4:187 VOL. 6:285;
  VOL. 7:375

FOOD PLANTS
- APIUM GRAVEOLENS LINN. (CELERY)
- ASTER ERICOIDES LINN.
- BRASSICA OLERACEA LINN. (CABBAGE)
- BRASSICA RAPA LINN. (TURNIP)
- CABOMA CAROLINIANA GRAY
- CONRINGIA ORIENTALIS (LINN.)
- CUCUMIS MELO LINN. (MELON)
- FRAGARIA CHILOENSIS (LINN.) (STRAWBERRY)
- IPOMOEA BATATAS (LINN.) (SWEET POTATO)
- LACTUCA SATIVA LINN. (LETTUCE)
- LUPINUS SP. (LUPINES)
- MALUS PUMILA MILL. (APPLE)
- NICOTIANA TABACUM LINN. (TOBACCO)
- PHASEOLUS VULGARIS LINN. (BEAN)
- PISUM SATIVUM LINN. (PEAS)
- PLANTAGO SP.

Poa pratensis Linn.
Rumex crispua Linn.
Secale cereale Linn. (rye)
Solanum lycopersicum Linn. (tomato)
Solanum tuberosum Linn. (potato)
Solidago sp.
Stellaria media Linn.
Taraxacum officinale Weber
Trifolium sp. (clovers)
Triticum aestivum Linn. (wheat)
Vaccinium angustifolium Ait.
Vaccinium myrtilloides Michx.
Verbascum sp.
Zea mays Linn. (corn)
General feeder-Field Crops, Fruit Trees, Low Plants, Most Anything, Shrubs, Vegetables

ducens Walker
syn. see ipomoeae Doubleday

dumeti Behr
syn. see mormo Felder & Felder

dumetorum (Boisduval), Callophrys Lycaenidae
syn. viridis Edwards
a perplexa Barnes & Benjamin
LIFE HISTORY
Amer. Ent. Soc. Trans. 50:329, 1925 Coolidge egg, larva, pupa
Comstock, J. A.-Butter California p. 168, 1927 egg, larva, chrysalis
FOOD PLANTS
Eriogonum sp.
Lotus argophyllus ornithopus Greene
Lotus grandiflorus (Benth.)

duncani Gunder
ssp. see sara Boisduval

duncani Gunder
ssp. see wheeleri Henry Edwards

dunni Gunder
ssp. see baroni Edwards (Euphydryas)

duovata Bird, Papaipema-Noctuidae
LIFE HISTORY
FOOD PLANTS
Solidago sp.
Solidago sempervirens Linn.

duplicata (Bethune), Zale-Noctuidae
LIFE HISTORY
FOOD PLANTS
Pinus banksiana Lamb
Pinus strobus Linn.

duplicata Bird, Papaipema-Noctuidae
ab obsolescens Strand
LIFE HISTORY
Can. Ent. 40:25, 1908 Bird larva, pupa
FOOD PLANTS
Collinsonia canadensis Linn.

duplicata Neumoegen & Dyar
syn. see colona Hubner

duplicata Packard
syn. see ocellinata Guenee

duplicata Worthington
syn. see ilia Cramer

duryi Edwards
syn. see mormo Felder & Felder

duscalis Franclemont
ssp. see bethunei Grote & Robinson

dwinellei Henry Edwards
ssp. see chalcedona Doubleday & Hewitson

dyari Cassino & Swett
syn. see honestarius Walker

dyari Gunder

SSP. SEE THAROS DRURY

DYARI HULST
SYN. SEE SUBCROCEATA WALKER

DYARI RIVERS
SSP. SEE AMERICANA HARRIS (EPICNAPTERA)

DYARI TAYLOR, GABRIOLA-GEOMETRIDAE
LIFE HISTORY
FOOD PLANTS
PSEUDOTSUGA MUCRONATA RAF.
PSEUDOTSUGA MENZIESII (MIRB.)

DYARII MERRICK
SSP. SEE LECONTEI GUERIN-MENEVILLE

DYAUS GROTE
SYN. SEE OO CRAMER

DYMAR EDWARDS, MELITAEA-NYMPHALIDAE
SYN. LARUNDA STRECKER
SYN. MALE SENRABII BARNES
A CHARA EDWARDS
AB JACINTOI GUNDER
AB NITELA COMSTOCK
LIFE HISTORY
SOUTHERN CALIFORNIA ACAD. SCI. BULL. 28:50, 1929 COMSTOCK EGG* LARVA* CHRYSALIS* (CHARA)
LOC. CIT. 33:35, 1934 COMSTOCK LARVA* PL. 15
FOOD PLANTS
BELOPERONE CALIFORNICA BENTH.

E

EAVESII HENRY EDWARDS
SSP. SEE LORQUINI BOISDUVAL (BASILARCHIA)

EAVESII STRETCH
SSP. SEE FULVA STRETCH

EBORATA HULST
SYN. SEE VIRIDATA PACKARD (ACASIS)

ECHINOCYSTIDIS STRECKER
SYN. SEE NI HUBNER

ECHO (ABBOT & SMITH), SEIRARCTIA-ARCTIIDAE
SYN. NIOBE STRECKER
LIFE HISTORY
BOSTON SOC. NAT. HIST. PROC. 27: 135, 1896 DYAR LARVAL SETAE
ENT. AMER. 5:154, 1889 SLOSSON EGG, LARVA
NEW YORK ACAD. SCI. ANNALS 8: 80, 1893 PACKARD EGG, LARVA
PSYCHE 5:349, 1890 PACKARD LARVA
UNITED STATES DEPT. AGRI. MISC. PUB. 657:355, 388, 1950 CRAIGHEAD LARVA
ABBOT & SMITH-LEP. INS. GEORGIA 2:135, 1797 LARVA* PL. 68 COLOR
HOLLAND-MOTH BOOK 1905 LARVA* PL. 1, F. 10 COLOR
FOOD PLANTS
DIOSPYROS VIRGINIANA LINN.
LACTUCA SATIVA LINN. (LETTUCE)
2 SABAL PALMETTO (WALT.)
SABAL TEXANA (COOK.)
GROUND OAK, GENERAL FEEDER ON MOST ANYTHING.

ECHO EDWARDS
SSP. SEE PSEUDARGIOLUS BOISDUVAL & LECONTE

ECHO ROTHSCHILD
SYN. SEE VESTALIS PACKARD (MAENAS)

ECHO ROTHSCHILD
SYN. SEE ACREA DRURY

ECTYPA (MORRISON), HADENA-NOCTUIDAE
SYN. BELLA GROTE
LIFE HISTORY
ENT. NEWS 40:98, 1929 HERZ LARVAL HABITS ONLY
FOOD PLANTS

SILENE STELLATA(LINN.)(LEAVES, FLOWERS AND SEED PODS)

EDITHA(BOISDUVAL), EUPHYDRYAS-NYMPHALIDAE
FOR THE MANY SYNS. AND SSP. OF THIS SPECIES SEE MCDUNNOUGH'S LIST
LIFE HISTORY
CALIFORNIA ACAD. SCI. PROC. 5: 167, 1873 HENRY EDWARDS LARVA, CHRYSALIS
CAN. ENT. 17:155, 1885 LARVA (RUBICUNDA) EDWARDS (IT IS CLAIMED THAT THIS IS REALLY THE SPECIES TAYLORI OF EDWARDS AND NOT THE VAR. RUBICUNDA OF EDITHA)
COMSTOCK, J. A.-BUTTER, CALIFORNIA P. 102, 1927 LARVA, CHRYSALIS
FOOD PLANTS
CHELONE GLABRA LINN.
ERODIUM CICUTARIUM(LINN.)
ORTHOCARPUS PUSILLUS BENTH.
TRIFOLIUM SP. (CLOVERS)
VIOLA SP. (VIOLETS)

EDITHA EDWARDS
SSP. SEE AMATRIX HUBNER

EDITHA(MEAD), LYCAENA-LYCAENIDAE
AB VANDUZEEI GUNDER
A MONTANA FIELD
AB FEMALE MEADI FIELD
LIFE HISTORY
FOOD PLANTS
HORKELIA TENUILOBA(TORR.)
POTENTILLA SP.
POTENTILLA DOUGLASII GREENE

EDMANDSI PACKARD
SYN. SEE UNICORNIS ABBOT & SMITH

EDNA BEUTENMULLER
SYN. SEE ANGUSI GROTE

EDONIS GUNDER
SSP. SEE EDWARDSII REAKIRT

EDUCTALIS(WALKER), LOMANALTES-NOCTUIDAE
S SYN. LAETULUS GROTE
LIFE HISTORY
FOOD PLANTS
ALNUS SP. (ALDERS)

EDUSA DRURY
SYN. SEE LUNATA DRURY

EDWARDSI BOISDUVAL
SYN. SEE MACCULLOCHI KIRBY

EDWARDSI GRINNELL
SSP. SEE ATALANTA LINNAEUS

EDWARDSI KUSNEZOV
SSP. SEE CALIFORNICA EDWARDS (CATOCALA)

EDWARDSI SMITH, TRICHOCLEA-NOCTUIDAE
A DESERTICOLA HILL
LIFE HISTORY
SOUTHERN CALIFORNIA ACAD. SCI. BULL. 40:165, 1941 COMSTOCK & HENNE LARVA* PL. 19; PUPA* PL. 20
FOOD PLANTS
ATRIPLEX LEUCOPHYLLA (MOQ.)

EDWARDSII BEHR
SYN. SEE LANCEOLATA BOISDUVAL

EDWARDSII BUTLER
SYN. SEE ALOPE DRURY

EDWARDSII EDWARDS
SSP. SEE ALEXANDRA EDWARDS

EDWARDSII(GROTE), LYMIRE-AMATIDAE
LIFE HISTORY
BOSTON SOC. NAT. HIST. PROC. 27: 139, 1896 DYAR LARVAL SETAE
ENT. AMER. 3:167, 1887 HENRY EDWARDS LARVA
INSECT LIFE 2:361, 1890 DYAR EGG, LARVA, PUPA
FOOD PLANTS
FICUS ELASTICA ROXB.

EDWARDSII (PACKARD), HEMIHYALEA-ARCTIIDAE
SYN. TRANSLUCIDA WALKER
SYN. QUERCIA BOISDUVAL
SYN. ARGILLACEA ROTHSCHILDS
FORM OCHREOUS MEADOWS
LIFE HISTORY
CALIFORNIA ACAD. SCI. PROC. 7:21, 1876 HY. EDW. EGG, LARVA
SOUTHERN CALIFORNIA ACAD. SCI. BULL. 43:113, 1944 COMSTOCK-DAMMERS EGG*PL. 23A; LARVA* PL. 23B; PUPA* PL. 23C
STRETCH-ZYGAEN. BOMBYC. NORTH AMERICA P. 88, 1873 LARVA* PL. 10, F. 2 COLOR
FOOD PLANTS
QUERCUS SP. (OAKS)
QUERCUS CHRYSOLEPIS LIEBM.

EDWARDSII REAKIRT, ARGYNNIS-NYMPHALIDAE
AB EDONIS GUNDER
LIFE HISTORY
CAN. ENT. 20:3, 1888 EDWARDS EGG, LARVA, CHRYSALIS
FOOD PLANTS

EDWARDSII (SAUNDERS), STRYMON-LYCAENIDAE
SYN. FABRICII KIRBY
LIFE HISTORY
NEW YORK ENT. SOC. JOURN. 48:62 1940 COMSTOCK LARVAL HABITS
NEW YORK STATE MUS. MEM. 8: 704, 1905 FELT LARVA (BRIEF)
HOLLAND-BUTTERFLY BOOK 1905 CHRYSALIS*PL. 5, F. 29
SCUDDER-BUTTER. EAST. UNITED STATES & CANADA PP. 894-897, 1889 LIFE HISTORY*PL. 65, F. 2; PL. 68, F. 2; PL. 75, F. 25; PL. 84, F. 29
SCUDDER-GUIDE TO COMMON BUTTER. P. 121, 1893 LARVA, CHRYSALIS
FOOD PLANTS
CARYA ILLINOENSIS (WANG.)
QUERCUS SP.
QUERCUS ALBA LINN.
QUERCUS COCCINEA MOENCH
QUERCUS ILICIFOLIA WANG.
"IN ANTS' NESTS AT ROOTS OF WHITE OAK"

EDWARDSII STRETCH
SSP. SEE ORNATA PACKARD

EFFECTA (WALKER), EUCHLAENA-GEOMETRIDAE
AB CLEMENTINA GUMPPENBERG
LIFE HISTORY
PSYCHE 13:117, 1906 DYAR EGG LARVA
FOOD PLANTS
COMPTONIA PEREGRINA (LINN.)
PRUNUS SEROTINA EHRH.
ROSA SP. (ROSES)
SALIX SP. (WILLOWS)
WISTERIA SINENSIS SWEET

EFFUSALIS WALKER
SYN. SEE AEMULA HUBNER

EFFUSALIS WALKER
SYN. SEE METONALIS WALKER

EFFUSARIA PROUT
SSP. SEE AMATURARIA WALKER

EGENA (GUENEE), AUTOGRAPHA-NOCTUIDAE
SYN. EGENOIDES STRAND
LIFE HISTORY
PAN PACIFIC ENT. 21:13, 1945 TILDEN LARVA
INDEX AMER. ECON. VOL. 7:117
FOOD PLANTS
PHASEOLUS VULGARIS LINN. (BEANS)

EGENARIA WALKER
SYN. SEE MELSHEIMERI HARRIS

EGENOIDES STRAND
SYN. SEE EGENA GUENEE

EGEREMET SCUDDER
SSP. SEE OTHO ABBOT & SMITH

EGLANTERINA (BOISDUVAL), PSEUDOHAZIS-SATURNIIDAE

SYN. NORMALIS DYAR
FORM SHASTAENSIS BEHRENS
FORM DENUDATA NEUMOEGEN
SYN. BOISDUVALII OBETTHUR
A NUTTALLI STRECKER
FORM ARIZONENSIS STRECKER
SYN. UNIFORMIS COCKERELL

LIFE HISTORY

CALIFORNIA ACAD. SCI. PROC. 7: 22, 1876 EDWARDS EGG, LARVA
ENT. SOC. AMERICA ANNALS 7:296, 1914 MOSHER PUPA
NAT. ACAD. SCI. MEM. 12:134, 1914 PACKARD LARVA*PL. 24, FF. 5, 6; PL. 25, FF. 1, 2 COLOR
LOC. CIT. P. 136, 1914 PACKARD EGG, LARVA*PL. 25, F. 3 COLOR (SHASTAENSIS); PUPA; COCOON
PSYCHE 5:325, 1890 PACKARD LARVA
PSYCHE 7:91, 1894 DYAR EGG, LARVA, PUPA, COCOON(SHASTAENSIS(
PSYCHE 15:103, 1908 COOLIDGE EGG
UNITED STATES NAT. MUS. PROC. 25:372, 1903 DYAR LARVA
ESSIG-INJ. BENEF. INS. P. 388, 1915 EGG*LARVA*PUPA*ALL F. 388
ESSIG-INS. WEST. NORTH AMERICA P. 672, 1926 EGG, LARVA
PEALE-LEPIDOPTERA AMERICANA 1833 LARVA*PL. 2 COLOR

FOOD PLANTS

ARCTOSTAPHYLOS GLAUCA LINDL.
ARCTOSTAPHYLOS UVA-URSI (LINN.)
CEANOTHUS SP.
CEANOTHUS THYRISFLORUS ESCHS.
HOLODISCUS DISCOLOR (PURSH.)
PRUNUS CERASUS LINN. (CHERRY)
PRUNUS DOMESTICA LINN. (PLUM)
PRUNUS DOMESTICA V. GELATENSIS EX HOOK (PRUNE)
PRUNUS SEROTINA EHRH.
RHAMNUS ALNIFOLIA L'HER
RHAMNUS CALIFORNICA V. TOMENTELLA (BENTH. & WOLF)
RHAMNUS PURSHIANA DC
ROSA SP. (ROSES)
ROSACEAE
RUBUS ALLEGHENIENSIS PORTER
SALIX SP. (WILLOWS)
VITIS CALIFORNICA BENTH.
VITIS VULPINA LINN.

EGLE (DRURY), EUCHAETIAS-ARCTIIDAE
FORM CYCLICA HENRY EDWARDS

LIFE HISTORY

AMER. MUS. NAT. HIST. BULL. 10: 368, 1898 BEUTEN. LARVA
CAN. ENT. 12:230, 1880 JEWETT EGG, LARVA
LOC. CIT. 16:221, 1884 FRENCH LARVA
CAN. JOURN. INDUSTRY, SCI. & ART 8:366, 1863 SAUNDERS LARVA
MUS. COMP. ZOOL. BULL. 87, NO. 6, 1941 DETHIER LARVAL ANTENNA* F. 75
NEW YORK STATE CABIN. NAT. HIST. ANN. REPT. NO. 24:136, 1871 LINTNER LARVA
PAPILIO 3:147, 1883 EDWARDS LARVA
PSYCHE 2:271, 1879 ANDREWS LARVA
SMITHSONIAN MISC. COLLECTIONS 4, ART. 2:344, 1862 MORRIS LARVA
CRAGEN-OUR INS. FRIENDS & FOES P. 149, 1899 LARVA*F. 88
DICKERSON-MOTHS & BUTTERFLIES P. 197, 1901 LARVA*F. 174
EMMONS-NAT. HIST. NEW YORK 5:227, 1854 LARVA
HARRIS-ENT. CORRESP. P. 288, 1869 LARVA*
HARRIS-INS. INJ. VEG. 1ST EDIT. P. 256, 1841 LARVA, PUPA
HARRIS-LOC. CIT. 2ND EDIT. P. 276, 1852 LARVA, PUPA, COCOON
HARRIS-LOC. CIT. FLINT EDIT. P. 359, 1862 LARVA*PUPA COCOON
HART-INS. COLLECTIONS ILLINOIS

High Schools p.23,1903 egg, cocoon
Holland-Moth Book 1904 larva* Pl.1,f.5 COLOR
Morris-Synop.Lep.North America p.344,1862 larva (brief)
FOOD PLANTS
Asclepias syriaca Linn.
Asclepias tuberosa Linn.

EGLEIS Behr
syn.see MONTIVAGA Behr

EGLENENSIS(Clemens),PYGARCTIA-Arctiidae
LIFE HISTORY
Amer.Mus.Nat.Hist.Bull.10: 369,1898 Beuten.larva
Papilio 3:147,1883 Hy.Edw. larva
Psyche 8:70 1897 Dyar egg, larva
FOOD PLANTS
Asclepias sp.
Asclepias tuberosa Linn.

EHRMANNI Chermock
ssp.see PHILODICE Godart

EHRMANNI Ehrmann
ssp.see ASTERIUS Cramer

EHRAMANNI Dhrmann
ssp.see GLAUCUS Linnaeus

EILEENA Dyar
syn.see PERSPICUA Grote & Robinson

EJECTARIA Walker
syn.see SIGNATARIA Walker

ELATA Smith
syn.see SCANDENS Riley

ELDA Behr
ssp.see RELICTA Walker

ELDORA Smith
ssp.see AMERICANA Harris(Acronicta)

ELDORADENSIS Beutenmuller
syn.see CALIFORNICA Edwards (Catocala)

ELDORADO Strand
ssp.see EPITHORE Edwards

ELEANORA Barnes & McDunnough
ssp.see SUBJUNCTA Grote & Robinson

ELECTILELLA Strand
syn.see ELECTILIS Walker

ELECTILIS Walker,CATOCALA-Noctuidae
?syn.ASPASIA Strecker
syn.CASSANDRA Henry Edwards
?syn.ELECTILELLA Strand
LIFE HISTORY
New York Ent.Soc.Journ6:250 1898 Coquillett larva
Southern California Acad.Sci. Bull37:113,1938 Comstock-Dammers egg* Pl.36;larva* Pl.37;pupa*Pl.38
FOOD PLANTS
Eriogonum sp.
Eriogonum fasciculatum Benth.
Polygonum scandens Linn.

ELEGANS Hubner
syn.see HIEROGLYPHICA Cramer

ELEGANS(Strecker),ODONTOSIA-Notodontidae
a GRISEA Strecker
syn.NOTARIA Henry Edwards
LIFE HISTORY
New York State Agri.Exp.Sta. (Cornell),Mem.274:220, 1948 Forbes larva
United States Nat.Mus.Proc. 25:393,1903 Dyar egg, larva
FOOD PLANTS
Populus tremuloides Michx.

ELEGANS(Strecth),EUCHAETIAS-Arctiidae
LIFE HISTORY

SOUTHERN CALIFORNIA ACAD.SCI.
BULL.35:99,1936 COMSTOCK
&DAMMERS EGG*PL.24;
LARVA*PLS.24,25,26;PUPA*
PL.25
FOOD PLANTS
ASCLEPIAS SP.

ELEGANTALIS FITCH
SYN.SEE DECORALIS HUBNER

ELEGANTARIA(HENRY EDWARDS),PHILTRAEA
GEOMETRIDAE
A PAUCIMACULA BARNES & MCDUNNOUGH
LIFE HISTORY
SOUTHERN CALIFORNIA ACAD.SCI.
BULL.33:35,1934 COMSTOCK
EGG*PL.16;PUPA*JPL.17
FOOD PLANTS

ELENAE BARNES & BENJAMIN
SSP.SEE ASTRICTA MORRISON

ELEUCHEA(HUBNER),ATHENA-NYMPHALIDAE
SYN.PELLENIS GODART
LIFE HISTORY
FOOD PLANTS
FICUS SP.(FIGS)

ELIAS BRYK
SSP.SEE SMINTHEUS DOUBLEDAY &
HEWITSON

ELIMATA(GUENEE),ANOMOGYNA-NOCTUIDAE
A GRISATRA SMITH
B BADICOLLIS GROTE
LIFE HISTORY
CAN.ENT.74:21,1942 BROWN-
MCGUFFIN LARVA*
GEOL.SURVEY TERR.BULL.6,NO.
3:566,1882 GROTE LARVA
UNITED STATES NAT.MUS.BULL.
38:52,1890 SMITH LARVA
PUPA
INDEX AMER.ECON.ENT.
VOL.7:43
FOOD PLANTS
ABIES BALSAMEA(LINN.)
CHRYSANTHEMUM SP.
PICEA ENGELMANNI(PARRY)
PICEA GLAUCA(MOENCH)
PICEA MARIANA(MILL.)
PINUS BANKSIANA LAMB.
PINUS RESINOSA AIT.
TSUGA CANADENSIS(LINN.)

ELIS STRECKER
SSP.SEE MEADII EDWARDS(COLIAS)

ELIVATA BARNES & BENJAMIN
SSP.SEE SISYMBRII BOISDUVAL

ELIZA FISCHER
SSP.SEE CEROGAMA GUENEE

ELIZABETA SMITH
SSP.SEE INTERRUPTA GUENEE

ELIZABETH CASSINO
SSP.SEE CALIFORNICA EDWARDS
(CATOCALA)

ELKO EDWARDS
SSP.SEE AMPELOS EDWARDS

ELLENENSIS REIFF
SSP.SEE AHOLIBAH STRECKER

ELLO(LINNAEUS),ERINNYIS-SPHINGIDAE
SYN.VIRIDIS COCKERELL
LIFE HISTORY
AMER.JOURN.SCI.36:297,1839
HARRIS LARVA
AMER.MUS.NAT.HIST.BULL.7:
296,1895 BEUTEN.LARVA
PUPA
CAN.ENT.18:103,1886 HOLLAND
LARVA(GREEN,BROWN)
CONNECTICUT AGRI.EXP.STA.
BULL.247:370,1923 BRITON
LARVA* PL.14
ENT.AMER.3:166,1887 HY.EDW.
LARVA,PUPA
LOC.CIT.6:143,1890 DYAR EGG,
LARVA,PUPA,COCOON
ENT.SOC.AMERICA,ANNALS 11:

EMIGDIONIS(GRINNELL), PLEBEIUS-LYCAENIDAE
SYN. MELIMONA WRIGHT
LIFE HISTORY
SOUTHERN CALIFORNIA ACAD. SCI. BULL. 29:23, 1930 COMSTOCK EGG*
LOC. CIT. 31:92, 1932 COMSTOCK-DAMMERS EGG* PL. 27; LARVA* PL. 27; PUPA* PL. 28
FOOD PLANTS
ATRIPLEX SP.
LOTUS AMERICANUS(NUTT.)

EMILIA EDWARDS
SSP. SEE ALEXANDRA EDWARDS

EMILIA HENRY EDWARDS
SYN. SEE LACRYMOSA GUENEE

EMISSA EDWARDS
SYN. SEE CAMPESTRIS BEHR.

EMMEDONIA(CRAMER), NEPHELODES-NOCTUIDAE
SYN. MINIANS GUENEE
SYN. EXPANSA WALKER
SYN. SUBNOTATA WALKER
FORM VIOLANS GUENEE
SYN. SUBDOLENS WALKER
A TERTIALIS SMITH
B PECTINATA SMITH
LIFE HISTORY
AMER. MUS. NAT. HIST. BULL. 16: 425, 1902 BEUTEN. LARVA
AMER. NAT. 15:576, 1881 LARVA, PUPA
CAN. ENT. 10:61, 1878 FRENCH LARVA
CONNECTICUT AGRI. EXP. STA. BULL. 461:542, 1942 FRIEND LARVA* F. 8
ENT. SOC. AMERICA, ANNALS 23: 396, 1930 HAWKINS LARVAL CLAWS* F. 6
ILLINOIS AGRI. EXP. STA. BULL. 95 1904 FORBES LARVA*
ILLINOIS DEPT. AGRI. TRANS. 15:99 220, 1877 FRENCH-THOMAS LARVA
LOC. CIT. 18, APPEND:139, 1880 MARTEN LARVA
KENTUCKY AGRI. EXP. STA. BULL. 227, 1920 GERMAN LARVAL CUTICLE* F. 6
MINNESOTA ACAD. NAT. SCI. BULL. 1:169, 1875 MINDENHALL LARVA
NEBRASKA AGRI. EXP. STA. BULL. 81:22, 1935 WHELAN LARVA TECH. DESCR. FF. 1 B, 8 A
NEW YORK STATE CABIN. NAT. HIST. ANN. REPT. NO. 41:175, 1888 LINTNER LARVA*
WASHINGTON ENT. SOC. PROC. 28: 201, 1926 CRUMB EGG, LARVA PUPA
CROSBY & LEONARD-MANUAL VEG. GARD. INS. P. 286, 1918 LARVA
HART-INS. COLLECTIONS ILLINOIS HIGH SCHOOLS P. 35, 1903 LARVA* F. 48
METCALF & FLINT-DESTR. & USEFUL INS. P. 348, 1939 LARVA* F. 206
INDEX AMER. ECON. ENT.
VOL. 1:220 MINIANS
VOL. 2:250 MINIANS
VOL. 4:325 MINIANS
VOL. 5:447; VOL. 6:531; VOL. 7:706
FOOD PLANTS
GRAMINEAE(GRAINS, CEREALS)
PHELEUM PRATENSE LINN.
POLYGONUM SP.
POLYGONUM HYDROPIPER LINN.
TRIFOLIUM SP. (CLOVERS)
ZEA MAYS LINN. (CORN)
FRUIT TREES, GRASSES

EMPETRI FREE
SSP. SEE SCUDDERII EDWARDS

ENDROPIARIA GROTE & ROBINSON
ELLOPIA-GEOMETRIDAE
SYN. FATUARIA STRECKER
LIFE HISTORY
CAN. ENT. 11:194, 1879 GOODELL

430,1918 MOSHER PUPA* PL. 36,F.6
NEW YORK STATE AGRI.EXP.STA. (CORNELL),MEM.274-199, 1948 FORBES LARVA
NOVITATES ZOOLOGICAE 9,SUPPLEMENT P.366,1903 ROTHSCHILD & JORDAN LARVA,PUPA
ONTARIO ENT.SOC.ANN.REPT.P. 77,1899 MOFFATT LARVA
PHILADELPHIA ACAD.NAT.SCI. JOURN.5:176,1859 CLEMENS LARVA
SMITHSONIAN MISC,COLLECTIONS 4,ART:2:201,1862 MORRIS LARVA
SOUTHERN CALIFORNIA ACAD.SCI. BULL.37:105,1938 COMSTOCK-DAMMERS LARVAE*PLS.30,31, 32; PUPA* PLS.32,33
LOC.CIT.42:133,1943 COMSTOCK LARVA* PLS.14,15
BOISDUVAL-SPECLGEN.LEP.SPHING P.120,1874 LARVA
GUNDLACH-ENTOM.CUBANA P.233, 1886 LARVA,PUPA
MERIAN-INS.SURINAM P.61,1719 LARVA*PUPA*PL.61 COLOR
MORRIS-SYNOP.LEP.NORTH AMERICA P.201,1862 LARVA,PUPA
INDEX AMER.ECON.ENT. VOL.6:270;VOL.7:360

FOOD PLANTS
CARICA PAPAYA LINN.
EUPHORBIA SP.
EUPHORBIA HETEROPHYLLA LINN.
MANIHOT SP.
MANIHOT ESCULENTA CRANTZ
PSIDIUM GUAJAVA LINN.
VITIS SP.(GRAPES)
YUCCA SP.

ELONGATA GROTE & ROBINSON
SYN.SEE MARTHESIA CRAMER

ELONYMPHA(HUBNER),ALLOTRIA-NOCTUIDAE
LIFE HISTORY
BROOKLYN ENT.SOC.BULL.7:33, 1884 HULST LARVA
NEW YORK STATE MUS.MEM.8: 720,1905 FELT LARVA
GLOVER-ILLUST.NORTH AMER.ENT. 1878 LARVA*PL.2,F.9
GUENEE-SPEC.GEN.LEP.NOCT. 3:38,1852 LARVA

FOOD PLANTS
GLYCINE SP.
JUGLANS NIGRA LINN.
NYSSA SYLVATICA MARSH

ELSA BEUTENMULLER
SYN.SEE CREUSA DOUBLEDAY & HEWITSON

ELSA(STRECKER),DICTYOSOMA-SPHINGIDAE
LIFE HISTORY
ENT.AMER.4:61,1888 HY.EDW. LARVA
FOOD PLANTS

ELYMI RAMBUR
SSP.SEE CARDUI LINNAEUS

EMACULATA SMITH
SSP.SEE IMPRESSA WALKER

EMARGATARIA GUENEE
SYN.SEE FERVIDARIA HERRICH-SCHAEFFER

EMARGINATA GROTE
SSP.SEE EXSERTISTIGMA MORRISON

EMARGINATA SMITH
SSP.SEE GEORGII GROTE

EMASCULATA(DYAR),ANACAMPTODES-GEOMETRIDAE
LIFE HISTORY
FOOD PLANTS
SALIX HOOKERIANA BARR.
SALIX SCOULERIANA BARR.

EMENDATA PEARSALL
SSP.SEE DESIGNATA HUFNAGEL

LARVA
NEW YORK ENT. SOC. JOURN. 52: 331, 1944 RUPERT EGG, LARVA* PL. 11, F. 2; PUPA
LOC. CIT. 53:202, 1945 FORBES PUPA
NEW YORK STATE AGRI. EXP. STA. (CORNELL), MEM. 274:102, 1948 FORBES LARVA
NEW YORK STATE MUS. MEM. 8: 706, 1905 FELT LARVA
PSYCHE 8:407, 1899 DYAR EGG LARVA
FOOD PLANTS
ACER SP. (MAPLES)
CARPINUS CAROLINIANA WALT.
CASTANEA DENTATA (MARSH.)
QUERCUS SP.
QUERCUS ALBA LINN.
QUERCUS RUBRA LINN.

ENERVIS GUENEE
SYN. SEE VECORS GUENEE (ORTHODES)

ENGELHARDTI BIRD, PAPAIPEMA-NOCTUICAE
LIFE HISTORY
CAN. ENT. 71:75, 1939 BIRD LARVA PUPA
FOOD PLANTS
RUDBECKIA LACINIATA LINN.

ENGELHARDTI CLARK
SSP. SEE SEQUOIAE BOISDUVAL

ENGELHARDTI LEMMER
SSP. SEE SORDIDA GROTE (CATOCALA)

ENIGMATA BARNES & MCDUNNOUGH
SSP. SEE COORTARIA HULST

ENIXA GROTE
SSP. SEE XANTHIOIDES GUENEE

ENNIUS SCUDDER & BURGESS
SYN. SEE JUVENALIS FABRICIUS

ENOPTES (BOISDUVAL), PHILOTES-LYCAENITIDAE
A ANCILLA BARNES & MCDUNNOUGH
B DAMMERSI COMSTOCK & HENNE
C MOHAVE WATSON & COMSTOCK
LIFE HISTORY
FOOD PLANTS
ERIOGONUM SP.

ENOTATA GUENEE
SYN. SEE AEMULATARIA WALKER

ENUCLEATA (GUENEE), SCOPULA-GEOMETRIDAE
SYN. RESTRICTATA WALKER
SYN. RECONDITARIA WALKER
SYN. CONTINUARIA WALKER
FORM MENSURATA WALKER
FORM RELEVATA SWETT
FORM ADORNATA PROUT
ALABASTARIA HULST
LIFE HISTORY
CAN. ENT. 11:194, 1879 GOODELL LARVA
LOC. CIT. 12:236, 1890 GOODELL EGG, LARVA
NEW YORK STATE AGRI. EXP. STA. (CORNELL), MEM. 274:123, 1948 FORBES LARVA
NEW YORK STATE MUS. MEM. 8: 733, 1905 FELT LARVA
PSYCHE 9:165, 1901 DYAR EGG, LARVA
GUENEE-SPEC. GEN. LEP. PHAL. 1: 505, 1857 LARVA* F. 3
GUMPPENBERG SYSTEMA GEOMET. 4:492, 1890 LARVA, PUPA
PACKARD-MONOGRAPH GEOMET. P. 348, 1876 LARVA* PUPA* PL. 13, FF. 26, 26A
FOOD PLANTS
GALIUM SP.
MALUS PUMILA MILL. (APPLE)
PRUNUS SEROTINA EHRH.
RHEXIA LUTEA WALT.
VACCINIUM SP.

ENYS SCUDDER
SYN. SEE PECKIUS KIRBY

EPHELIDARIA (HULST), NEOTERPES-GEOMETRIDAE
FORM KUNZEI HULST

LIFE HISTORY
Southern California Acad.Sci. Bull.41:44,1942 Comstock &Dammers larva* Pl.10, pupa* Pl.10
FOOD PLANTS
Argeomone platyceras v.hispida A Gray

EPHESTIAENA Hubner
syn.see ASTYANAX Fabricius

EPHESTION Stoll
syn.see ASTYANAX Fabricius

EPHYRARIA(Walker),ANACAMPTODES-Geometridae
syn.EXPRESSARIA Walker
syn.TAKENARIA Pearsall
LIFE HISTORY
New york State Agri.Exp.St. (Cornell),Mem.274:57, 1948 Forbes larva
United States Dept.Agri.Misc. Pub.657:361,430,1950 Craighead larva
FOOD PLANTS
Tsuga canadensis(Linn.)

EPIGENATA Barnes & McDunnough
ssp.BITACTATA Walker

EPILAIS(Walker),SYNTOMEIDA-Amatidae
a JUCUNDISSIMA Dyar
LIFE HISTORY
Boston Soc.Nat.Hist.Proc.27: 138,1896 Dyar larval setal structure
Florida Ent.15:57,1931 Bratley egg* Pl.1,f.12;larva*Pl. 1,ff.11-19;pupa*Pl.1,ff. 20-21
Insect Life 2:360,1890 Dyar egg,larva,pupa,cocoon
New YorkEnt.Soc.Journ.4:72, 1896 Dyar larva* Pl.3
FOOD PLANTS
Bougainvillea sp.
Carissa grandiflora DC.
Echites umbellata Jacq.
Nerium indicum Mill.

EPILOBII Harris
syn.see GALLI Rottenburg

EPIMENIS(Drury),PSYCHOMORPHA-Noctuidae
LIFE HISTORY
Amer.Mus.Nat.Hist.Bull.10: 357,1898 Beuten.larva
Illinois Dept.Agri.Trans.15: 177,1877 French larva
Loc.cit.18,Append:111,172 1880 Coquillett-Marten larva* pupa*
Minnesota Agri.Exp.Sta.Bull. 61:122,1899 Lugger larva* f.67; pupa
Missouri,Rept.Ins. of 2:83, 1870 larva as EPIMENIS in error:really UNIO Hubner
Loc.cit.3:64,1874 Riley larva* pupa
New York State Cabin.Nat.Hist. Ann.Rept.No.26:121,1874 Lintner larva*
Cragen-Our Ins.Friends & Foes p.145,1899 larva* f.83
Harris-Ent.Corresp.p.130, 1869 larva
Holland-Moth Book p.232,1905 larva* f.83
Saunders-Ins.Inj.Fruits p.264, 1883 larva*pupa*
Saunders-loc.cit.p.264,1889 larva
Stretch-Zygaen.&Bombyc.North America p.235,1873 larva* Pl.10,f.12 COLOR; pupa
FOOD PLANTS
Campsis radicans(Linn.)
Vitis sp.(grapes)

EPIONE(Drury),CATOCALA-Noctuidae
LIFE HISTORY
Amer.Mus.Nat.Hist.Bull.38:

149,1918 Barnes & McDunnough egg,larva
Amer.Mus.Nat.Hist.Mem.3, New Series,part 1,1918 Barnes & McDunnough egg, larva* Pl.13,f.3; Pl.15, f.7;Pl.16,f.1 COLOR
Brooklyn Ent.Soc.Bull.7:49, 1884 Hulst larva
Can.Ent.33:225,1901 Dodge larva,pupa
New York State Mus.Mem.8: 705,1905 Felt larva(brief)
Guenee-Spec.Gen.Lep.Noct. 3:93,1852 larva
FOOD PLANTS
Carya sp.(hickories)
Juglans cinera Linn.
Quercus sp.(oaks)

epionoides Guenee
syn.see decoralis Hubner

epipsodea Butler,Erebia-Satryidae
syn.rhodia Edwards
ab brucei Elwes
syn.sineocellata Skinner
LIFE HISTORY
Can.Ent.22:48,1890 Edwards note on life hist.
Loc.cit.28:274,1896 Lyman egg, larva, chrysalis
Ontario Ent.Soc.Ann.Rept.p. 58,1899 Fletcher habits and chrysalis
Edwards-Butter.North America 3rd series 1897 life hist* (Erebia 3,ff.1-g)
Loc.cit.3rd series Supplementary Notes egg(correction)
Loc.cit.3rd series life hist.* (Erebia 3 brucei)
FOOD PLANTS
Grasses

epithore(Boisduval),Brenthis-Nymphalidae
ab eldorado Strand
ab wawonae Gunder
ab obscuripennis Gunder
LIFE HISTORY
Can.Ent.24:108,1892 Edwards larva
Loc.cit.26:45,1894 Lembert egg
FOOD PLANTS
Viola ocellata T. & G.

epixanthe(Boisduval & Leconte), Lycaena-Lycaenidae
syn.hypoxanthe Kirby
a amicetus Scudder
syn.phaedrus Hall
LIFE HISTORY
Can.Ent.1:57,1869 Saunders egg
French-Butter.East.United States p.282,1886 egg
Scudder-Butter.East.United States & Canada pp.987-989,1889 egg* larva*Pl.65, ff.16,23;Pl.68,f.11;Pl. 79,f.41
Scudder-Guide to Common Butter.p.129,1893 larva, chrysalis
FOOD PLANTS
Laureceae sp.
Menyanthes trifoliata Linn.
Oxalis sp.
Polygonum sp.
Polygonum persicaria Linn.
Rumex patientia Linn.
Rumex verticillatus Linn.
Vaccinium macrocarpon Ait.

epula Boisduval
syn.see mylitta Edwards

epyaxa Strecker
syn.see terlootii Behr

erastrioides(Guenee),Tarachidia-Noctuidae
LIFE HISTORY
Can.Ent.26:21,1894 Dyar egg, larva

PAPILIO 1:8,1881 COQUILLETT
LARVA
PAPILIO 3:84,1883 COQUILLETT
LARVA COMPARED WITH THAT
OF CANDFACTA HUBNER
FOOD PLANTS
AMBROSIA ARTEMISIIFOLIA LINN.
ARCTIUM LAPPA LINN.

ERATO BOISDUVAL
SYN. SEE PHAETON GROTE & ROBINSON

ERECHTEA(CRAMER), CAENURGINA-NOCTUIDAE
SYN. PATIBILIS WALKER
SYN. NARRATA WALKER
SYN. MUNDULA GROTE & ROBINSON
SYN. AGRICOLA GROTE & ROBINSON
FORM PARVA BLACKMORE
LIFE HISTORY
CAN. ENT. 1:4, 1868 SAUNDERS
LARVA
ILLINOIS DEPT. AGRI. TRANS. 15:
233, 1877 FRENCH LARVA
LOC. CIT. 18, APPEND. P. 148, 1880
COQUILLETT LARVA
INSECT LIFE 5:87, 1892 SLINGER-
LAND LARVA COMPARED WITH
THAT OF CRASSIUSCULA
HAWORTH
JOURN. ECON. ENT. 17:313, 1924
R.C. SMITH EGG*F. 11, NO. 8;
LARVA*F. 11 NOS. 3,4,7,11;
PUPA
NEW YORK ENT. SOC. JOURN. 6:27,
1898 WEBSTER EGG* PL. 4,
F. 7; LARVA*PLS. 4,5, F. 8;
COCOON* PLS. 4,5
ONTARIO ENT. SOC. ANN. REPT. P.
37, 1875 SAUNDERS LARVA
LOC. CIT. P. 47, 1881 SAUNDERS
LARVA
PAPILIO 4:149, 1884 FRENCH EGG,
LARVA, PUPA
FERNALD-IN KINGSLEY'S NATURAL
HISTORY 2:449, 1885 LARVA
GLOVER-ILLUST. NORTH AMER. ENT.
1878 LARVA* PL. 9, FF. 5, 20;
PL. 16, F. 2
HARRIS-ENT. CORRESP. PP. 175,
318, 1869 LARVA*
O'KANE-INJ. INS. P. 179, 1912
LARVA* F. 210
PACKARD-GUIKD STUDY INS. P.
317, 1869 LARVA
PEAIRS-PESTS FARM, GARD.,
ORCHARD P. 199, 1941 LARVA*
F. 229
INDEX AMER. ECON. ENT.
VOL. 1:113 DRASTERIA
VOL. 2:123 DRASTERIA
VOL. 3:66; VOL. 6:114;
VOL. 7:149
FOOD PLANTS
AMBROSIA TRIFIDA LINN.
MEDICAGO SATIVA LINN.
TRIFOLIUM SP. (CLOVERS)
GRASS

ERECTALIS GUENEE
SYN. SEE SCABRA FABRICIUS

EREMIATA(GUENEE), SEMIOTHISA-GEOMETRIDAE
SYN. RETECTATA WALKER
SYN. GRADATA WALKER
SYN. RETENTATA WALKER
SYN. SUBCINCTARIA WALKER
LIFE HISTORY
NEW YORK ENT. SOC. JOURN. 53:
189, 1945 FORBES PUPA
FOOD PLANTS
TEPHROSIA AMBIGUA CURTIS

EREMITA WRIGHT
SSP. SEE PALLA BOISDUVAL

EREMITOIDES STRECKER, SPHINX-SPHINGIDAE
LIFE HISTORY
KANSAS ACAD. SCI. TRANS. 4:28
1875 SNOW LARVA, PUPA
NOVITATES ZOOLOGICAE(TRING) 9
SUPPL. P. 125, 1903 ROTHS-
CHILD, & JORDAN LARVA, PUPA
FOOD PLANTS
SALVIA AZUREA LAM.
SALVIA REFLEXA HORNEM.

EREMITUS(HUBNER), SPHINX–SPHINGIDAE
SYN. SORDIDA HARRIS
A MCCREARYI CLARK
LIFE HISTORY
AMER. MUS. NAT. HIST. BULL. 7: 304, 1895 BEUTEN. LARVA, PUPA
LOC. CIT. 16:396, 1902 BEUTEN. EGG, LARVA
CAN. ENT. 11:59, 1879 FYLES LARVA
ENT. NEWS 4:310, 1893 DYAR-DOLL LARVA
ENT. SOC. AMERICA, ANNALS 11: 422, 1918 MOSHER PUPA
NEW YORK STATE AGRI. EXP. STA. (CORNELL), MEM. 274:190, 1948 FORBES LARVA
NOVITATES ZOOLOGICAE(TRING) 9 SUPPL. P. 124, 1903 ROTHSCHILD & JORDAN LARVA, PUPA
ONTARIO ENT. SOC. ANN. REPT. 3: 652, 1864 LINTNER LARVAL DESCR. (FEEDING ON WILD BERGAMONT PROBABLY THIS SP)
UNITED STATES DEPT. AGRI. REPT. P. 80, 1870 GLOVER LARVA*
FERNALD–SPHING. NEW ENGLAND P. 47, 1886 LARVA, PUPA
FOOD PLANTS
MENTHA ARVENSIS V. VILLOSA (BENTH.)
MENTHA PIPERITA LINN.
MONARDA DIDYMA LINN.
MONARDA FISTULOSA LINN.
PYCNANTHEMUM VIRGINIANUM(LINN)
SALVIA AZUREA LAM.
SALVIA REFLEXA HORNEM.
SALVIA OFFICINALIS LINN.

EREPTA(GROTE), APAMEA–NOCTUIDAE
A RYENSIS BIRD
LIFE HISTORY
ENT. NEWS 24:224, 1913 BIRD LARVA, PUPA, LIFE HISTORY
FOOD PLANTS
TRIPSACUM DACTYLOIDES LINN.

ERICETORUM(BOISDUVAL), HELOPETES–HESPERIIDAE
SYN. ALBA EDWARDS
LIFE HISTORY
ENT. NEWS 34:142, 1923 COOLIDGE LIFE HISTORY
COMSTOCK, J. A.–BUTTERFLIES OF CALIFORNIA P. 209, 1927 EGG* (A 69); LARVA* (A 70); CHRYSALIS* (A 70)
FOOD PLANTS
ALTHAEA ROSEA CAV.
AMARANTHACEAE
SPHARERALCEA THRUBERI(GRAY)

ERICHTO GUENEE
SYN. SEE CRASSIUSCULA HAWORTH

ERIDANIA(CRAMER), PRODENIA–NOCTUIDAE
SYN. EXTERNA WALKER
SYN. NIGROFASCIA HULST
FORM LINEA FABRICIUS
SYN. PHYTOLACCAE ABBOT & SMITH
SYN. DERUPTA MORRISON
LIFE HISTORY
BROOKLYN ENT. SOC. BULL. 22:51, 1927 CRUMB MATURE LARVA* PL. 5, F
LOC. CIT. 27:93, 1932 CRUMB LARVA
ENT. SOC. AMERICA, ANNALS 33: 57, 1940 HOWLAND EGG
LOC. CIT. 37:214, 1944 MAYER-BAKER LARVAL HEAD
UNITED STATES DEPT. AGRI. BUR. ENT. BULL. 66, PART 5:55, 1909 NEW SERIES CHITTENDEN-RUSSELL EGG* F. 8, A, B, C; LARVA* F. 8 E, F, G.
UNITED STATES DEPT. AGRI. TECH. BULL. 762:3, 1941 WOKE LARVAL INTERNAL STRUCTURES
ABBOT & SMITH–LEP. INS. GEORGIA 2:193, 1797 LARVA* PL. 97 COLOR
CROSBY & LEONARD–MANUAL VEG. GARD. INS. P. 297, 1918 EGG, LARVA, PUPA

Peairs–Pests Farm. Gard. Orchard p. 251, 1941 larva
Index Amer. Econ. Ent.
Vol. 1:257: Vol. 2:299;
Vol. 2:387; Xylomyges;
Vol. 3:355 and 438 Xylomyges
Vol. 4:514 Xylomyges;
Vol. 5:516; Vol. 6:616;
Vol. 7:825, 1059

FOOD PLANTS

Althaea rosea Cav.
Amaranthus spinosus Linn.
Apium graveolens Linn. (celery)
Arachis hypogaea Linn. (peanut)
Beta vulgaris Linn. (beet)
Brassica oleracea Linn. (cabbage)
Brassica oleracea v. viridis Lam. (collards)
Capsicum annum Linn. (peppers)
Citrullus vulgaris Schrad
Citrus grandis (Linn. (grapefruit)
Citrus limon (Linn.) (lemon)
Citrus sinensis (Linn.) (orange)
Cucurbita maxima Dusch. (squash)
Daucus carota v. sativus Hoff. (carrot)
Digitaria ischaemum (Schreb.)
Digitaria sanguinalis (Linn.)
Glycine max (Linn.) (soy bean)
Gossypium herbacium Linn.
Helianthus sp.
Hibiscus esculentus Linn. (okra)
Ipomoea batatas (Linn.) (sweet potato)
Nerium indicum Mill.
Nicotiana tabacum Linn.
Persea americana (Linn.) (avacado)
Phaseolus polystachios (Linn.)
Phaseolus vulgaris Linn. (bean)
Phytolacca americana Linn.
Rheum rhaponticum Linn. (rhubarb)
Ricinus communis Linn.
Rumex sp.
Salix sp. (willows)
Sanguinaria canadensis Linn.
Solanum sp.
Solanum lycopersicum Linn. (tomato)
Solanum melongena Linn.
Solanum tuberosum Linn. (potato)
Sonchus oleraceus Linn.
Stizolobium deeringianum Bort.
Trifolium sp.
Vaccinium macrocarpon Ait.
Vigna sinensis (Linn.)
Zea mays Linn. (corn)
Field crops, low plants, vegetables

ERINACEA Retzius
syn. see CAJA Linnaeus

ERINNA Edwards
ssp. see MORMONIA Boisduval

ERIPHYLE Edwards
ssp. see RUTYTHEME Boisduval

ERIS Igel
syn. see EURYNOME Edwards

ERITIOSA Boisduval
syn. see MELISSA Fabricius

ERNESTINAE Bryk & Eisner
syn. see SMINTHEUS Doubleday & Hewitson

EROS Edwards
syn. see ARCHIPPUS Cramer

EROSA Hubner, ANOMIS–Noctuidae

LIFE HISTORY

Amer. Nat. 16:327:1882 Riley egg (brief)
Florida Agri. Exp. Sta. Bull. 151:171, 1919 Watson larva* f. 95
Insecutor Inscitiae Menstruus 1:11, 1913 Dyar larva (brief)
Journ. Econ. Ent. 10:541, 1917 Dozier egg* larva* pl. 27,

F.3;PUPA* PL.26,F.2
UNITED STATES DEPT.AGRI.BUR. ENT.BULL.126:7,1913 NEW SERIES CHITTENDEN EGG; LARVA*PL.4 B;PUPA*PL.4 C
UNITED STATES DEPT.AGRI. FARMER'S BULL.1495:11,1926 WIEGEL & MIDDLETON LARVA* F.7
UNITED STATES DEPT.AGRI.REPT. ENTOM.COMMISSION 4:345, 1895 RILEY LIFE HISTORY* PL.2,FF.1,2,3 COLOR
CROSBY & LEONARD-MANUAL VEG. GARDEN INS.P.254,1918 EGG, LARVA,PUPA
GLOVER-ILLUST. NORTH AMER. ENT. 1878 LARVA* PL.15,F.6
GUENEE-SPEC.GEN.LEP.NOCT.2: 395,1852 LARVA
GUNDLACH-ENTOM.CUBANA P.321, 1886 LARVA,PUPA
INDEX AMER.ECON.ENT. VOL.5:32; VOL.6:34; VOL.7:42

FOOD PLANTS

ABUTILON STRIATUM DICKS.
ABUTILON THEOPHRASTI MEDIC.
ALTHAEA ROSEA CAV.
CAPISCUM SP.(PEPPERS)
GOSSYPIUM HERBACEUM LINN.
HIBISCUS ESCULENTUS LINN.
HIBISCUS MOSCHEUTOS LINN.
HIBISCUS MUTABILIS LINN.
HIBISCUS ROSA-SINENSIS LINN.
HIBISCUS TRIONUM LINN.
MALVA SP.
MALVA ROTUNDIFOLIA LINN.
MALVAVISCUS DRUMMONDI T.&G.
URENA LOBATA LINN.

EROSA STRETCH
SSP.SEE DISSTRIA HUBNER

EROSIATA PACKARD
SYN.SEE DRYOPTERATA GROTE

EROSIATA WALKER
SSP.SEE PAMPINARIA GUENEE

ERRANS(SKINNER), PANOQUINA-HESPERIIDAE

LIFE HISTORY

ENT.NEWS 3:175,1892 DYAR LIFE HISTORY
SOUTHERN CALIFORNIA ACAD.SCI. BULL.29:139,1930 COMSTOCK LIFE HISTORY PUPA*
COMSTOCK,J.A.-BUTTER CALIFORNIA P.227,1927 LARVA

FOOD PLANTS

CYNODON DACTYLON (LINN.)
DISTICHLIS SPICATA(LINN.) GRASSES

ERRATICA PLOETZ
SYN.SEE ZABULON BOISDUVAL & LECONTE

ERRUCATA DYAR, SCHIZURA-NOTODONTIDAE

LIFE HISTORY

BROOKLYN MUS.ARTS,ACI.,BULL. 1:198,1906 DYAR LARVA

FOOD PLANTS

ERUBESCENS BIRD
SYN.SEE SAUZALITAE GROTE

ERUBESCENS WARREN
SSP.SEE RECTIFASCIA HULST

ERUTARIA DUPONCHEL
SYN.SEE CAMBRICA CURTIS

ERYNGII BIRD, PAPAIPEMA-NOCTUIDAE

LIFE HISTORY

CAN.ENT.49:127,1917 BIRD LARVA*PL.8,FF.8,9; PUPA

FOOD PLANTS

ERYNGIUM AQUATICUM LINN.
ERYNIGIUM YUCCIFOLIUM MICHX.
HIERACIUM VENOSUM LINN.

ERYNGII HENRY EDWARDS
SSP.SEE CALIFORNIA WESTWOOD & HEWITSON

ERYPHON(BOISDUVAL),INCISALIA-LYCAENIDAE
LIFE HISTORY
FOOD PLANTS
RIBES SP.?

ERYTHEMARIA(GUENEE),DEILINIA-GEOMETRIDAE
SYN.SIMILARIA WALKER
LIFE HISTORY
NEW YORK ENT.SOC.JOURN.53: 191,1945 FORBES LARVA, PUPA
NEW YORK STATE AGRI.EXP.STA. (CORNELL),MEM.274:71, 1948 FORBES EGG,LARVA
NEW YORK STATE MUS.MEM.8: 743,1905 FELT LARVA(BRIEF)
PSYCHE 10:195,1903 DYAR EGG,LARVA
UNITED STATES DEPT.AGRI.MISC. PUB.657:363,428,1950 CRAIGHEAD LARVA
FOOD PLANTS
PUPULUS SP.(POPLARS)
SALIX SP.(WILLOWS)

ESMERALDA OBERTHUR
SSP.SEE MONETA FABRICIUS

E STESI CLENCH
SSP.SEE HALESUS CRAMER

ESTHER BARNES,PARAPHIA-GEOMETRIDAE
FORM LEMMERI BARNES
LIFE HISTORY
FOOD PLANTS
CARYA SP.(HICKORIES)
PINUS VIRGINIANA MILL.

ESULA DRUCE,CATABENS-NOCTUIDAE
LIFE HISTORY
SOUTHERN CALIFORNIA ACAD.SCI. BULL.33:145,1934 COMSTOCK-DAMMERS LIFE HISTORY;EGG* PL.51;LARVA*PL.52;PUPA* PL.53; COCOON*PL.52
FOOD PLANTS
LANTANA SP.
VERBENA SP.

ETHELA(HULST),DYSSTROMA-GEOMETRIDAE
A KASLOATA TAYLOR
LIFE HISTORY
NEW YORK STATE AGRI.EXP.STA. (CORNELL),MEM.274:136, 1948 FORBES LARVA(KASLOATA?)
FOOD PLANTS
RIBES SANGUINEUM PURSH.
RIBES SATIVUM SYME(CURRANT)

ETHLIUS (CRAMER),CALPODES-HESPERIIDAE
SYN.CHEMNIS FABRICIUS
SYN.OLYNTHUS BOISDUVAL & LECONTE
LIFE HISTORY
ENT.AMER.3:163,1887 HY.EDW. CHRYSALID
ENT.NEWS 2:78,1892 COCKERELL LIFE HISTORY
LOC.CIT.9:163,1898 DYAR LIFE HISTORY
FIELD & FOREST 1:73,1876 DODGE LARVA
JAMAICA INSTITUTE,JOURN.1:29 1891 COCKERELL LARVA, CHRYSALID
MUS.COMP.ZOOLOGY,BULL.87, NO.6,1941 DETHIER LARVAL ANTENNA* FF.43,44,49,50, 53,57
PSYCHE 3:322,1882 KING EGG, LARVA INCLUDES INTERNAL STRUCT.
UNITED STATES DEPT.AGRI.BUR. ENT.BULL.54:55,1905 NEW SERIES CHITTENDEN EGG, LARVA*F.18 B;PUPA* F.18 D
UNITED STATES DEPT.AGRI.DIV. ENT.CIRC.145:4,1912 CHITTENDEN EGG,LARVA*FF.1, 6 B; CHRYSALID*F.6 C,D
UNITED STATES DEPT.AGRI. FARMER'S BULL.1495:18, 1926 WEIGEL-MIDDLETON LARVA*F.27
COMSTOCK,J.A.-BUTTER CALIFORNIA P.227,1927 LARVA, CHRYSALIS
ESSIG-INS.WEST.NORTH AMERICA

P.651,1926 LARVA*F.530 B;
CHRYSALIS* F.530 C,D
GUNDLACH-ENTOM.CUBANA P.160
1881 LARVA, CHRYSALIS
HOLLAND-BUFFERFLY BOOK 1905
CHRYSALIS* PL.6,F.48
SCUDDER-BUTTER.EAST.UNITED
STATES & CANADA PP.1752
1756,1889 LIFE HISTORY* PL.
66,F.34;PL.69,F.15;PL.77,
F.20;PL.80,FF.72-74;PL.85,
F.48
INDEX AMER.ECON.ENT.
VOL.5:101; VOL.6:117;
VOL.7:155

FOOD PLANTS
CANNA FLACCIDA SALISB.
CANNA INDICA LINN.
COLOCASIA ESCULENTA(LINN.)

ETHNE HEMMING
SYN.SEE APHRODITE FABRICIUS

ETOLUS BOISDUVAL
SYN.SEE THYSBE FABRICIUS

EUBULE LINNAEUS
SSP.S E SENNAE LINNAEUS

EUCALYPTULA DYAR
SSP.SEE MINUSCULA ZELLER

EUDIOPTA GUENEE
SSP.SEE ORNITHOGALLI GUENEE

EUDIOPTOIDES BARNES & BENJAMIN
SSP.SEE PRAEFICA GROTE

EUFALA(EDWARDS),LERODEA-HESPERIIDAE
SYN.FLORIDAE MABILLE
SYN.DISPERSUS GUNDLACH

LIFE HISTORY
ENT.NEWS 33:305,1922 COOLIDGE
LIFE HISTORY
SOUTHERN CALIFORNIA ACAD.SCI.
BULL.28:29,56,1929 COOLIDGE
EGG*LARVA*CHRYSALIS*
COMSTOCK,J.A.-BUTTER CALIFORNIA P.226,1927 EGG,LARVA,
CHRYSALIS

FOOD PLANTS
POA PRATENSIS LINN.
GRASSES

EULALIA DOUBLEDAY & HEWITSON
SYN.SEE BREDOWII GEYER

EUMELUSALIS(WALKER),PHILOMETRA-NOCTUIDAE
SYN.SERRATICORNIS GROTE

LIFE HISTORY
ILLINOIS DEPT.AGRI.TRANS.15:
246,1877 FRENCH LARVA
(SERRATICORNIS)

FOOD PLANTS
GRASS ROOTS

EUNOMIA FIELD
SSP.SEE AMPELOS EDWARDS

EUNUS (EDWARDS),PSEUDOCOPAEODES-HESPERIIDAE
SYN.WRIGHTII EDWARDS

LIFE HISTORY
SOUTHERN CALIFORNIA ACAD.SCI.
BULL.30:88,1931 COMSTOCK
EGG* P.87

FOOD PLANTS

EUPATORII(LYMAN),PAPAIPEMA-NOCTUIDAE

LIFE HISTORY

FOOD PLANTS
EUPATORIUM PURPUREUM LINN.

EUREKA DYAR
SSP.SEE MACULATA HARRIS

EUROPOME HAWORTH
SEE PHILODICE GODART

EURYALUS(BOISDUVAL),PLATYSAMIA-SATURNIIDAE
SYN.CALIFORNICA GROTE
SYN.CEANOTHI BEHR
SYN.RUBRA NEUMOEGEN & DYAR
AB PARVIMACULA GROTE

A KASLOENSIS COCKERELL

LIFE HISTORY

CALIFORNIA ACAD. SCI. PROC. 5: 370, 1873 EDWARDS LARVA, PUPA (CEANOTHI)

CAN. ENT. 12:227, 1880 WAILLY LARVA (CEANOTHI)

DAVENPORT ACAD. SCI. PROC. 2: 277, 1878 STRECKER COMPARES LARVAE OF GLOVERI & CEANOTHI

ENT. SOC. AMERICA, ANNALS. 9:151 1916 MOSHER PUPAL STRUC. CALIFORNICA

NAT. ACAD. SCI. MEM. 12:225, 1914 PACKARD LIFE HISTORY*EGG, LARVA*PL. 10, FF. 2-5; PL. 11, F. 1 COLOR CALIFORNICA

ESSIG-INS. WEST. NORTH AMERICA P. 667, 1926 LARVA, COCOON

STRECKER-LEPID, RHOP. ET HETER. 12:103, 1876 LARVA, PUPA, COCOON

INDEX AMER. ECON. ENT.
VOL. 2:321 SAMIA RUBRA
VOL. 6:653 SAMIA RUBRA

FOOD PLANTS

AMELANCHIER FLORIDA LINDL.
ARCTOSTAPHYLOS GLAUCA LINDS.
BETULA SP. (BIRCHES)
CEANOTHUS INTEGERRIMUS H. &A.
CEANOTHUS THYRSIFLORUS ESCHS.
LARIX LARICINA(DUROI.)
MALUS PUMILA MILL. (APPLE)
PHOTINA ARBUTIFOLIA LINDL.
PICEA SP. (SPRUCE)
PRUNUS VIRGINIANA LINN.
PRUNUS SEROTINA EHRH.
PSEUDOTSUGA MENZIESII (MIRB.)
RHAMNUS CALIFORNICA V. TOMENTELLA(BENTH. & WOLF)
RHAMNUS CROCEA NUTT.
RHAMNUS PURSHIANA DC.
RIBES SANGUINEUM PURSH(KASLOENSIS)
SALIX SP. (WILLOWS)

EURYDICE(BOISDUVAL), ZERENE-PIERIDAE

FOR THE MANY SSP. AND SYNS. OF THIS SP. SEE MCDUNNOUGH'S LIST

LIFE HISTORY

CALIFORNIA ACAD. SCI. PROC. 7:60 1876 HY. EDW. LARVA CHRYSALIS

CAN. ENT. 15:224, 1883 W. H. EDW. EGG, LARVA, CHRYSALIS

COMSTOCK, J. A.-BUTTER. CALIFORNIA P. 47, 1927 EGG, LARVA* F. A 20; CHRYSALIS

EDWARDS-BUTTER. NORTH AMERICA 3RD SER. 1897 LIFE HIST* (COLIAS 1 FF. A-K)

KELLOGG-AMER. INS. P. 446, 1908 LARVA

SCUDDER-EVERYDAY BUTTERFLIES P. 286, 1899 EGG, LARVA, CHRYSALIS

SCUDDER-GUIDE TO COMMON BUTTER. P. 109, 1893 LARVA, CHRYSALIS

FOOD PLANTS

AMORPHA CALIFORNICA NUTT. EX T. &G.

EURYDICE(JOHANSSEN), SATYRODES-SATYRIDAE

SYN. CANTHUS LINNAEUS
SYN. CANTHUS GODART
A TRANSOMONTANA GOSSE
SYN. BOISDUVALII HARRIS
AB BOWERI CHERMOCK
FORM FEMALE RAWSONI FIELD
B FUMOSUS LEUSSLER
C APPALACHIA CHERMOCK

LIFE HISTORY

AMER. MUS. NAT. HIST. BULL. 5: 274, 1893 BEUTEN. LARVA, CHRYSALIS

BUFFALO SOC. NAT. SCI. BULL. 19: 21, 1939 WILD LARVA

CAN. ENT. 15:64, 1883 EDWARDS EGG, LARVA

LOC. CIT. 17:112, 1885 EDWARDS CHRYSALIS

ONTARIO ENT. SOC. ANN. REPT. P. 98, 1896 FYLES LARVA(BRIEF)

COMSTOCK-HOW TO KNOW BUTTER

p.192,1904 larva
Fernald-Butterflies Maine p. 68,1884 egg,larva
French-Butter.East.United States p.232,1886 life history (canthus)
Holland-Butterfly Book 1905 larva* chrysalis* Pl.3, f.9; Pl.4,f.9
Maynard-Butter.New England p.6,1886 larva
Scudder-Butter.East.United States & Canada pp.195-198,1889 life history*Pl. 64,f.10;Pl.67,f.2;Pl.70, f.1;Pl.74,f.9;Pl.78,ff.9, 11;Pl.86,ff.19,38;Pl.83, f.9

FOOD PLANTS
Carex bromoides Schkuhr.
Scirpus rubricosus Fern
Grasses

eurymedon Lucas,Papilio-Papilionidae
syn.lewisii Kirby
form albanus Felder & Felder
ab cocklei Gunder
ab columbiana Gunder

LIFE HISTORY
Butterfly Park Nature Club 2:no page number,1931 egg, larva
California Acad.Sci.Proc.5:164, 1873 Hy.Edw.larva,chrysalis
Ent.News 4:243,1893 Dyar egg, larva, chrysalis
Montana Univ.,Bull.30:21,1906 Elrod larva(brief)
Southern California Acad.Sci. Bull.25:65,1926 Comstock larva* pupa*
Comstock,J.A.-Butter California p.24,1927 egg,larva* chrysalis*
Edwards-Butter.North America 2nd series 1874 larva* chrysalis* Pl.Papilio 1 COLOR
Holland-Butterfly Book p.309, 1905 larva; 3rd edit:320, 1931
Peale-Lepid.Americana 1833 larva* Pl.1 COLOR

eurynome Edwards,Argynnis-Nymphalidae
form clio Edwards
syn.eris Igel
ab gunderi Field
a bischoffii Edwards
b opis Edwards
c washingtonia Barnes &McDunnough
form rainierensis Gunder
d artonis Edwards
form rubyensis Gunder
e luski Barnes & McDunnough
ab brucei Gunder

LIFE HISTORY
Edwards-Butter.North America 2nd series,1875 egg* Pl. Argynnis 1 COLOR
Mead-Rept.Lepid.Wheeler Exped.West 100 Merid p. 756,1875 egg

FOOD PLANTS

eurytheme Boisduval,Colias-Pieridae
For the many ssp.and syns. of this sp.see McDunnough's List

LIFE HISTORY
Amer.Mus.Nat.Hist.Bull.5: 250,1893 Beuten.larva
Buffalo Soc.Nat.Sci.Bull.19: 20,1939 Wild larva
California Acad.Sci.Proc.5: 166,1873 Hy.Edw.egg, larva,chrysalis
Illinois Dept.Agri.Trans.18: 78,1880 Middleton larva (Append.)
Journ.Econ.Ent.36:666,1943 Smith&McLeon larva*ff. 1-2;chrysalis*f.1 no descr.
Louisiana Agri.Exp.Sta.Bull. 323:19,1940 Floyd egg*f.1; larva* f.2; chrysalis f.3
Montana Univ.Bull.30:38,1906

Elrod egg,larva,chrysalis
United States Dept.Agri.Bull. 124:9,1914 Wildermuth egg* f.5; chrysalis*f.6;larva* f.2
United States Dept.Agri.Div. Ent.Circ.133:3,1911 Wildermuth egg*f.4;larva*f.1; chrysalis* f.5
United States Dept.Agri. Farmer's Bull.1094:6,1920 Wildermuth egg*f.3;larva* f.4;chrysalis* f.5
Comstock,J.A.-Butter California p.50,1927 egg*Pl.A 21; larva*Pl.A22;chrysalis* Pl.A22
Comstock,J.H.-How Know Butter p.95 1904 larva
Edwards,W.H.-Butter.North America 2nd series 1869 egg*larva*chrysalis* Pl. Colias No.3
Edwards,W.H.-loc.cit.Suppl. Notes 1872 egg,larva, chrysalis
Edwards,W.H.-loc.cit.2nd series 1878 egg*larva* Pl.Colias No.4
Essig-Inj.& Benef.Ins.p.458, 1915 egg,larva*f.469; chrysalis
Essig-Ins.West.North America p.638,1926 larva* f.515; chrysalis
Fernald-Applied Entom.p.299, 1926 larva*f.316
Fernald & Shepard-Applied Entom.p.294,1942 larva* f.310
French-Butter.East.United States p.131,1886 life history
Holland-Butterfly Book 1905 larva*chrysalis*Pl.2, f.1;Pl.5,f.53
Kellogg-Amer.Ins.P.446,1908 larva
Metcalf & Flint-Destr.&Use ful Ins.p.426,1939 larva* chrysalis* f.265
Scudder-Butter.East.United States & Canada pp.1129-
1 1130,1889 Life history*Pl. 65,f.22;Pl.76,f.1;Pl.79, f.61;Pl.84,f.53
Scudder-Guide to Common Butter.p.136,1893 larva, chrysalis'
Index Amer.Econ.Ent. Vol.1:129:Vol.2:142; Vol.3:159;Vol.4:182; Vol.5:149;Vol.6:171; Vol.7:206 Kootenae

FOOD PLANTS

Astragalus sp.
Astragalus carassicarpus Nutt.
Astragalus crotalariae(Benth)
Citrullus vulgaris Schard. (watermelon)
Glycine max(Linn.)
Gossypium herbaceum Linn.
Leguminosaw
Lotus sp.
Lotus americanus(Nutt.)
Lotus grandiflorus(Benth.)
Lupinus perennis Linn.
Medicago hispida Gaertn.
Medicago sativa Linn.(alfalfa)
Melilotus alba Desr.
Pisum sativum Linn.(peas)
Trifolium sp.(clovers)
Trifolium reflexum Linn.
Trifolium repens Linn.
Trifolium stoloniferum Eat.
Trifolium tridentatum Lindl.
Vicia sativa Linn.

eurytion Mead
ssp.see anicia Doubleday&Hewitson

eurytulus Hubner
syn.see columella Fabricius

eurytus(Fabricius),Megisto-Satyridae
syn.cymela Cramer
syn.viola Maynard

LIFE HISTORY

AMER. MUS. NAT. HIST. BULL. 5: 273, 1893 BEUTEN. LARVA, CHRYSALIS
BUFFALO SOC. NAT. SCI. BULL. 19: 21, 1939 WILD LARVA
CAN. ENT. 2:139, 1870 SAUNDERS EGG, LARVA* F. 9
LOC. CIT. 10:105, 1878 EDWARDS EGG, LARVA, CHRYSALIS
ILLINOIS DEPT. AGRI. TRANS. 15: 156, 1877 FRENCH LARVA
LOC. CIT. 18, APPEND: 90, 184, 1880 COQUILLETT & MIDDLETON EGG, LARVA, CHRYSALIS
MUS. COMP. ZOOLOGY, BULL. 87, NO. 6, 1941 DETHIER LARVAL ANTENNA* F. 22
NATURALISTE CANADIEN P. 189, 1928 HUARD LARVA, CHRYSALIS
BOISDUVAL&LECONTE-HIST. LEPID NORTH AMERICA 1833 LARVA* CHRYSALIS* PL. 61 COLOR
COMSTOCK-HOW KNOW BUTTER. P. 197, 1904 LARVA
FERNALD-BUTTER. MAINE P. 67, 1884 LARVA, CHRYSALIS
FRENCH-BUTTER. EAST. UNITED STATES P. 238, 1886 LIFE HISTORY
HOLLAND-BUTTERFLY BOOKP. 203, 1905 LARVA* CHRYSALIS* PL. 3, FF. 3. 6. 10, 13, 14; PL. 4, F. 28
HOLLAND-LOC. CIT. 3RD EDIT: 180, 1931 LIFE HISTORY LARVA*' PL. 3; CHRYSALIS* PL. 4
SCUDDER-BUTTER. EAST. UNITED STATES & CANADA PP. 216-221, 1889 LIFE HISTORY* PL. 64, FF. 5, 11; PL. 70, F. 4; PL. 75, FF. 3, 6, 10, 13, 14; PL. 78, FF. 25-29; PL. 86, FF. 40, 42; PL. 83, F. 28
SCUDDER-EVERYDAY BUTTER. P. 177, 1899 EGG, LARVA, CHRYSALIS
SCUDDER-GUIDE TO COMMON BUTTER. P. 108, 1893 LARVA, CHRYSALIS

FOOD PLANTS

TRIFOLIUM SP. (CLOVER)
XYRIS TORTA J.E. SMITH
GRASSES

<u>EUTERPE</u> HERRICH-SCHAEFFER
SYN. SEE <u>IPOMOEAE</u> HARRIS

<u>EVANIDA</u> GROTE
SSP. SEE <u>HENRICI</u> GROTE

<u>EVANIDALIS</u> ROBINSON
SYN. SEE <u>HUMULI</u> HARRIS (HYPENA)

<u>EVELINA</u> FRENCH
SSP. SEE <u>LACRYMOSA</u> GUENEE

<u>EVELINA</u>(FRENCH), <u>FISHIA</u>-NOCTUIDAE
A <u>HANHAMI</u> SMITH
LIFE HISTORY
CAN. ENT. 26:20, 1894 DYAR LARVA
FOOD PLANTS
ACER MACROPHYLLUM PURSH (<u>HANHAMI</u>)
ALNUS RUBRA BONG. (<u>HANHAMI</u>)
LUPINUS SP.
OSMARONIA CERASIFORMIS (T&G) (<u>HANHAMI</u>)
PRUNUS EMARGINATA (DOUGL.) (<u>HANHAMI</u>)
RIBES AMERICANUM MILL.
RIBES SANGUINEUM PURSH (<u>HANHAMI</u>)
SALIX HOOKERIANA BARR. (<u>HANHAMI</u>)

<u>EVIUS</u> BOISDUVAL
SSP. SEE <u>ICARIOIDES</u> BOISDUVAL

<u>EXACTA</u> HUBNER, <u>ANOMIS</u>-NOCTUIDAE
SYN. <u>CONDUCTA</u> WALKER
SYN. <u>HOSTIA</u> HARVEY
LIFE HISTORY
FOOD PLANTS
MALVAVISCUS SP.

<u>EXAUSPICATA</u>(WALKER), <u>ITAME</u>-GEOMETRIDAE
LIFE HISTORY
FOOD PLANTS
SALIX SP. (WILLOWS)

EXCAECATA(ABBOT&SMITH)PAONIAS-SPHINGIDAE

SYN. PAVONINA GEYER

A PECOSENSIS COCKERELL

B BOREALIS CLARK

LIFE HISTORY

AMER.JOURN.SCI.36:290,1839 HARRIS LARVA

AMER.MUS.NAT.HIST.BULL.7: 314,1895 BEUTEN.LARVA, PUPA

CAN.ENT.1:40,1869 REED LARVA

LOC.CIT.16:10,1884 SAUNDERS LARVA*

LOC.CIT.23:14,1891 BEUTEN. EGG,LARVA

ENT.NEWS 9:150,1898 SOULE DURATION OF EGG STAGE

ENT.SOC.AMERICA,ANNALS 11: 426,1918 MOSHER PUPAL STRUCT.

MINNESOTA AGRI.EXP.STA.,BULL 61:102,1899 LUGGER LARVA* F.52

LOC.CIT.BULL.84,5,1903 WASHBURN LARVA* F.2 COLOR

NAT.ACAD.SCI.MEM.9,1905 PACKARD PUPA* PL.56,F.8

NEW YORK STATE AGRI.EXP.STA. (CORNELL),MEM.274:194, 1948 FORBES LARVA

NEW YORK STATE MUS.MEM.8: 724,1905 FELT LARVA(BRIEF)

NEW YORK STATE,REPT.INS.OF P.341,1856 FITCH LARVA ALSO P.23, SEC.35

NOVITATES ZOOLOGICAE(TRING)9 SUPPL.P.329,1903 ROTHSCHILD & JORDAN LARVA,PUPA

ONTARIO ENT.SOC.ANN.REPT.P. 62,1881 REED LARVA,PUPA

LOC.CIT.P.26,1884 SAUNDERS LARVA,PUPA

PHILADELPHIA ACAD.NAT.SCI. JOURN.5:182,1859 CLEMENS LARVA,PUPA

PHILADELPHIA ENT.SOC.PROC.3: 665,1864 LINTNER LARVA

PSYCHE 3:368,1882 GOODELL LARVA (BRIEF)

SMITHSONIAN MISC.COLLECTIONS 4:ART.2:209,1862 MORRIS LARVA,PUPA

UNITED STATES DEPT.AGRI.MISC. PUB.657:358,375,1950 CRAIGHEAD LARVA

ABBOT & SMITH-LEP.INS.GEORGIA 1:49,1797 LARVA*PUPA*PL.25

BOISDUVAL-SPEC.GEN.LEP.SPHING. P.38,1874 LARVA,PUPA

ELIOT & SOULE-CATER.&THEIR MOTHS P.184,1902 EGG, LARVA* PUPA

FERNALD-IN KINGSLEY'S STANDARD NAT.HIST.2:468,1885 LARVA

FERNALD-SPHING.NEW ENGLAND P.74,1885 LARVA*PUPA*

GLOVER-ILLUST.NORTH AMER.ENT. 1878 LARVA* PL.2,F.14

HARRIS-INS.INJ.VEG.2ST EDIT. P.230,1841 LARVA,PUPA

MORRIS-SYNOP.LEP.NORTH AMERICA P.209,1862 EGG (REALLY EGG OF POLYPHEMUS) LARVA,PUPA

PACKARD-HALF HOURS WITH INSECTS P.180, 1877 LARVA

SAUNDERS-INS.INJ.FRUITS P.85, 1883 LARVA*

SAUNDERS-LOC,COT.P.85,1889 LARVA* F.83L

STRECKER-RHOP.ET.HETER.7:55 1873 LARVA

INDEX AMER.ECON.ENT.

VOL.1:229;VOL.1:275 SATURNIA PAVONIA

VOL.2:324 SATURNIA PAVONIA

FOOD PLANTS

AMELANCHIER ALNIFOLIA NUTT.

BETULA SP.(BIRCHES)

BETULA PAPYRIFERA MARSH.

CARPINUS CAROLINIANA WALT.

CORYLUS SP.(HAZELS)

CRATAEGUS SP.(THRONS)

CRATAEGUS DOUGLASII LINDL.

CYDONIA JAPONICA(THUNB.)

Fraxinus sp. (ashes)
Malus pumila Mill. (apple)
Ostrya virginiana (Mill.)
Physocarpus opulifolius(Linn.)
Populus alba Linn.
Populus balsamifera Linn.
Prunus cerasus Linn.
Prunus domestica Linn. (plum)
Prunus serotina Ehrh.
Pyrus communis Linn. (pear)
Quercus sp. (oaks)
Rosa carolina Linn.
Rubus allegheniensis Porter
Rubus idaeus v. strigosus(Michx.)
Salis sp. (willows)
Spiraea salicifolia Linn.
Tilia americana Linn.
Ulmus sp. (elms)
Wisteria frutescens(Linn.)
Wisteria sinensis Sweet

excavaria Morrison
syn. see fervidaria Herrich-Schaeffer

excellens(Grote), Euxoa-Noctuidae
syn. perexcellens Grote
a infelix Smith
LIFE HISTORY
Can. Ent. 49:401, 1917 Gibson larva, pupa
Washington Ent. Soc. Proc. 4: 318, 1898 Dyar larva (perexcellens)
Index Amer. Econ. Ent. Vol. 1:133 perexcellens Vol. 2:144; Vol. 7:372
FOOD PLANTS
Garden crops-vegetables

excelsa Neumoegen
ssp. see figurata Drury

excurvaria Packard
syn. see denticularia Walker

excurvata Grote
syn. see gueneata Packard

exemplaris Smith
syn. see kelloggi Henry Edwards

exigua(Hubner), Laphygma-Noctuidae
syn. flavimaculata Harvey
LIFE HISTORY
Brooklyn Ent. Soc. Bull. 22:50 1927 Crumb mature larva* Pl. 5 J, K
California Agri. Exp. Sta. Bull. 660:35, 1942 Smith larva* f. 27
Loc. cit. Bull. 707:10, 1948 Michelbacher etc. larva* f. 6A
Can. Ent. 26:65, 1894 Dyar egg larva, pupa
Florida Agri. Exp. Sta. Bull. 271:6, 1934 Wilson egg* f. 2; larva* f. 4; pupa* f. 5
Florida Ent. 16:36, 1932 Wilson egg, larva* (see Vol. 17:2, Pl. f. 7); pupa
Illinois Agri. Exp. Sta. Bull. 60:498, ; 1900 Forbes larva
Southern California Acad. Sci. Bull. 41:176, 1942 Comstock-Dammers three larval forms, green, brown, olive black larva* and pupa* Pl. 34
United States Dept. Agri. Bur. Ent. Bull. 33 new series p. 38, 1902 Chittenden egg* f. 8 e, f; larva* f. 8 b, c, d
Loc, cit. 43 new series p. 37, 1903 Chittenden egg* f. 32 e, f; larva* f. 32 b, c, d; f. 33
Loc. cit. 57 new series p. 36, 1906 Sanderson egg* f. 19 e, f; larva* f. 19 b, c, d
United States Dept. Agri. Farmer's Bull. 223:15, 1905 Sanderson egg* f. 13 e, f; larva* f. 13 b, d
Loc. cit. 856:30, 1917 Orton-Chittenden egg* f. 30 e, f;

LARVA*F.13 B,C,D
WASHINGTON ENT.SOC.PROC.32:
48,1930 CAMPBELL & DURAN
EGG*
CHITTENDEN-INS.INJ.VEG.P.125,
1907 EGG* F.79; LARVA* F.69
CROSBY & LEONARD-MANUAL VEG.
GARD.INS.P.294,1918 EGG,
LARVA,PUPA
ESSIG-INJ.&BENEF.INS.P.400,
1915 EGG*LARVA* BOTH F.400
FLAVIMACULATA
ESSIG-INS.WEST.NORTH AMERICA
P.686,1926 EGG,LARVA
O'KANE-INJ.INS.P.174,1912
LARVA
PEAIRS-PESTS FARM.GARD.&
ORCHARD P.257,1941 EGG*
LARVA* BOTH F.296
INDEX AMER.ECON.ENT.
VOL.1:178;VOL.2:197 EXIGUA
AND FLAVIMACULATA
VOL.3:226;VOL.4:260;
VOL.5:373;VOL.6:442;
VOL.7:590

FOOD PLANTS

ALLIUM CEPA LINN.(ONION)
AMARANTHUS RETROFLEXUS LINN.
ASPARAGUS OFFICINALIS LINN.
(ASPARAGUS)
ATRIPLEX SP.
BETA VULGARIS LINN.(BEET)
BRASSICAE OLERACEA LINN.
(CABBAGE)
BRASSICAE OLERACEA V.BOTYRTIS
D.C.(CAULIFLOWER)
CAPSICUM ANNUUM LINN.
CARDAMINE SP.
CHENOPODIUM ALBUM LINN.
CLEOME SP.
GLADIOLUS SP.
GOSSYPIUM HERBACIUM LINN.
HELIANTHUS SP.
HELIANTHUS ANNUUS LINN.
HIBISCUS SP.
IPOMOEA BATATAS(LINN.)(SWEET
POTATO)
LACTUCA SATIVA LINN.(LETUCE)
MALUS PUMILA MILL.(APPLE)
MALVA SP.
MALVA ROTUNDIFOLIA LINN.
MEDICAGO SATIVA LINN.
NICOTIANA GLAUCA GRAHAM
NICOTIANA RUSTICA LINN.
PHASEOLUS VULGARIS LINN.(BEAN)
PISUM SATIVUM LINN.(PEAS)
PLANTAGO SP.
RICINUS COMMUNIS LINN.
SALSOLA KALI V.TENUIFOLIA
TAUSCH
SOLANUM LYCOPERSICUM LINN.
(TOMATO)
SOLANUM TUBEROSUM LINN.
(POTATO)
ZEA MAYS LINN.(CORN)
GENERAL FEEDER-FIELD
CROPS,GARDEN CROPS,
VEGETABLES
GRASSES
FOREST TREES

EXILIS(BOISDUVAL),BREPHIDIUM-LYCAENIDAE
SYN.FEA EDWARDS
AB COOLIDGEI GUNDER

LIFE HISTORY

CAN.ENT.27:37,1894 EDWARDS
LARVA,CHRYSALIS
ENT.NEWS 35:115,1924 COLLIDGE
LIFE HISTORY EGG,LARVA,
CHRYSALIS
NEW YORK ENT.SOC.JOURN.7:211,
1899 COQUILLETT LARVA
COMSTOCK,J.A.-BUTTER.CALIF-
ORNIA P.179,1927 EGG*
F.A61A;LARVA*F.A 60;F.A
61 B;CHRYSALIS* F.A 61C

FOOD PLANTS

ATRIPLEX CANESCENS (PURSH)
ATRIPLEX COULTERI(MOQ.)
ATRIPLEX LENTIFORMIS V.
BREWERI(WATS.)
ATRIPLEX LEUCOPHYLLA(MOQ.)
ATRIPLEX PATULA LINN.
ATRIPLEX PATULA V.HASTATA
(LINN.)
ATRIPLEX SEMIBACCATA R.BR.

Atriplex serenana Nels.
Chenopodium sp.
Chenopodium album Linn.
Chenopodium leptophyllum Nutt.
Petunia parviflora Juss.

EXIMIA Grote
SYN. SEE SEMIRUFESCENS Walker

EXNOTATA Walker
SSP. SEE GRANITATA Guenee

EXOLETA Henry Edwards
SSP. SEE NELSONI Boisduval

EXPANSA Walker
SYN. SEE EMMEDONIA Cramer

EXPANSA Walker
SYN. SEE ANTICARIA Walker
(Eupithecia)

EXPOSITA Dyar
SSP. SEE APERA Drury

EXPRESSARIA Walker
SYN. SEE EPHYRARIA Walker

EXPRIMENS Walker
SSP. SEE UMBRA Hufnagel

EXPULSA Walker
SYN. SEE MESSORIA Harris

EXPULTRIX Grote
SSP. SEE CYMATOPHOROIDES Guenee

EXPUNCTA Walker
SYN. SEE CHALCEDONIA Hubner

EXPUNCTARIA Grote
SSP. SEE LIMBATA Haworth

EXQUISITA Hampson
SYN. SEE HOWLANDI Grote

EXSANGUIS Dyar
SSP. SEE BILINEATA Packard
(Heterocampa)

EXERTISTIGMA (Morrison), RHYNCHAGROTIS-NOCTUIDAE
For the many ssp. of this sp. see McDunnough's List
LIFE HISTORY
Washington Ent. Soc. Proc. 4: 316, 1898 Dyar larva
FOOD PLANTS
Vitis sp. (grapes)
"Many plants" "Low plants"

EXSINUARIA Guenee
SSP. SEE AMICARIA Herrich-Schaeffer

EXSUPERATA Walker
SYN. SEE SUBATOMARIA Wood

EXTERNA Walker
SYN. SEE ERIDANIA Cramer

EXTERSA Walker
SYN. SEE BINOTATA Walker

EXTIMA (Walker), PLATYPERIGEA-NOCTUIDAE
SYN. CIVICA Grote
LIFE HISTORY
United States Nat. Mus. Proc. 25:375, 1903 Dyar egg, larva
Washington Ent. Soc. Proc. 4: 322, 1898 Dyar larva
FOOD PLANTS
Polygonum sp.
Taraxacum officinale Weber

EXTINCTOANALIS Bryk & Eisner
SYN. SEE CLODIUS Menetries

EXTRANEA Guenee
SYN. SEE UNIPUNCTA Haworth

EXUSTA Guenee
SYN. SEE PICTA Harris

F

FABRICII EDWARDS
SSP. SEE INTERROGATIONIS FABRICIUS

FABRICII GROTE
SYN. SEE IO FABRICIUS

FABRICII KIRBY
SYN. SEE DWARDSII SAUNDERS

FABULA GROTE
SYN. SEE ARCUATA WALKER

FACETA HENRY EDWARDS
SSP. SEE GRAPHICA HUBNER

FACTIOSALIS (WALKER), RENIA–NOCTUIDAE
SYN. CLITOSALIS WALKER
SYN. CENTRALIS GROTE
SYN. PLENILINEALIS GROTE
SYN. ALUTALIS GROTE
SYN. TILOSALIS SMITH
LIFE HISTORY
ENT. AMER. 14:167, 1934 CRUMB
LARVA* PL. 24 E
FOOD PLANTS
DEAD LEAVES

FACTORIS SMITH
SYN. SEE DIVERGENS (WALKER (EUXOA)

FACULA GROTE
SSP. SEE EXSERTISTIGMA MORRISON

FACULANA STRAND
SYN. SEE EXSERTISTIGMA MORRISON

FACULELLA STRAND
SSP. SEE EXSERTISTIGMA MORRISON

FACULOIDES STRAND
SSP. SEE EXSERTISTIGMA MORRISON

FALACER (GODART), STRYMON–LYCAENIDAE
SYN LORATA GROTE & ROBINSON
SYN. INORATA GROTE & ROBINSON
LIFE HISTORY
CAN. ENT. 1:57, 1869 SAUNDERS
EGG
LOC. CIT. 2:61, 1870 SAUNDERS
LARVA, CHRYSALIS (CALANUS HUBNER?)
BOISDUVAL & LECONTE–HIST. LEP. NORTH AMERICA P. 93, 1833
LARVA* CHRYSALIS* PL. 29
COLOR PROBABLY CALANUS HUBNER
FOOD PLANTS
CARYA SP. (HICKORIES)
CRATAEGUS SP. (THORNS)
JUGLANS CINEREA LINN.
QUERCUS SP. (OAKS)

FALCATA (PACKARD), APICIA–GEOMETRIDAE
LIFE HISTORY
PACKARD–MONOGRAPH GEOMETRIDAE P. 558, 1876 EGG, (BRIEF)
FOOD PLANTS

FALCIFERA (KIRBY), AUTOGRAPHA–NOCTUIDAE
FORM SIMPLEX GUENEE
LIFE HISTORY
CAN. ENT. 13:21, 1881 COQUILLETT
EGG, LARVA, PUPA (AS PRECATIONIS IN ERROR CORRECTED ON PAGE 60 OF SAME ISSUE)
UNITED STATES DEPT. AGRI. BUR. ENT. BULL. 33 NEW SERIES 0.73, 1902 CHITTENDEN
LARVA* F. 16
CHITTENDEN–INS. INJ. VEG. P. 182, 1907 LARVA* F. 117
CROSBY & LEONARD–MANUAL VEG. GARD. INS. P. 192, 1918 EGG, LARVA
ESSIG–INS. WEST. NORTH AMERICA P. 687, 1926 LARVA
O'KANE–INJ. INS. P. 179, 1912 LARVA (SIMPLEX)
PEAIRS–PESTS FARM, GARD., ORCHARD P. 265, 1941 LARVA* F. 306
INDEX AMER. ECON. ENT.
VOL. 1:249 PLUSIA SIMPLEX
VOL. 2:47 SIMPLEX
VOL. 3:51 SIMPLEX
VOL. 4:57; VOL. 5:68;

VOL. 6:79;VOL. 7:35

FOOD PLANTS

ALTHAEA ROSEA CAV.
APIUM GRAVEOLENS LINN.
ARCTIUM LAPPA LINN.
BETA VULGARIS LINN. (BEET)
BRASSICA OLERACEA LINN. (CABBAGE)
CIRSIUM SP. (THISTLES)
DAUCUS CAROTA V. SATIVUS HOFF.
LACTUCA SATIVA LINN. (LETTUCE)
PHASEOLUS VULGARIS LINN. (BEANS)
PLANTAGO SP.
TARAXACUM OFFICINALE WEBER
VACCINIUM MACROCARPON AIT.
GRASS LOW WEEDS

FALCULA (GROTE), ACRONICTA-NOCTUIDAE

LIFE HISTORY

AMER. MUS. NAT. HIST. BULL. 14: 248, 1901 BEUTEN. LARVA
NEW YORK STATE MUS. MEM. 8: 723, 1905 FELT LARVA (BRIEF)
PAPILIO 1:6, 1881 COQUILLETT LARVA
UNITED STATES NAT. MUS. PROC. 21:110, 1898 SMITH & DYAR LARVA

FOOD PLANTS

CORYLUS SP. (HAZELS)

FALLACIALIS WALKER
SYN. SEE DISCOLORALIS GUENEE

FALLAX (HERRICH-SCHAEFFER), AGRIOPODES-NOCTUIDAE

LIFE HISTORY

NEW YORK ENT. SOC. JOURN. 7:67 1899 DYAR EGG, LARVA, PUPA, COCOON
NEW YORK STATE MUS. MEM. 8: 738, 1905 FELT LARVA (BRIEF)

FOOD PLANTS

POPULUS SP. (POPLARS)
VIBURNUM DENTATUM LINN.

FAMULATA HULST
SSP. SEE NOTATARIA WALKER

FANNIAE GUNDER
SSP. SEE EURYDICE BOISDUVAL

FANNYAE GUNDER
SSP. SEE RUTULUS LUCAS

FANTASIA BUTLER
SSP. SEE LYSIDE GODART

FARCTA (GROTE), LEUCANIS-NOCTUIDAE
A ROSEOLA SMITH

LIFE HISTORY

UNITED STATES NAT. MUS. PROC. 25:376, 1903 DYAR EGG, LARVA
LOC. CIT. 27:863, 1904 DYAR EGG LARVA (ROSEOLA)

FOOD PLANTS

GRASS (ROSEOLA)

FASCIATA (BARNES & MCDUNNOUGH), CYMATOPHORA-THYATIRIDAE

LIFE HISTORY

BRITISH COLUMBIA, ENT. SOC. PROC. 1 NEW SERIES P.32, 1911 G. O. DAY EGG, LARVA

FOOD PLANTS

AMELANCHIER FLORIDA LINDL.
PRUNUS VIRGINIANA LINN.
PRUNUS LYONII (EASTW.)

FASCIATA BEUTENMULLER
SYN. SEE NUBILIS HUBNER

FASCIATA COCKERELL
SSP. SEE BELLONA FABRICIUS

FASCIATA SMITH
SSP. SEE STYRACIS GUENEE

FASCIATA STRECKER
SSP. SEE HYPOPHLAEAS BOISDUVAL

FASCIATA SWAINSON
SYN. SEE ALOPE DRURY

FASCIATA TAYLOR
SYN. SEE LUTEATA PACKARD

FASCIATA WALKER
SYN. SEE IMPRESSA¼WALKER

FASCIATUS(SULZER), PHOLUS-SPHINGIDAE
SYN. JUSSIEUAE HUBNER
SYN. STRIGILIS VOGEL
LIFE HISTORY
FLORIDA ENT. 15:49,1931 MILLER LARVA, PUPA
NEW YORK STATE AGRI. EXP. STA. (CORNELL), MEM. 274:200, 1948 FORBES LARVA
NOVITATES ZOOLOGICAE(TRING)9 SUPPL. P. 495,1903 ROTHSCHILD & JORDAN LARVA, PUPA
INDEX AMER. ECON. ENT. VOL. 5:488
FOOD PLANTS
JUSSIAEA DECURRENS(WALT.)
VITIS SP. (GRAPES)

FASCIOFERARIA HULST
SSP. SEE RIBEARIA FITCH

FASTUOSA GUENEE
SYN. SEE HIEROGLYPHICA CRAMER

FATUARIA STRECKER
SYN. SEE ENDROPIARIA GROTE & ROBINSON

FAUNUS(EDWARDS), POLYGONIA-NYMPHALIDAE
SYN. VIRESCENS SCUDDER
A SMYTHI CLARK
SYN. FEMALE SILVIUS EDWARDS
C ARCTICUS LEUSSLER
LIFE HISTORY
AMER. MUS. NAT. HIST. BULL. 5: 265,1893 BEUTEN. LARVA CHRYSALIS
BUFFALO SOC. NAT. SCI. BULL. 19:25,1939 WILD LARVA
CALIFORNIA ACAD. SCI. PROC. 5: 169,1873 H. Y. EDWARDS LARVA, CHRYSALIS(ZEPHYRUS IN ERROR?)
CAN. ENT. 7:49,1875 CAULFIELD LARVA
ILLINOIS DEPT. AGRI. TRANS. 18:APPEND:84,1880 MIDDLETON LARVA
MONTANA UNIV. BULL. 30:85, 1906 ELROD LARVA, CHRYSALIS
NEW YORK STATE MUS. MEM. 8: 742,1905 FELT LARVA (BRIEF)
COMSTOCK, J. A.-BUTTER CALIFORNIA P. 122,1927 LARVA, CHRYSALIS (RUSTICUS)
COMSTOCK, J. H.-HOW TO KNOW BUTTER. P. 138,1904 LARVA
W. H. EDWARDS-BUTTER. NORTH AMERICA 2ND SERIES 1879 LARVA*CHRYSALIS*PL. GRAPTA 3 COLOR RUSTICUS
FERNALD-BUTTER. MAINE P.52, 1884 EGG, LARVA, CHRYSALIS*
FRENCH-BUTTER. EAST. UNITED STATES P. 188,1886 LARVA, CHRYSALIS
HOLLAND-BUTTERFLY BOOK 1905 LARVA*CHRYSALIS*PL.3, F.32:PL.4, FF.31,33-35 COLOR
MAYNARD-BUTTER. NEW ENGLAND P. 13,1886 LARVA
SCUDDER-BUTTER. EAST. UNITED STATES & CANADA PP. 351-356,1889 LIFE HISTORY*PL. 64, F.21;PL.67, F.10;PL. 70, F.8;PL.74, F.32;PL.78, FF.41-47, PL.86, F.82;PL. 86, FF.31,33-35
SCUDDER-BUTTER. THEIR STRUCTURE ETC. P.42,1881 CHRYSALIS*
SCUDDER-EVERYDAY BUTTER. P.291,1899 EGG, LARVA, CHRYSALIS
SCUDDER-GUIDE TO COMMON BUTTER. P.94,1893 LARVA, CHRYSALIS
INDEX AMER. ECON. ENT. VOL. 7:799

FOOD PLANTS
ALNUS CRISPA(AIT.)
BETULA PAPYRIFERA MARSH.
RHODODENDRON OCCIDENTALE (T.& G.) GRAY
RIBES CYNOSBATI LINN.
RIBES GROSSULARIA LINN.
RIBES SATIVUM SYME(CURRANT)
SALIX DISCOLOR MUHL.
ULMUS SP.(ELMS)
URTICA SP.(NETTLES)

FAUSTINA STRECKER, CATOCALA-NOCTUIDAE
FORM RUBRA CASSINO
FORM CARLOTA BEUTENMULLER
FORM ZILLAH STRECKER
FORM LYDIA BEUTENMULLER
LIFE HISTORY
AMER. MUS. NAT. HIST. BULL. 38: 161, 1918 BARNES & MCDUNNOUGH LARVA
ENT. NEWS 25:165, 1914 ROWLEY EGG, LARVA
PSYCHE 20:199, 1913 BARNES & MCDUNNOUGH EGG, LARVA
FOOD PLANTS
SALIX SP. (WILLOWS)

FAUSTINARIA STRECKER
SYN. SEE MACULARIA HARRIS

FAUSTINULA(BOISDUVAL), CISTHENE-ARCTIIDAE
FORM FUSCA STRETCH
LIFE HISTORY
ENT. NEWS 16:257, 1905 WILLIAMS NOTE ON LIFE HIST.
FOOD PLANTS
RAMALINA MENZIESII TUCK AND OTHER LICHENS

FAVILLIFERA WALKER
SYN. SEE OSSULARIA HUBNER

FAVONIUS(ABBOT & SMITH), STRYMON-LYCAENIDAE
LIFE HISTORY
UNITED STATES DEPT. AGRI. BUR. ENT. BULL. 13, OLD SERIES P. 31, 1887 PACKARD LARVA, CHRYSALIS
ABBOT & SMITH-LEP. INS. GEORGIA 1:27, 1797 LARVA* CHRYSALIS* PL. 14 COLOR
BOISDUVAL & LECONTE-HIST. LEP. NORTH AMERICA P. 96, 1833 LARVA*PUPA*PL. 30
FRENCH-BUTTER. EAST UNITED STATES P. 258, 1886 LARVA, CHRYSALIS
FOOD PLANTS
QUERCUS SP. (OAKS)
QUERCUS VELUTINA LAM.
QUERCUS VIRGINIANA MILL.

FAXONI MINOT
SYN. SEE AMITARIA GUENEE

FEA EDWARDS
SYN. SEE EXILIS BOISDUVAL

FEBRUALIS GROTE, FERALIA-NOCTUIDAE
LIFE HISTORY
NEW YORK ENT. SOC. JOURN. 7: 210, 1899 COQUILLETT LARVA
SOUTHERN CALIFORNIA ACAD. SCI. BULL. 43:117, 1944 COMSTOCK-DAMMERS LARVA* PL. 25
FOOD PLANTS
PRUNUS ILICIFOLIA(NUTT. EX H.&A.)
SAMBUCUS COERULEA RAF.

FEGEUS CRAMER
SYN. SEE LUGUBRIS LINNAEUS

FEILDENI MCLACHLAN
SSP. SEE HYPOPHLAEAS BOISDUVAL

FELINA GROTE
SSP. SEE LEPUSCULINA GUENEE

FENESTRA PERRY
SYN. SEE POLYPHEMUS CRAMER

FENNICA(TAUSCHER), ACTEBIA-NOCTUIDAE
SYN. INTRACTA WALKER
LIFE HISTORY
AMER. MUS. NAT. HIST. BULL. 14: 276, 1901 BEUTEN. LARVA
BROOKLYN ENT. SOC. BULL. 27: 84, 1933 CRUMB LARVA* PL. 5 B. Q
CAN. ENT. 16:204, 1884 FLETCHER LARVA
ENT. SOC. AMERICA, ANNALS 23: 396, 1930 HAWKINS LARVAL CLAWS
MAINE AGRI. EXP. STA. BULL. 340, 204, 1927 PHIPPS LARVA* F. 29 A, D; PUPA* F. 29E
LOC. CIT. 356:158, 1930 PHIPPS LARVA* F. 23C; PUPA
ONTARIO ENT. SOC. ANN. REPT. P. 15, 1884 SAUNDERS LARVA
LOC. CIT. P. 100, 1896 FYLES LARVA
CROSBY & LEONARD-MANUAL VEG. GARD. INS. P. 275, 1918 LARVA
INDEX AMER. ECON. ENT.
VOL. 1:221 NOCTUA
VOL. 2:252 NOCTUA
VOL. 4:14 AGROTIS
VOL. 5:5; VOL. 6:61
VOL. 7:8
FOOD PLANTS
ACER SP. (MAPLES)
ACER NEGUNDO LINN.
ALLIUM CEPA LINN. (ONION)
ALNUS SP. (ALDER)
ARONIA MELANOCARPA (MICHX.)
ASPARAGUS OFFICINALIS LINN.
ASTER SP.
BETULA SP. (BIRCHES)
CHENOPODIUM ALBUM LINN.
COMTONIA PEREGRINA (LINN.)
CORNUS CANADENSIS LINN.
DENNSTAEDTIA PUNCTILOBULA (MICHX.)
FAGUS SP. (BEECH)
FRAGARIA CHILOENSIS (LONN.)
JUGLANS NIGRA LINN.
KALMIA SP.
LEPIDIUM VIRGINICUM LINN.
MALUS PUMILA MILL. (APPLE)
MELILOTUS ALBA DESR.
ORYZA SATIVA LINN. (RICE)
PEDICULARIS CANADENSIS LINN.
PHLEUM PRATENSE LINN.
PISUM SATIVUM LINN. (PEAS)
POA PRATENSIS LINN.
POPULUS SP. (POPLARS)
PRUNUS CERASUS LINN. (CHERRY)
PRUNUS SEROTINA EHRH.
QUERCUS SP. (OAKS)
QUERCUS ALBA LINN.
RHUS SP.
RIBES SANGUINEUM PURSH.
ROSA SP.
ROSA BLANDA AIT.
RUBUS ALLEGHENIENSIS PORTER
RUBUS IDAEUS V. STRIGOSUS (MICHX.)
RUMEX SP.
RUMEX CRISPUS LINN.
SALIX SP. (WILLOWS)
SALIX ALBA LINN.
SOLIDAGO SP.
SONCHUS OLERACEUS LINN.
SPIRAEA SALICIFOLIA LINN.
TRIFOLIUM SP. (CLOVER)
TRIFOLIUM PRATENSE LINN.
ULMUS SP. (ELMS)
VACCINIUM ANGUSTIFOLIUM AIT.
VACCINIUM MYRTILLOIDES MICHX.
VERBASCUM SP.
VERBASCUM THAPSUS LINN.
VICIA CRACCA LINN
FERNS, FOREST TREES, FRUIT TREES, SHRUBS

FERMATA BRYK
SYN. SEE SMINTHEUS DOUBLEDAY & HEWITSON

FERNALDARIA HULST
SYN. SEE CREPUSCULARIA SCHIFFER-MULLER

FERNALDI GROTE

SYN. SEE ATROPUNCTATA PACKARD

FERONIA STEPHENS
SYN. SEE MONUSTE LINNAEUS

FEROX WALKER
SYN. SEE IPOMOEAE HARRIS

FERREA GROTE
SYN. SEE STRICTA WALKER

FERRICOSTA SMITH
SSP. SEE ARCIGERA GROTE

FERRUGATA(LINN. (, XANTHORHOE-GEOMETRIDAE
SYN. INCLINATARIA WALKER
FORM UNIDENTARIA HAWORTH
A INFUMATA BARNES & MCDUNNOUGH
B ALASKAE CASSINO & SWETT
LIFE HISTORY
NEW YORK STATE AGRI. EXP. SAT. (CORNELL), MEM. 274:151, 1948 FORBES LARVA
ROYAL SOC. CANADA, TRANS. VOL. 1 SERIES 3:66, 1907 FLETCHER-GIBSON EGG, LARVA, PUPA
GUMPPENBERG-SYSTEMA GEOMETRARUM 3:365, 1890 LARVA, PUPA
NEWMAN-BRITISH MOTHS P.168, 1869
PACKARD-MONOGRAPH GEOMETRIDAE P.148, 1876 LARVA
STAINTON-BRITISH BUTTER. & MOTHS 2:105, 1869 LARVA
FOOD PLANTS
ARMORACIA LAPATHIFOLIA GILIB.
GLECHOMA HEDERACEA LINN.
POLYGONUM SP.
STELLARIA MEDIA(LINN.)
LOW PLANTS

FERRUGINARIA PACKARD
SYN. SEE FULVARIA VILLERS

FERRUGINEA PACKARD
SSP. SEE AMERICANA HARRIS(EPICNAPTERA)

FERRUGINEA PACKARD, LOPHODONTA-NOTODONTIDAE
A REDUCTA MCDUNNOUGH
LIFE HISTORY
AMER. MUS. NAT. HIST. BULL. 10: 412, 1898 BEUTEN. LARVA
BOSTON SOC. NAT. HIST. PROC. 26:394, 1894 DYAR EGG, LARVA
NAT. ACAD. SCI. MEM. 7:151, 1895 PACKARD EGG, LARVA* PL. 18, F. 1 COLOR
NEW YORK STATE AGRI. EXP. STA., (CORNELL), MEM. 274, 219, 1948 FORBES LARVA
NEW YORK STATE MUS. MEM. 8: 749, 1905 FELT LARVA (BRIEF)
FOOD PLANTS
BETULA PAPYRIFERA MARSH.

FERRUGINEA SMITH
SYN. SEE REPUGNALIS HUBNER

FERRUGINEOIDES GUENEE
SSP. SEE BICOLORAGO GUENEE

FERVIDARIA(HUBNER), ELLOPIA-GEOMETRIDAE
LIFE HISTORY
UNITED STATES DEPT. AGRI. REPT. P.329, 1886 LARVA, PUPA
GUMPPENBERG-SYSTEMA GEOMETRARUM 7:400, 1895 LARVA, PUPA
PACKARD-MONOGRAPH GEOMETRIDAE P.494, 1876 LARVA* PL. 13, F.34; PUPA* PL. 13, F. 13A
INDEX AMER. ECON. ENT. VOL. 5:210
FOOD PLANTS
ABIES BALSAMEA(LINN.)
BETULA SP. (BIRCHES)
FAGUS SP.
HALESIA CAROLINA LINN.
PICEA SP.
PRUNUS CERASUS LINN. (CHERRY)
PSEUDOTSUGA MENZIESII(MIRB.)

QUERCUS VIRGINIANA MILL.
TSUGA CANADENSIS(LINN.)
ULMUS SP.

FERVIDARIA HERRICH-SCHAEFFER, PLAGODIA-GEOMETRIDAE
SYN. EMARGATARIA GUENEE
SYN. EXCAVARIA MORRISON
LIFE HISTORY
NEW ;YORK STATE MUS. MEM. 8: 706, AND 754, 1905 FELT LARVA
FOOD PLANTS
ACER SP. (MAPLES)
BETULA SP. (BIRCHES)
FAGUS SP. (BEECHES)
FRAXINUS SP. (ASHES)
PICEA SP. (SPRUCES)
PRUNUS CERASUS LINN. (CHERRY)
QUERCUS SP. (OAKS)

FESTIVA FABRICIUS
SYN. SEE HIEROGLYPHICA CRAMER

FESTUS GEYER
SYN. SEE CELLUS BOISDUVAL & LECONTE

FICUS(LINNAEUS), PACHYLIA-SPHINGIDAE
SYN. CRAMERI MENETRIES
SYN. LYNCEA CLEMENS
LIFE HISTORY
ENT. NEWS3:4, 1892 COCKERELL LARVAL NOTE
NOVITATES ZOOLOGICAL(TRING)9 SUPPL. P. 374, 1903 ROTHSCHILD & JORDAN LARVA, PUPA
BOISDUVAL-SPEC. GEN. LEP. SPHING. P. 137, 1874 LARVA
GUNDLACH-ENTOM. CUBANA P. 201, 1886 LARVA, PUPA
MERIAN-HIST. INS. SURINAM 1719 LARVA* PL. 33
SMITH, J. B.-MONOGRAPH SPHINGIDAE P. 139, 1888
FOOD PLANTS
FICUS SP. (FIGS)

FIDELISSIMA HERRICH-SCHAEFFER, COMPOSIA PERICOPIDAE
SYN. OLYMPIA BUTLER
A VAGRANS BATES
LIFE HISTORY
FLORIDA ENT. 28:29, 1945 WATSON LARVA(BRIEF)
INSECUTOR INSCITIAE MENSTRUUS 2:62, 1914 DYAR LARVA(BRIEF)
NEW YORK ENT. SOC. JOURN. 4:70 1896 DYAR EGG, LARVA, PUPA, COCOON
FOOD PLANTS
CANAVALIA ROSA(SW.)
ECHITES UMBELLATA JACQ.
STEPHANOTIS SP.

FIDONIARIA WALKER
SYN. SEE SUBATOMARIA WOOD

FIDONIATA WALKER
SYN. SEE NOTATARIA WALKER

FIELDI GUNDER
SSP. SEE EDITHA BOISDUVAL

FIGURATA(DRURY), APANTESIS-ARCTIIDAE
SYN. CERAMICA HUBNER
FORM SNOWI GROTE
FORM EXCELSA NEUMOEGEN
AB LUGUBRIS HULST
FORM PRECIOSA NIXON
LIFE HISTORY
BROOKLYN ENT. SOC. BULL. 1:3, 1878 GRAEF LARVA(BRIEF)
CAN. ENT. 35:152, 1903 GIBSON LARVA
NEW YORK ENT. SOC. JOURN. 8:42, 1900 DYAR LARVA(BRIEF)
UNITED STATES NAT. MUS. PROC. 25:372, 1903 DYAR LARVA
WASHINGTON ENT. SOC. PROC. 13: 232, 1911 NIXON LARVA
FOOD PLANTS
MEDICAGO SATIVA LINN. (ALFALFA)

FILAMENTARIA GUENEE
SYN. SEE LIMBATA HAWORTH

FILARIA WALKER

SYN. SEE PORCELARIA GUENEE

FILMATA PEARSALL, EUPITHECIA-GEOMETRIDAE
LIFE HISTORY
CAN. ENT. 74:202, 1942 MCDUNNOUGH EGG, LARVA
LOC. CIT. 77:54, 1945 MCGUFFIN LARVA
FOOD PLANTS
ABIES BALSAMEA(LINN.)
PICEA SP. (SPRUCE)
TSUGA CANADENSIS(LINN.)

FIMBRIAGO STEPHENS
SSP. SEE FLAVA FABRICIUS

FINIS SMITH
SYN. SEE TESSELLATA HARRIS

FINITIMA(GUENEE), SEPTIS-NOCTUIDAE
A CERIVANA SMITH
LIFE HISTORY
CAN. ENT. 52:121, 1920 CRIDDLE LARVA
MUS. COMP. ZOOL. BULL. 87; NO. 6, 1941 DETHIER LARVAL ANTENNA* F. 110 HADENA BASILINEA
INDEX AMER. ECON. ENT. VOL. 7:875
FOOD PLANTS
BROMUS INERMIS LEYSS.
CAREX SP. (SEDGES)
PHLEUM PRATENSE LINN. (TIMOTHY)
TRITICUM AESTIVUM LINN. (WHEAT KERNELS)
ZEA MAYS LINN. (CORN)

FISCELLARIA GUENEE, ELLOPIA-GEOMETRIDAE
FORM FLAGITARIA GUENEE
FORM JOHNSONI SWETT
FORM PECCATARIA SWETT
A LUGUBROSA HULST
LIFE HISTORY
CAN. ENT. 74:229, 1942 DETHIER EGG, LARVA*PUPA*PL. 15, FF. 6,7,8,10; PL. 17, F. 2
JOURN. ECON. ENT. 20:288, 1927 FRACKER & GRANOVSKY EGG.
MICHIGAN AGRI. EXP. STA. CIRC. 141:19, 1932 MCDANIEL EGG* F. 22; LARVA*F. 21; PUPA*F. 23
MICHIGAN STATE COLLEGE, SPECIAL BULL. 243:34, 1933 MCDANIEL LIFE HISTORY LARVA*PUPA
NEW YORK ENT. SOC. JOURN. 53: 202, 1945 FORBES PUPA
PSYCHE 10:13, 1903 DYAR EGG, LARVA, PUPA
UNITES STATES DEPT. AGRI. MISC. PUB. 657:362, 434, 1950 CRAIGHEAD LARVA* F. 91E
GRAHAM-PRINCIPLES FOREST ENT. P. 197, 1939 LARVA* F. 79
HERRICK-INS. ENEMIES SHADE TREES P. 355, 1935 LARVA
INDEX AMER. ECON. ENT. VOL. 4:160; VOL. 5:210; VOL. 6:244; VOL. 7:588
FOOD PLANTS
ABIES SP.
ABIES BALSAMEA(LINN.)
ACER CIRCINATUM PURSH.
ACER SACCHARINUM LINN.
ALNUS SP.
BETULA ALLEGHENIENSIS BRITT.
BETULA PAPYRIFERA MARSH.
FAGUS SP.
PICEA ENGELMANNI ENGELM.
PINUS RESINOSA AIT.
PINUS STROBUS LINN.
PRUNUS SEROTINA EHRH.
PRUNUS VIRGINIANA LINN.
PSEUDOTSUGA MENZIESII(MIRB.)
QUERCUS SP.
THUJA OCCIDENTALIS LINN.
TSUGA CANADENSIS(LINN.)
TSUGA HETEROPHYLLA(RAF.)
ULMUS AMERICANA LINN.
VACCINIUM SP.

FISSINOTATA WALKER
SYN. SEE GRANITATA GUENEE

FLAGELLUM(WALKER), AUTOGARPHA-NOCTUIDAE
SYN. MONODON GROTE
SYN. INSOLITA SMITH

LIFE HISTORY
FOOD PLANTS
HELIANTHUS SP.
LIATRIS SP.

FLAGITIARIA GUENEE
SSP. SEE FISCELLARIA

FLAMMEA NEUMOEGEN
SSP. SEE PLACENTIA ABBOT & SMITH

FLAMMIFERA WALKER
SYN. SEE HERSILIATA GUENEE

FLAVA BARNES & BENJAMIN
SSP. SEE OPELLA GROTE

FLAVA EDWARDS
SSP. SEE SISYMBRII BOISDUVAL

FLAVA EDWARDS
SSP. SEE NAPI LINNAEUS

FLAVA(FABRICIUS), ANOMIS-NOCTUIDAE
A FIMBRIAGO STEPHENS
SYN. SERRATA BARNES&MCDUNNOUGH
LIFE HISTORY
JOURN. ECON. ENT. 20:95, 1936
RAINWATER DIFFERENCE BETWEEN LARVA OF THIS SPECIES AND OF ALABAMA ARGILLACEA HUBNER
FOOD PLANTS
GOSSYPIUM HERBACIUM LINN

FLAVA GROTE
SSP. SEE POLYPHEMUS CRAMER

FLAVA STRECKER
SSP. SEE EURYTHEME BOISDUVAL

FLAVA STRECKER
SSP. SEE NICIPPE CRAMER

FLAVICANS GOEZE
SYN. SEE ALOPE DRURY

FLAVICARIA(PACKARD), ITAME-GEOMETRIDAE
SYN. SUBFALCATA HULST
A HELENA CASSINO
LIFE HISTORY
FOOD PLANTS
RIBES LACUSTRE(PERS.)
RIBES SATIVUM SYME(CURRANT)

FLAVICOLLIS(SMITH), EUXOA-NOCTUIDAE
LIFE HISTORY
FOOD PLANTS
MELILOTUS ALBA DESR.

FLAVICOLORIS GUNDER
SSP. SEE SARA BOISDUVAL

FLAVIDA SKINNER
SYN. SEE MIDEA HUBNER

FLAVIDALIS COMSTOCK
SSP. SEE AUSONIDES BOISDUVAL

FLAVIDALIS GROTE
SSP. SEE INNUBENS GUENEE

FLAVILINEATA RILEY
SYN. SEE AERATA FABRICIUS

FLAVIMACULATA HARVEY
SYN. SEE EXIGUA HUBNER

FLAVIMEDIA HARVEY
SYN. SEE ORNITHOGALLI GUENEE

FLAVIPUNCTALIS(GEYER), RENIA-NOCTUIDAE
SYN. PHALEROSALIS WALKER
SYN. HELIUSALIS WALKER
SYN. PASTORALIS GROTE
SYN. BELFRAGEI GROTE
LIFE HISTORY
ENT. AMER. 14:169, 1934 CRUMB
LARVA*PL. 23 G
FOOD PLANTS
DEAD LEAVES

FLAVISCAPULA SMITH
SSP. SEE TESSELLATA HARRIS

FLAVISTRIARIA(HUBNER), PTICHODIS-NOCTUIDAE

SYN. PERPLEXA GUENEE
SYN. PERSIPCUA WALKER
SYN. GLANS GROTE
LIFE HISTORY
GUENEE-SPEC. GEN. LEP. NOCT, 3:302,1852 LARVA*PL. 2
FOOD PLANTS
SCUTELLARIA SP.

FLAVITINCTA COMSTOCK
SSP. SEE SISYMBRII BOISDUVAL

FLAVITINCTA NIXON
SSP. SEE JAMAICENSIS DRURY

FLAVOOCELLATA BRYK
SYN. SEE CLODIUS MENETRIES

FLAVOFACIATA PACKARD
SYN. SEE NEPTARIA GUENEE

FLAVOFASCIATA(WALKER), PROSERPINUS-SPHINGIDAE
A ULALUME STRECKER
B RACHEL BRUCE
LIFE HISTORY
CAN. ENT. 23:42,1891 BRUCE LARVA
ENT. NEWS 2:87,109,1891 BRAUN. NOTE ON LIFE HIST.
NEW YORK STATE AGRI. EXP. STA. (CORNELL), MEM. 274:197,1948 FORBES LARVA
OTTAWA NATURALIST 28:143,1915 GIBSON LARVA (ULALUME)
UNITED STATES NAT. MUS. PROC. 27:790,1904 DYAR EGG, LARVA
LIFE HISTORY
FOOD PLANTS
EPILOBIUM SP.
EPILOBIUM ANGUSTIFOLIUM(LINN.)

FLAVOTINCTA COCKERELL
SSP. SEE SCUDDERI REAKIRT

FLEBILIS, GROTE, CATOCALA-NOCTUIDAE
SYN. CAROLINA HOLLAND
LIFE HISTORY
AMER. MUS. NAT. HIST. MEM. 3 NEW SERIES PART 1,1918 BARNES & MCDUNNOUGH LARVA* PL. 12, F. 19 COLOR
CAN. ENT. 51:226,1919 ROWLEY LARVA
ENT. AMER. 2:46,1886 KELLICOTT LARVA AS FLEBILIS IN ERROR, REALLY RETECTA GROTE
ENT. NEWS 23:209,1912 ROWLEY BERRY LARVA
NEW YORK STATE MUS. MEM. 8: 716,1905 FELT LARVA(BRIEF)
PAPILIO 1:141,1881 KELLICOTT LARVA
FOOD PLANTS
CARYA SP. (HICKORIES)
CARYA OVATA(MILL.)

FLEBILIS HULST
SYN. SEE GELIDATA MOESCHLER

FLETCHERATA TAYLOR, EUPITHECIA-GEOMETRIDAE
SYN. PROMULGATA PEARSAL
LIFE HISTORY
CAN. ENT. 77:53,1945 MCGUFFIN LARVA
FOOD PLANTS
LARIX LARICINA(DUROI.)
PICEA GLAUCA(MOENCH.)
PICEA MARIANA(MILL.)
PICEA RUBENS SARG.

FLETCHERI BEUTENMULLER
SSP. SEE UNIJUGA WALKER

FLETCHERI GROTE
SSP. SEE PATALIS GROTE

FLETCHERI KEMP
SSP. SEE GLAUCUS LINNAEUS

FLETCHERI SMITH
SSP. SEE GEORGII GROTE

FLEXUOSA WALKER
SYN. SEE FRATER GROTE

FLORA EDWARDS
SSP. SEE CLYTON BOISDUVAL&LECONTE

FLORA WRIGHT
SSP. SEE SARA BOISDUVAL

FLORENCIA CLEMENS
SYN. SEE GYAS EDWARDS

FLORIDA GUENEE, RHODOPHORA-NOCTUIDAE
LIFE HISTORY
BROOKLYN ENT. SOC. BULL. 4:28, 1881 SMITH LARVA
CAN. ENT. 2:6, 1869 SAUNDERS LARVA
LOC. CIT. 3:76, 1871 SAUNDERS EGG, LARVA
MUS. COMP. ZOOL., BULL. 87, NO. 6, 1941 DETHIER LARVAL ANTENNA* F. 67
NEW YORK ENT. SOC. JOURN. 9: 90, 1901 BEUTEN. LARVA
LOC. CIT. 28:35, 1920 WEISS-DICKERSON EGG, LARVA
NEW YORK STATE AGRI. SCO. REPT. 12:900, 1867 FITCH LARVA
FERNALD-IN KINGSLEY'S STANDARD NAT. HIST. 2:450, 1885 EGG, LARVA
GLOVER-ILLUST. NORTH. AMER. ENT. 1878 LARVA* PL. 14, F. 14
FOOD PLANTS
OENOTHERA BIENNIS LINN.
OENOTHERA BIENNIS V. GRANDIFLORA (SOL.)

FLORIDA MABILLE
SYN. SEE EUFALA EDWARDS

FLORIDALIS JOHSON & COMSTOCK
SSP. SEE AIDEA GUERIN-MENEVILLE

FLORIDANA CASSINO
SYN. SEE RADIANS WALKER

FLORIDANA GRAEF
SYN. SEE RANAECEPS GUERIN-MENEVILLE

FLORIDANA HENRY EDWARDS
SSP. SEE TORREFACTA ABBOT & SMITH

FLORIDARIA GUENEE
SYN. SEE CONFUSARIA HUBNER

FLORIDENSIS CLARK
SSP. SEE NESSUS CRAMER

FLORIDENSIS CLARK
SYN. SEE HYLAEUS DRURY

FLORIDENSIS GROTE & ROBINSON
SYN. SEE THYSBE FABRICIUS

FLORIDENSIS (GUENEE), CALLOPISTRIA-NOCTUIDAE
SYN. STRENA GROTE
LIFE HISTORY
MICHIGAN STATE AGRI. EXP. STA. BULL. 134:55, 1924 PETTIT LARVA* F. 34
MICHIGAN STATE COLLEGE, SPECIAL BULL. 214:80, 1931 MCDANIEL LARVA* F. 56
NEW YORK ENT. SOC. JOURN. 8: 33, 1900 SWAINSON LARVA, COCOON
NEW YORK STATE MUS. BULL. 198 74, 1918 FELT LARVA
ONTARIO ENT. SOC. ANN. REPT. P. 111, 1916 GIBSON EGG, LARVA* PUPA
SOUTHERN CALIFORNIA ACAD. SCI. BULL. 38:34, 1939 COMSTOCK-DAMMERS LARVA*
UNITED STATES DEPT. AGRI. BRU. ENT. BULL. NEW SERIES 125: 7, 1913 CHITTENDEN LARVA* F. 1; PUPA
UNITED STATES DEPT. AGRI. FARMER'S BULL. 1362:43, 1922 WEIGEL-SASSCER EGG, LARVA* F. 49; PUPA
LOC. CIT. 1495:21, 1926 WEIGEL-MIDDLETON LARVA* F. 30
METCALF & FLINT-DESTR. &USE FUL INS. P. 735, 1939 LARVA,

PUPA
INDEX AMER. ECON. ENT.
VOL. 1:58; VOL. 2:64;
VOL. 3:68; VOL. 5:100
VOL. 6:116; VOL. 7:154
FOOD PLANTS
ADIANTUM SP.
ADIANTUM PEDATUM LINNAEUS
ASPARAGUS SPRENGERI REG EL
BLECHNUM SP.
CYRTOMIUM SP.
NEPHROLEPIS SP.
NEPHROLEPIS EXALTATA V.
WHITMANI (BARROWS)
POLYPODIUM SP.
PTERIS SP.
OTHER FERNS
SMILAX ROTUNDIFOLIA LINN.
?WOODWARDIA VIRGINICA (LINN.)

FLORIDENSIS HOLLAND
SSP. SEE MARCELLUS CRAMER

FLORIDENSIS MORRISON
SSP. SEE THEONUS LUCAS

FLORIDENSIS PLOETZ
SYN. SEE PALATKA EDWARDS

FLORIDENSIS ROEBER
SEE AGARITHE BOISDUVAL

FLORIDENSIS STRECKER
SSP. SEE ARCHIPPUS CRAMER

FLORIDUM GROTE
SSP. SEE INVOLUTUM SEPP.

FLORUS EDWARDS
SSP. SEE HELLOIDES BOISDUVAL

FLOSCULARIA GROTE
SYN. SEE SERINARIA HERRICH-SCHAEFFER

FLUCTUATA LINNAEUS
SYN. SEE IDUATA GUENEE

FLUTEA SMITH
SSP. SEE QUADRIDENTATA GROTE & ROBINSON

FLUVIATA HUBNER
SYN. SEE OBSTIPATA FABRICIUS

FLUXA BIRD
SSP. SEE CATAPHRACTA GROTE

FLCINA SMITH
SYN. SEE TESSELLATA HARRIS

FOEDARIA WALKER
SYN. SEE HYPOCHRARIA HERRICH-SCHAEFFER

FORBESI BENJAMIN, ROTHSCHILDIA-SATURNIIDAE
FORM DRAUDTI BENJAMIN
LIFE HISTORY
PAN PACIFIC ENT. 10:14, 1934
BENJAMIN EGG, LARVA
FOOD PLANTS
LIGUSTRUM SP. IN CAPTIVITY
SALIX SP. (WILLOWS)

FORBESI FRANCLEMONT
SSP. SEE PARTA GUENEE

FORBESI FRENCH, EPIZEUXIS-NOCTUIDAE
SYN. MERRICKI SMITH
LIFE HISTORY
ENT. AMER. 14:160, 1934 CRUMB
LARVA* PL. 24 L
FOOD PLANTS
DEAD LEAVES

FORFICARIA (GUENEE), PROCHOERODES-GEOMETRIDAE
SYN. NUBILIATA PACKARD
A CATENULATA GROTE
SYN. ANFRACTATA HULST
LIFE HISTORY
UNITED STATE S SEPT. AGRI. BUR.
ENT. BULL. OLD SERIES
32:23, 1894 COQUILLETT
LARVA
QUALYE-INSECT OF CITRUS &

SUBTROPICAL FRUITS P.387, 1938 LARVA(NUBILATA)
FOOD PLANTS
JUGLANS REGIA LINN.
MALUS PUMILA MILL. (APPLE)
SALIX SP. (WILLOWS)

FORFICULARIA GUENEE, METANEMA-GEOMETRIDAE
LIFE HISTORY
NEW YORK ENT. SOC. JOURN. 53: 193, 1945 FORBES PUPA
FOOD PLANTS
JUGLANS SP.
JUGLANS REGIA LINN.

FORMALIS GROTE
SSP. SEE EXSERTISTIGMA MORRISON

FORMOSA FISCHER
SSP. SEE ZELICAON LUCAS

FORMOSA(GROTE), CHRYSANYMPHA-NOCTUIDAE
LIFE HISTORY
ENT. NEWS 15:301, 1904 KEARFOOT LARVA, PUPA
FOOD PLANTS
GAYLUSSACIA DUMOSA(ANDR.)
VACCINIUM SP.

FORMOSA HULST
SYN. SEE PANNARIA GUENEE

FORMOSA HENRY EDWARDS
SSP. SEE SEPTENTRIONALIS WALKER (GLUPHISIA)

FORMOSUS ELLSWORTH
SSP. SEE ARCTICA BOISDUVAL

FORMULA GROTE
SYN. SEE ROSEA WALKER(ORETA)

FORMULA GROTE & ROBINSON
SYN. SEE SIMILIS EDWARDS

FORNAX HUBNER, AGERONIA-NYMPHALIDAE
LIFE HISTORY
AMER. ACAD. SCI. PROC. 27:250, 1892 SCUDDER LIFE HIST.
FOOD PLANTS
DALECHAMPIA SP.

FORNICA SMITH
SYN. SEE PURPUREA GROTE

FORSYTHEA GUNDER
SSP. SEE CRESPHONTES CRAMER

FORSYTHEA WOOD
SSP. SEE ASTERIUS CRAMER(AJAX LINN.)

FORTIS(GROTE), HOMONCOCNEMIS-NOCTUIDAE
SYN. VORAX BEHR
A PICINA GROTE
LIFE HISTORY
PAPILIO 4:21, 1884 BEHRENS LARVA(VORAX)
SOUTHERN CALIFORNIA ACAD. SCI. BULL. 38:31, 1939 COMSTOCK-DAMMERS LIFE HISTORY LARVA* PL. 18
FOOD PLANTS
FRAXINUS SP. (ASHES)
PENSTEMON ANTIRRHINOIDES BENTH.

FORTITARIA BARNES & MCDUNNOUGH
SSP. SEE COGNATARIA GUENEE

FORTUNATA PEARSALL
SYN. SEE PYGMAEATA HUBNER

FOXI GUNDER
SSP. SEE NUBIGENA BEHR

FRACTILINEA(GROTE), OLIGIA-NOCTUIDAE
FORM VULGIVAGA MORRISON
FORM MODIOLA GROTE
SYN. MACTATOIDES BARNES & MCDUNNOUGH
FORM UNA STRECKER
FORM MISERA GROTE
A ALBESCENS BARNES & MCDUNNOUGH

LIFE HISTORY
INSECT LIFE 6:155,1893 WEBSTER LARVA* PUPA*
NEW YORK STATE MUS. BULL. 175:14,1915 FELT LARVA
ONTARIO ENT. SOC. ANN. REPT. P.92,1893 WEBSTER LARVAL HABITS
OSBORN-MEADOW PASTURE INSECTS P.226,1939 LARVA* F.94
INDEX AMER. ECON. ENT. VOL.2:257;VOL.3:285; VOL.4:332:VOL.5:547; VOL.6:544;VOL.7:721

FOOD PLANTS
PHLEUM PRATENSE LINN. (TIMOTHY)
ZEA MAYS LINN. (CORN)

FRAGILARIA(GROSSBECK), ANACAMPTODES-GEOMETRIDAE

LIFE HISTORY
NEW YORK ENT. SOC. JOURN. 53: 189,1945 FORBES PUPA (THIS SP.?)
SOUTHERN CALIFORNIA ACAD. SCI. BULL. 45:17,1946 COMSTOCK & DAMMERS EGG* PL.3; LARVA* PL.304;PUPA*PL.5

FOOD PLANTS
ACACIA MELANOXYLON R. BR.
ACACIA NERIIFOLIA CUNN.
ACACIA PRUINOSA CUNN.
ALBIZIA JULIBRISSIN DURAZ.
ANETHUM GRAVEOLENS LINN.
CEANOTHUS SP.
CERCOCARPUS SP.
ERIOGONUM SP.
JACARANDA SP.
PELARGONIUM DOMESTICUM BAILEY
PROSOPIS SP.
PRUNUS AMERICANA MARSH.
PRUNUS ARMENIACA LINN. (APRICOT)
PRUNUS TRIFLORA RAFIN (SATSUMA PLUM)
SALIX SP. (WILLOWS)
THUJA OCCIDENTALIS LINN.
ULMUS PARVIFOLIA JACQ.
VERBENA SP.
LEMON VERBENA

FRAGILIS(GUENEE), ACRONICTA-NOCTUIDAE
SYN. SPECTANS WALKER
A MINELLA DYAR
B FRAGILOIDES BARNES &BENJAMIN
C ATRIOR FRANCLEMONT

LIFE HISTORY
AMER. MUS. NAT. HIST. BULL. 14:246,1901 BEUTEN. LARVA
UNITED STATES NAT. MUS. PROC. 21:96,1898 SMITH & DYAR EGG, LARVA
INDEX AMER. ECON. ENT. VOL.7:8

FOOD PLANTS
AMELANCHIER FLORIDA LINDL.
BETULA SP. (BIRCHES)
CRATAEGUS DOUGLASII LINDL.
MALUS PUMILA MILL(APPLE)
PRUNUS DOMESTICA LINN. (PLUM)
SORBUS AMERICANA MARCH

FRAGILIS(STRETCH), MALACOSOMA-LASIOCAMPIDAE
FORM LUTESCENS NEUMOEGEN & DYAR
A INCURVA HENRY EDWARDS
FORM CONSTRICTINA NEUMOEGEN & DYAR
B MUS NEUMOEGEN
FORM DISCOLORATA NEUMOEGEN
C PERLUTEA NEUMOEGEN & DYAR

LIFE HISTORY
AMER. PHIL. SOC. PROC. 31:178, 1893 PACKARD LARVA
CAN. ENT. 25:41,1893 DYAR LARVA
LOC. CIT. 71:76,1939 SWAIN EGG, LARVA
ENT. AMER. 4:62,1888 HY. EDWARDS COCOON
NEW YORK ENT. SOC. JOURN. 3: 23,1895 DYAR LARVAL STRUCT.
PSYCHE 7:189,1895 DYAR EGG,

LARVA
PSYCHE 7:260,1895 DYAR
COMPARES LARVA WITH
PLUVIALIS DYAR
ESSIG-INJ.&BENEF.INS.P.411,
1915 LARVA
ESSIG-INS.WEST.NORTH AMER-
ICA P.697,1926 LARVA
HERRICK-INS.ENEMIES SHADE
TREES P.34,1935 LARVA
SLINGERLAND & CROSBY-MANUAL
FRUIT INS.P.118,1914
LARVA* F.128
INDEX AMER.ECON.ENT.
VOL.1:198;VOL.2:224;
VOL.3:287;VOL.5:403;
VOL.6:481

FOOD PLANTS

AMELANCHIER ALNIFOLIA NUTT.
ARCTOSTAPHYLOS GLAUCA LINDL.
ARONIA MELANOCARPA(MICHX.)
BIGELOVIA SP.
CEANOTHUS SP.
FRAXINUS SP.(ASHES)
MALUS PUMILA MILL.(APPLE)
POPULUS FREMONTI S.WATS.
POPULUS TREMULOIDES MICHX.
PRUNUS SP.
PRUNIS ILICIFOLIA(NUTT.EX
H.&A.)
PRUNUS SEROTINA EHRH.
PRUNUS SUBCORDATA BENTH.
PRUNUS VIRGINIANA LINN.
QUERCUS SP.
RIBES CYNOSBATI LINN.
ROSA SP.
ROSE CAROLINA LINN.
SALIX SP.(WILLOWS)
SHEPHERDIA ARGENTEA(PURSH)

FREGILOIDES BARNES & BENJAMIN
SSP.FRAGILIS GUENEE

FRANCISCA PLOETZ
SYN.SEE SYLVANOIDES BOIDVUAL

FRANCKII NEUMOEGEN,SPHINX-SPHINGIDAE

LIFE HISTORY

ENT.NEWS 23:9,1912 SMITH
LARVA,PUPA
NEW YORK STATE AGRI.EXP.STA.
(CORNELL),MEM.274:190,
1948 FORBES LARVA

FOOD PLANTS

FRAXINUS SP.(ASHES)
ULMUS SP.(ELMS)

FRANCLEMONTI RUPERT,METARRANTHIS-
GEOMETRIDAE

LIFE HISTORY

NEW YORK ENT.SOC.JOURN.51:
143,1943 RUPERT EGG,
LARVA,PUPA
NEW YORK STATE AGRI.EXP.
STA.(CORNELL),MEM.274:
94,1948 FORBES LARVA

FOOD PLANTS

ARONIA MELANOCARPA(MICHX.)
PRUNUS SEROTINA EHRH.
VACCINIUM SP.

FRANCONIA HENRY EDWARDS
SSP.SEE CELIA SAUNDERS

FRANCONIA SLOSSON
SSP.SEE ASSIMILANS WALKER

FRANKLINII CURTIS
SYN.SEE AQUILO BOISDUVAL

FRATELLA GROTE
SYN.SEE OU GUENEE

FRATER GROTE,RAPHIA-NOCTUIDAE
SYN.PERSONATA WALKER
SYN.FLEXUOSA WALKER

LIFE HISTORY

AMER.MUS.NAT.HIST.BULL.14:
235,1901 BEUTEN.LARVA
MUS.COMP.ZOOL.BULL.87,NO.
6,1941 DETHIER LARVAL
ANTENNA* F.109
NEW YORK STATE MUS.MEM.8:
747,1905 FELT LARVA
PAPILIO 3:13,1883 THAXTER
LARVA

UNITED STATES DEPT. AGRI. MISC. PUB. 657:358,395,1950 CRAIGHEAD LARVA
FOOD PLANTS
POPULUS SP. (POPLARS)
POPULUS DELTOIDES MARSH.
POPULUS HETEROPHYLLA LINN.
POPULUS TREMULOIDES MICHX.
SALIX SP.

FRATERCULA GROTE & ROBINSON
SSP. SEE MICRONYMPHA GUENEE

FRATERNALIS SMITH, RENIS-NOCTUIDAE
LIFE HISTORY
ENT. AMER. 14:168,1934 CRUMB LARVA
FOOD PLANTS
DEAD LEAVES

FRADULENTARIA ZELLER
SYN. SEE PAMPINARIA GUENEE

FREEMANI STALLINGS
SSP. SEE GEMMA HUBNER

FREIJA WARREN
SSP. SEE CENTAUREAE RAMBUR

FRETCHINI CHERMOCK
SSP. SEE SCUDDERI EDWARDS

FRIABILIS GROTE
SSP. SEE MESSORIA HARRIS

FRIDAYI CHERMOCK
SSP. SEE MELISSA EDWARDS

FRIDAYI GUNDER
SSP. SEE WEIDENMEYERII EDWARDS

FRIDAYI GUNDER
SYN. SEE NUBIGENA BEHR.

FRIDAYI GUNDER
SSP. SEE NEUMOEGENI SKINNER

FRIEDLEI WATSON
SSP. SEE HOBOMOK HARRIS

FRIGGA (THUNBERG), BRENTHIS-NYMPHALIDAE
A SAGA STAUDINGER
B SAGATA BARNES & BENJAMIN
C IMPROBA BUTLER
SYN. YOUNGI HOLLAND
D GIBSONI BARNES & BENJAMIN
ALASKENSIS LEHMAN
SYN. LEHMANNI HOLLAND
LIFE HISTORY
THE ENTOMOLOGIST 46:80,81, 1913 SHELDON LARVA
FOOD PLANTS
SALIX SP.
VIOLA SP.

FRIGIDA SCUDDER
SSP. SEE NAPI LINNAEUS

FRIGIDA SMITH
SYN. SEE LUPUSCULINA GUENEE

FRIGIDA (SMITH), PAPAIPEMA-NOCTUIDAE
SYN. PEROBSOLETA LYMAN
FORM THALICITRI LYMAN
FORM TERMINALIS STRAND
LIFE HISTORY
CAN. ENT. 37:308,1905 LYMAN LARVA, PUPA (THALICTRI)
INDEX AMER. ECON. ENT. VOL. 6:558
FOOD PLANTS
THALICTRUM POLYGAMUM MUHL.

FRIGIDANA WALKER
SYN. SEE REVAYANA SCOPOLI

FRISIA)POEY), ANTHANASSA-NYMPHALIDAE
SYN. GYGES HEWITSON
A TULCIS BATES
SYN. GENIGUEH REAKIRT
SYN. ARCHESILEA R. FELDER
SYN. PUNCTATA EDWARDS
LIFE HISTORY
CAN. ENT. 79:142,1947 CHERMOCK-CHERMOCK EGG*F. F; LARVA* F. G; CHRYSALIS*F. H, I

NEW YORK ENT.SCO.JOURN.9:80, 1901 SWAINSON LARVA, CHRYSALIS
FOOD PLANTS
BELOPERONE GUTTATA BRANDEGEE

FRITILLARIA(GUENEE),HOMOCHLODES-GEOMETRIDAE
SYN.DISCONVENTA WALKER
SYN.LACTISPARGARIA WALKER
SYN.IMMERSATA WALKER
LIFE HISTORY
NEW YORK STATE AGRI.EXP.STA. (CORNELL),MEM.274:89, 1948 FORBES LARVA,PUPA
WASHINGTON ENT.SOC.PROC.51: 143,1949 RUPERT LARVA
FOOD PLANTS
FERNS
PTERIDIUM AQUILINUM V. LATIUSCULUM (DESV.)

FROSTIATA SWETT
SSP.SEE MISTURATA HULST

FRUGALIARIA GUENEE
SYN.SEE PAMPINARIA GUENEE

FRUGIPERDA(ABBOT & SMITH),LAPHYGMA-NOCTUIDAE
SYN.MACRA GUENEE
SYN.INEPTA WALKER
SYN.SIGNIFERA WALKER
SYN.AUTUMNALIS RILEY
FORM FULVOSA RILEY
FORM OBSCURA RILEY
LIFE HISTORY
AMER.ENT.2:328,365,1870 RILEY LARVA,PUPA
AMER.MUS.NAT.HIST.BULL.16: 421,1902 BEUTEN.LARVA
BROOKLYN ENT.SOC.BULL.22:49 1927 CRUMB MATURE LARVA* PL.4H;PL.5 E,L.O
LOC.CIT.27:91,1932 CRUMB LARVA
ILLINOIS AGRI.EXP.STA.BULL. 60:497,1900 FORBES LARVA
ILLINOIS DEPT.AGRI.TRANS.15: 97,219,1877 FRENCH-THOMAS LARVA
LOC.CIT.18,APPEND:138,1880 MARTIN LARVA*
JOURN.ECON.ENT.6:361,1913 DEW LIFE HISTORY
LOC.CIT.37:113,1944 BISSELL LARVA FRUGIPERDA COMPARED WITH THAT OF UNIPUNCTA HAWORTH
KENTUCKY AGRI.EXP.STA.BULL. 227:70,1920 GARMAN LARVA* F.7 CUTICLE;F.15 A LARVAL PATTERN
MASSACHUSETTS AGRI.EXP.SAT. BULL.239,35,1928 FRAN-KLIN LARVA* PL.2,FF.8A COLOR
MICHIGAN AGRI.EXP.STA. SPECIAL BULL.183:62, 1931 PETTIT LARVA*F.51
MINNESOTA AGRI.EXP.STA. BULL.61:218,1899 LUGGER LARVA* F.163
MISSOURI,REPT.INS.OF 3:112, 1871 RILEY EGG,LARVA*
LOC.CIT.8:48,1876 RILEY LARVA*
NEBRASKA AGRI.EXP.STA.RE-SEARCH BULL.81:25,1935 WHELAN LARVAL STRUCTURES* FF.3A,4 TECH.
NEW YORK STATE MUS.BULL.6: 14,1888 LINTNER LARVA*
LOC.CIT.165:42,1913 FELT EGG*LARVA*PUPA*
UNITED STATES DEPT.AGRI.BUR. ENT.BULL.23:81,1900 NEW SERIES CHITTENDEN EGG, LARVA* F.19 D,E;PUPA*F. 19E
LOC.CIT.29:16,1901 CHITTENDEN EGG* F.2A;LARVA*F.1C,F.2 C; PUPA* F.1E
UNITED STATES DEPT.AGRI. FARMER'S BULL.132:34, 1901 MARLATT LARVA*F.21

c,d;pupa*f.21 e
Loc.cit.752:6,1916 Walton-Luginbill egg* f.3;larva* ff.4,5;pupa*f.6
Loc.cit.890:9,1917 Pierce larva* f.6 c;pupa*f.6d
Loc.cit.1688:21,1932 Folsom larva*f.27 e;pupa*f.27d
United States Dept.Agri.Tech. Bull.34:32;1928 Lugenbill egg*ff.2,8,9;larva*ff.5, 7,10-21;pupa*ff.22-24
Loc.cit.138:3,1929 Vicery life history but no descr.
Abbot & Smith-Lep.Ins.Georgia 2:191,1797 larva*pl.96

COLOR

Chittenden-Ins.Inj.Veg.p.56, 1907 larva*ff.30,30 a; pupa* f.30
Crosby & Leonard-Manual Veg. Gard.Ins.p.293,1918 egg, larva,pupa
Fernald-Applied Entom.p.277u, 1926 larva* f.282
Fernald & Shepard-Applied Ent.p.274,1942 larva*f. 277
Glover-Illust.North Amer.Ent. 1878 larva*pl.18,f.5;l pl.19,f.18
Guenee-Spec.Gen.Lep.Noct. 1:159,1852 larva
Gundlach-Ent.Cubana p.288, 1885 larva,pupa
Hart-Ins.Collections Illinois High Schools p.28,1903 larva*pupa*
Holland-Moth Book p.174, 1905 larva* f.97
Hubbard-Ins.Affect.Oranges p.150,1885 egg,larva*
Metcalf & Flint-Destr.& Useful Ins.pp.134,344, 1939 egg*f.78 d;larva* f.201
O'Kane-Inj.Ins.p.173,1912 larva* f.198
Peairs-Pests Farm.Gard. Orchard p.108,1941 egg* larva* f.116
Index Amer.Econ.Ent. Vol.1:178;Vol.2:197; Vol.3:226;Vol.4:260; Vol.5:373;Vol.6:443; Vol.7:590

FOOD PLANTS

Agrostis alba Linn.
Agrostis hyemalis(Walt.)
Allium cepa Linn.
Althaea rosea Cav.
Amaranthus sp.
Andropogon virginicus Linn.
Arachis hypogaea Linn.
Asclepias sp.(not favored)
Asparagus officinalis Linn. (asparagus)
Atropa belladonna Linn.
Avena sativa Linn.(oats)
Beta vulgaris Linn.(beet)
Brassica oleracea Linn. (cabbage)
Brassica oleracea v.viridis Lam.(kale)
Brassica rapa Linn.(turnip)
Capsicum annuum Linn.
Carex sp.(sedges)
Carya sp.(hickories)
Carya illinoensis(Wang.)
Cenchrus pruciflorus Benth.
Chenopodium album Linn.
Chloris gayana Kunth.
Chrysanthemum sp.
Cicer arietinum Linn.
Citrullus vulgaris Schard. (watermelon)
Citrus limin(Linn.)(lemon)
Citrus nobilis v.unshiu Swingle
Citrus sinensis(Linn.)(orange)
Convolvulus sp.
Croton capitatus Michx.
Cucumis sativus Linn.(cucumber)
Cyperus rotundus Linn.
Dactyloctenium aegyptium(Linn)
Dahlia variabilis Desf.
Digitaria ischaemum(Schreb.)

DIGITARIA SANGUINALIS(LINN.)
EUCHLAENA MEXICANA SCHRAD.
FAGOPYRUM SAGITTATUM GILIB.
FRAGARIA CHILOENSIS(LINN.) (STRAWBERRY)
GERANIUM SP.
GLADIOLUS SP.
GLYCINE MAX(LINN.)(SOY BEAN)
GOSSYPIUM HERBACIUM LINN. (COTTON)
GRAMINEAE SP. (GRAINS, CEREALS)
HORDEUM VULGARE LINN. (BARLEY)
IPOMOEA BATATA(LINN.)(SWEET POTATO)
IPOMOEA PURPUREA (LINN.)
MALUS PUMILA MILL. (APPLE)
MEDICAGO SATIVA LINN. (ALFALFA)
NICOTIANA TABACUM LINN. (TOBACCO)
ONOSMODIUM VIRGINIANUM(LINN.)
ORYZA SATIVA LINN. (RICE)
PANICUM MILIACEUM LINN.
PANICUM PURPURASCENS RADDI
PANICUM TEXANUM BUCKL.
PENNISETUM GLAUCUM(LINN.)
PHASEOLUS VULGARIS LINN. (BEAN)
PHLEUM PRATENSE LINN. (TIMOTHY)
PISUM SATIVUM LINN. (PEAS)
PLATANUS OCCIDENTALIS LINN.
POA ANNUA LINN.
POA PRATENSIS LINN.
PORTULACA OLERACEA LINN.
PRUNUS PERSICA(LINN.)(PEACH)
SACCHARUM OFFICINARUM LINN. (SUGAR)
SECALE CEREALE LINN. (RYE)
SETARIA ITALICA(LINN.)(MILLET)
SOLANUM SP.
SOLANUM DULCAMARA LINN.
SOLANUM LYCOPERSICUM LINN. (TOMATO)
SOLANUM TUBEROSUM LINN. (POTATO)
SORGHUM BICOLOR(LINN.)
SORGHUM CAFFRORUM RETZ.
SORGHUM HALEPENSE(LINN.)
SORGHUM SACCHARATUM(LINN.)
SORGHUM SUDANENSIS(PIPER)
SPINACIA OLERACEA MILL. (SPINACH)
STIZOLOBIUM DEERINGIANUM BORT.
TRIFOLIUM SP. (CLOVERS)
TRIFOLIUM PRATENSE LINN.
TRIFOLIUM REPENS LINN.
TRITICUM AESTIVUM LINN. (WHEAT)
VACCINIUM MACROCARPON AIT. (CRANBERRY)
VIGNA SINENSIS (LINN.)
VIOLA SP. (VIOLETS)
VITIS SP. (GRAPES)
VITIS VINIFERA LINN.
WISTERIA HISPIDA(MAXIUM)
XANTHIUM STRUMARIUM LINN.
ZEA MAYS LINN. (CORN)
GENERAL FEEDER ON FOREST TREES AND SHRUBS
GRASSES

FRUTETORUM BOISDUVAL
SYN. SEE AMERICANA FABRICIUS

FOCOSA HUBNER, HYPOPREPIA-ARCTIIDAE
SYN. TRICOLOR FITCH
FORM PLUMBEA HENRY EDWARDS
A SUBORNATA NEUMOEGEN & DYAR
SYN. INORNATA OTTOLENGUI
FORM DOLLI DYAR

LIFE HISTORY

AMER. MUS. NAT. HIST. BULL. 10: 362, 1898 BEUTEN. LARVA
CAN. ENT. 12:45, 1880 COQUILLETT LARVA
ILLINOIS DEPT. AGRI. TRANS. 18, APPEND: 184, 1880 COQUILLETT LARVE
NEW YORK ENT. SOC. JOURN. 3: 176, 1895 PACKARD FIRST LARVAL STAGE
HARRIS-INS. INJ. VEG. 1ST EDIT. P. 241, 1841 LARVA, PUPA
HARRIS-LOC. CIT. 2ND EDIT. P. 262, 1852 LARVA

HARRIS-LOC.CIT.FLINT EDIT
P.342,1862 LARVA,COCOON
MORRIS-SYNOP.LEP.NORTH
AMERICA P.256,1862 LARVA
STRETCH-ZYGAEN.&BOMBYC.
NORTH AMERICA P.47,1873
LARVA,COCOON
FOOD PLANTS
LICHENS ON TREES
MOSSES ON TREES

FULDA SMITH
SYN.SEE MESSORIA HARRIS

FULIGINOSA(EDWARDS),SATYRIUM-LYCAENIDAE
SYN SUASA BOISDUVAL
AB IMMACULATA GUNDER
A SEMILUNA KLOTS
LIFE HISTORY
SOUTHERN CALIFORNIA ACAD.SCI.
BULL.32:114,1933 COMSTOCK
& DAMMERS EGG
FOOD PLANTS
LUPINUS PS.(LUPINES)

FULIGINOSA(LINNAEUS),PHRAGMATOBIA-
ARCTIIDAE
A RUBRICOSA HARRIS
?SYN.BOREALIS STAUDINGER
LIFE HISTORY
AMER.MUS.NAT.HIST.BULL.10:
374,1898 BEUTEN.LARVA
CAN.JOURN.INDUST.SCI.&ART.
8:372,1863 SAUNDERS LARVA
CAN.NAT.P.195,1840 GOSSE
LARVA(BRIEF)
ENT.AMER.3:168,1887 NY.EDW.
LARVA(RUBRICOSA)
INSECT LIFE 2:236,1890 LUGGER
LARVA* PUPA*
ONTARIO ENT.SOC.ANN.REPT.
P.12,1896 FYLES LARVA
(BRIEF)
PAPILIO 3:127,1883 EDWARDS-
ELLIOT LARVA,PUPA,COCOON'
HOLLAND-MOTH BOOK P.126,1905
LARVA*PUPA*F.70(POOR)
INDEX AMER.ECON.ENT.
VOL.4:359
FOOD PLANTS
EUPATORIUM SP.
FRAXINUS SP.(ASHES)
SOLIDAGO SP.
SYMPLOCARPUS FOETIDUS(LINN.)
VERNONIA NOVEBORACENSIS
(LINN.)
GENERAL FEEDER LOW PLANTS

FULLA EDWARDS
SYN.SEE ICARIOIDES BOISDUVAL

FULLIOLUS HULST
SSP.SEE HYPOPHLAEAS BOISDUVAL

FULVA STRETCH,KODIOSOMA-ARCTIIDAE
FORM NIGRA STRETCH
FORM EAVESII STRETCH
FORM TRICOLOR STRETCH
LIFE HISTORY
SOUTHERN CALIFORNIA ACAD.
SCI.BULL.34:131,1935
COMSTOCK-DAMMERS LIFE
HISTORY EGG,LARVA*PL.25;
PUPA*PL.25
STRETCH-ZYGAEN.&BOMBYC.
NORTH AMERICA P.68,1873
LARVA(BRIEF)
FOOD PLANTS
ESCHSHOLTZIA CAESPITOSA V.
HYPECOIDES GRAY
FESTUCA MYUROS LINN.
GILIA ACHILLEAEFOLIA BENTH.
LACTUCA CANADENSIS LINN.
TRIFOLIUM CILIOLATUM BENTH.
GENERAL FEEDER ON LOW
PLANTS

FULVARIA(VILLERS),ITAME-GEOMETRIDAE
SYN.BRUNNEATA THUNBERG
SYN.FERRUGINARIA PACKARD
LIFE HISTORY
NEW YORK STATE AGRI.EXP.STA.
(CORNELL),MEM.274:41,
1948 FORBES LARVA
PACKARD-MONOGRAPH GEOMETRI-
DAE P.258,1876 LARVA

(BRUNNEATA)
SPULER-DIE RAUPEN DER SCHMETTERLINGE EUROPAS 1904 LARVA*TAFEL 44,F.7A
STAINTON-BRITISH BUTTER.MOTHS 2:62,1859 LARVA
FOOD PLANTS
VACCINIUM SP.
VACCINIUM ANGUSTIFOLIA AIT.

FULVESCENS HENRY EDWARDS
SSP.SEE SAEPIUM BOISDUVAL

FULVIA DODGE
SSP.SEE VIRGINIENSIS DRURY (VANESSA)

FULVIA(EDWARDS),MELITAEA-NYMPHALIDAE
AB SINEFASCIA WILLIAMS
LIFE HISTORY
SOUTHERN CALIFORNIA ACAD.SCI. BULL.32:102,1933 COMSTOCK-SPERRY LARVA* PL.44;CHRYSALIS*PL.45
FOOD PLANTS
CASTILLEJA LANATA GRAY

FULVICOLLIS(HUBNER),CISSEPS-AMATIDAE
SYN.SEMIDIAPHANA HARRIS
A PALLENS HENRY EDWARDS
LIFE HISTORY
AMER.MUS.NAT.HIST.BULL.10: 353,1898 BEUTEN.LARVA
CAN.ENT.12:44,1880 COQUILLETT LARVA
ILLINOISE DEPT.AGRI.TRANS.18, APPEND:171,1880 COQUILLETT LARVA
ONTARIO ENT.SOC.ANN.REPT.P. 99,1896 FYLES LARVA
SOUTHERN CALIFORNIA ACAD.SCI. BULL.36:111,1937 COMSTOCK LARVA*PL.44;PUPA*PL.45
UNITED STATES NAT.MUS.PROC. 23:264,1900 DYAR EGG, LARVA
HART-INSECT COLLECTIONS ILLINOIS HIGH SCHOOLS P. 20,1903 COCOON
HART-INSECT COLLECTIONS ILLINOIS HIGH SCHOOLS P.20,1903 COCOON
FOOD PLANTS
ELEOCHARIS SP.
GRASSES
LICHENS

FULVICOSTA CLEMENS
SSP.SEE COLONA HUBNER

FULVOFASCIATA BUTLER
SYN.SEE CALIFORNIAE WALKER

FULVOFLAVA WALKER
SYN.SEE MACULATA HARRIS

FULVOSA RILEY
SSP.SEE FRUGIPERDA ABBOT&SMITH

FULVUS RUMMEL
SSP.SEE HYPOPHLAEAS BOISDUVAL

FUMATA BARNES & MCDUNNOUGH
SSP.SEE ANTICARIA WALKER (AETHALURA)

FUMATA TAYLOR
SSP.SEE SATYRATA HUBNER

FUMATARIA MINOT
SSP.SEE CREPUSCULARIA SCHIFFER-MULLER

FUMEOLA(HAMPSON),ASEPTIS-NOCTUIDAE
SYN.PROBATA BARNES&MCDUNNOUGH
LIFE HISTORY
SOUTHERN CALIFORNIA ACAD.SCI. BULL.39:106,1940 COMSTOCK LARVA*PL.17;PUPA*PL.18
FOOD PLANTS
ARCTOSTAPHYLOS GLAUCA LINDL.

FUMIDA SCUDDER
SSP.SEE PSEUDARGIOLUS BOISDUVAL & LECONTE

FUMIDUS HENRY EDWARDS
SYN. SEE ALBICOSTA WALKER

FUMIPENNIS MCDUNNOUGH
SSP. SEE VOLUBILIS HARRIS

FUMOSA DRAUDT
SSP. SEE DERIDENS GUENEE

FUMOSA FIELD
SYN. SEE CLAUDIA CRAMER

FUMOSA HULST
SYN. SEE LITARIA HULST

FUMOSA HULST
SYN. SEE COAGULATA GUENEE

FUMOSA MORRISON
SSP. SEE HENRICI GROTE

FUMOSA REVERDIN
SYN. SEE SYRICHTUS FABRICIUS

FUMOSA STRECKER
SSP. SEE EURYTHEME BOISDUVAL

FUMOSA STRECKER
SSP. SEE VIRGINICA FABRICIUS

FUMOSA STRECKER
SYN. SEE DIFFINIS BOISDUVAL

FUMOSARIA STRECKER
SYN. SEE SEMICLUSARIA WALKER

FUMOSUS GUNDER
SSP. SEE VANILLAE LINNAEUS

FUMOSUS HULST
SSP. SEE PLEXIPPUS LINNAEUS

FUMOSUS LEUSSLER
SSP. SEE EURYDICE JOHANSSON

FUMOSUS WURSTER
SSP. SEE POLYPHEMUS CRAMER

FUNERALIS GROTE & ROBINSON, ACRONICTA-NOCTUIDAE
FORM FUSCALIS BARNES&BENJAMIN
LIFE HISTORY
AMER. MUS. NAT. HIST. BULL. 14: 247, 1901 BEUTEN. LARVA
BOSTON SOC. NAT. HIST. PROC. 27:132, 1896 DYAR LARVAL SETAE
BRITISH COLUMBIA ENT. SOC., PROC. 43:21, 1947 DENNIS LARVAL PICTURE ONLY*
NATURALISTE CANADIEN P.6, 1936 DAVIAULT LARVA
NEW YORK STATE MUS. ME. 8: 716, 1905 FELT LARVA
UNITED STATES NAT. MUS. PROC. 21:103, 1898 SMITY&DYAR LARVA, COCOON
INDEX AMER. ECON. ENT. VOL. 1:20 APATELA
FOOD PLANTS
ACER SP. (MAPLES)
ALNUS RUBRA BONG.
BETULA SP. (BIRCHES)
CARYA SP. (HICKORIES)
CORNUS SERICEA LINN.
CRATAEGUS SP.
CRATAEGUS DOUGLASII LINDL.
MALUS PUMILA MILL. (APPLE)
POPULUS CANADENSIS MOENCH
PRUNUS CERASIFERA EHRH.
PRUNUS DOMESTICA LINN. (PLUM)
PRUNUS MARITIMA MARSH.
PRUNUS SEROTINA EHRH.
SALIX SP. (WILLOWS)
SALIX SCOULERIANA BARR.
ULMUS SP. (ELMS)

FUNERALIS SCUDDER & BURGESS, ERYNNIS HESPERIIDAE
LIFE HISTORY
NEW YORK ENT. SOC. JOURN. 7: 211, 1899 COQUEILETT LARVA
LOC. CIT. 31:175, 1923 COOLIDGE LIFE HISTORY
SOUTHERN CALIFORNIA ACAD. SCI. BULL. 29:24, 1930 COMSTOCK

EGG*LARVA*CHRYSALIS*
COMSTOCK, J.A.-BUTTER CALIFORNIA P.215,1927 EGG,LARVA.CHRYSALIS
FOOD PLANTS
LOTUS GRANDIFLORUS(BENTH.)
MEDICAGO HISPIDA GAERTN.
MEDICAGO SATIVA LINN.(ALFALFA)
PHOLISTOMA MEMBRANACEA (BENTH)

FURCATA(SMITH), PAPAIPEMA-NOCTUIDAE
LIFE HISTORY
CAN.ENT.47:150,1915 BIRD LARVA* PL.1,FF.3,4;PUPA
INDEX AMER.ECON.ENT. VOL.1:230;VOL.2:273; VOL.7:740
FOOD PLANTS
FRAXINUS SP.(ASHES)

FURCATA(THUNBERG), HYDRIOMENA-GEOMETRIDAE
FORM FUSCOUNDATA DONOVAN
LIFE HISTORY
NEW YORK STATE AGRI.EXP.STA.(CORNELL),MEM.274:141, 1948 FORBES LARVA
FOOD PLANTS
ALNUS SP.(ALDERS)
ALNUS RUGOSA(DUROI)
ALNUS SERRULATA(AIT.)
BETULA SP.(BIRCHES)
POPULUS SP.(POPLARS)
SALIX SP.(WILLOWS)
VACCINIUM SP.(BILLBERRIES)

FURCIFASCIA WALKER
SYN.SEE HASTATA LINNAEUS

FURCIFERA GUENEE, ACRONICTA-NOCTUIDAE
A MANITOBA SMITH
LIFE HISTORY
AMER.MUS.NAT.HIST.BULL.14: 244,1901 BEUTEN.LARVA
NEW YORK STATE MUS.MEM.8: 733,1905 FELT LARVA(BRIEF)
UNITED STATES NAT.MUS.PROC. 21:86,1898 SMITH&DYAR EGG,LARVA*PL.6,F.10 COLOR;COCOON
INDEX AMER.ECON.ENT. VOL.7:8
FOOD PLANTS
PRUNUS PENSYLVANICA LINN.
PRUNUS SEROTINA EHRH.
PRUNUS VIRGINIANA LINN.

FURCIFERATA PACKARD
SSP.SEE ARCASARIA WALKER

FURCILLA(PACKARD), PANTHEA-NOCTUIDAE
A PALLESCENS MCDUNNOUGH
B CENTRALIS MCDUNNOUGH
LIFE HISTORY
AMER.MUS.NAT.HIST.BULL.4: 233,1901 BEUTEN.LARVA
AMER.NAT.8:691,1874 LARVA* F.102 NO DESCR.
ILLINOIS DEPT.AGRI.TRANS.18: APPEND:168,1880 COQUILLETT LARVA
NEW YORK STATE CABIN.NAT. HIST.REPT.NO.26:131,1875 LINTNER LARVA*
NEW YORK STATE MUS.MEM.8: 679,1905 FELT LARVA
PAPILIO 3:11,1883 THAXTER LARVA
UNITED STATES NAT.MUS.PROC. 21:16,1898 SMITH & DYAR LARVA* PL.8,F.37 COLOR
FERNALD-IN KINGSLEY'S STANDARD NAT.HIST.2:453,1885 LARVA
PACKARD-INS.INJ.FOREST TREES P.203,1881 LARVA
FOOD PLANTS
LARIX LARICINA(DUROI)
PINUS ECHINATA MILL.
PINUS STROBUS LINN.
TILIA SP.(LINDENS)

FURCILLATA SAY
SSP.SEE MILBERTI GODART

FURTIVE SMITH

SYN. SEE COSTATA GROTE

FURTIVA SMITH
SSP. SEE JOCOSA GUENEE

FUSCA GROTE & ROBINSON
SYN. SEE L'HERMINIERI GODART(LERODEA)

FUSCA HARVEY
SSP. SEE HYDROMELI HARVEY

FUSCA STRETCH
SSP. SEE FAUSTINULA BOISDUVAL

FUSCALIS BARNES & BENJAMIN
SSP. SEE FUNERALIS GROTE & ROBINSON

FUSCARIA BARNES & BENJAMIN
SSP. SEE CANOSARIA WALKER

FUSCICAUDIS WALKER
SSP. SEE THYSBE FABRICIUS

FUSCOBRUNNEA STRAND
SSP. SEE MARGARITOSA HAWORTH

FUSCOSA NEUMOEGEN
SSP. SEE BEANII NEUMOEGEN

FUSCOSTIGMA COCKERELL
SYN. SEE TEXANA RILEY

FUSCOUNDATA DONOVAN
SSP. SEE FURCATA THUNBERG

FUCUS LUTHER
SYN. SEE IO FABRICIUS

FUSIFASCIATA WALKER
SYN. SEE LIMITARIA WALKER

FUSIMACULA BARNES
SSP. SEE CHALCEDONIA DOUBLEDAY & HEWITSON

FUSIMACULA SMITH
SSP. SEE DIVERGENS WALKER(EUXOA)

FUSISECUNDA COMSTOCK
SYN. SEE CHALCEDONIA DOUBLEDAY & HEWITSON

FUTILIS(GROTE & ROBINSON)LITOPROSOPUS-NOCTUIDAE
LIFE HISTORY
ENT. AMER. 14:150, 1934 CRUMB LARVA
FOOD PLANTS
SABAL PALMETTO (S. WALT.)
SEREONA REPENS(BARTR.)

GABBII BEHR. MELITAEA-NYMPHALIDAE
SYN. SONORAE BOISDUVAL
AB NEWCOMBI COMSTOCK
AB GUNDERI COMSTOCK
AB PASADENAE GUNDER
A SABINA WRIGHT
LIFE HISTORY
BROOKLYN ENT. SOC. BULL. 19: 46, 1924 COOLIDGE EGG, YOUNG LARVA
ENT. NEWS 13:158, 1902 MINOT CHRYSALIS
SOUTHERN CALIFORNIA ACAD. SCI. BULL. 30:15, 89, 1931 COMSTOCK EGG, LARVA*CHRYSALIS*
FOOD PLANTS
CORETHROGYNE FILAGINIFOLIA V. BERNARDINA(ABRAMS)
HAPLOPAPPUS SQUARROSUS H.&A.
ORTHOCARPUS PURPURASCENS BENTH.

GALACTINUS BOISDUVAL
SSP. SEE CALIFORNIA WESTWOOD & HEWITSON

GALBINA(CLEMENS), AGAPEMA-SATURNIIDAE
AB INTERRUPTA DRAUDT
A ANONA OTTOLENGUI
LIFE HISTORY
ENT. AMER. 4:61, 1888 HY. EDW. EGG, LARVA PUPA, COCOON
ENT. SOC. AMERICA, ANNALS 9:143, 1916 MOSHER PUPAL STRUCT.
NAT. ACAD. SCI. MEM. 12:158, 1914

PACKARD LARVA*PL.14,F.6
COLOR
PSYCHE 80:151,1898 DYAR EGG, LARVA,PUPA
FOOD PLANTS
SARCOBATUS VERMICULATUS(HOOK)

GALBINEATA ZELLER
SYN.SEE GISIGNATA WALKER

GALLATINUS STICHEL
SSP.SEE CLODIUS MENETRIES

GALLII(ROTTENBURG),CELERIO-SPHINGIDAE
A INTERMEDIA KIRBY
SYN.CHAMAENERII HARRIS
SYN.OXYBAPHI CLEMENS
SYN.CANADENSIS GUENEE
LIFE HISTORY
AMER.JOURN.SCI.36:305,1839 HARRIS LARVA,PUPA (CHAMAENERII)
AMER.MUS.NAT.HIST.BULL.7: 286,1895 BEUTEN.LARVA,PUPA
CAN.ENT.3:145,1871 BOWLES PUPS
LOC.CIT.9:64,1877 SAUNDERS LARVA (CHAMAENERII)
LOC.CIT.35:109,1903 FLETCHER LARVA
ENT.SOC.AMERICA,ANNALS 11: 441,1918 MOSHER PUPA
ILLINOIS DEPT.AGRI.TRANSS18, APPEND:159,1880 COQUILLETT LARVA
MAINE AGRI.EXP.STA.BULL.162: 356,1908 PATCH LARVA*FF. 47,50;PUPA* F.49
MINNESOTA AGRI.EXP.STA.BULL. 61:91,1899 LUGGER LARVA
NEW YORK STATE AGRI.EXP.STA. (CORNELL),MEM.274:202, 1948 FORBES LARVA
NEW YORK STATE MUS.MEM.8: 736,1905 FELT LARVA(BRIEF)
ONTARIO ENT.SOC.ANN.REPT.P. 36,1876 SAUNDERS LARVA
LOC.CIT.P.55,1881 REED LARVA
PHILADELPHIA ACAD.NAT.SCI. JOURN.5:144,1859 CLEMENS LARVA
PHILADELPHIA ENT.SOC.PROC.3: 661,1864 LINTNER LARVA PUPA(CHAMAENERII)
PSYCHE 15:110,1909 COX EGG, LARVA,PUPA
SMITHSONIAN MISC.COLLECTIONS 4,ART.2:165,1862 MORRIS LARVA
AGASSIZ-LAKE SUPERIOR P.338, 1850 LARVA
BOISDUVAL-SPEC.GEN.LEP. SPHING.P.170,1874 LARVA
FERNALD-SPHINGIDAE NEW ENGLAND P.55,1886 LARVA
MORRIS-SYNOP.LEP.NORTH AMERICA P.165,1862 LARVA
SAUNDERS-INS.INJ.FRUITS P.257, 1883 LARVA,PUPA,COCOON
SAUNDERS-LOC.CIT.P.257,1889 LARVA
SMITH-MONOGRAPH SPHINGIDAE P.133,1888 LARVA(BRIEF)
INDEX AMER.ECON.ENT.
VOL.1:103 DEILEPHILA
VOL.2:72
FOOD PLANTS
EPILOBIUM ANGUSTIFOLIUM(LINN.)
OENOTHERA BIENNIS LINN.
PORTULACA OLERACEA LINN.
ROSA SP. (ROSES)
VITIS SP. (GRAPES)

GAMMA COCKERELL
SSP.SEE JAMAICENSIS DRURY

GAOSALIS WALKER
SYN.SEE METONALIS WALKER

GARGAMELLE(STRECKER),GLOVERIA-LASIOCAMPIDAE
A MEDUSA STRECKER
LIFE HISTORY
NEW YORK ENT.SOC.JOURN.7: 210,1899 COQUILLETT LARVA, PUPA(MEDUSA)

PACIFIC COAST ENT. SOC. PROC.
P.71,1909 RICKSECKER LARVA
FOOD PLANTS
ERIOGONUM FASCICULATUM BENTH.

GARGANTUA(DYAR), ARZAMA-NOCTUIDAE
LIFE HISTORY
ASOUTHERN CALIFORNIA ACAD. SCI. BULL.43:84,1944 COMSTOCK= DAMMERS LIFE HISTORY EXCEPT EGG; LARVA* PUPA*
FOOD PLANTS
TYPHA SP.(CAT TAILS)

GARITA(REAKIRT), OARISMA-HESPERIIDAE
SYN. HYLAX EDWARDS
LIFE HISTORY
CAN. ENT.42:145,1910 GIBSON EGG, LARVA, CHRYSALIS
FOOD PLANTS
POA PRATENSIS LINN.

GARTHI GUNDER
SSP. SEE SAEPIOLUS BOISDUVAL

GASPEENSIS MCDUNNOUGH
SSP. SEE BREVICAUDA SAUNDERS

GAURAE(ABBOT & SMITH), PROSPERPINUS-SPHINGIDAE
SYN. CIRCAE HENRY EDWARDS
LIFE HISTORY
AMER. MUS. NAT. HIST. BULL.16: 396,1902 BEUTEN. LARVA
NOVITATES ZOOLOGICA(TRING)9 SUPPL. P.610,1903
PHILADELPHIA ACAD. NAT. SCI. JOURN.5:133,1859 CLEMENS LARVA, PUPA
SMITHSONIAN MISC. COLLECTIONS 4, ART.2:154,1862 MORRIS LARVA
ABBOT & SMITH-LEP. INS. GEORGIA 1:61,1797 LARVA*PUPA*PL.31
BOISDUVAL-SPEC. GEN. LEP. SPHING P.315,1874 LARVA
MORRIS-SYNOP. LEP. NORTH AMERICA P.154,1862 LARVA, PUPA
FOOD PLANTS
EPILOBIUM SP.
GAURA BIENNIS LINN.

GAURAE(ABBOT & SMITH), RHODOPHORA-NOCTUIDAE
SYN. MATUTINA HUBNER
LIFE HISTORY
ABBOT & SMITH-LEP. INS. GEORGIA 2:197,1797 LARVA*PL.99 COLOR
DUNCAN-NATUR. LIBRARY 22:200 1858 LARVA*PL.24, F.5 COLOR
GUENEE-SPEC. GEN. LEP. NOCT. 2:170,1852 LARVA
FOOD PLANTS
GUARA BIENNIS LINN.

~~GEBA~~ SCUDDER
SYN. SEE BOREALIS GROTE & ROBINSON

GEDDESI NEUMOEGEN
SSP. SEE PLANTAGINIS LINNAEUS

GELIDA GROTE
SYN. SEE CANIPLAGA WALKER

GELIDA MOESCHLER
SSP. SEE QUENSELII PAYKULL

GELIDATA MOESCHLER, EUPITHECIA-GEOMETRIDAE
SYN. HYPERBOREATA STAUDINGER
SYN. FLEBILIS HULST
SYN. LAGGANATA TAYLOR
?SYN. COMPACTA TAYLOR
?SYN. YOUNGATA TAYLOR
SYN. NORDEGGENSIS CASSINO&SWETT
LIFE HISTORY
FOOD PLANTS
BETULA SP.(BIRCHES)
LEDUM PLAUSTRE LINN.

GEMINATUS SAY
SSP. SEE JAMAICENSIS DRURY

GEMMA HUBNER, NEONYMPHA-SATYRIDAE
?SYN. CORNELIUS FABRICUS
LIFE HISTORY
CAN. ENT. 11:31, 1879 W.H.EDW. EGG, LARVA* CHRYSALIS
ILLINOIS DEPT. AGRI. TRANS. 18, APPEND:91, 1880 MIDDLETON LARVA, CHRYSALIS
MUS. COMP. ZOOL. BULL. 87, NO. 6, 1941 DETHIER LARVAL ANTENNA* F.32
BOISDUVAL & LECONTE-HIST. LEP. NORTH AMERICA 1833 LARVA* CHRYSALIS* PL.62 COLOR
COMSTOCK-HOW KNOW BUTTER. P. 194, 1904 LARVA
EDWARDS-BUTTER. NORTH AMERICA 3RD SERIES, 1897 LIFE HIST* NEONYMPHA 1 FF. A-H COLOR
FRENCH-BUTTER. EAST. UNITED STATES P.235, 1881 LIFE HISTORY
HOLLAND-BUTTERFLY BOOK P.202, 1905
HOLLAND-LOC. CIT. P.179, 1931 LIFE HISTORY
SCUDDER-BUTTER. EAST. UNITED STATES & CANADA PP. 1783-1785, 1889 LIFE HISTORY (CORNELIUS)
FOOD PLANTS
GRASSES

GEMMATILIS HUBNER, ANTICARSIA-NOCTUIDAE
LIFE HISTORY
JOURN. ECON. ENT. 9:522, 1916 WATSON EGG, LARVA* PL.39, FF. 1, 2; PUPA
LOC. CIT. 23:687, 1930 DOUGLAS EGG, LARVA, PUPA
LOC. CIT. 23:714, 1930 HINDS EGG* PL.25, FF.2, 3; LARVA* PL.25, F.4; PL.26, F.7; PUPA* PL.25, F.6
LOC. CIT. 33:636, 1940 GUYTON LARVAL VARIATIONS
LOUISIANA AGRI. EXP. STA. BULL. 350;17 1942 ELLISOR EGG, LARVA, PUPA
INDEX AMER. ECON. ENT. VOL.1:19; VOL.2:25; VOL.3:31; VOL.4:35; VOL.5:41; VOL.6:46; VOL.7:66
FOOD PLANTS
ARACHIS HYPOGAEA LINN. (PEANUT)
CANAVALIA SP.
CANAVALIA ROSEA (SW.)
GLYCINE MAX (LINN.) (SOY BEAN)
GOSSYPIUM HERBACEUM LINN. (COTTON)
MEDICAGO SATIVA LINN. (ALFALFA)
PHASEOLUS SEMIERECTUS LINN.
PUERARIA LOBATA (WILLD.)
ROBINIA PSEUDO-ACACIA LINN.
SESBANIS EXALTATA (RAF.)
STIZOLOBIUM DERINGIANUM BORT.
VIGNA SINENSIS (LINN.)

GENERALIS WALKER
SYN. SEE SIDCOLORALIS GUENEE

GENETRIX (GROTE), ASEPTIS-NOCTUIDAE
LIFE HISTORY
WASHINGTON ENT. SCO. PROC. 4: 320, 1898 DYAR LARVA
FOOD PLANTS
ADENOSTOMA FASCICULATUM H.&A.

GENEURA STRECKER
SSP. SEE NEVADENSIS GROTE&ROBINSON

GENICULA GROTE
SSP. SEE ARCUATA WALKER (DREPANA)

GENIGUAH REAKIRT
SYN. SEE FRISIA POEY

GENOA PLOETZ
SYN. SEE SABULETI BOISDUVAL

GENOVEVA (CRAMER), JUNONIA-NYMPHALIDAE
LIFE HISTORY
NEW YORK ENT. SOC. JOURN. 9:79 1901 SWAINSON LARVA, CHRYSALIS

FOOD PLANTS
Verbena hastata Linn.

genutia Fabricius
see midea Hubner

geometrica Grote
syn. see plantaginis Linn.

geometroides Guenee
syn. see decoralis Hubner

geometroides Walker, Melanchroia-Geometridae
syn. mors Lucas
LIFE HISTORY
Washington Ent. Soc. Proc. 6:77 1904 Dyar larva
Gundlach-Entom. Cubana p. 410, 1886 larva
FOOD PLANTS
Phyllanthus acidus (Reed)
Rhyllanthus racemosa Lour.

georgeana Beutenmuller
syn. see grotiana Bailey

georgei Gunder
ssp. see chalcedona Doubleday & Hewitson

georgiana Dyar
syn. see varia Walker (Heterocampa)

georgica (Herrich-Schaeffer) Hyperaeschra-Notodontidae
LIFE HISTORY
Amer. Mus. Nat. Hist. Bull. 10: 414, 1898 Beuten. larva
Nat. Acad. Sci. Mem. 7:154, 1895 Packard larva, pupa
New York State Agri. Exp. Sta. (Cornell), Mem. 274:220, 1948 Forbes larva
New York State Mus. Mem. 8:705 1904 Felt larva (brief)
FOOD PLANTS
Fraxinus sp. (ashes)
Prunus serotina Ehrh.
Quercus sp. (oaks)
Prunus serotina Ehrh.
Quercus sp. (oaks)
Quercus alba Linn.

georgii Grote, Lithophane-Noctuidae
a emarginata Smith
b holocinerea Smith
c ancilla Smith
d vertina Smith
e fletcheri Smith
f. oregonensis Harvey
LIFE HISTORY
Can. Ent. 26:67, 1894 Dyar larva
FOOD PLANTS
Cornus sericea Linn.
Malus pumila Mill. (apple)
Quercus kelloggii Newb.
Salix hookeriana Barr.
Salix scouleriana Barr.

gerhardi Barnes & Benjamin
ssp. see herodias Strecker

gerhardi Gunder
ssp. see glaucus Linnaeus

germana Henry Edwards, Annaphila-Noctuidae
LIFE HISTORY
FOOD
A specimen in the United States Nat. Mus. bears the following label "Larvae feeding in numbers on larvae of wasps" Sacramento, California, Harry Smith Feb. 16, 1915

germana Morrison
ssp. see solidagenis Hubner

germanalis Walker
syn. see humuli Harris (Hypena)

gertana Smith
ssp. see cuneata Grote

gibboni Gunder
ssp. see dione Scudder

GIBBOSA(ABBOT & SMITH), NADATA-
NOTODONTIDAE
A DOUBLEDAYI PACKARD
B RUBRIPENNIS NEUMOEGEN & DYAR
C OREGONENSIS BUTLER
D BEHRENSI HENRY EDWARDS
LIFE HISTORY
AMER. MUS. NAT. HIST. BULL. 10: 411, 1898 BEUTEN. LARVA
AMER. NAT. 8:891, 1874 PACKARD LARVA*F. 103
ENT. NEWS 9:150, 1898 SOULE DURATION EGG STAGE
NAT. ACAD. SCI. MEM. 7:143, 280, 1895 PACKARD EGG, LARVA* PL. 17, FF. 1, 2 COLOR: PUPA*
NEW YORK ENT. SOC. JOURN. 1:57 1893 PACKARD EGG, LARVA
NEW YORK STATE CABIN. NAT. HIST ANN. REPT. NO. 26:150, 1874 LINTNER LARVA*
NEW YORK STATE MUS. MEM. 8: 521, 1905 FELT LARVA
PSYCHE 6:197, 1891 SOULE EGG, LARVA
PSYCHE 6:338, 1892 DYAR EGG, LARVA, PUPA, COCOON (OREGONENSIS)
UNITED STATES DEPT. AGRI. MISC. PUB. 657:363, 401, 1950 CRAIGHEAD LARVA
ABBOT & SMITH-LEP. INS. GEORGIA 2:163, 1797 LARVA* PL. 82 COLOR
DOANE ETC-FOREST INSECTS P. 295 1936 LARVA
GLOVER-ILLUST. NORTH AMER. ENT. 1878 LARVA* PL. 6, F. 9
HARRIS-ENT. CORRESP. P. 308, 1869 LARVA
FOOD PLANTS
ACER SP. (MAPLES)
ACER MACROPHYLLUM PURSH
ALNUS SP. (ALDERS)
AMELANCHIER FLORIDA LINDL.
BETULA PAPYRIFERA MARSH.
CASTANEA SATIVA MILL.
FAGUS SP. (BEECH)
PRUNUS DOMESTICA LINN.
QUERCUS SP.)OAKS)
QUERCUS KELLOGGII NEWB.
ROSA SP. (ROSES)
SALIX SP. (WILLOWS)

GIBSONATA TAYLOR
SYN. SEE CHAGNONI SWETT(EUPITHECIA)

GIBSONI BARNES & BENJAMIN
SSP. SEE FRIGGA THUNBERG

GIBSONI HOLLAND
SSP. SEE MELISSA FABRICIUS

GIGANS MCDUNNOUGH
SSP. SEE OCCIDENTALIS LINTNER

GIGANTARIA SWETT
SSP. SEE DEFENSARIA GUENEE

GIGANTEA STRECKER, COLIAS-PIERIDAE
SYN. PELIDNEIDES STAUDINGER
LIFE HISTORY
FOOD PLANTS
SALIX SP. (WILLOWS)

GIGANTEUS GROSSBECK, PERO-GEOMETRIDAE
A PROVOENSIS CASSINO & SWETT
LIFE HISTORY
FOOD PLANTS
ALNUS RUBRA BONG.
HOLODISCUS DISCOLOR (PURSH)
RIBES SANGUINEUM PURSH

GIGAS BUTLER
SYN. SEE NEVADENSIS FELDER & FELDER

GILLETTI (BARNES), EUPHYDRYAS-NYMPHALIDAE
SYN. GLACIALIS SKINNER
AB HERRI GUNDER
LIFE HISTORY
AOUTHERN CALIFORNIA ACAD. SCI. BULL. 39:111, 1940 COMSTOCK EGG, LARVA* PL. 20
FOOD PLANTS
LONICERA INVOLUCRATA(RICH.)
SYMPHORICARPOS ALBUS(LINN.)

GISELA MEYER
SSP. SEE MICRONYMPHA GUENEE

GLACIALIS McLACHLAN
SSP. SEE HECLA LEFEBRE

GLACIALIS SKINNER
SYN. SEE GILLETTII BARNES

GLADIARIA MORRISON, AGROTIS-NOCTUIDAE
SYN. MORRISONIANA RILEY
LIFE HISTORY
BROOKLYN ENT. SOC. BULL. 22:44 1927 CRUMB MATURE LARVA* PL. 4 D, F
ENT. SOC. AMERICA, ANNALS 23: 396, 1930 HAWKINS LARVAL CLAES
ILLINOIS AGRI. EXP. STA. BULL. 95:358, 1904 FORBES LARVA*
JOURN. ECON. ENT. P. 394, 1915 CRUMB KEY TO LARVA-LARVAL SKIN* PL. 20, F. 11
NEBRASKA AGRI. EXP. STA. RESEARCH BULL. 81:16, 1935 WHELAN TECH. DESC. LARVA-LARVAL SKINN* F. 6D; CLAWS * F. 9A
UNITED STATES DEPT. AGRI. FARMER'S BULL. 1494:5, 1926 CRUMB LARVA* F. 6 B, C
UNITED STATES DEPT. AGRI. TECH. BULL. 88:70, 1929 CRUMB EGG, LARVA* PL. 6 C, I; PUPA* PL. 19 A
CROSBY & LEONARD-MANUAL VEG. GARD. INS. P. 274, 1918 LARVA
HART-INSECT COLLECTIONS ILLINOIS HIGH SCHOOLS P. 32, 1903 LARVA* F. 42
INDEX AMER. ECON. ENT. VOL. 1:134; FELTIA
Vo VOL. 6:17; VOL. 7:24
FOOD PLANTS
ALLIUM CEPA LINN. (ONION)
ASTER SP.
ASTER ERICOIDES LINN.
AVENA SATIVA LINN. (OATS)
BRASSICA OLERACEA LINN. (CABBAGE)
CANNABIS SATIVA LINN. (HEMP)
FRAGARIA CHILOENSIS (LINN) (STRAWBERRY)
IPOMOEA BATATAS (LINN. (SWEET) POTATO)
NICOTIANA TABACUM LINN. (TOBACCO)
PHASEOLUS VULGARIS LINN. (BEAN)
RUBUS ALLEGHENIENSIS PORTER
RUBUS IDAEUS V. STRIGOSUS (MICHX.)
RUMEX CRISPUS LINN.
SOLANUM LYCOPERSICUM LINN. (TOMATO)
SOLANUM TUBEROSUM LINN. (POTATO)
SOLIDAGO SP.
TRIFOLIUM SP. (CLOVER)
VIOLA TRICOLOR LINN. (PANSY)
ZEA MAYS LINN. (CORN)
GENERAL FEEDER-FIELD CROPS, BERRIES, GARDEN CROPS
GRASS

GLANS GROTE
SYN. SEE FLAVISTRIARIA HUBNER

GLAUCARIA GUENEE
SYN. SEE AERATA FABRICIUS

GLAUCATA BARNES & McDUNNOUGH
SSP. SEE KENTARIA GROTE & ROBINSON

GLAUCOPIS HAMPSON
SSP. SEE LUTRA GUENEE

GLAUCOVARIA WALKER
SYN. SEE TRIFOLII ROTTENBURG

GLAUCUS LINNAEUS, PAPILIO-PAPILIONIDAE
FORM TURNUS LINNAEUS
SYN. AJAX LINNAEUS (PARTIM)
SYN. ANTILOCHUS LINNAEUS
SYN. ALCIDAMUS CRAMER

AB EHRAMANNI EHRMANN
AB DIETZI GUNDER
AB GERHARDI GUNDER
A CANADENSIS ROTHSCHILD & JORDAN
SYN. ARCTICUS SKINNER
AB MALE FLETCHERI KEMP
B AUSTRALIS MAYNARD

LIFE HISTORY

AMER. MUS. NAT. HIST. BULL. 5: 244, 1893 BEUTEN. LARVA, CHRYZALIS
BUFFALO SOC. NAT. SCI. BULL. 19: 18, 1919 WILD LARVA
CAN. ENT. 1:51, 1859 SAUNDERS EGG, LARVA
CAN. ENT. 1:74, 1869 SAUNDERS LARVA
LOC. CIT. 6:3, 1874 SAUNDERS EGG, LARVA*
LOC. CIT. 21:201, 1889 FLETCHER EGG, LARVA*CHRYSALIW
CAN. NAT. &GEOL. 2:223, 1857 LARVA, CHRYSALIS
ENT. NEWS 16:113, 1905 JEHEBER GLAUCUS AND TURNUS LARVAE COMPARED
ILOINOIS DEPT. AGRI. TRANS. 16: 139, 1877 FRENCH LARVA
LOC. CIT. 18:74, 1880 MIDDLETON LARVA
LOC. CIT. 18:173, 1880 COQUILLETT LARVA
MARYLAND ACAD. SCI. BULL. 3(2): 8, 1923 MOORE EGG, LARVA, CHRYSALIS
MARYLAND NAT. HIST. SOC. PROC. P. 9, 1933 HAYDON EGG, LARVA CHRYSALIS
MINNESOTA AGRI. EXP. STA. BULL. 61:81, 1899 LUGGER EGG* LARVA* CHRYSALIS* ALL F. 28
MONTANA UNIV. BULL. 30:21, 1906 ELROD EGG(BRIEF)
MUST. COMP. ZOOL. BULL. 82, NO. 6, 1941 DETHIER LARVAL ANTENNA*FF. 19, 24
NATURALISTA CANADIEN P. 43, 1927 HUARD LARVA, CHRYSALIS
NATURE STUDY REVIEW P. 223, SEPT. 1917 LARVA*
NEW YORK STATE AGRI. EXP. STA. (CORNELL), MEM. 56:1072, 1922 WELLHOUSE LARVA (BRIEF)
NEW YORK STATE MUS. MEM. 8: 804, 1904 FELT LARVA (BRIEF)
NEW YORK STATE REPT. INS. 3: 341, 376, ALSO P. 24, SEC. 36, 1856 FITCH LARVA
ONTARIO ENT. SOC. ANN. REPT. P. 20, 1873 SAUNDERS LARVA* CHRYSALIS
LOC. CIT. P. 73, 1879 SAUNDERS LARVA*
LOC. CIT. P. 39, 1889 HARRINGTON EGG, LARVA* CHRYSALIS
LOC. CIT. P. 39, 1894 BETHUNE LARVA*
LOC. CIT. P. 76, 1902 GIBSON LARVA
PAPILIO 3:2, 1883 EDWARDS GLAUCUS & DAUNUS LARVA COMPARED
PAPILIO 4:86, 88, 1884 GRUBER LARVA*
PHILADELPHIA ENT. SOC. PROC. 1:352, 1863 WALSH LARVA
LOC. CIT. 3:50, 1864 LINTNER CHRYSALIS
PSYCHE 8:208, 1898 SCUDDER LARVA
SMITHSONIAN MISC. COLLECTIONS 4, ART. 2:2, 1862 MORRIS LARVA, CHRYSALIS
UNITED STATES DEPT. AGRI. MISC. PUB. 657:351, 368, 1950 CRAIGHEAD LARVA*F. 70 B
UNITED STATES NAT. MUS. BULL. 157:189, 1932 A. CLARD LARVA, CHRYSALIS
VERMONT STATE BOARD AGRI. REPT. 2:589, 1876 PERKINS LARVA
BOISDUVAL–SPEC. GEN. LEP.

DIURNAE 1:339,1936 LARVA, CHRYSALIS

BOISDUVAL & LECONTE-HIST.LEP. NORTH AMERICA PP.7,22, 1833 LARVA*PLS.7,9COLOR

COMSTOCK-HOW KNOW BUTTER.P. 52,1904 LARVA*

CRAGEN-OUR INSECE FRIENDS & FOES P.102,1899 LARVA

DICKERSON-MOTHS & BUTTER.P. 57,1901 LIFE HISTORY LARVA* FF.48-53,56,57;CHRYSALIS* FF.54,55

EDWARDS-BUTTER.NORTH AMERICA 2ND SERIES 1877 LIFE HIST* PL.PAPILIO 3,4 COLOR

FERNALD-BUTTER.MAINE P.24, 1884 EGG*LARVA*CHRYSALIS*

FRENCH-BUTTER.EAST.UNITED STATES P.100,1886 LIFE HIST.EGG,LARVA*CHRYSALIS*

GLOVER-ILLUST.NORTH AMER.ENT. 1878 LARVA,PUPA*PL.2,F4

GOSSE-LETTERS FROM ALABAMA P.123,1859 LARVA(BRIEF)

HARRIS-INS.INJ.VEG.FLINT EDIT. P.268,1862 LARVA*CHRYSALIS*

HOLLAND-BUTTERFLY BOOK 1905 LARVA *CHRYSALIS*PL.2,FF. 15,26,28;PL.6,FF.1-4 COLOR

KELLOGG-AMER.INSECTS P.449, 1908 LARVA

MAYNARD-BUTTER.NEW ENGLAND P.51,1886 LARVA,CHRYSALIS

MORRIS-SYNOP.LEP.NORTH AMERICA P.2,1861 LARVA

PACKARD-ONE HALF HOURS WITH INS.P.180,1877 LARVA

SAUNDERS-LOC.CIT.P.83,1889 EGG,LARVA*F.81 1;CHRYSALIS* F.82 P

SCUDDER-BUTTER.EAST.UNITED STATES & CANADA PP.1291-1294;1296-1300,1889 LIFE HISTORY* PL.66,F.1;PL.68, F.18;PL.76,FF.15,26;PL. 80,FF.7-10;PL.85,FF.1-4

SCUDDER-BUTTER.THEIR STRUCT. P.7,1881 EGG*

SCUDDER-EVERDAY BUTTER.P.161, 1899 EGG,LARVA,PUPA (AJAX)

SCUDDER-GUIDE TO COMMON BUTTER.P.148,1893 LARVA, CHRYSALIS

VANCE-LIFE AMONG BUTTER.P. 50,1925 LARVA

INDEX AMER.ECON.ENT. VOL.1:230;VOL.2:264; VOL.3:294;VOL.4:342; VOL.5:469;VOL.6:559; VOL.7:741

FOOD PLANTS

ACER SP.
ALNUS SP.
ALNUS RUGOSA (DUROI)
AMELANCHIER CANADENSIS(LINN.)
BETULA SP.(BIRDH)
CARYA SP.(HICKORIES)
CATALPA BIGNONOIDES WALT.
CINNAMOMUM CAMPHORA(LINN.)
CRATAEGUS SP.(HAWS)
CYDONIA OBLONGA MILL.(QUINCE)
FRAXINUS SP.(ASHES)
FRAXINUS CAROLINIANA MILL.
FRAXINUS NIGRA MARSH
HUMULUS LUPULUS LINN.(HOPS)
LINDERA BENZOIN(LINN.)
LIRIODENDRON TULIPIFERA LINN.
MAGNOLIA ACUMINATA LINN.
MALUS PUMILA MILL.(APPLE)
POPULUS SP.(POPLARS)
POPULUS TREMULOIDES MICHX.
PRUNUS CERASUS LINN.(CHERRY)
PRUNUS DOMESTICA LINN.(PLUM)
PRUNUS PERSICA(LINN.)(PEACH)
PRUNUS SEROTINA EHRH.
PRUNUS VIRGINIANA LINN.
PYRUS COMMUNIS LINN.(PEAR)
QUERCUS SP.(OAKS)
SALIX SP.(WILLOWS)
SASSAFRAS ALBIDUM VAR.MOLLE (RAF.)
STYRAX AMERICANA LAM.
SYRINGA VULGARIS LINN.

TILIA AMERICANA LINN.
ULMUS SP. (ELMS)
VACCINIUM SP.
VITIS SP. (GRAPES)

GLENWOODI BARNES, TOLYPE-LASIOCAMPIDAE
LIFE HISTORY
BROOKLYN MUS. ARTS&SCI. BULL. 1:195, 1906 DYAR LARVA
SOUTHERN CALIFORNIA ACAD. SCI. BULL. 41:86, 1942 COMSTOCK-HENNE. EGG*PL. 16; EGG MASS* PL. 17; LARVA*PL. 18; PUPA* AND COCOON* PL. 19
FOOD PLANTS
CERCOCARPUS BETULOIDES NUTT.

GLOVERI (GROTE & ROBINSON), COPIDRYAS-AGARISTIDAE
LIFE HISTORY
ILLINOIS AGRI. EXP. STA. BULL. 60:592, 1900 FORBES EGG* LARVA*PUPA*
INDUSTRIALIST MANHATTAN KANSAS OCT. 1, 1887 POPENOE LIFE HISTORY EGG*LARVA* PUPA*
INSECT LIFE 1:104, 1888 RILEY-HOWARD LIFE HISTORY EGG* LARVA* PUPA*
HOLLAND-MOTH BOOK 1905 P. 141, EGG*LARVA*PUPA*FF. 78, 79, 80
FOOD PLANTS
BETA VULGARIS LINN. (BEET)
PORTULACA GRANDIFLORA HOOK
PORTULACA OLERACEA LINN.

GLOVERI STRECKER, PLATYSAMIA-SATURNIIDAE
A REDUCTA NEUMOEGEN
LIFE HISTORY
AMER. ACAD. ARTS. &SCI. PROC. 28:65, 1893 PACKARD YOUNG LARVA
BOSTON SOC. NAT. HIST. PROC. 26: 164, 1893 DYAR EGG, LARVA
BROOKLYN ENT. SOC. BULL. 1:75, 1879 GRAEF CODOON
CAN. ENT. 72:89, 1940 DUNCAN HIBERNATION
DAVENPORT ACAD. SCI. PROC. 2: 277, 1878 STRECKER LARVAE OF CECROPIA, COLUMBIA AND GLOVERI COMPARED
ENT. SOC. AMERICA, ANNALS 9;154 1916 MOSHER PUPAL STRUCT.
MONTANA AGRI. EXP. STA. BULL. 75:110, 1908 COOLEY EGG, LARVA* PL. 3; PUPA COCOON* PL. 2
NAT. ACAD. SCI. MEM. 12:219, 1914 PACKARD LIFE HISTORY* PL. 8, FF. 2-7; PL. 9, FF1, 2
COLOR
LOC. CIT. P. 267 HYBRID LARVAE CROSSES WITH PROMETHEA AND CECROPIA
SOUTHERN CALIFORNIA ACAD. SCI. BULL. 29:152, 1930 COMSTOCK LARVA*
ELIOT & SOULE-CATER AND MOTHS P. 246, 1902 EGG, LARVA* COCOON
ESSIG-INS. WEST. NORTH AMERICA P. 668, 1926 LARVA, COCOON
INDEX AMER. ECON. ENT. VOL. 7:862
FOOD PLANTS
ACER SP. (MAPLES)
ACER NEGUNDO LINN.
LARIX LARICINA (DUROI)
PRUNUS VIRGINIANA LINN.
RIBES LACUSTRE (PERS.)
RIBES SATIVUM SYME. (CURRANT)
SALIX SP. (WILLOWS)
ULMUS SP. (ELMS)

GNOPHARIA GUENEE
SYN. SEE UMBROSARIA HUBNER

GNOPHOSARIA (GUENEE), SEMIOTHISA-GEOMETRIDAE
SYN. INFECTATA WALKER
SYN. REDUCTARIA WALKER
SYN. CAESIARIA HULST
SYN. SEDUCTATA WALKER

LIFE HISTORY
FOOD PLANTS
SALIX SP. (WILLOWS)

GOLOVINUS HOLLAND
SSP. SEE SMINTHEUS DOUBLEDAY & HEWITSON

GONIATA GUENEE
SYN. SEE TRANSVERSATA DRURY

GOPHERI SMITH, EPIZEUXIS-NOCTUIDAE
LIFE HISTORY
ENT. AMER. 14:158, 1934 CRUMB LARVA
FOOD
EXCREMENT OF GOPHERUS POLYPHEMUS (GOPHER TURTLE) LIVES UNDERGROUND

GORDIUS CRAMER, SPHINX-SPHINGIDAE
SYN. POECILA STEPHENS
A COXEYI CADBURY
B BOREALIS CLARK
C CAMPESTRIS MCDUNNOUGH
D OSLARI ROTHSCHILD & JORDAN
LIFE HISTORY
AMER. JOURN. SCI. 36:295, 1839 HARRIS LARVA, PUPA
AMER. MUS. NAT. HIST. BULL. 7: 303, 1895 BEUTEN. LARVA, PUPA
ENT. SOC. AMERICA, ANNALS 11: 420, 1918 MOSHER PUPA
ILLINOIS DEPT. AGRO. TRANS. 18: APPEND:105, 1880 MIDDLETON LARVA
MICHIGAN STATE COLLEGE SPEC. BULL. 243:49, 1933 MCDANIEL LARVA
MINNESOTA AGRI. EXP. STA. BULL. 61:102, 1899 LUGGER LARVA
NEW YORK ENT. SOC. JOURN. 9:87 1901 BEUTEN. EGG, LARVA
NEW YORK STATE AGRI. EXP. STA. (CORNELL), MEM. 274:191, 1948 FORBES LARVA
NEW YORK STATE MUS. MEM. 8: 727, 1905 FELT LARVA (BRIEF)
NEW YORK STATE, REPT. INS. 3: 341, ALSO SEC. 34, P. 23, 1856 FITCH LARVA
NOVITATES ZOOLOGICA (TRING) 9 SUPPL. P. 136, 1903 ROTHSCHILD & JORDAN LARVA, PUPA
PHILADELPHIA ACAD. NAT. SCI. JOURN. 5:173, 1859 CLEMENS LARVA, PUPA
PSYCHE 5:228, 1889 ELIOT & SOULE LARVAL VARIATIONS
SMITHSONIAN MISC. COLLECTIONS 4, ART. 2:198, 1862 MORRIS LARVA, PUPA
UNITES STATES DEPT. AGRI. MISC. PUB. 657:358, 374, 1950 CRAIGHEAD LARVA
ELIOT & SOULE-CATER. AND THEIR MOTHS P. 143, 1902 EGG, LARVA*PUPA
FERNALD-SPHINGIDAE NEW ENGLAND P. 45, 1886 LARVA
MORRIS-SYNOP. LEP. NORTH AMERICA P. 198, 1862 LARVA, PUPA
PACKARD-HALF HOURS WITH INSECTS P. 180, 1877 LARVA
SAUNDERS-INS. INJ. FRUITS P. 86, 1883 LARVA, PUPA (BRIEF)
SAUNDERS-LOC. CIT. P. 86, 1889 LARVA
INDEX AMER. ECON. ENT. VOL. 5:580; VOL. 6:678
FOOD PLANTS
BETULA SP. (BIRCHES)
COMPTONIA PEREGRINA (LINN.)
FRAXINUS SP. (ASHES)
GAYLUSSACIA DUMOSA (ANDR.)
LIGUSTRUM SP. (PRIVET)
MALUS PUMILA MILL. (APPLE)
MYRICA CERIFERA LINN.
MYRICA PENSYLVANICA LOISEL
PYRUS COMMUNIS LINN. (PEAR)
ROSA CAROLINA LINN.
VACCINIUM CORYMBOSOUM LINN.
VACCINIUM MACROCARPON AIT. (CRANBERRY)

VACCINIUMMYRTILLUS LINN.
VACCINIUM ULIGINOSUM LINN.
XOLISMA LIGUSTRINA(LINN.)

GORGON(BOISDUVAL),LYCAENA-LYCAENIDAE
LIFE HISTORY
CAN. ENT. 40:346,1908 COOLIDGE EGG,LARVA,CHRYSALIS
ENT. NEWS 15:97,1904 GRUNDEL LIFE HISTORY LARVA
PSYCHE 14:119,1907 COOLIDGE EGG
SOUTHERN CALIFORNIA ACAD. SCI. BULL. 33:25,1934 COMSTOCK-DAMMERS LIFE HISTORY EGG* PL. 3;LARVA*PL. 4;CHRYSALIS* PL. 5
FOOD PLANTS
ERIOGONUM ELONGATUM BENTH.
ERIOGONUM LATIFOLIUM V. NUDUM (DOUGL. EX BENTH.)

GORGONE(HUBNER),PHYCIODES-NYMPHALIDAE
SYN. ISMERIA BOISDUVAL & LECONTE
SYN. CARLOTA REAKIRT
AB NIGRA CARY
AB NOX GUNDER
LIFE HISTORY
CAN. ENT. 25:93,1893 DYAR LARVA CHRYSALIS(CARLOTA)
LOC. CIT. 26:3,1894 EDWARDS EGG, LARVA,CHRYSALIS
LOC. CIT. 65:95,1933 HANDFORD HIBERNATING LARVA
SMITHSONIAN MISC. COLLECTIONS 4,ART. 2:51,1862 MORRIS LARVA,CHRYSALIS(ISMERIA)
BOISDUVAL&LECONTE-HIST. LEP. NORTH AMERICA P. 168,1833 LARVA* CHRYSALIS*PL. 46 ISMERIA
HOLLAND-BUTTERFLY BOOK P. 152, 1905 AND 3RD EDIT. P. 139, 1931 LARVA,CHRYSALIS
MORRIS-SYNOP. LEP. NORTH AMERICA P. 50,1862 LARVA, CHRYDALIS (ISMERIA)
SCUDDER-BUTTER. EAST. UNITED STATES & CANADA P. 1811, 1889 LARVA (ISMERIA)
FOOD PLANTS
ERIOGONUM SP.
HELIANTHUS SP. (SUNFLOWERS)
HELIANTHUS LAETIFLORUS PERS.
HELIANTHUS TRACHELIFOLIUS MILL.

GORGONIOI GUNDER
SSP. SEE PIASUS BOISDVVAL

GORTYNOIDES WALKER,BELLURA-NOCTUIDAE
SYN. VULNIFICA GROTE
LIFE HISTORY
ENT. SOC. AMERICA, ANNALS 16: 375,1923 ROBERTSON-MILLER LIFE HISTORY EGG* EGG NESTS,LARVA*PUPA
FOOD PLANTS
NUPHAR ADVENA(AIT.)
PONTEDERIA CORDATA LINN.
TYPHA SP.

GOTHICATA GUENEE
SSP. SEE HASTATA LINNAEUS

GOZORA BOISDUVAL
SSP. SEE PSEUDARGIOLUS BOISDUVAL & LECONTE

GRACILARIA PACKARD
SYN. SEE AERATA FABRICIUS

GRACILENTA GRAEF
SYN. SEE DISPARATA HENRY EDWARDS

GRACILENTA HUBNER,SCHINIA-NOCTUIDAE
LIFE HISTORY
BROOKLYN ENT. SOC. BULL. 26: 203,1941 WYATT LARVA
FOOD PLANTS
KUHNIA EUPATORIOIDES LINN.

GRACILENEATA GUENEE
SSP. SEE DIVERSILINEATA HUBNER

GRACILIS EDWARDS,CATOCALA-NOCTUIDAE

SYN. TELA STRAND
FORM CINEREA MAYFIELD
FORM LEMMERI MAYFIELD
LIFE HISTORY
BROOKLYN ENT. SOC. BULL. 17:116, 1922 MAYFIELD EGG, LARVA
FOOD PLANTS
QUERCUS SP. (OAKS)
VACCINIUM SP. (HUCKLEBERRIES)

GRACILIS (GROTE & ROBINSON), HEMARIS SPHINGIDAE
LIFE HISTORY
FOOD PLANTS
VACCINIUM SP. (HUCKLEBERRY)

GRACILIS (GROTE & ROBINSON), POLYGONIA NYMPHALIDAE
LIFE HISTORY
ONTARIO ENT. SOC. ANN. REPT. P. 30, 1892 FYLES EGG
SCUDDER-BUTTER. EAST. UNITED STATES & CANADA P.360, 1889 EGG
FOOD PLANTS
AMBROSIA TRIFIDA LINN.

GRADATA WALKER
SYN. SEE EREMIATA GUENEE

GRAEFI GROTE
SYN. SEE INNOTATA GUENEE

GRAEFIANA (GROTE), PYREFERRA-NOCTUIDAE
LIFE HISTORY
FOOD PLANTS
HAMAMELIS VIRGINIANA LINN.

GRANDIOSA HULST
SSP. SEE IMPLICATA GUENEE

GRANDIPUNCTA GUENEE
SYN. SEE ARGILLACEA HUBNER

GRANDIS (BOISDUVAL), POLIA-NOCTUIDAE
SYN. LIBERA WALKER
LIFE HISTORY
AMER. MUS. NAT. HIS. BULL. 14: 293, 1901 BEUTEN. LARVA
NEW YORK STATE CABIN. NAT. HIST. ANN. REPT. NO. 42:210 1889 LINTNER EGG, LARVA, PUPA
PAPILIO 3:17, 1883 THAXTER LARVA
ROYAL SOC. CANADA, TRANS. 1 SERIES 3:57, 1907 FLETCHER-GIBSON EGG, LARVA, PUPA
UNITED STATES NAT. MUS. PROC. 27:848, 1904 DYAR EGG, LARVA
FOOD PLANTS
ARCTIUM LAPPA LINN.
CHENOPODIUM ALBUM LINN.
POPULUS SP. IN CAPTIVITY

GRANDIS EHRMANN
SSP. SEE ANTIOPA LINNAEUS

GRANDIS SPEYER
SYN. SEE HARUSPICA GROTE

GRANDIS (STRECKER), COPABLEPHARON-NOCTUIDAE
SYN. SUBFLAVIDENS GROTE
LIFE HISTORY
PSYCHE 27:82, 1920 STRICKLAND LARVA, PUPA (BRIEF)
FOOD PLANTS

GRANITATA (GUENEE), SEMIOTHIS GEOMETRIDAE
A IRREGULATA WALKER
B QUADRISIGNATA WALKER
C DISPUNCTA WALKER
D RETINOTATA WALKER
E EXNOTATA WALKER
F SUBMARMORATA
FOR SYNS. SEE MCDUNNOUGH'S LIST
LIFE HISTORY
CAN. ENT. 62:112, 1930 MCDUNNOUGH LARVA (DISPUNCTA)
LOC. CIT. 75:134, 1943 MCGUFFIN LARVA* PL. 12, FF. A AND 1 A MANDIBLE
ENT. NEWS 5:63, 1894 DYAR LARVA

PUPA
New York State Agri. Exp. Sta. (Cornell), Mem. 274:46, 1948 Forbes larva
Psyche 11:105, 1904 Dyar egg, larva
United States Dept. Agri. Misc. Pub. 657:361, 429, 1950 Craighead larva
Index Amer. Econ. Ent. Vol. 7:875

FOOD PLANTS
Abies amabilis (Dougl.)
Abies balsamea (Linn.)
Abies lasiocarpa (Hook)
Larix laricina (DuRoi.)
Larix occidentalis Nutt.
Picea engelmanni (Parry)
Picea glauca (Moench)
Picea mariana (Mill.)
Picea sitchensis (Bong.)
Pinus strobus Linn.
Pseudotsuga menziesii (Mirb.)
Ribes sp.
Tsuga canadensis (Linn.)
Jack Fir

GRAPHICA Hubner, DRASTERIA-Noctuidae
SYN. CAPTICOLA Walker
SYN. MEDIA Morrison
SYN. FACETA Henry Edwards
A ATLANTICA Barnes & McDunnough
LIFE HISTORY
Brooklyn Ent. Soc. Bull. 27:80 1932 Crumb larva*Pl. 10G
Maine Agri. Exp. Sta. Bull. 356; 177, 1930 Phipps egg, larva, pupa (ATLANTICA)
Guenee-Spec. Gen. Lep. Noct. 3: 72, 1852 larva*Pl. 2, f. 10 color
FOOD PLANTS
Vaccinium sp.

GRASSATA Hulst
SYN. SEE BICOLORATA Fabricius

GRATA (Fabricius), EUTHISANOTIA-Noctuidae
SYN. ASSIMILIS Boisduval
LIFE HISTORY
Amer. Ent. 2:152, 1870 larva
Amer. Mus. Nat. Hist. Bull. 10: 358, 1898 Beuten. larva
Amer. Nat. 4:615, 1871 larva (brief note)
Can. Dept. Agri. Rept. Ent. p. 30 1888 Fletcher egg, larva*
Can. Ent. 7:41, 1875 Saunders egg* larva*
Essex Institute, Proc. 4:27 1864 Packard larva, pupa
Illinois Dept. Agri. Trans. 15: 178, 1877 French larva
Loc. cit. 18, Append: 111, 173, 1880 Coquillett-Marten egg* larva*
Minnesota Agri. Exp. Sta. Bull. 61:123, 1899 Lugger egg* f. 68; larva* f. 68 and Pl. 12, f. 68; pupa
Missouri, Rept. Ins. 2:83, 1870 Riley larva* pupa*
Missouri, loc. cit. 6:89, 1874 Riley egg, *larva*pupa*
New York State Cabin. Nat. Hist. Rept. No. 26:123, 1874 Lintner egg* larva*
Loc. cit. No. 42:182, 1889 Lintner egg* larva*
Loc. cit. No. 45:170, 1891 Lintner egg* larva*
New York State Mus. Mem. 8: 740, 1905 Felt larva
New York State, Rept. Ins. 3: 399, also sec. 123:81, 1865 Fitch larva
Ontario Ent. Soc. Ann. Rept. p. 33, 1871 Saunders egg, larva, pupa
Loc. cit. p. 38, 1875 Saunders egg* larva*
Loc. cit. p. 10, 1893 Fletcher larva*
Smithsonian Misc. Collections 4, art. 2:246, 1862 Morris larva

CRAGEN-OUR INS. FRIENDS & FOES P. 146, 1899 LARVA
FERNALD-IN KINGSLEY'S STANDARD NAT. HIST. 2:462, 1885 EGG, LARVA
GLOVER-ILLUST. NORTH AMER. ENT. 1878 LARVA* PL. 14, F. 13
HARRIS-ENT. CORRESP. P. 306, 1869 LARVA
HARRIS-INS. INJ. VEG. 1ST EDIT. P. 310, 1841 LARVA
HARRIS-LOC. CIT. 2ND EDIT. P. 329, 1852 LARVA
HARRIS-LOC. CIT. FLINT EDIT. P. 427, 1862 LARVA*
HART-INS. COLLECTIONS ILLINOIS HIGH SCHOOLS P. 38, 1903 EGG* LARVA* F. 52
MOTH BOOK P. 233, 1905 EGG* LARVA* F. 145
MORRIS-SYNOP. LEP. NORTH AMERICA P. 246, 1862 LARVA
PACKARD-GUIDE TO STUDY INSECTS P. 281, 1869 LARVA
SAUNDERS-INS. INJ. FRUITES P. 259, 1883 EGG* LARVA* PUPA
SAUNDERS-LOC. CIT. P. 258, 1889 EGG* LARVA* F. 269; PUPA
STRETCH-ILLUST. ZYGAEN. & BOMBYC. NORTH AMERICA P. 148, ADDENDA P. 233, 1873 LARVA
INDEX AMER. ECON. ENT. VOL. 1:124 EUDRYAS

FOOD PLANTS

CEPHALANTHUS OCCIDENTALIS LINN.
HUMULUS LUPULUS LINN. (HOPS)
PARTHENOCISSUS QUINQUEFOLIA (LINN.)
VITIS SP. (GRAPES)
VITIS VINIFERA LINN. (GRAPE)

GRATA (HUBNER), ELAPHRIA - NOCTUIDAE

SYN. RASILIS MORRISON
SYN. SUBUSTA DRUCE
SYN. ORINA GUENEE

LIFE HISTORY

CAN. ENT. 5:206, 1873 SAUNDERS LARVA (ORINA)
WASHINGTON ENT. SOC. PROC. 4: 321, 1898 DYAR LARVA
PACKARD-INS. INJ. FOREST TREES P. 47, 1881 LARVA

FOOD PLANTS

QUERCUS SP. (OAKS)
TARAXACUM OFFICINALE WEBER
VIOLA SP.

GRATARIA (FABRICIUS), HAEMATOPIS - GEOMETRIDAE

SYN. SANIARIA HUBNER
SYN. SUCCESSARIA WALKER
FORM ANNETTEARIA HAIMBACH

LIFE HISTORY

MISSOURI, REPT. INS. 1:179, 1859 RILEY LIFE HISTORY*
NEW YORK STATE AGRI. EXT. STA. (CORNELL), MEM. 274:119, 1948 FORBES LARVA, PUPA
PSYCHE 9:155, 1900 DYAR EGG, LARVA
GUMPPENBERG-SYSTEMA GEOMETRARUM 5:297, 1892 LARVA, PUPA
HART-INS. COLLECTIONS ILLINOIS HIGH SCHOOLS P. 50, 1903 EGG, LARVA, COCOON
PACKARD-MONOGRAPH GEOMETRIDAE P. 220, 1876 LARVA, PUPA

FOOD PLANTS

CERASTIUM ARVENSE LINN.
CERASTIUM VISCOSUM LINN.
POLYGONUM AVICULARE LINN.
POLYGONUM HYDROPIPER LINN.
POLYGONUM RAYII BAB.
STELLARIA MEDIA (LINN.)
TRIFOLIUM SP. (CLOVERS)

GRATATA PACKARD

SYN. SEE INCERTATA WALKER

GRATULATA (WALKER), MESOLEUCA - GEOMETRIDAE

SYN. BRUNNEICILIATA PACKARD

A LATIALBATA BARNES & McDUNNOUGH
LIFE HISTORY
PSYCHE 10:191,1903 DYAR EGG, LARVA
FOOD PLANTS
QUERCUS SP. (OAKS)

GRAVENOTATA KLOTS
SSP. SEE HETERONEA BOISDUVAL

GRAVILINEARIA ANDREWS
SYN. SEE VIRGINARIA CRAMER

GRAVIS HENRY EDWARDS
SSP. SEE WRIGHTII STRETCH

GREGONSONI GUNDER
SSP. SEE HYDASPE BOISDUVAL

GRETA SMITH
SYN. SEE PALLIGERA SMITH

GREYI FIELD
SSP. SEE NYCTEIS DOUBLEDAY & HEWITSON

GRISATRA SMITH
SSP. SEE ELIMATA GUENEE

GRISEA BARNES & McDUNNOUGH
SSP. SEE VAGANS BARNES & McDUNNOUGH

GRISEA COCKERELL
SSP. SEE OLIVIAE COCKERELL

GRISEA PACKARD
SYN. SEE NEXA BOISDUVAL

GRISEA STRECKER
SSP. SEE ELEGANS STRECKER

GRISEA WALKER, ACRONICTA-NOCTUIDAE
SYN. PUDORATA MORRISON
A REVELLATA SMITH
LIFE HISTORY
AMER. MUS. NAT. HIST. BULL. 4: 249, 1901 BEUTEN. LARVA
NEW YORK STATE MUS. MEM. 8: 724, 1905 FELT LARVA (BRIEF)
PAPILIO 3:131, 1883 EDWARDS-ELLIOTT LARVA
UNITED STATES NAT. MUS. PROC. 21:144, 1898 SMITH & DYAR LARVA
LOC. CIT. 27:802, 1904 DYAR EGG, LARVA (REVELLATA)
INDEX AMER. ECON. ENT. VOL. 7:8
FOOD PLANTS
ALNUS RUBRA BONG. (REVELLATA)
BETULA SP. (BIRCHES)
MALUS PUMILA MILL. (APPLE)
POPULUS SP. (POPLARS)
SALIX HOOKERIANA BARRATT
SALIX SCOULERIANA BARR.
ULMUS SP. (ELMS)
VIBURNUM DENTATUM LINN.

GRISEATA CASSINO & SWETT
SSP. SEE DESTINATA MOESCHLER

GRISEFACTA DYAR, OLENE-LIPARIDAE
LIFE HISTORY
FOOD PLANTS
ABIES SP. (FIRS)
CHAMAECYPARIS THYOIDES (LINN.)
JUNIPERUS SP. (JUNIPERS)
PICEA SP. (SPRUCE)
PINUS SP. (PINES)
TAXODIUM DISTICHUM (LINN.)
THUJA SP.
TSUGA CANADENSIS (LINN)

GRISEOCINCTA HARVEY
SSP. SEE VECORS GUENEE (ORTHODES)

GRISEOR DYAR
SSP. SEE INNOTATA GUENEE

GRISEOR McDUNNOUGH
SSP. SEE PENDULINARIA GUENEE

GROENLANDICA (HOMEYER), BYRDIA-LIPARIDAE
LIFE HISTORY
PSYCHE 8:153, 1897 DYAR

COMPARES PUPS WITH THAT OF ROSSI CURTIS AND DESCRIBES LARVA, LARVA DESCRIBED AS THAT OF ROSSI BY PACKARD IS PROBABLY THIS SPECIES

FOOD PLANTS

GROENLANDICA STAUDINGER
SYN. SEE INTERROGATIONIS LINNAEUS

GROSSBECKI STRAND
SYN. SEE BELLA LINNAEUS

GROSSBECKIATA SWETT
SYN. SEE MISERULATA GROTE

GROSSULA HERRICH-SCHAEFFER
SYN. SEE MACULATA EDWARDS

GROTEI BARNES & BENJAMIN
SSP. SEE ARCUATA WALKER

GROTEI BUTLER
SYN. SEE DIFFINIS BOISDUVAL

GROTEI HENRY EDWARDS, CAUTETHIA-SPHINGIDAE

LIFE HISTORY
ENT. AMER. 3:164, 1887 HW. EDW. PUPA
PSYCHE 7:385, 1896 DYAR LARVA
UNITED STATES NAT. MUS. PROC. 23:255, 1900 DYAR EGG, LARVA
GUNDLACH-ENT. CUBANA P. 179, 1881 LARVA

FOOD PLANTS
RUBIACEAE
SYMPHORICARPOS ALBUS (LINN.)

GROTEI PLOETZ
SYN. SEE HIANNA SCUDDER

GROTEI RILEY
SYN. SEE CINEROSA GROTE

GROTIANA BAILEY, CATOCALA-NOCTUIDAE
SYN. GEORGEANA BEUTENMULLER

LIFE HISTORY
AMER. MUS. NAT. HIST. BULL. 38: 163, 1918 BARNES & McDUNNOUGH EGG, LARVA
AMER. MUS. NAT. HIST. MEM. NEW SERIES 3, PART 1, 1918 BARNES & McDUNNOUGH LARVA* PL. 14, F. 4; PL. 15, F. 4; PL. 16, F. 15; PL. 17, F. 2 COLOR

FOOD PLANTS
POPULUS SP. (POPLARS)
SALIX SP. (WILLOWS)

GRUNDELI COOLIDGE
SYN. SEE CHALCEDONA DOUBLEDAY & HEWITSON

GRUNUS (BOISDUVAL), HABRODAIS-LYCAENIDAE

LIFE HISTORY
CAN. ENT. 25:94, 1893 DYAR LARVA CHRYSALIS
SOUTHERN CALIFORNIA ACAD. SCI. BULL. 34:81, 1935 COMSTOCK-DAMMERS LARVA*
HOLLAND-BUTTERFLY BOOK P. 238, 1905 LARVA, CHRYSALIS
HOLLAND-LOC. CIT. 3RD EDIT. P. 234, 1931 LARVA, CHRYSALIS

FOOD PLANTS
QUERCUS CHRYSOLEPIS LIEBM.

GRYNEA (CRAMER), CATOCALA-NOCTUIDAE
SYN. POLYGAMA GUENEE
SYN. NUPTULA WALKER
AB CONSTANS HULST

LIFE HISTORY
AMER. MUS. NAT. HIST. BULL. 26: 382, 1902 BEUTEN. LARVA* F. 2 HEAD
AMER. MUS. NAT. HIST. MEM. NEW SERIES 3, PART 1, 1918 BARNES & McDUNNOUGH LARVA* PL. 10, F. 37; PL. 13, F. 8 COLOR
BROOKLYN ENT. SOC. BULL. 4: 22, 1881 KOEBLE LARVA
LOC. CIT. BULL. 7:36, 1884 HULST

LARVA
CAN.ENT.2:30,1869 REED LARVA (POLYGAMA)
LOC.CIT.8:74,1876 SAUNDERS LARVA
ENT.NEWS 21:450,1910 ROWLEY LARVA
ILLINOIS DEPT.AGRI.TRANS.18, APPEND:183,1880 COQUILLETT LARVA
MINNESOTA AGRI.EXP.STA.BULL. 61:228,1899 LUGGER LARVA
NEW YORK STATE MUS.MEM.8: 705,743,1905 FELT LARVA (BRIEF)

FOOD PLANTS
CRATAEUGS SP. (THORNS)
MALUS CORONARIA(LINN.)
MALUS PUMILA MILL. (APPLE)
PRUNUS DOMESTICA LINN. (PLUM)
PYRUS SP.
QUERCUS SP. (OAKS)
SALIX SP. (WILLOWS)

GRYNEUS HUBNER
SYN. SEE DAMON CRAMER

GUENEATA PACKARD, CERATODALIS-GEOMETRIDAE
SYN. EXCURVATA GROTE
LIFE HISTORY
PSYCHE 10:194,1903 DYAR EGG, LARVA
FOOD PLANTS
POLYGONUM SP.

GUENEEATA PACKARD
SYN. SEE CUSTODIATA GUENEE

GUENEEI AVINOFF
SSP. SEE NASTES BOISDUVAL

GUENEI GROTE
SYN. SEE MAESTOSA HULST(CATOCALA)

GULARIS GROTE
SSP. SEE OCHROGASTER GUENEE

GULOSA HENRY EDWARDS
SSP. SEE VETUSTA BOISDUVAL

GUNDERI COMSTOCK
SSP. SEE GABBII BEHR

GUNDERI FIELD
SSP. SEE EURYNOME EDWARDS

GUNDERI FIELD
SSP. SEE LORQUINI BOISDUVAL (BASILARCHIA)

GUNDERI HILL
SSP. SEE HERA HARRIS

GUNDERI INGHAM
SSP. SEE SARA BOISDUVAL

GUNDERI INGHAM
SSP. SEE BECKERII EDWARDS

GUNDERI RUDKIN
SSP. SEE HELLOIDES BOISDUVAL

GUNDERIAE HOLLAND
SSP. SEE POLA BOISDUVAL

GUSTATA DYAR
SYN. SEE BUXEA GROTED

GUTTATA(BOISDUVAL), PLATYPREPIA-ARCTIIDAE
FORM VIRGINALIS BOISDUVAL
FORM OCHRACEA STRETCH
LIFE HISTORY
SOUTHERN CALIFORNIA ACAD.SCI. BULL.34:149,1935 COMSTOCK LARV A*PL.41 OCHRACEA
PACKARD-HAYDEN'S SURVEY TERRIT. P.559,1873 LARVA (BRIEF)
STRETCH-ZYGAEN.&BOMBYC NORTH AMERICA P.71,1873 LARVA* PUPA*PL.10,F.1 COLOR
FOOD PLANTS
LUPINUS ARBOREUS SIMS

GUTTATA HULST
SYN. SEE PINIATA PACKARD(PARAPHIA)

GUTTIFERA HERRICH-SCHAEFFER
SYN. SEE MACULATA HARRIS

GUTTIVITTA(WALKER), HETEROCAMPA-
NOTODONTIDAE
SYN. ALBIPLAGA WALKER
SYN. UCOREA HERRICH-SCHAEFFER
SYN. CINEREUS PACKARD
SYN. HARRISI PACKARD
SYN. INDETERMINATA WALKER
SYN. DOUBLEDAYI SCUDDER

LIFE HISTORY

AMER. MUS. NAT. HIST. BULL. 10: 425,1898 BEUTEN. LARVA
BOSTON SOC. NAT. HIST. PROC. 24: 544,1890 PACKARD LARVA*
JOURN. AGRI. RESEARCH 32:690 1926 COLLINS EGG, LARVA
MAINE AGRI. EXP. SAT. BULL. 259:72,1917 MOSHER PUPA* F. 6 C, D
MAINE-LOC. CIT. BULL. 161:321 1908 PATCH EGG, LARVA* FF. 14, 15-22(VARIATIONS); PUPA
MICHIGAN STATE COLLEGE, SPEC. BULL. 243:20,1933 MCDANIEL LIFE HISTORY LARVA*
MINNESOTA AGRI. EXP. STA. BULL. 61:171,1899 LUGGER LARVA* F. 118
NAT. ACAD. SCI. MEM. 7:231,1895 PACKARD LARVA* PL. 31, FF. 1-7; PL. 33, FF. 1-3 COLOR; PUPA*
LOC. CIT. 9,1905 PACKARD LARVA* PL. 44, F. 11 HEAD
NEW YORK ENT. SOC. JOURN. 3:29 1895 PACKARD LARVA* PL. 1, F. 3
NEW YORK STATE AGRI. EXP. STA. (CORNELL), MEM. 274:232, 1948 FORBES LARVA
NEW YORK STATE MUS. BULL. 124; 21,1908 FELT LARVA, PUPA
NEW YORK STATE MUS. MEM. 8: 727,1905 FELT LARVA
ONTARIO ENT. SOC. ANN. REPT. P. 48,1917 CORCORAN LARVA
PSYCHE 6:178,1891 DYAR LARVA, PUPA, COCOON
UNITED STATES DEPT. AGRI. MISC. PUB. 657;363,403,1950 CRAIGHEAD LARVA* F. 81 B
DOANE ETC-FOREST INSECTS P. 295,1936 LARVA
HERRICH-INS. ENEMIES SHADE TREES P. 43,1935 LARVA
KELLOGG-AMER. INSECTS P. 304, 1908 LARVA* F. 564
SLINGERLAND-CROSBY--MANUEL FRUIT INS. P. 127,1914 EGG, LARVA, PUPA
INDEX AMER. ECON. ENT. VOL. 1:154; VOL. 2:174; VOL. 3:194, VOL. 4:219; VOL. 5:298, VOL. 6:350; VOL. 7:461

FOOD PLANTS

ACER SP. (MAPLES)
ACER RUBRUM LINN.
ACER SACCHAROPHORUM KOCH
ACER SPICATUM LAM.
BETULA SP. (BIRCHES)
CARYA SP. (HICKORIES)
CASTANEA DENTATA(MARSH)
CORYLUS SP. (HAZELS)
FAGUS SP. (BEECHES)
HAMAMELIS VIRGINIANA LAM.
JUGLANS SP.
MALUS PUMILA MILL. (APPLE)
OSTRYA VIRGINIANA(MILL.)
POPULUS SP. (POPLARS)
PRUNUS CERASUS LINN. (CHERRY)
PRUNUS SEROTINA EHRH.
PYRUS SP.
PYRUS COMMUNIS LINN. (PEAR)
QUERCUS SP. (OAKS)
RUBUS ALLEGHENIENSIS PORTER
SPIRAEA SP.
ULMUS SP. (ELMS)
VIBURNUM SP.

GUYASUTA CHERMOCK & CHERMOCK
SSP. SEE HAMAMELIS GUENEE

GYAS(EDWARDS), HEMIARGUS-LYCAENIDAE
SYN. ASTRAGALA WRIGHT
SYN. FLORENCIA CLEMENS
A ZACHAEINA BUTLER & DRURY
LIFE HISTORY
SOUTHERN CALIFORNIA ACAD. SCI. BULL. 33:139, 1934 COMSTOCK-DAMMERS LIFE HISTORY EGG* PL. 42; LARVA* PL. 43; PUPA* PLS. 43, 44
COMSTOCK, J. A.-BUTTERFLIES OF CALIFORNIA P. 181, 1827 EGG
FOOD PLANTS
ASTRAGALUS CROTALARIAE(BENTH.)
MEDICAGO SATIVA LINN.
PROSOPIS SP.

GYASALIS WALKER
SYN. SEE PYRAMUSALIS WALKER

GYGES HEWITSON
SYN. SEE FRISIA POEY

GYNDNDRATA PEARSALL
SSP. SEE DESIGNATA HUFNAGEL

H

HABILIS GROTE, CATOCALA-NOCTUIDAE
SYN. FEMALE BASALIS GROTE
LIFE HISTORY
AMER. MUS. NAT. HIST. BULL. 16: 389, 1902 BEUTEN. LARVA* PL. 52, F. 12 AS SERENA EDWARDS
LOC. CIT. 38:150, 1918 BARNES & MCDUNNOUGH EGG, LARVA
AMER. MUS. NAT. HIST. MEM. NEW SERIES 3, PART 1, 1918 BARNES & MCDUNNOUGH EGG, LARVA* PL. 10, F. 24; PL. 12, F. 1 COLOR
CAN. ENT. 33:226, 1901 DODGE LARVA
ENT. AMER. 2:46, 1886 KELLICOTT LARVA
ENT. NEWS 2:94, 1891 BUNKER LARVA(BRIEF)
LOC. CIT. 20:134, 1909 ROWLEY LARVA(LIKE ANGUSI GROTE)
NEW YORK STATE MUS. MEM. 8: 716, 1905 FELT LARVA
FOOD PLANTS
CARYA SP. (HICKORIES)
JUGLANS NIGRA LINN.

HADENIFORMIS BEHR
SSP. SEE JUCUNDA HUBNER

HAESITATA(GUENEE), TRIPHOSA-GEOMETRIDAE
FORM PUSTULARIA HENRY EDWARDS
A AFFIRMARIA WALKER
SYN. INDUBITATA GROTE
SYN. DUBITATA PACKARD
LIFE HISTORY
PSYCHE 9:59, 1900 DYAR EGG, LARVA
FOOD PLANTS
BERBERIS VULGARIS LINN.
CRATAEGUS SP. (THORNS)
PISONIA ACULEATA LINN.
PRUNUS DOMESTICA LINN. (PLUM IN CAPTIVITY)
RHAMNUS SP. (IN CAPTIVITY)
RHAMNUS CATHARTICA LINN.

HAGENII EDWARDS
SYN. SEE EURYTHEME BOISDUVAL

HAGENI(GROTE), ISOGRAMMA-SPHINGIDAE
LIFE HISTORY
KANSAS ENT. SOC. JOURN. 17:29 1944 STALLINGS-TURNER LARVA(BRIEF0
STRECKER=LEP. RHOP. ET HETER. 14:127, 1877 LARVA
FOOD PLANTS
MACLURA POMIFERA(RAF.)

HALESUS(CRAMER), ATLIDES-LYCAENIDAE
SYN. DOLICHOS HUBNER
SYN. JUANITA SCUDDER
AB CORCORANI GUNDER
LIFE HISTORY

California Acad. Sci. Proc. 5: 331, 1873 Hy. Edw. larva, chrysalis
Ent. Amer. 3:162, 1887 Hy. Edw. chrysalis
Ent. News 44:71-74, 1933 Haskins life history
Smithsonian Misc. Collections 4, art. 2:92, 1862 Morris larva, chrysalis
Southern California Acad. Sci. Bull. 30:88, 1931 Comstock egg* see page 87 for figure
Boisduval-Leconte-Hist. Lep. North America p. 86, 1883 larva*chrysalis*Pl. 25 COLOR
Comstock, J. A-Butter. California P. 154, 1927 larva* f. 48; Cbrysalis* f. 49o
Comstock, J. H.-How Know Butter. p. 220, 1904 larva
French-Butter. East. United States p. 256, 1886 larva, chrysalis
Morris-Synop. Lep. North America p. 92, 1862 larva, chrysalis
Scudder-Butter. East. United States & Canada p. 1828 1889 larva, chrysalis

FOOD PLANTS
Phoradendron californicum Nutt.
Phoradendron serotinum (Raf.)
Phoradendron villosum Nutt.
Quercus sp. (oaks)
Quercus incana Bartr.

HALIATA Guenee
?syn. see GRANITATA Hubner

HALIATA Walker
syn. see GRANITATA Hubner

HALLI Cook & Watson
ssp. see ARCHIPPUS Cramer

HAMADRYAS Harris
syn. see INFANS Moeschler

HAMAMAELIDIS Scudder
syn. see ICELUS Scudder & Burgess

HAMAMELIS Guenee, ACRONICTA-Noctuidae
a GUYSAUTA Chermock & Chermock
LIFE HISTORY
Amer. Mus. Nat. Hist. Bull. 14: 256, 1901 Beuten. larva
Boston Soc. Nat. Hist. Proc. 27:132, 1896 Dyar larval structure
Can. Ent. 9:61, 1877 Goodell larva, pupa
Illinois Dept. Agri. Trans. 18: Append: 131, 1880 Marten larva
New York State Mus. Mem. 8: 704, 1905 Felt larva
United States Nat. Mus. Proc. 21:143, 1898 Smith & Dyar larva
Guenee-Spec. Gen. Lep. Noct. 1:53, 1852 larva*Pl. 2 COLOR
Hart-Ins. Collections Illinois High Schools p. 25, 1903 cocoon

FOOD PLANTS
Aesculus hippocastanum Linn.
Betula sp. (birches)
Castanea dentata (Marsh.)
Hamamelis virginiana Linn.
Quercus sp. (oaks)

HAMARIA Guenee
ssp. see DUARIA Guenee

HAMPSONI Dyar
ssp. see PICTA Packard

HANHAMI Hulst
syn. see ROTUNDOPENNATA Packard

HANHAMI Smith
ssp. see EVELINA French (Fishis)

HANNO (STOLL), HEMIARGUS-LYCAENIDAE
SYN. ANTIBUBASTUS HUBNER
SYN. PSEUDOPTILETA BOISDUVAL & LECONTE
LIFE HISTORY
ENT. NEWS 44:156, 1933 HASKIN EGG, LARVA, CHRYSALIS
FOOD PLANTS
ABRUS PRECATORIUS LINN.
CASIA ALATA LINN.
CASSIS NICTITANS LINN.

HARFORDII HENRY EDWARDS, COLIAS-PIERIDAE
FORM FEMALE MARTINI GUNDER
AB WEARERAE GUNDER
LIFE HISTORY
PAPILIO 4:2, 1884 EDWARDS EGG, LARVA, CHRYSALIS
SOUTHERN CALIFORNIA ACAD. SCI. BULL. 25:63, 64, 1926 COMSTOCK EGG*LARVA* CHRYSALIS*
COMSTOCK, J. A.-BUTTER. CALIFORNIA P. 53, 1927 EGG*LARVA* CHRYSALIS* PL. A23
EDWARDS-BUTTER. NORTH AMERICA 3RD SERIES, 1897 LIFE HIST* PL. COLIAS 2, FF. A-G
FOOD PLANTS
ASTRAGALUS CROTALARIAE (BENTH.)
TRIFOLIUM REPENS LINN

HARPERI GUNDER
SSP. SEE NASTES BOISDUVAL

HARPERI GUNDER
SSP. SEE BATESII REAKIRT

HARISI (GROTE), PAPAIPEMA-NOCTUIDAE
AB RUBIGINOSA BIRD
AB MULIERIS STRAND
LIFE HISTORY
CAN. ENT. 32:282, 1900 BIRD LARVA
LOC. CIT. 43:37, 1911 BIRD LARVA
LOC. CIT. 43:40, 1911 BIRD LARVA (RUBIGINOSA)
FOOD PLANTS
ANGELICA ATROPURPUREA LINN. (RUBIGINOSA)
ARCTIUM LAPPA LINN. (RUBIGINOSA)
HERACLEUM MAXIMUM BARTR.

HARRISI OBERTHUR
SYN. SEE EGLANTERINA BOISDUVAL

HARRISI PACKARD
SYN. SEE GUTTIVITTA WALKER

HARRISII CLEMENS
SYN. SEE BOMBYCOIDES WALKER

HARRISII DYAR
SYN. SEE LECONTEI GUERIN-MENEVILLE

HARRISII EDWARDS
SYN. SEE COMMA HARRIS

HARRISII SCUDDER, MELITAEA-NYMPHALIDAE
A ALBIMONTANA AVINOFF
B LIGGETTI AVINOFF
LIFE HISTORY
AMER. MUS. NAT. HIST. BULL. 5: 262, 1893 BEUTEN. LARVA CHRYAALIS
CAN. ENT. 2:163, 1869 EDWARDS LARVA
LOC. CIT. 4:161, 1872 SAUNDERS EGG, LARVA (?NYCTEIS)
LOC. CIT. 9:165, 1877 EDWARDS EGG, LARVA, CHRYSALIS
ILLINOIS DEPT. AGRI. TRANS. 18, APPEND: 162, 1880 COQUILLETT LARVA
NATURALISTE CANADIEN P. 133, 1927 HUARD LARVA
PAPILIO 4:58, 1884 EDWARDS LARVAL VARIETY
COMSTOCK-HOW KNOW BUTTER. P. 126, 1904 LARVA
FRENCH-BUTTER. EAST. UNITED STATES & CANADA P. 170, 1886 LARVA* CHRYSALIS*
HOLLAND-BUTTERFLY BOOK P. 144, L 1904 LARVA, CHRYSALIS* PL. 5, FF. 17, 18

Holland-loc.cit.3rd.edit.p. 127,1932 life history, chrysalis* Pl.5
Maynard-Butter.New England p.29,1886 larva,chrysalis
Scudder-Butter.East.United States & Canada pp.677-682,1889 life history*Pl. 64,f.41;Pl.67,f.15;Pl.72, f.1;Pl.75,ff.13,14,Pl.79, ff.20-24;Pl.82,f.8;Pl.86, ff.54-57;Pl.84,ff.17,18
Scudder-Butter.Their Structure etc.p.32,1881 chrysalis
Scudder-Guide to Common Butter.p.68,1893 larva, chrysalis

FOOD PLANTS
Actinomeris alternifolia(Linn.)
Aster sp.
Aster umbellatus Mill.
Helianthus sp.
Verbesina helianthoides Michx.

harrisii Walsh,Halisidota-Arctiidae
LIFE HISTORY
Nantucket Marie Mitchell Assoc. Publication Vol.3,No.2:87, 1930 Johnson larva(brief)
New York Acad.Sci.Annals 8: 71,1893 Packard larva
New York State Mus.Mem.8: 71,1903 Felt larva
Psyche 6:163,1891 Dyar egg, larva,pupa, cocoon
United States Dept.Agri.Misc, Pub.657:356,387,1950 Craighead larva
Herrick-Ins.Enemies Shade Trees p.125,1935 larva

FOOD PLANTS
Carum carvi Linn.
Carya sp.(hickories)
Platanus occidentalis Linn.

hartmanni French
syn.see laetifica Lintner

haruspica(Grote) Pseudospaelotis-Noctuidae
unimacula Morrison
syn.grandis Speyer
a sierrae Harvey
b inopinatus Smith
LIFE HISTORY
United States Nat.Mus.Proc. 7:828,1904 Dyar egg, larva (sierrae)
FOOD PLANTS
Malus pumila Mill.(apple) (inopinatus)

harveyata Cassino & Swett
ssp.see incursata Hubner

harveyata Taylor
ssp.see destinata Moeschler

harveyi Grote
syn.see albilinea Hubner

hasdrubal Cramer
syn.see tetrio Linnaeus

hastata(Linnaeus),Eulype-Geometridae
a gothicata Guenee
syn.furcifascia Walker
b sughastata Nolcken
ab moestata Nolcken
LIFE HISTORY
New York State Agri.Exp.Sta. (Cornell),Mem.274:146, 1948 Forges larva
New York State Mus.Mem.8: 749,1905 Felt larva
United States Dept.Agri.Misc. Publ657:353,428,1950 Craighead larva
Duncan-Natural.Library 4:256 1836 larva(brief)
Fernald-Kingsley's Stand.Nat. Hist.2:447,1885 larva*
Humphrey-Genera British Moths. p.113,1859 larva*P.43 color
Kirby-European Butter.&Moths

P. 380,1882 LARVA
NEWMAN-BRITISH MOTHS P.157, 1869 LARVA
PACKARD-MONOGRAPH GEOMETRIDAE P.165,1876 LARVA
STAINTON-BRITISH BUTTER.& MOTHS 2:100,1859 LARVA
STEPHENS-ILLUST.BRITISH ENT. HAUSTELL.3:249 LARVA
WOOD-INSECTS AT HOME P. 458,1872 LARVA
INDEX AMER.ECON.ENT. VOL.7:366

FOOD PLANTS
BETULA SP. (BIRCHES)
COMPTONIA PEREGRINA(LINN.)
MYRICA CERIFERA LINN.
MYRICA GALE LINN.
RHODODENDRON HIRSUTUM LINN.
SALIX SP. (WILLOWS)
VACCINIUM ULIGINOSUM LINN.

HASTINGSI HENRY EDWARDS
SSP. SEE MIRIFICA HENRY EDWARDS

HASTULIFERA(ABBOT & SMITH), ACRONICTA-NOCTUIDAE

LIFE HISTORY
AMER.MUS.NAT.HIST.BULL.14: 239,1901 BEUTEN. LARVA
AMER.PHIL.SOC.PROC.31,1893 PACKARD LARVA*PL.11,F.20 (NO DESCR.)
NEW YORK STATE CABIN.NAT. HIST.ANN.REPT.NO.26:158 1874 LINTNER LARVA
NEW YORK STATE MUS.MEM.8: 750,1905 FELT LARVA
UNITED STATES NAT.MUS.PROC. 21:49,1898 SMITH & DYAR LARVA* PL.6,FF.12,13 COLORED OR UNCOLORED; PUPA, COCOON
ABBOT & SMITH-LEP.INS.GEORGIA 2:183,1797 LARVA*PL.92 COLOR
ELIOT & SOULE-CATER.AND MOTHS P.298,1902 LARVA*
GUENEE-SPEC.GEN.LEP.NOCT. 1:47,1852 LARVA

FOOD PLANTS
ACER SP. (MAPLES)
AESCULUS SP.
AESCULUS HIPPOCASTANUM LINN.
ALNUS SP. (ALDERS)
CASTANEA DENTATA(MARSH.)
SALIX SP. (WILLOWS)
TILIA AMERICANA LINN.
ULMUS SP. (ELMS)

HATUI BARNES & BENJAMIN
SSP. SEE ALEXANDRA EDWARDS

HAUSTA GROTE
SYN. SEE SEMICANA WALKER

HAYDENATA(PACKARD), SOMATOLOPHIA-GEOMETRIDAE
SYN. UMBRIPENNIS HULST

LIFE HISTORY
UNITED STATES NAT.MUS.PROC. 25:390,1903 DYAR LARVA

FOOD PLANTS
POLYGONUM SP.
PRUNUS SEROTINA EHRH.

HAYHURSTII(EDWARDS), PHOLISORA-HESPERIIDAE

LIFE HISTORY
INSECT LIFE 3:389,1891 DYAR LARVA, CHRYSALIS
SCUDDER-BUTTER.EAST.UNITED STATES & CANADA PP.1857 1858,1889 LIFE HISTORY

FOOD PLANTS
ALTERNANTHERA FLAVESCENS HBK

HERBRAEICUM HUBNER, POLYGRAMMATE-NOCTUIDAE

LIFE HISTORY
UNITED STATES NAT.MUS.PROC. 21:9,1898 DYAR LARVA, PUPA

FOOD PLANTS
NYSSA SYLVATICA MARSH.

HECLA LEFEBRE, COLIAS-PIERIDAE
A GLACIALIS McLACHLAN
SYN. CHRYSOTHEMOIDES VERITY
B HELA STRECKER
C PALLIDA SKINNER
LIFE HISTORY
ENT. NEWS 46:58, 1935 KLOTS OVOVIVPARITY
ENTOMOLOGIST 47:82-85, 1914 SHELDON EGG, LARVA
FOOD PLANTS
ASTRAGALUS ALPINUS LINN.

HEDYSARUM SCUDDER
SYN. SEE LYCIDAS ABBOT & SMITH

HEGESIA (CRAMER), EUPTOIETA-NYMPHALIDAE
LIFE HISTORY
NEW YORK ENT. SOC. JOURN. 9:79 1901 SWAINSON LARVA
FOOD PLANTS
OENOTHERA SPECIOSA NUTT.

HEGON (SCUDDER), AMBLYSCIRTES-HESPERIIDAE
SYN. SAMOSET SCUDDER
SYN. NEMORIS EDWARDS
LIFE HISTORY
HOLLAND-BUTTERFLY BOOK 1905 CHRYSALIS* PL. 6, F. 45
COLOR SAMOSET
SCUDDER-BUTTER. EAST. UNITED STATES & CANADA PP. 1590 1592, 1889 LARVA*CHRYSALIS* PL. 77, F. 29; PL. 85, F. 45
FOOD PLANTS
SORGHUM BICOLOR (LINN.)

HEILIGBRODTI (HARVEY), ADELOCEPHALA-CITHERONIIDAE
A HUBBARDI DYAR
LIFE HISTORY
NEW YORK ENT. SOC. JOURN. 14: 131, 1906 KUNZE LARVA
SOUTHERN CALIFORNIA ACAB. SCI. BULL. 46:72, 1947 COMSTOCK EGG, LARVA* PL. 16; PL. 17A; PUPA* PL. 17 B, C, D HUBBARDI
WASHINGTON ENT. SOC. PROC. 4: 428, 1900 DYAR LARVA
FOOD PLANTS
CERCICIUM FLORIUM BENTH.
CERCIDIUM MICROPHYLLUM (TORR.)
PROSOPIS GLAUNDULOSA TORR.

HELA STRECKER
SSP. SEE HECLA LEFEBRE

HELCITA BOISDUVAL
SYN. SEE HOFFMANNI BEHR.

HELENA CASSINO
SYN. SEE UNIJUGA WALKER

HELENA CASSINO
SSP. SEE FLAVICARIA PACKARD

HELENA HULST
SSP. SEE LIBERARIA WALKER

HELENA REAKIRT
SYN. SEE EURYDICE BOISDUVAL

HELENA TAYLOR
SSP. SEE TOGATA HUBNER

HELENAE DOS PASSOS
SSP. SEE AUGUSTUS KIRBY

HELENE PILATE
SSP. SEE MICRONYMPHA GUENEE

HELICTA HUBNER
SYN. SEE PHOCION FABRICIUS (NEONYMPHA)

HELIOS EDWARDS
SSP. SEE ICARIOIDES BOISDUVAL

HELIOTHIDATA AUCT.
SEE NIGROCOMINA WARREN

HELIUSALIS WALKER
SYN. SEE FLAVIPUNCTALIS GEYER

HELLOIDES (BOISDUVAL), LYCAENA-LYCAENIDAE

SYN. CASTRO BEAKIRT
AB GUNDERI RUDKIN
AB WILLIAMSI GUNDER
AB STERNITZKYI GUNDER
A FLORUS EDWARDS
FORM FEMALE HULBIRTI FIELD
FORM FEMALE STERNITZKYI FIELD
LIFE HISTORY
ENT. NEWS 35:306,1924 COOLIDGE LIFE HISTORY
SOUTHERN CALIFORNIA ACAD. SCI. BULL. 28:55,1929 COMSTOCK LARVA* PUPA*
COMSTOCK, J. A.-BUTTER CALIFORNIA P. 174,1927 EGG*PL. A 56; LARVA, CHRYASLIA
MEAD-REPT. LEP. WHEELER'S EXPED. WEST 100 MERID. P. 780,1875 LARVA MAY BE THIS SPECIES
FOOD PLANTS
ERIOGONUM SPERGULINUM GRAY
GAYOPHYTUM DIFFUSUM T. & G.
OENOTHERA BINNIS LINN.
POLYGONUM AVICULARE LINN.
POLYGONUM AMPHIBUM LINN
RUMEX SP.

HELVIA SCUDDER
SSP. SEE ANICIA DOUBLEDAY & HEWITSON

HEMIFLAVA FIELD
SSP. SEE AUSONIDES BOISDUVAL

HEMIFUSA GUNDER
SSP. SEE PALLA BOISDUVAL

HEMILUTEOFUSCUS GUNDER
SYN. SEE CHALCEDONIA DOUBLEDAY & HEWITSON

HEMIMELANICA COMSTOCK
SYN. SEE CHALCEDONIA DOUBLEDAY & HEWITSON

HENNEI GUNDER
SSP. SEE CHALCEDONIA DOUBLEDAY & HEWITSON

HENRICI (GROTE), SIMYRA-NOCTUIDAE
FORM EVANIDA GROTE
FORM FUMOSA MORRISON
NOTE: THIS SPECIES IF OFTEN LISTED AS ARSILONCHE ALBOVENOSA GOEZE
LIFE HISTORY
AMER. MUS. NAT. HIST. BULL. 14: 261,1901 BEUTEN LARVA ALBOVENOSA
BOSTON SOC. NAT. HIST. PROC. 27: 133,1896 DYAR LARVAL SETAE ALBOVENOSA
CAN. ENT. 12:45,1880 COQUILLETT LARVA
ENT. AMER. 3:171,1887 HY. EDW. LARVA
ILLINOIS DEPT. AGRI. TRANS. 18, APPEND:171,1880 COQUILLETT' LARVA
NEW YORK STATE AGRI. EXP. STA. (CORNELL), MEM. 47:484, 1921 CLASSEN EGG*PL. 46, F. 61; LARVA*PL. 46, F. 62; PUPA*PL. 46, F. 63 ALBOVENOSA
NEW YORK STATE MUS. MEM. 8: 743,1905 FELT LARVA (BRIEF)
PSYCHE 1:188,1876 THAXTER LARVA (BRIEF-ALBOVENOSA)
SOUTHERN CALIFORNIA ACAD. SCI. BULL. 33:143,1934 COMSTOCK & DAMMERS LARVA*PLS. 48, 49,50, PUPA*PL. 50
UNITED STATES NAT. MUS. PROC. 21.176,1898 SMITH & DYAR LARVA, PUPA, COCOON
INDEX AMER. ECON. ENT.
VOL. 2:41 ARSILONCHE ALBOVENOSA
VOL. 3:45 ARSILONCHE ALBOVENOSA
VOL. 7:880
FOOD PLANTS
CAREX SP. (SEDGES)
POLYGONUM HYDROPIPER LINN.
POLYGONUM LAPATHIFOLIUM LINN.

POPULUS SP. (POPLARS)
SALIX SP. (WILLOWS)
TYPHA SP.
TYPHA LATIFOLIA LINN.
GRASSES

HENRICI (GROTE & ROBINSON), INCISALIS-LYCAENIDAE
A SOLATUS COOK & WATSON
LIFE HISTORY
CAN. ENT. 39:181, 229, 1907 COOK LIFE HISTORY EGG* LARVA, CHRYSALIS*
PAPILIO 1:150, 1881 HY. EDW. EGG, LARVA, CHRYSALIS
FERNALD-BUTTER. MAINE P. 81 1884 LIFE HISTORY
FRENCH-BUTTER. EAST. UNITED STATES P. 274, 1886 LIFE HISTORY
FOOD PLANTS
CERCIS CANDENSIS LINN.
CYRILLA RACEMIFLORA LINN.
GAYLUSSACIA BOACCATA (WANG.)
ILEX SP.
LEUCOTHOE RACEMOSA (LINN.)
LUPINUS PERENNIS LINN.
PRUNUS SP.
PRUNUS DOMESTICA LINN. (PLUM FRUIT)
PRUNUS SEROTINA EHRH.
VACCINIUM CORYMBOSUM LINN.
VACCINIUM VACILLANS TORR.

HENRIETTA (SMITH), EUXOA-NOCTUIDAE
FORM ADUSTA BARNES & MCDUNNOUGH
LIFE HISTORY
UNITED STATES NAT. MUS. PROC. 22:453, 1900 SMITH LARVA
FOOD PLANTS
ROOTS OF MANY PLANTS

HENSHAWI (EDWARDS), NEONYMPHA-SATYRIDAE
LIFE HISTORY
EDWARDS-BUTTER. NORTH AMERICA 3RD SERIES 1897 EGG* (NEONYMPHA 1, F. 1)
HOLLAND-BUTTERFLY BOOK P. 202 1905 EGG
HOLLAND-LOC. CIT. 3RD EDIT. P. 179, 1931 EGG
FOOD PLANTS

HENSHAWI SWETT
SSP. SEE AUTUMNATA GUENEE

HEPBURNI GODMAN & SALVIN
SSP SEE PALMERII EDWARDS

HERA (HARRIS), PSEUDOHAZIS-SATURNIIDAE
SYN. PICA WALKER
SYN. CHRYSOCARENA HARRIS
A MARCATA NEUMOEGEN
AB GUNDERI HILL D
LIFE HISTORY
NAT. ACAD. SCI. MEM. 12:138 1914 PACKARD LARVA* PL. 25, F. 4 COLOR
FOOD PLANTS
ARTEMISIA SP.

HERBACEA GUENEE
SYN. SEE PRASINA SCHIFFENMULLER

HERBARIA (FABRICIUS), RACHEOSPILA-GEOMETRIDAE
SYN. SITELLARIA GUENEE
SYN. CONGRUATA WALKER
A HULSTIANA DYAR
LIFE HISTORY
WASHINGTON ENT. SOC. PROC. 12: 177, 1910 RUSSELL EGG, LARVA, PUPA (SITELLARIA)
FOOD PLANTS
PHYTOLACCA AMERICANA LINN.
PISONIA ACULEATA LINN.

HERBIMACULA GUENEE
SYN. SEE RENIGERA STEPHENS

HERILIS (GROTE), FELTIA-NOCTUIDAE
LIFE HISTORY
ENT. SOC. AMERICA, ANNALS 23: 396 1930 HAWKINS LARVAL CLAWS
ILLINOIS DEPT. AGRI. TRANS. 15:

90,205,1877 FRENCH & THOMAS LARVA
ILLINOIS DEPT.AGRI.TRANS.18, APPEND:133,1880 MARTEN LARVA
INDEX AMER.ECON.ENT. VOL.1:134
FOOD PLANTS
VEGETABLE AND FIELD CROPS

HERMES(EDWARDS),LYCAENA-LYCAENIDAE
SYN.DEL-SUD WRIGHT
LIFE HISTORY
SOUTHERN CALIFORNIA ACAD.SCI. BULL.34:124,1935 COMSTOCK-DAMMERS LIFE HISTORY EGG* PL.19; LARVA*PL.20; CHRYSALIS* PL.21
FOOD PLANTS
RHAMNUS CROCEA NUTT.

HERMES(FABRICIUS),MEGISTO-SATYRIDAE
SYN.CANTHE HUBNER
FORM SOSYBIUS FABRICIUS
LIFE HISTORY
CAN.ENT.9:229,1877 EDWARDS EGG,LARVA, CHRYSALIS
ILLINOIS DEPT.AGRI.TRANS.18, APPEND:91,1880 MIDDLETON EGG,LARVA, CHRYSALIS
MUS.COMP.ZOOL.BULL.87,NO. 6,1941 DETHIER LARVAL ANTENNA* F.35 SOSYBIUR
BOISDUVAL & LECONTE-HIST.LEP. NORTH AMERICA 1833 LARVA* CHRYZALIS* PL.63 COLOR
COMSTOCK-HOW TO KNOW BUTTER. P.199,1904 LARVA(SOSYBIUS)
FRENCH-BUTTER.EAST.UNITED STATES P.241,1886 LIFE HIST. SOSYBIUS
SCUDDER-BUTTER.EAST.UNITED STATES & CANADA PP.1787-1788,1889 LIFE HISTORY
FOOD PLANTS
GRASSES

HERMIA HENRY EDWARDS,CATOCALA-NOCTUIDAE
FORM VESTA BARNES & MCDUNNOUGH
A VERECUNDA HULST
FORM ROSA BEUTENMULLER
FORM DIANTHA BEUTENMULLER
FORM RITANA BEUTENMULLER
LIFE HISTORY
AMER.MUS.NAT.HIST.BULL.16: 388,1902 BEUTEN.LARVAL HEAD* F.10
AMER.MUS.NAT.HIST.MEM.NEW SERIEZ 3,PART 1,1918 BARNES & MCDUNNOUGH LARVA* PL.13,F.11;PL.15, F.15;PL.16,F.21;PL.17, F.7 COLOR VERECUNDA
WASHINGTON ENT.SOC.PROC.4: 327,1898 DYAR LARVA
FOOD PLANTS
SALIX SP. (WILLOWS)

HERMINIATA(GUENEE),EUDEILINIA-DREPANIDAE
SYN.ALBATA GUENEE
A BISERIATA PACKARD
LIFE HISTORY
AMER.MUS.NAT.HIST.BULL.10: 387,1898 BEUTEN.LARVA
NEW YORK ENT.SOC.JOURN 3:67, 1985 DYAR EGG,LARVA
HART-INS.COLLECTIONS ILLINOIS HIGH SCHOOLS P.49,1903 LARVAL AND PUPAL HABITS
FOOD PLANTS
CORNUS FLORIDA LINN.
CORNUS RACEMOSA LAM.
CORNUS SERICEA LINN.

HERMINIOIDES WALKER
SYN.SEE AEMULA HUBNER

HERMODUR HENRY EDWARDS
SSP.SEE SMINTHEUS DOUBLEDAY & HEWITSON

HERMOSA COMSTOCK
SSP.SEE NOKOMIS EDWARDS

HERO HENRY EDWARDS
SSP.SEE MICRONYMPHA GUENEE

HERODIAS STRECKER, CATOCALA-NOCTUIDAE
A GERHARDI BARNES & BENJAMIN
LIFE HISTORY
BROOKLYN ENT. SOC. BULL. 17: 138, 1922 MAYFIELD LARVA
FOOD PLANTS
QUERCUS SP. (OAKS)
QUERCUS ILICIFOLIA WANG.

HERRI FIELD
SSP. SEE CALIFORNICA BOISDUVAL (NYMPHALIS)

HERRI GUNDER
SSP. SEE GILLETTII BARNES

HERRII GRINNELL
SSP. SEE COMYNTAS GODART

HERSCHELENSIS BENJAMIN
SSP. SEE INTERROGATIONIS LINNAEUS

HERSCHELI LEUSSLER, EREBIA-SATYRIDAE
LIFE HISTORY
FOOD PLANTS
SARCOSTEMMA CRISPUM BENTH.

HERSE HALL
SSP. SEE THAROS DRURY

HERSILIATA (GUENEE), DYSSTROMA-GEOMETRIDAE
SYN. FLAMMIFERA WALKER
A CERVINIFASCIA WALKER
FORM MIRANDATA TAYLOR
B MANITOBA MCDUNNOUGH
LIFE HISTORY
NEW YORK STATE AGRI. EXP. STA. (CORNELL), MEM. 274:136, 1948 FORBES ?LARVA
UNITED STATES NAT. MUS. PROC. 27:897, 1904 DYAR LARVA
FOOD PLANTS
RIBES SANGUINEUM PURSH

HESPERIDA SMITH, ACRONICTA-NOCTUIDAE
LIFE HISTORY
UNITED STATES NAT. MUS. PROC. 27:797, 1904 DYAR EGG, LARVA, COCOON
FOOD PLANTS
ALNUS RUBRA BONG.

HESPERIS EDWARDS, ARGYNNIS-NYMPHALIDAE
LIFE HISTORY
CAN. ENT. 20:67, 1888 W.H. EDWARDS EGG, LARVA, CHRYSALIS
FOOD PLANTS

HETERODOXA SMITH
SSP. SEE INSUETA GUENEE

HETERONEA BOISDUVAL, LYCAENA-LYCAENIDAE
AB COLORADENSIS GUNDER
A CLARA HENRY EDWARDS
B GRAVENOTATA KLOTS
C KLOTSI FIELD
LIFE HISTORY
ENT. NEWS 21:37, 1910 WILLIAMS LARVA, CHRYSALIS
SOUTHERN CALIFORNIA ACAD. SCI. BULL. 26:68, 1927 COMSTOCK EGG
FOOD PLANTS
ERIOGONUM FASCICULATUM BENTH
ERIOGONUM MICROTHECUM NUTT.

HEWITSONI FIELD
SSP. SEE NYCTEIS DOUBLEDAY & HEWITSON

HEWLETTAE GUNDER
SSP. SEE VANILLAE LINNAEUS

HEWLETTARIA WRIGHT
SSP. SEE CITRATA LINAEUS

HEWLETTI BARNES & MCDUNNOUGH
SSP. SEE ORNATA PACKARD

HEXASPILATA WALKER
SYN. SEE TRIGUTTARIA HERRICH-SCHAEFFER

HIANNA (SCUDDER), ATRYTONOPSIS-

HESPERIIDAE
SYN. GROTEI PLOETZ
LIFE HISTORY
FOOD PLANTS
GRASSES

HIBISCI (GUENEE), ORTHOSIA-NOCTUIDAE
ALIA AUCT.
SYN CONFLUENS MORRISON
FORM INSCIENS WALKER
A BRUCEI SMITH
SYN. MALORA SMITH
B NUBILATA SMITH
C QUINQUEFASCIATA SMITH
SYN. LATIRENA DOD
SYN. PROBA SMITH
SYN. INHERITA SMITH
SYN. INFLAVA SMITH
LIFE HISTORY
JOURN. ECON. ENT. 42:754, 1949
RASKI-BORDEN LARVA* F. 1;
PUPA
THEY SAY THAT SLINGERLAND
(NEW YORK STATE AGRI. EXP.
STA. (CORNELL), BULL. 123:
502, 1896) DESCRIBES THE
LIFE HISTORY OF HIBISCI UNDER
XYLINA LATICINEREA GROTE
BY MISTAKE
NEW YORK STATE, REPT. INS. 3:
343, ALSO P. 25, SEC. 41, 1856
FITCH LARVA (INSTABILIS)
UNITED STATES DEPT. AGRI. MISC.
PUB. 657:363, 394, 1950
CRAIGHEAD LARVA
GUENEE-SPEC. GEN. LEP. NOCT.
1:355, 1852 LARVA (HIBISCI)
STAINTON-BRITISH BUTTER. &
MOTHS 1:243, 1857 LARVA
INDEX AMER. ECON. ENT.
VOL. 2:163 GRAPHIPHORA ALIA
FOOD PLANTS
ACER NEGUNDO LINN.
ACER SACCHARINUM LINN.
AESCULUS HIPPOCASTANUM LINN.
ALNUS RUBRA BONG.
AMELANCHIER FLORIDA LINDL.
BETULA SP. (BIRCHES)
CARYA SP. (HICKORIES)
FRAXINUS SP. (ASH)
HOLODISCUS DISCOLOR (PURSH)
MALUS PUMILA MILL. (APPLE)
POPULUS SP. (POPLARS)
PRUNUS ARMENIACA LINN.
(APRICOT)
PRUNUS EMARGINATA (DOUGL.)
PRUNUS PERSICA LINN. (PEACH)
PRUNUS SEROTINA EHRH.
PYRUS COMMUNIS LINN. (PEAR)
QUERCUS SP. (OAKS)
QUERCUS ALBA LINN.
RHODODENDRON SP.
RIBES SANGUINEUM PURSH
SALIX HOOKERIANA BARRATT
SALIX SCOULERIANA BARR.

HIEMALIS EDWARDS
SSP. SEE PHAON EDWARDS

HIEMALIS EDWARDS
SYN. SEE VESTA EDWARDS

HIEMALIS GROTE, XYLOMYGES-NOCTUIDAE
SYN. CALIFORNICA STRECKER
LIFE HISTORY
FOOD PLANTS
ALNUS RUBRA BONG.
AMELANCHIER FLORIDA LINDL.
HOLODISCUS DISCOLOR (PURSH)
PRUNUS EMARGINATA (DOUBL.)
QUERCUS GARRYANA DOUGL.
RIBES SANGUINEUM PURSH
SALIX HOOKERIANA BARRATT
SALIX SCOULERIANA BARR.

HIEROGLYPHICA (CRAMER), NOROPSIS-
NOCTUIDAE
SYN. FESTIVA FABRICIUS
SYN. ELEGANS HUBNER
SYN. FASTUOSA GUENEE
LIFE HISTORY
ENT. NEWS 33:277, 1922
BENJAMIN EARLY STAGES
FOOD PLANTS

HILDA GRINNELL

ssp. see saepiolus Boisduval

hilli Grote
ssp. see concumbens Walker

hilli Gunder
ssp. see chalcedona Doubleday & Hewitson

hilliana Harvey
syn. see oblata Morrison

hinda French
syn. see innubens Guenee

hippolyta Lyman
ssp, see antiopa Linnaeus

hircina Morrison, Homoglaea-Noctuidae
LIFE HISTORY
FOOD PLANTS
Vaccinium sp.

hirsuta Skinner
ssp. see philenor Linnaeus

hirtipes Grote, Zosteropoda-Noctuidae
LIFE HISTORY
United States Nat. Must. Proc. 27:866, 1904 Dyar egg, larva
FOOD PLANTS

hiseri Cassino
ssp. see minuta Edwards (Catocala)

histrio Grote
syn. see inficita Walker (Marathyssa)

hobomok (Harris), Poanes-Hesperiidae
ab pallida Watsono
ab alfaratta Holland
form female pocahontas Scudder
syn. quadaquina Scudder
ab friedlei Watson
ab ridingsii Chermock
LIFE HISTORY
Buffalo Soc. Nat. Sci. Bull. 19: 34, 1939 Wild larva
Can. Ent. 1:66, 1869 Saunders egg, larva
Ent. News 16:245, 1905 Laurent hibernation of larva in all stages
Loc. cit. 19:410, 497, 1908 Laurent larva almost identical with that of zabulon Boiaduval & Leconte
Mus. Comp. Zool., Bull. 87, No. 6, 1941 Dethier larval antenna* f. 61
Southern California Acad. Sci. Bull. 41:41, 1942 Dethier compares larva with that of themistocles Latreille
FOOD PLANTS
Grass

hochenwarthi (Hochenwarth), Syngrapha-Noctuidae
a ignea Grote
LIFE HISTORY
Kirby-European Butter. & Moths p. 266, 1882 larva
FOOD PLANTS
Antennaria dioica (Linn.)
Carex filifolia Nutt.
Castilleja lemmonii Gray
Gentiana newberryi Gray
Mimulus primuloides Benth.
Penstemon confertus Doubl.
Salix sp. (willows)
Vaccinium sp.

hoffmanni Behr. Melitaea, Nymphalidae
syn. helcita Boisduval
ab abnorma Wright
ab hollandae Gunder
a segregata Barnes & McDunnough
syn. bridgei Comstock
LIFE HISTORY
FOOD PLANTS
Chrysopsis brewerii Gray (probably)

HOLLANDAE GUNDER
SSP. SEE HOFFMANNI BEHR

HOLLANDI BRYK & EISNER
SYN. SEE SMINTHEUS DOUBLEDAY & HEWITSON

HOLLANDI CHERMOCK
SSP. SEE ATLANTIS EDWARDS

HOLLANDI GUNDER
SSP. SEE SATYRUS EDWARDS

HOLLANDI ROEBER
SYN. SEE ILAIRE GODART

HOLLANDII EDWARDS
SSP. SEE BAIRDII EDWARDS

HOLOCINEREA SMITH
SSP. SEE GEORGII GROTE

HOMOPTEROIDES HULST
SYN. SEE PACKARDARIA HULST

HOMURARIA GROTE
SSP. SEE HYPOCHRARIA HERRICH-SCHAEFFER

HONESTA WALKER
SYN. SEE RUFAGO HUBNER

HONESTARIUS (WALKER), PERO-GEOMETRIDAE
SYN. STYGIARIUS WALKER
SYN. DYARI CASSINO & SWETT
LIFE HISTORY
NEW YORK ENT. SOC. JOURN. 53: 192, 1945 FORBES PUPA
NEW YORK STATE AGRI. EXP. STA. (CORNELL), MEM. 274:83, 1948 FORBES LARVA
UNITED STATES DEPT. AGRI. MISC. PUB. 657:362,426, 1950 CRAIGHEAD LARVA
FOOD PLANTS
LARIX SP. (LARCHES)
PRUNUS SEROTINA EHRH.
ROBINIA PSEUDO-ACACIA LINN.

HOPFFERI GROTE & ROBINSON
SYN. SEE LATIPENNIS BOISDUVAL

HOPFINGERI GUNDER
SSP. SEE ANICIA DOUBLEDAY & HEWITSON

HORATIUS (SCUDDER & BURGESS), ERYNNIS-HESPERIIDAE
SYN. VIRGILIUS SCUDDER & BURGESS
SYN. PETRONIUS LINTNER
LIFE HISTORY
SCUDDER-BUTTER. EAST. UNITED STATES & CANADA PP. 1488-1489, 1889 LARVA, CHRYSALIS
FOOD PLANTS
LEGUMINOSAE
QUERCUS SP. (OAKS)
WISTERIA FRUTESCENS (LINN.)

HORMOS HUBNER, HYPSOROPHA-NOCTUIDAE
LIFE HISTORY
CAN. ENT. 31:289, 1899 DYAR EGG, LARVA, PUPA, COCOON
GLOVER-ILLUST. NORTH AMER. ENT. 1878 LARVA*PL. 13, FF. 1,2; PL. 15, F. 7
GUENEE-SPEC. GEN. LEP. NOCT. 2:403, 1852 LARVA
FOOD PLANTS
DIOSPYROS VIRGINIANA LINN.
GOSSYPIUM HERBACEUM LINN. (COTTON)

HORNBECKIANA HARRIS
SYN. SEE VITIS LINNAEUS

HORNI STRAND
SYN. SEE APPASIONATA HARVEY

HORTARIA FABRICIUS
SYN. SEE VIRGINIARIA CRAMER

HORTULANA MORRISON
SYN. SEE VANCOUVERENSIS GROTE

HOSPITA DENNIS & SCHIFFENMULLER
SSP. SEE PLANTAGINIS LINNAEUS

HOSPITONINA LeCerf
SSP. SEE RUTULUS Lucas

HOSTIA Harvey
SYN. SEE EXACTA Hubner

HOWARDI (Dyar), GLOVERIA–Lasiocampidae
LIFE HISTORY
New York Ent. Soc. Journ. 4:24, 1896 Dyar egg, larva, pupa, cocoon
FOOD PLANTS
Quercus emoryii Torr.

HOWARDI Skinner
SSP. SEE AARONI Skinner

HOWLANDI (Grote), SYNEDOIDA–Noctuidae
SYN. EXQUISITA Hampson
LIFE HISTORY
United States Nat. Mus. Proc. 25:382, 1903 Dyar egg, larva
FOOD PLANTS
Eriogonum sp.

HOWLANDI Stallings
SSP. SEE ANICIA Doubleday & Hewitson

HOYI Grote
SYN. SEE BREPHOIDES Walker

HUACHUCA Beutenmuller
SSP. SEE ARIZONAE Grote

HUALAPAI Neumoegen
SSP. SEE SORORIUS Henry Edwards

HUBBARDI Dyar
SSP. SEE HEILIGBRODTI Harvey

HUBBARDIANA Dyar
SSP. SEE SEMICROCEA Guenee

HUDSONIANUS Clark
SSP. SEE MACHAON Linnaeus

HUDSONICA Henry Edwards
SYN. SEE LANGTONI Couper

HUGHI Clark
SSP. SEE MASSASOIT Scudder

HUGUENINI Gunder
SSP. SEE XERCES Boisduval

HULBIRTI Field
SSP. SEE HELLOIDES Boisduval

HULDA Edwards
SSP. SEE NAPI Linnaeus

HULSTI Reiff
SSP. SEE ILIA Cramer

HULSTIANA Dyar
SSP. SEE HERBARIA Fabricius

HULSTII Edwards
SYN. SEE ARCHIPPUS Cramer

HUMARIA (Guenee), ANACAMPTODES–Geometridae
SYN. INTRARIA Guenee
SYN. INTRACTARIA Walker
SYN. ILLAUDATA Walker
LIFE HISTORY
Ent. Amer. 14:145, 1934 Crumb larva
Illinois Dept. Agri. Trans. 18, Append: 148, 1880 Coquillett larva
New York Ent. Soc. Journ. 7:70 1899 Dyar larval structure Pl. 2, f. 6
New York State Agri. Exp. Sta. (Cornell), Mem. 15:190, 1918 Hawley egg* f. 42; larva* ff. 43, 46, 47; pupa* f. 48
New York State, Rept. of Ins. 2:234, 1856 Fitch larva, pupa
Ontario Ent. Soc. Ann. Rept. p. 31, 1872 Bethune larva*

PUPA*
LOC.CIT.P.19,1888 FYLES LARVA PUPA
UNITED STATES DEPT.AGRI.BUR. ENT.BULL,4 OLD SERIES P. 39,1884 SMITH LARVA,PUPA
UNITED STATES NAT.MUS.PROC. 27:882,1904 DYAR EGG, LARVA
GLOVER-ILLUST.NORTH AMER. ENT.1878 LARVA* PL.9,F.2; PL.21,F.10;PL.22,F.8
HOLLAND-MOTH BOOK P.287, 1905 LARVA* PUPA*F.178
O'KANE-INJ.INS.P.179,1912 LARVA* F.212
INDEX AMER.ECON.ENT. VOL.1:159;VOL.7:480
FOOD PLANTS
HUMULUS LUPULUS LINN.(HOPS)
URTICA DIOICA LINN

HUMULI HARRIS
SYN.SEE MELINUS HUBNER

HUNTERA FABRICIUS
SYN.SEE VIRGINIENSIS DRURY(VANESSA)

HUNTERI HUBNER
SYN.SEE VIRGINIENSIS DRURY(VANESSA)

HURON
SYN.SEE CAMPESTRIS BOISDUVAL

HUTCHINSI GUNDER
SSP.SEE ATLANTIS EDWARDS

HYALINA HULST
SYN.SEE UNICOLOR ROBINSON

HYALINOPUNCTA PACKARD
SYN.SEE ANGELICA GROTE

HYANTIS EDWARDS
SSP.SEE CREUSA DOUBLEDAY & HEWITSON

HYBRIDA BUTLER
SSP.SEE BELLA LINNAEUS

HYDASPE(BOIASUVAL),ARGYNNIS-NYMPHALIDAE
AB CALIGINOSA COMSTOCK
A VIRIDICORNIS COMSTOCK
B PURPURASCENS HENRY EDWARDS
C RHODOPE EDWARDS
AB GREGSONI GUNDER
F.ALT.MINOR MCDUNNOUGH
D SAKUNTALA SKINNER
SYN.SKINNERI HOLLAND
LIFE HISTORY
SOUTHERN CALIFORNIA ACAD.SCI.D BULL.39:75,1940 COMSTOCK EGG* PL.8A
FOOD PLANTS
VIOLA SP.

HYDASPUS CRAMER
SSP.SEE ANTAEUS DRURY

HYDROMELI HARVEY,LITODONTA-NOTODONTIDAE
AB FUSCA HARVEY
A CONTRASTA BARNES &MCDUNNOUGH
LIFE HISTORY
WASHINGTON ENT.SOC.PROC.6: 3,1904 DYAR EGG,LARVA
FOOD PLANTS
BUMELIA CELASTRINA H.B.K.

HYEMALIS EDWARDS
SYN.SEE NAPI LINNAEUS

HYEMALIS STRETCH
SYN.SEE MINNA BUTLER

HYGIAEA HEYDENREICH
SSP.SEE ANTIOPA LINNAEUS

HYLAEUS(DRURY),DOLBA-SPHINGIDAE
A PRINI ABBOT & SMITH
SYN.FLORIDENSIS CLARK
LIFE HISTORY
AMER.JOURNSCI.36:296,1839 HARRIS LARVA,PUPA
AMER.MUS.NAT.HIST.BULL.7:

309,1895 Beuten. larva pupa
Can.Ent.16:55,1884 Harrington larva?
Ent.News 9:150,1898 Soule duration egg stage
Ent.Soc.America,Annals 11: 414,1918 Mosher pupa
Naturalista Canadien p.153, 1929 Huard larva
New York State Agri.Exp.Sta. (Cornell),Mem.274:188, 1948 Forbes larva
New York State Mus.Mem.8: 739,1905 Felt larva(brief)
Novitates Zoologica 9 Suppl. p.103,1903 Rothschild & Jordan larva,pupa(Tring Mus)
Ontarion Ent.Soc.Ann.Rept.p. 70,1881 Reed larva,pupa
Philadelphia Acad.Nat.Sci. Proc.5:178,1859 Clemens larva,pupa
Psyche 2:77,1877 Scudder undetermined larva this sp.?
Psyche 5:267,1889 Eliot & Soule egg,larva,pupa
Psyche 6:436,1893 Scudder larval habits,pupa
Smithsonian Misc.Collections 4,art 2:204,1862 Morris larva, pupa
Abbot & Smith-Lep.Ins.Georgia 1:69,1797 larva*pupa*Pl. 35 color
Boisduval-Spec.Gen.Lep. Sphing.p.99,1874 larva, pupa
Eliot & Soule-Cater.&Moths p.158,1902 egg,larva,pupa
Fernald-Sphing.New England p.49,1886 larva
Glover-Illust.North Amer.Ent. 1878 larva*Pl.2,f.18
Morris-Synop.Lep.North America p.204,1862 larva pupa
Westwood-In Drury's Illust. Exot.Ent.2:49,1837 larva,pupa

FOOD PLANTS
Asimina triloba(Linn.)
Comptonia peregrina(Linn.)
Ilex glabra(Linn.)
Vaccinium angustifolium Ait.
Vaccinium myrtillus Linn.
Vaccinium uliginosum Linn.

hylax Edwards
syn.see garita Reakirt

hyperbola Slosson
ssp.see juglandis Abbot & Smith

hyperborea(Hulst), Anthelia-Geometridae
LIFE HISTORY
FOOD PLANTS
Alnus rubra Bong.

hyperborea Seitz
ssp.see antiopa Linnaeus

hyperboreata Staudinger
syn.see gelidata Moeschler

hyperici Boisduval & Leconte
syn.melinus Hubner

hypochraria(Herrich-Schaeffer), Metarranthis-Geometridae
syn.refractaria Guenee
syn.mestusata Walker
a lateritiaria Guenee
syn.foedaria Walker
syn.indeclinata Walker
b homuraria Grote
syn.amethystaria Strecker
LIFE HISTORY
Ent.News 5:61,1894 Dyar larva pupa(according to Rupert this is distinct sp.broweri Rupert)
New York Ent.Soc.Journ.51: 149,1943 Rupert egg,larva

Loc.cit.41:150,1943 Rupert
see note under browerI
Rupert
New York Ent.Soc.Journ.53:
200,1945 Forbes pupa
New York State Agri.Exp.Sta.
(Cornell),Mem.274:94,
1948 Forbes larva
New York State Mus.Mem.8:
732 1905 Felt larva(brief)
Psyche 8:415,1899 Dyar egg,
cocoon(browerI Rupert)

FOOD PLANTS
Diospyrus virginiana Linn.
Malus pumila Mill.(apple)
Prunus serotina Ehrh.
Prunus virginiana Linn.
Rosaceae sp.
Sassafras albidum v.molle Raf.
Vaccinium sp.

hypophlaeeas(Boisduval),Lycaena-
Lycaenidae
For the many syns. and ssps. of
this species see McDunnough's
List

LIFE HISTORY
Amer.Mus.Nat.Hist.Bull.5:
283,1893 Beuten.larva,
chrysalis
Bufalo Soc.Nat.Sci.Bull.19
29,1939 Wild larva
Can.Ent.1:4,1868 Saunders
larva,chrysalis
Can.Nat.P.221,1840 Gosse
egg
Illinois Dept.Agri.Trans.18,
Append:95,1880 Middleton
larva
Montana Univ.Bull.30:134,1906
Elrod egg,larva,chrysalis
Smithsonian Misc.Collections
4,art.2:91,1862 Morris
larva,chrysalis
Comstock-How to Know Butter
p.242,1904 larva
Edwards,Hy.-In Kingsley's
Stand.Nat.Hist.2:476,
1885 larva
Fernald-Butter.Maine p.88,
1884 larva* chrysalis*
French-Butter.East.United
States p.283,1886 egg*
larva* chrysalis*
HarrisOIns.Inj.Veg.Flint edit.
p.274,1862 larva,chrysalis
Holland-Butteffly Book.p.
255,1905 chrysalis* Pl.5,
f.49,3rd edit.p.251,
1931 larva,chrysalis
Maynard-Butter.New England
p.41,1885 larva,chrysalis
(americana)
Packard-Second Rept.Nat.Hist.
& Geol.Maine p.166,1862
larva
Scudder-Butter.East.United
States & Canada pp.1002-
1006,1889 life history*
Pl.65,f.21;Pl.68,f.10;
Pl.71,f.1;Pl.75,f.42;
Pl.79,ff.39,40;Pl.84,
f.49;Pl.86.f.33
Scudder-Butter.Their Struc-
ture pp.7,20,31,& 43,
1882 egg*larva*chrysalis*
Scudder-Everyday Butter.p.
90,1899 larva,chrysalis

FOOD PLANTS
Rumex acetosella Linn.
Rumes crispus Linn.
Trifolium sp.(clovers)

hypoxanthe Kirby
syn.see epixanthe Boisduval &
Leconte

I

ianthe Edwards
syn.see nivalis Boisduval

iberidis Boisduval
syn.see napi Linnaeus

ICARIOIDES(BOIADUVAL), PLEBIUS-LYCAENIDAE
FOR THE MANY SYNS. AND SSP.S OF THIS SPECIES SEE MCDUNNOUGH'S LIST
LIFE HISTORY
CAN. ENT. 1:12,1868 SAUNDERS LARVA(V. PEMBINA)
CAN. ENT. 43:87,1911 NEWCOMER EGG, LARVA, CHRYSALIS (FULLA EDWARDS)
PSYCHE 15:84,1908 COOLIDGE EGG, YOUHG LARVA
SOUTHERN CALIFORNIA ACAD. SCI. BULL. 34:82,1935 COMSTOCK-DAMMERS LIFE HISTORM(EVIUS) EGG*PL. 8;LARVA*PLS. 9,10 CHRYSALIS* PL. 10
MEAD-REPT. LEP. WHEELER EXPED. WEST 100 MERID. 5:785, 1875 LARVA (LYCEA)
FOOD PLANTS
LUPINUS LATIFOLIUS V. COLUMBIANUS (HELLER)
LUPINUS MICRANTHUS DOUGL.
TRIFOLIUM MONATHUM GRAY
VICIA CRACCA LINN.

ICELUS(SCUDDER & BURGESS), ERYNNIS-HESPERIIDAE
SYN. BAUTISTA PLOETZ
SYN. HAMAMAELIDIS SCUDDER
LIFE HISTORY
AMER. MUS. NAT. HIST. BULL. 5: 299,1893 BEUTEN. LARVA, CHRYSALIS
BUFFALO SOC. NAT. SCI. BULL. 19: 31,1939 WILD LARVA
CAN. ENT. 17:98,1885 EDWARDS EGG, LARVA, CHRYSALIS
NEW YORK STATE CABIN. NAT. HIST. REPT. NO. 23, P. 163, 1873 LINTNER EGG
NEW YORK STATE MUS. MEM. 8: 742,1905 FLINT LARVA (BRIEF)
ONTARIOENT. SOC. ANN. REPT. P. 88,1888 FLETCHER EGG
OTTAWA NATURALIST 6:103,1892 HARRINGTON LARVA
PSYCHE 7:403,1896 FOLSOM EGG (PROBABLY THIS SPECIES)
UNITED STATES DEPT. AGRI. MISC. PUB. 657:352,353,1950 CRAIGHEAD LARVA
FLETCHER-A TRIP TO NEPIGON P. 15,1888 EGG
FRENCH-BUTTER. EAST. UNITED STATES P. 355,1886 EGG
HOLLAND-BUTTERFLY BOOK 1905 CHRYSALIS*PL. 6, F. 27
SCUDDER-BUTTER. EAST. UNITED STATES & CANADA PP. 1509-1514,1889 LIFE HISTORY* PL. 77, F. 1; PL. 85, F. 27
SCUDDER-EVERYDAY BUTTER. P 74,1899 EGG, LARVA
SCUDDER-GUIDE TO COMMON BUTTER. P. 163,1893 LARVA, CHRYSALIS
FOOD PLANTS
BAPTISIA TINCTORIA(LINN.)?
CORYLUS SP. (HAZELS)
HAMAMELIS VIRGINIANA LINN.
POPULUS SP. (POPLARS)
QUERCUS SP.)OAKS)
QUERCUS ILICIFOLIA WANG.
QUERCUS RUBRA LINN.
QUERCUS VELUTINA LAM.
SALIX CORDATA MICHX.

IDAHO EDWARDS
SSP. SEE COMMA LINNAEUS

IDAHOENSIS BRYK & EISNER
SSP. SEE SMINTHEUS DOUBLEDAY & HEWITSON

IDAHOENSIS GROTE
SSP. SEE COSTATA GROTE

IDALIS(DRURY), ARGYNNIS-NYMPHALIDAE
AB ASHTAROTH FISCHER
ASTARTE FISCHER
SYN. INFUMATA OBERTHUS
AB DOLLI GUNDER
LIFE HISTORY

AMER. MUS. NAT. HIST. BULL. 5: 254, 1893 BEUTEN. LARVA, CHRYSALIS
CAN. ENT. 7:151, 1875 EDWARDS LARVA (BRIEF)
LOC. CIT. 11:217, 1879 EDWARDS EGG, LARVA, CHRYSALIS
ILLINOIS DEPT. AGRI. TRANS. 18, APPEND: 81, 162, 1880 COQUILLETT-MIDDLETON LARVA
MONTANA UNIV. BULL. 30:52, 1906 ELROD EGG, LARVA, CHRYSALIS
COMSTOCK-HOW TO KNOW BUTTER. P. 113, 1904 LARVA
FERNALD-BUTTER. MAINE P. 37, 1884 EGG, LARVA
FRENCH-BUTTER. EAST. UNITED STATES P. 150, 1886 LIFE HISTORY
HOLLAND-BUTTERFLY BOOK P. 103, 1905 CHRYSALIS* PL. 5, F. 4
HOLLAND-LOC. CIT. 3RD EDIT. P. 84, 1931 LARVA CHRYSALIS* PL. 5, F. 4
KELLOGG-AMER. INS. P. 456, 1908 LARVA
MAYNARD-BUTTER. NEW ENGLAND P. 22, 1886 LARVA, CHRYSALIS
SCUDDER-BUTTER. EAST. UNITED STATES & CANADA PP. 540-543, 1889 LIFE HIST* PL. 64, F. 34; PL. 57, F. 19; PL. 72, F. 10; PL. 75, F. 10; PL. 79, F. 2; PL. 84, F. 4
SCUDDER-EVERYDAY BUTTER. P. 252, 1899 EGG, LARVA, CHRYSALIS
SCUDDER-GUIDE TO COMMON BUTTER. P. 80, 1893 LARVA, CHRYSALIS

FOOD PLANTS
SERICOCARPUS ASTEROIDES (LINN.) (EGG LAID ON THIS PLANT-SCUDDER)
VIOLA SP. (VIOLETS)
VIOLA TRICOLOR LINN.

IDEATA WALKER
SYN. SEE UNANGULATA HAWORTH

IDUATA (GUENEE), XANTHORHOE-GEOMETRIDAE
SYN. PLANATA TAYLOR
?SYN. FLUCTUATA LINNAEUS

LIFE HISTORY
KIRBY-EUROPEAN BUTTER. & MOTHS P. 378, 1882 LARVA
NEWMAN-BRITISH MOTHS P. 164, 1869 LARVA
PACKARD-MONOGRAPH GEOMETRIDAE P. 157, 1876 LARVA
STAINTON-BRITISH BUTTER. MOTHS 2:102, 1859 LARVA

FOOD PLANTS
BRASSICA OLERACEA LINN. (CABBAGE)
TROPAEOLUM MAJUS LINN.
GENERAL FEEDER ON VEGETABLES

IDUNA EDWARDS
SSP. SEE NEVADENSIS FELDER & FELDER

IGNA BARNES & BENJAMIN
SSP. SEE SIGNATA WALKER

IGNEA GROTE
SSP. SEE HOCHENWARTHI HOCHENWARTH

IGNOTARIA WALKER
SYN. SEE MYRTARIA GUENEE

ILAIRE (GODART), APPIAS-PIERIDAE
SYN. MYSIA GODART
SYN. MARGARITA HUBNER
SYN. MOLPODIA HUBNER
A POEYI BUTLER
SYN. NEUMOEGENII SKINNER
SYN. HOLLANDI ROEBER

LIFE HISTORY
CAN. ENT. 79:142, 1947 CHERMOCK & CHERMOCK EGG, LARVA, CHRYSALIS* FF. A, B
MUS. COMP. ZOOL., BULL. 87, NO. 6, 1941 DETHIER LARVAL ANTENNA* FF. 34, 47

FOOD PLANTS

DRYPETES LATERIFLORA(S.W.)

ILAVIA BEUTENMULLER
SSP.SEE AUTOLYCUS EDWARDS

ILIA CRAMER ,CATOCALA-NOCTUIDAE
FOR THE MANY SSPS.AND SYNS. OF THIS SPECIES SEE MCDUNNOUGH'S LIST
LIFE HISTORY
AMER.MUS.NAT.HIST.MEM.NEW SERIES 3,PART 1,1918 BARNES & MCDUNNOUGH LARVA*PL. 10,F.26;PL.12,F.9COLOR
LOC.CIT.LARVA 5 ZOE* PL.16, F.5;PL.15,FF.10,42;PL. 16,F.10 COLOR
BROOKLYN ENT.SOC.BULL.4: 22,1881 KOEBLE LARVA
LOC.CIT. 7:43,1884 HULST LARVA
CAN.ENT.7:208,1875 CAULFIELD LARVA
LOC.CIT.16:12,1884 FRENCH EGG,LARVA,PUPA
ENT.NEWS 20:18,128,1909 ROWLEY EGG,LARVA
LOC.CIT.21:442,1910 ROWLEY LARVA
NEW YORK STATE MUS.MEM.8: 705,1905 FELT LARVA
PSYCHE 20:189,1913 BARNES & MCDUNNOUGH EGG,LARVA
GLOVER-ILLUST.NORTH AMER. ENT.1878 LARVA* PL.12,F. 14
GUENEE-SPEC.GEN.LEP.NOCT. 3:92,1852 LARVA
FOOD PLANTS
QUERCUS SP. (OAKS)
QUERCUS MACROCARPA MICHX.

ILIANA STRAND
SYN.SEE ILIA CRAMER

ILICIFOLIA AUCT.
SEE AMERICANA HARRIS(EPICNAPTERA)

ILIONEUS ABBOT & SMITH
SSP.SEE TROILUS LINNAEUS

ILLATA WALKER
SYN.SEE TESSELLATA HARRIS

ILLAUDABILIS(GROTE) ,LACINIPOLIA-NOCTUIDAE
A ALBOGUTTATA GROTE
B RESTORA SMITH
LIFE HISTORY
UNITED STATES NAT.MUS.PROC. 27:857,1904 DYAR EGG, LARVA
FOOD PLANTS

ILLAUDATA WALKER
SYN.SEE HUMARIA GUENEE

ILLECTA WALKER ,CATOCALA-NOCTUIDAE
SYN.MAGDALENA STRECKER
LIFE HISTORY
AMER.MUS.NAT.HIST.BULL.16: 390,1902 BEUTEN.LARVAL HEAD* F.16
LOC.CIT.38:169,1918 BARNES & MCDUNNOUGH EGG
AMER.MUS.NAT.HISTMEM.NEW SERIES 3 PART 1,1918 BARNES & MCDUNNOUGH LARVA* PL.10,F.30;PL.12, F.10 COLOR
CAN.ENT.24:307,1892 FRENCH LARVA,PUPA
LOC.CIT.33:300,1901 DODGE LARVA(BRIEF)
ENT.NEWS 20:129,1909 ROWLEY LARVA
NEW YORK ENT.SOC.JOURN.9: 189,1901 BEUTEN.LARVA
NEW YORK STATE MUS.MEM.8: 729,1905 FELT LARVA (BRIEF)
WASHINGTON ENT.SOC.PROC.4: 327,1898 DYAR LARVA
FOOD PLANTS
GLEDITSIA TRIACANTHOS LINN.
ROBINIA PSEUDO-ACACIA LINN.

ILLINOIS DODGE
SYN. SEE BIMACULA GROTE & ROBINSON

ILLINOISENSIS FRENCH
SYN. SEE AURANTIAGO GUENEE

ILLITA SMITH
SSP. SEE IMPLETA WALKER (ACRONICTA)

ILLITERATA GROTE
SYN. SEE AURANTIAGO GUENEE

ILLOCATA HULST
SYN. SEE MUNITATA HUBNER

ILMAE WALKER
SSP. SEE BURNSI WATSON

IMBRARIA GUENEE
SYN. SEE CONFUSARIA HUBNER

IMBRIFERA (GUENEE), POLIA-NOCTUIDAE
LIFE HISTORY
ENT. SOC. AMERICA, ANNALS 23: 396, 1930 HAWKINS LARVAL CLAWS
ONTARIO ENT. SOC. ANN. REPT. P. 14, 1896 FYLES LARVA
FOOD PLANTS
PRUNUS VIRGINIANA LINN.

IMITATA (WALKER), MELANOLOPHIA GEOMETRIDAE
SYN. SUBGENERICATA DYAR
LIFE HISTORY
CAN. ENT. 76:126, 1944 MCGUFFIN LARVA, LARVAL MANDIBLE* F,
NEW YORK ENT. SOC. JOURN. 53: 210, 1945 FORBES PUPS (BRIEF NOTE)
NEW YORK STATE AGRI. EXP. STA. (CORNELL), MEM. 274:60, 1948 FORBES EGG, LARVA
INDEX AMER. ECON. ENT. VOL. 7:647
FOOD PLANTS
ALNUS RUBRA BONG.
PICEA ENGELMANNI (PARRY)
PICEA SITCHENSIS (BONG.)
PSEUDOTSUGA MENZIESII (MIRB.)

IMMACULARIA CASSINO & SWETT
SSP. SEE PLUMOGERARIA HULST

IMMACULARIA SWETT
SSP. SEE CATENARIA DRURY

IMMACULATA BARNES & MCDUNNOUGH
SSP. JUTURNARIA GUENEE

IMMACULATA BENJAMIN
SSP. SEE SEMICROCEA GUENEE

IMMACULATA CHERMOCK
SSP. SEE LUPINI BOISDUVAL

IMMACULATA COCKLE
SSP. SEE IROIDES BOISDUVAL

IMMACULATA FIELD
SYN. SEE IOLE BOISDUVAL

IMMACULATA FOLOGNE
SYN. SEE RAPAE LINNAEUS

IMMACULATA GRAEF
SYN. SEE INVOLUTUM SEPP

IMMACULATA GUNDER
SSP. SEE FULIGINOSA EDWARDS

IMMACULATA JEWETT
SSP. SEE BICOLOR HARRIS

IMMACULATA SKINNER
SSP. SEE CLODIUS MENETRIES

IMMACULATA SKINNER
SSP. SEE PYLADES SCUDDER

IMMACULATA SKINNER & AARON
SEE RAPAE LINNAEUS

IMMACULATA WHITTAKER
SSP. SEE LISA BOISDUVAL & LECONTE

IMMACULATUS WILLIAMS
SSP. SEE RURICOLA BOISDUVAL

IMMACULOSUS COMSTOCK
SSP. SEE TITUS FABRICIUS

IMMACULSECUNDA GUNDER
SSP. SEE CAESONIA STOLL

IMMANATA HAWORTH
SSP. SEE CITRATA LINNAEUS

IMMANIS GUENEE, HYDROECIA-NOCTUIDAE
LIFE HISTORY
NEW YORK STATE AGRI. EXP. STA. (CORNELL), MEM. 15:159, 1918 ROWLEY EGG* FF. 18, 12 LARVA* FF. 19, 20; PUPA* F. 21
NEW YORK STATE, REPT. STATE ENT. 2:42, 1885 LINTNER EGG, LARVA
ONTARIO ENT. SOC. ANN. REPT. P. 49, 1928 GILBERT LIFE HISTORY, LARVA*
UNITED STATES SEPT. AGRI. BUR. ENT. BULL. 4, OLD SERIED P. 34, 1884 SMITH EGG, LARVA, PUPA
LOC. CIT. BULL. 7 NEW SERIES P. 40, 1897 HOWARD EGG, LARVA* F. 35 A, B; PUPA* F. 35 C
EDGE-AGRI. OF PENNSYLVANIA REPT. P. 106, 1884 LARVA, PUPA
HOLLAND-MOTH BOOK P. 212, 1905 LARVA* PUPA* F. 210
O'KANE-INJ. INS. P. 137, 1912 LARVA
INDEX AMER. ECON. ENT. VOL. 1:158; VOL. 2:162 GORTYNA VOL. 6:359; VOL. 7:475
FOOD PLANTS
HUMULUS LUPULUS LINN. (HOPS)
LUPINUS MICROCARPUS SIMS.
SILPHIUM SP. (ROOTS)
ZEA MAYS LINN. (CORN)

IMMERSATA WALKER
SYN. SEE FRITILLARIA GUENEE

IMPARATA WALKER
SYN. SEE INSULSARIA GUENEE

IMPASTA GUENEE, ANOMIS-NOCTUIDAE
SYN. DOCTORIUM DYAR
LIFE HISTORY
FOOD PLANTS
GOSSYPIUM HERBACEIUM LINN. (COTTON)

IMPAUPERATA WALKER
SYN. SEE STELLATA GUENEE

IMPECUNIOSA (GROTE), PAPAIPEMA-NOCTUIDAE
LIFE HISTORY
CAN. ENT. 32:279, 1900 BIRD EGG, LARVA, PUPA
LOC. CIT. 37:230, 1905 DOD LARVA
FOOD PLANTS
ASTER PUNICEUS LINN.
ASTER UMBELLATUS MILL.
HERACLEUM MAXIMUM BARTR.

IMPERATOR STRECKER
SSP. SEE MODESTA HARRIS

IMPERATORIA ABBOT & SMITH
SYN. SEE IMPERIALIS DRURY

IMPERIALIS (DRURY), EACLES-CITHERONIIDAE
SYN. IMPERATORIA ABBOT & SMITH
AB PUNCTATISSIMA NEUMOEGEN
A DIDYMA BEAUVOIS
B NOBILIS NEUMOEGEN
C OSLARI ROTHSCHILD
D PINI MICHNER
LIFE HISTORY
AMER. MUS. NAT. HIST. BULL. 10: 442, 1898 BEUTEN. LARVA
AMER. NAT. 19:716, 1885 PACKARD PUPA (NOTE ON)
AMER. PHI. SOC. PROC. 31:157, 1893 PACKARD EGG, LARVA*

PL. 6, FF. 3, 4
CAN. ENT. 6:87, 1874 GENTRY LARVA(NORMAL AND THREE VARIETIES)
ENT. NEWS 9:150, 1898 SOULE DURATION OF EGG
ENS. SOC. AMERICA, ANNALS 7:281 1914 MOSHER PUPAL STRUCTS.
ILLINOIS DEPT. AGRI. TRANS. 15: 196, 1877 FRENCH LARVA
LOC. CIT. 18, APPEND:121, 1880 MARTEN LARVA
MAINE AGRI. EXP. STA. BULL. 162, 1908 PATCH LARVA* F. 51
MARYLAND NAT. HIST. SOC. PUBLICATION NO PAGE 1930 HAYDON EGG, LARVA
MICHIGAN STATE COLLEGE, SPEC. BULL243:55, 1933 MCDANIEL LARVA*
NAT. ACAD. SCI. MEM. 9:120, 1905 PACKARD LARVA* (PLS. 9, 15 COLOR; PL. 27 EGG AND LARVA UNCOLORED; PL. 28 LARVA UNCOLORED; PL. 47, F. 2 AND PL. 48, FF. 1-3 LARVAL STRUCTURE) PUPA* PL. 56, F. 5
NATURALISTE CANADIEN P. 251, 1929 HUARD LARVA
NEW YORK STATE CABIN, NAT. HIST. REPT. NO. 24:150, 1871 LINTNER EGG, LARVA
LOC. CIT. NO. 51:344, 1897 LINTNER EGG, LARVA, PUPA SPINES OF LARVA* PL. 2
NEW YORK STATE MUS. MEM. 8: 677, 1905 FELT LARVA* PL. 19, F. 1 COLOR
NEW YORK STATE REPT. INS. OF 4, 742, ALSO P. 56, SEC 271 FITCH LARVA (BRIEF)
OHIO AGRI. EXP. STA. BULL. 332: 265, 1918 HOUSER-FYLES LARVA
ONTARIO ENT. SOC. ANN. REPT. P. 31, 1910 FYLES LARVA
PHILADELPHIA ACAD. NAT. SCI. PROC. 25:274, 1872 GENTRY LARVAL NOTE
PSYCHE 9:299, 1902 SOULE HATCHING OF EGG
SMITHSONIAN MISC. COLLECTIONS 4, ART. 2:230, 1862 MORRIS LARVA, PUPA
SOCIETE ACCLIM. BULL. FRANCE SERIES 3, VOL. 9:268, 1882 EGG, PUPA
UNITED STATES DEPT. AGRI. MISC. PUB. 657:357, 384, 1950 CRAIGHEAD LARVA* F. 74 BO
ABGOT & SMITH LEP. INS. GEORGIA 2:109, 1797 LARVA* PL. 55 COLOR
CRAGEN-OUR INS. FRIENDS & FOES P. 118, 1899 LARVA
DUNCAN-NATURALISTA LIBRARY 22:160, 1858 LARVA
ELIOT & SOULE-CATER. AND THEIR MOTHS P. 271, 1902 EGG, LARVA* PUPA*
EMMONS-NAT. HIST. NEW YORK 5:234, 1854 LARVA
FERNALD-APPLIED ENT. P. 284, 1926 LARVA* F. 294
FERNALD & SHEPHARD-APPLIED ENT. P. 281, 1942 LARVA * F. 289
GLOVER-ILLUST. NORTH AMER. ENT. 1878 LARVA* PL. 2, F. 7; PL. 6, F. 28
GOSSE-LETTERS FROM ALABAMA P. 172, 1859 LARVA, PUPA (BRIEF)
HARRIS-INS. INJ. VEG. 2ST EDIT. P. 290, 1841 LARVA, PUPA
HARRIS-LOC. CIT. 2ND EDIT. P. 310, 1852 LARVA
HARRIS-LOC. CIT. FLINT EDIT. P. 404, 1862 LARVA*
HERRICK-INS. ENEMIES SHADE TREES P. 158, 1935 LARVA
KELLOGG-AMER. INS. P. 426, 1908 LARVA
MORRIS-SYNOP. LEP. NORTH AMERICA P. 230, 1862 LARVA, PUPA

PackardOIns. Inj. Forest Trees
p. 203, 1881 larva
Westwood-On Drury's Illust.
Exotic Ent. 1:17, 1837
larva, pupa
Index Amer. Econ. Ent.
Vol. 1:47 Basilona
Vol. 2:49 Basilona
Vol. 3:53 Basilona
Vol. 4:49 Basilona
Vol. 5:72 Basilona
Vol. 5:206; Vol. 6:241;
Vol. 7:316

FOOD PLANTS

Abies grandis (Dougl.)
Acer sp. (maples)
Acer negundo Linn
Acer pseudoplatanus Linn.
Acer rubrum Linn.
Acer saccharinum Linn.
Acer sacharophorum Koch
Aesculus hippocastanum Linn.
Alnus sp. (alders)
Alnus serrulata (Ait.)
Betula alba Linn.
Betula alba v. laciniata DeVos
Betula papyrifera Marsh.
Betula populifolia Marsh.
Carpinus carokinaiana Walt.
Carya sp. (hickories)
Castanea dentata (Marsh. 0
Castanea pumila (Linn.)
Castanea sativa Mill.
Chamaecyparis thyoides (Linn.)
Cornus sp.
Crataegus sp. (thrrorns)
Fagus grandifolia Ehrh.
Fagus sylvatica Linn.
Gleditsia triacanthos Linn.
Gossypium herbaceum Linn. (cotton)
Juglans cinerea Linn.
Juniperus commonus Linn.
Juniperus virginiana Linn.
Koelreuteria paniculata Laxm.
Laris laricina (DuRoi)
Lindera benzoin Linn.
Liquidambar styraciflua Linn
Myrica cerifera Linn.
Ostrya virginiana (Mill.)
Picea sp. (spruces)
Pinus sp. (pines)
Pinus banksiana Lamb.
Pinus echinata Mill.
Pinus griffithii McCl.
Pinus strobus Linn.
Platanus occidentalis Linn.
Platanus orientalis Linn.
Prunus serotina Ehrh.
Prunus virginiana Linn.
Quercus coccinea Muench.
Quercus laevis Walt.
Quercus macrocarpa Michx.
Quercus palustris Muench
Quercus rubra Linn.
Rhus glabra Linn.
Rosa sp. (roses)
Salix sp. (willows)
Sambucus canadensis Linn.
Sassafras albidum v. molle (Raf.)
Taxodium distichum (Linn.)
Tilis sp. (lindens)
Tilia americana Linn.
Tilia europaea Linn.
Tilia heterophylla Vent.
Tsuga canadensis (Linn.)
Ulmus alata Michx.
Ulmus americana Linn.
Ulmus procera Salisb.
Ulmus rubra Muhl.
Ulmus thomasii Sarg.
Viburnum dentatum Linn.

imperturbata Bird
syn. see necopina Grote

impingens Dyar
syn. see manifesta Morrison

impleta Walker, Acronicta-Noctuidae
syn. luteicoma Grote & Robinson
ab krautwormi Chermock
a illita Smith
LIFE HISTORY
Amer. Mus. Nat. Hist. Bull. 14:

257,1901 BEUTEN. LARVA (LUTEICOMA)
CAN. ENT. 27:18,1894 DYAR LARVA
NEW YORK STATE AGRI. EXP. STA. (CORNELL), MEM. 56:1073, 1922 WELLHOUSE LARVA (BRIEF)
NEW YORK STATE MUS. MEM. 8: 749,1905 FELT LARVA(BRIEF)
PAPILIO 3:16,1883 THAXTER LARVA
UNITED STATES NAT. MUS. PROC. 21:153,1898 SMITH & DYAR LARVA*PL. 6,F. 16 COLOR OR UNCOLORED;PUPA;COCOON (LUTEICOMA)
LOC. CIT. 27:807,1904 DYAR EGG, LARVA (ILLITA)
INDEX AMER. ECON. ENT. VOL. 7:8

FOOD PLANTS
ACER SP. (MAPLES)
ALNUS SP. (ALDER)
ALNUS RUBRA BONG.
BETULA SP. (BIRCHES)
CRATAEGUS SP. (THORNS)
FRAXINUS SP. (ASHES)
HAMAMELIS VIRGINIANA LINN.
JUGLANS NIGRA LINN.
MALUS FLORIBUNDA(LINDL),
MALUS PUMILA MILL. (APPLE)
POPULUS SP. (POPLARS)
PRUNUS CERASUS LINN.
PRUNUS VIRGINIANA LINN.
QUERCUS SP. (OAKS)
QUERCUS ROBUR LINN.
SALIX SP. (WILLOWS)
TILIS SP. (LINDENS)
ULMUS SP. (ELMS)

IMPLICATA(GUENEE), EUPHYIA-GEOMETRIDAE
A MULTILINEATA PACKARD
SYN. WILLIAMSI SWETT
B GRANDIOSA HULST
LIFE HISTORY
SOUTHERN CALIFORNIA ACAD. SCI. BULL. 38:198,1939 COMSTOCK LARVA* PL. 52;PUPA* PL. 53

FOOD PLANTS
ABRONIA UMBELLATA LAM.

IMPLICATA WALKER
SYN. SEE ANTICARIA WALKER (EUPITHECIA)

IMPOLITA MORRISON
SYN. SEE CRISTIFERA WALKER

IMPRESSA WALKER, ACRONICAT-NOCTUIDAE
SYN. FASCIATA WALKER
SYN. VERRILLI GROTE & ROBINSON
A EMACULATA SMITH
B LEMMERI CHERMOCH & CHERMOCK
LIFE HISTORY D
AMER. MUS. NAT. HIST. BULL. 14; 259,1901 BEUTEN. LARVA
CAN. ENT. 47:174,1915 BARNES & MCDUNNOUGH EGG, LARVA
MINNESOTA AGRI. EXP. STA. BULL. 61:187,1899 LUGGER LARVA* F. 134
NATURALISTA CANADIEN P. 7,1936 DAVIAULT LARVA, PUPA
NEW YORK STATE MUS. MEM. 8:743 1905 FELT LARVA(BRIEF)
ONTARIO ENT. SOC. ANN. REPT. P. 13,1873 SAUNDERS LARVA, PUPA(VERRILLI)
UNITED STATES NAT. MUS. PROC. 21:160,1898 SMITH & DYAR LARVA* PL. 8,FF. 33,34 COLOR OR UNCOLORED;PUPA; COCOON
INDEX AMER. ECON. ENT. VOL. 7:8

FOOD PLANTS
CORYLUS SP. (HAZELS)
HAMAMELIS VIRGINIANA LINN.
MALUS PUMILA MILL. (APPLE)
POPULUS SP. (POPLARS)
PRUNUS DOMESTICA LINN. (PLUM)
RIBES SATIVUM SYME(CURRANT)
ROSA SP. (ROSES)
RUBUS ALLEGHENIENSIS PORTER
RUBUS IDAEUS V. STRIGOSUS(MICHX) (RASPBERRY)

SALIX SP. (WILLOWS)
FOREST TREES

IMPROBA BUTLER
SSP. SEE FRIGGA THUNBERG

IMPROPRIATA WALKER
SYN. ZEE SUBATOMARIA WOOD

IMPROVISA HENRY EDWARDS, CYMATOPHORA-THYATIRIDAE
LIFE HISTORY
BRITISH COLUMBIA ENT. SOC. PROC. 1 NEW SERIES P. 32, 1911 G. O. DAY EGG, LARVA
CALIFORNIA ACAD. SCI. PROC. 5: 190, 1873 EDWARDS CHRYSALIS
FOOD PLANTS
PRUNUS VIRGINIANA LINN.

IMPUNCTATA FISCHER
SSP. SEE ZELICAON LUCAS

INAEQUALIS TUTT
SSP. SEE PSEUDARGIOLUS BOISDUVAL & LECONTE

INATOMARIA GUENEE, METANEMA-GEOMETRIDAE
LIFE HISTORY
NEW YORK ENT. SOC. JOURN. 53: 198, 1945 FORBES PUPA
NEW YORK STATE AGRI. EXP. STA. (CORNELL), MEM. 274:91, 1948 FORBES LARVA (BRIEF)
FOOD PLANTS
POPULUS SP. (POPLARS)
POPULUS DELTOIDES MARSH.

INATTENATA SMITH
SSP. SEE PERATTENTA GROTE

INCALLIDA SMITH
SSP. SEE QUINQUELINEA SMITH

INCANA EDWARDS
SSP. SEE ALOPE FABRICIUS

INCANDESCENS GROTE
SYN. SEE DEDUCTA MORRISON

INCARCERATA BOISDUVAL
SYN. SEE APICALIS WALKER

INCARNATA RILEY
SSP. SEE VANILLAE LINNAEUS

INCARNATA STRETCH
SSP. SEE PHALERATA HARRIS

INCARNATA WALKER
SYN. SEE AULAEA GEYER (ARACHNIS)

INCARNATA WALKER, LERINA-ARCTIIDAE
SYN. ROBINSONII BOISDUVAL
LIFE HISTORY
WASHINGTON ENT. SOC. PROC. 10: 16, 1908 DYAR LARVA
FOOD PLANTS
ASCLEPIAS SP. (MILKWEEDS)

INCARNATORUBRA GOEZE
SYN. SEE ARGE DRURY

INCEPTARIA WALKER
SYN. SEE METONALIS WALKER

INCERTATA (WALKER), MESOTHEA-GEOMETRIDAE
SYN. OPORARIA ZELLER
SYN. GRATATA ZELLER
LIFE HISTORY
FOOD PLANTS
VACCINIUM CORYMBOSUM LINN.

INCINCTA (MORRISON), ORTHODES-NOCTUIDAE
LIFE HISTORY
JOURN. ECON. ENT. 13:148, 1920 WADLEY LARVA, PUPA
LOC. CIT. 30:695-699, 1937 WALKDEN EGG, LARVA, PUPA
FOOD PLANTS
HORDEUM PUSILLUM NUTT.
MEDICAGO SATIVA LINN. (ALFALFA)
TRITICUM AESTIVUM LINN. (WHEAT)
GENERAL FEEDER VEG. CROPS

INCISARIA WALKER
SYN. SEE OBTUSARIA HUBNER

INCIVIS GUENEE
SYN. SEE INFECTA OCHSENHEIMER

INCLINATARIA WALKER
SYN. SEE FERRUGATA LINNAEUS

INCLUDENS (WALKER), SPARTINIPHAGA-NOCTUIDAE
SYN. NORMA MORRISON
SYN. PENITA MORRISON
SYN. MARIAE GROTE
FORM LUNARIS STRAND
LIFE HISTORY
INSECTUR INSCITIAE MENSTRUUS 1: 124,1913 BIRD LARVA
FOOD PLANTS
CAREX STRICTA LAM.

INCLUSA HUBNER, ICHTHYURA-NOTODONTIDAE
SYN. AMERICANA HARRIS
A INVERSA PACKARD
SYN. PALLA FRENCH
B JOCOSA HENRY EDWARDS
LIFE HISTORY
AMER. MUS. NAT. HIST. BULL. 10: 409,1898 BEUTEN. LARVA
BOSTON SOC. NAT. HIST. PROC. 24: 516,1890 PACKARD LARVA
CAN. ENT. 14:34,1882 FRENCH LARVA (PALLA)
LOC. CIT. 17:41,1885 FRENCH EGG, LARVA, PUPA (PALLA)
ENT. NEWS 26:227,1915 GIRAULT LIFE HISTORY
MAINE AGRI. EXP. STA. BULL. 259: 61,1917 MOSHER PUPA* FF. 5 C, I
MINNESOTA AGRI. ESP. STA. BULL. 61:201,1899 LUGGER LARVA* F. 134
NAT. ACAD. SCI. MEM. 7:133,1895 PACKARD LARVA* PL. 16, FF. 1, 2 COLOR: PUPA* COCOON
LOC. CIT. 9,1905 PACKARD LARVAL HEAD* PL. 44, F. S AMERICANA
NEW YORK ENT. SOC. JOURN 1:23, 1893 PACKARD EGG, LARVA
NEW YORK STATE AGRI. EXP. STA. (CORNELL), MEM. 274:211, 1948 FORBES LARVA
NEW YORK STATE MUS. MEM. 8: 560,1905 FELT LARVA* PL. 16, F. 1 COLOR
NEW YORK STATE REPT. INS. OF 5: 845, ALSO P. 65, SEC. 351, 1858 FITCH LARVA
OHIO AGRI. EXP. STA. BULL. 332: 256,1918 HOUSER LARVA* PL. 38, F. 4
PAPILIO 3:24,1883 EDWARDS LARVA
PSYCHE 5:262,1889 ELLIOT & SOULE EGG, LARVA, PUPA
UNITED STATES DEPT. AGRI. MISC. PUB. 657:354,396,1950 CRAIGHEAD LARVA* F. 79
ABBOT & SMITH-LEP. INS. GEORGIA 2:143,1797 LARVA* PL. 72 COLOR
CRAGEN-OUR INS. FRIENDS & FOES P. 163,1899 LARVA
GLOVER-ILLUST. NORTH AMER. ENT. 1878 LARVA, PUPA, COCOON PL. 6, F. 13
HARRIS-ENT. CORRESP. P. 310,1869 LARVA* PL. 3
HARRIS-INS. INJ. VEG. 1ST EDIT. P. 314,1841 LARVA
HARRIS-LOC. CIT. 2ND EDIT. P. 333 1852 LARVA
HARRIS-LOC. CIT. FLINT EDIT. P. 431 1862 LIFE HISTORH* COLOR
HERRICH-INS. ENEMIES SHADE TREES P. 277,1935 LARVA* F. 235
PACKARD-INS. INJ. FOREST TREES P. 122,1881 LARVA, PUPA
INDEX AMER. ECON. ENT.
VOL. 1:203; VOL. 2:228; VOL. 3:254; VOL. 4:295; VOL. 6:367; VOL. 7:647
FOOD PLANTS
ACER SP. (MAPLES)
POPULUS SP. (POPLARS)

Populus tremuloides Michx.
Salix sp. (willows)
Ulmus sp. (elms)

inclusaria Walker
syn. see lixaria Guenee

incoloraria Walker
syn. see basiaria Walker

incolorata Dyar
ssp. see sexmaculata Packard

incommodata Walker
syn. see magnoliata Guenee

incompleta Butler
syn. see radians Walker

incongruaria Hulst
ssp. see quercivoraria Guenee

inconstans (Geyer), Eutrepsia-Geometridae
syn. cephisaria Grote
LIFE HISTORY
Ent. News 43:16, 1932 Wickwire
egg, larva, pupa
Southern California Acad. Sci. Bull. 34:151, 1935 Comstock
egg* Pl. 44 cephisaria
FOOD PLANTS
Breynia nivosus v. roseo pictua Hort.
Monarda fistulosa Linn.

incorrupta Henry Edwards
syn. see nevadensis Grote & Robinson

incredibilis Bryk
ssp. see clodius Menetries

incursata (Hubner), Xanthorhoe-Geometridae
a lagganata Swett & Cassino
b harveyata Swett & Cassino
LIFE HISTORY
Gumppenberg-Systema Geometrarum 3:350, 1890 larva, pupa
FOOD PLANTS
Vaccinium sp.
Vaccinium myrtillus Linn.
Vaccinium uliginosum Linn.

incurva Henry Edwards
ssp. see fragilis Stretch

incurvata Guenee
ssp. see transversata Drury

indeclinata Walker
syn. see hypochraria Herrich-Schaeffer

indentata Harvey
ssp. see latipes Guenee

Indentata Packard
syn. see apicalis Walker

indeterminata Walker
syn. see guttivitta Walker

indicans Walker
syn. see laudabilis Guenee

indicataria Walker
ssp. see porcelaria Guenee

indigens Walker
syn. see videns Guenee

indirecta Walker
syn. see messoria Harris

indirecta (Walker), Pyreferra-Noctuidae
syn. see moffatiana Grote
LIFE HISTORY
Connecticut Agri. Exp. Sta. Bull. 234:195, 1921 Britton larva* f. 13 a
Ent. Amer. 4:59, 1888 Pearsall
larva, pupa (moffatiana)
New York State Mus. Mem. 8: 723, 1905 Felt larva (brief)
FOOD PLANTS
Hamamelis virginiana Linn.

INDISTINCTA HENRY EDWARDS
SSP. SEE ANGELICA GROTE

INDISTINCTA SMITH, HYPPA-NOCTUIDAE
LIFE HISTORY
UNITED STATES NAT. MUS. PROC. 27:816, 1904 DYAR EGG, LARVA
FOOD PLANTS

INDISTINCTA STRAND
SSP. SEE PENSILIS GROTE

INDOMITA (WALKER), MELIPOTIS-NOCTUIDAE
SYN. NIGRESCENS GROTE & ROBINSON
SYN. MALE OCHREIPENNIS HARVEY
LIFE HISTORY
BROOKLYN INST. ARTS. SCI. BULL. 1:201, 1906 DYAR LARVA
PAN PACIFIC ENT. 1:94, 1924 BRISLEY LARVA
FOOD PLANTS
PROSOPIS GLANDULOSA TORR.

INDRA REAKIRT, PAPILIO-PAPILIONIDAE
A PERGAMUS HENRY EDWARDS
B MINORI CROSS
LIFE HISTORY
SOUTHERN CALIFORNIA ACAD. SCI. BULL. 27:82, 1928 COMSTOCK LIFE HISTORY EGG* LARVA* CHRYSALIS* (PERGAMUS)
EDWARDS-BUTTER. NORTH AMERICA 3RD SERIES SUPPL. NOTES 1897 EGG
INDEX AMER. ECON. ENT. VOL. 2:264
FOOD PLANTS
ARTEMISIA GLAUCA PALL.
SANICULA GRAVEOLENS POEPP. EX DC.
TAUSCHIA PARISHII C. & R.
TORILIS JAPONICA (HOUTT.)

INDUBITATA GROTE
SYN. SEE HAESITATA GUENEE

INDUCTA DYAR
SSP. SEE TIBRIS DYAR

INDUCTATA (GUENEE), SCOPULA-GEOMETRIDAE
SYN. ANTICARIA WALKER
FORM SOBRIA WALKER
SYN. CONSECUTARIA WALKER
SYN. SUPPRESSARIA WALKER
A DELICATA CASSINO
B OLIVEATA CASSINO
LIFE HISTORY
NEW YORK STATE AGRI. EXP. STA. (CORNELL), MEM. 274:123, 1948 FORBES LARVA
PSYCHE 9:239, 1901 DYAR EGG, LARVA
HART-INS. COLLECTIONS ILLINOIS HIGH SCHOOLS P. 51, 1903 COCOON
FOOD PLANTS
AMBROSIA TRIFIDA LINN. (IN CAPTIVITY)
TRIFOLIUM SP. (CLOVERS) (IN CAPTIVITY)

INDUCTURA STALLINGS
SSP. SEE MACHAON LINNAEUS

INDURATA DYAR
SYN. SEE MONICARIA GUENEE

INDURATA SMITH
SSP. SEE CURIALIS GROTE

INEFFUSARIA GUENEE
SYN. SEE CONFUSARIA HUBNER

INELEGANS SMITH
SYN. SEE EXSERTISTIGMA MORRISON

INEPTA WALKER
SYN. SEE FRUGIPERDA ABBOT & SMITH

INERMIS HARRIS
SYN. SEE MARGARITOSA HAWORTH

INES EDWARDS
SSP. SEE LEDA EDWARDS

INEXTRICATA WALKER
SYN. SEE MESSORIA HARRIS

INFANS MOESCHLER, BREPHOS-GEOMETRIDAE
SYN. HAMADRYAS HARRIS
A OREGENENSIS SWETT
LIFE HISTORY
CAN. ENT. 29:272, 1897 BRAINERED EGG, LARVA, PUPA
NATURALISTE CANADIEN P.329, 1935 DAVIAULT LARVA
NEW YORK STATE AGRI. EXP. STA. (CORNELL), MEM. 274:14, 1948 FORBES LARVA
OTTAWA NATURALIST 11:12, 1898 FLETCHER LARVAL HABITS
UNITED STATES DEPT. AGRI. MISC. PUB. 657:360, 422, 1950 CRAIGHEAD LARVA
FOOD PLANTS
BETULA PAPYRIFERA MARSH.
BETULA POPULIFOLIA MARSH.
POPULUS SP.

INFECTA (OCHSENHEIMER), ANICLA-NOCTUIDAE
SYN. INCIVIS GUENEE
SYN. ALABAMAE GROTE
LIFE HISTORY
CAN. ENT. 26:18, 1894 DYAR EGG, LARVA
ILLINOIS AGRI. EXP. STA. BULL. 60: 499, 1900 FORBES LARVA
JOURN. ECON. ENT. 8:393, 1915 CRUMB KEY TO LARVA
UNITED STATES DEPT. AGRI. TECH. BULL. 88:104, 1929 CRUMB EGG* PL. 5 B, C; LARVA* PL. 7, G. L, M; PUPA* PL. 18 C, D
UNITED STATES NAT. MUS. PROC. 23:273, 1901 DYAR EGG, LARVA PUPA
INDEX AMER. ECON. ENT. VOL. 4:282 LYCOPHOTIA
FOOD PLANTS
BETA VULGARIS LINN. (BEET)
CENCHRUS PAUCIFLORUS BENTH.
DIGITARIA SANGUINALIS (LINN.)
MUHLENBERGIA SP.
NICOTIANA TABACUM LINN. (TOBACCO)
PLANTAGO SP.
POA PRATENSIS LINN.
PORTULACA OLERACEA LINN.
TRIFOLIUM SP. (CLOVERS)
GRASSES

INFECTA WALKER
SYN. SEE RENIGERA STEPHENS

INFECTATA WALKER
SYN. SEE GNOPHOSARIA GUENEE

INFELIX SMITH
SSP. SEE EXCELLENS GROTE

INFERIOR GROTE, BLEPTINA-NOCTUIDAE
LIFE HISTORY
ENT. AMER. 14:171, 1934 CRUMB LARVA* PL. 23C; PL. 24 I, MB
FOOD PLANTS
DEAD LEAVES

INFERNALIS STRECKER
SSP. SEE REGALIS FABRICIUS

INFICITA (WALKER), MARATHYSSA-NOCTUIDAE
SYN. HISTRIO GROTE
A MINUS DYAR
LIFE HISTORY
FOOD PLANTS
RHUS SP.

INFIXARIA WALKER
SYN. SEE COGNATARIA HUBNER

INFLAVA SMITH
SYN. SEE HIBISCI GUENEE

INFRACTA (MORRISON), EUXOA-NOCTUIDAE
SYN. RUBEFACTALIS GROTE
LIFE HISTORY
FOOD PLANTS
POA PRATENSIS LINN.
GRASSES

INFRAPICTA STRAND
SSP. SEE PACIFICA HARVEY

INFRUCTUOSA WALKER
SYN. SEE CONFUSA HUBNER

INFUMATA BARNES & MCDUNNOUGH
SSP. FERRUGATA LINNAEUS

INFUMATA GROTE
SSP. SEE DECOLOR WALKER

INFUMATA OBERTHUR
SYN. SEE IDALIA DRURY

INFUSA DYAR
SSP. SEE PERSPICUA GROTE & ROBINSON

INGENS HENRY EDWARDS, HALISIDOTA-ARCTIIDAE
SYN. SCAPULARIS STRETCH
LIFE HISTORY
SOUTHERN CALIFORNIA ACAD. SCI. BULL. 36:115, 1937 COMSTOCK LARVA, PUPA* COCOON* PL. 48
FOOD PLANTS
IRIS SP.?
PINUS ECHINATA MILL.
PINUS EDULIS ENGELM

INGHAMI GUNDER
SSP. SEE SARA BOISDUVAL

INGHAMI GUNDER
SSP. SEE LACINIA GEYER

INGHAMI GUNDER
SSP. SEE PHILENOR LINNAEUS

INHERITA SMITH
SYN. SEE HIBISCI GUENEE

INNOTATA GUENEE, ACRONICTA-NOCTUIDAE
SYN. GRAEFI GROTE
AB RISEOR DYAR
LIFE HISTORY
AMER. MUS. NAT. HIST. BULL. 14:241, 1901 BEUTEN. LARVA
ENT. NEWS 2:153, 1891 BEUTEN. LARVA
NEW YORK STATE MUS. MEM. 8:716, 1905 FELT LARVA
UNITED STATES NAT. MUS. PROC. 27: 801, 1904 DYAR EGG, LARVA (GRISEOR)
LOC. CIT. 21:74, 1898 SMITH DYAR LARVA COCOON
FOOD PLANTS
ABIES BALSAMEA (LINN.)
BETULA SP. (BIRCHES) GRISEOR
CARYA SP. (HICKORIES)

INNOTATA (HUFNAGEL), EUPITHECIA-GEOMETRIDAE
A WINNATA TAYLOR
B ALBERTA TAYLOR
C KOOTENAIATA DYAR
LIFE HISTORY
NEW YORK STATE AGRI. EXP. STA. (CORNELL), MEM. 274:171, 1948 FORBES LARVA
FOOD PLANTS
ARTEMISIA SP.
FRAXINUS SP. (ASHES)
TAMARIX CHINENSIS LOUR.

INNUBENS GUENEE, CATOCALA-NOCTUIDAE
SYN. FEMALE HINDA FRENCH
SYN. INNUBENTA STRAND
AB FLAVIDALIS GROTE
FORM SCINTILLANS GROTE & ROBINSON
LIFE HISTORY
AMER. MUS. NAT. HIST. BULL. 16: 388, 1902 BEUTEN. LARVAL HEAD* F. 11
LOC. CIT. 28:147, 1918 BEUTEN. EGG, LARVA
AMER. MUS. NAT. HIST. MEM. NEW SERIES 3, PART 1, 1918 BARNES & MCDUNNOUGH EGG, LARVA* PL. 10, F. 35; PL. 11, F. 10 COLOR
CAN. ENT. 20:170, 1888 FRENCH LARVA, PUPA
LOC. CIT. 21:107, 1910 ROWLEY-BERRY LARVA
NEW YORK STATE MUST. MEM. 8: 729, 1905 FELT LARVA
HOLLAND-MOTH BOOK 1905 LARVA* PL. 1, F. 7 COLOR
FOOD PLANTS
GLEDITSIA TRIACANTHOS LINN.

JUGLANS NIGRA LINN.

INNUBENTA STRAND
SYN. SEE INNUBENS GUENEE

INO HALL
SYN. SEE ALOPE FABRICIUS

INOPINATUS SMITH
SSP. SEE HARUSPICA GROTE

INORATA GROTE
SYN. SEE MARGANA FABRICIUS

INORATA GROTE & ROBINSON
SYN. SEE FALACER GODART

INORDINARIA WALKER
SYN. SEE GRANITATA GUENEE

INORNATA BEUTENMULLER
?SYN. SEE DETRITA GUERIN-MENEVILLE

INORNATA EDWARDS, COENONYMPHA-SATYRIDAE
SYN. QUEBECENSIS BARNES & BENJAMIN
A MCSAACI DOS PASSOS
B BENJAMINI McDUNNOUGH
C INSULANA McDUNNOUGH
D COLUMBIANA McDUNNOUGH
LIFE HISTORY
MUS. COMP. ZOOL. BULL. 87, NO. 4: 263, 1941 DAVENPORT LIFE HISTORY EGG, LARVA* F. 32; CHRYSALIS* F. 33
FOOD PLANTS
POA PRATENSIS LINN. (IN CAPTIVITY)
PROBABLY MANY SPECIES OF GRASS

INORNATA GROTE
SSP. SEE PYRAMIDOIDES GUENEE

INORNATA NAKAHARA
SSP. SEE ASTYANAX FABRICIUS

INORNATA NEUMOEGEN, ICHTHYURA-NOTODONTIDAE
LIFE HISTORY
FOOD PLANTS

POPULUS FREMONTI S. WATS

INORNATA OTTOLENGUI
SYN. SEE FUCOSA HUBNER

INORNATA WOLLCOT
SSP. SEE CELTIS BOISDUVAL & LECONTE

INQUAESITA (GROTE & ROBINSON), PAPAIPEMA-NOCTUIDAE
FORM WYATTI BARNES & BENJAMIN
LIFE HISTORY
CAN. ENT. 33:61, 1901 BIRD LARVA, PUPA
FOOD PLANTS
ONOCLEA SENSIBILIS LINN. (ROOTS)

INQUIETA WALKER
SYN. SEE TRIFOLII ROTTENBURG

INQUINARIA HULST
SYN. SEE QUADRILINEARIA PACKARD

UNSARIA DYAR
SYN. SEE COGNATARIA HUBNER

INSCIENS WALKER
SSP. SEE HIBISCI GUENEE

INSCRIPTUM (HARRIS), DEIDAMIA-SPHINGIDAE
LIFE HISTORY
AMER. MUS. NAT. HIST. BULL. 7: 285, 1895 BEUTEN. LARVA, PUPA
LOC. CIT. 16:395, 1902 BEUTEN. EGG, LARVA
ENT. NEWS 4:310, 1893 DYAR-DOLL LARVA
ENT. SOC. AMERICA, ANNALS 11:436, 1918 MOSHER PUPAL STRUCTURE* PL. 36, F. 8
MINNESOTA AGRI. EXP. STA. BULL. 61:88, 1899 LUGGER LARVA
NEW YORK STATE AGRI. EXP. STA. (CORNELL), MEM. 274:198 1948 FORBES LARVA
NEW YORK STATE MUS. MEM. 8: 740, 1905 FELT LARVA (BRIEF(

NOVITATES ZOOLOGICAE(TRING)9
SUPPL. P.604,1903 ROTHSCHILD
JORDAN LARVA, PUPA
PSYCHE 5:397,1890 PACKARD EGG, LARVA
PSYCHE 6:116,149,1891 SOULE LARVA
PSYCHE 7:317,1895 SOULE LARVA
PSYCHE 8:105,1897 SOULE EGG, LARVA
SMITHSONIAN MISC. COLLECTIONS 4, ART 2:159,1862 MORRIS PUPS
UNITED STATES DEPT. AGRI. MISC. PUB. 657:359,1950 CRAIGHEAD LARVA
ELIOT & SOULE-CATER. &MOTHS P. 89,1902 LARVA* PUPA
MORRIS-SYNOP. LEP. NORTH AMERICA P. 159,1862 PUPA
STRECKER-LEP. RHOP. ET HETER. 13: 112,1876 LARVA
INDEX AMER. ECON. ENT. VOL. 2:110

FOOD PLANTS
OXYDENDRUM ABOREUM (LINN.)
PARTHENOCISSUS QUINQUEFOLIA (LINN.)
PARTHENOCISSUS TRICUSPIDATA (S. & Z.)
VITIS SP. (GRAPES)
VITIS VINIFERA LINN. (GRAPE)

INSIGNALIS WALKER
SYN. SEE ASOPIALIS GUENEE

INSIGNATA STRAND
SSP. SEE RENIFORMIS GROTE

INSIGNATA WALKER
SYN. SEE DUBITANS WALKER

INSIGNATA WALKER
SYN. SEE TESSELLATA HARRIS

INSIGNATA WALKER
SSP. SEE OCHROGASTER GUENEE

INSIGNIFICATA TAYLOR
SYN. SEE MISTURATA HULST

INSIGNIS WALKER
SYN. SEE COMPRESSIPALPIS GUENEE

INSINUARIA GUENEE
SYN. SEE AMICARIA HERRICH-SCHAEFFER

INSOLABILELLA STRAND
SYN. SEE INSOLABILIS GUENEE

INSOLABILIS GUENEE, CATOCALA-NOCTUIDAE
SYN. INSOLABILELLA STRAND

LIFE HISTORY
AMER. MUS. NAT. HIST. BULL. 19: 509,1903 BEUTEN. LARVA
AMER. MUS. NAT. HIST. MEM. NEW SERIES 3, PART 1,1918 BARNES & MCDUNNOUGH LARVA* PL. 10, F. 23; PL. 12, F. 11 COLOR

FOOD PLANTS
CARYA SP. (HICKORIES)

INSOLITA GROTE
SSP. SEE OBLINITA ABBOT & SMITH

INSOLITA SMITH
SYN. SEE FLAGELLUM WALKER

INSPERATA GROTE
SYN. SEE VALENS HENRY EDWARDS

INSPINOSA GUENEE
SYN. SEE MALEFIDA GUENEE

INSTABILIS MARTYN
SYN. SEE JUGLANDIS ABBOT & SMITH

INSUETA GUENEE, LEUCANIA-NOCTUIDAE
SYN. ADONEA GROTE
SYN. MIMICA STRECKER
A HETERODOXA SMITH
B DIA GROTE
C MEGADIA SMITH

LIFE HISTORY
UNITED STATES NAT. MUS. PROC. 27L865,1904 DYAR EGG,

LARVA(HETERODOXA)

FOOD PLANTS

GRASSES

INSULANA McDUNNOUGH

SSP. SEE INORNATA EDWARDS(COENONYMPHA)

INSULANUS BLACKMORE

SSP. SEE SAEPIOLUS BOISDUVAL

INSULARIS CLARK

SSP. SEE PICTA PACKARD

INSULARIS VOLL

SYN. SEE WALKERI FELDER & FELDER

INSULATA(WALKER), CYCNIA-ARCTIIDAE

SYN. CADAVEROSA GROTE
SYN. AFFINIS GROTE
SYN. AURATA BUTLER

LIFE HISTORY

NEW YORK ENT. SOC. JOURN. 8:33, 1900 SWAINSON LARVA, COCOON

FOOD PLANTS

HERNANDIA SONORA LINN.

INSULIDENS, BIRD, PAPAIPEMA-NOCTUIDAE

LIFE HISTORY

CAN. ENT. 53:79, 1921 BIRD, LARVA, PUPA

FOOD PLANTS

SENECIO HYDROPHILUS NUTT.

INSULSA WALKER

SYN. SEE TESSELLATA HARRIS

INSULSARIA(GUENEE), PLEUROPRUCHA-GEOMETRIDAE

SYN. ASTHENARIA WALKER
SYN. INVARIATA WALKER
SYN. IMPARATA WALKER
SYN. PERSIMILATA GROTE

LIFE HISTORY

CAN. ENT. 44:369, 1912 PROUT LARVA

ENT. AMER. 3:175, 1887 HULST LARVA, PUPA

OHIO JOURN. SCI. 23:95, 1923 AINSLIE EGG, LARVA, PUPA

HART-INS. COLLECTIONS ILLINOIS HIGH SCHOOLS P.50, 1903 PUPA

PACKARD-MONOGRAPH GEOMETRIDAE P.336, 1876 PUPA*PL.13, F.32

FOOD PLANTS

ACACIA RIPARIA H.BK.
CASSIA FASCICULATA MICHX.
CASTANEA DENTATA(MARSH.)
CELASTRUS SCANDENS LINN.
GALIUM SP.
PROSOPIS SP.
QUERCUS SP. (OAKS)
SOLIDAGO SP. (FLOWERS)
ZEA MAYS LINN. (CORN)

INTEGERRIMA GROTE & ROBINSON, DATANA-NOTODONTIDAE

A COCHISE DYAR

LIFE HISTORY

AMER. MUS. NAT. HIST. BULL. 10: 408, 1898 BEUTEN. LARVA

BOSTON SOC. NAT. HIST. PROC. 24: 518, 1890 PACKARD LARVA

CAN. ENT. 20:135, 1888 BEUTEN. KEY TO DATANA LARVA DESCR. OF LARVA

ILLINOIS AGRI. EXP. STA. BULL. 151: 470, 1911 FORBES LARVA* PUPA*

MAINE AGRI. EXP. STA. BULL. 259: 67, 1917 MOSHER PUPA*F.5G

MICHIGAN STATE COLLEGE, SPECL BULL. 243:29, 1933 McDANIEL LIFE HISTORY

MINNESOTA AGRI. EXP. STA. BULL. 61:163, 1899 LUGGER LARVA

MISSOURI AGRI. EXP. STA. BULL. 418:5, 1940 HASEMAN EGG* F.3; LARVA*FF.4-10; PUPA*F.11

NAT. ACAD. SCI. MEM. 7:120, 1895 PACKARD EGG, LARVA*PL.13, FF.106 COLOR

NEW YORK STATE AGRI. EXP. STA. (CORNELL), MEM. 274:216, 1948 FORBES LARVA

New York State Mus. Mem. 8:304 1905 Felt egg, larva* Pl. 15, ff. 1,2 color
Ohio Agri. Exp. Sta. Bull. 332: 226, 1918 Houser larva* Pl. 29, f. 5
Oklahoma Agri. Exp. Sta. Bull. 246:11, 1941 Huxon egg* f. 3; larva* ff. 4,5; pupa* f. 6
Philadelphia Ent. Soc. Proc. 6:13 1866 Grote & Robinson mature larva
Psyche 6:573, 1893 Dyar key to larva
United States Dept. Agri. Farmer's Bull. 843:39, 1917 Gill egg* f. 35; larva* f. 36; pupa* f. 38
Loc. cit. 1364:31, 1924 Gill egg* f. 38; larva* f. 39; pupa f. 41
Loc. cit. 1654:34, 1931 Moznett egg* f. 45; larva* f. 46; pupa* f. 48
United States Dept. Agri. Misc. Pub. 657:355, 399, 1950 Craighead larva* f. 80 B
Doane etc.-Forest Insects p. 294 1936 larva(f. 172
Hereich-Ins. Enemies Shade Trees p. 301, 1935 larva* ff. 254-256
Metcalf & Flint-Destr. & Useful Ins. p. 696, 1939 egg* f. 459; larva* f. 460; pupa
Quayle-Citrus & Subtropical Fruit p. 386, 1938 larva
Index Amer. Econ. Ent. Vol. 1:102; Vol. 2:109; Vol. 3:115; Vol. 4:133; Vol. 5:174; Vol. 6:198; Vol. 7:246

FOOD PLANTS
Carya sp. (hickories)
Carys illinoensis (Wang.)
Crataegus sp. (thorns)
Fagus sp. (beeches)
Gleditsia triacanthos Linn.
Juglans cinerea Linn.
Juglans nigra Linn.
Juglans regia Linn.
Juglans cordifolia v. ailanthifolia (Carr.)
Malus pumila Mill. (apple)
Quercus sp. (oaks)
Robinis pseudo-acacia Linn.
Salix sp. (willows)

<u>integerrima</u> Harris
syn. see <u>astylus</u> Drury

<u>integerrima</u> Walker
syn. see <u>amella</u> Guenee

<u>integraria</u> Walker
syn. see <u>liberaria</u> Walker

<u>intensalis</u> Smith
ssp. see <u>lubricalis</u> Geyer

<u>intentata</u> Walker
syn. see <u>variolaria</u> Guenee

<u>interalbicans</u> Warren
syn. see <u>aequaliaria</u> Grote

<u>interalis</u> Ottolengui
ssp. see <u>alias</u> Ottolengui

<u>interfaunus</u> Neumoegen
ssp. see <u>astylus</u> Drury

<u>interior</u> Scudder, <u>Colias</u>-Pieridae
form female <u>nepi</u> Barnes & Benjamin
a <u>laurentina</u> Scudder
LIFE HISTORY
Can. Ent. 29:249, 1897 Lyman life history egg, larva, chrysalis
New York State Mus. Mem. 8: 742, 1905 Felt larva
Scudder-Butter. East. United States & Canada pp. 1107-1110, 1889 egg, larva
FOOD PLANTS
Desmodium sp.
Salix sp. (willows)
Vaccinium crassifolium Andr.

VACCINIUM MYRTILLOIDES MICHX.

INTERJACENS GROTE
SYN. SEE PHLOXIPHAGA GROTE&ROBINSON

INTERLINEARIA GUENEE
SYN. SEE CONFUSARIA HUBNER

INTERMEDIA BUTLER
SSP. SEE BELLA LINNAEUS

INTERMEDIA CHERMOCK
SYN. SEE XERCES BOISDUVAL

INTERMEDIA COCKERELL
SYN. SEE EURYTHEME BOISDUVAL

INTERMEDIA CLARK
SSP. SEE SATELLITIA LINNAEUS

INTERMEDIA FITCH
SYN. SEE LEUCOSTIGMA ABBOT & SMITH

INTERMEDIA GRINNELL
SSP. SEE CARYE HUBNER

INTERMEDIA KIRBY
SSP. SEE GALLI ROTTENBERG

INTERMEDIA SPEYER, CUCILLIA-NOCTUIDAE
A CINDERELLA SMITH
LIFE HISTORY
NEW YORK STATE CABIN. NAT. HIST. ANN. REPT. NO. 23:213, 1873 LINTNER LARVA* PUPA* PL. 8, FF. 6,7
LOC. CIT. NO. 30:237, 1878 LINTNER LARVA(BRIEF)
FOOD PLANTS
COMPOSITAE SP.
LACTUCA BIENNIS (MOENCH)

INTERMEDIA STRECKER
SSP. SEE PSEUDARGIOLUS BOISDUVAL & LECONTE

INTERMEDIA (STRETCH) APANTESIS-ARCTIIDAE
FORM STRETCHII GROTE
LIFE HISTORY
NEW YORK ENT. SOC. JOURN. 9:25 1901 DYAR LARVA
FOOD PLANTS
GRASSES

INTERMEDIARIA BARNES & McDUNNOUGH, PLAGODIS-GEOMETRIDAE
LIFE HISTORY
CAN. ENT. 65:124, 1933 McD. EGG, LARVA, PUPA
FOOD PLANTS
TILIA SP. (LINDENS)

INTERMEDIATA GUENEE
SSP. SEE UNANGULATA HAWORTH

INTERNA PACKARD
SYN. SEE THYATIROIDES WALKER

INTERPOSITA DYAR
SSP. SEE MANTO STRECKER

INTERROGATIONIS (FABRIXIUS), POLYGONIA-NYMPHALIDAE
SYN. UMBROSA LINTNER
SYN. CRAMERI SCUDDER
FORM FABRICII EDWARDS
LIFE HISTORY
AMER. ENT. SOC. TRANS. 3:8, 1870 EDWARDS EGG, LARVA, CHRYSALIS
AMER. ENT. 1:39, 1869 LARVA
D AMER. MUS. NAT. HIST. BULL. 5:263 1893 BEUTEN. LARVA, CHRYSALIS
BUFFALO SOC. NAT. SCI. BULL. 19: 24, 1839 WILDE LARVA
CAN. ENT. 1:76, 1896 SAUNDERS LARVA
LOC. CIT. 14:201, 1882 EDWARDS EGG, LARVA, CHRYSALIS
ILLINOIS DEPT. AGRI. TRANS. 15: 152, 1877 FRENCH LARVA
LOC. CIT. 18, APPEND: 84, 164, 1880 COQUILLETT-MIDDLETON LARVA
NATURALISTE CANADIEN P. 135, 1927 HUARD LARVA, CHRYSALIS
NEW YORK STATE AGRI. EXP. STA. (CORNELL), MEM. 15:213, 1918

HAWLEY LARVA* FF.66,67;
CHRYSALIS* F.65 NO DESCR.
NEW YORK STATE MUS.MEM.8:
544,1905 FELT LARVA
ONTARION ENT.SOC.ANN.REPT.P.
31,1872 BETHUNE LARVA,
CHRYSALIS
LOC.CIT.P.76,1890 CAULFIELD
LARVA
LOC.CIT.P.3,1897 LYMAN EGGS
PSYCHE 4:184,1884 MURTFELDT
CHRYSALIS
PSYCHE 20:161,1913 RAU DURATION
EGG STAGE
SMITHSONIAN MISC.COLLECT.4,ART
2:54,1862 MORRIS LARVA,
CHRYSALIS
UNITED STATES DEPT.AGRI.BUR.
ENT.BULL.7 NEW SERIES P.
48,1897 HOWARD EGG*F.37A;
LARVA*F.37B;CHRYSALIS*F.37C
UNITED STATES DEPT.AGRI.MISC.
PUB.657:356,369,1950 CRAIG-
HEAD LARVA* F.70 D
ABBOT & SMITH-LEP.INS.GEORGIA
1:32,1797 LARVA*PUPA*PL.11
COLOR
BOISDUVAL & LECONTE HIST.LEP.
NORTH AMERICA P.192,1883
LARVA* CHRYSALIS* PL.51
COMSTOCK-HOW KNOW BUTTER.P.
135,1904 LARVA
DICKERSON-MOTHS & BUTTER.P.89,
1901 EGG* FF.79,80;LARVA*
F.78; CHRYSALIS* F.81
EDWARDS,W.H.-BUTTER.NORTH
AMERICA 2ND SERIES 1871
LIFE HISTORY* PL.S GRAPTA
NO.4,5 3RD SERIES PL.I A-D
EMMONS-NATURAL HIST.NEW YORK
5:207,1854 LARVA* PL.35
COLOR
FERNALD-BUTTER.MAINE P.49,1884
EGG,LARVA,CHRYSALIS
FRENCH-BUTTER.EAST.UNITED
STATES P.184,1886 LIFE
HISTORY CHRYSALIS*
HARRIS-ENT.CORRESP.P.280,1869
LARVA
HARRIS-INS.INJ.VEG.1ST EDIT.
P.220,1841 LARVA,CHRYSALIS
HARRIS-LOC.CIT.2ND EDIT.P.240,
1852 LARVA,CHRYSALIS
HARRIS-LOC.CIT.FLINT EDIT.P.
300,1862 LARVA,CHRYSALIS
GLOVER-ILLUST.NORTH AMER.ENT.
1878 LARVA,PUPA* PL.1,F.19;
PL.2,F.1
HOLLAND-BUTTERFLY BOOK 1905
LARVA* PL.3,FF.23,27;
CHRYSALIS*PL.4,FF.21,22,
24,26,40 COLOR
KELLOGG-AMER.INS.P.453,1908
LARVA* FF.643,644
MAYNARD-BUTTER.NEW ENGLAND
P.11,1885 LARVA
MORRIS-SYNOP.LEP.NORTH AMERICA
P.54,1862 LARVA,CHRYSALIS
PACKARD-INS.INJ.FOREST TREES
P.65,1881 LARVA,CHRYSALIS
SCUDDER-BUTTER.EAST.UNITED
STATES & CANADA PP.325-
329,1889 LIFE HISTORY* PL.
64,FF.16,17,PL.74,FF.23,
27;PL.78,FF.37-41;PL.83,
FF.21,22,24,25,26,40
SCUDDER-BUTTER.THEIR STRUCTURE
ETC P.258,1881 CHRYSALIS*
SCUDDER-EVERYDAY BUTTER.P.54,
1899 LARVA,CHRYSALIS
SCUDDER-GUIDE TO COMMON BUTTER.
P.97 ,1893 LARVA,CHRYSALIS
VANCE-LIFE AMONG BUTTER.P.42,
1925 EGG,LARVA,CHRYSALIS
INDEX AMER.ECON.ENT.
VOL.1:251;VOL.2:293;
VOL.5:604;VOL.7:709

FOOD PLANTS

AMBROSIA ARTEMISIIFOLIA LINN.
AMBROSIA TRIFIDA LINN.
ANSOSTICHUS CAPREOLATUA(LINN.)
BOEHMERIA CYLINDRICA(LINN.)
CELTIS OCCIDENTALIS LINN.
HUMULUS LUPULUS LINN.(HOPS)
LIRIODENDRON TULIPIFERA LINN.
TILIA SP.

TILIA AMERICANA LINN.
TILIA CAROLINIANA MILL.
ULMUS SP. (ELMS)
URITCA SP. (NETTLES)

INTERROGATIONIS(LINNAEUS), AUTOGRAPHA-NOCTUIDAE
A U-AUREUM GUENEE
SYN. GROENLANDICA STAUDINGER
SYN. ARCTICA MOESCHLER
B VACCINII HENRY EDWARDS
C HERSCHELENSIS BENJAMIN
LIFE HISTORY
FOOD PLANTS
VACCINIUM SP.

INTERRUPTA DRAUDT
SSP. SEE GALBINA CLEMENS

INTERRUPTA GUENEE, ACRONICTA-NOCTUIDAE
SYN. OCCIDENTALIS GROTE & ROBINSON
FORM ELIZABETA SMITH
LIFE HISTORY
AMER. MUS. NAT. HIST. BULL. 14: 242, 1901 BEUTEN. LARVA
CAN. ENT. 4:49, 1872 SAUNDERS LARVA(OCCIDENTALIS)
ILLINOIS DEPT. AGRI. TRANS. 18, APPEND:139, 1880 MARTEN LARVA
MINNESOTA AGRI. EXP. STA. BULL. 61:207, 1899 LUGGER LARVA* F. 152
NEW YORK STATE AGRI. EXP. STA. (CORNELL), MEM. 56:1074, 1922 WELLHOUSE LARVA(BRIEF)
NEW YORK STATE MUS. MEM. 8:724 1905 FELT LARVA(BRIEF)
PAPILIO 2:181, 1882 PACKARD LARVA, PUPA
PAPILIO 3:132, 1883 EDWARDS-E ELLIOTT LARVA
UNITED STATES NAT. MUS. PROC. 21:80, 1898 SMITH-DYAR LARVA* PL. 4, FF. 7, 8 COLOR OR UNCOLORED; PUPA; COCOON
GLOVER-ILLUST. NORTH AMER. ENT. 1878 LARVA* PL. 10, F. 5
HARRIS-ENT. CORRESP. P. 311, 1869 LARVA
KELLOGG-AMER. INS. P. 404, 1908 LARVA* F. 585; OCCIDENTALIS
SAUNDERS-INS. INJ. FRUITS P. 166, 1883 LARVA
SAUNDERS-LOC. CIT. P. 165, 1889 LARVA
INDEX AMER. ECON. ENT.
VOL. 1:20 APATELA
VOL. 7:8
FOOD PLANTS
BETULA SP. (BIRCHES)
CRATAEGUS SP. (THORNS)
MALUS PUMILA MILL. (APPLE)
PRUNUS SP.
PRUNUS CERASUS LINN. (CHERRY)
PRUNUS DOMESTICA LINN. (PLUM)
PRUNUS MARITIMA WANG.
PRUNUS SEROTINA EHRH.
PYRUS COMMUNIS LINN. (PEAR)
QUERCUS SP. (OAKS)
SORBUS AMERICANA MARSH.
ULMUS SP. (ELMS)

INTERRUPTOFASCIATA PACKARD
SSP. SEE SOBRINATA HUBNER

INTERRUPTOMARGINATA BEAUVOIS
SYN. CLYMENE BROWN

INTIMATA PEARSALL
SSP. SEE SATYRATA HUBNER

INTRACTA WALKER
SYN. SEE FENNICA TAUSCHER

INTRACTARIA WALKER
SYN. SEE HUMARIA GUENEE

INTRARIA GUENEE
SYN. SEE HUMARIA GUENEE

INTRATARIA WALKER
SYN. SEE CREPUSCULARIA SCHIFFENMULLER

INTRITA(MORRISON), EUXOA-NOCTUIDAE
FORM STRIGILIS GROTE

SYN. TITUBATIS SMITH
FORM REUDA STRECKER
A ALTICOLA SMITH
LIFE HISTORY
FOOD PLANTS
BRASSICA OLERACEA LINN. (CABBAGE)
TARAXACUM OFFICINALE WEBER

INTROFERENS GROTE
SYN. SEE AUXILIARIS GROTE

INULTA(GROTE), METAXAGLAEA-NOCTUIDAE
LIFE HISTORY
WASHINGTON ENT. SOC. PROC. 4:324, 1898 DYAR LARVA
FOOD PLANTS
VIBURNUM NUDUM LINN.

INVALIDA(SMITH), POLIA-NOCTUIDAE
LIFE HISTORY
UNITED STATES NAT. MUS. PROC. 27:849, 1904 DYAR EGG, LARVA
FOOD PLANTS

INVARIATA WALKER
SYN. SEE INSULSARIA GUENEE

INVERSA PACKARD
SSP. SEE INCLUSSA HUBNER

INVERSATA NAKAHARA
SSP. SEE PHILODICE GODART

INVEXATA WALKER
SYN. SEEPULTARIA GUENEE

INVOLUTA DYAR
SYN. SEE APERA DRURY

INVOLUTA WALKER
SYN. SEE MINEREA GUENEE

INVOLUTUM(SEPP), EUPSEUDOSOMA-ARCTIIDAE
SYN. NIVEA HERRICH-SCHAEFFER
SYN. NIVEUM GROTE
SYN. BICOLOR ROTHSCHILD(MALE)
A FLORIDUM GROTE
SYN. IMMACULATA GRAEF
LIFE HISTORY
UNITED STATES NAT. MUS. PROC. 23:259, 1900 DYAR EGG, LARVA
FOOD PLANTS
EUGENIA BUXIFOLIA WILLD.
EUGENIA PROCERA(SW.)
PSIDIUM GUAJAVA LINN.

INYOENSIS GUNDER
SSP. SEE MELISSA EDWARDS

IO(FABRICIUS), AUTOMERIS-SATURNIIDAE
AB ARGUS NEUMOEGEN & DYAR
AB CAECA IGEL
A LUTHERI COCKERELL
B COLORADENSIS COCKERELL
C NEOMEXICANA BARNES & BENJAMIN
D LILITH STRECKER
E TEXANA BARNES & BENJAMIN
FOR SYNS. SEE MCDUNNOUGH'S LIST
LIFE HISTORY
AMER. ENT. 1:39, 1868 LARVA (BRIEF)
LOC. CIT. 2:31, 1869 LARVA(BRIEF)
AMER. MUS. NAT. HIST. BULL. 10: 438, 1898 BEUTEN. LARVA
AMER. NAT. 7:477, 1873 LARVA* F. 126 NO DESCR.
AMER. PHIL. SOC. PROC. 31:168, 1893 PACKARD EGG, LARVA* PL. 7, FF. 6, 7; PL. 8, FF. 8, 9, 10; PL. 9, F. 11
CAN. ENT. 2:19, 1869 BETHUNE LARVA
LOC. CIT. 5:109, 1873 RILEY EGG
LOC. CIT. 6:227, 1874 REED EGG, LARVA*
ENT. SOC. AMERICA, ANNALS 7:299 1914 MOSHER PUPAL STRUCTURE
LOC. CIT. 9, 1916 PUPA* PL. 5, FF. 1, 2
ILLINOIS DEPT. AGRI. TRANS. 18, APPEND: 127, 169, 1880 COQUILLETT-MARTEN LARVA*
MARYLAND NAT. HIST. SOC. BULL. 8:51, 1938 COCOON*
MARYLAND NAT HIST. SOC. PUBLICATION NO PAGE 1930 HAYDON

EGG, LARVA, PUPA
MICHIGAN STATE COLLEGE, SPEC. BULL. 243:52, 1934 MCDANIEL LARVA
MINNESOTA AGRI. EXP. STA. BULL. 61:182, 1899 LUGGER EGG; LARVA* F. 130; PUPA
LOC. CIT. 84:71, 1903 WASHBURN LARVA* F. 57
MISSOURI, REPT. INS. OF 5:133, 1873 LIFE HISTORH*
NAT. ACAD. SCI. MEM. 12:99, 1914 PACKARD LIFE HISTORY EGG, LARVA* PL. 19, F. 7; PL. 20, F. 1 COLOR; PUPA
LOC. CIT. P. 102, LARVA OF <u>LILITH</u>
NATURALISTE CANADIEN P. 231, 1929 HUARD LARVA* F. 31
NEW YORK ENT. SCO. JOURN. 3:138, 1895 BEUTEN. LARVA OF <u>LILITH</u> LIKE THAT OF <u>IO</u>
NEW YORK ENT. SOC. JOURN 15:131, 1907 LUTHER EGG, LARVA (<u>FUSCUS</u>)
NEW YORK STATE AGRI. EXP. STA. (CORNELL), MEM. 56:1073, 1922 WELLHOUSE LARVA (BRIEF)
NEW YORK STATE AGRI. SOC. TRANS. P. 81, 1855 GLOVER LARVA
NEW YORK STATE CABIN. NAT. HIST. ANN. REPT. NO. 24:146, 1871 LINTNER EGG, LARVA, COCOON
NEW YORK STATE MUS. MEM. 8:521 1905 FELT LARVA
NEW YORK STATE REPT. INS. OF 3: 379, ALSO P. 61, SEC. 81, 1856 FITCH LARVA
LOC. CIT. 5:832, ALSO P. 52 SEC. 329, 1858 FITCH LARVA
OHIO AGRI. EXP. STA. BULL. 332:270, 1918 HOUSER EGG, LARVA, PUPA
ONTARIO ENT. SOC. ANN. REPT. P. 33, 1872 BETHUNE LARVA (<u>VARIA</u>)
LOC. CIT. P. 11, 1874 REED LARVA*
LOC. CIT. P. 75, 1879 SAUNDERS LARVA*
LOC. CIT. P. 27, 1886 HAGUE LARVA*
LOC. CIT. P. 48, 1889 HARRINGTON LARVA*
LOC. CIT. P. 78, 1902 GIBSON LARVA
LOC. CIT. P. 54, 1903 GIBSON LARVA*
PSYCHE 3:352, 1882 DIMMOCK LARVAL DEFENSE ORGANS
PSYCHE 5:318, 1890 HINKLER LARVA
SMITHSONIAN MISC. COLLECTIONS 4, ART. 2:220, 1862 MORRIS LARVA
UNITED STATES DEPT. AGRI. FARMER'S BULL. 890:15, 1917 PIERCE LARVA (BRIEF)
UNITED STATES DEPT. AGRI. MISC. PUB. 657:357, 380, 1950 CRAIGHEAD LARVA* F. 72 F
ABBOT & SMITH–LEP. INS. GEORGIA P. 97, 1797 LARVA* PL. 49 COLOR
BALLARD–INSECT LIVES P. 87, 1879 LARVA* PUPA*
CRAGEN–OUR INS. FRIENDS & FOES P. 115, 1899 LARVA* F. 49
DICKERSON–MOTHS & BUTTER. P. 146, 1901 EGG* F. 128; LARVA* FF. 127–129; PUPA* F. 130
DUNDAN–NATURALIST LIBRARY 22: 157, 1858 LARVA
ELIOT & SOULE–CATER. & MOTHS. P. 267, 1902 EGG; LARVA* PUPA; COCOON
EMMONS–NAT. HIST. NEW YORK 5: 232, 1854 LARVA
FERNALD–C. H.–KINGSLEY'S STANDARD NAT. HIST. 2:455, 1885 EGG, LARVA, COCOON
FERNALD, H. T.–APPLIED ENT. P. 289, 1926 LARVA
FERNALD & SHEPARD–APPLIED ENT. PL284, 1942 LARVA
GLOVER–ILLUST. NORTH AMER. ENT. 1878 LARVA, PUPA, COCOON* PL. 5, F. 6
GOSSE–LETTERS FROM ALABAMA P. 281, 1859 LARVA
HARRIS–ENT. CORRESP. P. 295, 1869 LARVA
HARRIS–INS. INJ. VEG. 2ST EDIT. P. 284, 1841 LARVA, PUPA
HARRIS–LOC. CIT. 2ND EDIT. P. 303

1852 LARVA,COCOON
HARRIS-LOC.CIT.FLINT EDIT.P.393
1862 LARVA* COCOON*
HART-INS.COLLECTIONS ILLINOIS
HIGH SCHOOLS P.19,1903
LARVA* F.15
HERRICK-INS.ENEMIES SHADE TREES
P.157,1935 LARVA* F.132
HOLLAND-MOTH BOOK P.90,1905
LARVA* F.43
JAEGER-LIFE NORTH AMER.INS.P.
124,1864 LARVA*
KELLOGG-AMER.INS.P.424,1908
LARVA* F.611
MORRIS-SYNOP.LEP.NORTH AMERICA
P.220,1862 LARVA
PACKARD-HALF HOURS WITH INS.
P.85,1877 LARVA*
PAC KARD-INS.INJ.FOREST TREES
P.111,1881 LARVA*
PEARIS-PESTS FARM,GARD.ORCHARD
P.22,1941 LARVA* F.28
SAUNDERS-INS.INJ.FRUITS P.210
1883 LIFE HISTORY*
SAUNDERS-LOC.CIT.P.210,1889
EGG; LARVA* F.217; PUPA
INDEX AMER.ECON.ENT.
VOL.1:45; VOL.2:47;
VOL.3:51; VOL.4:57;
VOL.5:69 VOL.6:79;
VOL.7:117

FOOD PLANTS

ABIES BALSAMEA(LINN.)
ACER SP.(MAPLES)
AMORPHA FRUTICOSA LINN.
BAPTISIA TINCTORIA(LINN.)
BETULA SP.(BIRCHES)
CARPINUS CAROLINIANA WALT.
CARYA SP.(HICKORIES)
CASSIA FASCICULATA MICHX.
CERCIS CANADENSIS LINN.
COMPTONIA PEREGRINA(LINN.)
CORNUS FLORIDA LINN.
CORYLUS SP.(HAZELS)
CORYLUS AVELLANA LINN.
FAGUS SP.(BEECHES)
FRAXINUS SP.(ASHES)
GOSSYPIUM HERBACEUM LINN.(COTTON)
HIBISCUS LASIOCARPOS CAV.
HUMULUS LUPULUS LINN.(HOPS)
ILEX VERTICILLATA(LINN.)
LESPEDEZA SP.
LIRIODENDRON TULIPIFERA LINN.
MAGNOLIA VIRGINIANA LINN.
MALUS PUMILA MILL.(APPLE)
MELILOTUS ALBA DESV.
MELILOTUS OFFICINALIS(LINN.)
MYRICA PENSYLVANICA LOISEL
OSTRYA VIRGINIANA(MILL.)
PLATANUS OCCIDENTALIS LINN.
POPULUS GILEADENSIS ROULEAU
POPULUS TREMULOIDES MICHX.
PRUNUS CERASUS LINN.(CHERRY)
PRUNUS DOMESTICA LINN.(PLUM)
PRUNUS PENSYLVANICA LINN.
PRUNUS SEROTINA EHRH.
PRUNUS VIRGINIANA LINN.
PYRUS COMMUNIS LINN.(PEAR)
QUERCUS SP.(OAKS)
RHIZOPHORA MANGLE LINN.
RHODODENDRON SP.
RIBES SATIVUM SYME(CURRANT)
ROBINIA PSEUDO-ACACIA LINN.
ROSA SP.(ROSES)
RUBUS ALLEGHENIENSIS PORTER
RUBUS CANADENSIS LINN.
RUBUS IDAEUS V.STRIGOSUS(MICHX.)
SALIX SP.(WILLOWS)
SASSAFRA ALBIDUM V.MOLLE(RAF.)
SERENOA REPENS(BARTR.)
SPIRAEA SP.
SYMPHORICARPOS ALBUS(LINN.)
TILIA SP.(LINDENS)
TRIFOLIUM SP.(CLOVERS)
TRIFOLIUM REPENS LINN.
ULMUS SP(ELMS)
VIBURNUM DENTATUM LINN.
WISTERIA FRUTESCENS(LINN.)
WISTERIA SINENSIS SWETT
ZEA MAYS LINN.(CORN)

<u>IO</u> GRAY
SYN.SEE <u>ASTYLUS</u> DRURY

<u>IOLE</u> BOISDUVAL, <u>NATHALIS</u>-PIERIDAE
SYN.<u>IRENE</u> FITCH

SYN. LUTEOLUS REAKIRT
SYN. PALLIDA FIELD
SYN. IMMACULATA FIELD
LIFE HISTORY
ENT. NEWS 7:15, 1896 BRUCE EGG, LARVA
LOC. CIT. 45:252, 1934 BERRY LARVA
SOUTHERN CALIFORNIA ACAD. SCI. BULL. 27:87, 1928 COMSTOCK LIFE HISTORY EGG* LARVA* CHRYSALIS*
COMSTOCK, J. A. - BUTTER. CALIFORNIA P. 36, 1927 LARVA
COMSTOCK, J. H. - HOW TO KNOW BUTTER. P. 86, 1904 LARVA
SCUDDER - GUIDE TO COMMON BUTTER P. 140, 1893 LARVA, CHRYSALIS
INDEX AMER. ECON. ENT. VOL. 7:699
FOOD PLANTS
BIDENS PILOSA LINN.
DYSSODEA PAPPOSA (VENT.)
ERODIUM CICUTARIUM (LINN.)
HELENIUM BIGELOVII GRAY
STELLARIA MEDIA (LINN.)

IOLE CRAMER
SYN. SEE VIRGINIENSIS DRURY (VANESSA)

IPOMOEAE DOUBLEDAY, SCHIZURA - NOTODONTIDAE
SYN. BIGUTTATUS PACKARD
SYN. CONFUSA WALKER
SYN. DUCENS WALKER
SYN. CORTICEA WALKER
SYN. COMPTA WALKER
SYN. NIGROSIGNATA WALKER
FORM TELIFER GROTE
FORM CINEREOFRONS PACKARD
SYN. USTIPENNIS WALKER
LIFE HISTORY
AMER. MUS. NAT. HIST. BULL. 10: 420, 1898 BEUTEN. LARVA
AMER. NAT. 4:229, 1872 LARVA* PL. 2, FF. 2A, 2B; PUPA
BOSTON SOC. NAT. HIST. PROC. 24: 535, 1890 PACKARD EGG, LARVA* PL. 4, F. 1
ENT. NEWS 4:158, 1893 TOWNSEND LARVA (BRIEF)
ENT. NEWS 9:150, 1898 SOULE DURATION OF EGG STAGE
ENTOMOLOGIST 4:60, FEB. 1841 DOUBLEDAY LARVA
INSECT LIFE 3:62, 1890 DYAR LARVA
MAINE AGRI. EXP. STA. BULL. 259: 70, 1917 MOSHER PUPA* F. 5B
MINNESOTA AGRI. EXP. STA. BULL. 61:169, 1899 LUGGER LARVA* PL. 19, F. 115
MUS. COMP. ZOOL. BULL. 87, NO. 6, 1941 DETHIER LARVAL ANTENNA* FF. 81, 84
NAT. ACAD. SCI. MEM. 7:195, 1895 PACKARD EGG, LARVA* PL. 25, FF. 2, 3 COLOR; PUPA* COCOON
NEW YORK ENT. SOC. JOURN. 1:69 1893 PACKARD LARV
NEW YORK STATE AGRI. EXP. STA. (CORNELL), MEM. 274:229, 1948 FORBES LARVA
NEW YORK STATE MUS. MEM. 8:705 729, 1905 FELT LARVA (BRIEF)
OHIO AGRI. EXP. STA. BULL. 45:168, 1893 WEBSTER LARVA
DOANE ETC - FOREST INSECTS P. 293 1936 LARVA
FOOD PLANTS
ACER SP. (MAPLES)
ACER RUBRUM LINN.
AMELANCHIER FLORIDA LINDL.
BETULA SP. (BIRCHES)
CEANOTHUS SP.
CERCIS OCCIDENTALIS TORR.
CONVOLVULUS SP.
FAGUS SP. (BEECHES)
GLEDITSIA TRIACANTHOS LINN.
HAMAMELIS VIRGINIANA LINN.
IPOMOEA COCCINA LINN.
QUERCUS SP, (OAKS)
ROSA SP. (ROSES)
RUBUS ALLEGHENNSIS PORTER
RUBUS IDAEUS V. STRIGOSUS (MICHX)
TILIA AMERICANA LINN.
ULMUS SP. (ELMS)

VACCINIUM SP.

IPOMOEAE HARRIS, SYNTOMEIDA-AMATIDAE
SYN. FEROX WALKER
SYN. EUTERPE HERRICH-SCHAEFFER
LIFE HISTORY
BOSTON SOC. NAT. HIST. PROC. 27: 138, 1896 DYAR LARVAL SETAE
FOOD PLANTS

IRENE BEHR, CATOCALA-NOCTUIDAE
FORM VOLUMNIA HENRY EDWARDS
FORM VIRGILIA HENRY EDWARDS
A VALERIA HENRY EDWARDS
LIFE HISTORY
AMER. MUS. NAT. HIST. BULL. 38: 159, 1918 BARNES & MCDUNNOUGH EGG, LARVA
AMER. MUS. NAT. HIST. MEM. NEW SERIES 3, PART 1, 1918 BARNES & MCDUNNOUGH LARVA* PL. 13 F. 14; PL. 15, F. 21; PL. 16, FF. 6, 18 COLOR
PSYCHE 20:202, 1913 BARNES & MCDUNNOUGH EGG, LARVA
FOOD PLANTS
POPULUS SP. (POPLARS)
SALIX SP. (WILLOWS)

IRENE BOISDUVAL, ARGYNNIS-NYMPHALIDAE
SYN. COTTLEI COMSTOCK
A DODGEI GUNDER
LIFE HISTORY
FOOD PLANTS
VIOLA SP. (VIOLETS)

IRENE FITCH
SYN. SEE IOLE BOISDUVAL

IRIDARIA (GUENEE), DICHORDA-GEOMETRIDAE
SYN. REMOTARIA WALKER
SYN. CONSEQUARIA HENRY EDWARDS
SYN. LATIPENNIS HULST
LIFE HISTORY
NEW YORK STATE AGRI. EXP. STA. (CORNELL), MEM. 274:114, 1948 FORBES LARVA
FOOD PLANTS
RHUS SP. (SUMACS)

IROIDES (BOISDUVAL), INCISALIA-LYCAENIDAE
AB IMMACULATA COCKLE
B ANNETEAE DOS PASSOS
LIFE HISTORY
CALIFORNIA ACAD. SCI. PROC. 1 NEW SERIES P. 104, JUNE, 1888 EDWARDS LARVA, CHRYSALIS
SOUTHERN CALIFORNIA ACAD. SCI. BULL. 32:77, 1933 COMSTOCK-DAMMERS LIFE HISTORY EGG* PL. 19; LARVA* PLS. 20, 21; CHRYSALIS* PL. 22
FOOD PLANTS
CUSCUTA GRONOVII WILLD.
ERIOGONUM SP.
HOLODISCUS DISCOLOR (PURSH)
MALUS PUMILA MILL. (APPLE)
SEDUM SPATHULIFOLIUM HOOK QUESTIONABLE FOOD PLANT

IRRARIA BARNES & MCDUNNOUGH, EUCHLAENA-GEOMETRIDAE
MAJORARIA PACKARD
LIFE HISTORY
FOOD PLANTS
ACER SP. (MAPLES) (NOT FAVORED)
QUERCUS SP. (OAKS)

IRREGULATA WALKER
SSP. SEE GRANTITATA GUENEE

IRRESOLUTA WALKER
SYN. SEE CHALCEDONIA HUBNER

IRRORATA HENRY EDWARDS
SYN. SEE ALBATA PACKARD

IRRORATA PACKARD
SSP. SEE ROSEA WALKER

IRRORATA (PACKARD), SEMIOTHISA-GEOMETRIDAE
LIFE HISTORY
ENT. NEWS 5:63, 1894 DYAR LARVA, PUPA
NEW YORK ENT. SOC. JOURN. 53: 189, 1945 FORBES PUPA

Psyche 9:310,1902 Dyar egg, larva

FOOD PLANTS

Populus fremonti S. Wats
Populus wislizenii (S. Wats.)
Salix sp. (willows)

irrorata Smith, Orthordes-Noctuidae

LIFE HISTORY

FOOD PLANTS

Acer macrophyllum Pursh
Alnus rubra Bong.
Amelanchier florida Lindl.
Plantago lanceolata Linn.
Plantago major Linn.
Prunus emarginata (Doubl.)
Ribes sanguineum Pursh
Salix hookeriana Barratt
Salix scouleriana Barr.

irus (Godart), Incisalia-Lycaenidae

syn. arsace Boisduval & Leconte
ab balteata Scudder

LIFE HISTORY

Amer. Mus. Nat. Hist. Bull. 5:280 1893 Beuten. larva, chrysalis
California Acad. Sci. Proc. 5:331 1873 Henry Edwards larva, chrysalis (arsace)
Can. Ent. 38:142, 1906 Cook life history very complete
Illinois Dept. Agri. Trans. 18, Append: 94, 1880 Middleton larva, chrysalis
New York State Mus. Mem. 8:732, 1905 Felt larva (brief)
Smithsonian Misc. Collections 4 art. 2:97, 1862 Morris larva, chrysalis (irus, arsace)
Boisduval & Leconte-Hist. Lep. North America p. 101, 1833 larva* chrysalis* Pl. 31 COLOR
Boisduval & Leconte-loc. cit. p. 103, 1833 larva* (arsace Pl. 32 chrysalis also in COLOR)
Comstock-How to Know Butter. p. 233, 1904 larva
Holland-Butterfly Book 1905 chrysalis* Pl. 5, ff. 32-34 COLOR
Maynard-Butter. New England p. 37, 1886 larva, chrysalis
Morris-Synop. Lep. North America p. 97, 1862 larva, chrysalis (irus and arsace)
Scudder-Butter. East. United States & Canada pp. 838-841 1889 life history* Pl. 65, ff. 9, 10; Pl. 68, f. 8; Pl. 75, ff. 22, 23, 28, Pl. 79, f. 42; Pl. 84, ff. 26, 32. 34
Scudder-Guide to Common Butter. p. 115, 1893 larva, chrysalis

FOOD PLANTS

Baptisia tinctoria (Linn.)
Cercis canadensis Linn.
Cyrilla racemiflora Linn.
Ilex sp.
Leucothoe racemosa (Linn.)
Lupinus perennis Linn.
Prunus sp.
Prunus domestica Linn. (plum)
Prunus ilicifolia (Nutt. ex H. & A)
Vaccinium corymbosum Linn.
Vaccinium myrtillus Linn.

isabella (Abbot & Smith), Isia-Arctiidae

syn. californica Packard

LIFE HISTORY

Amer. Ent. 1:148, 1869 larva (brief)
Loc. cit. 2:182, 1870 larva* cocoon*
Loc. cit. 3:133, 1880 life history larva* pupa*
Amer. Mus. Nat. Hist. Bull. 10: 373, 1898 Beuten. larva
Boston Soc. Nat. Hist. Proc. 27: 135, 1896 Dyar larval setae
Boston Zoological Soc. Journ1: 29, 54, 1882 Coleman larva and larval varieties
California Acad. Sci. Proc. 5:370 1873 Hy. Edw. larva, pupa

CAN. ENT. 5:75, 1873 SAUNDERS LIFE HISTORY*
LOC. CIT. 16:67, 1884 BEAN LARVA (BRIEF)
CAN. JOURN. INDUSTRY, SCIENCE & ART 8:364, 1863 SAUNDERS LARVA
CAN. NATURALIST PP. 78, 308, 1840 GOSSE LARVA
ILLINOIS AGRI. EXP. STA. BULL. 60: 505, 1900 FORBES LARVAL HEAD MARKINGS* F. 84C
ILLINOIS DEPT. AGRI. TRANS. 15: 182, 1877 FRENCH LARVA
LOC. CIT. 18, APPEND: 115, 169, 1880 COQUILLETT-MARTEN LARVA, PUPA, COCOON (FF.)
MINNESOTA AGRI. EXP. STA. BULL. 61:131, 1899 LUGGER LARVA* COCOON
MISSOURI, REPT. INS. OF 4:143, 1872 RILEY LARVA* PUPA*
MUS. COMP. XOOL. BULL. 87, NO. 6, 1941 DETHIER LARVAL ANTENNA* F. 87
NATURALISTE CANADIEN P. 41, 1929 HUARD LARVA* F. 37
NEW YORK ACAD. SCI. ANNALS 8:76 1893 PACKARD EGG, LARVA
NEW YORK ENT. SOC. JOURN 17:139 1909 LUTZ LARVAL VARIATION DUE TO MOISTURE
NEW YORK STATE CABIN. NAT. HIST. ANN. REPT. NO. 44:226, 1890 LINTNER LARVA* PUPA* COCOON*
ONTARIO ENT. SOC. ANN. REPT. P. 22, 1873 SAUNDERS LARVA* PUPA*
LOC. CIT. P. 46, 1877 GOTT LARVA* PUPA*
LOC. CIT. P. 117, 1914 GIBSON LARVAL COLOR VARIATION
OTTAWA NATURALIST 19:159, 1905 GIBSON LARVA* PUPA*
PSYCHE 2:270, 1879 MANN LARVA
UNITED STATES DEPT. AGRI. BUR. ENT. BULL. NEW SERIES 43:44 1903 CHITTENDEN LARVA* F. 42
ABBOT & SMITH-LEP. INS. GEORGIA 2:131, 1797 LARVA* PL. 66 COLOR
CRAGEN-OUR INS. FRIENDS & FOES P. 147, 1899 LARVA
DICKERSON-MOTHS & BUTTER. P. 191, 1901 LARVA* F. 169
EMMONS-NAT. HIST. NEW YORK 5: 229, 1854 LARVA
ESSIG-INS. WEST. NORTH AMERICA P. 676, 1926 LARVA
FERNALD-APPLIED ENT. P. 280, 1926 LARVA
FERNALD & SHEPARD-APPLIED ENT. P. 277, 1942 LARVA
GLOVER-ILLUST. NORTH AMER. ENT. 1878 LARVA* PL. 8, FF. 9, 10
HART-INS. COLLECTIONS ILLINOIS HIGH SCHOOLS P. 22, 1903 LARVA* FF. 23, 24; PUPA* F. 23
HOLLAND-MOTH BOOK P. 125, 1905 LARVA* PUPA* F. 69
KELLOGG-AMER. INSECTS P. 412 1908 LARVA
O'KANE-INJ. INS. P. 182, 1912 LARVA* F. 219
SMITH, J. B.-INS. NEW JERSEY P. 400, 1900 LARVA* PUPA* F. 187
TENNEY-MANUAL NAT. HIST. P. 397, 1867 LARVA
WOOD-INSECTS ABROAD P. 664, 1874 LARVA*
INDEX AMER. ECON. ENT. VOL. 1:169; VOL. 6:434; VOL. 7:581

FOOD PLANTS

ACER SACCHAROPHORUM KOCH
AGROPYRON REPENS (LINN.)
ARCTIUM MAJUS BERNH.
ASTER SP.
ATROPA BELLADONNA LINN.
BETA VULGARIS LINN. (BEET)
BETULA ALBA LINN.
BETULA PAPYRIFERA MARSH.
CLETHRA ALNIFOLIA LINN.
DIANTHUS CARYOPHYLLUS LINN.
FRAGARIA CHILOENSIS (LINN.)
HELIANTHUS ANNUUS LINN.
MARTYNIA LOUSIANICA MILL.

PARTHENOCISSUS QUINQUEFOLIA(LINN)
PISUM SATIVUM LINN. (PEAS)
PLANTAGO MAJOR LINN.
POLYGONUM PERSICARIA LINN.
PRUNUS SEROTINA EHRH.
RIBES AUREUM PURSH
RICINUS COMMUNIS LINN.
RUBUS ALLEGHENIENSIS PORTER
RUBUS IDAEUS V.STRIGOSUS(MICHX.)
SORBARIA SORBIFOLIA(LINN.)
SYRINGA VULGARIS LINN.
TARAXACUM OFFICINALE WEBER
TRIFOLIUM SP. (CLOVERS)
TROPAEOLUM MAJUS LINN.
ULMUS AMERICANA LINN.
VIBURNUM DENTATUM LINN.
VITIS LABRUSCA LINN.
ZANTEDESCHIA AETHIOPICA SPRENG.
ZEA MAYS LINN. (CORN)
GRASS
LOW PLANTS

ISABELLA HENRY EDWARDS
SSP. SEE SIMILIS EDWARDS

ISIAS BOISDUVAL, ADELOCEPHALA-CITHERONIIDAE
LIFE HISTORY
NAT. ACAD. SCI. MEM. 9:80, 1905
PACKARD LARVA* PL. 3, FF. 2,3
COLOR; PL. 49, F. 3 STRUCTURE
FOOD PLANTS
ACACIA FLEXICAULIS(MEXICAN)
CERCIDIUM MICROPHYLLUM(TORR.)
PITHECELLOBIUM FLEXICAULE)BENTH)
PROSOPIS GLANDULOSA TORR.

ISMERIA BOISDUVAL & LECONTE
SYN. GORGONE HUBNER

ISTAPA REAKIRT
SYN. SEE COLUMELLA FABRICIUS

ITYS EDWARDS
SSP. SEE SYLVINUS BOISDUVAL

IVALLDA MEAD
SSP. SEE CHRYZUS DOUBLEDAY & HEWITSON

IXION LINNAEUS
SYN. SEE TANTALUS LINNAEUS

J

JACINTOI GUNDER
SSP. SEE DYMAS EDWARDS

JACKI STALLINGS-TURNER
SSP. SEE LYGDAMUS DOUBLEDAY

JACKSONI CLARK
SSP. SEE CARDUI LINNAEUS

JACQUENETTA HENRY EDWARDS
SSP. SEE MICRONYMPHA GUENEE

JACULIFERA GUENEE
SYN. SEE SUBGOTHICA HAWORTH

J-ALBUM(BOIASUVAL&LECONTE), NYMPHALIS
NYMPHALIDAE
SYN. POCAHONTAS SCUDDER
AB AUREOMARGINATA COCKERELL
A WATSONI HALL.
LIFE HISTORY
AMER. MUS. NAT. HIST. BULL. 5:266
1893 BEUTEN. LARVA, CHRYSALIS
BUFFALO SOC. NAT. SCI. BULL. 19:
25, 1939 WILD LARVA
CALIFORNIA ACAD. SCI. PROC. 3:123,
1864 BEHR LARVA(BRIEF)
CAN. ENT. 21:164, 1889 DENTON
LARVA
LOC. CIT. 21:179, 1889 DAWSON
LARVA, CHRYSALIS
LOC. CIT. 32:273, 1900 FLETCHER
LARVA
MONTANA UNIV. BULL. 30:90, 1906
ELROD LARVA, CHRYSALIS
NATURALISTE CANADIEN P. 138, 1927
HUARD LARVA, CHRYSALIS
NEW YORK STATE MUS. MEM. 8:749
1905 FELT LARVA(BRIEF)
PHILADELPHIA ENT. SOC. PROC. 3:58
1864 LINTNER LARVA, CHRYSALIS
PSYCHE 9:85, 1900 DURAND VARIATIONS IN LARVA & CHRYSALIS

UNITED STATES DEPT. AGRI. MISC. PUB. 657:357,369,1950 CRAIGHEAD LARVA

COMSTOCK-HOW TO KNOW BUTTER. P. 146,1904 LARVA

FERNALD-BUTTER. MAINE P. 55,1884 LARVA, CHRYSALIS

FRENCH-BUTTER. EAST. UNITED STATES P. 193,1886 LARVA, CHRYSALIS

MAYNARD-BUTTER. NEW ENGLAND P. 15,1886 LARVA

SCUDDER-BUTTER. EAST. UNITED STATES&CANADA PP. 383-386, 1889 LIFE HISTORY* PL. 83, FF. 36,44,45

SCUDDER-GUIDE TO COMMON BUTTER. P. 92,1893 LARVA, CHRYSALIS

INDEX AMER. ECON. ENT. VOL. 7:711

FOOD PLANTS

BETULA PAPYRIFERA MARSH.
BETULA POPULIFOLIA MARSH.
HUMULUS LUPULUS LINN. (HOPS)
MALUS PUMILA MILL. (APPLE)
POPULUS SP. (POPLARS)
SALIX SP. (WILLOWS)
ULMUS SP. (ELMS)

JAMAICENSIS (DRURY), SMERINTHUS-SPHINGIDAE

AB CLARKII FRANCK
AB TRIPARTITUS GROTE
FORM NORM. GEMINATUS SAY.
AB FLAVITINCTA NIXON
A GEMMA COCKERELL

LIFE HISTORY

AMER. MUS. NAT. HIST. BULL. 7:313 1895 BEUTEN. LARVA, PUPA

CAN. NAT. PP. 280,295,1840 GOSSE LARVA, PUPA

ENT. SOC. AMERICA, ANNALS 11: 425,1918 MOSHER PUPA

MINNESOTA AGRI. EXP. STA. BULL. 61:105,1899 LUGGER LARVA

NEW YORK ENT. SOC. JOURN. 9:89 1901 BEUTEN. LARVA

NEW YORK STATE AGRI. EXP. STA. (CORNELL), MEM. 274:193,1948 FORBES LARVA

NEW YORK STATE CABIN. NAT. HIST. ANN. REPT. NO. 24:119,1871 LINTNER EGG, LARVA, PUPA

NEW YORK STATE MUS. MEM. 8:724 1905 FELT LARVA(BRIEF)

NOVITATES ZOOLOGICAE(TRING) 9 SUPPL. P. 326,1903 ROTHSCHILD & JORDAN LARVA, PUPA

ONTARIO ENT. SOC. ANN. REPT. P. 62,1881 REED LARVA

PHILADELPHIA ENT. SOC. PROC. 3: 666,1864 LINTNER LARVA

UNITED STATES DEPT. AGRI. MISC. PUB. 657:359,375,1950 CRAIGHEAD LARVA

BOISDUVAL-SPEC. GEN. LEP. SPHING. P. 39,1874 LARVA

DICKERSON-MOTHS & BUTTER. P. 232 1901 LARVA* F. 197,199; PUPA* F. 198

ELIOT & SOULE-CATER. & MOTHS. P. 179,1902 EGG, LARVA*PUPA

FERNALD-SPHINGIDAE NEW ENGLAND P. 78,1886 EGG, LARVA

KELLOGG-AMERICAN INSECTS P. 437, 1908 LARVA* F. 626

STRECKER-LEP. RHOP. ET HETER., 7:57,1873 LARVA

FOOD PLANTS

BETULA PAPYRIFERA MARSH.
CARPINUS CAROLINIANA WALT
CORYLUS SP. (HAZELS)
FRAXINUS SP. (ASHES)
MALUS PUMILA MILL. (APPLE)
OSTRYA VIRGINIANA (MILL.)
POPULUS SP. (POPLARS)
PRUNUS CERASUS LINN. (CHERRY)
PRUNUS DOMESTICA LINN. (PLUM)
PRUNUS SEROTINA EHRH.
QUERCUS SP. (OAKS)
SALIX SP. (WILLOWS)
SPIRAEA SP.
TILIA SP. (LINDENS)
ULMUS SP. (ELMS)

JAMAICENSIS MOESCHLER

SYN. SEE JATROPHAE JOHANSSEN

JANAIS(DRURY), CHLOSYNE-NYMPHALIDE
LIFE HISTORY
ENT. AMER. 3:161, 1887 HY. EDW. CHRYSALIS
PAPILIO 3:188, 1883 SCHAUS JR. LARVA CHRYSALIS
WASHINGTON ENT. SOC. PROC. 13: 228, DYAR LARVA
FOOD PLANTS

JANUALIS GROTE
SYN. SEE DILUCIDA MORRISON

JASMINEARUM(GUERIN-MENEVILLE) CHALAENOGRAMMA-SPHINGIDAE
SYN. ROTUNDATA ROTHSCHILD
LIFE HISTORY
AMER. MUS. NAT. HIST. BULL. 7:306 1895 BEUTEN. LARVA, PUPA
ENT. SOC. AMERICA, ANNALS 11: 413, 1918 MOSHER PUPA
ILLINOIS DEPT. AGRI. TRANS. 18, APPEND:103, 1880 MARTEN LARVA
NEW YORK STATE AGRI. EXP. STA. (CORNELL), MEM. 274:187, 1948 FORBES LARVA
NEW YORK STATE MUS. MEM. 8:727, 1905 FELT LARVA
NOVITATES ZOOLOGICA(TRING) 9 SUPPL. P. 96, 1903 ROTHSCHILD & JORDAN LARVA, PUPA
PHILADELPHIA ACAD. NAT. SCI. JOURN 5:174, 1859 CLEMENS PUPA
SMITHSONIAN MISC. COLLECTIONS 4, ART. 2:199, 1862 MORRIS PUPA
BOISDUVAL-SPEC. GEN. LEP. SPHING P. 114, 1874 LARVA, PUPA
CUVIER-ANIMAL KINGDOM 15, 1832 LARVA* PL. 83
MORRIS-SYNOP. LEP. NORTH AMERICA P. 199, 1862 PUPA(BRIEF)
STRECKER-RHOP. ET HETER. 13:116 1876 LARVA
FOOD PLANTS
FRAXINUS SP. (ASHES)
SYRINGA SP.

JATROPHAE FABRICIUS
SYN. SEE ANTAEUS DRURY

JATROPHAE(JOHANSSON), ANARTIA-NYMPHALIDAE
A SATURATA STAUDINGER
SYN. JAMAICENSIS MOESCHLER
LIFE HISTORY
AMER. ACAD. SCI. PROC. 27:239, 1892 SCUDDER LARVA, CHRYSALIS
FLORIDA ENTOMOLOGIST 7:42, 1924 BATES LARVA, CHRYSALIS
INDEX AMER. ECON. ENT. VOL. 4:21; VOL. 7:36
FOOD PLANTS
JATROPHA SP.
LIPPIA SP.
MANIHOT SP.

JEMEZENSIS BREHME
SSP. SEE PICTA EDWARDS

JEMEZENSIS GUNDER
SSP. SEE COMYNTAS GODART

JENISTAE STALLINGS
SSP. SEE BELLONA FABRICIUS

JENNINGSAE HOLLAND
SSP. SEE MYRINA CRAMER

JESSICA BARNES
SSP. SEE CLIO PACKARD

JOANNISI VERITY
SYN. SEE MACHAON LINNAEUS

JOCOSA(GUENEE), FERALIA-NOCTUIDAE
A FURTIVA SMITH
LIFE HISTORY
AMER. MUS. NAT. HIST. BULL. 14: 236, 1901 BEUTEN. LARVA
NEW YORK ENT. SOC. JOURN. 6:182 1898 SEIFERT EGG, LARVA, PUPA, COCOON
NEW YORK STATE MUS. MEM. 8:754 1905 FELT LARVA
UNITED STATES DEPT. AGRI. MISC. PUB. 657:363, 394, 1950

CRAIGHEAD LARVA
INDEX AMER. ECON. ENT.
VOL. 7:376
FOOD PLANTS
ABIES SP. (FIRS)
ABIES BALSAMEA(LINN.)
LARIX LARICINA(DUROI)
PICEA SP. (SPRUCES)
TSUGA CANADENSIS(LINN.)

JOCOSA HENRY EDWARDS
SSP. SEE INCLUSA HUBNER

JOHNSONARIA(FITCH), EUCHLAENA-GEOMETRIDAE
SYN. BILLINEARIA PACKARD
A MINORARIA HULST
LIFE HISTORY
PACKARD-INS. INJ. FOREST TREES
P. 49, 1881 LARVA(BRIEF)
FOOD PLANTS
CRATAEGUS SP. (THORNS)
PRUNUS SP.
PRUNUS VIRGINIANA LINN.
QUERCUS SP. (OAKS)

JOHNSONI(SKINNER), MITOURA-LYCAENIDAE
LIFE HISTORY
ENT. NEWS 15:298, 1904 SKINNER
LARVA
FOOD PLANTS
ARCEUTHOBIUM DOUGLASSII ENGEL.

JOHNSONI SWETT
SSP. SEE FISCELLARIA GUENEE

JORDANI BARNES & BENJAMIN
SYN. SEE DIFFINIS BOISDUVAL

JORULLA(WESTWOOD), ROTHSCHILDIA-SATURNIIDAE
A CINCTUS TEPPER
LIFE HISTORY
AMER. ACAD. ARTS. SCI. PROC. 39:
552, 1904 PACKARD LARVA,
COCOON
CAN. ENT. 16:131, 1884 MURTFELDT
COCOON
ENT. SOC. AMER. ANNALS 9:149, 1916
MOSHER PUPAL STRUCT.
NAT. ACAD. SCI. MEM. 12:155, 1914
PACKARD LARVA* PL. 4, FF.
4, 5; PL. 5, FF. 1, 2 COLOR:
COCOON
FOOD PLANTS
FRAXINUS SP. (ASH)(IN CAPTIVITY)
PRUNUS SEROTINA EHRH. (IN CAPTIVITY)

JUANITA SCUDDER
SYN. SEE HALESUS CRAMER

JUANITA(STRECKER), PROSERPINUS-SPHINGIDAE
A OSLARI ROTHSCHILD & JORDAN
LIFE HISTORY
AMER. MUS. NAT. HIST. BULL. 16:
396, 1902 BEUTEN. LARVA
ENT. SOC. AMER. ANNALS 11:438,
1918 MOSHER PUPAL STRUCTURE
FOOD PLANTS
EPILOBIUM SP.

JUBA(SCUDDER), HESPERIA-HESPERIIDAE
AB OGDENENSIS HOLLAND
LIFE HISTORY
DENISON UNIV. SCIENFITIC LABOR.
JOURN. 20:121, 1923 LINDSEY
EGG, LARVA* PL. 16 STRUCTURE
FOOD PLANTS
GRASSES

JUBARARIA(HULST), SYNAXIS-GEOMETRIDAE
A SERICEATA BARNES & MCDUNNOUGH
LIFE HISTORY
FOOD PLANTS
ACER MACROPHYLLUM PURSH
CRATAEGUS DOUGLASII LINDL.
HOLODISCUS DISCOLOR(PURSH)
JUNIPERUS VIRGINIANA LINNL.
OSMARONIA CERASIFORMIS(T.&G.)
POPULUS SP. (POPLARS)
RIBES SANGUINEUM PURSH
SALIX HOOKERIANA BARR.
SALIX SCOULERIANA BARR.

JUCUNDA HUBNER. MEKIPOTIS-NOCTUIDAE
SYN. CINIS GUENEE

SYN. VERSABILIS HARVEY
SYN. FEMALE TETRICA HENRY EDWARDS
A HADENIFORMIS BEHR
LIFE HISTORY
FOOD PLANTS
ACACIA SP.
QUERCUS PRINOIDES WILLD.
SALIX AMYGDALOIDES ANDERS.
SALIX BONPLANDIANA H. BK.

JUCUNDARIA STRECKER
SYN. SEE DECOLORARIA HULST

JUCUNDISSIMA DYAR
SSP. SEE EPILAIS WALKER

JUDITY STRECKER, CATOCALA-NOCTUIDAE
SYN. LEVETTEI GROTE
LIFE HISTORY
AMER. MUS. NAT. HIST. BULL. 16: 390, 1902 BEUTEN. LARVAL HEAD* F. 15
AMER. MUS. NAT. HIST. MEM. NEW SERIES 3, PART 1, 1918 BARNES & MCDUNNOUGH LARVA* PL. 10, F. 28; PL. 11, F. 15 COLOR
CAN. ENT. 33:224, 1901 DODGE LARVA PUPA
NEW YORK STATE MUS. MEM. 8:716 1905 FELT LARVA (BRIEF)
FOOD PLANTS
CARYA SP. (HICKORIES
JUGLANS SP.

JUGLANDIS (ABBOT & SMITH), CRESSONIA-SPHINGIDAE
SYN. INSTABILIS MARTYN
SYN. ROBINSONII BUTLER
AB PALLENS STRECKER
AB HYPERBOLA SLOSSON
A ALPINA CLARK
B MANITOBAE CLARK
LIFE HISTORY
AMER. JOURN. SCI. 36:292, 1839 HARRIS LARVA, PUPA
AMER. MUS. NAT. HIST. BULL. 7:317 1895 BEUTEN. LARVA, PUPA
ENT. NEWS 9:150, 1898 SOULE DURATION EGG STAGE
ENT. SOC. AMER. ANNALS 11:428, 1918 MOSHER PUPA
MINNESOTA AGRI. EXP. STA. BULL. 61:106, 1899 LUGGER LARVA
NATURALISTE CANADIEN P. 185, 1929 HUARD LARVA
NEW YORK STATE AGRI. EXP. STA. (CORNELL), MEM. 274:192, 1948 FORBES LARVA
NEW YORK STATE MUS. MEM. 8:518 720, 1905 FELD LARVA
NEW YORK STATE, REPT. INS. OF 3: 462, ALSO P. 144, SEC. 186, 1856 FITCH LARVA
NOVITATES ZOOLOGICA (TRING) 9 SUPPL. P. 345, 1903 ROTHSCHILD & JORDAN LARVA, PUPA
ONTARIO ENT. SOC. ANN. REPT. P. 63 1881 REED LARVA, PUPA
OUTLOOK 74, NO. 14, APRIL 7, P. 831. 1900 SOULE LARVA
PAPILIO 3:127, 1883 EDWARDS-ELLIOTT YOUNG LARVA
PHILADELPHIA ACAD. NAT. SCI. JOURN. 5:186, 1859 CLEMENS LARVA, PUPA
PHILADELPHIA ENT. SOC. PROC. 3: 668, 1864 LINTNER LARVA, PUPA
PSYCHE 5L264, 1889 ELIOT & SOULE EGG, LARVA, PUPA
UNITED STATES DEPT. AGRI. MISC. PUB. 657:349, 376, 1950 CRAIGHEAD LARVA* F. 71E
ABBOT & SMITH-LEP. INS. GEORGIA 1:57, 1797 LARVA* PUPA* PL. 29
BOISDUVAL-SPEC. GEN. LEP. SPHING. P. 27, 1874 LARVA
ELIOT & SOULE-CATER. & MOTHS. P. 201, 1902 EGG, LARVA* PUPA
FERNALD-SPHINGIDAE NEW ENGLAND P. 81, 1886 LARVA, PUPA
GLOVER-ILLUST. NORTH AMER. ENT 1878 LARVA* PL. 2, F. 19
HARRIS-ENT. CORRESP. P. 281, 1869 LARVA
MORRIS-SYNOP. LEP. NORTH AMERICA

P.213,1862 LARVA
STRECKER-LEP.RHOP.ET HETER.
7:54, 1873 LARVA,PUPA
INDEX AMER.ECON.ENT.
VOL.1:94;VOL.4:123
FOOD PLANTS
CARYA ILLINOENSIS(WANG.)
CARYA TOMENTOSA NUTT.
JUGLANS CINEREA LINN.
JUGLANS NIGRA LINN.
OSTRYA VIRGINIANA(MILL.)
PRUNUS PENSYLVANICA LINN.
PRUNUS SEROTINA EHRH.

JULIA BARNES & MCDUNNOUGH,EPIZEUXIS-NOCTUIDAE
LIFE HISTORY
ENT.AMER.14:160,1934 CRUMB
LARVA* PL.24,R
FOOD PLANTS
DEAD WOOD

JULIA EDWARDS
SSP.SEE SARA BOISDUVAL

JULIA HULST
SYN.SEE LORICARIA EVERSMANN

JUNCIMACULA SMITH
SSP.SEEPURPURISSATA GROTE(POLIA)

JUNCTURANA STRAND
SYN.SEE ARIZONAE GROTE

JUNCTURELLA STRAND
SYN.SEE ARIZONAE GROTE

JUNIPERARIA COMSTOCK
SSP. SEE SIVA EDWARDS

JUNIPERARIA PACKARD
SYN.SEE PUBER GROTE & ROBINSON

JUNO PACKARD,HEMILEUCA-SATURNIIDAE
SYN.YAVAPAI NEUMOEGEN
LIFE HISTORY
AMER.PHILSOC.PROC.31,NO PAGE
1893 PACKARD LARVA*PL.10
F.14 YAVAPAI NO DESCR.
ENT.AMER.3:167,1887 HENRY
EDWARDS LARVA(YAVAPAI)
NAT.ACAD.SCI.MEM.12:116,1914
PACKARD LARVA* PL.23,F.6
COLOR YAVAPAI
PAPILIO 2:102,1882 GROTE EGG
SOUTHERN CALIFORNIA ACAD.SCI.
BULL.37:116,1938 COMSTOCK
DAMMERS LARVA* PL.38;PUPA*
P.40
FOOD PLANTS
POPULUS FREMONTI S.WATS.
PROSOPIS GLANDULOSA TORR.
SALIX SP.(WILLOWS)

JUSSIEUAE HUBNER
SYN.SEE FASCIATUS SULZER

JUTTA(HUBNER),OENEIS-SATYRIDAE
SYN.BALDER GUERIN-MENEVILLE
A TERRAE-NOVAE DOSPASSOS
B ALASKENSIS HOLLAND
C LEUSSLERI BRYANT
D REDUCTA MCDUNNOUGH
E RIDINGSII CHERMOCK & CHERMOCK
LIFE HISTORY
CAN.ENT.20:131,1888 FYLES EGG,
LARVA,CHRYSALIS
ENT.TIDSCH.7:151,1886 HOLMGREN
EGG*LARVA*
NATURALISTE CANADIEN P.203,
1928 HUARD LARVA,CHRYSALIS
EDWARDS-BUTTER.NORTH AMERICA
3RD SERIES 1897 LIFE HIST*
PL.CHIONOBAS 5,FF.A-G
SCUDDER-BUTTER.EAST.UNITED
STATES & CANADA PP.151-
152,1889 LIFE HISTORY*
PL.64,F.2;PL.70,F.2;PL.
74,F.11;PL.78,F.14;PL.84,
FF.17,18
FOOD PLANTS
CAREX OLIGOSPERMA MICHX.
JUNCUS ARTICULATUS LINN.

JUTURNARIA(GUENEE),SERICOSEMA-GEOMETRIDAE

A CALIFORNIARIA PACKARD
B IMMACULATA BARNES & MCDUNNOUGH
LIFE HISTORY
ENT. NEWS 5:63,1894 DYAR, LARVA PUPA
UNITED STATES NAT. MUS. PROC. 25:392,1903 DYAR EGG
FOOD PLANTS
RHAMNUS SP.

JUVENALIS ABBOT & SMITH
SYN. SEE BRIZO BOISDUVAL

JUVENALIS(FABRICIUS)ERYNNIS-HESPERIIDAE
SYN. JUVENIS HUBNER
SYN. COSTALIS WESTWOOD & HEWITSON
SYN. ENNIUS SCUDDER & BURGESS
SYN. PLAUTUS SCUDDER & BURGESS
LIFE HISTORY
AMER. MUS. NAT. HIST. BULL. 5: 301,1893 BEUTEN. LARVA, CHRYSALIS
BUFALO SOC. NAT. SCI. BULL. 19: 32,1939 WILD LARVA
ILLINOIS DEPT. AGRI. TRANS. 18, APPEND:97,1880 MIDDLETON LARVA, CHRYSALIS
NEW YORK STATE MUS. MEM. 8:704 1905 FELT LARVA (BRIEF)
PSYCHE 7:362,1896 FOLSOM EGG AND ITS HATCHING
ABBOT & SMITH-LEP. INS. GEORGIA 1:41 1797 LARVA* CHRYSALIS* PL. 21 COLOR
BOISDUVAL & LECONTE-HIST. LEP. NORTH AMER. 1833 LARVA* PL. 65, F. 4 COLOR: CHRYSALIS*
COMSTOCK-BUTTER. CALIFORNIA P. 213,1927 EGG, LARVA, CHRYSALIS
FRENCH-BUTTER. EAST. UNITED STATES P.364,1886 LARVA, CHRYSALIS
GLOVER-ILLUST. NORTH AMER. ENT. 1878 LARVA, PUPA* PL. 2, F. 2
HARRIS-INS. INJ. VEG. FLING. EDIT. P.309,1862 LARVA, CHRYSALIS
HOLLAND-BUTTERFLY BOOK 1905 CHRYSALIS* PL. 6, F. 33
MAYNARD-BUTTER. NEW ENGLAND P.56,1886 LARVA
SCUDDER-BUTTER. EAST. UNITED STATES & CANADA PP. 1480-1485,1889 LIFE HISTORY* PL. 77, FF. 2-6,10,11, PL. 80, FF. 30,36,41; PL. 82, FF. 1-2; PL. 86, FF. 48-51, PL. 85, F. 33
SCUDDER-EVERYDAY BUTTEFFLIES P.45,1899 EGG, LARVA, CHRYSALIS
SCUDDER-GUIDE TO COMMON BUTTER. P.161,1893 LARVA, CHRYSALIS
FOOD PLANTS
APIOS AMERICANA MEDIC.
BAPTISIA SP.
CORYLUS SP. (HAZELS)
GALACTIA VOLUBILIS(LINN.)
LATHYRUS SP.
LEGUMINOSAE SP. (LESS COMMONLY)
QUERCUS SP. (OAKS)
QUERCUS ALBA LINN.
QUERCUS ILICIFOLIA WANG.
QUERCUS PHELLOS LINN.
QUERCUS RUBRA LINN.
QUERCUS VELUTINA LAM.

JUVENIS HUBNER
SYN. SEE JUVENALIS FABRICIUS

K

KALI STRECKER
SYN. SEE BEHRII EDWARDS(CALLIPSYCHE)

KALLIAS EHRMAN
SYN. SEE CLODIUS MENETRIES

KALMIAE ABBOT & SMITH, SPHINX-SPHINGIDAE
LIFE HISTORY
AMER. JOURN. SCI. 36:295,1839 HARRIS LARVA, PUPA
AMER. MUS. NATHIST. BULL. 7:301 1895 BEUTEN. LARVA, PUPA
LOC. CIT. 8:292,1896 BEUTEN. EGG, LARVA, PUPA
CAN. NAT. P.260,1840 GOSSE EGG

LARVA
ENT. NEWS 9:150, 1898 SOULE
DURATION OF EGG
ENT. SOC. AMER. ANNALS 11:421, 1918
ILLINOIS DEPT. AGRI. TRANS. APPEND 18:104, 1880 MARTEN LARVA
NEW YORK STATE AGRI. EXP. STA. (CORNELL), MEM. 274:191, 1948 FORBES LARVA
NEW YORK STATE MUS. MEM. 8:548 1905 FELT LARVA
NOVITATES ZOOLOGICAE (TRING) 9 SUPPLP. 136, 1903 ROTHSCHILD & JORDAN LARVA, PUPA
ONTARIO ENT. SOC. ANN. REPT. P. 68 1881 REED LARVA, PUPA
PHILADELPHIA ACAD. NAT. SCI. JOURN. 5:171, 1859 CLEMENS LARVA, PUPA
PHILADELPHIA ENT. SOC. PROC. 3: 657, 1864 LINTNER LARVA
SMITHSONIAN MISC. COLLECTIONS 4, ART. 2:196, 1862 MORRIS LARVA, PUPA
UNITED STATES DEPT. AGRI. MISC. PUB. 657, 1950 CRAIGHEAD LARVA* F. 71, G
ABBOT & SMITH-LEP. INS. GEORGIA 1:7, 1797 LARVA*PUPA*PL. 37 COLOR
BOISDUVAL-SPEC. GEN. LEP. P. 92, 2874 LARVA, PUPA
ELIOT & SOULE-CATER. &MOTHS P. 135, 1902 EGG, LARVA*PUPA
FERNALD-SPHINGIDAE NEW ENGLAND P. 41, 1886 LARVA, PUPA
HOLLAND-MOTH BOOK P. 7, 1905 LARVA* F. 4 COLOR
MORRIS-SYNOP. LEP. NORTH AMER. P. 196, 1862 LARVA, PUPA

FOOD PLANTS
CHIONANTHUS VIRGINICUS LINN.
FRAXINUS NIGRA MARSH.
ILEX GLABRA (LINN.)
KALMIA SP. (LAURELS)
LIGUSTRUM SP. (PRIVETS)
PRUNUS MARITIMA MARSH.
RHODODENDRON SP.
SYRINGS SP.

KANAWAHENSIS CHERMOCK
SSP. SEE CATALPAE BOISDUVAL

KAPPA BARNES & BENJAMIN
SEE STRICTA WALKER

KASLOA DYAR
SSP. SEE VAGANS BOISDUVAL

KASLOATA TAYLOR
SSP. SEE ETHELA HULST

KASLOATA (DYAR), EUPITHECIA-GEOMETRIDAE
LIFE HISTORY
UNITED STATES NAT. MUS. PROC. 27:891, 1904 DYAR EGG, LARVA
FOOD PLANTS

KASLOENSIS COCKERELL
SSP. SEE EURYALUS BOISDUVAL

KATAHDIN NEWCOMB
SSP. SEE POLIXENES FABRICIUS

KEDEMA BUTLER
SYN. SEE CAMPESTRIS BOISDUVAL

KEEWAYDIN EDWARDS
SYN. SEE EURYTHEME BOISDUVAL

KELLOGGI (HENRY EDWARDS), ONCOCNEMIS-NOCTUIDAE
SYN. EXEMPLARIS SMITH
LIFE HISTORY
FOOD PLANTS
CAREX FILIFOLIA NUTT.
GENTIANA NEWBERRYI GRAY

KELSEYI WRIGHT
SSP. SEE ACMON WESTWOOD & HEWITSON

KENTARIA (GROTE & ROBINSON), SELENIA-GEOMETRIDAE
A. GLAUCATA BARNES & MCDUNNOUGH

LIFE HISTORY
ENT. AMER. 3:47, 1887 BRUCE EGG
NEW YORK STATE AGRI. EXP. STA. (CORNELL), MEM. 274:91, 1948 FORBES EGG, LARVA
FOOD PLANTS
ACER SP. (MAPLES)
BETULA SP. (BIRCHES)
FAGUS SP. (BEECHES)
QUERCUS SP. (OAKS)
TILIA AMERICANA LINN.
GENERAL FEEDER FOREST TREES

KERRI COMSTOCK
SSP. SEE BERENICE CRAMER

KERRVILLARIA CASSINO & SWETT
SYN. SEE SCOLOPACINARIUS GUENEE

KEUTZINGARIA PACKARD, PLAGODIS-GEOMETRIDAE
SYN. ALTRURARIA PACKARD
LIFE HISTORY
NEW YORK ENT. SOC. JOURN. 53:205 210, 1945 FORBES LARVA, PUPA
FOOD PLANTS
MALUS PUMILA MILL. (APPLE)

KEUTZINGI GROTE
SYN. SEE NIGRESCARIA HULST

KIOWAH REAKIRT
SYN. SEE RURICOLA EDWARDS

KLAGESII EHRMANN
SYN. SEE ACREA DRURY

KLEENEI WATSON
SYN. SEE BELLONA FABRICIUS

KLOTSI FIELD
SSP. SEE HETERONEA BOISDUVAL

KLOTSI RICHARDS
SSP. SEE MIRIFICA HENRY EDWARDS

KODIAK EDWARDS
SSP. SEE SCUDDERII EDWARDS

KODIAKATA PACKARD
SYN. SEE LUCTUATA SCHIFFERMULLER

KOHLSAATI GUNDER
SSP. SEE AQUILO BOISDUVAL

KOOTENAI COCKLE
SSP. SEE EURYTHEME BOISDUVAL

KOOTENAIATA DYAR
SSP. SEE INNOTATA HUFNAGEL

KRAUTHII KOTS
SSP. SEE CHRISTINA EDWARDS

KRAUTWORMI CHERMOCK
SSP. SEE IMPLETA WALKER (ACRONICTA)

KRAUTWORMI HOLLAND
SSP. SEE CYBELE FABRICIUS

KROMBEINI FRANCLEMONT
SSP. SEE MESKEI GROTE

KUNZEI HULST
SSP. SEE EPHELIDARIA HULST

KUNZEI ROTHSCHILD & JORDAN
SSP. MODESTA HARRIS

L

LABECULA WATSON & COMSTOCK
SSP. SEE ACMON DOUBLEDAY & HEWITSON

LABESARIA GROTE
SSP. SEE VARIARIA GROTE

LABRADORIATA MOESCHLER
SYN. SEE SEXMACULATA PACKARD

LABRADORIENSIS SOMMER
SSP. SEE PALUDATA THUNBERG

LABRUSCAE (LINNAEUS), PHOLUS-SPHINGIDAE
CLOTHO FABRIDIUS
LIFE HISTORY
ENTOMOLOGIST'S RECORD & JOURN.

VARIATION 41:143,1929
HAYWARD LIFE HISTORY
NEW YORK ENT. SOC. JOURN. 15,1907
VERRILL LARVA*PL.1,F.5
NEAR P.50 COLOR NO DESCR,
NEW YORK STATE AGRI. EXP. STA.
(CORNELL), MEM. 274:200,1948
FORBES LARVA(BRIEF)
NOVITATES ZOOLOGICAE(TRING)9
SUPPL. P.497,1903 ROTHSCHILD
& JORDAN LARVA, PUPA
PHILADELPHIA ACAD. NAT. SCI.
JOURN. 5:156,1859 CLEMENS
LARVA
BOISDUVAL-SPEC. GEN. LEP. SPHING
P.193,1874 LARVA
BURMEISTER-DESCR. REP. ARGENT.
F.347,1878 LARVA* ATLAS
PL.15,F.36
GUNDLACH-ENT. CUBANA P.187,1886
MERIAN-METAM. INS. SURINAM P.34,
1705 LARVA*PUPA*F.34 COLOR
SEPP. SURINAMSCHE VLINDERS 1855
LARVA* PL.32

FOOD PLANTS

PARTHENOCISSUS SP.
VITIS SP.

LACHRYMOSA HULST
SYN. SEE PACKARDARIA HULST

LACINARIAE BIRD
SSP. SEE BEERIANA BIRD

LACINIA(GEYER), CHLOSYNE-NYMPHALIDAE
A ADJUSTRIX SCUDDER
SYN. NIGRA COCKERELL
SYN. BICOLOR COCKERELL
SYN. RUFA COCKERELL
AB INGHAMI GUNDER
B RUFESCENS EDWARDS
C CROCALE EDWARDS
D NIGRESCENS COCKERELL

LIFE HISTORY

CAN. ENT. 25:286,1893 EDWARDS
EGG, LARVA, CHRYSALIS
SOUTHERN CALIFORNIA ACAD, SCI.
BULL. 45:144,1946 COMSTOCK
CHRYSALIS* PL.18
WASHINGTON ENT. SOC. PROC. 13:228
1911 DYAR LARVA
COMSTOCK-BUTTER. CALIFORNIA
P.119,1927 EGG, LARVA,
CHRYSALIS

FOOD PLANTS

HELIANTHUS ANNUUS LINN.
HELIANTHUS CILIARIS DC
PALAFOXIA SPHACELATUM(NUTT.
EX. TORR.)
VERBESINA ENCELIOIDES(CAV.)
XANTHIUM STRUMARIUM V.
CANADENSE(MILL.)

LACINIOSA ZELLER
SYN. SEE BALTIMORALIS GUENEE

LACRYMOSA GUENEE, CATOCALA-NOCTUIDAE
FORM EVELINA FRENCH
SYN. EMILIA HENRY EDWARDS
FORM ZELICA FRENCH
FORM PAULINA HENRY EDWARDS
FORM ALBOMARGINATA CASSINO

LIFE HISTORY

AMER. MUS. NAT. HIST. BULL. 38:
155,1918 BARNES & MCDUNNOUGH
EGG, LARVA
CAN. ENT. 47:338,1915 ROWLEY EGG,
LARVA
LOC. CIT. 51:227,1919 ROWLEY
LARVA
ENT. NEWS 33:235,1922 BROWER
EGG, LARVA

FOOD PLANTS

CARYA SP. (HICKORIES)
CARYA ILLINOENSIS(WANG.)
CARYA OVATA(MILL.)
JUGLANS REGIA LINN.
QUERCUS SP. (OAKS)

LACTATA SMITH
SYN. SEE COLONA HUBNER

LACTEUS GUNDER
SSP. SEE NYCTEIS DOUBLEDAY&HEWITSON

LACTISPARGARIA WALKER
SYN. SEE FRITILLARIA GUENEE

LACUSTRATA(GUENEE), XANTHORHOE-GEOMETRIDAE
LIFE HISTORY
UNITED STATES NAT. MUS. PROC. 27:.897, 1904 DYAR EGG, LARVA
FOOD PLANTS
BETULA SP. (BIRCHES)
CRATAEGUS SP. (THORNS)
IMPATIENS SP.
RUBUS SP.
RUBUS ALLEGHENIENSIS PORTER
RUBUS IDAEUS V. STRIGOSUS (MICHX.)
SALIX SP. (WILLOWS)
GENERAL FEEDER. MOST ANYTHING

LACUSTRUS FREEMAN
SSP. SEE AQUILO BOISDUVAL

LAETA(EDWARDS), ERORA-LYCAENIDAE
SYN. FEMALE CLOTHILDE EDWARDS
LIFE HISTORY
CAN. ENT. 55:27, 1923 MOUSLEY EGG, LARVA
SCUDDER-BUTTER. EAST. UNITED STATES & CANADA P. 831, 1889 EGG* PL. 65, F. 8
FOOD PLANTS
FAGUS SP. (BEECHES)
GENISTA SP. (?(
HEDYSARUM SP. (?)
RUBUS IDAEUS V. STRIGOSUS (MICHX.)

LAETA(MORRISON), ARCHANARA-NOCTUIDAE
LIFE HISTORY
NEW YORK ENT. SOC. JOURN. 33:185, 1925 BIRD LARVA
FOOD PLANTS
SPARGANIUM EURYCARPUM ENGELM.

LAETIFICA LINTNER, CUCILLIA-NOCTUIDAE
SYN. CITA GROTE
SYN. HARTMANNI FRENCH
LIFE HISTORY
UNITED STATES NAT. MUS. PROC. 25:381, 1903 DYAR LARVA
FOOD PLANTS
CHRYSOTHAMNUS SP.

LAETIFICANS(SMITH), EUXOA-NOCTUIDAE
SYN. MASCULINUS SMITH
LIFE HISTORY
FOOD PLANTS
BETA VULGARIS LINN. (BEET)
LACTUCA SATIVA LINN. (LETTUCE)
MEDICAGO SATIVA LINN. (ALFALFA)
MELILOTUS ALBA DESR.
SALSOLA KALI V. TENUIFOLIA TAUSCH

LAETULA BARNES & MCDUNNOUGH
SSP. SEE MACULARIA HARRIS

LAETULA GROTE
SSP. SEE EXSERTISTIGMA MORRISON

LAETULUS GROTE
SSP. SEE EDUCATLIS WALKER

LAGGANATA CASSINO & SWETT
SSP. SEE INCURSATA HUBNER

LAGGANATA TAYLOR
SYN. SEE GELIDATA MOESCHLER

LAGUS EDWARDS
SSP. SEE LOGAN EDWARDS

LAMINA FABRICIUS
SYN. SEE ARTHEMIS DRURY

LAMINIS SMITH
SYN. SEE TESSELLATA HARRIS

LANCEOLARIA(GROTE), ACRONICTA-NOCTUIDAE
LIFE HISTORY
NEW YORK STATE MUS. MEM. 8:742 1905 FELT LARVA(BRIEF)
ONTARIO ENT. SOC. ANN. REPT. P. 115, 1909 WILLING LARVA
UNITED STATES NAT. MUS. PROC. 21:172, 1898 SMITH & DYAR LARVA, PUPA, COCOON
FOOD PLANTS
COMPTONIA SP.
GAILLARDIA SP.
PLANTAGO SP.

PRUNUS CERASUS LINN. (CHERRY
PRUNUS SEROTINA EHRH.
SALIX SP. (WILLOWS)

LANCEOLATA BOISDUVAL, ANTHOCHARIS- PIERIDAE
SYN. EDWARDSII BEHR
A AUSTRALIS GRINNELL
LIFE HISTORY
CAN. ENT. 40:130, 1908 COOLIDGE PUPA
LOC. CIT. 42:315, 1910 COOLIDGE EGG, LARVA (AUSTRALIS)
PSYCHE 2:183, 1878 MEAD LARVA WHICH MAY BE THIS SPECIES
LOC. CIT. 15:81, 1908 COOLIDGE PUPA*
SOUTHERN CALIFORNIA ACAD. SCI. BULL. 39:22, 1930 COMSTOCK EGG* AUSTRALIS
COMSTOCK-BUTTER. CALIFORNIA P. 39, 1927 EGG, LARVA* PL. A FF. 14, 15; CHRYSALIS* PL. 63, F. 10 COLOR
EDWARDS-BUTTER. NORTH AMERICA 3RD SERIES 1897 LARVA, CHRYSALIS PL. ANTHOCARIS 1, F. X
HOLLAND-BUTTER. BOOK P. 285, 1905 LARVA 3RD EDIT. P. 288, 1931
FOOD PLANTS
ARABIS GLABRA (LINN.)
ARABIS GRACILIPES GREENE
CRUCIFERAE
SISYMBRIUM OFFICINALE (LINN.)

LANGEI COMSTOCK
SSP. SEE MORMO FELDER & FELDER

LANGTONI COUPER, ALYPIA-AGARISTIIDAE
SYN. SACRAMENTO GROTE & ROBINSON
SYN. HUDSONICA HENRY EDWARDS
LIFE HISTORY
CAN. ENT. 27:278, 1895 DYAR EGG, LARVA
FOOD PLANTS
EPILOBIUM ANGUSTIFOLIUM LINN.
?PARTHENOCISSUS QUINQUEFOLIA (LINN.)
?VITIS SP. (GRAPES)

LANICE LINTNER
SYN. SEE LYSIDE GODART

LANTHANIS COOK & WATSON
SYN. SEE ARCHIPPUS CRAMER

LAPIDARIA GROTE
SYN. SEE MULTILINEA WALKER

LAQUEATA HENRY EDWARDS
SYN. SEE STRIGOSA WALKER

LARENTIOIDES GROTE, PHALAENOSTOLA- NOCTUIDAE
A CITIMA GROTE
LIFE HISTORY
ENT. AMER. 14:176, 1934 CRUMB LARVA* PL. 24 A
FOOD PLANTS
DEAD LEAVES, DEAD GRASS

L-ARGENTEUM SCUDDER
SSP. SEE PROGNE CRAMER

LARGERA SMITH
SSP. SEE BENESIGNATA HARVEY

LARICIS (FITCH), TOLYPE-LASIOCAMPIDAE
SYN. MINUTA GROTE
LIFE HISTORY
AMER. MUS. NAT. HIST. BULL. 10: 445, 1898 BEUTEN. LARVA
NEW YORK STATE MUS. MEM. 8: 680 1905 FELT LARVA
NEW YORK STARE REPT. INS. OF 2:264 ALSO P. 66, SEC. 287, 1856 FITCH LIFE HISTORY
PAPILIO 3:25, 1883 GILBERT COCOON ON WHITE PINE
PAPILIO 3:131, 1883 EDWARDS ELLIOTT LARVA
DOANE ETC-FOREST INSECTS P. 282 1936 LARVA (BRIEF)
PACKARD-INS. INJ. FOREST TREES P. 254, 1881 LARVA, COCOON

Index Amer. Econ. Ent.
Vol. 4:495
FOOD PLANTS
Larix laricina (DuRoi)
Pinus sp. (pines)

LARUNDA Strecker
syn. see DYMAS Edwards

LARVARIA (Guenee), ANACAMPTODES-Geometridae
LIFE HISTORY
Can. Ent. 6:32, 1874 Saunders larva
New York State Agri. Exp. Sta. (Cornell), Mem. 274:57, 1948 Forbes larva
New York Stare Mus. Mem. 8:743, 1905 Felt larva (brief)
Gumppenberg-Systema Geometrarum 5:332, 1892 larva, pupa
Packard-Monograph Geometridae p. 438, 1876 larva*Pl. 13, f. 17; pupa* Pl. 13, f. 36
FOOD PLANTS
Populus sp. (poplars)
Prunus serotina Ehrh.
Salix sp. (willows)

LATEBRA Benjamin
ssp. see BRUNNEIGERA Grote

LATEBRICOLA Grote
syn. see DORMITANS Gurnee

LATERITIA (Hufnagel), AGROPERINA-Noctuidae
LIFE HISTORY
Amer. Mus. Nat. Hist. Bull. 14: 304, 1901 Beuten. larva
FOOD PLANTS

LATERITIARIA Guenee
ssp. see HYPOCHRARIA Herrich-Schaeffer

LATIA Strecker
syn. see BUFFALOENSIS Grote

LATIALBATA Barnes & McDunnough
ssp. GRATULATA Walker

LATIARIA Packard
syn. see MIMOSARIA Guenee

LATICINCTA Walker
syn. see AMICARIA Herrich-Schaeffer

LATICINEREA Grote, LITHOPHANE-Noctuidae
a WINNIPEG Smith
LIFE HISTORY
New York State Agri. Exp. Sta. (Cornell), Bull. 123:157, 1896 Slingerland larva*pupa* Pl. 2 Note-This may be HIBISCI Guenee
Papili 3:135, 1883 Edwards-Elliott larva
Index Amer. Econ. Ent.
Vol. 1:321 Xylena
Vol. 2:164; Vol. 3:180;
Vol. 3:437 Xylena
Vol. 5:282; Vol. 6:326
FOOD PLANTS
Acer sp.
Acer saccharinum Linn.
Prunus sp.
Prunus cerasus Linn. (cherry)
Prunus serotina Ehrh.

LATIFASCIA Barnes & McDunnough
ssp. LUCINA Henry Edwards

LATIFACICA Walker, PRODENIA-Noctuidae
LIFE HISTORY
Florida Ent. 6:44, 1922 Mason larva
FOOD PLANTS
Citrus grandis (Linn.) (grapefruit)

LATIFASCIATA Butler
syn. see CALIFORNIAE Walker

LATIFERRUGATA (Walker), ITAME-Geometridae
syn. BRUNNEATA Packard
LIFE HISTORY
New York State Agri. Exp. Sta.

(CORNELL),MEM.274:37,1948
FORBES LARVA
FOOD PLANTS
MALUS PUMILA MILL.(APPLE)
PRUNUS DOMESTICA LINN.(PLUM)
PRUNUS PENSYLVANICA LINN.

LATIORELLA(WALKER),ARGUISA-NOCTUIDAE
SYN. LUTEA SMITH
LIFE HISTORY
ENT.AMER.14:155,1934 CRUMB
LARVA* PL.21 D;PL.24 K
FOOD PLANTS
POA PRATENSIS LINN.

LATIPENNIS(BOISDUVAL),GNOPHAELA-PERICOPIDAE
SYN.HOPFFERI GROTE & ROBINSON
A DISCRETA STRETCH
SYN.ARIZONAE FRENCH
SYN.MORRISONI DRUCE
B CERMICULATA GROTE & ROBINSON
SYN.CONTINUA HENRY EDWARDS
LIFE HISTORY
CAN.ENT.21:220,1889 COCKERELL
EGG(VERMICULATA)
LOC.CIT.53:101,1921 WHITEHOUSE
EGG,LARVA(VERMICULATA)
ENT.AMER.4:24,1888 BRUCE LARVA
PUPA
LOC.CIT.5:57,1889 COCKERELL
LARVA,PUPA
ENT.NEWS 6:74,1895 BRUCE LARVA
(BRIEF) VERMICULATA
INSECUTOR INSCITIAE MENSTRUUS
2:62,1914 DYAR LARVA
NEW YORK ACAD.SCI.ANNALS 80:
228,1894 DYAR LARVAL SETAE
(VERMICULATA)
PAPILIO 2:82,1882 STRETCH LARVA
(BRIEF) HOPFFERI
WASHINGTON ENT.SOC.PROC.4:408
1899 DYAR LARVE
STRETCH-ILLUST.ZYGAEN.&MOMBYC
NORTH AMERICA P.236,1873
LARVA
FOOD PLANTS
CYNOGLOSSUM OCCIBENTALE GRAY
HACKELIA FLORIBUNDA (LEHM.)
MERTENSIA PANICULATA(AIT.)
MERTENSIA VIRGINICA(LINN.)
MYOSOTIS

LATIPENNIS HULST
SYN.SEE CASTICATA HUBNER

LATIPENNIS HULST
SSP.SEE OCCIDENTALIS HULST
(OPOROPHTERA)

LATIPENNIS HULST
SYN.SEE IRIDARIA GUENEE

LATIPENNIS(STRETCH),DIACRISIA-ARCTIIDAE
LIFE HISTORY D
AMER.MUS.NAT.HIST.BULL.10:
374,1898 BEUTEN.LARVA
ENT.NEWS 2:115,1891 DYAR LARVA
NEW YORKSTATE MUS.MEM.8:727
1905 FELT LARVA
PSYCHE 7:71,1894 SOULE EGG,
LARVA,PUPA
FOOD PLANTS
ARISAEMA TRIPHYLLUM(LINN.)
FRAXINUS SP.(ASHES)YOUNG LARVA
IMPATIENS SP.
PLANTAGO SP.
PLANTAGO MAJOR LINN.
PRUNUS SEROTINA EHRH.
GENERAL FEEDER LOW PLANTS

LATIPES(GUENEE),MOCIS-NOCTUIDAE
AB IDENTATA HARVEY
LIFE HISTORY
JOURN.ECON.ENT.17:404,1924
VICKERY NO DESCR,EGG*PL.5;
LARVA*PL.5,F.1;PUPA*PL.5,
FF,2,3 MOCIS REPANDA
FABRICIUS
NEW YORK ENT.SOC.JOURN.8:33
1900 SWAINSON LARVA,PUPA
UNITED STATES NAT.MUS.PROC.
23:276,1901 DYAR EGG,LARVA
GUENEE-SPEC.GEN.LEP.NOCT.
3:315,1852 LARVA
GUNDLACH-ENT.CUBANA P.354,1886

LARVA, PUPA

FOOD PLANTS

CENCHRUS SP.

CENCHRUS BROWNII ROEM.

DIGITARIA SANGUINALIS(LINN.)

ERIOCHLOA PUNCTATA(LINN.)

HYPERICUM SP.

LEPTOCHLOA NEALLEYI VASEY

PANICUM SP.

PANICUM FASCICULATUM SWARTZ

SACHARUM OFFICINARUM LINN. (SUGAR)

TRICHLORUS PLURIFLORA FOURN.

ZEA MAYS LINN. (CORN)

GRASSES

LATIRENA DOD

SYN. SEE HIBISCI GUENEE

LATIREPTANA WALKER

SYN. SEE SEMICANA WALKER

LATIRUPTA WALKER

SYN. SEE CENTROSTRIGARIA WOLLASTON

LATIUSCULA HERRICH-SCHAEFFER, LEUCANIA-NOCTUIDAE

SYN. SUBPUNCTATA HARVEY

SYN. COMPLICATA STRECKER

LIFE HISTORY

JOURN. AGRI. RESEARCH 23:1104, 1926 VICKERY EGG, LARVA* F. 2 DIAGRAMMATIC SEG.

FOOD PLANTS

AVENA SATIVA LINN. (OATS)

DIGITARIA SANGUINALIS(LINN.)

HORDEUM VULGARE LINN. (BARLEY)

SACCHARUM OFFICINARUM LINN (SUGAR)

SORGHUM BICOLOR(LINN.)

ZEA MAYS LINN. (CORN)

LATREILLANA KIRBY

SYN. SEE VIRGINICA CHARPENTIER

LAUDABILIS(GUENEE), LACINIPOLIA-NOCTUIDAE

SYN. INDICANS WALKER

FORM RUFOIRRORATA STRAND

FORM MEDIOSUFFUSA STRAND

LIFE HISTORY

CAN. ENT. 35:273, 1903 DYAR EGG, LARVA

GUENEE-SPEC. GEN. LEP. NOCT. 2:30, 1852 LARVA, PUPA

FOOD PLANTS

LEGUMINOSAE SP.

GENERAL FEEDER LOW PLANTS

LAURAE CHERMOCK

SSP. ZEE EURYTHEME BOISDUVAL

LAURENTINA LYMAN

SYN. SEE COMMA LINNAEUS

LAURENTINA SCUDDER

SSP. SEE INTERIOR SCUDDER

LAURINA WRIGHT

SSP. SEE MACARIA EDWARDS

LAVINIA CARMER

SEE NOTE UNDER COENIA HUBNER

LEANIRA FELDER & FELDER, MELITAEA-NYMPHALIDAE

FORM LEONA WRIGHT

FORM OBSOLETA HENRY EDWARDS

A WRIGHTII EDWARDS

AB CAROLYNAE GUNDER

FORM PELONA GUNDER

B CERRITA WRIGHT

LIFE HISTORY

SOUTHERN CALIFORNIA ACAD. SCI. BULL. 31:9, 1923 COMSTOCK DAMMERS LIFE HISTORY EGG, LARVA* PUPA*

COMSTOCK-BUTTER. CALIFORNIA P. 114, 1927 EGG(WRIGHTII)

FOOD PLANTS

CASTILLEJA AFFINIS H. & A.

CASTILLEJA FOLIOLOSA H. & A.

CORDYLANTHUS PILOSA GRAY

CORYDYLANTHUS TENUIS GRAY

LECONTEI(GUERIN-MENEVILLE), HAPLOA-ARCTIIDAE

SYN. LEUCOMELAS HERRICH-SCHAEFFER
FORM MILITARIS HARRIS
SYN. CONFINIS WALKER
SYN. HARRISII DYAR
FORM SMITHII DYAR
FORM DYARII MERRICK
FORM VESTALIS PACKARD
SYN. OCHROLEUCA STRAND

LIFE HISTORY

AMER. MUS. NAT. HIST. BULL. 10; 366, 1898 BEUTEN. LARVA
CAN. ENT. 1:20, 1868 SAUNDERS LARVA
ENT. NEWS 16:90, 1905 WASMUTH LARVA HIBERNATES?
ILLINOIS DEPT. AGRI. TRANS. 15:181 1877 FRENCH LARVA
LOC. CIT. 18, APPEND: 115, 1880 MARTEN LARVA
ILLINOIS, REPT. INS. OF 1:47, 1871 LEBARON LARVA* PUPA*
KENTUCKY AGRI. EXP. STA. CIRC. 25, 1921 GARMAN LARVA* FF. 1, 2
MINNESOTA AGRI. EXP. STA. BULL. 61:129, 1899 LUGGER LARVA
NATURALISTE CANADIEN P. 17, 1929 HUARD LARVA
NEW YORK ENT. SOC. JOURN. 3:59 1895 DYAR EGG
LOC. CIT. 3:176, 1895 PACKARD EGG, LARVA
NEW YORK STATE CABIN. NAT. HIST. ANN. REPT. NO. 26:142, 1874 LINTNER LARVA (BRIEF)
PSYCHE 13:32, 1906 FOSTER TABLE OF HAPLOA LARVAE
SAUNDERS-INS. INJ. FRUITS P. 198, 1883 LARVA, PUPA*
SAUNDERS-LOC. CIT. P. 198, 1889 LARVA* F. 206; PUPA
INDEX AMER. ECON. ENT. VOL. 3:187

FOOD PLANTS

CYDONIA JAPONICA (THUNB.)
CYNOGLOSSUM SP.
MALUS ANGUSTIFOLIA (AIT.)
MALUS PUMILA MILL. (APPLE)
MENTHA SPICATA LINN.
ONOSMODIUM VIRGINIANUM (LINN.)
PRUNUS PERSICA (LINN. ((PEACH)
PRUNUS SEROTINA EHRH.
PYRUS COMMUNIS LINN. (PEAR)
RUBUS ALLEGHENSIS PORTER
RUBUS IDAEUS V. STRIGOSUS (MICHX.)
TRIOSTEUM PERFOLIATUM LINN.

LECONTEI ROTHSCHILD & JORDAN
SSP. MARCELLUS CRAMER

LEDA (EDWARDS), STRYMON-LYCAENIDAE
A INES EDWARDS

LIFE HISTORY

SOUTHERN CALIFORNIA ACAD. SCI. BULL. 35:6, 1936 COMSTOCK-DAMMERS EGG* PL. 2; LARVA* PL. 3; PUPA* PL. 3

FOOD PLANTS

ACACIA GREGGII A. GRAY
PROSOPIS JULIFLORA TORR.

LEGITIMA (GROTE), POLIA-NOCTUIDAE

LIFE HISTORY

AMER. MUS. NAT. HIST. BULL. 4: 77, 1892 BEUTEN. LARVA
ENT. SOC. AMERICA, ANNALS 23:396 1930 HAWKINS LARVAL CLAWS
JOURN. ECON. ENT. 8:393, 1915 CRUMB KEY TO LARVA
OKLAHOMA AGRI. EXP. STA. BULL. 100:11, 1912 SANBORN LARVA* PUPA* BOTH F. 1
UNITED STATES DEPT. AGRI. BUR. ENT. BULL. 66 NEW SERIES PART 3:28, 1907 CHITTENDEN LARVA* F. 7 B, C, D; PUPA* F. 7E
UNITED STATES DEPT. AGRI. FARMER'S BULL. 120:25, 1900 HOWARD LARVA* F. 20 B, C, D; PUPA* F. 20E
UNITED STATES DEPT. AGRI. YEAR-BOOK FOR 1898 P. 143 HOWARD LARVA* F. 26 B, C, D; PUPA* F. 26E
O'KANE-INJ. INS. P. 171, 1912 LARVA
INDEX AMER. ECON. ENT.
VOL. 1:198 MAMESTRA
VOL. 2:225 MAMESTRA

VOL. 4:378:VOL. 6:602

FOOD PLANTS

AGROSTIS HYEMALIS(WALT.)
ASCLEPIAS INCARNATA LINN.
ASPARAGUS OFFICINALIS LINN.
ASTER SP.
BRASSICA OLERACEA LINN. (CABBAGE)
BRASSICA OLERACEA V. VIRIDIS LAM; (TURNIP)
CALENDULA OFFICINALIS LINN.
COMPOSITAE SP.
DIGITARIA SANGUINALIS(LINN.)
GNAPHALIUM SP.
MUHLENBERGIA TORREYI(KUNTH)
NICOTIANA TABACUM LINN. (TOBACCA)
PHASEOLUS VULGARIS LINN. (BEAN)
PISUM SATIVUM LINN. (PEAS)
PRUNUS SEROTINA EHRH.
RUBUS ALLEGHENSIS PORTER
SALIX HUMILIS MARSH
SOLIDAGO SP.
TRIFOLIUM SP. (CLOVERS)
VIOLA SP. (VIOLETS)
GRASSES

LEHMANI GUNDER
SSP. SEE NUBIGENA BEHR

LEHMANNI HOLLAND
SYN. SEE FRIGGA THUNBERG

LEILAE CLARK
SSP. SEE CELLUS BOISDUVAL & LECONTE

LEMMERI BARNES
SSP. SEE ESTHER BARNES

LEMMERI BARNES & BENJAMIN
SSP. SEE ATOMARIA WALKER (OLENE)

LEMMERI CHERMOCK & CHERMOCK
SSP. IMPRESSA WALKER

LEMMERI MCDUNNOUGH
SSP. SEE BENESIGNATA HARVEY

LEMMERI MAYFIELD
SSP. SEE GRACILIS EDWARDS

LENA BOISDUVAL
SYN. SEE CALIFORNIAE WALKER

LEONA WRIGHT
SSP. SEE LEANIRA FELDER & FELDER

LEONARDUS HARRIS, HESPERIA-HESPERIIDAE
A STALLINGSI FREEMSN

LIFE HISTORY

CAN. ENT. 71:118, 1939 DETHIER EGG, LARVA
ENT. NEWS 19:411, 1908 LAURENT EARLY STAGES EGG
SCUDDER-BUTTER. EAST. UNITED STATES & CANADA PP. 1676-1677 1889 EGG*LARVA*PL. 66, F. 33; PL. 69, F. 11; PL. 73, F. 10 PL. 77, FF. 32, 33; PL. 80, FF. 66, 67
SCUDDER-GUIDE TO COMMON BUTTER P. 107, 1893 LARVA, CHRYSALIS

FOOD PLANTS

AGROSTIS SP.
TARAXACUM OFFICINALE WEBER
GRASSES

LEPIDULA(GROTE), LEUCONYCTA-NOCTUIDAE
A AVIRIDA SMITH

LIFE HISTORY

FOOD PLANTS

RUMEX SP.
TARAXACUM OFFICINALE WEBER

LEPORINA(LINNAEUS), ACRONICTA-NOCTUIDAE
A VULPINA GROTE
SYN. SANCTA HENRY EDWARDS
B CRETATOIDES BENJAMIN
CRETATA AUCT.
C MOESTA DYAR
D CASSINOI BARNES & BENJAMIN

LIFE HISTORY

NEW YORK STATE MUS. MEM. 8:747 1905 FELT LARVA
PAN PACIFIC ENT. 11:152, 1935 BENJAMIN LARVAL NOTE
PAPILIO 3:14, 1883 THAXTER LARVA (VULPINA)
UNITED STATES DEPT. AGRI. MISC.

PUB.657:356,391,1950
CRAIGHEAD LARVA(VULPINA)
UNITED STATES NAT.MUS.PROC. 21:60,1898 SMITH & DYAR LARVA*PL.8,F.26 COLOR OR UNCOLORED; PUPA,COCOON
LOC.CIT.27:799,1904 DYAR EGG, LARVA (MOESTA)
GROTE-DIE APATELIDEN(MITTHEILUNGER AUS DEM ROEMER MUSEUM) HILDESHEIM 1896 LARVA* PL. 1,F.2

FOOD PLANTS
BETULA SP.(BIRCHES)
POPULUS SP.(POPLARS)
SALIX SP.(WILLOWS)

LEPTINOIDES(GROTE),SCHIZURA-NOTODONTIDAE
SYN. MUSTELINA PACKARD

LIFE HISTORY
AMER.MUS.NAT.HIST.BULL.10: 421 1898 BEUTEN.LARVA
BOSTON SOC.NAT.HIST.PROC.24: 540,1890 PACKARD LARVA
CAN.ENT.18:92,1886 FRENCH LARVA
ENT.AMER.6:230,1890 DYAR EGG, LARVA,PUPA,COCOON
NAT.ACAD.SCI.MEM.7:200,1895 PACKARD EGG,LARVA*PL.26, FF.1-4 COLOR PUPA8COCOON
NEW YORK ENT.SOC.JOURN.1:71 1893 PACKARD EGG,LARVA
NEW YORK STATE AGRI.EXP.STA. MEM.274:228,1948 FORBES LARVA
NEW YORK STATE MUS.MEM.8:716 736,1905 FELT LARVA(BRIEF)

FOOD PLANTS
CARPINUS CAROLINANA WALT.
CARYA SP.(HICKORIES)
CARYA ILLINOENSIS(WANG.)
CASTANEA DENTATA(MARSH.)
FAGUS SP.(BEECHES)
HELIANTHUS ANGUSTIFOLIUS LINN.
JUGLANS CINEREA LINN.
JUGLANS NIGRA LINN.
MALUS PUMILA MILL.(APPLE)
NYSSA SYLVATICA MARSH.
OSTRYA VIRGINIANA(MILL.)
POPULUS SP.(POPLARS)
PRUNUS SEROTINA EHRH.
QUERCUS SP.(OAKS)
ROSA SP.(ROSES)
SALIX SP.(WILLOWS)

LEPUSCULINA GUENEE,ACRONICTA-NOCTUIDAE
FORM TRANSVERSATA SMITH
FORM CANADENSIS SMITH
FORM SIMILANA SMITH
A FELINA GROTE
B CYANESCENS HAMPSON

LIFE HISTORY
AMER.MUS.NAT.HIST.BULL.14: 240,1901 BEUTEN.LARVA (POPULI)
CALIFORNIA ACAD.SCI.PROC.7:23 1876 HENRY EDWARDS LARVA
CAN.ENT.14:221,1882 SAUNDERS LARVA*
LOC.CIT.19:48,1887 FRENCH LARVA,PUPA
LOC,CIT.54:138,1922 MCDUNNOUGH LARVA(BRIEF) CANADENSIS
ILLINOIS AGRI.EXP.STA.BULL.151: 468,1911 FORBES LARVA*
ILLINOIS DEPT.AGRI.TRANS.15:201, 1877 FRENCH LARVA
LOC.CIT.18,APPEND:130,1880 MARTEN LARVA*
MISSOURI,REPT.NOX.INS.2:119, 1870 RILEY LARVA* PUPA
NEW YORK STATE MUS.MEM.8:747 1905 FELT LARVA
ONTARIO ENT.SOC.ANN.REPT.P.25 1883 SAUNDERS LARVA
PAN PACIFIC ENTOMOLOGIST 11:152 1935 BENJAMIN LARVAL NOTE
LOC.CIT.11:151,1935 BENJAMIN-DYARS DESCR.OF LARVA OF FELINA IS REALLY LARVA CYANESCENS
UNITED STATES DEPT.AGRI. FARMER'S BULL.1169:47,1921 KOTINSKY LARVA* F.28
UNITED STATES DEPT.AGRI.MISC.

PUB.657:356,391,1950 CRAIGHEAD LARVA* F.78
UNITED STATES NAT.MUS.PROC. 21:55,1898 SMITH & DYAR LARVA,PUPA,COCOON (FRIGIDA)
LOC.CIT.21:62,1898 SMITH & DYAR LARVA* PL.7,F.25 COLOR OR UNCOLORED POPULI
LOC.CIT.27:798,1904 DYAR EGG, LARVA,COCOON (FELINA)
DOANE ETC.-FOREST INSECTSP.284 1936 LARVA (BRIEF)
HART-INS.COLLECTIONS ILLINOIS HIGH SCHOOLS P.24,1903 COCOON
HOLLAND-MOTH BOOK P.154,1905 LARVA* F.86 POPULI
PACKARD-INS.INJ.FOREST TREES P.116,1881 LARVA* PUPA
INDEX AMER.ECON.ENT.
VOL.1:20 APATELA POPULI RILEY
VOL.2:3 POPULI;VOL.3:4;
VOL.3:35 APATELA POPULI
VOL.4:4;VOL.5:46 POPULI
VOL.6:5;VOL.7:8

FOOD PLANTS
ABIES BALSAMEA(LINN.)
POULUS SP.(POPLARS)
POPULUS DELTOIDES MARSH.
POPULUS TREMULOIDES MICHX.
SALIX SP.(WILLOWS)
SALIX BARCLEYI ANDERS
SALIX SCOULERIANA BARR.

LETCHERI GRINNELL
SSP.SEE CARYE HUBNER

LETHE GUNDER
SSP.SEE LETO BEHR

LETIS WRIGHT
SSP.SEE LETO BEHR

LETO BEHR.ARGYNNIS-NYMPHALIDAE
AB LETIS WRIGHT
AB LETHE GUNDER
A CHARLOTTII BARNES

LIFE HISTORY
SOUTHERN CALIFORNIA ACAD.SCI. BULL.39:75,1940 COMSTOCK EGG, LARVA* PL.8D

FOOD PLANTS
VIOLA SP.(VIOLETS)

LEUCANIDIA HAMPSON
SYN.SEE COLORADA SMITH

LEUCITYS FRANCLEMONT,SYMMERISTA-NOTODONTIDAE

LIFE HISTORY
CAN.ENT.78:97,1946 FRANCLEMONT LARVA
NEW YORK STATE AGRI.EXP.STA. (CORNELL),MEM.274:224, 1948 FORBES LARVA

FOOD PLANTS
ACER SP.(MAPLES)

LEUCOGRAPHA GEYER
SYN.SEE LEUCOSTIGMA ABBOT & SMITH

LEUCOMELANA MORRISON
SYN.SEE ACRONYCTOIDES WALKER

LEUCOMELAS HERRICH-SCHAEFFER
SYN.SEE LECONTEI GUERIN-MENEVILLE

LEUCOPHAEA(ABBOT & SMITH)OLENE-LIPARIDAE

LIFE HISTORY
AMER.MUS.NAT.HIST.BULL.5:92 1893 BEUTEN. LARVA
LOC.CIT.10:385,1898 BEUTEN LARVA
NEW YORK STATE AGRI.EXP.STA. (CORNELL),MEM.274:243, 1948 FORBES LARVA
NEW YORK STATE MUS.MEM.8:717 1905 FELT LARVA(BRIEF)
PSYCHE 7:135,1894 DYAR LARVA, COCOON
PSYCHE 18:158,1911 BARNES & MCDUNNOUGH LARVA
WASHINGTON ENT.SOC.PROC.1:88 1887 RILEY LARVA(MAY BE

THIS SP.)
LOC. CIT. 13:17, 1911 DYAR LARVAL NOTE
ABBOT&SMITH-LEP. INS. GEORGIA 2:155, 1797 LARVA*PL. 78 COLOR
HARRIS-INS. INJ. VEG. 1ST EDIT. P. 264, 1841 LARVA

FOOD PLANTS
BETULA SP. (BIRCHES)
CARYA SP. (HICKORIES)
DIOSPYROS VIRGINIANA LINN.
MALUS PUMILA MILL. (APPLE)
QUERCUS SP. (OAKS)

LEUCOSTIGMA (ABBOT & SMITH), HEMEROCAMPA
LIPARIDAE
SYN. LEUCOGRAPHA GEYER
SYN. INTERMEDIA FITCH
SYN. BOREALIS FITCH
SYN. OBLIVIOSA HENRY EDWARDS

LIFE HISTORY
AMER. ENT. 1:79, 1868 LARVA*F. 67
LOC. CIT. 2:306, 1870 LARVA*
AMER. MUS. NAT. HIST. BULL. 10: 383, 1898 BEUTEN. LARVA
AMER. NATURALIST 20:814, 1886 PACKARD LARVAL GLANDS
LOC. CIT. 28:326, 1894 HANCOCK LARVA*PUPA*COCOON*
BOSTON ZOOLOGICAL SOC. JOURN. 1:39, 1882 COLEMAN LARVA
CAN. ENT. 3:14, 1871 SAUNDERS EGG, LARVA*
LOC. CIT. 15:186, 1883 SAUNDERS EGG, LARVA
ENT. NEWS 16:67, 1905 GIRAULT NUMBER OF EGGS
HOVEY'S MAGAZINE OF HORT. 1:52 1833 IVES EGG
ILLINOIS AGRI. EXP. STA. BULL. 151: 472, 1911 FORBES LARVA*
ILLINOIS DEPT. AGRI. TRANS. 15:185 1877 FRENCH LARVA*
LOC. CIT. 18, APPEND:117, 166, 1880 COQUILLETT & MARTEN LARVA*
ILLINOIS, REPT. ON INS. OF 1:13, 1871 LEBARON LIFE HISTORY*
ILLINOIS, REPT. NORMAL UNIV. 6:44 1880 FRENCH LARVA
INSECT LIFE 6:390, 1891 DYAR CONTRASTS EGGS OF LEUCOSTIGMA DEFINITA PACKARD AND ANTIQUA LINNAEUS
IOWA ACAD. SCI. BIO. SURVEY. 1936 NO. 1 HOW TO KNOW INS. JAQUES LARVA P. 96
IOWA ACAD. SCI. PROC. 29:165, 1922 POULTER EGG*LARVA*PUPA*
MICHIGAN AGRI. EXP. STA. CIRC. 137:27, 1935 PETIT LARVA
MICHIGAN STATE COLLEGE, SPEC. BULL. 243:15, 1933 MCDANIEL LIFE HISTORY LARVA* COCOON*
MINNESOTA AGRI. EXP. STA. BULL. 61:143, 1899 LUGGER EGG, LARVA* F. 87; PUPA*F. 87
LOC. CIT. BULL. 84ALSO 6:85 1903 WASHBURN LARVA* F. 9COLOR AND F. 73
MISSOURI, REPT. ONINS. OF 1:144, 1869 RILEY LIFE HISTORY*
NATURALISTE CANADIEN P. 164, 1918 CHAPAIS LARVA*F. 3 COCOON
NEW YORK ACAD. SCI. ANNALS 8:59 1893 PACKARD LARVA
NEW YORK STATE AGRI. EXP. STA. (CORNELL), MEM. 56:1075, 1922 WELLHOUSE LARVA
LOC. CIT. 274:245, 1948 FORBES LARVA
NEW YORK STATE AGRI. EXP. STA. (GENEVA), BULL. 312:44, 1909 SCHOENE EGG*PL. I; LARVA* PLS. I, IIPUPA* PL. III
NEW YORK STATE CABIN. NAT. HIST. ANN. REPT. NO. 41:168, 1888 LINTNER LARVA*
NEW YORK STATE, REPT. ENT. 2:210 ALSO 3:20, SEC, 32, 1856 FITCH LARVA, PUPA, COCOON
NEW YORK STATE MUS. BULL. 27:42, 1899 FELT EGG* LARVA* PUPA*
LOC, CIT. 37:17, 1900 FELT EGG* LARVA* PUPA*

Loc.cit.109:7,1907 Felt egg* larva*pupa* COLOR
Loc,cit.155:43,1912 Felt egg magnified* Pl.9
Loc.cit.156:15,1912 Felt egg* larva*pupa* COLOR
New York State Mus.Mem.8:134 1905 Felt egg*Pl.7,f.3; larva*Pl.7,ff.1,5;pupa* Pl.6,f7;cocoon*Pl.7,f.4 all in COLOR
New York State Rept.State Ent. 2:70,1885 life history* Lintner
Loc.cit.4:48,1888 Lintner larva*
Nova Scotia Ent.Soc.Proc.1:93, 1915 Saunders life history (brief)
Loc.cit.3:62,1917 Payne life history* Pl.13 all stages
Loc.cit.8:110,1922 Dustan all stages* Pl.1
Ohio Agri.Exp.Sta.Bull.332:207 1918 Houser larva* Pl.17, ff.1,2
Ontario Ent.Soc.Ann.Rept.p.17, 1871 Bethune egg,larva*pupa
Loc.cit.p.19,1874 Saunders larva*pupa*
Loc.cit.p.29,1886 Hague larva*
Loc.cit.p.49,1889 Harrington larva*
Loc.cit.p.106,1894 Southwick larva* pupa
Loc.cit.p.55,1903 Gibson larva*
Loc.cit.p.11 ,1905 Bethune larva*
Loc.cit.p.16,1905 Fyles egg* larva* pupa*
Loc.cit.p.3,1908 Hahn egg*larva* pupa*
Loc.cit.p.55,1911 Gibson larva*
Loc.cit.p.41,1917 Caesar larva*
Loc.cit.p.40,1941 Caesar egg mass
Papilio 2:164,1882 Coleman larva*
Psyche 6:324,1892 Dyar egg, larva,pupa,cocoon
Psyche 7:341,1896 Dyar larva and key to characters
Smithsonian Misc,Collections 4,art.2:249,1862 Morris larva
United States Dept.Agri.Bur. Ent.Bull.4 old series p. 41,1884 Smith larva,pupa
Loc.cit.Bull.10:29,1888 Riley egg,larva*ff.13,14;pupa* f.13 c,d
Loc.cit.Bull.34 new series p. 33,1902 Banks egg* f.22 K,H.;larva*f.22A;pupa; cocoon* f.22 B,C
United States Dept.Agri.Div. Ent.Bull.10:23028,1887 Riley life history*
United States Dept.Agri. Farmer's Bull.99:13,1899 Howard egg*f.4h;larva*f.4 a,5,6;pupa*f.4b,c;cocoon* f.24 i,k
Loc.cit.1169:41,1921 Kotinsky egg* f.24h;larva*f.24a; pupa*f.24b,c;cocoon*f.24 i,k
Loc,cit.1270:46,1922 Quaintance & Siegler egg*f.91;larva* f.93;pupa
Loc.cit.1495:13,1926 Weigel & Middleton egg*f.15;larva* f.15
United States Dept.Agri.Misc. Pub.657:356,406,1950 Craig-head larva
United States Dept.Agri.Year-book for 1895 p.369 Howard egg*f.86;larva*ff.86,87,88; pupa* f.86
Washington Ent.Soc.Proc.1:88 1887 Riley larval glands
Abbot &Smith-Lep.Ins.Georgia 2:157,1797 larva*Pl.79 COLOR
Cragen-Our Ins.Friends & Foes p.152,1899 larva*f.93
Dickerson-Moths & Butter.p.211, 1901 egg*f.183;larva*ff. 182,184;pupa* f.185

DOANE ETC-FOREST INSECTS P.271, 1936 EGG, LARVA, PUPA
EMMONS-NAT. HIST. NEW YORK 5: 230, 1854 LARVA*PL.37 COLOR
ESSIG-INS. WEST. NORTH AMERICA P.693 1926 LARVA
FERNALD-APPLIED ENT. P.264, 1926 EGG*F.265; LARVA*F.266
FERNALD & SHEPARD-APPLIED ENT. P.261, 1942 EGG*F.262; LARVA* F.263
GLOVER-ILLUST. NORTH AMER. ENT. 1878 EGG, LARVA, PUPA*PL.10, FF.6-10
HARRIS-INS. INJ. VEG. 1ST EDIT. P. 262, 1841 LARVA, PUPA
HARRIS-LOC. CIT. 2ND EDIT. P.284, 1852 EGG, LARVA, COCOON
HARRIS-LOC. CIT. FLINT EDIT. P.366 1862 LIFE HISTORH*PL.7 COLOR
HART-INS. COLLECTIONS ILLINOIS HIGH SCHOOLS P.47, 1903 EGG* LARVA* FF.55,57 D
HERRICH-INS. ENEMIES SHADE TREES P.90, 1935 LARVA* F.60
HOLLAND-MOTH BOOK P.307, 1905 LARVA* F.183; PUPA*F.184
JAEGER-LIFE NORTH AMER. INS. P. 169, 1864 LARVA*
KELLOGG-AMER. INSECTS P.405, 1908 LARVA*F.588
METCALF&FLINT-DESTR.&USEFUL INS. P.699, 1939 EGG*F.463; LARVA* F.461
MORRIS-SYNOP. LEP. NORTH AMER. P.249, 1862 LARVA
O'KANE-INJ. INS. P.295, 1912 EGG* F.437; LARVA
PACKARD-INS. INJ. FOREST TREES P.239, 1881 LARVA*PUPA* COCOON
PEAIRS-PESTS FARM, GARDEN, ORCHARD P.376, 1951 LARVA* F.446
SAUNDERS-INS. INJ. FRUITS P.58, 1883 LIFE HISTORY*
SAUNDERS-LOC. CIT. P.57, 1889 EGG, LARVA* F.50; PUPA*F.51
SLINGERLAND&CROSBY-MANUAL FRUIT INS. P.100, 1914 EGG* F.113; LARVA*F.111; PUPA*F. 112
INDEX AMER. ECON. ENT. VOL.1:153; VOL.2:172; VOL.3:191; VOL.4:217; VOL.5:296; VOL.6:347; VOL.7:458

FOOD PLANTS

ABIES SP. (FIRS)
ACER SP. (MAPLES)
ACER NEGUNDO LINN.
ACER PLATANOIDES LINN.
ACER RUBRUM LINN.
ACER SACCHARINUM LINN.
ACER SACCHAROPHORUM KOCH
AESCULUS GLABRA WILLD.
AESCULUS HIPPOCASTANUM LINN.
AESCULUS OCTANDRA MARSH.
AILANTHUS ALTISSIMA(MILL.)
ALNUS SP. (ALDERS)
AMELANCHIER CANADENSIS(LINN.)
ASIMINA TRILOBA(LINN.)
BERBERIS CANDENSIS MILL.
BERBERIS VULGARIS LINN.
BETULA ALBA LINN.
BETULA LENTA LINN.
BETULA NIGRA LINN.
BUXUS SEMPERVIRENS LINN.
CAMPSIS RADICANS(LINN.)
CARPINUS CAROLINIANA WALT.
CARYS SP. (HICKORIES)
CASTANEA DENTATA(MARSH.)
CASTANEA PUMILA(LINN.)
CATALPA SP.
CATALPA BIGNONIODES WALT.
CATALPA SPECIOSA WARDER
CELTIS OCCIDENTALIS LINN.
CEPHALANTHUS OCCIDENTALIS LINN.
CERCIS CANADENSIS LINN.
CHAMAECYPARIS THYOIDES(LINN.)
CHENOPODIUM ALBUM LINN.
CHIONANTHUS VIRGINICUS LINN.
CLEMATIS SP.
CONVALLARIA MAJALIS LINN.
CORNUS ALTERNIFOLIA LINN.

Cornus florida Linn.
Corylus sp. (hazels)
Corylus americana Walt.
Cotinus coggyria Scopoli
Crataegus sp.
Cupressus sp. (cypress)
Cydonia oblonga Mill. (quince)
Diospyros kaki Linn.
Diospyros virginiana Linn.
Euonymus atropurpureus Jacq.
Fagus sp. (beeches)
Fagus grandifolia Ehrh.
Ficus carica Linn. (fig)
Fraxinus sp. (ashes)
Fraxinus americana Linn.
Fraxinus excelsior Linn.
Geranium maculatum Linn.
Gleditsia triacanthos Linn.
Gossypium herbaceum Linn. (cotton)
Gymnocladus dioica(Linn.)
Hamamelis virginiana Linn.
Helianthus sp.
Hibiscus syriacus Linn.
Hibiscus trionum Linn.
Humulus lupulus Linn.
Ilex opaca Ait.
Ipomoea purpurea(Linn.)
Iris sp.
Jasminium sp.
Juglans cinerea Linn.
Juglans Nigra Linn.
Juglans regia Linn.
Juniperus virginiana Linn.
Kalmis sp.
Larix decidua Mill.
Larix laricina(DuRoi)
Leptochloa nealleyi Vasey
Ligustrum vulgare Linn.
Liquidambar styraciflua Linn.
Liriodendron tulipifera Linn
Lonicera sp. (honeysuckles)
Maclura pomifera(Raf.)
Magnolia sp.
Malus pumila Mill. (apple)
Malva sp.
Morus sp.
Morus rubra Linn.
Nyssa sylvatica Marsh.
Ostrya virginiana (Mill.)
Oxydendrum arboreum (Linn.)
Parthenocissus Quinquefolia (Linn.)
Paulownia tomentosa(Thun.)
Picea sp. (spruces)
Pinus sp. (pines)
Plantago sp.
Platanus occidentalis Linn.
Poa pratensis Linn.
Populus sp. (poplars)
Populus alba Linn.
Populus balsamifera Linn.
Populus deltoides Marsh.
Populus fremonti S. Wats.
Populus nigra v. italica Muench.
Populus tremuloides Michx.
Prunus americana Marsh.
Prunus armeniaca Linn. (apricot)
Prunus cerasus Linn. (cherry)
Prunus domestica Linn. (plum)
Prunus ilicifolia(Nutt. ex. H. &A.)
Prunus persica(Linn.)
Prunus virginiana Linn.
Pyrus communis Linn. (pear)
Quercus sp. (oaks)
Quercus alba Linn.
Quercus coccinea Muench
Quercus phellos Linn.
Quercus prinus Linn.
Quercus rubra Linn.
Rhamnus alnifolia L'Her
Rhododendron sp.
Ribes sp.
Ricinus communis Linn.
Robinia pseudo-acacia Linn.
Rosa sp. (roses)
Rubus sp.
Salix xp. (willows)
Sambucus canadensis Linn.
Sassafras albidum v. molle(Raf.)
Spiraea sp.
Staphylea trifolia Linn.
Symphoricarpos albus(Linn.)
Syringa vulgaris Linnl
Taraxacum officinale Weber.
Taxus sp.
Tilia sp. (lindens)

TILIA AMERICANA LINN.
TILIA EUROPAEA LINN.
TRIFOLIUM SP. (CLOVERS)
ULMUS SP. (ELMS)
ULMUS AMERICANA LINN
ULMUS RUBRA MUHL.
VIBURNUM SP.
WISTERIA FRUTESCENS (LINN.)
WISTERIA SINENSIS SWETT
ZANTHOXYLUM CLAVA-HERCULIS LINN.
ZEA MAYS LINN. (CORN)

LEUSSLERI BRYANT
SSP. SEE JUTTA HUBNER

LEUSSLERI GUNDER
SSP. SEE ANICIA DOUBLEDAY&HEWITSON

LEUSSLERI GUNDER
SSP. SEE LYGDAMUS DOUBLEDAY

LEUSSLERI GUNDER
SSP. SEE SAEPIOLUS BOISDUVAL

LEUSSLERI LINDSAY
SSP. SEE COMMA LINN.

LEVETTEI GROTE
SYN. SEE JUDITH STRECKER

LEVIS HUDSON
SSP. SEE BILINEATA PACKARD (DREPANA)

LEWISI SWETT
SSP. SEE MACULARIA HARRIS

LEWISII KIRBY
SYN. SEE EURYMEDON LUCAS

L'HERMINIERI GODART
SSP. SEE MIDEA HUBNER

L'HERMINIERI (GODART), LERODEA-HESPERIIDAE
SYN. FUSCA GROTE & ROBINSON
LIFE HISTORY
ENT. NEWS 19:416, 1908 LAURENT
EGG, LARVA FIRST INSTAR (FUSCA)
FOOD PLANTS
GRASSES

LIBATRIX (LINNAEUS), SCOLIOPTERYX-NOCTUIDAE
LIFE HISTORY
NEW YORK STATE CABIN. NAT. HIST. ANN. REPT. NO. 45:171, 1891 LINTNER LARVA
NEW YORK STATE MUS. MEM. 8:743 1905 FELT LARVA (BRIEF)
PAPILIO 1:56, 1881 COQUILLETT LARVA
UNITED STATES DEPT. AGRI. BUR. ENT. BULL. 13:27, 1887 OLD SERIES RILEY LARVA
DUNCAN-NATURALIST'S LIBRARY 4:231, 1836 LARVA (BRIEF)
GUENEE-SPEC. GEN. LEP. NOCT. 2: 406, 1852 LARVA
HUMPHREYS-GENERA BRITISH MOTHS P. 61, 1859 LARVA*PL. 25 COLOR
KIRBY-EUROPEAN BUTTER. &MOTHS P. 185, 1882 LARVA, PUPA (BRIEF)
STAINTON-BRITISH BUTTER. &MOTHS 1:309, 1857 LARVA
STEPHENS-ILLUST. BRITISH ENT. HAUST. 3:50, 1829 LARVA
WOOD-INSECTS AT HOME P. 484, 1872 LARVA*
INDEX AMER. ECON. ENT. VOL. 1:280; VOL. 6:662
FOOD PLANTS
POPULUS SP. (POPLARS)
POPULUS NIGRA V. ITALICA MUENCH.
POPULUS TRICHOCARPA T. &G.
SALIX SP. (WILLOWS)
SALIX LASIANDRA BENTH.
SALIX SCOULERIANA BARR.

LIBERA WALKER
SYN. SEE GRANDIS BOISUDUVAL

LIBERARIA (WALKER), DREPANULATRIX-GEOMTRIDAE
SYN. INTEGRARIA WALKER
SYN. LINTNERARIA PACKARD

AB HELENA HULST
LIFE HISTORY
CAN. ENT. 65:121, 1933 MCDUNNOUGH LARVA, PUPA
NEW YORK STATE AGRI. EXP. STA. (CORNELL), MEM. 274:71, 1948 FORBES LARVA
FOOD PLANTS
CEANOTHUS AMERICANUS LINN.

LIBURNA(GEYER), SCOLECOCAMPA-NOCTUIDAE
SYB. LIGNI GUENE
LIFE HISTORY
AMER. MUS. NAT. HIST. BULL. 16: 447, 1902 BEUTEN. LARVA
ENT. AMER. 14:149, 1934 CRUMB LARVA* PL. 19A, B; PL. 22 C, D: PL. 21 B
NEW YORK STATE MUS. MEM. 8:705 1905 FELT LARVA
PAPILIO 3:134, 1883 EDWARDS & ELLIOTT LARVA
PSYCHE 2:272, 1879 ANDREWS LARVA
UNITED STATES DEPT. AGRI. MISC. PUB. 657:366, 395, 1950 CRAIGHEAD LARVA
GLOVER-ILLUST. NORTH AMER. ENT. 1878 LARVA* PL. 18, F. 18
GUENEE-SPEC. GEN. LEP. NOCT. 1: 131, 1852 LARVA* PL. 2 COLOR LIGINI
FOOD PLANTS
IN DEAD STUMPS OF
CARYA SP. (HICKORIES)
CASTANEA DENTATA(MARSH)
PLATANUS OCCIDENTALIS(LINN.)
PRUNUS CERASUS LINN. (CHERRY)
QUERCUS SP. (OAKS)
PROBABLY IN DEAD LOGS OF MANY OTHER SPECIES: FUNGI

LIBYA SCUDDER, PHOLISORA-HESPERIIDAE
LIFE HISTORY
SOUTHERN CALIFORNIA ACAD. SCI. BULL. 31:96, 1932 COMSTOCK & DAMMERS LIFE HISTORY EGG* PL. 31; LARVA* PL. 32: PUPA* PL. 32

FOOD PLANTS
ATRIPLEX CANESCENS (PURSH)

LIGATA WALKER
SYN. SEE LOREA GUENEE

LEGGETTI AVINOF
SSP. SEE HARRISII SCUDDER

LIGNI GUENEE
SYN. SEE LIBURNA GEYER

LIGNICOLOR WALKER, DICENTRIA-NOTODONTIDAE
SYN. VIRGATA PACKARD
SYN. LIGNIGERA WALKER
A DELICATA DYAR
LIFE HISTORY
AMER. MUS. NAT. HIST. BULL. 10: 419, 1898 BEUTEN. LARVA
BOSTON SOC. NAT. HIST. PROC. 24: 542, 1890 PACKARD LARVA
ENT. AMER. 5:91, 1889 DYAR EGG, LARVA, PUPA
MINNESOTA AGRI. EXP. STA. BULL. 61-187, 1899 LUGGER LARVA* PL. 134, F. 1 NO DESCR.
NAT. ACAD. SCI. MEM. 7:190, 282, 1895 PACKARD EGG LARVA* PL. 25, F. 1 COLOR; PUPA; COCOON
NEW YORK STATE AGRI. EXP. STA. (CORNELL), MEM. 274:226, 1948 FORBES LARVA
NEW YORK STATE MUS. MEM. 8:705 1905 FELT LARVA (BRIEF)
UNITED STATES DEPT. AGRI. MISC. PUB. 657:360, 404 1950 CRAIGHEAD LARVA
GLOVER-ILLUST. NORTH AMER. ENT. 1878 LARVA* PL. 13, F. 28
FOOD PLANTS
BETULA ALBA LINN.
BETULA PAPYRIFERA MARSH.
FAGUS SP. (BEECHES)
QUERCUS SP. (OAKS)
QUERCUS ALBA LINN.

LIGNIGERA WALKER

SYN. SEE LIGNICOLOR WALKER

LILIANA HENRY EDWARDS, ARGYNNIS-
NYMPHALIDAE

AB BARONI EDWARDS

LIFE HISTORY

EDWARDS-BUTTER. NORTH AMER. 3RD SERIES 1897 EGG*LARVA* (ARGYNNIS PL. 5, FF. A, B)

HOLLAND-BUTTERFLY BOOK P. 120 1905, EGG, LARVA

HOLLAND-LOC. CIT. 3RD EDIT. PP. 98, 99, 1931 EGG, LARVA

FOOD PLANTS

VIOLA SP.

LILITH STRECKER

SSP. SEE IO FABRICIUS

LIMBATA(HAWORTH), NEMATOCAMPA-
GEOMETRIDAE

SYN. VESTITARIA HERRICH-SCHAEFFER

SYN. RESISTARIA HERRICH-SCHAEFFER

SYN. FILAMENTARIA GUENEE

FORM EXPUNCTARIA GROTE

AB CHAGNONI SWETT

A ORDORDENSIS CASSINO & SWETT

LIFE HISTORY

ILLINOIS DEPT. AGRI. TRANS. 15:242 1877 FRENCH LARVA

LOC. CIT. 18, APPEND:149, 1880 COQUILLETT LARVA

MAINE AGRI. EXP. STA. BULL. 259: 42, 1917 MOSHER LARVA, PUPA* F. 2H

MINNESOTA AGRI. EXP. STA. BULL. 61:234, 1899 LUGGER LARVA* F. 182

NEW YORK ENT. SOC. JOURN. 53:197, 1945 FORBES PUPA(FILAMENTARIA)

NEW YORK STATE AGRI. EXP. STA. (CORNELL) MEM. 15:197, 1918 HAWLEY EGG*F. 51; LARVA*F. 53; PUPA* F. 54

LOC. CIT. 274:100, 1948 FORBES EGG, LARVA(FILAMENTARIA)

NEW YORK STATE CABIN. NAT. HIST. ANN. REPT. NO. 26:165, 1874 LINTNER LARVA

NEW YORK STATE MUS. MEM. 8:530 1905 FELT LARVA

UNITED STATES DEPT. AGRI. MISC. PUB. 657:360, 433, 1950 CRAIGHEAD LARVA

WISCONSIN AGRI. SOC. TRANS. 13:5, 1883 FORBES LARVA

GUMPPENBERG-SYSTEMA GEOMETRARUM 7:426, 1895 LARVA, PUPA(FILAMENTARIA)

HARRIS-ENT. CORRESP. P. 322, 1869, LARVA*PUPA*PL. 8, FF. 7, 7A

PACKARD-MONOGRAPH GEOMETRIDAE P. 471, 1876 LARVA*PUPA* PL. 13, FF. 8, 8A

SAUNDERS-INS. INJ. FRUITS P. 167, 1883 LARVA*PUPA*COCOON

SAUNDERS-LOC, CIT. P. 167, 1889 LARVA*F. 176(FILAMENTARIA

INDEX AMER. ECON. ENT.
VOL. 2:18 ANIA
VOL. 2:249 NEMATOCAMPA

FOOD PLANTS

ACER SP. (MAPLES)
ACER SACCHAROPHORUM KOCH
AESCULUS HIPPOCASTANUM LINN
BETULA SP. (BIRCHS)
BETULA POPULIFOLIA MARSH.
CARYS SP. (HICKORIES)
CASTANEA DENTATA(MARSH.)
CORYLUS SP.
CRATAEGUS SP.
FRAGARIA CHILOENSIS(LINN.) STRAWBERRY)
HUMULUS LUPULUS LINN. (HOPS)
MALUS IOENSIS(WOOD)
MALUS PUMILA MILL. (APPLE)
PRUNUS DOMESTICA LINN. (PLUM)
PYRUS COMMUNIS LINN. (PEAR)
QUERCUS SP. (OAKS)
RIBES AMERICANUM MILL.
RIBES LACUSTRE(PERS.)(GOOSEBERRIES)
RIBES SATIVUM SYME(CURRANTS)
ROBINIA PSEUDO-ACACIA LINN.
ROSA SP. (ROSE PETALS)

RUBUS ALLEGHENIENSIS PORTER
RUBUS IDAEUS V.STRIGOSUS(MICHX.)
SEDUM SP.
TILIA SP. (LINDENS)
LOW PLANTS

LIMBOPUNCTATA NOLCKEN
SYN.SEE TRISTATA LINNAEUS

LIMITARIA(WALKER), NYCTOBIA-GEOMETRIDAE
SYN. LOBOPHORATA WALKER
SYN. FUSIFASCIATA WALKER
SYN. LONGIPENNIS WALKER
FORM REIFFI SWETT
A NIGROANGULATA STRECKER
LIFE HISTORY
CAN.ENT.77:197,1945 MCGUFFIN EGG,LARVA*FF.12,1 MANDIBLE
NEW YORK STATE AGRI.EXP.STA. (CORNELL),MEM.274:173,1948 FORBES LARVA
FOOD PLANTS
ABIES BALSAMEA(LINN.)
AMELANCHIER ALNIFOLIA NUTT.
AMELANCHIER CANADENSIS(LINN.)
LARIX LARICINA(DUROI)
PICEA ENGELMANNI BARRY
PICEA GLAUCA(MOENCH)
PINUS BANKSIANA LAMB.
TSUGA CANADENSIS(LINN.)

LIMITATA SMITH
SSP.SEE ALBILINEA HUBNER

LIMPIDA(GUENEE), PAPAIPEMA-NOCTUIDAE
SYN. AERATA LYMAN
LIFE HISTORY
CAN.ENT.33:64,1901 BIRD ERROR REALLY SCIATA BIRD SEE CAN. ENT.11:28,1908
LOC.CIT.33:320,1901 LYMAN LARVA PUPA(AERATA)
LOC.CIT.40:252,1908 LYMAN LARVA (AERATA)
FOOD PLANTS
ARCTIUM LAPPA LINN.

LINDA BIRD
SSP.SEE NELITA STRECKER

LINEA FABRICIUS
SSP.SEE ERIDANIA CRAMER

LINEAINITA GUNDER
SSP.SEE EURYDICE BOISDUVAL

LINEATA(FABRICIUS), CELERIO-SPHINGIDAE
SYN. DAUCUS CRAMER
LIFE HISTORY
AMER.ENT.1:206,1869 LAVA,PUPA
LOC.CIT.2:257,1870 LARVA*
AMER.JOURNSCI.36:304,1839 HARRIS LARVA,PUPA
AMER.MUS.NAT.HIST.BULL.7:286 1895 BEUTEN.LARVA,PUPA
AMER.NATURALIST 6:294,1872 LARVA*F.99 NO DESCR.
CALIFORNIA ACAD.SCI.PROC.7:20, 1876 HY.EDW.LARVA,PUPA (DAUCUS)
CALIFORNIA AGRI.EXP.STA.BULL. 660:33,1942 SMITH LARVA* F.26 A
CAN.ENT. 9:64,1877 SAUNDERS LARVA*FF.4,5
ENT.NEWS 8:49,1897 LARVA*
LOC.CIT.9:150,1898 SOULE DURATION OF EGG STAGE
ILLINOIS AGRI.EXP.STA.BULL.60: 503,1900 FORBES LARVA*
ILLINOIS DEPT.AGRI.TRANS.15:164 1877 FRENCH LARVA
LOC.CIT.18:100,159 APPEND.1880 COQUILLETT-MARTEN LARVA*
MICHIGAN STATE COLLEGE,SPEC. BULL.243:48,1933 MCDANIEL LARVA
MINNESOTA AGRI.EXP.STA.BULL. 61:90,1899 LUGGER LARVA* FF.34,35,AND PL.VI,F.36
MISSOURI,REPT.ON INS.OF 3:141, 1871 RILEY LARVA*
MONTANA AGRI.EXP.STA.BULL.62: 225,1905 COOLEY EGG*LARVA* PUPA*
NEW YORK STATE AGRI.EXP.STA.

(CORNELL), MEM. 274:201, 1948 FORBES LARVA

NEW YORK STATE MUS. MEM. 8:724, 1905 FELT LARVA(BRIEF)

NOVITATES ZOOLOGICAE(TRING)9 SUPPL. P. 731, 1903 ROTHSCHILD & JORDAN LARVA, PUPA

ONTARIO ENT. SOC. ANN. REPT. P. 28 1874 SAUNDERS LARVA*

LOC. CIT. P. 36, 1876 SAUNDERS LARVA*

LOC. CIT. P. 55, 1881 REED LARVA* PUPA

LOC. CIT. P. 25, 1888 FYLES LARVA*

PHILADELPHIA ACAD. NAT. SCI. JOURN 5:143, 1859 CLEMENS LARVA, PUPA

PHILADELPHIA ENT. SCO. PROC. 3:662 1864 LINTNER LARVA

PSYCHE 7:458, 1896 SOULE EGG, LARVA PUPA

SMITHSONIAN MISC. COLLECTIONS 4, ART. 2:164, 1862 MORRIS LARVA, PUPA

UNITED STATES DEPT. AGRI. BUR. ENT. BULL. 43 NEW SERIES P. 41, 1903 CHITTENDEN LARVA*FF. 38B, C; PUPA* F. 38D

LOC. CIT. BULL. 57:15, 1906 SANDERSON LARVA* FF. 5 B, C; PUPA*F. 5D

UNITED STATES DEPT. AGRI. FARMER'S BULL. 223:9, 1905 SANDERSON LARVA* F. 4 B, C; PUPA* F. 4D

LOC. CIT. 1220:34, 1921 QUAINTANCE LARVA

ABBOT&SMITH–LEP. INS. GEORGIA 1:77, 1797 LARVA* PUPA PL. 39

BOISDUVAL–SPEC. GEN. LEP. SPHING. P. 173, 1874 LARVA

CHENU–ENCYC. HIST. NAT. PAPILLONS 1:262, 1856 LARVA, PUPA

CRAGEN–OUR INS. FRIENDS&FOES P. 136, 1899 LARVA

DICKERSON–MOTHS & BUTTER. P. 239, 1901 LARVA* F. 207; PUPA*F. 208

DUNCAN–NATURALISTS LIBRARY 4:152 1836 LARVA

ELIOT&SOULE–CATER. &MOTHS P. 95 1902 EGG, LARVA* PUPA

ESSIG–INJ. &BENEF. INS. P. 383, 1915 LARVA*F. 382; PUPA*F. 382

ESSIG–INS. WEST. NORTH AMER. P. 666, 1926 LARVA* F. 540; PUPA* F. 540

FERNALD–SPHINGIDAE NEW ENGLAND P. 57, 1886 LARVA*

GLOVER–ILLUST. NORTH AMER. ENT. 1878 LARVA*PL. 3, FF. 2, 3

GUNDLACH–ENT. CUBANA P. 196, 1886 LARVA, PUPA

HART–INS. COLLECTIONS ILLINOIS HIGH SCHOOLS P. 15, 1903 LARVA* F. 9

HOLLAND–MOTH BOOKP. 76, 1905 LARVA* FF. 34, 35

KELLOGG–AMER. INS. P. 436, 1908 LARVA

MORRIS–SYNOP. LEP. NORTH AMER. P. 146, 1862 LARVA, PUPA

SAUNDERS–INS. INJ. FRUITS P. 225, 1883 LARVA*PUPA

SAUNDERS–LOC, CIT. P. 255, 1889 LARVA* FF. 265, 266; PUPA

STEPHENS–ILLUST. BRITISH ENT. HAUSTELL. 1:127, 1828 LARVA, PUPA

WOOD–INSECTS ABROAD P. 637, 1874 LARVA*

INDEX AMER. ECON. ENT.
VOL. 1:103 DEILEPHILA
VOL. 2:72; VOL. 3:75;
VOL. 4:89; VOL. 5:580 SPHINX
VOL. 6; 134; VOL. 7:172

FOOD PLANTS

AMARANTHUS SP.
AZALEA NUDIFLORA LINN.
BETA VULGARIS LINN. (BEET)
BOISDUVALIA SP.
BRASSICA RAPA LINN. (TURNIP)
CHENOPODIUM SP.
CITRULLUS VULGARIS SCHRAD. (WATERMELON)
CLARKIS SP.
CLARKIS RHOMBOIDEA DOUGLE
CUCUMIS MELO LINN. (MELON)
EPILOBIUM GLANDULOSUM V. ADENOCAULON (HAUSSK.)

EUCHARIDIUM SP.
FAGOPYRUM SAGITTATUM GILIB. (BUCKWHEAT)
FUCHSIA MAGELLANICA V.RICCARTONII LEBAS
GOSSYPIUM HERBACEUM LINN(COTTON)
LACTUCA SATIVA LINN. (LETTUCE)
MALUS PUMILA MILL. (APPLE)
OENOTHERA BIENNIS LINN.
ONAGRACEAE
PARTHENOCISSUS QUINQUEFOLIA (LINN.)
PELARGONIUM SP.
PLANTAGO SP.
POLYGONUM SP.
PORTULACA GRANDIFLORA HOOK
PORTULACA OLERACEA LINN.
PRUNUS DOMESTICA LINN. (PLUM)
PYRUS COMMUNIS LINN. (PEAR)
RIBES LACUSTRE(PERS.)(GOOSEBERRY)
RIBES SATIVUM SYME(CURRANT)
ROSA SP. (ROSES)
ROSA WOODSII LINDL.
RUMEX SP.
RUMEX OBTUSIFOLIUS LINN.
SALIX INTERIOR ROWLEE
SOLANIUM LYCOPERSICUM LINN. (TOMATO)
STELLARIA MEDIA(LINN.)
ULMUS SP. (ELMS)
VITIS SP. (GRAPES)

LINEATA WALKER
SYN. SEE TRIFASCIA HUBNER

LINEATELLA HARVEY
SYN. SEE ORNITHOGALLI GUENEE

LINEELLA GROTE
SSP. SEE AMICA HUBNER

LINEOLATA WALKER, CATABENA-NOCTUIDAE
SYN. PETRAEA WALKER
SYN. MISCELLUS GROTE
LIFE HISTORY
AMER. MUS. NAT. HIST. BULL. 4:69 1892 BEUTEN. LARVA, PUPA. (MISCELLUS)
LOC. CIT. 16:450, 1902 BEUTEN. LARVA
FOOD PLANTS
VERBENA STRICTA VENT.

LINEOSA WALKER
SYN. SEE LUNIFERA HUBNER

LINNEI GROTE & ROBINSON
SYN. SEE VITIS LINN.

LINTNERANA SPEYER
SSP. SEE REVAYANA SCOPOLI

LINTNERARIA PACKARD
SYN. SEE LIBERARIA WALKER

LINTNERI COCKERELL
SSP. SEE MAIA DRURY

LINTNERI(GROTE), GLUPHISIA-NOTODONTIDAE
FORM AVIMACULA HUDSON
FORM SLOSSONI PACKARD
A SEVERA HENRY EDWARDS
FORM DANBYI NEUMOEGEN
FORM NORMALIS DYAR
LIFE HISTORY
AMER. MUS. NAT. HIST. BULL. 10: 404, 1898 BEUTEN. LARVA (SEVERA)
ENT. AMER. 2:167, 1886 HY. EDW. EGG(SEVERA)
NAT. ACAD. SCI. MEM. 7:98, 1895 PACKARD EGG, LARVA, PUPA, COCOON
PSYCHE 6:503, 1893 DYAR EGG, LARVA, PUPA, COCOON
ROYAL SOC. CANADS. TRANS. VOL. 1, SERIES 3:63, 1907 FLETCHER & GIBSON EGG, PUPA(SEVERA)
FOOD PLANTS
POPULUS BALSAMIFERA LINN.
POPULUS GRANDIDENTATA MICHX.
POPULUS TREMULOIDES MICHX.
RHIZOPHORA MANGLE LINN.

LINTNERI GROGE, OMMATOSTOLA-NOCTUIDAE
LIFE HISTORY

FOOD PLANTS
ARENARIA SP.

LINTNERII FITCH
SYN. SEE ANTIOPA LINNAEUS

LIPAROPS(BOISDUVAL&LECONTE), STRYMON-LYCAENIDAE
A STRIGOSA HARRIS
AB PRUINA SCUDDER
LIFE HISTORY
AMER. MUS. NAT. HIST. BULL. 5:279 1893 BEUTEN. LARVA, CHRYSALIS
BUFFALO SOC. NAT. SCI. BULL. 19:28 1939 WILD LARVA
CAN. ENT. 1:99, 1869 SAUNDERS LARVA
ILLINOIS DEPT. AGRI. TRANS. 18, APPEND:92, 1880 MIDDLETON LARVA
MINNESOTA AGRI. EXP. STA. BULL. 61:78, 1899 LUGGER LARVA
NEW YORK STATE CABIN. NAT. HIST. ANN. REPT. NO. 41:257, 1888 LINTNER LARVA(BRIEF)
NEW YORK STATE MUS. MEM. 8:704 1905 FELT LARVA (BRIEF)
SMITHSONIAN MISC. COLLECTIONS 4, ART. 2:96, 1862 MORRIS LARVA CHRYSALIS(MAY BE MELINUS HUBNER INSTEAD)
COMSTOCK-HOW KNOW BUTTER. P. 228 1904 LARVA
EDWARDS-BUTTER. NORTH AMER. 1ST SERIES 1869 LARVA* CHRYSALIS*(PL. THECLA NO. 2)
FERNALD-BUTTER. MAINE P. 78, 1884 LARVA
FRENCH-BUTTER. EAST. UNITED STATES P. 267, 1886 LARVA (STRIGOSA)
HOLLAND-BUTTERFLY BOOK 1905 CHRYSALIS*PL. 5, F. 28 COLOR
SAUNDERS-INS. INJ. FRUITS P. 176, 1883 LARVA, CHRYSALIS
SAUNDERS-LOC. CIT. P. 176, 1889 LARVA, CHRYSALIS (STRIGOSA)
SCUDDER-BUTTER. EAST. UNITED STATES & CANADA PP. 880-883, 1889 LIFE HISTORY*PL. 65 F. 17; PL. 71, F. 2; PL. 75, FF. 27 32; PL. 79, F. 26; PL. 86, F. 27; PL. 84, F. 28
SCUDDER-GUIDE TO COMMON BUTTER. P. 119, 1893 LARVA, CHRYSALIS
FOOD PLANTS
AMELANCHIER CANADENSIS(LINN.)
ARONIA MELANOCARPA(MICHX.)
CASTANEA DENTATA(MARSH)
CRATAEGUS SP. (THORNS)
ERICACEAE SP.
ILEX SP.
ILEX OPACA AIT.
MALUS PUMILA MILL. (APPLE)
PODPHYLLUM PELTATUM LINN.
PRUNUS CERASUS LINN. (CHERRY)
PRUNUS DOMESTICA V. CEREOLA POIR.
PRUNUS ILICIFOLIA(NUTT. EX. H. &A)
QUERCUS SP. (OAKS)
ROSACEAE SP.
SALIX SP. (WILLOWS)
VACCINIUM CORYMBOSUM LINN.

LIQUIDA(GROTE), POLIA NOCTUIDAE
LIFE HISTORY
UNITED STATES NAT. MUS. PROC. 27:843, 1904 DYAR EGG, LARVA
FOOD PLANTS

LIQUORARIA GUENEE, SYNCHLORA-GEOMETRIDAE
SYN. TRICOLOARARIA PACKARD
SYN. RUBRIFRONTARIA DYAR
LIFE HISTORY
SOUTHERN CALIFORNIA ACAD. SCI. BULL. 36:71, 1937 COMSTOCK & DAMMERS LARVA*PLS. 33, 34, 35; PUPA*PLS. 35, 36
FOOD PLANTS
CHILOPSIS LINEARIS(CAV.) (BLOSSOMS)
COREOPSIS SP. (BLOSSOMS)
ERIOGONUM FASCICULATUM BENTH. (BLOSSOMS)

LIRIODENDRARIA Abbot & Smith
syn. VIRGINARIA Cramer

LISA(Boisduval & Leconte), EUREME
Pieridae

form CLAPPII Maynard
form female ALBA Strecker
LIFE HISTORY
Amer. Mus. Nat. Hist. Bull. 5:251, 1893 Beuten. larva, chrysalis
Buffalo Soc. Nat. Sci. Bull. 19:19 1939 Wild larva
California Acad. Sci. Proc. 5:327 1873 Edwards larva, chrysalis (brief)
Ent. News 1:8, 1890 Skinner egg laying
Illinois Dept. Agri. Trans. 18, Append. 79, 1880 Middleton larva
Smithsonian Misc, Collections 4, art. 2:34, 1862 Morris larva, chrysalis
Washington Biological Soc. Proc. 33:35, 1920 A. A. Saunders duration egg, larval stages
Boisduval-Spec. Gen. Lep. Diurnae 1:662, 1836 larva* chrysalis* Pl. 2 COLOR
Boisduval & Leconte-Hist. Lep. North America p. 54, 1833 larva* chrysalis* Pl. 19 COLOR
Comstock-How Know Butter. p. 99, 1904 larva (EUTERPE Menetries)
French-Butter. East. United States p. 139, 1886 larva
Holland-Butterfly Book 1905 larva* Pl. 2, f. 3; chrysalis* Pl. 5, f. 56 both in COLOR
Morris-Synop. Lep. North Amer. p. 34, 1862 larva, chrysalis
Scudder-Butter. East. United States & Canada pp. 1089-1090, 1889 life history* Po. 65, f. 32; Pl. 76, f. 3; Pl. 79, f. 55; Pl. 84, f. 56
Scudder-Everyday Butter. p. 346 1899 egg, larva, chrysalis
Scudder-Guide to Common Butter. p. 138, 1893 larva, chrysalis
FOOD PLANTS
Cassia sp.
Cassia fasciculata Michx.
Cassia marilandica Linn.
Cassia nictitans Linn.
Glycine sp.
Trifolium sp. (clovers)

LITARIA(Hulst), DREPANULATRIX Geometridae

syn. FUMOSA Hulst
LIFE HISTORY
Psyche 10:199, 1903 Dyar egg, larva
United States Nat. Mus. Proc. 27:904, 1904 Dyar larva (FUMOSA)
FOOD PLANTS
Ceanothus sp.

LITHOSPILA Grote, ACRONICTA-Noctuidae
LIFE HISTORY
Amer. Mus. Nat. Hist. Bull. 14: 246, 1901 Benten. larva
Ent. News 17:23, 1906 Pearsall larvapupates in wood
New York State Mus. Mem. 8: 721, 1905 Felt larva
Papilio 3:132, 1883 Edw. & Elliott larva
United States Nat. Mus. Proc. 21:99, 1898 Smith & Dyar egg, larva
Index Amer. Econ. Ent. Vol. 7:8
FOOD PLANTS
Carya sp. (hickories)
Castanea dentata (Marsh.)
Juglans sp.
Quercus sp. (oaks)

LITOPHORA(Grote), HORMISA-Noctuidae
LIFE HISTORY

ENT. AMER. 14:179, 1934 CRUMB
LARVA*PL. 21G; PL. 24C
FOOD PLANTS
POA PRATENSIS LINN. (DEAD GRASS)
TRIFOLIUM SP. (CLOVERS)

LITURALIS (HUBNER), ZANCLOGNATHA-NOCTUIDAE
LIFE HISTORY
ENT. AMER. 14:181, 1934 CRUMB
LARVA
WASHINGTON ENT. SOC. PROC. 4:328
1898 DYAR LARVA
FOOD PLANTS
DEAD LEAVES

LITURATA SMITH
SSP. SEE BRUMOSA GUENEE

LIVIDUS HUBNER
SYN. SEE AMYNTAS FABRICIUS

LIXARIA GUENEE, RACHEOSPILA-GEOMETRIDAE
SYN. INCLUSARIA WALKER
LIFE HISTORY
FOOD PLANTS
MYRICA PENSYLVANICA LOISEL

LOBELIAE GUENEE, ACRONICTA-NOCTUIDAE
SYN. GROTEI BUTLER
LIFE HISTORY
AMER. MUS. NAT. HIST. BULL. 14:
243, 1901 BEUTEN. LARVA
CAN. ENT. 18:118, 1886 FRENCH
LARVA
NEW YORK STATE MUS. MEM. 8:704
1905 FELT LARVA
PAPILIO 1:6, 1881 COQUILLETT
LARVA
UNITED STATES NAT. MUS. PROC.
21:83, 1898 SMITH&DYAR EGG,
LARVA*PL. 7, F. 25 COLOR OR
UNCOLORED; PUPA, COCOON
GUENEE-SPEC. GEN. LEP. NOCT. 1:44
1852 LARVA
HART-INS. COLLECTIONS ILLINOIS
HIGH SCHOOLS P. 25, 1903
COCOON

FOOD PLANTS
PRUNUS SEROTINA EHRH.
QUERCUS SP. (OAKS)
QUERCUS MACROCARPA MICHX.

LOBOPHORATA WALKER
SYN. SEE LIMITARIA WALKER

LOCULATA MORRISON
SYN. SEE PASSER GUENEE

LOGAN (EDWARDS), ATRYTONE-HESPERIIDAE
SYN. MALE DELAWARE EDWARDS
A LAGUS EDWARDS
LIFE HISTORY
AMER. MUS. NAT. HIST. BULL. 5:289,
1893 BEUTEN. LARVA, CHRYSALIS
CAN. ENT. 11:192, 1879 EDWARDS
LARVA, CHRYSALIS
ILLINOIS DEPT. AGRI. TRANS. 18,
APPEND: 96, 174, 1880
COQUILLETT & MIDDLETON
LARVA
FRENCH-BUTTER. EAST. UNITED
STATES & CANADA PP. 1615-
1616, 1889 LARVA, CHRYSALIS
FOOD PLANTS
ERIANTHUS ALOPECUROIDES (LINN.)
GRASSES

LOIPERDA DYAR
SYN. SEE PANDORA BLAKE

LOKI (SKINNER), MITOURA-LYCAENIDAE
LIFE HISTORY
ENT. NEWS 35:199, 1925 COOLIDGE
LIFE HISTORY IN ERROR, REALLY
SIVA EDWARDS
SOUTHERN CALIFORNIA ACAD. SCI.
BULL. 31:40, 1932 COMSTOCK-
DAMMERS EGG, LARVA*PL. 20;
CHRYSALIS* PL. 21
FOOD PLANTS
JUNIPERUS SP. (JUNIPERS)
JUNIPERUS CALIFORNICA CARR.

LONGA GUENEE, ACRONICTA-NOCTUIDAE
SYN. XYLINOIDES GUENEE

SYN. XYLINIFORMIA GUENEE
SYN. PALLIDICOMA GROTE
FORM SET BARNES & BENJAMIN
LIFE HISTORY
AMER. MUS. NAT. HIST. BULL. 14: 260, 1901 BEUTEN. LARVA (XYLINIFORMIS)
MISSOURI, REPT. NOX. INS. 5:126, 1873 RILEY LARVA (XYLINOIDES)
NEW YORK STATE MUS. MEM. 8:749 1905 FELT LARVA XYLINIFORMIA (BRIEF)
OHIO AGRI. EXP. STA. BULL. 45:174 1893 WEBBER LARVA
UNITED STATES NAT. MUS. PROC. 21: 168, 1898 SMITH & DYAR LARVA* PL. 8, F. 35 COLOR OR UNCOLORED PUPA
INDEX AMER. ECON. ENT. VOL. 6:5; VOL. 7:8
FOOD PLANTS
BETULA SP. (BIRCHES)
DIOSPYROS VIRGINIANA LINN.
GOSSYPIUM HERBACEUM LINN (COTTON)
PRUNUS PERSICA (LINN.) (PEACH)
QUERCUS SP. (OAKS)
RHUS TOXICODENDRON LINN.
ROSA SP. (ROSES)
RUBUS ALLEGHENIENSIS PORTER
RUBUS IDAEUS V. STRIGOSUS MICHX.
SALIX SP. (WILLOWS)

LONGILABRIS GROTE
SYN. SEE METONALIS WALKER

LONGIOR (SMITH), LITHOPHANE-NOCTUIDAE
LIFE HISTORY
SOUTHERN CALIFORNIA ACAD. SCI. BULL. 42:132, 1943 COMSTOCK & HENNE LARVA
FOOD PLANTS
JUNIPERUS OSTEOSPERMA (TORR.)

LONGIPALPATA HULST, MYCETEOPHORA-NOCTUIDAE
LIFE HISTORY
UNITED STATES NAT. MUS. PROC. 27: 877, 1904 DYAR EGG, LARVA
FOOD PLANTS
LICHENS

LONGIPENNIS GROTE, COPABLEPHARON-NOCTUIDAE
LIFE HISTORY
CAN. ENT. 57:289, 1925 SEAMANS LARVA, PUPA
FOOD PLANTS
ROSA SP. (ROSES)

LONGIPENNIS WALKER
SYN. SEE LIMITARIA WALKER

LONGLEYI FRENCH
SSP. SEE NUMITOR FABRICIUS

LORATA GROTE & ROBINSON
SYN. SEE FALACER GODART

LORATA GROTE, TETRACIS-GEOMETRIDAE
?SYN. CACHEXIATA GUENEE
LIFE HISTORY
AMER. NAT. 18:935, 1884 PACKARD LARVA, PUPA
CAN. ENT. 2:28, 1869 MINOT EGG, YOUNG LARVA
LOC. CIT. 9:62, 1877 GOODELL LARVA, PUPA
MAINE AGRI. EXP. STA. BULL. 259: 46, 1917 MOSHER LARVA, PUPA F. 4, B, E
NATURALISTE CANADIEN P. 328, 1935 DAVIAULT LARVA, PUPA
NEW YORK ENT. SOC. JOURN. 53:194 1945 FORBES PUPB
NEW YORK STATE AGRI. EXP. STA. (CORNELL), MEM. 274:107, 1948 FORBES LARVA, PUPA (CACHEXIATA GN)
NEW YORK STATE MUS. MEM. 8:754 1905 FELT LARVA (BRIEF)
UNITED STATES DEPT. AGRI. MISCH. PUB. 657:361, 437, 1950 CRAIGHEAD LARVA
FOOD PLANTS
BETULA ALLEGHENIENSIS BRITT.
COMPTONIA PEREGRINA (LINN.)

IRUS SP.
PINUS SP. (PINES)
PRUNUS SEROTINA EHRH.
TILIA SP.
TSUGA CANADENSIS(LINN.)

LOREA(GUENEE), LACINIPOLIA-NOCTUIDAE
SYN. LIGATA WALKER
SYN. DODGEI MORRISON
LIFE HISTORY
AMER. MUS. NAT. HIST. BULL. 14:300 1901 BEUTEN. LARVA
CAN. ENT. 22:16, 1890 BEUTEN. LARVA
MAINE AGRI. EXP. STA. BULL. 356: 171, 1930 PHIPPS LARVA*F. 24D PUPA
UNITED STATES NAT. MUS. PROC. 27: 858, 1904 DYAR EGG, LARVA
INDEX AMER. ECON. ENT. VOL. 5:505; VOL. 7:587
FOOD PLANTS
BETULA POPULIFOLIA MARSH
COMPTONIA PEREGRINA(LINN.)
FRAGARIA VIRGINIANA DUSCH.
GERANIUM MACULATUM LINN.
MEDICAGO SATIVA LINN. (ALFALFA)
PRUNUS PENSYLVANICA LINN.
SPIRAEA LATIFOLIA(AIT.)
TRIFOLIUM SP.
VACCCINIUM CASSINOIDES LINN.
DEAD LEAVES

LORETTA BARNES & MCDUNNOUGH
SSP. SEE NEOGAMA ABBOT & SMITH

LORICARIA(EVERSMANN), DYSMIGIA-GEOMETRIDAE
SYN. JULIA HULST
SYN. NUBILATA WARREN
LIFE HISTORY
CAN. ENT. 79:115, 1947 MCGUFFIN LARVA
NEW YORK STATE AGRI. EXP. STA. (CORNELL), MEM. 274:42, 1948 FORBES LARVA
FOOD PLANTS
BETULA SP. (BIRCHES)
POPULUS SP. (POPLARS)
SALIX SP. (WILLOWS)

LORQUINI BEHR
SEE PIASUS BOISDUVAL

LORQUINI(BOISDUVAL), BASILARCHIA-NYMPHALIDAE
FORM EAVESII HENRY EDWARDS
AB COMSTOCKI GUNDER
AB POWELLI FIELD
A BURRISONII MAYNARD
FORM MAYNARDI FIELD
AB GUNDERI FIELD
LIFE HISTORY
CALIFORNIA ACAD. SCI. PROC. 5:171 330, 1873 EDWARDS LARVA, CHRYSALIS
CAN. ENT. 23:172, 1891 DYAR EGG, LARVA, CHRYSALIS
SOUTHERN CALIFORNIA ACAD. SCI. BULL. 30:89, 1931 COMSTOCK EGG*F. ON P. 87
COMSTOCK-BUTTER. CALIFORNIA P. 143, 1927 EGG, LARVA, CHRYSALIS* A 43
EDWARDS, W. H.-BUTTER. NORTH AMER. 1ST SERIES, 1869 LARVA* PL. LIMENITIS NO. E
EDWARDS-LOC. CIT. 3RD SERIES, 1897 SUPPL. NOTES EGG, LARVA
ESSIG-INS. WEST. NORTH AMER. P. 647, 1926 LARVA*F. 527; CHRYSALIS
FOOD PLANTS
CRATAEUGUS SP. (THORNS)
MALUS BACCATA(LINN.)
MALUS DIVERSIFOLIA(BONG.)
MALUS PUMILA MILL. (APPLE)
POPULUS HETEROPHYLLA LINN.
POPULUS TREMULOIDES MICHX.
POPULUS TRICHOCARPA T. &G.
PRUNUS DOMESTICA V. GALATENSIS EX HOOK(PRUNE)
PRUNUS EMARGINATA DOUGL. EX. HOOK
PRUNUS VIRGINIANA LINN.
QUERCUS SP. (OAKS)
SALIX SP. (WILLOWS)
SPIRAEA DOUGLASII HOOK

LORQUINI BOISDUVAL
SYN. SEE EURYDICE BOISDUVAL

LORQUINI FIELD
SSP. SEE GRUNUS BOISDUVAL

LORQUINI GROTE & ROBINSON
SSP. SEE MAC-CULLOCHI KIRBY

LORQUINI OBERTHUS
SYN. SEE CHALCEDONIA DOUBLEDAY & HEWITSON

LORQUINI OBERTHUR
SSP. SEE CLODIUS MENETRIES

LOTIS LINTNER
SSP. SEE MELISSA EDWARDS

LOTTA BEUTENMULLER
SSP. SEE CREUSA DOUBLEDAY & HEWITSON

LOUISA HULST
SYN. SEE CUPEDINARIA GROTE

LOUISA STALLINGS
SSP. SEE CLYTON BOISDUVAL&LECONTE

LOYA SMITH
SSP. SEE SPONSA GUENEE

LUBENS GROTE
SYN. SEE LUTRA SMITH

LUBRICALIS GEYER, EPIZEUXIS-NOCTUIDAE
SYN. BISTRIGALIS STEPHENS
SYN. PHAEALIS GUENEE
SYN. SURRECTALIS WALKER
A OCCIDENTALIS SMITH
FORM INTENSALIS SMITH
A PARTITALIS SMITH
C COBETA SMITH
LIFE HISTORY
CAN. ENT. 12:44, 1880 COQUILLETT LARVA
ENT. AMER. 14:161, 1934 CRUMB LARVA* PL. 20 C, D; PL. 21 A; PL. 22E; PL. 23F; PL. 24ND P9
ILLINOIS DEPT. AGRI. TRANS. 18, APPEND: 138:182, 1880 COQUILLETT & MARTEN LARVA
GLOVER-ILLUST. NORTH AMER. ENT. 1878 LARVA* PL. 12, FF. 15, 16
INDEX AMER. ECON. ENT. VOL. 1:120
FOOD PLANTS
GRASS
ROTTEN WOOD

LUBRICANS(GUENEE), EUAGROTIS NOCTUIDAE
SYN. SPRETA SMITH
SYN. ASSOCIANS WALKER
LIFE HISTORY
CAN. ENT. 12:14, 1880 FRENCH LARVA
ILLINOIS DEPT. AGRI. TRANS. 18:135, 1880 APPEND. MARTEN LARVA
ILLINOIS STATE NORMAL UNIV. REPT. 6:45, 1880 FRENCH LARVA
FOOD PLANTS

LUCETTA HENRY EDWARDS
SSP. SEE ANGUSI GROTE

LUCIANA STRECKER, CATOCALA-NOCTUIDAE
FORM NEBRASKAE DODGE
FORM SOMNUS DODGE
LIFE HISTORY
AMER. MUS. NAT. HIST. BULL. 38: 158, 1918 BARNES & MCDUNNOUGH EGG, LARVA
ENT. NEWS 24:197, 1913 ROWLEY LIFE HISTORY D
FOOD PLANTS
SALIX SP. (WILLOWS)

LUCIDUS BOISDUVAL, ARCTONOTUS-SPHINGIDAE
A CLARKI BARNES & BENJAMIN
LIFE HISTORY
SOUTHERN CALIFORNIA ACAD. SCI. BULL. 41:167, 1942 COMSTOCK & HENNE EGG, LARVA* PL. 31; PUPA* PL. 32
FOOD PLANTS
OENOTHER DENTATA V. CAMPESTRIS (GREENE) (IN CAPTIVITY)

LUCILIUS(SCUDDER & BURGESS)ERYNNIS-HESPERIIDAE

LIFE HISTORY

AMER.MUS.NAT.HIST.BULL.5:300 1893 BEUTEN.LARVA,CHRYSALIS

CAN.ENT.7:163,1875 MEAD EGG?

DENISON UNIV.SCI.LABORATORIES, JOURN.22:111,1927 LINDSEY LARVAL HEAD CAPSULE*F.1C

ILLINOIS DEPT.AGRI.TRANS.18, APPEND:97,1880 MIDDLEDON LARVA

NEW YORK STATE CABIN.NAT.HIST. ANN.REPT.NO.24:164,1871 LINTNER LARVA

LOC.CIT.NO.30:179,1878 LINTNER EGG,LARVA,CHRYSALIS

FRENCH-BUTTER.EAST.UNITED STATES P.359,1886 EGG,LARVA CHRYSALIS

HOLLAND0BUTTERFLY BOOK 1905 CHRYSALIS*PL.6,FF.30-32

SCUDDER-BUTTER.EAST.UNITED STATES&CANADA PP.1461-1467 1889 LIFE HISTORY*PL.69,FF. 3,6;PL.73,F.2;PL.77,FF.8,9; PL.80,FF.31-35;PL.82,F.10; PL.86,FF.28-30,46,47;PL.85, FF.30-32

SCUDDER-GUIDE TO COMMON BUTTER. P.158,1893 LARVA,CHRYSALIS

INDEX AMER.ECON.ENT. VOL.7:1005

FOOD PLANTS

AQUILEGIA CANADENSIS LINN.
CHENOPODIUM ALBUM LINN.
POPULUS SP.(POPLARS)
SALIX SP.(WILLOWS)

LUCILLA WORTHINGTON

SYN.SEE UNIJUGA WALKER

LUCINA HENRY EDWARDS,HEMILEUCA-SATURNIIDAE

AB OBSELETA REIFF
AB LUTEA REIFF
A LATIFASCIA BARNES & MCDUNNOUGH

LIFE HISTORY

NAT.ACAD.SCI.MEM.7:120,1914 PACKARD LARVA,PUPA

PSYCHE 17:29,1910 REIFF UNITED STATES DEPT.AGRI.MISC. PUB.657:357,380,1950 CRAIGHEAD LARVA

FOOD PLANTS

BETULA POPULIFOLIA MARSH.
PRUNUS SEROTINA EHRH.
QUERCUS SP.(OAKS)
SPIRAEA SALICIFOLIA LINN.

LUCINA SMITH

SSP.SEE OLIVACEA MORRISON

LUCINDA BEUTENMULLER

SSP.SEE ULTRONIA HUBNER

LUCIPARA AUCT

SEE BENESIMILIS MCDUNNOUGH

LUCTUATA(SCHIFFERMULLER),SPARGANIA-GEOMETRIDAE

SYN.LUGUBRATA STAUDINGER
SYN.KODIAKATA PACKARD
A OBDUCTATA MOESCHLER
SYN.CONCORDATA WALKER

LIFE HISTORY

NEW YORK STATE AGRI.EXP.STA. (CORNELL),MEM.274:152,1948 FORBES LARVA?

GUMPPENBERG-SYSTEMA GEOMETRARUM 3:279,1890 LARVA

FOOD PLANTS

CHAMAENERION ANGUSTIFOLIUM LINN.
GALIUM SP.

LUCTUOSA HULST

SSP.SEE RETECTA GROTE

LUCTUOSA WATSON & COMSTOCK

SSP.XANTHOIDES BOISDUVAL

LUCTUOSUS BOISDUVAL

SYN.SEE LUGUBRIS LINN.

LUCULENTA HENRY EDWARDS

SSP.SEE STRIGOSA GROTE

LUGENS WALKER, SPHINX-SPHINGIDAE
SYN. ANDROMEDAE BOISDUVAL
LIFE HISTORY
KANSAS ACAD. SCI. TRANS. 4:28 1875 SNOW LARVA, PUPA
OBSERVER OF NATURE 3, NO. 1, 1875 LARVA
STRECKER-LEP. RHOP. ET HETER. 13:115, 1876 LARVA
FOOD PLANTS

LUGUBRATA MOESCHLER
SSP. SEE DESTINATA MOESCHLER

LUGUBRATA STAUDINGER
SYN. SEE LUCTUATA SCHIFFENMULLER

LUGUBRIS HULST
SSP. SEE FIGURATA DRURY

LUGUBRIS (LINNAEUS), ENYO-SPHINGIDAE
SYN. FEGEUS CRAMER
SYN. LUCTUOSUS BOISDUVAL
LIFE HISTORY
AMER. JOURN. SCI. 36:307, 1839 HARRIS LARVA, PUPA
AMER. MUS. NAT. HIST. BULL. 7:281 1895 BEUTEN. LARVA
NEW YORK ENT. SOC. JOURN. 15, 1907 LARVA* PL. 1 COLOR NO DESCR.
NEW YORK STATE MUS. MEM. 8:740 1905 FELT LARVA
PHILADELPHIA ACAD. NAT. SCI. JOURN 5:140, 1859 CLEMENS LARVA
SMITHSONIAN MISC. COLLECTIONS 4 ART. 2:162, 1862 MORRIS LARVA
WASHINGTON ENT. SOC. PROC. 4:451 1900 DYAR LARVA (THIS SP.?)
ABBOT & SMITH-LEP. INS. GEORGIA 1:59, 1797 LARVA*PUPA*PL. 30
BOISDUVAL-SPEC. GEN. LEP. SPHING. P. 297, 1874 LARVA
MORRIS-SYNOP. LEP. NORTH AMER. P. 162, 1862 LARVA
WESTWOOD-DRURY'S ILLUST. EXOTIC ENT. 1:56, 1837 LARVA, PUPA
FOOD PLANTS
PARTHENOCISSUS QUINQUEFOLIA (LINN.)
RANDIA ACULEATA LINN. (?)
VITIS SP.
VITIS VINIFERA LINN. (GRAPE)

LUGUBROSA HULST
SSP. SEE FISCELLARIA GUENEE

LUMBONIGERA DYAR
SSP. SEE CONTIGUA WALKER (HAPLOA)

LUMENARIA HUBNER
SYN. SEE PENDULINARIA GUENEE

LUNA (LINNAEUS), ACTIAS-SATURNIIDAE
AB ROSSI ROSS
AB RUBROSUFFUSA COCKERELL
FORM VERN. RUBROMARGINATA DAVIS
A DICTYNNA WALKER
FORM VERN. MARIAE BENJAMIN
LIFE HISTORY
AMER. ACAD. ARTS. SCI. PROC. 28: 87, 1893 PACKARD EGG, LARVA
AMER. MUS. NAT. HIST. BULL. 5:90 1893 BEUTEN. HYBRID LARVA BETWEEN LUNA AND SELENS
LOC. CIT. 10:437, 1898 BEUTEN LARVA
CAN. ENT. 2:27, 1869 MINOT. EGG
LOC. CIT. 6:86, 1874 GENTRY LARVAL VARIATIONS
LOC. CIT. 7:141, 1875 ROGERS LARVA
LOC. CIT. 7:199, 1875 ROGERS LARVA
LOC. CIT. 9:33, 1877 SAUNDERS EGG, LARVA
LOC. CIT. 64:263, 1932 BIRDSONG EGG, LARVAL VARIATIONS
LOC. CIT. 65:56, 1933 CALALE & WICKWIRE EGG
LOC. CIT. 65:266, 1933 CALALE & WICKWIRE EGG, LARVA AND HYBRID LARVA WITH A. SELENE
LOC. CIT. 69:137, 1937 WICKWIRE LARVAL VARIATIONS
ENT. NEWS 9:150, 1898 SOULE DURATION OF EGG STAGE
ENT. SOC. AMER. ANNALS. 9:141,

1916 Mosher pupal structure
Illinois Dept. Agri. Trans. 15: 192,1877 French larva
Loc. cit. 18:124,178 Append. 1880 Coquillett & Marten larva
Maryland Nat. Hist. Soc. Bull. 8: 51,1938 cocoon*
Maryaldnd Nat. Hist. Soc. Bull. 15:39,1945 larva(brief)
Maryland Nat. Hist. Soc. Pub. no page 1930 Haydon egg, larva, pupa
Michigan State College, Spec. Bull. 253:52,1933 McDaniel larva
Minnesota Agri. Exp. Sta. Bull. 61:180,1899 Lugger larva* f.127
Missouri, Rept. Ins. of 4:123,1872 Riley life history*
Nat. Acad. Sci. Mem. 12:193,1914 Packard life history egg; larva* Pls. 15,16,17, ff. 1-5 COLOR
Naturaliste Canadien p.204,1929 Huard larva
New York State Cabin. Nat. Hist. Rept. No. 26:126,1874 Lintner larva
New York State Mus. Mem. 8:526 1905 Felt larva
New York State, Rept. Ins. of 3: 452 also p.134, sec. 180, 1856 Fitch larva
Ohio Agri. Exp. Sta. Bull. 332:268, 1918 Houser larva* Pl. 42, f. 2
Ontario Ent. Soc. Ann. Rept. p. 43, 1875 Rogers egg, larva, pupa
Loc. cit. p. 35,1876 Saunders compares larvae of <u>luna</u> and <u>polyphemus</u>
Loc. cit. p. 29,1882 Moffat larva*
Loc. cit. p. 48,1889 Harringtton larva*
Outlook 64, No. 14:832, April 7, 1900 Soule larva*
Psyche 7:275,1895 Soule larval variation
Smithsonian Misc. Collections 4, art. 2:226,1862 Morris larva, cocoon
Societe Acclim. Bull. France series 3, Vol. 9:264,1882 egg
United States Dept. Agri. Misc. Pub. 657:358,379,1950 Craighead larva* f. 72D
Abbot & Smith-Lep. Ins. Georgie 1:95,1797 larva* Pl. 48 COLOR
Cragen-Our Ins. Friends & Foes p. 114,1899 larva* f. 48
Dickerson-Moths & Butter. p. 179 1901 larva* cocoon* f. 162
Eliot & Soule-Cater. & Moths p. 258,1902 egg, larva* coooon*
Emmons-Nat. Hist. New York 5: 233,1854 larva* pupa* COLOR
Fernald-Kingsley's Stand. Nat. Hist. 2:458,1885 egg, larva cocoon
Glover-Illust. North Amer. Ent. 1878 larva* Pl. 7, f. 3
Grote-Amer. Ent. p. 13,1865 Cocoon(brief)
Harris-Ent. Corresp. p. 293,1869 larva* Pl. 4
Harris-Ins. Inj. Veg. 2st edit. p. 277,1841 larva, pupa
Harris-loc. cit. 2nd edit. p. 297 1852 larva
Harris-loc. cit. Flint edit. p. 382,1862 larva, cocoon
Hart-ins. Collections Illinois High Schools p. 18,1903 larva* f. 14
Herrick-Ins. Enemies Shade Trees p. 156,1935 larva
Holland-Moth Book p. 87,1905 larva* f. 40
Kellogg-Amer. Ins. P. 423,1908 coco on* f. 608
Morris-Synop. North Amer. Lep. p. 225,1862 larva
Packard-Ins. Inj. Forest Trees p. 76,1881 larva(brief)

TENNEY-NATURAL HISTORY P.401
1867 LARVA
WESTWOOD-DRURY'SILLUST. EXOTIC
ENT. 1:45, 1837 LARVA
WOOD-INSECTS ABROAD P.674
1874 LARVA*
INDEX AMER. ECON. ENT.
VOL. 2:378 TROPEA
VOL. 4:502 TROPEA
VOL. 5:673 TROPEA
VOL. 6:795 TROPEA
VOL. 7:1036 TROPEA

FOOD PLANTS
BETULA SP. (BIRCHES)
BETULA PAPYRIFERA MARSH.
CARYA SP. (HICKORIES)
CASTANEA DENTATA(MARSH.)
DIOSPYROS VIRGINIANA LINN.
FAGUS SP. (BEECHES)
HUMULUS LUPULUS LINN. (HOPS_
JUGLANS CINEREA LINN.
JUGLANS NIGRA LINN.
LIQUIDAMBAR STYRACIFLUA LINN.
NYSSA SYLVATICA MARSH.
OSTRYA VIRGINIANA(MILL.)
PRUNUS DOMESTICA LINN.
QUERCUS SP. (OAKS)
SALIX SP. (WILLOWS)

LUNARIS STRAND
SSP. SEE INCLUDENS WALKER

LUNATA(DRURY), ZALE-NOCTUIDAE
SYN. MALE EDUSA DRURY
SYN. PUTRESCENS GUERIN-MENEVILLE
FORM SAUNDERSI BETHUNE
FORM LUNATOIDES STRAND
A SALICIS BEHR
SYN. ROSAE BEHR

LIFE HISTORY
AMER. ENT. SOC. TRANS. 3:28, 1870
BEHR LARVA(SALICIS& ROSAE)
AMER. NAT. 4:229, 1871 PACKARD
LARVA*PL. 2, F. 5A; PUPA*PL. 2,
F. 5B
CAN. ENT. 14:131, 1882 FRENCH EGG,
LARVA, PUPA
CAN. JOURN. IND. SCI. &ART 10:253,
255, 1865 BETHUEN LARVA
(LUNATA & EDUSA)
NEW YORK ENT. SOC. JOURN. 19:192
1902 BEUTEN. LARVA (EDUSA$
NEW YORK STATE CABIN. NAT. HIST.
REPT. NO. 41:179, 1888
LINTNER LARVA
NEW YORK STATE MUS. MEM. 8:538
743, 1905 FLINT LARVA
NEW YORK STATE ENT. REPT. INS.
OF 4:58, 1888 LINTNER LARVA
GUENEE-SPEC. GEN. LEP. NOCT. 3:
13, 14, 1852 LARVA*PL. 2
COLOR
HART-INS. COLLECTIONS ILLINOIS
HIGH SCHOOLS P.43, 1903
LARVAL MIMICRY

FOOD PLANTS
ACER SP. (MAPLES)
HYPERICUM SP.
PRUNUS DOMESTICA LINN. (PLUM)
QUERCUS SP. (OAKS)
ROSA SP. (ROSES)
SALIX SP. (WILLOWS)
SOLIDAGO SP.
GENERAL FEEDER ON SHRUBS

LUNATA HENRY EDWARDS, HETEROCAMPA-
NOTODONTIDAE
SYN. PLUMOSA HENRY EDWARDS

LIFE HISTORY
INSECUTOR INSCITIAE MENSTRUUS
13:47, 1925 DYAR LARVA

FOOD PLANTS
POPULUS TREMULOIDES MICHX.

LUNATA STRETCH
SSP. SEE MARIPOSA GROTE&ROBINSON

LUNATOIDES GUENEE
SSP. SEE LUNATA DRURY

LUNIFERA(HUBNER), ZALE-NOCTUIDAE
SYN. LINEOSA WALKER

LIFE HISTORY
NEW YORK STATE MUS. MEM. 8:
753, 1905 FELT. LARVA
UNITED STATES DEPT. AGRI. BUR.

ENT. BULL. 12 OLD SERIED P. 22
1886 PACKARD LARVA, PUPA
FOOD PLANTS
PINUS SP. (PINES)

LUNILINEA (GROTE), STRENOLOMA-NOCTUIDAE
LIFE HISTORY
FOOD PLANTS
GLEDITSIA TRIACANTHOS LINN.

LUPINI (BOISDUVAL), PLEBIUS-LYCAENIDAE
AB IMMACULATA CHERMOCK
LIFE HISTORY
FOOD PLANTS
ASTRAGALUS SP.
ERIOGONUM SP.
LOTUS SP.

LURIDULA HULST
SYN. SEE COGNATARIA HUBNER

LUSCA SMITH
SYN. SEE AMERICANA SPEYER

LUSCA STICHEL
SYN. SEE CLODIUS MENETRIES

LUSCINATA ZELLER
SYN. SEE CENTROSTRIGARIA WOLLASTON

LUSCITIOSA CLEMENS, SPHINX-SPHINGIDAE
AB FEMALE UNA SKINNER
A BOMBAX BARNES & BENJAMIN
SYN. BOREALIS CLARK
LIFE HISTORY
AMER. MUS. NAT. HIST. BULL. 7:302
1895 BEUTEN. LARVA, PUPA
LOC. CIT. 8:292, 1896 BEUTEN. EGG
LARVA, PUPA
ENT. AMER. 5:189, 1889 DYAR LARVA
ENT. NEWS 5:265, 1894 KUNZE LARVA
PUPA (COMPARES WITH EXCAECATUS)
ENT. SOC. AMER. ANNALS 11:419,
1918 MOSHER PUPA
NATURALISTE CANADIEN P. 152,
1929 HUARD LARVA
NEW YORK STATE MUS. MEM. 8:742,
1905 FELT LARVA
NOVITATES ZOOLOGICAE (TRING) 9
SUPPLY P. 138, 1903 ROTHS-
CHILD & JORDAN LARVA, PUPA
ONTARIO ENT. SOC. ANN. REPT.
P. 104, 1899 FYLES LARVA
PSYCHE 5:85, 1888 SOULE LARVA
ELIOT & SOULE-CATER. & MOTHS.
P. 147, 1902 EGG, LARVA, PUPA
FOOD PLANTS
BETULA SP. (BIRCHES)
BETULA PAPYRIFERA MARSH.
ILEX LAEVIGATA (PURSH)
MYRICA PENSYLVANICA LOISEL
POPULUS TREMULOIDES MICHX.
SALIX SP. (WILLOWS)
SALIX DISCOLOR MUHL.
SALIX PURPUREA LINN.
TILIA SP.
ZOLISMA LIGUSTRINA (LINN.)

LUSKI BARNES & MCDUNNOUGH
SSP. SEE EURYNOME EDWARDS

LUSTRALIS (GROTE), LACINIPOLIA-NOCTUIDAE
SYN. CERVINA SMITH
A SUFFUSA SMITH
LIFE HISTORY
FOOD PLANTS
POPULUS FREMONTI S. WATS.

LUTARIUS WALKER
SYN. SEE MAGNARIUS GUENEE

LUTEA GROTE
SSP. SEE CRESSONANA GROTE

LUTEA REIFF
SSP. SEE LUCINA HENRY EDWARDS

LUTEA SMITH
SYN. SEE LATIORELLA WALKER

LUTEA PACKARD, EUPITHECIA-GEOMETRIDAE
SYN. CATSKILLATA PEARSALL
SYN. FASCIATA TAYLOR
A BIFASCIATA DYAR
LIFE HISTORY
CAN. ENT. 77:53, 1945 MCGUFFIN

LARVA
NEW YORK STATE AGRI .EXP.STA. (CORNELL),MEM.274:165,1948 FORBES LARVA
NEW YORK STATE MUS.MEM.8:754 1905 FELT LARVA
UNITED STATES DEPT.AGRI.BUR. ENT.BULL.3 OLD SERIES P.24, 1883 PACKARD LARVA(BRIEF)
INDEX AMER.ECON.ENT. VOL.4:180

FOOD PLANTS
ABIES BALSAMEA(LINN.)
LARIX LARICINA(DUROI)
PICEA ENGELMANNI ENGELM.
PICEA GLAUCA(MUENCH)
PINUS SP.(PINES)
PSEUDOTSUGA MENZIESII(MIRB)
TSUGA CANADENSIS(LINN.)

LUTEICOMA GROTE & ROBINSON
SYN.SEE IMPLETE WALKER(ACRONICTA)

LUTEIPICTA STRAND
SYN.SEE PURPURIFASCIA GROTE & ROBINSON

LUTEITINCTA WOOLCOT
SYN.SEE PHILODICE GODART

LUTEITINCTUS GROTE
SSP.SEE PHLOCIPHAGA GROTE &ROBINSON

LUTEOCOSTA FRANCLEMONT
SSP.SEE BETHUNEI GROTE & ROBINSON

LUTEOLA GROTE & ROBINSON
SYN.SEE CORDIGERA THUNBERG

LUTEOLA SMITH
SYN.SEE MOERENS GROTE

LUTEOLUS REAKIRT
SYN.SEE IOLE BOISDUVAL

LUTEOSCAPUS NEUMOEGEN & DYAR
SSP.MULTIFARIA WALKER(CTENUCHA)

LUTESCENS CLARK
SSP.SEE VERSICOLOR HARRIS

LUTESCENS CLARK
SSP.SEE MYRON CRAMER

LUTESCENS NEUMOEGEN & DYAR
SSP.SEE FRAGILIS STRETCH

LUTHERI COCKERELL
SSP.SEE IO FABRICIUS

LUTRA(GUENEE),POLIA-NOCTUIDAE
STN,LUBENS GROTE
SYN.RUFULA MORRISON
A GLAUCOPIS HAMPSON
LIFE HISTORY
FOOD PLANTS
BETULA SP.(BIRCHES)
PINUS CONTORTA DOUGL.
PINUS CONTORTA V.LATIFOLIA ENGELM.
RHUS SP.(SUMACS)
VACCINIUM SP.

LUTULENTA SMITH
SSP.SEE QUINQUELINEA SMITH

LYCEA EDWARDS
SSP.SEE ICARIOIDES BOISDUVAL

LYCIADES GEYER
SYN.SEE LYCIDAS ABBOT & SMITH

LYCIDAS(ABBOT & SMITH,ACHALARUS HESPERIIDAE
SYN.LYCIADES GEYER
SYN.HEDYSARUM SCUDDER
LIFE HISTORY
AMER.ENT.SOC.TRANS.37:188, 1911 SKINNER LARVA, CHRYSALIS
AMER.MUS.NAT.HIST.BULL.5:304 1893 BEUTEN.LARVA,CHRYSALIS
BUFFALO SOC.NAT.SCI.BULL.19:31 1939 WILD LARVA
UNITED STATES NAT.MUS.BULL. 157:204,1932 AUSTIN CLARK

LARVA
ABBOT & SMITH-LEP.INS.GEORGIA 1:39,1797 LARVA,PUPA PL.20 COLOR
GLOVER-ILLUST.NORTH AMER.ENT. 1878 LARVA,PUPA PL.1,F.13
HOLLAND-BUTTERFLY BOOK 1905 LARVA* PL.2,F.23;CHRYSALIS* PL.6,F.21
SCUDDER-BUTTER.EAST UNITED STATES&CANADA PP.1420-1422, 1889 LIFE HISTORY* PL.66,F. 16; PL.73,F.6;DPL.76,F.23; PL.80,FF.17,18,23,24;PL. 85,F.21
INDEX AMER.ECON.ENT. VOL.5:2

FOOD PLANTS
AMORPHA FRUTICOSA LINN.
CORYDALIS SEMPERVIRENS(LINN.)
DESMODIUM GLABELLUM(MICHX.)
DESMODIUM PANICULATUM(LINN.)
INDIGOFERA CAROKINA WALT.
IPOMOEA PANDURATA(LINN.)
LESPEDEZA HIRTA(LINN)

LYCOPERSICI BOISDUVAL
SYN.SEE SEXTA JOHANSSON

LYDIA BEUTENMULLER
SSP.SEE FAUSTINA STRECKER

LYGDAMUS(DOUBLEDAY),GLAUCOPSYCHE-LYCAENIDAE
A COUPERI GROTE
SYN.FEMALE AFRA EDWARDS
AB MACDUNNOUGHI GUNDER
AB LEUSSLERI GUNDER
B ORO SCUDDER
C BEHRII EDWARDS
AB STERNITZKYI GUNDER
D COLUMBIA SKINNER
E AUSTRALIS GRINNELL
AB SINEPUNCTA COMSTOCK
F ARIZONENSIS MCDUNNOUGH

LIFE HISTORY
ENT.NEWS 19:482,1908 WILLIAMS, LARVA,CHRYSALIS
LOC.CIT.22:359,1911 BOWER EGG* LARVA*CHRYSALIS PL.12
PAN PACIFIC ENT.23:42,1947 TILDEN CHRYSALIS AND NEST
SOUTHERN CALIFORNIA ACAD.SCI. BULL28:6,1929 COMSTOCK EGG* P.11 AUSTRALIS
LOC.CIT.34:127,1935 COMSTOCK& DAMMERS LARVA* PL.22; CHRYSALIS* PL.22
COMSTOCK,J.A.-BUTTER.CALIFORNIA P.198,1927 LARVA CHRYSALIS (BEHRII)
EDWARDS-BUTTER.NORTH AMER. 3RD SERIES 1897 SUPPL.NOTES LARVA
SCUDDER-BUTTER.EAST.UNITED STATES&CANADA P.955,1889 LARVA(COUPERI)

FOOD PLANTS
ASTRAGALUS SP.
LATHYRUS JAPONICUS V.GLABER (SER.)
LATHYRUS OCHROLEUCUS HOOK.
LA THYRUS PALUSTRIS LINN.
LOTUS GRANDIFLORUS(BENTH.)
LUPINUS SP.
VICIA CAROLINIANA WALT
VICIA CRACCA LINN.

LYMANI DYAR
SSP.SEE CONFUSA LYMAN

LYMAX DRUCE
SYN.SEE VARIARIA GROTE

LYNCEA CLEMENS
SYN.SEE FICUS LINNAEUS

LYSIDE(GODART),KRICOGONIA-PIERIDAE
SYN.LANICE LINTNER
FORM MALE TERISSA LUCAS
FORM FEMALE UNICOLOR GODMAN & SALVIN
SYN.XANTHOPILA ROEBER
FORM FEMALE FANTASIA BUTLER

LIFE HISTORY
JOURN.INSTITUTE OF JAMAICA 1:29

1891 COCKERELL LARVA, CHRYSALIS
FOOD PLANTS

LYSIMACHIAE BIRD, PAPAIPEMA-NOCTUIDAE
LIFE HISTORY
CAN. ENT. 47:71, 1914 BIRD LARVA, PUPA
FOOD PLANTS
LYSIMACHIA QUADRIFOLIA LINN.

M

MACARIA EDWARDS, ARGYNNIS-NYMPHALIDAE
A LAURINA WRIGHT
LIFE HISTORY
SOUTHERN CALIF. ACAD. SCI. BULL. 30:40, 1931 COMSTOCK DAMMERS LARVA* CHRYSALIS*
LOC. CIT. 33:34, 1934 COMSTOCK EGG* PL. 14
FOOD PLANTS
VIOLA SP. (VIOLETS)

MAC-CULLOCHI (KIRBY), ANDROLOMA-AGARISTIDAE
A LORQUINI GROTE & ROBINSON
B SIMILIS STRETCH
SYN. EDWARDSII BOISDUVAL
AB CONJUNCTA HENRY EDWARDS
LIFE HISTORY
UNITED STATES NAT. MUS. PROC. 25:386, 1903 DYAR LARVA
FOOD PLANTS
EPILOBIUM SP.
EPILOBIUM ANGUSTIFOLIUM (LINN.)

MACDUNNOUGHI BLACKMORE
SSP. SEE NUBILATA PACKARD (EUSTROMA)

MACDUNNOUGHI CASSINO & SWETT
SSP. SEE PEPLARIOIDES HULST

MACDUNNOUGHI GUNDER
SSP. SEE LYGDAMUS DOUBLEDAY

MACDUNNOUGH GUNDER
SSP. SEE SNOWI EDWARDS (LYCAENA)

MACDUNNOUGHI GUNDER
SSP. SEE ZELICAON LUCAS

MACDUNNOUGHI GUNDER
SSP. SEE CAMPESTRIS BEHR

MACERATA SMITH
SSP. SEE MISELIOIDES GUENEE

MACHAON LINNAEUS, PAPILIO-PAPILIONIDAE
A ALIASKA SCUDDER
SYN. JOANNISI VERITY
FORM PETERSII CLARK
B HUDSONIANUS CLARK
FORM AVINOFFI CHERMOCK
LIFE HISTORY
CAN. ENT. 21:160, 1889 BEUTEN. EGG, LARVA
INSECT LIFE 2:45, 1889 FLETCHER LARVA*
ONTARIO ENT. SOC. ANN. REPT. P. 85 1888 FLETCHER EGG, LARVA
LOC. CIT. P. 19, 1889 FYLES LARVA
EDWARDS-BUTTER. NORTH AMER. 3RD SERIES 1897 EGG* LARVA* (CHIONOBAS 10, FF. A-G)
FLETCHER-A TRIP TO NEPIGON P. 12, 1888 EGG, LARVA
SCUDDER-BUTTER. EAST. UNITED STATES & CANADA PP. 1776-1777, 1889 EGG, LARVA
FOOD PLANTS
CAREX SP.
CAREX PEDUNCULATA MUHS.
POA PRATENSIS LINN.
ROSA BLANDA AIT.
ROSA CAROLINA LINN.
(EGG DEPOSITED)
GRASSES

MACRA GUENEE
SYN. SEE FRUGIPERDA ABBOT & SMITH

MACRA PLOETZ
SYN. SEE AURANTIACA HEWITSON

MACROCARPATA MCDUNNOUGH, EUPITHECIA-GEOMETRIDAE

LIFE HISTORY
FOOD PLANTS
CUPRESSUS GUADELUPENSIS S. WATS.

MACTATOIDES BARNES & MCDUNNOUGH
SYN. FRACTILINEA GROTE

MACULA REIFF
SSP. SEE CECROPIA LINNAEUS

MACULARIA HARRIS, SICYA-GEOMETRIDAE
SYN. SUBLIMARIA GUENEE
SYN. TRUNCATARIA GUENEE
SYN. CALIPUSARIA WALKER
SYN. FAUSTINARIA STRECKER
FORM SOLFATARIA GUENEE
A AGYLLARIA WALKER
B CROCEARIA PACKARD
C LEWISI SWETT
D LATULA BARNES & MCDUNNOUGH
E SUBCIRIDIS CASSINO & SWETT
F CRUZENSIS DYAR
LIFE HISTORY
MAINE AGRI. EXP. STA. BULL. 259: 44, 1917 MOSHER LARVA, PUPA F. 2 D, G
NEW YORK ENT. SOC. JOURN. 53:196 1945 FORBES PUPA
NEW YORK STATE AGRI. EXP. STA. (CORNELL), MEM. 274:104, 1948 FORBES PUPA
FOOD PLANTS
CEANOTHUS SP.
FRAXINUS SP. (ASHES)
SPIRAEA SP.
ULMUS SP. (ELMS)

MACULATA BRYK & EISNER
SYN. SEE CLODIUS MENETRIES

MACULATA (EDWARDS), OLIGORIA-HESPERIIDAE
SYN. DELETA HERRICH-SCHAEFFER
SYN. NORUS PLOETZ
LIFE HISTORY
CAN. ENT. 11:191, 1879 EDWARDS LARVA, CHRYSALIS
ILLINOIS DEPT. AGRI. TRANS. 18, APPEND. P. 176, 1880 COQUILLETT LARVA
FRENCH-BUTTER. EAST. UNITED STATES P. 330, 1886 LARVA, CHRYSALIS
SCUDDER-BUTTER. EAST. UNITED STATES & CANADA P. 1763 1889 LARVA, CHRYSALIS
FOOD PLANTS

MACULATZ (HARRIS), HALISIDOTA-ARCTIIDAE
SYN. FULVOFLAVA WALKER
SYN. GUTTIFERA HERRICH-SCHAEFFER
A ANGULIFERA WALKER
B ALNI HENRY EDWARDS
SYN. TEXANA ROSHSCHILD
C AGASSIZII PACKARD
SYN. CALIFORNICA WALKER
SYN. SALICIS BOISDUVAL
D EUREKA DYAR
LIFE HISTORY
AGRI. GAZETTE OF CANADA 4:1045, 1917 SWAINE LARVA, PUPA
AMER. MUS. NAT. HIST. BULL. 10: 371, 1898 BEUTEN. LARVA
BOSTON SOC. NAT. HIST. PROC. 9:289 1863 WALSH LARVA
LOC. CIT. 27:135, 1896 DYAR LARVAL SETAE
CALIFORNIA ACAD. SCI. PROC. 7:128 1876 HY. EDW. LARVA (ALNI)
CAN. ENT. 3:186, 1871 SAUNDERS LARVA
CAN. JOURN. IND. ART. SCI. 8:369 1863 SAUNDERS LARVA
NATURALISTE CANADIEN P. 19, 1930 HUARD LARVA
NEW YORK ACAD. SCI. ANNALS 8:67 1893 PACKARD LARVA
NEW YORK ENT. SOC. JOURN. 5:132 1897 DYAR LARVAL STRUCTURE* PL. 7, F. 4
NEW YORK STATE MUS. MEM. 8:523 1905 FELT LARVA
OHIO AGRI. EXP. STA. BULL. 332: 244, 1918 HOUSER LARVA* PL. 36, F. 1
ONTARIO ENT. SOC. ANN. REPT. P. 56 1903 GIBSON LARVA

Loc.cit.p.84,1904 Gibson larva
Ottawa Naturalist 19:50,1905 Gibson larva
Psyche 6:165,1891 Dyar larva, pupa,cocoon
Psyche 6:323,1892 Dyar larva (agassizii)
Southern California Acad.Sci. Bull.36:117,1937 Comstock larva*Pls.49,50;pupa*Pl.51, cocoon* Pl.50 Agassizii
United States Dept.Agri.Misc. Pub.657:356,386,1950 Craighead larva
United States Nat.Mus.Proc. 27:796,1904 Dyar larva
Washington Ent.Soc.Proc.6:19, 1904 Dyar larva(alni)
Loc.cit.6:20,1904 Dyar larva (angulifera)
Essig-Ins.West.North America p.673,1926 larva* f.547
Harris-Ent.Corresp.P.290, 1869 larva* Pl.9
Harris-Ins.Inj.Veg.2st edit.p. 259,1841 larva,pupa
Harris-loc,cit.2nd edit.p.279, 1852 larva (brief)
Harris-loc.cit.Flint edit.p.363 larva (brief)
Herrick-Ins.Enemies Shade Trees p.124,1935
Packard-Guide to Study Ins.p. 287,1869 larva
Packard-Ins.Inj.Forest Trees p.46,1881 larva
Stretch-Zygaen.Bombyc.North Amer.p.103,1873 larva * cocoon* Pl.10,f.7 COLOR
Index Amer.Econ.Ent. Vol.1:147;Vol.2:169

FOOD PLANTS

Acer so.(maples)
Acer negundo Linn.
Acer saccharophorum Koch.
Alnus sp.(alders)
Alnus rubra Bong.
Amelanchier florida Lindl.
Arbutus menziesii Pursh.
Betula sp.(birches)
Carya ovata(Mill.)
Cercis canadensis Linn.
Corylus americnaa Walt.
Fagus sp.(beeches)
Holodiscus discolor(Pursh.)
Malus pumila Mill.(apple)
Ostrya virginiana(Mill.)
Oxydendrum arboreum(Linn.)
Populus sp.(poplars)
Prunus illicifolia(Nutt.ex H&A)
Prunus serotina Ehrh.
Quercus sp.(oaks)
Quercus alba Linn.
Quercus prinus Linn.
Quercus velutina Lam.
Rhus toxicodendron Linn.
Rhus vernix Linn.
Ribes lacustre(Pers.)
Ribes sanguineum Pursh
Ribes sativum Syme(currant)
Robinia pseudo-acacia Linn.
Salix sp.(willows)
Salix hookeriana Barratt
Salix scouleriana Barr.
Spiraea douglasii Hook.
Tilia sp.
Ulmus rubra Muhl.
Vernonia fasciculata Michx
Vernonia noveboracensis(Linn.)

maculata-suffusa Cockle
syn.see pseudargiolus Boisduval & Leconte

maculinita Gunder
ssp.see cupreus Edwards

maculosa Stretch
syn.see ornata Packard

macyi Fender
ssp.see mylitta Edwards

madefactalis(Guenee),Bomolocha-Noctuidae
syn.achatinalis Zeller

SYN. MALE DAMNOSALIS WALKER
SYN. FEMALE CADUCALIS WALKER
SYN. PROFECTA GROTE
LIFE HISTORY
ENT. AMER. 14:143, 1934 CRUMB LARVA
NEW YORK STATE MUS. MEM. 8:720 1905 FELT LARVA
WASHINGTON ENT. SOC. PROC. 4:329 1898 DYAR LARVA
GUENEE-SPEC. GEN. LEP. DELT. & PYRAL. 8:35, 1854 LARVA
FOOD PLANTS
ALISMA TRIVIALE PURSH
JUGLANS NIGRA LINN.

MADUSARIA WALKER
SYN. SEE ASTYLUSARIA WALKER

MAESTOSA (HULST), CATOCALA-NOCTUIDAE
SYN. MODERNA GROTE
SYN. GUENEI GROTE
LIFE HISTORY
AMER. MUS. NAT. HIST. MEM. NEW SERIES 3, PART 1, 1918 BARNES & McDUNNOUGH LARVA* PL. 10, F. 25; PL. 11, F. 7 COLOR
INDEX AMER. ECON. ENT. VOL. 1:67
FOOD PLANTS
CARYA SP.
CARYA ILLINOENSIS (WANG.)
JUGLANS NIGRA LINN.

MAESTOSA HULST
SYN. SEE PORCELARIA GUENEE

MAESTUS GODMAN & SALVIN
SYN. SEE CLITUS EDWARDS

MAGDALENA BARNES & McDUNNOUGH EUPHYDRYAS-NYMPHALIDAE
LIFE HISTORY
SOUTHERN CALIFORNIA ACAD. SCI. BULL. 34:145, 1935 COMSTOCK LIFE HISTORY* EGG* PL. 38; LARVA* PL. 39; CHRYSALIS* PL. 40
FOOD PLANTS
PLANTAGO SP.

MAGDALENA STRECKER
SYN. SEE ILLECTA WALKER

MAGDALENA STRECKER, EREBIA-SATYRIDAE
LIFE HISTORY
EDWARDS-BUTTER. NORTH AMER. 3RD SERIES 1897 EGG* LARVA* (EREBIA FF. A-C)
FOOD PLANTS
CAREX SP. (?)
GRASSES (?)

MAGDELENAE GUNDER
SSP. SEE SIERRA WRIGHT

MAGNARIUS (GUENEE), DEUTERONOMOS-GEOMETRIDAE
SYN. ALNIARIA PACKARD
SYN. LUTARIUS WALKER
A OCHRAETUS HULST
LIFE HISTORY
CAN. ENT. 40:170, 1908 FLETCHER EGG
ENT. AMER. 4:13, 1888 CRANE COCOON (BRIEF)
NATURALISTE CANADIEN P. 326, 1935 DAVIAULT LARVA, PUPA
NEW YORK ENT. SOC. JOURN. 3:137 1895 BEUTEN. EGG, LARVA
LOC. CIT. 53:203, 1945 FORBES LARVA, PUPA
NEW YORK STATE AGRI. EXP. STA. (CORNELL), MEM. 56:1076, 1922 WELLHOUSE LARVA (BRIEF)
LOC. CIT. 274:86, 1948 FORBES EGG, LARVA, PUPA
NEW YORK STATE MUS. BULL. 155: 54, 1912 FELT EGG* PL. 18; LARVA, PUPA
NEW YORK STATE MUS. MEM. 8:725 1905 FELT LARVA
UNITED STATES DEPT. AGRI. MISC. PUB. 657:363, 437, 1950 CRAIGHEAD LARVA
GUMPPENBERG-SYSTEMA GEOMET-

RARUM 7:374,1895 LARVA,PUPA (ALNIARIA)
HART-INS.COLLECTIONS ILLINOIS HIGH SCHOOLS P.52,1903 EGG, LARVA,COCOON
HERRICK-INS.ENEMIES SHADE TREES' P.153,1935 EGG* F.126
HUMPHREYS-GENERA BRITISH MOTHS 1:84,1859 LARVA
KIRBY-EUROPEAN BUTTER.&MOTHS. P.299,1882 LARVA
NEWMAN-BRITISH MOTHS P.56,1869 LARVA
PACKARD-INS.INJ.FOREST TREES P.92,1881 LARVA,PUPA
PACKARD-MONOGRAPH GEOMETRIDAE P.630,1876 PUPA*PL.13,F.1A
STAINTON-BRITISH BUTTER.&MOTHS 2:15,1859 LARVA
INDEX AMER.ECON.ENT.
VOL.4:162 ENNOMOS
VOL.5:213 ENNOMOS ALNIARIA

FOOD PLANTS
ACER SP.(MAPLES)
ALNUS SP.(ALDERS)
BETULA SP.(BIRCHES)
BETULA ALLEGHENIENSIS BRITT.
BETULA NIGRA LINN.
BETULA POPULIFOLIA MARSH.
CARYA SP.
CASTANEA DENTATA(MARSH.)
COMPTONIA PEREGRINA(LINN.)
CRATAEGUS SP.(THORNS)
FAGUS SP.(BEECHES)
FRAXINUS AMERICANA LINN.
LIQUIDAMBAR STYRACIFLUA LINN.
MALUS PUMILA MILL.(APPLE)
POPULUS SP.
POPULUS TREMULOIDES MICHX.
SYRINGS SP.
TILIA SP.(LINDENS)
ULMUS SP.(ELMS)
GENERAL FEEDER DECIDUOUS TREES.

MAGNIFERARIA WALKER
SYN.SEE OSSULARIA HUBNER

MAGNIFICA CLARK
SYN.SEE PHAETON DRURY

MAGNOLIATA GUENEE,SPARGANIA-GEOMETRIDAE
SYN.INCOMMODATA WALKER
SYN.PLACIDATA WALKER
SYN.ZICZACATA WALKER
SYN.CUMATILIS GROTE & ROBINSON
A PERNOTATA HULST
B QUADRIPUNCTATA PACKARD
C RUPTATA BARNES & MCDUNNOUGH

LIFE HISTORY
CALIFORNIA ACAD.SCI.PROC.7:24 1876 HENRY EDWARDS LARVA; PUPA (QUADRIPUNCTATA)
NEW YORK STATE AGRI.EXP.STA. (CORNELL),MEM.274:145, 1948 FORBES LARVA
PSYCHE 10:191,1903 DYAR EGG, LARVA

FOOD PLANTS
EPILOBIUM ANGUSTIFOLIUM(LINN.)
FUCHSIA SP.
FUCHSIA MAGELLANICA V.RICCARTONII LEBAS
GERANIUM SP.
OENOTHERA BIENNIS LINN.
PELARGONIUM SP.

MAGNUS WRIGHT
SSPSEE SMINTHEUS DOUBLEDAY& HEWITSON

MAIA(DRURY),HEMILEUCA-SATURNIIDAE
SYN.PROSERPINA FABRICIUS
AB LINTNERI COCKERELL

LIFE HISTORY
AMER.ENT.1:186,252,1869 LARVA (BRIEF)
AMER.MUS.NAT.HIST.BULL.10: 438,1898 BEUTEN.LARVA
AMER.NAT.7:475,1873 EGG* F.122;LARVA*F.123
AMER.PHIL.SOC.PROC.31:171 1893 PACKARD LARVA
CAN.ENT.17:243,1885 GOODHUE LARVA
ENT.SOC.AMER.ANNALS.P.294.

1914 Mosher pupa
Loc. cit. 9, 1918 pupa*Pl. 5, ff. 4-5
Illinois Dept. Agri. Trans. 15:106, 1877 French larva
Loc. cit. 18, Append: 127, 163, 1880 Coquillett & Marten larva; pupa*
Minnesota Agri. Exp. Sta. Bull. 61:183, 1899 Lugger egg* larva* pupa* all in f. 131
Missouri, Rept. Ins. of p. 127, 1873 Riley life history*
Nat. Acad. Sci. Mem. 12:118, 1914 Packard life history larva* Pl. 21, ff. 2, 3; Pl. 22, ff. 1-4 COLOR
New York State Cabin. Nat. Hist. Rept. No. 23:153, 1870 Lintner egg*larva*pupa*Pl. 8, ff. 2, 3
New York State Mus. Mem. 8:310 1905 Felt egg*Pl. 17, f. 10; larva*Pl. 17, f. 8 all COLOR
Psyche 8:299, 1898 Scudder larva
Smithsonian Misc. Collections 4 art. 2:21, 1862 Morris larva
United States Dept. Agri. Misc. Pub. 657:357, 380, 1950 Craighead larva* f. 72 G
United States Nat. Must. Proc. 25:371, 1903 Dyar larva
Abbot & Smith-Lep. Ins. Georgia 1:99, 1797 larva*Pl. 50 COLOR
Cragen-Our Ins. Frienda & Foes p. 123, 1899 larva
Duncan-Natur. Lib. 23:154, 1858 larva*Pl. 16, f. 2 COLOR
Emmons-Nat. Hist. New York 5:232 1854 larva*Pl. 39 COLOR
Essig-Ins. West. North America p. 670, 1926 egg*f. 544; larva* f. 544
Glover-Illust. North Amer. Ent. 1878 larva, pupa Pl. 5, f. 7
Harris-Ins. Inj. Veg. 1st edit. p. 285, 1841 larva, pupa
Harris-Loc, cit. 2nd edit. p. 306 1852 larva
Harris-Loc. cit. Flint edit. p. 396, 1862 larva
Herrick-Ins. Enemies Shade Trees p. 228, 1935 egg*larva*pupa* all f. 184
Holland-Moth Book p. 92, 1905 egg*larva*pupa* ff. 44, 45
Morris-Synop. Lep. North Amer. p. 221, 1862 larva
Packard-Half Hours with Ins. p. 86, 1877 larva*
Westwood-Drury's Illust. Exotic. Ent. 2:45, 1837 larva
Index Amer. Econ. Ent.
Vol. 1:154; Vol. 2:174
Vol. 3:192; Vol. 6:348
Vol. 7:458

FOOD PLANTS

Artemisia filifolia Torr.
Aster sp.
Corylus sp. (hazels)
Populus sp. (populars)
Populus tremuloides Michx.
Prosopis glandulosa Torr
Prunus cerasus Linn. (cherry)
Prunus serotina Ehrh.
Quercus sp. (oaks)
Quercus ilicifolia Wang.
Quercus laevus Walt.
Quercus rubra Linn.
Quercus stellata Wang.
Quercus velutina Lam.
Rubus allegheniensis Porter
Salix sp. (willows)
Salix humilis Marsh.
Spiraea salicifolia Linn.
Vaccinium macrocarpon Ait.

<u>maia</u> Ottolengui
ssp. see <u>picta</u> Packard

<u>maimes</u> Smith
syn. see <u>ridingsiana</u> Grote

<u>maisi</u> Fitch
syn. see <u>tessellata</u> Harris

<u>major</u> Grote & Robinson, <u>Datana</u>-Notodontidae

LIFE HISTORY

AMER. MUS. NAT. HIST. BULL. 4:76, 1892 BEUTEN. LARVA

LOC. CIT. 10:406, 1898 BEUTEN. LARVA

CAN. ENT. 21:34, 1889 DYAR LARVA

MAINE AGRI. EXP. STA. BULL. 259; 66, 1917 MOSHER PUPA*F. 5F

LOC. CIT. 162:257, 1908 PATCH LARVA

NAT. ACAD. SCI. MEM. 7:115, 1895 PACKARD EGG, LARVA*PL. 12, FF. 1-6 COLOR; PUPA

NEW YORK STATE MUS. MEM. 8:739 1905 FELT LARVA

OHIO AGRI. EXP. STA. BULL. 332: 213 1918 HOUSER LARVA

PSYCHE 2:272, 1879 ANDREWS LARVA

PSYCHE 5:415, 1890 DYAR EGG, LARVA, PUPA

PSYCHE 6:573, 1893 DYAR KEY TO LARVAE OF DATANA

UNITED STATES DEPT. AGRI. MISC. PUB. 657:355, 398, 1950 CRAIGHEAD LARVA

ELIOT & SOULE-CATER. & MOTHS P. 295, 1902 LARVA*NEXT TO P. 294

INDEX AMER. ECON. ENT. VOL. 1:102; VOL. 2:109; VOL. 7:246

FOOD PLANTS

ANDROMEDA SP.
AZALEA SP.
HAMAMELS VIRGINIANA LINN.
MALUS PUMILA MILL. (APPLE)
RHODODENDRON SP.
VACCINIUM SP.
VACCINIUM STAMINEUM LINN.
VACCINIUM ULIGINOSUM LINN.
XOLISMA LIGUSTRINA (LINN.)
XOLISMA MARIANA (LINN.)

MAJOR SPEYER
SYN. SEE TRIFOLII ROTTENBURG

MAJORARIA GUENEE
SYN. SEE CONFUSARIA HUBNER

MAJORARIA PACKARD
SEE IRRARIA BARNES & MCDUNNOUGH

MALANA (FITCH), BALSA-NOCTUIDAE
SYN. OBLIQUIFERA WALKER

LIFE HISTORY

AMER. MUS. NAT. HIST. BULL. 16: 449, 1902 BEUTEN. LARVA

MINNESOTA AGRI. EXP. STA. BULL. 61:222, 1899 LUGGER LARVA* PL. 23, F. 170

NEW YORK STATE, REPT. INS. OF 2:242, 1855 FITCH LARVA

LOC. CIT. 3:343, 1856 FITCH LARVA ALSO P. 25, SEC. 40

PACKARD-HALF HOURS WITH INS. P. 181, 1877 LARVA

SAUNDERSOINS. INJ. FRUITS P. 101 1883 LARVA, COCOON

SAUNDERS-LOC. CIT. P. 101, 1889 LARVA

SMITH-INS. NEW JERSEY P. 419 1900 LARVA (BRIEF)

INDEX AMER. ECON. ENT. VOL. 1:46; VOL. 7:120

FOOD PLANTS

AMELANCHIER CANADENSIS (LINN.)
MALUS PUMILA MILL. (APPLE)
POPULUS SP.
PRUNUS CERASUS LINN. (CHERRY)
PRUNUS DOMESTICA LINN. (PLUM)
PYRUS SP.
PYRUS COMMUNIS LINN. (PEAR)
ULMUS SP.
ULMUS AMERICANA LINN.

M-ALBUM (BOIADUVAL & LECONE) STRYMON
LYCAENIDAE
SYN. PSYCHE BOISDUVAL & LECONTE

LIFE HISTORY

ILLINOIS DEPT. AGRI. TRANS. 18, APPEND:92, 1880 MIDDLETON LARVA, CHRYSALIS

SMITHSONIAN MISC. COLLECTIONS 4, ART. 2:93, 1862 MORRIS LARVA, CHRYSALIS

BOISDUVAL & LECONTE-HIST. LEP. NORTH AMER. PP. 86, 88, 1883

LARVA*CHRYSALIS*PLS.26,
27 M-ALBUM & PSHCYE COLOR
COMSTOCK-HOW TO KNOW BUTTER.
P.222,1904 LARVA
FRENCH-BUTTER.EAST.UNITED
STATES P.256,1886 LARVA,
CHRYSALIS
MORRIS-SYNOP.LEP.NORTH AMER.
P.93,1862 LARVA,CHRYSALIS
SCUDDER-BUTTER.EAST.UNITED
STATES & CANADA P.1825,
1889 LARVA,CHRYSALIS

FOOD PLANTS

ASTRAGALUS SP.
CRATAEGUS SP.(THORN SEED PODS)
CRATAEGUS OXYACANTHA LINN.
HUMULUS LUPULUS LINN(HOPS)
(PODS)
HYPERICUM SP.
LOTUS ARGOPHYLLUS V.ORNITHOPUS
GREENE
QUERCUS SP.(OAKS)
TILIA AMERICANA LINN.
VICIA SATIVA LINN.

MALCOLMI COMSTOCK
SSP.SEE MONTIVAGA BEHR

MALCOLMI GUNDER
SSP.SEE CHALCEDONIA DOUBLEDAY &
HEWITSON

MALEFIDA GUENEE,AGROTIS-NOCTUIDAE
SYN.INSPINOSA GUENEE
SYN.CONSUETA WALKER

LIFE HISTORY

AMER.MUST.NAT.HIST.BULL.14:
281,1901 BEUTEN.LARVA
ENT.SOC.AMER.ANNALS.23:396,
1930 HAWKINS LARVAL CLAWS
JOURN.ECON.ENT.8:394,1915
CRUMB KEY TO AGROTIS LARVAE
NEW YORK STATE MUS.BULL.6:14
1888 LINTNER LARVA*
UNITED STATES DEPT.AGRI.BUR.
ENT.BULL.57,NEW SERIES P.
10,1906 SANDERSON EGG* F.E;
LARVA* FF.2A,B,C,D
UNITED STATES DEPT.AGRI.
FARMER'S BULL.47:25,1897
HOWARD EGG*,LARVA*F.13
UNITED STATES DEPT.AGRI.REPT.
OF ENT.P.292,1884 RILEY
LIFE HISTORY*PL.1
UNITED STATES DEPT.AGRI.TECH.
BULL.88:83,1929 CRUMB EGG
LARVA*PL.6 F,G.L;PUPA
CROSBY & LEONARD-MANUAL VEG.
GARDEN INS.P.272,1918
LARVA,PUPA
INDEX AMER.ECON.ENT.
VOL.1:134 FELTIA;
VOL.2:147 FELTIA;
VOL.4:187 FELTIA;'
VOL.6:17

FOOD PLANTS

ALLIUM CEPA LINN.(ONION)
BETA VULGARIS LINN.(BEET)
BRASSICA OLERACEA LINN.(CABBAGE)
BRASSICA OLERACEA V.VIRIDIS LAM.
(COLLARDS)
CAPSICUM ANNUUM LINN.
GOSSYPIUM HERBACEUM LINN.
(COTTON)
LACUTCA SATIVA LINN.(LETTUCE)
NICOTIANA TABACUM LINN.(TOBACCO)
PHASEOLUS VULGARIS LINN.(BEANS)
PISUM SATIVUM LINN.(PEAS)
PORTULACA GRANDIFLORA HOOK.
PORTULACA OLERACEA LINN.
SOLANUM LYCOPERSICUM LINN.
(TOMATO)
SOLANUM TUBEROSUM LINN.(POTATO)
SORGHUM HALEPENSE(LINN.)
STELLARIA MEDIA(LINN.)
TRAGOPOGON PORRIFOLIUS LINN.
TRIFOLIUM SP.(CLOVERS)
ZEA MAYS LINN.(CORN)
WEEDS AND LOW PLANTS

MALORA SMITH
SYN.SEE HIBISCI GUENEE

MAMMOTHI GUNDER
SSP.SEE MONTIVAGA BEHR

MAMMURARIA GUENEE
SYN. SEE SUBATOMARIA WOOD

MANATAAQUA(SCUDDER), POLITES-HESPERIIDAE
SYN. CERNES HARRIS
A RHENA EDWARDS
SYN. ALCINA SKINNER
LIFE HISTORY
BUFFALO SOC. NAT. SCI. BULL. 19:33 1939 WILD LARVA
ENT. NEWS 3:77, 1892 LARUENT LIFE HISTORY
LOC. CIT. 19:413, 1908 LAURENT EARLY STAGES INCLUDING EGG
SOUTHERN CALIF. ACAD. SCI. BULL. 40:109, 1941 DETHIER LARVA* PLS. 8, 9
LOC. CIT. 41;41, 1942 DETHIER COMPARES LARVAE OF MANATAAQUA AND THEMISTOCKES LATREILLE
SCUDDER-BUTTER. EAST. UNITED STATES & CANADA P. 1722, 1889 EGG*LARVA*PL. 66, F. 27; PL. 69, F. 14; PL. 73, F. 3; PL. 80, FF. 63, 64; PL. 86, F. 52
FOOD PLANTS
GRASSES

MANDAN EDWARDS
SYN. SEE PALAEMON PALLAS

MANIFESTA, MORRISON, AGROTIS-NOCTUIDAE
SYN. IMPINGENS DYAR
LIFE HISTORY
SMITH-INS. NEW JERSEY P. 408, 1900 LARVA(VERY BRIEF)
FOOD PLANTS
UNDER LEAVES IN SWAMP

MANITOBA MCDUNNOUGH
SSP. SEE HERILIATA GUENEE

MANITOBA SCUDDER
SSP. SEE COMMA LINNAEUS

MANITOBA SMITH
SSP. SEE FURCIFERA GUENEE

MANITOBAE CLARK
SSP. SEE JUGLANDIS ABBOT & SMITH

MANITOBAE STRAND
SSP. SEE PLEONECTUSA GROTE

MANITOBAENSIS BRYK
SYN. SEE SMINTHEUS DOUBLEDAY & HEWITSON

MANITOBENSIS CASSINO
SSP. SEE BLANDULA HULST

MANITOBENSIS MCDUNNOUGH, ANISOTA CITHERONIIDAE
LIFE HISTORY
CAN. ENT. 61:98, 1929 BRODIE EGG, LARVA*PUPA*
FOOD PLANTS
QUERCUS SP. (OAKS)

MANITOBENSIS MCDUNNOUGH
SSP. SEE MARTHESIA CRAMER

MANITOBENSIS SWETT
SSP. SEE SOSPETA DRURY

MANITOBOIDES FLETCHER
SSP. SEE SASSACUS HARRIS

MANITOU NEUMOEGEN & DYAR
SSP. SEE STRAGULA GROTE

MANTEO(DOUBLEDAY), HETEROCAMPA-NOTODONTIDAE
SYN. CINERASCENS WALKER
SYN. AUBALBICANS GROTE
LIFE HISTORY
BOSTONSOC. NAT. HIS. TPROC. 24: 545, 1890 PACKARD LARVA
ENT. NEWS 4:262, 1893 DYAR LARVA (BRIEF)COMPARES LARVA MANTEO AND BILINEATA PACKARD
MINNESOTA AGRI. EXP. STA. BULL. 61:160, 1899 LUGGER LARVA* F;116
NAT. ACAD. SCI. MEM. 7:224, 1895 PACKARD EGG, LARVA*PL. 29,

FF.2-10 COLOR;PUPA*
NEW YORK ACAD.SCI.ANNALS 8:226, 1894 DYAR LARVAL SETAE
NEW YORK STATE AGRI.EXP.STA. (CORNELL),MEM.56:1074,1922 WELLHOUSE LARVA (BRIEF)
LOC.CIT.274:231,1948 FORBES LARVA
NEW YORK STATE MUS.MEM.8:705 1905 FELT LARVA(BRIEF)
OHIO AGRI.EXP.STA.BULL.332: 247,1918 HOUSER LARVA* PL. 37,F.1
UNITED STATES DEPT.AGRI.MISC. PUB.657:363,404,1950 CRAIGHEAD LARVA
WASHINGTON ENT.SOC.PROC.2:257, 1891 PERGANDE LARVA*
DOANE ETC.-FOREST INSECTS P.295 1936 LARVA
INDEX AMER.ECON.ENT. VOL.2:175

FOOD PLANTS
BETULA SP.(BIRCHES)
BETULE LENTA LINN.
BETULA PAPYRIFERA MARSH.
CRATAEGUS SP.(THORNS)
DIOSPYRUS VIRGINIANA LINN.
FAGUS SP.(BEECHES)
JUGLANS NIGRA LINN.
JUGLAND REGIA LINN.'
MALUS PUMILA MILL.(APPLE)
PINCKNEYA PUBENS MICHX.
QUERCUS SP.(OAKS)
QUERCUS ALBA LINN.
QUERCUS IMBRICARIA MICHX.
QUERCUS MACROCARPA MICHX.
QUERCUS STELLATA WANG.
TILIA SP.(LINDENS)
ULMUS SP.(ELMS)

MANTO(STRECKER),OLENE-LIPARIDAE
A INTERPOSITA DYAR
B MONTANA BEUTENMULLER

LIFE HISTORY
AMER.MUS.NAT.HIST.BULL.19: 585,1903 BEUTEN.LARVA (MONTANA)
NEW YORK STATE AGRI.EXP.STA. (CORNELL),MEM.274:244 1948 FORBES LARVA
PSYCHE 18:158,1911 BARNES & MCDUNNOUGH LARVA
WASHINGTON ENT.SOC.PROC.13:19, 1911 DYAR LARVA(BRIEF)

FOOD PLANTS
ABIES FRASERI(PURSH.)
PICEA MARIANA(MILL.)
PINUS SP.(PINES)

MANZANITA TAYLOR,HYDRIOMENA-GEOMETRIDAE

LIFE HISTORY

FOOD PLANTS
ARBUTUS MENZIESII PURSH.

MARCATA NEUMOEGEN
SSP.SEE HERA HARRIS

MARCELLINA CRAMER
SSP.SEE SENNAE LINNAEUS

MARCELLUS CRAMER,PAPILIO-PAPILIONIDAE
SYN.AJAX LINNAEUS(PARTIM)
SYN.CAROLINIANUS G.EDWARDS
SYN.WALSHII EDWARDS
AB ABBOTTII EDWARDS
FORM FLORIDENSIS HOLLAND
FORM TELAMONIDES FELDER & FELDER
FORM AEST.LECONTEI ROTHSCHILD & JORDAN
AB BROWERI GUNDER
AB PRICEI FIELD
AB NIGROSUFFUSA FIELD

LIFE HISTORY
AMER.MUS.NAT.HIST.BULL.5:241 1893 BEUTEN.LARVA, CHRYSALIS
AMER.NATURALIST 8:257,1874 LIFE HISTORY SCUDDER
BUFFALO SOC.NAT.SCI.BULL.19: 19,1939 WILD LARVA
CAN.ENT.2:115,133,162,1870 EDWARDS REARING
LOC.CIT.3:70,1871 EDWARDS REARING
ENT.NEWS 17:175,1906 ROWLEY EGG

Loc. cit. 18:306, 1907 Rowley habits
Maryland Acad. Sci. Bull. 3(2):8 1923 Moore egg, larva, chrysalis
Maryland Nat. Hist. Sco. Proc. pp. 4, 12, 1933 Haydon egg, larva, chrysalis
New York State Mus. Mem. 8:735 1905 Felt larva(brief)
Papilio 4:88, 1884 Gruber larva*
Psyche 8:208, 1898 Scudder life history* Pl. 5
Smithsonian Misc. Collec. 4, Art 2:9, 1862 Morris larva, chrysalis
Abbot & Smith-Lep. Ins. Georgia 1:7, 1797 larva* chrysalis Pl. 4 COLOR
Boisduval-Spec. Gen. Lep. Dirun. 1:257, 259, 1836 larva
Boisduval & Leconte-Histlep North Amer. p. 9, 1883 larva* chrysalis* Pl. 1, 2 COLOR
Comstock-How to Know Butter. p. 50, 1904 larva
Edwards-Butter. North Amer. 1st series 1871 life hist* Pls. Papilio 1, 2, 3 COLOR
Edwards-Loc. cit. 3rd series 1897 Papilio 5 breeding exper.
French-Butter. East. United States p. 85, 1886 life hist.
Glover-Illust. North Amer. Ent. 1878 larva, pupa Pl. 2, f. 1
Gosse-Letters from Alabama p. 51, 1859 larva* chrysalis*
Jaeger-Life North Amer. Ins. p. 209, 1864 larva
Kellogg-Amer. Insects p. 449, 1908 larva
Scudder-Butter. East. United States & Canda pp. 1269-1278 1889 life hist* Pl. 66, f. 5, Pl. 73, f. 12; Pl. 76, f. 14; Pl. 80, ff. 13-16, Pl. 85, ff. 11, 12
Scudder-Guide to Common Butter p. 147, 1893 larva, chrysalis
Vance-Life Amont the Butter. p. 51, 1925 egg* larva* chrysalis*
Index Amer. Econ. Ent. Vol. 3:295; Vol. 7:741

FOOD PLANTS
Asimina angustifolia Gray
Asimina parviflora(Michx.)
Asimina incana(Bartr.)
Asimina triloba(Linn.)
Lindera benzoin Linn.
Persea palustris(Raf.)
Vaccinium sp.

<u>marcia</u> Edwards
ssp. see <u>tharos</u> Druty

<u>margana</u>(Fabricius), <u>Heliocontia</u>-Noctuidae
syn. <u>subapicana</u> Walker
syn. <u>male inorata</u> Grote
syn. <u>female sordida</u> Grote
LIFE HISTORY
FOOD PLANTS
Sphaeralcea incana Torr.

<u>margarita</u> Draudt
syn. see <u>brettus</u> Boisduval & Leconte

<u>margarita</u> Hubner
syn. see <u>ilaire</u> Godart

<u>margaritae</u> dos Passos
ssp. see <u>henrici</u> Grote & Robinson

<u>margaritosa</u>(Haworth), <u>Peridroma</u>-Noctuidae
syn. <u>inermis</u> Harris
form <u>saucia</u> Hubner
LIFE HISTORY
Amer. Ent. 1:188, 1869 egg*
Amer. Mus. Nat. Hist. Bull. 14: 272, 1901 Beuten. larva
Brooklyn Ent. Soc. Bull. 22:48 1927 Crumb larva* Pl. V, A. H
Loc. cit. 27:85, 1932 Crumb larva
California Agri. Exp. Sta. Bull. 660:31, 1942 Smith egg* f. 24b;

LARVA*F.24C
ENT.SOC.AMER.ANNALS 23:396, 1930 HAWKINS LARVAL CLAWS
ILLINOIS AGRI.EXP.STA.BULL.95: 353,1904 FORBES EGG*LARVA*
ILLINOIS DEPT.AGRI.TRANS.15: 94,211,1877 FRENCH&THOMAS LARVA
LOC.CIT.18,APPEND:134,1880 MARTEN LARVA
JOURN.ECON.ENT.8:393,1915 CRUMB LARVAL HEAD* PL.20,F.2 ALSO KEY TO LARVAE
LOC.CIT.14:273,1921 WADLEY EGG, LARVA,PUPA
MAINE AGRI.EXP.STA.BULL.356: 166,1930 PHIPPS EGG,LARVA, PUPA
MINNESOTA,AGRI.EXP.STA.BULL. 61:214,1899 LUGGER EGG*LARVA* BOTH F.158
MISSOURI,REPT.NOX.INS.OF 1:72, 1869 RILEY LIFE HISTORY*
MUS.COMP.ZOOL.BULL.87,NO.6, 1941 DETHIER LARVAL ANTENNA* F.107
NEBRASKA AGRI.EXP.STA.RESEARCH BULL.81:21,1935 WHELAN LARVA
NEW YORK STATE AGRI.EXP.STA. (CORNELL),BULL.104:580,1895 SLINGERLAND EGG*LARVA*BOTH PL.5
NEW YORK STATE AGRI.SCO.REPT. 44:59,1885 LINTNER EGG*
NEW YORK STATE CABIN.NAT.HIST. ANN.REPT.NO.42:201,1889 LINTNER EGG* F.16;LARVA* F.17
NEW YORK STATE MUS.BULL.6:8 16,1888 LINTNER EGG*LARVA*
LOC.CIT.37:29,1900 FELT LARVA*
NEW YORK STARE MUS.MEM.8:704 1905 FELT LARVA(BRIEF)
ONTARIO ENT.SOC.ANN.REPT.P.41, 1879 BOWLES LARVA (INERMIS)
LOC.CIT.P.100,1896 FYLES LARVA
LOC.CIT.PP.68,69,1900 WEBSTER EGG*PUPA
LOC.CIT.P.99,1907 BETHUNE EGG* LARVA*
LOC.CIT.P.120,1908 GIBSON EGG* LARVA
OREGON AGRI.EXP.STA.CIRC.111: 4,1935 THOMPSON EGG*F.3 (GOOD)
OREGON BIENN.CROP PEST.REPT. 11:144,1913 LARVA*PL.11
UNITED STATES DEPT.AGRI.BURN. ENT.BULL.27:51,1901 NEW SERIES CHITTENDEN EGG* PL. 17 E,F;LARVA*PL.17B,C,D; PUPA
LOC.CIT.29:47,1901 NEW SERIES CHITENDED EGG*F.9 E,F;F.10; LARVA*F.9 B,C,D;F.11;PUPA
LOC.CIT.43:30,1903 NEW SERIES CHITTENDEN EGG* F.25 E,F; LARVA*F.25 B,C,D
LOC.CIT.57:11,1906 NEW SERIES SANDERSON EGG*F.3 E,F; LARVA*F.3 B,C,D
UNITED STATES DEPT.AGRI.DIV. ENT.CIRC.123:2,1910 MORGAN EGG* PL.2 E,F;LARVA*PL. S, B,C,D NO DESCR.
UNITED STATES DEPT.AGRI.FARMER'S BULL.120:22,1900 HOWARD EGG* F.16 E,F;LARVA* F.16 B,C,D
LOC.CIT.890:5,1917 PIERCE EGG* F.1 E,F;LARVA*F.1 B,C,D
LOC.TIC.1306:17,1923 WEIGEL EGG* F.12 E,F;LARVA* F.13 B,C,D NO DESCR.
LOC.CIT.1362,20:1922 WEIGEL & SASSCER EGG* F.21 E,F; LARVA* F.21 B,C,D
LOC.CIT.1494:2,1926 CRUMB EGG* F.1;LARVA*F.3B,C;PUPA*F.2
UNITED STATES DEPT.AGRI.REPT ENT.P.297,1884 RILEY LIFE HIST* PL.3,FF.1,2
UNITED STATES DEPT.AGRI.TECH. BULL.88:110,1929 CRUMB EGG*PL.5A;LARVA* PL.7N,H; PUPA*PL.18 A,B
UNITED STATES DEPT.AGRI.YEAR-

book for 1898 p.140 Howard
egg* f.22 e,f;larva*f.22
b,c,d no descr.
Loc.cit. for 1910 p.284 Morgan
egg*f.7 e,f;larva*f.7 b,c,d
no descr.
Chittenden-Ins.Inj.Veg.p.54
1907 egg*and larva*f.29
Crosby&Leonard-Manual veg.
Garden Ins.p.277,1918 egg,
larva,pupa
Essig-Inj.Benef.Ins.p.400,1915
egg*f.401;larva*f.402
Essig-Ins.West.North Amer.p.
683,1926 egg* f.557;larva*
f.558
Fernald-Kingsley'sStand.Nat.
Hist.2:452,1885 egg,larva
Hart-Ins.Collections Illinois
High Schools p.29,1903 larva
Holland-Moth Book p.5,1905 egg*
f.2
Kirby-European Butter.Moths
p.194,1882 larva
Metcalf & Flint-Destr.&Usefu.
Ins.p.349,1939 larva
Peairs-Pests Farm,Gard.Orchard
p.112,1941 egg*larva*both
f.121
Saunders-Ins.Inj.Fruits p.106,
1889 egg*f.100;larva*f.101;
pupa*f.102
Stainton-British Butter.&Moths
1:224,1857 larva
Index Amer.Econ.Ent.
Vol.1:237:Vol.2:217 Lycophota
Vol.3:243 Lycophota
.Vol.3:304 Peridroma
Vol.4:282 Lycophota
Vol.4:353;Vol5:393 Lycophota
Vol.6:575;Vol.7:759

FOOD PLANTS

Abies balsamea(Linn.)(seedlings)
Abies grandis(Dougl.)
Acer sp.(maples)
Acer negundo Linn.
Acer saccharinum Linn.
Allium cepa Linn.(onion)
Alnus rubra Bong.
Althaea rosea Cav.
Ambrosia artemisiifolia Linn.
Amelanchier florida Lindl.
Anthemis cotula Linn.
Asparagus medeoloides Thunb.
Asparagus officinalis Linn.
(asparagus)
Asparagus plumosa Baker
Beta vulgaris Linn.(beet)
Brassica hirta Moench(mustard)
Brassica nigra(Linn.)(mustard)
Brassica oleracea Linn.(cabbage)
Brassica rapa Linn.(turnip)
1 Capsella bursa-pastoris(Linn.)
Capsicum annumLinn.(peppers)
Chamecyparis thyoides(Linn.)
(seedlings)
Chrysanthemum sp.
Cirsium sp.
Cirsium vulgare (Savi)
Citrus limon(Linn.)(Lemon)
Citrus sinensis(Linn.)(orange)
Conyza canadensis(Linn.)
Cucumis sativus Linn.(cucumber)
Datura stramonium Linn.
Daucus carota v.sativus Hoff.
(carrot)
Dianthus caryophyllus Linn.
Epilobium angustifolium(Linn.)
Eupatorium sp.
Fragaria chiloensis(Linn.)
Geranium sp.
Gleditsia triacanthos Linn.
Gossypium herbaceum Linn.
(cotton)
Helianthus annuus Linn.
Humulus lupulus Linn.(hops)
Lactuca sativa Linn.(lettuce)
Lathyrus odoratus Linn.
Maclura pomifera(Raf.)
Malus pumila Mill.(apple)
Medicago sativa Linn.
Melilotus alba desr.
Morus sp.(mulberry)
Nasturtium sp.
Nicotiana tabacum Linn.(tobacco)
Parthenium argentatum Gray

PERSEA AMERICANA MILL. (AVACADO)
PHLEUM PRATENSE LINN.
PISUM SATIVUM LINN. (PEAS)
PLANTAGO SP.
PLANTAGO LANCEOLATA LINN.
POLYGONUM AVICULARE LINN.
PORTULACA OLERACEA LINN.
PRUNUS AMERICANA MARSH.
PRUNUS ARMENIACA LINN. (APRICOT)
PRUNUS CERASUS LINN. (CHERRY)
PRUNUS DOMESTICA LINN. (PLUM)
PRUNUS DOMESTICA V. GALATENSIS EX. HOOK(PRUNE)
PRUNUS EMARGINATA DOUGL.
PRUNUS PERSICA(LINN.)(PEACH)
PTERIDIUM LATIUSCULUM(DESV.)
QUERCUS SP. (OAKS)
QUERCUS ALBA LINN.
RAPHANUS SATIVUS LINN. (RADISH)
RHEUM RHAPONTICUM LINN. (RHUBARB)
RHUS SP. (SUMACS)
RHUS COPALLINA LINN.
RIBES LACUSTRE(PERS.)GOOSEBERRY
RIBES SANGUINEUM PURSH
RIBES SATIVUM SYME(CURRANT)
ROSA SP. (ROSES)
RUBUS ALLEGHENIENSIS PORTER
RUBUS OCCIDENTALIS LINN.
RUBUS IDAEUS V. STRIGOSUS(MICHX.)
RUMEX CRISPUS LINN.
SALIX SP. (WILLOWS)
SALIX HOOKERIANA BARR.
SALIX XCOULERIANA BARR.
SALVIA OFFICINALIS LINN.
SMILAX ROTUNDIFOLIA LINN.
SOLANUM LYCOPERSICUM LINN. (TOMATO)
SOLANUM TUBEROSUM LINN. (POTATO)
SOLIDAGO LEAVENWORTHII T.&G.
STELLARIA MEDIA(LINN.)
TRIFOLIUM SP. (CLOVERS)
TRITICUM AESTIVUM LINN. (WHEAT)
TROPAELUM MAJUS LINN.
TSUGA CANADENSIS(LINN.)(SEEDLINGS)
URTICA SP.
VACCINIUM SP.
VACCINIUM ANGUSTIFOLIUM AIT.
VACCINIUM CORYMBOSUM LINN.
VACCINIUM MYRTILLOIDES MICHX.
VIOLA SP.
VIOLA TRICOLOR LINN.
VITIS SP. (GRAPES)
VITIS VINIFERA LINN.
XANTHIUM STRUMARIUM LINN.
ZEA MAYS LINN. (CORN)
FERNS, GENERAL FEEDER ON FIELD CROPS, FOREST TREES, FRUIT TREES; GRASS, LOW PLANTS, MOST ANYTHING, SHRUBS, VEGETABLES

MARGHETITA BEUTENMULLER
SSP. SEE STRETCHI BEHR

MARGINALIS CLARK
SSP. SEE DRUPIFERARUM ABBOT & SMITH

MARGINALIS GROTE
SYN. SEE DIFFINIS BOISDUVAL

MARGINALIS SCUDDER
SSP. SEE NAPI LINNAEUS

MARGINALIS SKINNER
SSP. SEE PALMERII EDWARDS

MARGINATA EDWARDS
SSP. SEE PSEUDARGIOLUS BOISDUVAL & LECONTE

MARGINATA(HAWORTH), SCHINIA-NOCTUIDAE
SYN. RIVULOSA GUENEE
SYN. CONTRACTA WALKER
SYN. DESIGNATA WALKER
AB CONSTRUCTA HENRY EDWARDS
LIFE HISTORY
NEW YORK ENT. SOC. JOURN. 10: 175, 1902 SEIFERT LARVA
WASHINGTON ENT. SOC. PROC. 5: 291, 1903 DYAR LARVA
FOOD PLANTS
AMBROSIA ARTEMISIIFOLIA LINN. (RIVULOSA)

MARGINATA WALKER
SSP. SEE ROSEA WALKER(ORETA)

MARGINATUS HARRIS
SYN. SEE NUMITOR FABRICIUS

MARGINEAPERTUS GUNDER
SSP. SEE VANILLAE LINNAEUS

MARGINIDENS (GUENEE), PAPAIPEMA-NOCTUIDAE
SYN. BIRDI DYAR
LIFE HISTORY
AMER. ENT. 1:226, 1869 LARVA
CAN. ENT. 32:229, 1900 BIRD LARVA, PUPA
FOOD PLANTS
ARCTIUM LAPPA LINN.
CICUTA MACULATA LINN.'
COSMOS BIPINNATUS CAV.
ROSA SP. (ROSES)
RUMEX SP.

MARIAE BARNES & BENJAMIN
SSP. SEE POLINGI SKINNER

MARIAE BENJAMIN
SSP. SEE LUNA LINNAEUS

MARIAE BRYK
SYN. SEE SMINTHEUS DOUBLEDAY & HEWITSON

MARIAE GROTE
SYN. SEE INCLUDENS WALKER

MARIANNE SCUDDER
SYN. SEE CLAUDIA CRAMER

MARIANA BARNES
SSP. SEE CHALCEDONIA DOUBLEDAY & HEWITSON

MARIANA STRECKER
SYN. SEE CALIFORNICA EDWARDS (CATOCALA)

MARICOPA REAKIRT
SYN. SEE ICARIOIDES BOISDUVAL

MARINA PACKARD
SYN. SEE UNICOLOR PACKARD

MARINA (REAKIRT), LEPTOTES-LYCAENIDAE
FORM FEMALE REAKIRTI FIELD
AB VIOLACEA GUNDER
AB BURDICKI HENNE
LIFE HISTORY
PSYCHE 18:32, 1911 COOLIDGE EGG
COMSTOCK, J. A.-BUTTERFLIES OF CAL. P. 177, 1927 EGG, LARVA, CHRYSALIS* F. A 58
FOOD PLANTS
ASTRAGALUS SP.
DOLICHOS SP.
LATHYRUS ODORATUS LINN.
MEDICAGO SATIVA LINN.
PLUMBAGO SP.
WISTERIA SINENSIS SWETT

MARINARIA STRECKER
SYN. SEE VIRIDIPENNATA HULST

MARIPOSA GROTE & ROBINSON, ALYPIA-AGARISTIDAE
FORM LUNATA STRETCH
LIFE HISTORY
CAN. ENT. 26:348, 1894 LEMBERT EGG, LARVA, PUPA
LOC. CIT. 27:31, 1895 DYAR EGG, LARVA
FOOD PLANTS
CLARKIA UNGUICULATA LINDL.
CLARKIS WILLIAMSONII (DUR. & HILG.)

MARIPOSA (REAKIRT), LYCAENA-LYCAENIDAE
SYN. ZEROE BOISDUVAL
A CHARLOTTENSIS HOLLAND
B PENROSEAE FIELD
LIFE HISTORY
CAN. ENT. 42:316, 1910 COOLIDGE EGG
LOC. CIT. 43:83, 1911 NEWCOMER EGG, LARVA, CHRYSALIS
PSYCHE 16:31, 1909 COOLIDGE EGG
FOOD PLANTS
POLYGONUM AVICULARE LINN.
POLYGONUM DOUGLASII GREENE
RUMEX SP. (IN CAPTIVITY)
VACCINIUM SP.

MARITIMA BIRD, PAPAIPEMA-NOCTUIDAE
LIFE HISTORY
CAN. ENT. 41:59, 1909 BIRD LARVA
LOC. CIT. 47:115, 1915 BIRD LARVA F.9
FOOD PLANTS
HELIANTHUS DIVARICATUS LINN.
HELIANTHUS GIGANTEUS LINN.
HELIANTHUS TUBEROSUS LINN.

MARITIMA EDWARDS
SSP. SEE ALOPE FABRICIUS

MARIUS CARMER
SYN. SEE CHIRON FABRICIUS

MARMORATA W.H. EDWARDS, CATOCALA-NOCTUIDAE
LIFE HISTORY
FOOD PLANTS
POPULUS SP. (?)
SALIS SP. (?)

MARMORATUS GROSSBECK, PERO-GEOMETRIDAE
SYN. BARNESI CASSINO & SWETT
LIFE HISTORY
FOOD PLANTS
ACER SP. (MAPLES)

MARS FABRICIUS
SYN. SEE ACIS DRURY

MARSYAS EDWARDS
SSP. SEE SATYRUS EDWARDS

MARTHESIS (CRAMER), FENTONIA-NOTODONTIDAE
SYN. TESELLA PACKARD
SYN. TURBIDA WALKER
SYN. FEMALE ELONGATA GROTE & ROBINSON
AB NIGRA CHERMOCK
A MANITOBENSIS MCDUNNOUGH
LIFE HISTORY
AMER. MUS. NAT. HIST. BULL. 10: 428, 1898 BEUTEN. LARVA
AMER. NAT. 18:1044, 1884 PACKARD LARVA, COCOON
LOC. CIT. 19:715, 1885 PACKARD NOTE ON PUPA
BOSTON SOC. NAT. HIST. PROC. 24: 552, 1890 PACKARD LARVA
LOC. CIT. 26:157, 1893 DYAR EGG, LARVA, COCOON
NAT. ACAD. SCI. MEM. 7:259, 1895 PACKARD EGG, LARVA*PL. 24, FF. 1-5; PL. 35, F. 4 COLOR: PUPA* PL. 34
LOC. CIT. 9, 1905 PACKARD LARVA* PL. 44, F. 12 HEAD
NEW YORK STATE AGRI. EXP. STA. (CORNELL), MEM. 274;235, 1948 FORBES LARVA
NEW YORK STATE MUS. MEM. 8:706 1905 FELT LARVA (BRIEF)
OTTOWA NAT. 16:161, 1902 GIBSON LARVA (BRIEF)
UNITED STATES DEPT. AGRI. MISC. PUBL. 657:360, 405, 1950 CRAIGHEAD LARVA
FOOD PLANTS
ACER SP. (MAPLES)
CASTANEA DENTATA (MARSH.)
FAGUS GRANDIFOLIA EHRH.
PLATANUS OCCIDENTALIS LINN.
POPULUS SP
QUERCUS SP.

MARTIALIS (HERRICH-SCHAEFFER), STRYMON-LYCAENIDAE
LIFE HISTORY
ENT. NEWS 12:180, 1901 SLOSSON LIFE HISTORY
FOOD PLANTS
TREMA MICRANTHA (LINN.)

MARTIALIS (SCUDDER), ERYNNIS-HESPERIIDAE
SYN. QUERCUS BUTERL
AB AUSONIUS LINTNER
LIFE HISTORY
AMER. MUS. NAT. HIST. BULL. 5:301 1893 BEUTEN. LARVA CHRYSALIS
BUFFALO SOC. NAT. SCI. BULL. 19: 32, 1939 WILD LARVA
HOLLAND-BUTTERFLY BOOK 1ST EDIT 1905 CHRYSALIS*PL. 67, F. 37
SCUDDER-BUTTER. EAST. UNITED

STATES & CANADA PP. 1496-;1948,1889 LIFE HISTORY*PL. 66,F.16;PL.69,F.9;PL.77, F.13;PL.85,F.37

FOOD PLANTS

ADENOSTOMS FASCICULATUM H.&A.
AMARANTHUS SP.
AMARANTHUS RETROFLEXUS LINN.
CEANOTHUS AMERICANUS LINN.
INDIGOFERA CAROLINA WALT.
LACHNANTHES TRINCTORIA(WALT.)
TREMA MICRANTHA(LINN.)

MARTINEAE COLEMAN
SSP.SEE PROGNE CRAMER

MARTINI GUNDER
SSP.SEE HARFORDII HENRY EDWARDS

MARYX(GUENEE),SIDERIDIS-NOCTUIDAE
SYN.RUBEFACTA MORRISON
SYN.VINDEMIALIS GROTE
LIFE HISTORY
WASHINGTON ENT.SOC.PROC.4:319 1898 DYAR LARVA(RUBEFACTA)
FOOD PLANTS

MASCULINUS SMITH
SYN.SEE LAETIFICANS SMITH

MASSACHUSETTENSIS GUNDER
SSP.SEE VIRGINIENSIS DRURY(VANESSA)

MASSASOIT(SCUDDER),POANES-HESPERIIDAE
FORM SUFFUSA LAURENT
A HUGHI CLARK
LIFE HISTORY
ENT.NEWS 19:409,1908 LARRENT EGG,EARLY STAGES
FOOD PLANTS
GRASSES

MASUMBROSUS GUNDER
SSP.SEE EURYDICE BOISDUVAL

MATA REAKIRT
SSP.SEE MYLITTA EDWARDS

MATILDA(DYAR),ITAME-GEOMETRIDAE
LIFE HISTORY
PSYCHE 13:22,1906 DYAR EGG, LARVA
FOOD PLANTS
POPULUS TREMULOIDES MICHX.

MATUTINA HUBNER
SSP.SEE GAURAE ABBOT & SMITH (RHODOPHORA)

MAXIMA NEUMOEGEN
SSP.SEE AGARITHE BOISDUVAL

MAXIMUS BRYK & EISNER
SSP.SEE SMINTHEUS DOUBLEDAY & HEWITSON

MAXWELLI FRANCK
SSP.SEE CRESPHONTES CRAMER

MAYAE GUNDER
SSP.SEE APHRODITE FABRICIUS

MAYI CHERMOCK & CHERMOCK
SSP.SEE AUSONIDES BOISDUVAL

MAYI CHERMOCK & CHERMOCK
SSP.SEE CHRISTINA EDWARDS

MAYI GUNDER
SSP.SEE ANICIA DOUBLEDAY&HEWITSON

MAYNARDI FIELD
SSP.SEE LORQUINI BOISDUVAL (BASILARCHIA)

MAYORIAE CHERMOCK
SSP.SEE CHRISTINA EDWARDS

MCCREARYI CLARK
SSP.SEE MYOPS ABBOT & SMITH

MCCREARYI CLARK
SSP.SEE EREMITUS HUBNER

MCDUNNOUGHI CLARK
SSP.SEE DIFFINIS BOISDUVAL

MCGLASHANII RIVERS
SSP. SEE CHALCEDONIA DOUBLEDAY & HEWITSON

MCISAACI DOSPASSOS
SSP. SEE INORNATA EDWARDS (COENONYMPHA)

MEADI FIELD
SSP. SEE EDITHA MEAD

MEADI (PACKARD), CORYPHISTA-GEOMETRIDAE
FORM BADIARIA HENRY EDWARDS
SYN. OPTIMATA STRECKER
LIFE HISTORY
CORNELL UNIV. THESIS 1947 TUNG SUN EGG*LARVA*PUPA*
JOURN. ECON. ENT. 34:387, 1941 NEISWANDER EGG, LARVA*F. 3; PUPA*F. 4; COCOON*F. 4
NEW YORK STATE AGRI. EXP. STA. (CORNELL), MEM. 274:144, 1948 FORBES LARVA
PSYCHE 9:396, 1902 DYAR EGG, LARVA
FOOD PLANTS
BERBERIS AQUIFOLIUM PURSH
BERBERIS THUNBERGII DC.
BERBERIS VULGARIS LINN.

MEADII EDWARDS, COLIAS-PIERIDAE
A ELIS STRECKER
FORM MEDI GUNDER
LIFE HISTORY
CAN. ENT. 21:41, 1889 EDWARDS EGG, LARVA, CHRYSALIS
FOOD PLANTS
ASTRAGALUS SP.
ASTRAGALUS ALPINUS LINN.
TRIFOLIUM SP. (CLOVERS)

MEADII (EDWARDS), MINOIS-SATYRIDAE
LIFE HISTORY
EDWARDS-BUTTER. NORTH AMER. 3RD SERIES 1897 LIFE HISTORH* (SATYRUS 2 FF. 1-D) AND SUPPL.
HOLLAND-BUTTERFLY BOOK PP. 216-217, 1905 LIFE HIST. 3RD EDIT. P. 190, 1931 LIFE HISTORY
MEAD-REPT. LEP. WHEELER'S EXPED. WEST 100TH MERID. P. 774, 1875 EGG
FOOD PLANTS
GRASSES

MEADOWSI COMSTOCK
SSP. SEE PICTA PACKARD

MEDI GUNDER
SSP. SEE MEADII EDWARDS (COLIAS)

MEDIA MORRISON
SSP. SEE GRAPHICA HUBNER

MEDIA MORRISON
SYN. SEE PITYOCHROMUS GROTE

MEDIALIS SMITH, BLEPTINA-NOCTUIDAE
LIFE HISTORY
ENT. AMER. 14:172, 1934 CRUMB LARVA
FOOD PLANTS
DEAD LEAVES

MEDIALIS SMITH, HYDROECIA-NOCTUIDAE
LIFE HISTORY
ONTARIO ENT. SOC. ANN. REPT. P. 51, 1908 GIBSON LARVA
FOOD PLANTS
GLYCERIA MAXIMA (HARTM.)
SOLANUM TUBEROSUM LINN. (POTATO)

MEDIONIGROOCELLATA BRYK
SYN. SEE CLODIUS MENETRIES

MEDIORUBROOCELLATA BRYK & EISNER
SYN. CLODIUS MENETRIES

MEDIOSUFFUSA STRAND
SSP. SEE LAUDABILIS GUENEE

MEDITATA (GROTE), LACINIPOLIA-NOCTUIDAE
A COLUMBIA SMITH
LIFE HISTORY
BROOKLYN ENT. SOC. BULL. 27:92 CRUMB LARVA

ENT.SOC.AMER.ANNALS 23:396, 1930 HAWKINS LARVAL CLAWS* F.5

JOURN.ECON.ENT.8:393,1915 CRUMB LARVAL KEY

UNITED STATES DEPT.AGRI.TECH. BULL.88:120,1929 CRUMB EGG, LARVA* PL.8 B,I;PUPA*F.17D

WASHINGTON ENT.SOC.PROC.4:318, 1898 DYAR LARVA

HART-INS.COLL.ILLINOIS HIGH SCHOOLS P.33,1903 LARVA*(F.44

INDEX AMER.ECON.ENT. VOL.4:387 POLIA

FOOD PLANTS

ANDROPOGON VIRGINICUS LINN.
ASTER ERICOIDES LINN.
MALUS PUMILA MILL.(APPLE)
NICOTIANA TABACUM LINN.(TOBACCO)
POA PRATENSIS LINN.
SOLIDAGO SP.
TRIFOLIUM SP.(CLOVERS)
VITIS SP.(GRAPES)
VITIS VINIFERA LINN.(GRAPE)
GENERAL FEEDER ON LOW PLANTS
GRASS(YOUNG LARVAE)

MEDOR STOLL
SYN.SEE ANTAEUS DRURY

MEDUSA STRECKER
SSP.SEE GARGAMELLE STRECKER

MEGADIA SMITH
SSP.SEE INSUETA GUENEE

MEGALO MCDUNNOUGH
SSP.SEE AQUILO BOISDUVAL

MEGARENA SMITH
SSP.SEE OLIVACEA MORRISON

MEGLASHANII RIVERS
SSP.SEE CHALCEDONIA DOUBLEDAY & HEWITSON

MEINERSI GUNDER
SSP.SEE MELINUS HUBNER

MEJICANUS BEHR
SSP.SEE MORMO FELDER & FELDER

MELANE(EDWARDS),POANES-HESPERIIDAE

LIFE HISTORY

SOUTHERN CAL.ACAD.SCI.BULL. 30:20,1931 COMSTOCK-DAMMERS LIFE HIST.EGG* LARVA*CHRYSALIS*

FOOD PLANTS

BROMUS INERMIS LEYSS.
CYNDON DACTYLON(LINN.)
DIGITARIA ISCHAEMUM(SCHREB.)
DIGITARIA SANGUINALIS(LINN.)

MELANIC EDWARDS
SYN.SEE PHILODICE GODART

MELANIPPE CRAMER
SSP.SEE PHILEA LINNAEUS

MELANIS BOISDUVAL
SYN.SEE CALIFORNIAE WALKER

MELANOCARPA(SWETT),PTEROTAEA-GEOMETRIDAE
SYN.TREMULARIA BARNES&MCDUNNOUGH

LIFE HISTORY

SOUTHERN CALIF.ACAD.SCI.BULL. 36:112,1937 COMSTOCK LARVA* PL.46;PUPA

FOOD PLANTS

QUERCUS SP.(OAKS)

MELANODISCA COMSTOCK
SSP.SEE ANICIA DOUBLEDAY &HEWITSON

MELANOPA ZELLER
SYN.SEE FORMOSALIS WALKER

MELANOPHORUS BRYK
SYN.SEE SMINTHEUS DOUBLEDAY & HEWITSON

MELANOPYGA(GROTE),BELLUAR-NOCTUIDAE

LIFE HISTORY
ENT. SOC. AMER. ANNALS 16:380 1923 ROBERTSON-MILLER EGG, LARVA, PUPA
MARINE BIOL. LABOR. WOODS HOLE, BULL. 27:97, 1914 WELCH LIFE HIST. AND HABITS*PL. 1
PAPILIO 1:147, 1881 COMSTOCK LARVA
FOOD PLANTS
NUPHAR ADVENA(AIT.)
NUPHAR VARIEGATUM(ENGELM.)
NYMPHAEA SP.

MELANOTICA REIFF
SSP. SEE AMICA HUBNER

MELANYMPHA GUENEE
SYN. SEE ANTINYMPHA HUBNER

MELIMONA WRIGHT
SYN. SEE EMIGDIONIS GRINNELL

MELINUS HUBNER, STRYMON-LYCAENIDAE
SYN. HYPERICI BOISDUVAL&LECONTE
SYN. HUMULI HARRIS
FORM YOUNGI FIELD
AB MEINERSI GUNDER
A PUDICA HENRY EDWARDS
B STROFASCIATA MCDUNNOUGH
C SETONIA MCDUNNOUGH
LIFE HISTORY
AMER. MUS. NAT. HIST. BULL. 5:278 1893 BEUTEN. LARVA, CHRYSALIS
CAL. AGRI. EXP. STA. BULL. 660:26 1942 SMITH EGG, LARVA*F. 20B; CHRYSALIS*F. 20C
ENT. NEWS 11:578, 1900 COCKERELL LARVA
ILLINOIS DEPT. AGRI. TRANS. 18, APPEND:92, 1880 MIDDLETON LARVA
INSECT LIFE 7:354, 1895 HOWARD LARVA*CHRYSALIS*
MONTANA UNIV. BULL. 30:126, 1906 ELROD LARVA(BRIEF)
SMITHSONIAN MISC. COLL. 4, ART, 2:94, 1862 MORRIS LARVA, CHRYSALIS (HYPERICI)
SOUTHERN CAL. ACAD. SCI. BULL 1:41, 1902 RIVERS LARVA, CHRYSALIS
UNITED STATES DEPT. AGRI. BUR. ENT. BULL. 33:101, 1902 NEW SERIES CHITTENDEN EGG*F. 24 E; LARVA*F. 24C; CHRYSALIS* F. 23 D
LOC. CIT. 890:19, 1917 PIERCE EGG, LARVA*F. 21C; CHRYSALIS*F. 21D
LOC. CIT. 1688:16, 1932 FOLSOM LARVA*F. 17C; CHRYSALIS*F. 17D
BOISDUVAL & LECONTE-HIST. LEP. NORTH AMER. P. 90, 1833 LARVA* CHRYSALIS* PL. 28 HYPERICI
BOISDUVAL & LECONTE-LOC. CIT. P. 100, 1883 LARVA* CHRYSALIS* PL. 31 LIPAROPS
CHITTENDEN-INS. INJ. VEG. P. 113, 1907 EGG* F. 69; LARVA*F. 69; CHRYSALIS* F. 69
COMSTOCK, J. A.-BUTTER CAL. P. 157, 1927 EGG*LARVA*CHRYSALIS* A 50
COMSTOCK, J. H.-HOW KNOW BUTTER P. 223, 1904 LARVA
CROSBY & LEONARD-MANUAL VEG. GARDEN INS. P. 84, 1918 EGG, LARVA
ESSIG-INS. WEST. NORTH AMER. P. 649, 1926 LARVA, PUPA*F. 529
HARRIS-INS. INJ. VEG. 2ND EDIT. P. 235, 1852 LARVA
HARRIS-LOC. CIT. FLINT EDIT. P. 276 1862 LARVA, CHRYSALIS
HOLLAND-BUTTERFLY BOOK 1905 CHRYSALIS* PL. 5, F. 39
MAYNARD-BUTTER. NEW ENGLAND P. 35, 1886 LARVA HUMULI
MORRIS-SYNOP. LEP. NORTH AMER. PP. 96, 97, 1862 LARVA, CHRYSALIS LIPAROPS
MORRIS-LOC. CIT. P. 94, 1862 LARVA, CHRYSALIS HYPERICI
O'KANE-INJ. INSECTS P. 213, 1912 LARVA
PEAIRS-PESTS FARM, GARDEN &

ORCHARD P.183,1941 LARVA*
CHRYSALIS*F.209
SCUDDER-BUTTER. EAST. UNITED STATES & CANADA PP.852-855, 1889 LIFE HIST*PL.65,FF.5,6 PL.68,F.3;PL.75,F.21;PL.84, F.39
SCUDDER-EVERYDAY BUTTER. P.59, 1899 EGG,LARVA,CHRYSALIS
VANCE-LIFE AMONT BUTTER. P.47, 1925 EGG,LARVA,CHRYSALIS
INDEX AMER. ECON. ENT.
VOL.1:317 URANOTES
VOL.2:381 URANOTES
VOL.3:400
VOL.3:443 URANOTES
VOL.4:472;VOL.5:627;
VOL.6:745;VOL.7:979

FOOD PLANTS

AMPHICARPA BRACTEATA(LINN.)
AMPHICARPA COMOSA(LINN.)
ASTRAGALUS CROTAL RIAE(BENTH.)
ASTRAGALUS MOLLISSIMUS TORR.
CARYA ILLINOENSIS(WANG.)(NUTS)
CRATAEGUS MARSHALII EGGL.
CRATAEGUS PEDICILLATA SARG.
CYNOGLOSSUM OFFICIALE LINN.
DESMODIUM CANADENSE(LINN.)
DESMODIUM NUDIFLORUM(LINN.)
ECHINOCACTUS INTERTEXTUS ENGELM. (BORES IN FLOWERS)
ERIOBOTRYA JAPONICA(THUNB.)
GOSSYPIUM HERBACEUM LINN(COTTON)
HIBISCUS ESCULENTUS LINN. (OKRA)
HUMULUS LUPULUS LINN. (HOPS)
HYPERICUM SP.
HYPTIS EMORYI TORR.
LESPEDEZA HIRTA(LINN.)
LUPINUS SP.
MALUS PUMILA MILL. (APPLE)
MALVA PARVIFLORA LINN.
MALVA ROTUNDIFOLIA LINN.
PHASEOLUS LIMENSIS MACFAD. (LIMA BEAN)
PHASEOLUS VULGARIS LINN. (BEAN PODS)
PINUS SP. (PINES
PISUM SATIVUM LINN. (PEAS)
POLYGONUM LAPATHIFOLIUM LINN.
ROSA VALIFORNICA CHAM.&SCH.
RUBUS IDAEUS V.STRIGOSUS(MICHX.) (ATROFASCIATA)
SPHAERALCEA AMBIGUA GRAY
ZEA MAYS LINN. (CORN)

MELISSA(EDWARDS), PLEBEIUS-LYCAENIDAE
AB INYOENSIS GUNDER
A LOTIS LINTNER
B PARADOXA CHERMOCK
C?SCUDDERII EDWARDS

LIFE HISTORY

PAPILIO 4:91,1884 W.H. EDWARDS EGG,LARVA,CHRYSALIS
SOUTHERN CAL. ACAD. SCI. BULL. 27:66,1928 COMSTOCK EGG
LOC. CIT. 28:24,1929 COMSTOCK YOUNG LARVA*(WITH TUBES ATTRACTING ANTS);PUPA*

FOOD PLANTS

ASTER SP
ASTRAGALUS SP.
GLYCYRRHIZA LEPIDOTA(NUTT.)
LOTUS SP.
LUPINUS SP.
PISUM SATIVUM LINN. (PEA BLOSSOMS)
TRIFOLIUM SP. (CLOVERS)

MELISSA(FABRICIUS), OENEIS-SATYRIDAE
SYN. OENO BOISDUVAL
A SEMIDEA SAY
SYN. ERITIOSA BOISDUVAL
FORM NIGRA EDWARDS
B GIBSONI HOLLAND
C ASSIMILIS BUTLER
SYN. ARCTICA GIBSON
D SEMPLEI HOLLAND

LIFE HISTORY

AMER. MUS. NAT. HIST. BULL. 5:87 1893 BEUTEN. EGG,LARVA
EDWARDS-BUTTER. NORTH AMER. 3RD SERIES 1897 LIFE HIST* (PL. CHIONOBAS 9 FF. A-E)
EDWARDS-LOC. CIT. LIFE HIST*PL. 13 FF I-O OENO
HOLLAND-BUTTERFLY BOOK 1905 LARVA*CHRYSALIS*PL.3,FF.

1,2,4,7,15;PL.4,FF.4,5
COLOR
MAYNARD-BUTTER.NEW ENGLAND
P.2,1886 LARVA(SEMIDEA)
SCUDDER-BUTTER.EAST.UNITED
STATES & CANADA PP.137-139
141-145,1889 LIFE HIST*PL.64,
F.8;PL.70,F.5;PL.74,FF.1,
2,4,7,15;PL.78,FF.6-8;PL.
83,F.45;PL.86,F.18 SEMIDEA
SCUDDER-CHARACTER INSECT.FAUNA
WHITE MTS.PP.623,625,1863
EGG,LARVA,CHRYSALIS
SCUDDER-GEOLOGY NEW HAMPSHIRE'
1874 LARVA*CHRYSALIS*

FOOD PLANTS
CAREX CONCOLOR R.BR.(SEMIDEA)
PELTIGERA CANINA(LINN.)(SEMIDEA)
LICHENS

MELITLA MOESCHLER
SYN.SEE AUGE LINNAEUS

MELLITULA HULST
SSP.SEE MINUTA EDWARDS(CATOCALA)

MELSHEIMERI(HARRIS),CICINNUS-LACOSOMIDAE
SYN.EGENARIA WALKER

LIFE HISTORY
AMER.MUS.NAT.HIST.BULL.10:434
1898 BEUTEN.LARVA
ENT.NEWS 3:18,1892 D.S.K.LARVA
LOC.CIT.9:120 1898 MOORE LARVA
ILLINOIS DEPT.AGRI.TRANS.18,
APPEND:156 1880 COQUILLETT
LARVA,CASE
MINNESOTA AGRI.EXP.STA.BULL.
61:158,1899 LUGGER LARVA*
F.134 NO.6
NEW YORK ACAD.SCI.ANNALS 8:50
1893 PACKARD EGG,LARVA,PUPA
NEW YORK ENT.SOC.JOURN.3:175
1895 PACKARD LARVA
LOC.CIT.4:92,1896 DYAR LARVAL
HEADC CAPSULE
NEW YORK STATE MUS.MEM.8:706
1905 FELT LARVA
UNITED STATES DEPT.AGRI.MISC.
PUB.657:438,1950 CRAIGHEAD
LARVA
GLOVER-ILLUST.NORTH AMER.ENT.
1878 LARVA*PL.13,F.19
HARRIS-ENT.CORRESP.P.151,1869
LARVA,PUPA,CASE
HARRIS-INS.INJ.VEG.1ST EDIT.P.
301,1841 LARVA,PUPA,CASE
HARRIS-LOC.CIT.2ND EDIT.P.319
1852 LARVA,PUPA,CASE
HARRIS-LOC.CIT.FLINT EDIT.P.
415,1862 LARVA*PUPA*
CASE* PL.6
PACKARD-GUIDE STUDY INSECTS P.
292,1869 LARVA,CASE

FOOD PLANTS
ILEX VERTICILLATA;(LINN.)
QUERCUS SP.(OAKS)

MENAPIA(FELDER & FELDER)NEOPHASIA-
PIERIDAE
SYN.TAU SCUDDER
SYN.NINONIA BOISDUVAL
SYN.FEMALE SUFFUSA STRETCH
AB NIGRACOSTA COMSTOCK

LIFE HISTORY
BUTTERFLY PARK NAT.CLUB NEWS
2:NO PAGE NUMBERS,1931 EGG,
LARVA
CAL.ACAD.SCI.PROC.5:165,1873
HY.EDW.CHRYSALIS
JOURN.AGRI.RESEARCH 33:340,
1926 EVENDEN EGG*F.2A;
LARVA*F.2B;CHRYSALIS*F.2
C,D
MUS.COMP.ZOOL.BULL.87,NO.6,
1941 DETHIER LARVAL
ANTENNA*FF.12,40
PAPILIO 2:105,1882 STRETCH EGG,
LARVA
SOUTHERN CAL.ACAD.SCI.BULL.25
85,1926 COMSTOCK EGG*
COMSTOCK,J.A.-BUTTER CAL.P.28
1927 EGG* A7;LARVA,CHRYSALIS
DOANE ETC.-FOREST INSECTS P.330,
1936 LARVA,CHRYSALIS
EDWARDS-BUTTER.NORTH AMER.
3RD SERIES 1897 LIFE HIST*

(NEOPHASIA 1 CUT. "A")
ESSIT-INS. WEST NORTH AMER. P. 635, 1926 LARVA, CHRYSALIS
MEAD-REPT. LEP. WHEELER EXPED. P. 743, 1875 CHRYSALIS
INDEX AMER. ECON. ENT. VOL. 1:220; VOL. 2:250; VOL. 3:276; VOL. 4:324; VOL. 5:531; VOL. 6:531; VOL. 6:592; PIERIS VOL. 7:705

FOOD PLANTS
ABIES SP. (FIRS)
ABIES BALSAMEA (LINN.)
PINUS CONTORTA DOUGL.
PINUS CONTORTA V. MURRAYANA (GREV. BALF.)
PINUS ECHINATA MILL.
PINUS PONDEROSA LAWS
PSEUDOTSUGA MENZIESII (MIRB.)

MENAS HARRIS
SYN. SEE ASTARTE DOUBLEDAY

MENDICA STICHLE
SYN. SEE SMINTHEUS DOUBLEDAY & HEWITSON

MENDICA (WALKER), EUDULE-GEOMETRIDAE
SYN. BISERIATA HERRICH-SCHAEFFER
LIFE HISTORY
CAN. ENT. 3:227, 1871 SAUNDERS EGG, LARVA
LOC. CIT. 30:248, 1898 LUBBOCK COCOON (BRIEF NOTE)
ENT. AMER. 1:167, 1886 HULST LARVA PUPA
NEW YORK STATE AGRI. EXP. STA. (CORNELL), MEM. 274:259, 1948 FORBES LARVA
PHILADELPHIA ACAD. NAT. SCI. PROC. 12:543, 1860 CLEMENS EGG
PSYCHE 9:189, 1901 DYAR EGG, LARVA
STRETCH-ZYGAEN. BOMBY. NORTH AMER. P. 53, 1873 EGG, LARVA
FOOD PLANTS
ACER SP. (MAPLES)
VIOLA SP. (VIOLETS)

MENDOCINENSIS STRAND
SSP. SEE PUTNAMI GROTE

MENDOCINO (BEHRENS), CALOSATURNIA-SATURNIIDAE
LIFE HISTORY
CAL. ACAD. SCI. PROC. JAN. 1879 EDWARDS EGG, LARVA
ENT. NEWS 16:153, 1905 WILLIAMS LARVA, PUPA (BRIEF)
JOURN. ENT. & ZOOLOGY 27 (POMONA COLLEGE) P. 49, 1935 BABSON LARVA, COCOON
PACIFIC COAST ENT. SOC. PROC. P. 27, 1904 WILLIAMS LARVA
PAN. PACIFIC ENT. 21:35, 1945 TILDEN LARVA
FOOD PLANTS
ARBUTUS MENZIESII PURSH
ARCTOSTAPHYLOS GLAUCA LINDL.

MENETRIESII HENRY EDWARDS
SSP. SEE CLODIUS MENETIRES

MENIPPE HUBNER
SYN. SEE PLEXIPPUS LINNAEUS

MENSURATA WALKER
SSP. SEE ENUCLEATA GUENEE

MENTHRASTRINA MARTYN
SYN. SEE ACREA DRURY

MEPHISTARIA SWETT
SSP. SEE DEFENSARIA GUENEE

MERCEDULATA STRECKER
SYN. SEE MONICARIA GUENEE

MERCENARIA GROTE
SYN. SEE AUXILIARIS GROTE

MERIDIONALIS BARNES & MCDUNNOUGH
SSP. SEE BASIFLAVA PACKARD

MERIDIONALIS DYAR
SSP. SEE NIVEA NEUMOEGEN

MERRICKATA BIRD, PAPAIPEMA-NOCTUIDAE
LIFE HISTORY
CAN. ENT. 39:270, 1907 BIRD LARVA
FOOD PLANTS
PODOPHYLLUM PELTATUM LINN.

MERRICKI SMITH
SYN. SEE FORBESI FRENCH

MERTENA SMITH
SSP. SEE CINERITIA GROTE

MERTILA EDWARDS
SSP. SEE XERCES BOISDUVAL

MESAPANO SCUDDER
SYN. SEE PALAEMON PALLAS

MESILLAE COCKERELL
SYN. SEE PERSPICUA GROTE&ROBINSON

MESKEI GROTE, CATOCALA-NOCTUIDAE
SYN. FEMALE BEANIANA GROTE
SYN. ROSALINDA HENRY EDWARDS
A ORION MCDUNNOUGH
FORM CONCOLORATA MCDUNNOUGH
FORM KROMBEINI FRANCLEMONT
LIFE HISTORY
CAN. ENT. 15:100, 1883 BUNKER LARVA
LOC. CIT. 54:288, 1922 MCDUNNOUGH LARVA(BRIEF)
NEW YORK STATE MUS. MEM. 8:747 1905 FELT LARVA(BRIEF)
FOOD PLANTS
POPULUS SP. (POPLARS)
POPULUS FREMONTI S. WATS.

MESKEI SPEYER
SYN. SEE VIDENS GUENEE

MESSORIA(HARRIS), EUXOA-NOCTUIDAE
AB FRIABILIS GROTE
A TERITORIALIS SMITH
SYB. FULDA SMITH
FORM ATRIFERA GROTE
SYN. PINDAR SMITH
FOR OTHER SYN. SEE MCDUNNOUGH'S LIST

LIFE HISTORY
AMER. MUS. NAT. HIST. BULL. 14: 284, 1901 BEUTEN. LARVA
AMER. NAT. 17:422, 1883 LARVA*
BROOKLYN ENT. SOC. BULL. 27:82 1932 CRUMB LARVA* PL. X, C
ENT. SOC. AMER. ANNALS 23:396, 1930 HAWKINS LARVAL CLAWS* F. 1
ILLINOIS AGRI. EXP. STA. BULL. 95: 363, 1904 FORBES LARVA
ILLINOIS DEPT. AGRI. TRANS. 15:92, 209, 1877 FRENCH-THOMAS LARVA*
LOC. CIT. 18, APPEND:134, 1880 MARTEN LARVA(
JOURN. ECON. ENT. 8:393, 1915 CRUMB KEY TO LARVA; LARVAL HEAD(PL. 20, F. 6
MICHIGAN STATE BD. AGRI. REPT. 13:110, 1874 COOK LARVA* PUPA*(COCHRANI)
MISSOURI, REPT. NOX. INS. OF 1:74 1869 RILEY LIFE HIST*
NEBRASKA AGRI. EXP. STA. RESEARCH BULL. 81:13, 1935 WHELAN TECH. DESCR. LARVA-ANTENNAL SEG* F. 8C
NEW YORK STATE MUS. BULL. 6:13 1888 LINTNER LARVA*
LOC. CIT. 97:407, 1904 FELT LARVA
LOC. CIT. 109:53, 1907 FELT LARVA
ONTARIO ENT. SOC. ANN. REPT. P. 42 1879 BOWLES LARVA
LOC. CIT. P. 8, 1888 FLETCHER LARVA* P. ONLY COCHRANI
LOC. CIT. P. 9, 1893 FLETCHER LARVA*
LOC. CIT. P. 76, 1895 SMITH LARVA*
LOC. CIT. P. 51, 1904 FLETCHER LARVA*
LOC. CIT. P. 12, 1912 GIBSON LARVA*
UNITED STATES DEPT. AGRI. FARMER'S BULL. 1494:6, 1926 CRUMB LARVA* F. 7, B, C
UNITED STATES DEPT. AGRL. REPT. ENT. P. 290, 1884 LARVA* PL. 2, F. 6 RILEY
LOC. CIT. 1886 RILEY LARVA*

UNITED STATES DEPT. AGRI. TECH. BULL. 88:91, 1929 CRUMB EGG, LARVA* PL. 7, B, J; PUPA* F. 19E
WISCONSIN AGRI. SCO. TRANS. 19:484 1881 CHASE LARVA*
CHITTENDEN-INS. INJ. VEG. P. 246, 1907 LARVA* F. 160
CROSBY & LEONARD-MANUAL VEG. GARDEN INS. P. 268, 1918 LARVA, PUPA
ESSIG-INS. WEST. NORTH AMER. P. 682, 1926 LARVA
HART-INS. COLL. ILLINOIS HIGH SCHOOLS P. 32, 1903 LARVA* F. 43
PACKARD-GUIDE STUDY INS. P. 308 1869 LARVA
PACKARD-HALF HOURS WITH INS. P 23, 1877 LARVA*
SAUNDERS-INS. INJ. FRUITS P. 107, 1883 LARVA* PUPA (BRIEF)
SAUNDERS-LOC. CIT. P. 107, 1889 LARVA* F. 104 COCHRANI
SMITH-INS. NEW JERSEY P. 410 1900 LARVA* F. 192
INDEX AMER. ECON. ENT. VOL. 1:132; VOL. 2:144; VOL. 3:161; VOL. 3:296 PARAGROTIS VOL. 4:185; VOL. 6:283; VOL. 7:372

FOOD PLANTS
ACER SACCHARINUM LINN.
ALLIUM CEPA LINN. (ONION)
BETA VULGARIS LINN. (BEET)
BRASSICA OLERACEA LINN. (CABBAGE)
BRASSICA RAPA LINN. (TURNIP)
FAGOPYRUM SAGITTATUM GILIB (BUCKWHEAT)
FRAGARIA CHILOENSIS (LINN.) (STRAWBERRY)
IPOMOEA BATATAS (LINN.) (SWEET POTATO)
LACTUCA SATIVA LINN. (LETTUCE)
LILIUM SP.
MALUS PUMILA MILL. (APPLE)
NICOTIANA TABACUM LINN. (TOBACCO)
PHASEOLUS VULGARIS LINN. (BEANS)
PISUM SATIVUM LINN. (PEAS)
PORTULACA OLERACEA LINN.
PRUNUS PERSICA (LINN.) (PEACH)
RAPHANUS SATIVUS LINN. (RADISH)
RIBES SATIVUM SYME (CURRANT)
SOLANUM LYCOPERSICUM LINN. (TOMATO)
SOLANUM TUBEROSUM LINN. (POTATO)
SPINACA OLERACIA MILL. (SPINACH)
TRIFOLIUM SP. (CLOVERS)
VITIS VINIFERA LINN. (GRAPE)
ZEA MAYS LINN. (CORN)
GENERAL FEEDER ON FLOWERS, FRUIT TREES, LOW PLANTS, SHRUBS AND VEGETABLES; GRASSES

MESTUSATA WALKER
SYN. SEE HYPOCHRARIA HERRICH-SCHAEFFER

META SMITH
SSP. SEE EXSERTISTIGMA MORRISON

META STRANS
SSP. SEE CINCTIPES HENRY EDWARDS

METACOMET HARRIS
SYN. SEE RURICOLA BOISDUVAL

METALLICA GROTE
SYN. SEE DELECTA WALKER

METALOMUS MAYFIELD
SSP. SEE SORDIDA GROTE (CATOCALA)

METATHETIS BUTLER
SYN. SEE DIFFINIS BOISDUVAL

METEA SCUDDER, HESPERIA-HESPERIIDAE
LIFE HISTORY
ENT. NEWS 19:410, 1908 LAURENT EARLY STAGES EGG
SCUDDER-BUTTER. EAST. UNITED STATES & CANADA PP. 1652-1653, 1889 EGG* LARVA* PL. 66 F. 30; PL. 80, F. 52

FOOD PLANTS
GRASSES

METONALIS(WALKER),PHILOMETRA-NOCTUIDAE
SYN.INCEPTARIA WALKER
SYN.EFFUSALIS WALKER
SYN.GAOSALIS WALKER
SYN.LONGILABRIS GROTE
LIFE HISTORY
CAN.ENT.29:49,1897 SLINGERLAND LARVA*PUPA
LOC.CIT.41:364,1909 LYMAN LARVA
ENT.AMER.14:177,1934 CRUMB LARVA*PL.24 B
INDEX AMER.ECON.ENT. VOL.1:100 INCEPTARIA
FOOD PLANTS
BETULA SP.(DEAD BIRCH LEAVES)
POA PRATENSIS LINN.(DEAD GRASS)
TARAXACUM OFFICINALE WEBER.(DEAD)
VACCINIUM ANGUSTIFOLIUM AIT.
VACCINIUM MACROCARPON AIT. (CRANBERRY)
DECAYED LEAVES

MÈTRA SMITH
SYN.SEE LEPUSCULINA GUENEE

METRA STEPHENS
SYN.SEE RAPAE LINNAEUS

METROCAMPARIA GUENEE
SSP.SEE CONFUSARIA HUBNER

MEXICANA(BOISDUVAL),EUREME-PIERIDAE
SYN.DAMARIS FELDER & FELDER
SYN.DUPUSIETI BOISDUVAL
AB BIEDERMANNI EHRMAN
LIFE HISTORY
FOOD PLANTS
CASSIA SP.
CASSIA MARILANDICA LINN.

MEXICANA GROTE & ROBINSON
SYN.SEE PROXIMA GUERIN-MENEVILLE

MEXICANA GROTE & ROBINSON CITHERONIA-CITHERONIIDAE
LIFE HISTORY
ENT.AMER.4:62,1888 HENRY EDWARDS PUPS
ENT.NEWS 4:311,1893 DYAR-DOLL LARVA
NAT.ACAD.SCI.MEM.9:138,1905 PACKARD PUPA
FOOD PLANTS

MIA STRAND
SSP.SEE DECOLOR WALKER

MICACEA(ESPER),HYDROECIA-NOCTUIDAE
LIFE HISTORY
NOVA SCOTIA ENT.SOC.PROC.1:96 1915 BRITTAIN EGG*LARVA* PUPA* PL.2
ONTARIO ENT.SOC.ANN.REPT.P.96 1917 BRITTAIN EGG*LARVA* PUPA
INDEX AMER.ECON.ENT.
VOL.2:162 GORTYNA
VOL.3:179 GORTYNA
VOL.4:206 GORTYNA
VOL.5:307;VOL.6:359
FOOD PLANTS
DAHLIA VARIABILIS DESF.
IRIS SP.
PISUM SATIVUM LINN.(PEAS)
RHEUM RHAPONTICUM LINN.(RHUBARB)
SOLANUM LYCOPERSICUM LINN.(TOMATO)
SOLANUM TUBEROSUM LINN.(POTATO)
ZEA MAYS LINN.(CORN)

MICHABO GROTE
SYN.SEE DORIS BOISDUVAL

MICHIGANENSIS RAWSON
SSP.SEE EPIXANTHE BOISDUVAL & LECONTE

MICRONYMPHA GUENEE,CATOCALA-NOCTUIDAE
FORM TIMANDRA HENRY EDWARDS
FORM HERO HENRY EDWARDS
FORM ATARAH STRECKER
FORM GISELA MEYER
FORM FRATERCULA GROTE & ROBINSON
FORM JACQUENATTA HENRY EDWARDS

FORM HELENE PILATE
FORM OUWAH POLING
LIFE HISTORY
AMER. MUS. NAT. HIST. BULL. 16:382, 1902 BEUTEN. LARVAL HEAD*F. 1
LOC. CIT. 38:176, 1918 B. &MCD. EGG, LARVA
AMER. MUS. NAT. HIST. MEM. NEW SERIES 3, PART 1, 1918 B. &MCD LARVA*PL. 14, F. 11; PL. 15, FF. 24, 41; PL. 16, F. 11 COLOR
BROOKLYN ENT. SOC. BULL. 7:37, 1884 HULST LARVA
NEW YORK STATE MUS. MEM. 8:705 1905 FELT LARVA
PAPILIO 1:7, 1881 COQUILLETT LARVA FRATERCULA
FOOD PLANTS
QUERCUS SP. (OAKS)
QUERCUS MACROCARPA MICHX.
QUERCUS VIRGINIANA MILL.

MICROSTRIATA COMSTOCK
SYN. SEE NAPI LI NAEUS

MIDEA (HUBNER), ANTHOCHARIS-PIERIDAE
SYN. GENUTIA FABRICIUS
A L'HERMINIERI GODART
SYN. FLAVIDA SKINNER
LIFE HISTORY
AMER. MUS. NAT. HIST. BULL. 5:248 1893 BEUTEN. LARVA, CHRYSALIS
LOC. CIT. 10:247, 1898 BEUTEN. LARVA
ENT. NEWS 11:465, 1900 SMYTH JR. LIFE HIST. HABITS
LOC. CIT. 14:252, 1903 HORNIG FEEDING HABITS OF LARVA
LOC. CIT. 16:131, 1905 GROSSBECK EGG, LARVA, CHRYSALIS
COMSTOCK-HOW TO KNOW BUTTER P. 82, 1904 LARVA
EDWARDS-BUTTER. NORTH AMER. 3RD SERIES 1897 LIFE HIST* ANTHOCHARIS 1, FF. A-H
HOLLAND-BUTTERFLY BOOK 1905 CHRYSALIS* PL. 2, F. 5; PL. 5, F. 59; PL. 4, F. 6; EGG*P. 4, F. B
SCUDDER-BUTTER. EAST. UNITED STATES &CANADA PP. 1148-1152, 1889 LIFE HIST*PL. 65, F. 29; PL. 73, F. 9; PL. 76, F. 5; PL. 79, F. 5; PL. 84, F. 59
SCUDDER-GUIDE TO COMMON BUTTER P. 140, 1893 LARVA, CHRYSALIS
FOOD PLANTS
ARABIDOPSIS THALIANA (LINN.)
ARABIS GLABRA (LINN.)
BARBAREA VULGARIS R. BR.
BERBERIS VULGARIS LINN.
CAPSELLA BURSA-PASTORIS (LINN.)
CARDAMINE SP.
CRUCIFERAE SP.
CENTARIA DIPHYLLA MICHX.
DENTARIA HETEROPHYLLA NUTT.
DENTARIA LACINIATA MUHL.

MIGRATA SMITH
SYN. SEE RELICINA MORRISON

MILBERTI (GODART), NYMPHALIS-NYMPHALIDE
AB ROTHKEI GUNDER
A FURCILLATA SAY
SYN. SUBPALLIDA COCKERELL
B VIOLA DOSPASSOS
LIFE HISTORY
AMER. MUS. NAT. HIST. BULL. 5:267 1893 BEUTEN. LARVA, CHRYSALIS
BUFFALO SOC. NAT. SCI. BULL. 19: 25, 1929 WILD LARVA
CAL. ACAD. SCI. PROC. 3:125, 1929 BEHR LARVAL NOTE
LOC. CIT. 5:170, 1873 HY. EDW. LARVA, CHRYSALIS
CAN. ENT. 1:76, 1869 SAUNDERS LARVA
LOC. CIT. 17:181, 1885 EDWARDS EGG, LARVA, CHRYSALIS
CAN. NAT. PP. 107, 221, 1840 GOSSE LARVA
ILLINOIS DEPT. AGRI. TRANS. 18, APPEND: 85, 164, 1880 COQUILLETT & MIDDLETON LARVA
MONTANA UNIV. BULL. 30:97, 1906 ELROD LARVA, CHRYSALIS
NAT. CANADIEN P. 142, 1927 HUARD LARVA, CHRYSALIS

New York State Mus. Mem. 8:742 1905 Felt larva
Philadelphia Ent. Soc. Proc. 2: 28, 1863 Saunders larva
Loccit. 3:61, 1864 Lintner larva chrysalis
Comstock, J. A. - Butter Cal. p. 125, 1927 egg, larva, chrysalis
Comstock, H. H. - How Know Butter. p. 151, 1904 larva
Cragen - Our Ins. Friends & Foes p. 88, 1899 larva
Essig - Ins. West. North Amer. p. 643, 1926 larva
Fernald - Butter. Maine p. 27, 1884 larva
French - Butter. East. United States p. 195, 1886 larva, chrysalis
Harris - Ins. Inj. Veg. Flint edit. p. 303, 1862 larva, chrysalis
Holland - Butterfly Book 1905 larva*chrysalis*Pl. 3, f. 36; Pl. 4, ff. 43, 49, 50
Kellogg - Amer. Ins. P. 456, 1908 larva
Maynard - Butter. New england p. 17, 1886 larva
Scudder - Butter. East. United States & Canada pp. 423-428, 1889 life hist. pl. 64, ff. 37, 40; Pl. 74, f. 36; Pl. 78, ff. 54-57; Pl. 83, ff, 3, 4; Pl. 86, f. 66; Pl. 83, ff. 43, 49, 50
Scudder - Butter. Their Structures p. 99, 1881 egg*
Scudder - Everyday Butter. p. 333 1899 egg, larva, chrysalis
Scudder - Guidae to Common Butter. p. 89, 1893 larva, chrysalis
Index Amer. Econ. Ent.
Vol. 3:12 Aglais
Vol. 5:15 Aglais
Vol. 7:711 Nymphalis

FOOD PLANTS
Helianthus annuus Linn.
Salix sp.
Urtica dioica Linn.
Urtica holosericea Nutt.
Urtica procera Muhl.

MILDEI Stretch
ssp. see AMERICANA Harris (Epicnaptera)

MILDREDAE Chermock
ssp. see LYGDAMUS Doubleday

MILDREDAE Franclemont
ssp. see NEOGAMA Abbot & Smith

MILITARIS Harris
ssp. see LECONTEI Guerin-Meneville

MILLBURNI Rummel
ssp. see NYCTEIS Doubleday & Hewitson

MILO Edwards
syn. see AGRICOLA Boisduval

MIMICA Strecker
syn. see INSUETA Guenee

MIMICATA Walker
syn. see AERATA Fabricius

MIMOSARIA (Guenee), NEMORIA - Geometridae
syn. TRACTARIA Walker
syn. VENUSTUS Walsh
syn. APPROXIMARIA Packard
syn. LATIARIA Packard
syn. CONFIRARIA Packard

LIFE HISTORY
Amer. Nat. 18:933, 1884 Packard larva, pupa
Boston Soc. Nat. Hist. Proc. 9:301 1863 Walsh larva pupa VENUSTUS
Connecticut Agri. Exp. Sta. Bull. 234:195, 1921 Britton larva
Maine Agri. Exp. Sta. Bull. 259: 40, 1917 Mosher larva pupa* f. 3E
New York State Agri. Exp. Sta. (Cornell), Mem. 274:113, 1948 Forbes larva
New York State Mus. Mem. 8:706

1905 FELT LARVA
PSYCHE 8:310,1899 DYAR EGG,LARVA
PACKARD-INS. INJ. FOREST TREES
P.49,1881 LARVA
PACKARD-MONOGRAPH GEOMETRIDAE
P.389,1876 LARVA,PUPA
FOOD PLANTS
ABIES SP. (FIRS)
ALNUS SP. (ALDERS)
MYRICA PENSYLVANICA LOISEL
QUERCUS SP. (OAKS)
TSUGA CANADENSIS(LINN.)

MIMULI BEHR. ANARTA-NOCTUIDAE
LIFE HISTORY
CAL. ACAD. SCI. BULL. 3:62,1885
BEHR LARVA
FOOD PLANTS
ANTIRRHINUM KINGII S. WATS.
MIMULUS NANUS H.&A.

MINEA SLOSSON
SSP. SEE DORIS BOISDUVAL

MINELLA DYAR
SSP. SEE FRAGILIS GUENEE

MINEREA(GUENEE), ZALE-NOCTUIDAE
SYN. INVOLUTA WALKER
AB ALBOFASCIATA BETHUNE
AB FEMALE MINEREOIDES STRAND
A NORDA SMITH
SYN. FEMALE MINEREANA STRAND
SYN. FEMALE MINEREELLA STRAND
LIFE HISTORY
CAN. JOURN. IND. SCI. ART. 10:255,
1865 BETHUNE LARVA
GUENEE-SPEC. GEN. LEP. NOCT. 3:
15,1852 LARVA,PUPA
FOOD PLANTS
MALUS PUMILA MILL. (APPLE)
OSTRYA VIRGINIANA(MILL.)
SALIX SP. (WILLOWS)

MINEREANA STRAND
SYN. SEE MINEREA GUENEE

MINEREELLA STRAND
SYN. SEE MINEREA GUENEE

MENEREOIDES STRAND
SSP. SEE MINEREA GUENEE

MINERVA CASSINO
SSP. SEE BRISEIS EDWARDS

MINIANS GUENEE
SYN. SEE EMMEDONIA CRAMER

MINIATA(KIRBY), HYPOPREPIA-ARCTIIDAE
SYN. VITTATA HARRIS
A MISSISSIPPIENSIS BARNES & BENJAMIN
LIFE HISTORY
CAN. ENT. 3:36,1871 SAUNDERS
LARVA VITTATA
ENT. NEWS 3:246,1892 DYAR EGG,
LARVA
NAT. CANADIEN P.274,1929 HUARD
LARVA
SMITHSONIAN MISC. COLL. 4, ART.
2:256,1862 MORRIS LARVA,
COCOON VITTATA
UNITED STATES NAT. MUS. PROC.
27:793,1904 DYAR EGG,
LARVA
CRAGEN-OUR INS. FRIENDS &FOES
P.144,1899 LARVA
FOOD PLANTS
LICHENS

MINIATA PACKARD
SSP. SEE PHOLUS DRURY

MINIMA (GROTE), PSEUDOCHARIS-AMATIDAE
LIFE HISTORY
NEW YORK ENT. SOC. JOURN. 5:100
1897 DYAR LARVA(BRIEF)
PSYCHE 8:42,1897 DYAR EGG,
LARVA,PUPA,COCOON
FOOD PLANTS
RHACEMA ILICIFOLIA(POIR.)

MINIMA(GUENEE), EUBLEMMA-NOCTUIDAE
SYN. CARMELITA MORRISON
LIFE HISTORY

WASHINGTON E.TSOC.PROC.4:326
1898 DYAR LARVA CARMELITA
FOOD PLANTS
GNAPHALIUM SP.

MINIMALIS GROTE
SSP.SEE PLACIDA GROTE

MINIMALIS GROTE
SYN.SEE PROTUMNUSALIS WALKER

MINISTRA(DRURY),DATANA-NOTODONTIDAE
A CALIFORNICA DYAR
LIFE HISTORY
AMER.ENT.2:306,1870 LARVA(BRIEF)
AMER.MUS.NAT.HIST.BULL.10:
405,1898 BEUTEN.LARVA
AMER.NAT.19:715,1885 PACKARD
PUPA(BRIEF)
BOSTON SOC.NAT.HIST.PROC.27:
140,1896 DYAR LARVAL SETAE
CAN.ENT.20:16,1888 BEUTEN.EGG,
LARVA,PUPA
LOC.CIT.20:135,1888BEUTEN.KEY
TO DANATA LARVAE
ENT.NEWS 3:129,1892 LARVA*
COLOR
ILLINOIS DEPT.AGRI.TRANS.15:189
1877 FRENCH LARVA*
LOC.CIT.18,APPEND:119,167,1880
COQUILLETT & MARTEN EGG,
LARVA*
ILLINOIS,REPT.NOX.INS.OF 3:186,
1873 LEBARON LIFE HIST*
MAINE AGRI.EXP.STA.BULL.259:
67,1917 MOSHER PUPA*F.5H
MICHIGAN AGRI.EXP.STA.CIRC.
137:17,1935 PETTIT LARVA
MICHIGAN STATE COLLEGE,SPEC.
BULL.243:30,1933 MCDANIEL
LIFE HISTORY
MINNESOTA AGRI.EXP.STA.BULL.
61:160,1899 LUGGER EGG *F.
104;LARVA*F.104AND F.135 5
LOC.CIT.84:65,1903 WASHBURN
LARVA*F.46
NAT.ACAD.SCI.MEM.7:107,1895
PACKARD EGG,LARVA* PL.10,
F.1 COLOR;PUPA
LOC.CIT.9,1905 PACKARD LARVAL
HEAD* PL.44,F.3
NEW YORK ENT.SOC.JOURN.1:62,
1893 PACKARD LARVA
NEW YORK STATE AGRI.EXP.STA.
(CORNELL),MEM.56:1075,1922
WELLHOUSE EGG,LARVA
LOC.CIT.274:214,1948 FORBES
LARVA
NEW YORK STATE MUS.MEM.8:535,
1905 FELT LARVA
NEW YORK STATE REPT.NOX.INX.
2:237,1855 FITCH LARVA*PUPA
LOC.CIT.3:337,1856 FITCH LARVA
ALSO P.19,SEC.29
OHIO AGRI.EXP.STA.BULL.332:229,
1918 HOUSER LARVA*PL.31,
F.3
ONTARIO ENT.SOC.ANN.REPT.P.19
1871 BETHUNE LARVA*
LOC.CIT.P.53,1903 GIBSON LARVA
PHILADELPHIA ENT.SOC.PROC.6:11
1866 GROTE & ROBINSON
MATURE LARVA
PSYCHE 6:573,1893 DYAR KEY TO
LARVAE MINISTRA AND CALIFORN-
ICA
UNITED STATES DEPT.AGRI.FARMER'S
BULL.1169:47,1921 KOTINSKY
LARVA*F.27
LOC.CIT.1270:41,1922 QUAINTANCE
&SIEGLER EGG*F.80;LARVA*
F.81
UNITED STATES DEPT.AGRI.MISC.
PUB.657:355,398,1950
CRAIGHEAD LARVA
ABBOT&SMITH-LEP.INS.GEORGIA
2:161,1797 LARVA*PL.81
COLOR
CRAGEN-OUR INS.FRIENDS&FOES
P.162,1899 LARVA*F.107
DICKERSON-MOTHS&BUTTERFLIES
P.220,1901 LARVA*FF.189,
190
DOANE ETC-FOREST INS.P.294,
1936 LARVA
ESSIG-INJ.&BENEF.INS.P.405,1915

EGG, LARVA*F.410; PUPA
ESSIG-INS. WEST NORTH AMER. P. 691, 1926 EGG, LARVA*F.567
FERNALD-APPLIED ENT. P.270, 1926 LARVA*F.271
FERNALD&SHEPARD-APPLIED ENT. P. 266, 1942 LARVA*F.268
GLOVER-ILLUST. NORTH AMER. ENT. 1878 LARVA*PL.10, F.17
HARRIS-ENT. CORRESP. P.308, 1869 LARVA* PL.2
HARRIS-INS. INJ. VEG. 2ST EDIT. P. 312, 1841 LARVA
HARRIS-LOC. CIT. 2ND EDIT. P.332, 1852 LARVA
HARRIS-LOC. CIT. FLINTEDIT. P.429 1862 LARVA*
HART-INS. COLL. ILLINOIS HIGH SCHOOLS P.45, 1903 EGG* LARVA* F.56
HERRICK-INS. ENEMIES SHADE TREES P.126, 1935 LARVA* F.96
HOLLAND-MOTH BOOK 1905 LARVA* PL.1, F.13 COLOR
KELLOGG-AMER. INS. P.393, 1908 EGG
METCALF&FLINT-DESTR. &USEFUL INS. P.564, 1939 EGG*F.350; LARVA* F.351
O'KANE-INJ. INS. P.270, 1912 LARVA* F.380
PEAIRS-PESTS FARM, GARD. ORCHARD P.379, 1941 LARVA
SAUNDERS-INS. INJ. FRUITS P.61, 1883 LARVA*
SAUNDERS-LOC. CIT. P.61, 1889 EGG, LARVA* F.55; PUPA
SLINGERLAND, CROSBY-MANUAL FRUIT INS. P.124, 1914 EGG, LARVA* FF.135, 136; PUPA*F.137
WALSH-PRACTICAL ENT. 2:7, 1866 LARVA
INDEX AMER. ECON. ENT. VOL.1:102: VOL.2:109; VOL.3:115; VOL.4:133; VOL.5:174; VOL.6:199

FOOD PLANTS

ACER SP. (MAPLES)
BETULA SP. (BIRCHES)
BETULA ALBA LINN.
BETULA NIGRA LINN.
BETULA PAPYRIFERA MARSH.
BETULA POPULIFOLIA MARSH.
CARPINUS CAROLINIANA WALT.
CARYA SP. (HICKORIES)
CARYA CORDIFORMIS (WANG.)
CARYA GLABRS (MILL.)
CARYA LACINIOSA (MICHX.)
CARYA OVALIS (WANG.)
CARYA TOMENTOSA NUTT.
CASTANEA DENTATA (MARSH.)
CASTANEA PUMILA (LINN.)
CASTANEA SATIVA MILL.
CORYLUS SP.
CORYLUS AMERICANA WALT.
CORYLUS AVELLANA LINN.
CRATAEGUS SP.
CYDONIA OBLONGA MILL. (QUINCE)
FAGUS SP. (BEECHES)
FAGUS SYLVATICA V. CUPREA LOUD.
FAGUS GRANDIFOLIA EHRH.
FAGUS SYLVATICA V. LACINIATA VIGN.
HAMAMELIS VIRGINIANA LINN.
JUGLANS CINEREA LINN.
JUGLANS NIGRA LINN.
MALUS PUMILA MILL. (APPLE)
OXYDENDRUM ARBOREUM (LINN.)
PRUNUS AMERICANA MARSH.
PRUNUS ARMENIACA LINN. (APRICOT)
PRUNUS CERASUS LINN. (CHERRY)
PRUNUS DOMESTICA LINN. (PLUM)
PRUNUS ILICIFOLIA (NUTT EX. H&A)
PRUNUS PERSICA (LINN.) (PEACH)
PYRUS COMMUNIS LINN. (PEAR)
QUERCUS SP. (OAKS)
QUERCUS ALBA LINN.
QUERCUS COCCINEA MUENCH
QUERCUS LAEVIS WALT
QUERCUS MACROCARPA MICHX.
QUERCUS PALUSTRIS MUENCH.
QUERCUS ROBUS LINN.
QUERCUS RUBRA LINN.
QUERCUS STELLATA WANG.
QUERCUS UNDULATA TORR.
QUERCUS VELUTINA LAM.
RHUS SP. (SUMACS)

ROBINIA PSEUDO-ACACIA LINN.
RUBUS ALLEGHENIENSIS PORTER
TILIA AMERICANA LINN
TILIA EUROPAEA LINN.
TILIA HETEROPHYLLA VENT.
ULMUS SP. (ELMS)
VACCINIUM SP.

MINNA(BUTLER),CELAMA-NOLIDAE
SYN. HYEMALIS STRETCH
LIFE HISTORY
PSYCHE 6:110,1891 DYAR LARVA, COCOON HYEMALIS
UNITED STATES NAT. MUS. PROC. 27: 914,1904 DYAR LARVA,COCOON
FOOD PLANTS
CEANOTHUS SP.
SALIX SP. (WILLOWS)

MINOR CHERMOCK
SSP. SEE PHILODICE GODART

MINOR COCKERELL
SSP. SEE CARDUI LINNAEUS

MINOR MCDUNNOUGH
SSP. SEE HYDASPE BOISDUVAL

MINOR VERITY
SYN. SEE SMINTHEUS DOUBLEDAY & HEWITSON

MINORALIS SMITH,ZANCLOGNATHA-NOCTUIDAE
LIFE HISTORY
CAN. ENT. 74:52,1942 BROWN & MCGUFFIN LARVA*PL.7,F.A AND LARVAL MAND.
FOOD PLANTS
PICEA GLAUCA(MUENCH)
PICEA MARIANA(MILL.)

MINORARIA HULST
SSP. SEE JOHNSONARIA FITCH

MINORATA(PACKARD),SEMIOTHISA-GEOMETRIDAE
LIFE HISTORY
FOOD PLANTS
PINUS RESINOSA AIT.
PINUS STROBUS LINN.

MINORATA TAYLOR
SYN. SEE MISTURATA HULST

MINORI CROSS
SSP. SEE INDRA REAKIRT

MINTHA EDWARDS
SYN. SEE ICARIOIDES BOISDUVAL

MINUS DYAR
SSP. SEE INFICITA WALKER(MARATHYSSA)

MINUSCULA(ZELLER),SARBENA-NOLIDAE
A PHYLLA DYAR
B EUCALYPTULA DYAR
LIFE HISTORY
CAN. ENT. 31:61,1899 DYAR LARVA, COCOON
NEW YORK ACAD. SCI. ANNALS 8:226 1894 DYAR LARVAL SETAE
NEW YORKENT. SOC. JOURN. 6:249 1898 COQUILLETT LARVA (INCORRECTLY IDENTIFIED AS INVOLUTA DYAR)
LOC. CIT. 7:210,1899 ABOVE CORRECTION
PSYCHE 6:248,1892 DYAR EGG, LARVA,PUPA,COCOON
FOOD PLANTS
QUERCUS SP. (OAKS)
SALIX SP. (WILLOWS)

MINUSCULUS BRYK
SYN. SEE SMINTHEUS DOUBLEDAY & HEWITSON

MINUTA EDWARDS,CATOCALA-NOCTUIDAE
FORM HISERI CASSINO
FORM PARVULA EDWARDS
FORM MELLITULA HULST
FORM EUREKA SCHWARZ
FORM OBLITERATA SCHWARZ
LIFE HISTORY
AMER. MUS. NAT. HIST. BULL. 16: 381,1902 BEUTEN. LARVA
AMER. MUS. NAT. HIST. MEM. NEW

SERIES 3, PART 1, 1918 BARNES & McDUNNOUGH LARVA*PL. 11, FF. 12, 13 COLOR
CAN. ENT. 27:331, 1901 DODGE EGG, LARVA, PUPA
ENT. NEWS 21:116, 1910 ROWLEY-BERRY LARVA
LOC. CIT. 24:200, 1913 ROWLEY LIFE HISTORY-LARVA
NEW YORK STATE MUS. MEM. 8:729 1905 FELT LARVA
FOOD PLANTS
GLEDITSIA TRIACANTHOS LINN.

MINUTA EDWARDS, MELITAEA-NYMPHALIDAE
SYN. APPROXIMATA STRECKER
A NYMPHA EDWARDS
LIFE HISTORY
CAN. ENT. 11:141, 1879 EDWARDS CHRYSALIS
FOOD PLANTS

MINUTA GROTE
SYN. SEE LARICIS FITCH

MIRA GROTE, CATOCALA-NOCTUIDAE
A DANA CASSINO
LIFE HISTORY
AMER. MUS. NAT. HIST. BULL. 38:172 1918 B. & McD. EGG, LARVA
AMER. MUS. NAT. HIST. MEM. NEW SERIES 3, PART 1, 1918 BARNES & McDUNNOUGH LARVA*PL. 13, F. 12; PL. 15, FF. 22, 36; PL. 16, F. 4 COLOR
FOOD PLANTS
CRATAEGUS SP. (THORNS)

MIRABELLE BARNES
SYN. SEE AUTOLYCUS EDWARDS

MIRABILIS (GROTE), ABAGROTIS-NOCTUIDAE
LIFE HISTORY
SOUTHERN CAL. ACAD. SCI. BULL. 43:115, 1944 COMSTOCK & DAMMERS LARVA*PL. 24
INDEX AMER. ECON. ENT. VOL. 7:1 1940
FOOD PLANTS
JUNIPERUS CALIFORNICA CARR.

MIRABILIS WRIGHT
SSP. SEE BARONI EDWARDS (EUPHYDRYAS)

MIRANDA (BROTE), PROXENUS-NOCTUIDAE
A NITENS DYAR
LIFE HISTORY
WASHINGTON ENT. SOC. PROC. 4, 322, 1898 DYAR LARVA
FOOD PLANTS

MIRANDATA TAYLOR
SSP. SEE HERSILIATA GUENEE

MIRIFICA (HENRY EDWARDS), DRASTERIA-NOCTUIDAE
A HASTINGSI HENRY EDWARDS
B PERPALLIDA HENRY EDWARDS
C KLOTSI RICHARDS
LIFE HISTORY
UNITED STATES NAT. MUS. PROC. 23:384, 1903 DYAR EGG, LARVA
LOC. CIT. 25:384, 1903 DYAR EGG, LARVA HASTINGSII
FOOD PLANTS
ERIOGONUM SP.

MIRIFICUM STRECKER
SYN. SEE PUERA CRAMER

MISCELLUS GROTE
SYN. SEE LINEOLATA WALKER

MISCELLUS SMITH
SSP. SEE MISCELIOIDES GUENEE

MISCIDICE SCUDDER
SSP. SEE PHILODICE GODART

MISELIOIDES (GUENEE), PHOSPHILA-NOCTUIDAE
FORM MISCELLUS SMITH
A MACERATA SMITH
LIFE HISTORY
CAN. ENT. 31:286, 1899 DYAR EGG, LARVA
ILLINOIS DEPT. AGRI. TRANS. 8:

Append:180,1880 Coquillett
larva
Coquillett-North Amer. Ent. 1:52
1880 larva
FOOD PLANTS
Smilax rotundifolia Linn.
Verbena sp.

misera Grote
ssp. see fractilinea Grote

miserulata(Grote), Eupithecia-Geometridae
syn. nebulosa Hulst
syn. plumbaria Hulst
syn. grossbeckiata Swett
a zela Swett & Cassino
LIFE HISTORY
New York State Agri. Exp. Sta.
1 (Cornell), Mem. 274:166,1948
Forbes larva
New York State Mus. Mem. 8:706
1905 Felt larva
Psyche 19:15,1912 B.&McD. larva,
pupa
United States Dept. Agri. Bur.
Ent. Bull. old series 3, pl24
1883 Packard larva, pupa
United States Nat. Mus. Proc. 23:
282,1901 Dyar larva, pupa
Glover-Illust. North Amer. Ent.
1878 larva*Pl. 14, f. 19
Packard-Ins. Inj. Forest Trees
248,1881 larva
Index Amer. Econ. Ent.
Vol. 1:127
FOOD PLANTS
Amelanchier canadensis(Linn.)
Aster sp.
Chamaecyparis thyoides(Linn.)
Eupatorium purpureum Linn.
Juniperus sp.
Juniperus communis Linn.
Juniperus virginiana Linn.
Laris laricina(DuRoi)
Myrica sp.
Prunus serotina Ehrh.
Ptelea trifoliata Linn.
Quercus sp. (oaks)
Rhexia virginica Linn.
Ribes sativum Syme(currant)
Rubus idaeus v. strigosus(Michx)
Rudbeckia sp.
Salix sp. (willows)
Solidago sp. (flowers)
Taraxacum officinale Weber
Taxus baccata Linn.
Vaccinium macrocarpon Ait.
Zizia aurea(Linn.)

misippus(Linnaeus), Hypolimnas-
Nymphalidae
LIFE HISTORY
FOOD PLANTS
Portulaca grandiflora Hook
(In captivity see Canadian
Ent. 20:128,1888)

missionensis Hovanitz
ssp. see icarioides Boisduval

mississippiensis Barnes & Benjamin
ssp. see miniata Kirby

missouriensis Schwarz
ssp. see robinsoni Grote

misturata(Hust), Eupithecia-Geometridae
syn. subfoveata Dyar
syn. minorata Taylor
syn. insignificata Taylor
syn. sublineata Taylor
syn. scelestata Taylor
syn. delzurata Swett & Cassino
a frostiata Swett
syn. conformata Pearsall
LIFE HISTORY
United States Nat. Mus. Proc. 27:
892,1904 Dyar egg, larva
subfoveata
FOOD PLANTS
Ceanothus sp.

moderata Walker
syn. see albilinea Hubner

moderna Grote

SYN.SEE MAESTOSA HULST(CATOCALA)

MODESTA BATES
SYN.SEE MONIMA CRAMER

MODESTA BEUTENMULLER,DATANA-NOTODONTIDAE
LIFE HISTORY
PSYCHE 6:573,1893 DYAR KEY TO DATANA LARVAE
FOOD PLANTS

MODESTA(HARRIS),PACHYSPHINX-SPHINGIDAE
SYN.PRINCEPS WALKER
SYN.POPULICOLA BOISDUVAL
SYN.CABLEI VON REIZENSTEIN
SYN.OCCIDENTALIS HENRY EDWARDS
A IMPERATOR STRECKER
FORM KENZEI ROTHSCHILD & JORDAN
B BOREALIS CLARK
LIFE HISTORY
AMER.MUS.NAT.HIST.BULL.7:312 1895 BEUTEN.LARVA,PUPA
BUFFALO SOC.NAT.SCI.BULL.4:29 1881 KELLICOTT LARVA,PUPA
CAN.ENT.9:210,1877 BUNKER EGG, LARVA
LOC.CIT.26:208,1894 HEATH LARVA (OCCIDENTALIS)
ENT.SOC.AMER.ANNALS 11:427, 1918 MOSHER PUPAL STRUCT* PL.36,F.5
ILLINOIS DEPT.AGRI.TRANS.18, APPEND:159,1880 COQUILLETT LARVA
INSECT LIFE 3:390,1891 DYAR LARVA
MICHIGAN STATE COL.SPEC.BULL. 243:48,1933 MCDANIEL LARVA
NAT.CANADIEN P.158,1929 HUARD LARVA
NEW YORK STATE MUS.MEM.8:742 1905 FELT LARVA(BRIEF)
ONTARIO ENT.SOC.ANN.REPT.63, PUPA
PSYCHE 2:72,1877 ANDREWS LARVA
PSYCHE 8:308,1899 SOULE EGG, LARVA,PUPA
SOUTHERN CAL.ACAD.SCI.BULL.42: 39,1943 COMSTOCK & DAMMERS LIFE HIST.EGG,LARVA* PL.1;PUPA*PL.1
CRAGEN-OUR INS.FRIENDS &FOES P.138,1899 LARVA
ELIOT&SOULE-CATER.AND MOTHS. P.174,1902 LARVA*PUPA
ESSIG-INS.WEST.NORTH AMER. P.665,1926 LARVA
FERNALD-SPHINGIDAE NEW ENGLAND P.72,1886 LIFE HIST.
KELLOGG-AMER.INS.P.437,1908 LARVA
INDEX AMER.ECON.ENT. VOL.4:340;VOL.5:466; VOL.7:735
FOOD PLANTS
MALUS PUMILA MILL.(APPLE)
POPULUS BALSAMIFERA LINN.
POPULUS DELTOIDES MARSH.
POPULUS FREMONTI S.WATS.
POPULUS GILEADENSIS ROUL.
OIYKYS HETEROPYYLLA LINN.
POPULUS NIGRA V.ITALICA MUENCH
POPULUS TREMULOIDES MICHX.
SALIX SP.(WILLOWS)

MODESTA HUDSON,CERURA-NOTODONTIDAE
LIFE HISTORY
NEW YORK STATE AGR .EXP.STA. (CORNELL),MEM.274:237,1948 FORBES LARVA
FOOD PLANTS
POPULUS SP.(POPLARS)
SALIX SP.(WILLOWS)

MODESTA MAYNARD
SYNSEE COLUMELLA FABRICIUS

MODESTA PACKARD
SSP.SEE PLANTAGINIS LINNAEUS

MODESTA SMITH,HYPENA-NOCTUIDAE
LIFE HISTORY -
ENT.AMER.14:146,1934 CRUMB LARVA
WASHINGTON ENT.SOC.PROC.4:329 1898 DYAR LARVA

FOOD PLANTS
URTICA SP. (NETTLES)
URTICA LYALLII S. WATS.

MODESTARIA HULST
SYN. SEE ARGILLACEARA PACKARD

MODICA WALKER, ACRONICTA-NOCTUIDAE
LIFE HISTORY
AMER. MUS. NAT. HIST. BULL. 14:254 1901 BEUTEN. LARVA
NEW YORK STATE MUS. MEM. 8:704, 1905 FELT LARVA
UNITED STATES NAT. MUS. PROC. 21: 139, 1898 SMITH&DYAR LARVA* PL. 5, F. 6 COLOR OR UNCOLORED; PUPA, COCOON
FOOD PLANTS
QUERCUS SP. (OAKS)

MODIOLA GROTE
SSP. SEE FRACTILINEA GROTE

MOERENS (GROTE), EUXOA-NOCTUIDAE
SYN. LUTEOLA SMITH
LIFE HISTORY
FOOD PLANTS
MELILOTIS ALBA DESR.

MOESERI BIRD
SYN. SEE NEPHELEPTENA DYAR

MOESTA DYAR
SSP. SEE LEPORINA LINNAEUS

MOESTATA NOLCKEN
SSP. SEE HASTATA LINNAEUS

MOFFATIANA GROTE
SYN. SEE INDIRECTA WALKER (PYREFERRA)

MOHAVE WATSON & COMSTOCK
SSP. SEE ENOPTES BOISDUVAL

MOILENA STRECKER
SYN. SEE VERSUTA SMITH

MOINA STRECKER
SSP. SEE NASTES BOISDUVAL

MOJAVE COMSTOCK
SSP. SEE PHAETON DRURY

MOLLICULARIA ZELLER
SSP. SEE AMICARIA HERRICH-SCHAEFFER

MOLLIFERA WALKER
SYN. SEE AEMULA HUBNER

MOLLIS WALKER
SYN. SEE BIUNDATA WALKER

MOLLIS WRIGHT
SYN. SEE SARA BOISDUVAL

MOLLISARIA (HULST), EUCHLAENA-GEOMETRIDAE
SYN. OCCANTARIA HULST
LIFE HISTORY
FOOD PLANTS
AMELANCHIER FLORIDA LIND.
CRATAEGUS DOUGLASII LINDL.
RHAMNUS CATHARTICA LINN.
RHAMNUS PURSHIANA DC.
RIBES SANGUINEUM PURSH
RUBUS VITIFOLIUS C.&S.
SALIX SP. (WILLOWS)
SALIX HOOKERIANA BARR.
SALIX SCOULERIANA BARR.

MOLPODIA HUBNER
SYN. SEE ILAIRE GODART

MONACHE COMSTOCK
SSP. SEE POLA BOISDUVAL

MONETA (FABRICIUS), CHRYSOPTERA-NOCTUIDAE
A ESMERALDA OBERTHUS
SYN. TRABEA SMITH
LIFE HISTORY
FOOD PLANTS
DELPHINIUM SP.

MONICARIA (GUENEE), DREPANULATRIX-GEOMETRIDAE
SYN. CELATARIA HULST
SYN. INDURATA DYAR

SYN. MERCEDULATA STRECKER
LIFE HISTORY
AMER. MUS. NAT. HIST. BULL. 94, (5):287, 1949 RINDGE EGG, LARVA, PUPA
FOOD PLANTS
CEANOTHUS SP.
CEANOTHUS SOREDIATUS H. &A.
CEANOTHUS THYRSIFLORUS ESCHS.

MONILIS (FABRICIUS), HYPSOROPHA-NOCTUIDAE
LIFE HISTORY
WASH. ENT. SOC. PROC. 5:291, 1903 DYAR LARVA
FOOD PLANTS
DIOSPYROS VIRGINIANA LINN.

MONIMA (CRAMER), EUNICA-NYMPHALIDAE
SYN. MYRTO GODART
SYN. MODESTA BATES
LIFE HISTORY
WAS. ENT. SOC. PROC. 14:54, 1912 DYAR LARVA
FOOD PLANTS
ZANTHOXYLUM PENTANOME DC.

MONO BURDICK
SSP. SEE AMPELOS EDWARDS

MONOCHROMATEA (MORRISON), HEMIPACHNOBIA NOCTUIDAE
LIFE HISTORY
ENT. NEWS 20:61, 1919 HOOKER LIFE HIST. LARVA*PUPA*PL. 4
FOOD PLANTS
DROSERA ROTUNDIFOLIA LINN.
VACCINIUM MACROCARPON AIT.

MONOCO SCUDDER
SYN. SEE ACCIUS ABBOT & SMITH

MONODIA DYAR
SYN. SEE TANTILLUS GROTE

MONODON GROTE
SYN. SEE FLAGELLUM WALKER

MONOENSIS GUNDER
SSP. SEE NIBIGENE BEHR

MONOTRAPA GROTE, SELENIS-NOCTUIDAE
LIFE HISTORY
ENT. NEWS 29:44, 1918 SKINNER NOTE ON LIFE HIST.
FOOD PLANTS
CASSIA NICTITANS LINN.

MONSTROSA STRAND
SYN. SEE ARELI STRECKER

MONTANA BEHR, PHYCIODES-NYMPHALIDE
SYN. ORSA BOISDUVAL
LIFE HISTORY
FOOD PLANTS
ASTER SP.
CHRYSOPSIS BREWERII GRAY

MONTANA BEUTENMULLER
SSP. SEE NEVADENSIS BEUTENMULLER

MONTANA BEUTENMULLER
SSP. SEE MANTO STRECKER

MONTANA FIELD
SSP. SEE EDITHA MEAD

MONTANA VERITY
SYN. SEE AUSONIDES BOISDUVAL

MONTANAE GROTE, CUCILLIA-NOCTUIDAE
LIFE HISTORY
CAN. ENT. 62:277, 1930 COOK LARVA (BRIEF)
FOOD PLANTS
CHRYSOTHAMNUS SP.

MONTANATA PACKARD
SYN. SEE DESTINATA MOESCHLER

MONTANENSIS WATSON & COMSTOCK
SSP. SEE ACADICA EDWARDS (STRYMON)

MONTANULA BRYK & EISNER
SYN. SEE SMINTHEUS DOUBLEDAY & HEWITSON

MONTANUS COOK
SSP. SEE AUXILIARIS GROTE

MONTANUS EHRMANN
SYN. SEE SMINTHEUS DOUBLEDAY & HEWITSON

MONTECLARA SMITH
SSP. SEE SPONSA SMITH

MONTICOLA BEHR
SSP. SEE ZERENE BOISDUVAL

MONTINUS (SCUDDER), BRENTHIS-NYMPHALIDAE
LIFE HISTORY
CAN. ENT. 68:239, 1936 DOS PASSOS EGG*LARVA
SCUDDER-BUTTER. EAST. UNITED STATES & CANADA PP. 604-607 1889 EGG*PL. 64, F. 39; PL. 67, F. 16
SCUDDER-GEOLOGY OF NEW HAMPSHIRE P. 380, 1874 LIFE HIST*PL. A
FOOD PLANTS
GEUM PECKII PURSH
SOLIDAGO CUTLERI FERN
VIOLA SP. (VIOLETS)

MONTIS BLACKMORE
SSP. SEE ICARIOIDES BOISDUVAL

MONTIS EDWARDS
SSP. SEE CELTIS BOISDUVAL & LECONTE

MONTIVAGA BEHR. ARGYNNIS-NYMPHALIDAE
SYN. EGLEIS BHR.
AB MAMMOTHI GUNDER
AB BOHARTI GUNDER
A MALCOLMI COMSTOCK
AB SINEARGENTATUS GUNDER
B TEHACHAPINA COMSTOCK
C OWENI EDWARDS
LIFE HISTORY
CAN. ENT. 11:177, 1879 EDWARDS EGG, LARVA, CHRYSALIS
LOC. CIT. 26:45, 1894 LEMBERT OVIPOSITION
ILLINOIS DEPT. AGRI. TRANS. 18: APPEND:164, 1880 COQUILLETT LARVA
COMSTOCK, J. A.-BUTTER. CAL. P. 83 1927 EGG*CHRYSALIS*(A 34)
EDWARDS-BUTTER. NORTH AMER. 3RD SERIES 1897 LIFE HIST* ARGYNNIS 9, FF. A-D
FOOD PLANTS
FESTUCA OVINA LINN.
POTENTILLA DOUGLASII GREENE
POTENTILLA GRACILLIPES PIPER
VIOLA SP.
VIOLA WALTERI HOUSE
EGGS ARE LAID ON PINE BURS, CONES, LEAVES AND STEMS. PROBABLY VIOLETS ARE THE ONLY FOOD PLANTS (COMSTOCK, J. A. LOC. CIT.)

MONTIVAGUS REAKIRT
SSP. SEE SYRICHTUS FABRICIUS

MONUSTE (LINNEUS), ASCIA-PIERIDAE
SYN. FERONIA STEPHENS
SYN. CLEOMES BOISDUVAL & LECONTE
FORM FEMALE PHILETA FABRICIUS
A CRAMERI HOLLAND
LIFE HISTORY
PENN. ACAD. SCI. PROC. 19:40, 1945 CHERMOCK CHRYSALIS* F. 1 A, B
SMITHSONIAN MISC, COLL. 4, ART. 2:17, 1862 MORRIS LARVA, CHRYSALIS
UNITED STATES DEPT. AGRI. REPT. ENT. PP. 117; 118, 1883 RILEY EGG, LARVA*CHRYSALIS PL. 10, F. 1
BOISDUVAL-SPEC. GEN. DIURNAE 1: 497, 1836 LARVA, CHRYSALIS
BOISDUVAL & LECONTE-HIST. LEP. NORTH AMER. P. 44, 1833 LARVA* PL. 16; PUPA* PL. 16
COMSTOCK-HOW TO KNOW BUTTER P. 72, 1904 LARVA
FRENCH-BUTTER. EAST. UNITED STATES P. 107, 1886 EGG, LARVA, CHRYSALIS

GUNDLACH-ENT. CUBANA P.101, 1881
LARVA, CHRYSALIS
MORRIS-SYNOP. LEP. NORTH AMER. P.17, 1862 LARVA, CHRYSALIS
INDEX AMER. ECON. ENT.
VOL. 2:294 PONTIA
VOL. 3:328 PONTIA
VOL. 4:381 PONTIA
VOL. 7:109
FOOD PLANTS
BRASSICA OLERACEA LINN. (CABBAGE)
CLEOME GYNANDRA LINN.
CRUCIFERAE SP.

MOPSA HENRY EDWARDS
SYN. SEE ULTRONIA HUBNER

MOPSUS HUBNER
SYN. SEE TITUS FABRICIUS

MORBIDALIS (GUENEE), CHYTOLIAT-NOCTUIDAE
LIFE HISTORY
CAN. ENT. 12:44, 1880 COQUILLETT LARVA
ENT. AMER. 14:182, 1934 CRUMB LARVA*PL. 24 G; PL. 25 D
ILLINOIS DEPT. AGRI. TRANS. 18, APPEND: 138, 182, 1880 COQUILLETT & MARTEN LARVA
UNITED STATES NAT. MUS. PROC. 23:279, 1900 DYAR EGG, LARVA
FOOD PLANTS
CORYLUS SP. (HAXELS)
TARAXACUM OFFICINALE WEBER
DEAD LEAVES GRASS

MORGANTA PLOETZ
SYN. SEE BRETTUS BOISDUVAL & LECONTE

MORMO (FELDER & FELDER), APODEMIA-RIODINIDAE
SYN. DUMETI BEHR
SYN. MORMONIA BOISDUVAL
A VIRGULTI BEHR.
SYN. SONORENSIS FELDER & FELDER
SYN. CYTHERA EDWARDS
B MEJICANUS BEHR
SYN. DURYI EDWARDS
C DESERTI BARNES & MCDUNNOUGH
LIFE HISTORY
AMER. ENT. SOC. TRANS. 50:234, 1925 COOLIDGE EGG, LARVA, CHRYSALIS
ENT. NEWS 16:86, 1905 GRUNDEL LIFE HIST. LARVA, CHRYSALIS (VIRGULTI) SEE NOTE BELOW
LOC. CIT. 17:140, 1906 COOLIDGE THE LARVA AND CHRYSALIS DEXCRIBED ABORE ARE REALLY MORMO
PACIFIC COAT ENT. SOC. PROC. P. 27, 1904 GUNDEL EGG, LARVA, CHRYSALIS
SOUTHERN CAL. ACAD. SCI. BULL. 29:22, 1930 COMSTOCK CHRYSALIS
LOC. CIT. 33:34, 1934 COMSTOCK & DAMMERS EGG OF DESERTI LIKE THAT OF VIRGULTI
COMSTOCK, J. A. BUTTER. CAL. P. 150, 1927 EGG*A 46; LARVA* A 47; CHRYSALIS
FOOD PLANTS
ERIOGONUM INFLATUM TORR. & FREM.
POLYGONUM SCANDENS LINN.

MORMONIA BOISDUVAL
SYN. SEE MORMO FELDER & FELDER

MORMONIA BOISDUVAL, ARGYNNIS-NYMPHALIDAE
SYN. ARGE STRECKER
AB BENJAMINI GUNDER
A ERINNA EDWARDS
AB CUNNINGHAMI OWEN
LIFE HISTORY
FOOD PLANTS
VIOLA SP. (VIOLETS)

MORMONICA NEUMOEGEN
SSP. SEE PROXIMA GUERIN-MENEVILLE

MORENSIS STERNITZKYI
SSP. SEE ICARIOIDES BOISDUVAL

MORPHEUS FABRICIUS
SYN. SEE THAROS DRURY

MORRISONARIUS(HENRY EDWARDS), PERO-GEOMETRIDAE
LIFE HISTORY
CAN. ENT. 75:186, 1943 MCGUFFIN LIFE HIST. LARVAL MANDIBLE* PL. 15, F. A
FOOD PLANTS
ABIES BALSAMEA(LINN.)
ASTER SP.
LARIX LARICINA(DUROI)
PICEA GLAUCA(MOENCH)
PINUS STROBUS LINN.

MORRISONI BARNES & BENJAMIN
SSP. SEE SOLIDAGINIS HUBNER

MORRISONI DRUCE
SYN. SEE LATIPENNIS BOISDUVAL

MORRISONI EDWARDS
SSP. SEE CETHURA FELDER & FELDER

MORRISONI(GROTE), EUPSILIA-NOCTUIDAE
SYN. CASTANEA STRAND
LIFE HISTORY
CAN. ENT. 16:30, 1884 THAXTER EGG, LARVA
NEW YORK STATE MUS. MEM. 8:705 1905 FELT LARVA(BRIEF)
FOOD PLANTS
PRUNUS SEROTINA EHRH.
QUERCUS SP. (OAKS)
ULMUS SP. (ELMS)

MORRISONI HOLLAND
SYN. SEE AIDEA GUERIN-MENEVILLE

MORRISONIANA RILEY
SYN. SEE GLADIARIA MORRISON

MORISONISTIGMA GROTE
SSP. SEE EXSERTISTIGMA MORRISON

MORS LUCAS
SYN. SEE GEOMETROIDES WALKER

MORTUORUM GUENEE
SSP. SEE RECTANGULA KIRBY

MORULA GROTE & ROBINSON, ACRONICTA-NOCTUIDAE
SYN. ULMI HARRIS
LIFE HISTORY
AMER. MUS. NAT. HIST. BULL. 14: 242, 1901 BEUTEN. LARVA
NEW YORK STATE CABIN. NAT. HIST. ANN. REPT. NO. 26:137, 1874 LINTNER LARVA
NEW YORK STATE MUS. MEM. 8:724 1905 FELT LARVA
PAPILIO 3:13, 1883 THAXTER EGG, LARVA
UNITED STATES NAT. MUS. PROC. 21:78, 1898 SMITH&DYAR EGG, LARVA* PL. 7, FF. 20, 21 COLOR OR UNCOLORED; COCOON
HARRIS-ENT. CORRESP. P. 312, 1869 LARVA*PL. 3 ULMI
INDEX AMER. ECON. ENT. VOL. 7:8
FOOD PLANTS
MALUS PUMILA MILL. (APPLE)
TILIA SP. (LINDENS)
ULMUS SP. (ELMS)

MOTYA BOISDUVAL & LECONTE
SYN. SEE BACHMANII KIRTLAND

MUCOREA HERRICH-SCHAEFFER
SYN. SEE GUTTIVITTA WALKER

MUELLERI LETCHER
SSP. SEE CARYA HUBNER

MUIRI HENRY EDWARDS
SSP. SEE NELSONI BOISDUVAL

MULIERCULA GUENEE, CATOCALA-NOCTUIDAE
AB PERAMANS HULZT
LIFE HISTORY
AMER. MUS. NAT. HIST. BULL. 16: 391, 1902 BEUTEN. LARVA
AMER. MUS. NAT. HIST. MEM. NEW SERIES 3, PART 1, 1918 BARNS & MCDUNNOUGH LARVA*PL. 12, F. 13 QUESTIONABLE COLOR
GUENEE-SPEC. GEN. LEP. NOCT. 3:

97,1852 LARVA* PL.2 COLOR
FOOD PLANTS
MYRICA CERIFERA LINN.

MULIERIS STRAND
SSP.SEE HARRISI GROTE

MULLEOLATA HULST
SSP.SEE CITRATA LINNAEUS

MULTICAUDATA KIRBY
SYN.SEE DAUNUS BOISDUVAL

MULTIFARIA(WALKER),CTENUCHA-AMATIDAE
FORM LUTEOSCAPUS NEUMOEGEN & DYAR
LIFE HISTORY
CAL.ACAD.SCI.PROC.5:344,1873
HENRY EDWARDS EGG,LARVA, PUPA
FOOD PLANTS
SYMPHORICARPOS ALBUS(LINN.)

MULTIFERATA(WALKER),EUPHYIA-GEOMETRIDAE
LIFE HISTORY
NEW YORK STATE AGRI.EXP.STA. (CORNELL),MEM.274:155,1948
FORBES LARVA
PSYCHE 10:193,1903 DYAR EGG, LARVA
FOOD PLANTS
EPILOBIUM ANGUSTIFOLIUM(LINN.)
POLYGONUM AVICULARE LINN.

MULTILINEA WALKER,LEUCANIA-NOCTUIDAE
SYN.SOLITA WALKER
SYN.LAPIDARIA GROTE
LIFE HISTORY
JOURN.ECON.ENT.8:389,1915
VICKERY LARVA LOOKS LIKE THAT OF LEUCANIA SUBPUNCTATA HARVEY(LATIUSCULA HERRICH-SCHAEFFER)
FOOD PLANTS
SACCHARUM OFFICINARUM LINN (SUGAR)
SORGHUM BICOLOR(LINN.)

MULTILINEATA PACKARD
SSP.SEE IMPLICATA GUENEE

MULTISCRIPTA RILEY
SSP.SEE SCITISCRIPTA WALKER

MULTNOMA DYAR
SSP.SEE BRUCEI HENRY EDWARDS (ICHTHYURA)

MULTOCONSPICUA REIFF
SSP.SEE ANTINYMPHA HUBNER

MUNDULA GROTE & ROBINSON
SYN.SEE ERECHTEA CRAMER

MUNIS(GROTE),EUXOA-NOCTUIDAE
SYN.DUBLATIS GROTE
LIFEHISTORY
FOOD PLANTS
BETA VULGARIS LINN.)SUGAR BEET)

MUNITATA(HUBNER),XANTHORHOE-GEOMETRIDAE
SYN.STRIGATA PACKARD
SYN.ANTICOSTIATA STRECKER
A CONVALLARIA GUENEE
B NEMORELLA HULST
SYN.ILLOCATA HULST
LIFE HISTORY
NEW YORK STATE AGRI.EXP.STA. (CORNELL),MEM.274:151,1948
FORBES LARVA
GUMPPENBERG-SYSTEMA GEOMETRARUM 3:360,1890 LARVA,PUPA
KIRBY-EUROPEAN BUTTER.MOTHS P.376,1882 LARVA
NEWMAN-BRITISH MOTHS P.168 1869 LARVA
PACKARD-MONOGRAPH GEOMETRIDAE' P.238,1876 LARVA
FOOD PLANTS
SENECIA SP.
GENERAL FEEDER ON LOW PLANTS

MURAENULA GROTE & ROBINSON
SYN.SEE VETUSTA WALKER(AGROTIS)

MURICOLOR(HULST)
SYN.SEE COGNATARIA HUBNER

MURINA(STRETCH), PYGARCTIA-ARCTIIDAE
SYN. POLYOCHROA HAMPSON
A OSLARI ROTHSCHILD
B ALBISTRIGATA BARNES & MCDUNNOUGH
LIFE HISTORY
SOUTHERN CAL. ACAD. SCI. BULL. 34:220, 1935 COMSTOCK-DAMMERS EGG, LARVA*PL. 59; PUPA* PLS. 59, 60
FOOD PLANTS
EUPHORBIA SP.

MUS NEUMOEGEN
SSP. SEE GRAGILIS STRETDH

MUSKOKA WATSON & COMSTOCK
SSP. SEE ACADICA EDWARDS(STRYMON)

MUSTELINA PACKARD
SYN. SEE LEPTINOIDES GROTE

MUSTELINA SMITH
SSP. SEE RUBICA HARVEY

MUTANS WALKER
SYN. SEE CUNEA DRURY

MUTATA BARNES & BENJAMIN
SSP. SEE VETUSTA WALKER(AGROTIS)

MUTATA PEARSALL
SSP. SEE TOGATA HUBNER

MUTICUM MCALPINE, CALEPHELIS-RIODINIDAE
LIFE HISTORY
BROOKLYN ENT. SOC. BULL. 33:111, 1938 MCALPINE EGG*LARVA* CHRYSALIS*
FOOD PLANTS
CIRSIUM MUTICUM MICHX.

MUZARIA WALKER
SYN. SEE OBTUSARIA HUBNER

MUZINA OBETHUR, ECPANTHERIA-ARCTIIDAE
SYN. ALBICOLLIS OBERTHUR
SYN. ABSCONDENS OBERTHUR
SYN. DEPAUPERATA OBERTHUR
SYN. SENNETTII LINTNER
LIFE HISTORY
BROOKLYN MUS. ARTS. &SCI. BULL. 1:194, 1906 DYAR LARVA
FOOD PLANTS

MYLITTA(EDWARDS), PHYCIODES-NYMPHALIDAE
SYN. COLLINA BEHR
SYN. EPULA BOISDUVAL
AB COLLINSI GUNDER
AB MACYI FENDER
A PALLIDA EDWARDS
AB MATA REAKIRT
B BARNESI SKINNER
C THEBAIS GODMAN & SALVIN
SYN. ARIDA SKINNER
LIFE HISTORY
CAL. ACAD. SCI. PRO. 5:168, 1873 HENRY EDWARDS LARVA, CHRYDSLIS
CAN. ENT. 23:203, 1891 DYAR EGG, LARVA, CHRYSALIS
MONTANA UNIV. BULL. 30:81, 1906 ELROD LARVA, CHRYSALIS
COMSTOCK, J. A. BUTTER. CAL. P. 188 1927, LARVA, CHRYSALIS
HOLLAND-BUTTERFLY BOOK P. 155 1905 LIFE HIST.
LOC. CIT. 3RD EDIT. P. 138, 1931 LIFE HISTORY
MEAD-REPT. LEP. WHEELER'S EXPED. WEST. 100 MERID., 5.764, 1875 LARVA, CHRYSALIS
FOOD PLANTS
CIRSIUM SP. (THISTLES)
CIRSIUM OCCIDENTALE(NUTT.)

MYOPS(ABBOT &SMITH), PAONIAS-SPHINGIDAE
SYN. ROSACEARUM BOISDUVAL
SYN. CERASI BOISDUVAL
SYN. SORBI BOISDUVAL
?SYN. TILIASTRI BOISDUVAL
A OCCIDENTALIS CLARK
B MCCREARYI CLARK
LIFE HISTORY
AMER. JOURN. SCI. 36:291, 1839 HARRIS LARVA, PUPA
AMER. MUS. NAT. HIST. BULL. 7:315

1895 Beuten. larva, pupa
Loc. cit. 8:297, 1896 Beuten. egg larva, pupa
Can. Ent. 8:239, 1876 Peck red blotched larva
Loc. cit. 16:11, 1884 Saunders larva
Loc. cit. 18:207, 1886 Bunker abnormal larva
Ent. news 9:150, 1898 Soule duration of egg stage
Ent. Soc. Amer. Annals 11:426, 1918 Mosher pupal structure
Minnesota Agri. Exp. Sta. Bull. 61:104,1899 Lugger larva
Nat. Acad. Sci. Mem. 9, 1905 Packard larva*Pl. 42, f. 3 structure; Pl. 44, f. 15 head
New York State Agri. Exp. Sta. (Cornell), Mem. 274L194, 1948 Forbes larva
New York State Mus. Mem. 8:733 1905 Felt larva(Brief)
New York State Rept. Nox. Ins. of 3:58, sec 79, 1856 Fitch larva
Novitates Zoo. (Tring.)9 Suppl. p. 330, 1903 Rothschild & Jordan larva, pupa
Ontario Ent. Soc. Ann. Rept. p. 61 1881 Reed larva
Loc. cit. p. 27, 1884 Saunders larva, pupa
Outlook 69no. 14, April1900 p. 827, Soule larva*
Philadelphia Acad. Nat. Sci. Journ. 5:182, 1859 Clemens larva, pupa
Psyche 5:266, 1889 Eliot &Soule egg, larva, pupa
Loc. cit. 8:155, 1897 Soule larval variety
Smithsonian Misc. Coll. 4, art, 2:208, 1862 Morris larva, pupa
United States Dept. Agri. Misc. pub. 657:358, 375, 1950 Craighead larva
Abbot & Smith-Lep. Ins. Georgie 1:51, 1797 larva*pupa*Pl. 26
Boisduval-Spec. Gen. Lep. Sphingidae p. 42, 1874 larva pupa
Eliot&Soule-Cater. &Moths p. 189, 1902 egg, larva*pupa
Fernald-Sphingidae New England p. 76, 1886 larva
Glover-Illust. North Amer. Ent. 1878 larva* Pl. 21, f. 7
Morris-Synop. Lep. NorthAmer. p. 208, 1862 larva
Saunders-Ins. Inj. Friits p. 208 1883 larva
Saunders-loc. cit. p. 208, 1889 larva
Strecker-Lep. Rhop. et. Heter. 7:56, 1873 larva

FOOD PLANTS

Betula sp. (birches)
Corylus sp. (haxels)
Crataegus sp. (thorns)
Populus sp. (poplars)
Prunus cerasus Linn. (cherry)
Prunus domestica Linn. (plum)
Prunus serotina Ehrh.
Salix sp. (willows)

myrina(Cramer), Brenthis-Nymphalidae
syn. myrissa Godart
ab nubes Dcudder
ab serratimarginata Gunder
ab nivea Gunder
a atrocostalis Huard
syn. albequina Holland
ab jenningsae Holland
ab baxteri Holland
b terrae-novae Holland
c nebraskensis Holland
d tollandensis Barnes & Benjamin

LIFE HISTORY

Amer. Mus. Nat. Hist. Bull. 5: 256, 1893 Beuten. larva, chrysalis
Buff. Soc. Nat. Sci. Bull. 19:23 1939 Wild larva
Cal. Acad. Sci. Proc. 5:328 1874 Edwards egg, larva, chrysalis

CAN. ENT. 1:55, 1869 SAUNDERS EGG, LARVA, CHRYSALIS
LOC. CIT. 7:189, 1875 EDWARDS NOTE ON LIFE HISTORY
LOC. CIT. 8:161, 1876 EDWARDS EGG, LARVA, CHRYSALIS
ILLINOIS DEPT. AGRI. TRANS. 18, APPEND:82, 161, 1880 COQUILLETT & MIDDLETONLARVA
MUS. COMP. ZOOL. BULL. 87, NO. 6, 1941 DETHIER LARVAL ANTENNA* F. 41
NAT. CAN. P. 117, 1927 HUARD LARVA
PAPILIO 4:91, 1884 BRUBER LARVA
COMSTOCK J.H. HOW KNOW BUTTER. P. 122, 1904 LARVA
FERNALD-BUTTER. MAINE P. 41, 1884 EGG, LARVA* CHRYSALIS
FRENCH-BUTTER. EAST. UNITED STATES P. 162, 1886 EGG*LARVA CHRYSALIS
HOLLAND-BUTTERFLY BOOK P. 129 1905 CHRYSALIS*PL. 5, FF. 12-14;3RD EDIT. P. 106, 1931 LIFE HISTORY
MAYNARD-BUTTER. NEW ENGLAND P. 25, 1886 LARVA
PACKARD-GUIDE STUDY INS. P. 254, 1869 LARVA, CHRYSALIS
SCUDDER-BUTTER. EAST. UNITED STATES & CANADA PP. 596-600 1889 LIFE HIST. *PL. 64, F, 28; PL. 67, F. 18; PL. 72. F. 2; PL. 75 FF. 2, 3; PL. 79, FF. 7-9; PL. 86, F. 81; PL. 84, FF. 12-14
SCUDDER-BUTTER. THEIR STRUCTURE ETC. P. 8, 1881 EGG*
SCUDDER-EVERYDAY BUTTERP./318, 1889 EGG, LARVA, CHRYSALIS
SCUDDER-GUIDE TO COMMON BUTTER. P. 75, 1893 LARVA, CHRYSALIS

FOOD PLANTS
VIOLA SP. (VIOLETS)
FERNS?(SCUDDER)

MYRINA MARTYN
SEE BELLONA FABRICIUS

MYRISSA GODART
SYN. SEE MYRINA CRAMER

MYRON(CRAMER), AMPELOECA-SPHINGIDAE
SYN. PAMPINATRIX ABBOT & SMITH
AB LUTESCENS CLARK
FORM CNOTUS HUBNER
A TEXANA CLARK

LIFE HISTORY
AMER. ENT. 2:22, 1869 LARVA*PUPA*
AMER. JOURN. SCI. 36:302, 1839 HARRIS LARVA, PUPA PAMPINATRIX
AMER. MUS. NAT. HIST. BULL. 7:292, 1895 BEUTEN. LARVA, PUPA
AMER. NAT. 4:612, 1871 LARVA* PUPA* F. 128 NO DESCR.
CAN. ENT. 3:66, 1871 SAUNDERS LARVA*PUPA*
ENT. NEWS 9:150, 1898 SOULD DURATION EGG STAGE
ENT. SOC. AMER. ANNSL 11:434, 1918 MOSHER PUPAL STRUCTURE
ILLINOIS DEPT. AGRI. TRANS. 15: 166, 1877 FRENCH LARVA
LOC. CIT. 18 APPEND:103, 1880 MARTEN LARVA*
MICHIGAN STATE COL. SPEC. BULL. 243, 49, 1933 MCDANIEL LARVA
MINNESOTA AGRI. EXP. STA. BULL. 61:97, 1899 LUGGER LARVA* F. 44; PUPA* F. 45
MISSOURI REPT. INS. OF 2:71, 73, 1870 RILEY LIFE HIST*LARVA* F. 45
NAT. CAN. P. 113, 1928 HUARD LARVA
NEW YORK STATE AGRI. EXP. STA. (CORNELL), MEM. 274:199, 1948 FORBES LARVA
NEW YORK STATE CABIN. NAT. HIST. ANN. REPT. NO, 42:175, 1889 LINTNER LARVA*
NEW YORK STATE MUS. MEM. 8:739 1905 FELT LARVA(BRIEF)
NEW YORK STATE REPT. INS. OF 3: 397, 1856 FITCH LARVA ALSO P. 79, SEC. 118
NOVITATES ZOOL. (TRING)9 SUPPL

P.524,1903 ROTHSCHILD & JORDAN LARVA
ONTARIO ENT.SOC.ANN.REPT.P.29 1871 SAUNDERS EGG*LARVA* PUPA*
LOC.CIT.P.58,1881 REED LARVA* PUPA*
LOC.CIT.P.10,1891 BETHUNE LARVA*
PHIL.ACAD.NAT.SCI.JOURN.5:147 1859 CLEMENS LARVA
PHIL.ENT.SO..PROC.3:663 1864 LINTNER LARVA,PUPA
PSYCHE 5:400,1890 PACKARD EGG, LARVA
SMITHSONIAN MISC.COLL.4,ART.2: 169,1862 MORRIS LARVA,PUPA
UNITED STATES DEPT.AGRI.FARMER'S BULL.1220:34,1921 QUAINTANCE LARVA
UNITED STATES DEPT.AGRI.MISC. PUBL657:359,1950 CRAIGHEAD LARVA
VERMONT BD.AGRI.REPT.5:277,1878 PERKINS LARVA*
ABBOT & SMITH-LEP.INS.GA.1:55 1797 LARVA*PL.28 PAMPINATRIX
BALLARD-INS.LIVES P.67,1879 LARVA*
BOISDUVAL-SPEC.GEN.LEP.SPHING. P.210,1874 LARVA,PUPA
CRAGEN-OUR INS.FRIENDS & FOES P.137,1899 LARVA* F.74
DICKERSON-MOTHS.BUTTER.P.238 1901 LARVA* F.205;PUPA*F.206
ELIOT&SOULE-CATER.&MOTHS P. 115,1902 EGG,LARVA*PUPA*
FERNALD-KINGSLEY'SSTAND.NAT. HIST.2:465,1885 EGG,LARVA
FERNALD-SPHINGIDAE NEW ENDLAND P.66,1886 EGG*LARVA*PUPA*
GLOVER-ILLUST.NORTH AMER.ENT. 1878 LARVA*PL.4,F.3;PL.5, F.4
HARRIS-ENT.CORRESP.P.283,1869 LARVA*PL.1,F.10
HARRIS-INS.INJ.VEG.1ST EDIT.P. 229,1841 LARVA,PUPA
HARRIS-LOC.CIT.2ND EDIT.P.249 1852 LARVA
HART-INS.COLL.ILLINOIS HIGH SCHOOLS P.16,1903 LARVA* F.11
HOLLAND-MOTH BOOK P.68,1905 LARVA*PUPA*FF.29,30,31
KELLOGG-AMER.INSECTS P.435 1908 LARVA
MORRIS-SYNOP.LEP.NO.AMER. P.169,1862 LARVA,PUPA
O'KANE-INJ.INSECTS P.272, 1912 LARVA
SAUNDERS-INS.INJ.FRUITS P.245 1883 EGG,LARVA*PUPA*
SAUNDERS-LOC.CIT.P.245,1889 EGG,LARVA*F.253;PUPA*F.254
INDEX AMER.ECON.ENT. VOL.1:9;VOL.1:101 DARAPSA VOL.2:15;VOL.3:18 DARAPSA VOL.5:24 DARAPSA;VOL.7:33

FOOD PLANTS

PARTHENOCISSUS QUINQUEFOLIA(LINN.)
VIBURNUM SP.
VITIS SP.(GRAPES)
VITIS VINIFERA LINNAEUS

MYRTARIA (GUENEE),COXYMBIA-GEOMETRIDAE
SYN.IGNOTARIA WALKER
FORM TRISERIATA PROUT

LIFE HISTORY

CAN.ENT.9:62,1877 GOODELL LARVA,PUPA
GUMPPENBERG-SYSTEMA GEOMETRARUM 5:292,1892 LARVA,PUPA

FOOD PLANTS

COMTPNIA SP.
COMPTONIA PEREGRINA(LINN.)
GAYLUSSACIA DUMOSA(ANDR.)
GAYLUSSACIA FRONDOSA(LINN.)
PSORALEA PSORALIOIDES(WALT.)
QUERCUS SP.(OAKS)

MYRTO GODART
SYN.SEE MONIMA CRAMER

MYSIA GODART
SYN.SEE ILAIRE GODART

MYSTIC(SCUDDER), POLITES-HESPERIIDAE
AB WEETAMOO SCUDDER
AB NUBS SCUDDER
A DECOTAH EDWARDS
SYN. PALLIDA SKINNER
LIFE HISTORY
AMER. MUS. NAT. HIST. BULL. 5:293, 1893 BEUTEN. LARVA
CAN. ENT. 1:65, 1869 SAUNDERS EGG LARVA
ENT. NEWS 19:412, 1908 LARU ENT EARLY STAGES INCLUDING EGG
ILLINOIS SEPT. AGRI. TRANS. 18, APPEND:97, 1880 MIDDLETON LARVA
SOUTHERN CAL. ACAD. SCI. BULL. 39: 108, 1940 COMSTOCK LARVA, CHRYSALIS*PL. 19
LOC. CIT. 41:41, 1942 DETHIER COMPARES LARVA WITH THAT OF THEMISTOCLES LATREILLE
FERNALD-BUTTER. MAINE P. 99, 1884 EGG, LARVA
FRENCH-BUTTER. EASTUNITED STATES P. 319, 1886 EGG, LARVA
MAYNARD-BUTTER. NEW ENGLAND P. 62, 1886 LARVA
SCUDDER-BUTTER. EAST. UNITED STATES & CANADA PP. 1707-1710 1889 EGG*LARVA*PL. 66, FF. 20, 24; PL. 69, F. 13; PL. 80, FF, 58, 59
SCUDDER-EVERYDAY BUTTER P. 186 1899 EGG, LARVA
SCUDDER-GUIDE TO COMMON BUTTER P. 172, 1893 LARVA, CHRYSALIS
FOOD PLANTS
AGROPYRON REPENS(LINN.)
CAREX SP.
ECHINOCHLOA CRUS-GALLI(LINN.)
PHELUM PRATENSE LINN.
GRASSES

N

NAEVIUS LINTNER
SYN. SEE ZARUCCO LUCAS

NAIDARIA SWETT
SYN. SEE PULCHERRIMA McDUNNOUGH

NAIDELLA STRAND
SSP. SEE NAIS DRURY

NAIS(DRURY), APANTESIS-ARCTIIDAE
SYN. CUNNEATA GOEZE
SYN. DEFLORIANA MARTYN.
SYN. OCHREATA BUTLER
AB NAIDELLA STRAND
AB SUBTERMINALIS STRAND
LIFE HISTORY
AMER. MUS. NAT. HIST. BULL. 10: 381, 1898 BEUTEN. LARVA
BROOKLYN ENT. SOC. BULL. 6:69 1883 HULST EGG, LARVA
CAN. ENT. 16:65, 1884 BEAN DURATION PUPA
ENT. NEWS 9:150, 1898 SOULE DURATION EGG STAGE
NEW YORK ENT. SOC. JOURN. 8:38 1900 DYAR LARVA(BRIEF)
LOC. CIT. 4:4, 1902 SEIFERT EGG, LARVA, PUPA
PAPILIO2:176, 1882 FRENCH EGG, LARVA, PUPA
FERNALD-KINGSLEY'S STAND. NAT. HIST. 2:460, 1885 EGG, LARVA
GLOVER-ILLUST. NORTH AMER. ENT. 1878 LARVA* PL. 8, F. 3
FOOD PLANTS
ARTEMISIA ABSINTHIUM LINN.
FRAGARIA CHILOENSIS(LINN.)
GRAMINEAE
PLANTAGO MAJOR LINN.
POLYGONUM ERECTUM LINN.
POLYGONUM PUNCTAUM ELL.
RUMEX ACETOSELLA LINN.
TARAXACUM OFFICINALE WEBER
TRIFOLIUM SP. (CLOVERS)
VIOLA SP. (VIOLETS)
LOW PLANTS

NAIS(EDWARDS)APODEMIA-RIODINAAE
LIFE HISTORY
EDWARDS-BUTTER. NORTH AMER. 2ND SERIES 1883 LIFE HIST*

Pl. 11 Lemonias No. 1 COLOR
Holland-Butterfly Book p. 231
1905 egg, larva, chrysalis
Loc. cit. 3rd edit. p. 215, 1932
life history
FOOD PLANTS
Prunus sp.
Prunus ilicifolia Walp. (Nutt. ex. H. &A.)

nanaimo Barnes, Ipimorpha-Noctuidae
LIFE HISTORY
FOOD PLANTS
Populus nigra v. italica Muench(?)

nanus Neumoegen
ssp. see smintheus Doubleday & Hewitson

napa Edwards
ssp. see sylvanoides Boisduval (Ochlodes)

napeae (Morrison), Litholomia-Noctuidae
a umbrifasciata Blackmore
LIFE HISTORY
FOOD PLANTS
Amelanchier florida Lindl.
Salix hookeriana Barratt
(both plants food for umbrifasciata)

napi (Linnaeus), Pieris-Pieridae
for the many ssps of this insect see McDunnough's List
LIFE HISTORY
Amer. Ent. 2:76, 1870 larva, pupa (oleracea Harris)
Amer. Mus. Nat. Hist. Bull. 5:247 1893 Beuten. larva, chrysalis
Can. Ent. 5:37, 1873 Bethune life hist. larva* (oleracea)
Can. Nat. &Geol. 2:347, 1857 egg, larva
Ent. News 19:314, 1908 Coolidge & Newcomer egg, larva, chrysalis (castoria Reakirt)
Illinois Dept. Agri. Trans. 15:143, 1877 French larva* (oleracea)
Loc. cit. 18, Append:77, 1880 Middleton larva* (oleracea)
Missouri Rept. Ins. of 2:105, 1870 Riley larva* chrysalis* (oleracea)
Nat. Can. p. 66, 1927 Huard larva (oleracea)
New England Farmer 7:402, 1829' Harris larva (oleracea)
New York State Cabin. Nat. Hist. Rept. No. 23:160, 1870 Linther egg, larva
Ontario Ent. Soc. Ann. Rept. p 37, 1894 Bethune larva, chrydalis
Phil. Ent. Soc. Proc. 3:54 1864 Lintner larva, chrysalis
Psyche 5:166, 1889 Scudder larva compared with that of European napi
Psyche 6:70, 1891 Scudder chrysalis
United States Dept. Agri. Rept. Ent. p. 79, 1807 Glover larva chrysalis*
Loc. cit. p. 108, 1883 Riley egg* larva* Pl. 1 oleracea
Agassiz-Lake Superior p. 386 larva (oleracea)
Boisduval-Spec. Gen. Lep. Diurnae 1:519, 1836 larva, chrysalis
Chittenden-Ins. Inj. Veg. p. 139 1907 larva* f. 92
Comstock, J. A. Butter-Cal. p. 33 1927 egg, larva, chrysalis
Comstock, J. H. How Know Butter. p. 76, 1904 larva
Crosby &Leonard-Manual Veg. Gard. Ins. p. 7, 1918 larva
Emmons-Nat. Hist. New York 5: 205, 1854 larva, chrysalis
Fernald-Butter. Maine p. 25, 1884 egg, larva* chrysalis
French-Butter. East. United States p. 112, 1886 life hist*
Harris-Ent. Corresp. p. 361, 1869

LARVA,CHRYSALIE*(OLERACEA)
HARRIS-INS.INJ.VEG.2ST EDIT. P.214,1841 LARVA CHRYSALIS (OLERACEA)
HARRIS-LOC.CIT.2ND EDIT.P.233 1852 LIFE HIST.)OLERACEA)
HARRIS-LOC.CIT.FLINT EDIT.P.270 1862 EGG,LARVA,CHRYSALIS (OLERACEA)
HOLLAND-BUTTERFLY BOOK 1905 LARVA*PL.2,FF 8,9:CHRYSALIS* PL.5,FF.57,63,64
MAYNARD-BUTTER.NEW ENGLAND P.47,1886 LARVA,CHRYSALIS
PACKARD-HALF HOURS INS.P.52, 1877 LARVA*(OLERACEA)
PACKARD-INS.INJ.WEST P.746,1877 EGG*LARVA*(OLERACEA)
SCUDDER-BUTTER.EAST.UNITED STATES &CANADA PP.1194-1195;1198-1203,1889 LIFE HIST*PL.65,F.27;PL.68,F.15; PL.76,FF.8.9;PL.79,FF.50-52 PL.86,FF.32,44;PL.84,FF.57, 63,64
SCUDDER-BUTTER.THEIR STRUCTURE ETC.P.165,1881 EGG*CHRYSALIS* (OLERACEA)
SCUDDER-GUIDE TO COMMON BUTTER P.143,1893 LARVE CHRYSALIS
STEPEHENS-ILLUST.BRITISH ENT. HAUST.1:20,1828 LARVA, CHRYSALIS
INDEX AMER.ECON.ENT.
VOL.1:252 PONTIA
VOL.2:294 PONTIA
VOL.7:780

FOOD PLANTS
ARABIS DRUMMONDI GRAY
ARABIS GLABRA(LINN.)
ARMORACIA LAPATHIFOLIA GILIB.
BARBAREA VULGARIS R.BR.
BERBERIS VULGARIS LINN.
BRASSICA HIRTA MOENCH(MUSTART)
BRASSICA NIGRA(LINN.)(MUSTARD)
BRASSICA OLERACEA LINN.(CABBAGE)
BRASSICA RAPA LINN.(TURNIP)
CRUCIFERAE SP.
DENTARIA DIPHYLLA MICHX.
DENTARIA INTEGRIFOLIA V. CALIFORNICA(NUTT.)
RAPHANUS SATIVUS LINN.(RADISH)

NARGENTA OTTOLENGUI
SSP.SEE RECTANGULA KIRBY

NARRATA WALKER
SYN.SEE ERECHTEA CRAMER

NASONI (GROTE), NATATA-NOTODONTIDAE
LIFE HISTORY
NEW YORK ENT.SOC.JOURN.7:63 1899 DYAR EGG,LARVA,COCOON
FOOD PLANTS
CARYS SP.(HICKORIES)
CASTANEA DENTATA(MARSH.)
FAGUS SP.(BEECHES)
OSTRYA VIRGINIANA(MILL.)
QUERCUS VELUTINA LAM.

NASTES BOISDUVAL,COLIAS-PIERIDAE
A ROSSII GUENEE
B GUENEEI AVINOFF
C MOINA STRECKER
SYN.COCANDICIDES VERITY
AB HARPERI GUNDER
D SUBARCTICA MCDUNNOUGH
E ALASKAE BANG-HAAS
F STRECKERI GRUM GRSCHIMAJLO
AB PALLIFLAVA MCDUNNOUGH
G OBSCURATA VERITY
LIFE HISTORY
ENT.45:122-125,1912 SHELDON LIFE HISTORY
FOOD PLANTS
ASTRAGALUS ALPINUS LINN.
HEDYSARUM ALPINUM V.AMERICANUM MICHX.
SALIX SP.(WILLOWS)
TRIFOLIUM REPENSLINN.

NASTURTII BOISDUVAL
SYN.SEE NAPI LINNAEUS

NASUTARIA ZELLER
SYN.SEE CALLITRICHOIDES GROTE

NAVAJO SKINNER
SSP.SEE YUCCAE BOISDUVAL & LEDONTE

NEBRASKAE DODGE
SSP.SEE LUCIANA LUCAS

NEBRASKENSIS HOLLAND
SSP.SEE MYRINA CRAMER

NEBRIS(GUENEE), PAPAIPEMA-NOCTUIDAE
FORM NITELA GUENEE
LIFE HISTORY
AMER.ENT.1:22,1868 LARVA*
LOC.CIT.P.206,1869 LARVA*
AMER.ENT.2ND SERIES 1:201,1880 FULLER LARVA*
AMER.MUS.NAT.HIST.BULL.5:93, 1893 BEUTEN.EGG,LARVA
CAN.ENT.30:128,1898 BIRD LARVA, PUPA
ENT.NEWS 1:137,1890 BLAKE LARVA
ILLINOIS DEPT.AGRI.TRANS.15:113 221,1877 FRENCH &SMITH LARVA PUPA
LOC.CIT.17,APPEND:142,1879 THOMAS LARVA*
LOC.CIT.18,APPEND:151,1880 COQUILLETT LARVA*
ILLINOIS,REPT.INS.OF 2:141 1872 LEBARON LARVA*
LOC.CIT.7:221,1878 FRENCH LARVA
JOURN.ECON.ENT.12:268,1919 MOSHER LARVA*SETAL MAPS AND CLAWS
MICHIGAN AGRI.EXP.STA.SPEC. BULL.183:28,1931 PETTIT LARVA* F.16
MINNESOTA AGRI.EXP.STA.BULL. 61:221,1899 LUGGER LARVA* F.167
LOC.CIT.84,7,1903 WASHBURN LARVA* F.10 COLOR
MISSOURI,REPT.INS.OF 1:92,1869 RILEY LARVAL STAGES*
NEW HAMPSHIRE AGRI.EXP.STA.TECH BULL34:7,1927 LOWERY EGG, LARVA,PUPA
NEW YORK ENT.SOC.JOURN.7:70 1899 DYAR LARVAL STRUCTURE ONLY* PL.2,F.5
NEW YORK STATE,AGRI.SOC.REPT. P.50,1879 LINTNER LARVA*
NEW YORK STATE CABIN.NAT.HIST. ANN.REPT.NO.45:191,1891 LINTNER LARVA*
NEW YORK STATE MUS.BULL.37:28 1900 FELT LARVA*
OHIO AGRI.EXP.STA.BULL.45:176 1893 WEBSTER LARVA*
ONTARIO ENT.SOC.ANN.REPT.P.83 1905 FLETCHER LARVA*PUPA*
LOC.CIT.P.49,1906 BETHUNE LARVA* PUPA*
UNITED STATES DEPT.AGRI.BUR. ENT.BULL.57 NEW SERIES P. 38,1906 SANDERSON LARVA* F.22 B,C,D;PUPA*F.22 E
UNITED STATES DEPT.AGRI.FARMER'S BULL.223:17,1905 SANDERSON LARVA* F.16 B,C,D;PUPA*F. 16 E
LOC.CIT.890:25,1917 PIERCE LARVA* F.32 B,C;PUPA* F.32E
LOC.CIT.1495:25 1926 WEIGEL & MIDDLETON LARVA* F.35
VERNOMT BD.OF AGRI.REPT.3:556 1876 PERKINS LARVA*
WISCONSIN AGRI.SOC.TRANS.13:11 1883 FORBES LARVA*
CHITTENDEN-INS.INJ.VEG.P.200, 1907 LARVA*F.127 PUPA*F.127
CROSBY & LEONARD-MANUAL VEG. GARD.INS.P.158,1918 EGG, LARVA*F.91;PUPA
FER;NALD-KINGSLEY'S STAND.NAT. HIST.2:451,1885 LARVA
HOLLAND-MOTH BOOK P.213,1905 LARVA* F.121
METCALF-FLINT-DESTR.&USEFUL INS.P.365,1939 EGG,LARVA* PUPA*BOTH F.248
O'KANE-INJ.INS.P.133,1912 LARVA* F.115
OSBORN-MEADOW&PASTURE INS. P.230,1939 LARVA
PACKARD-GUIDE STUDY INS.P.310

1869 LARVA*
PACKARD-INS. INJ. WEST P. 719, 1877 LARVA
PEAIRS-PESTS FARM, GARD. ORCHARD P. 110, 1941 LARVA* F. 120
SAUNDERS-INS. INJ. FRUITS P. 334, 1883 LARVA*
SAUNDERS-LOC. CIT. P. 334, 1889 LARVA* F. 347
INDEX AMER. ECON. ENT.
VOL. 1:230 NITELA
VOL. 2:263; VOL. 3:294 NITELA
VOL. 4:342; VOL. 5:468;
VOL. 6:555; PAIPEMA LAP. CAL. NITELA
VOL. 6:558; VOL. 7:740

FOOD PLANTS

ACER SACCHARIUM LINN.
ACHILLEA MILLEFOLIUM LINN.
ALTHAEA ROSEA CAV
AMBROSIA ARTEMISIIFOLIA LINN.
AMBROSIA TRIEIDA LINN.
APIUM GRAVEOLENS LINN. (CELERY)
ARCTIUM LAPPA LINN.
ASCLEPIAS SP.
ASPARAGUS OFFICINALIS LINN.
ASTER SP.
ATROPA BELLADONNA LINN.
AVENA SSATIVA LINN. (OATS)
BETA VULGARIS LINN. (BEETS)
BRASSICA OLERACEA LINN. (CABBAGE)
BRASSICA OLERACEA V. BOTYRTIS DC
CALENDULA OFFICINALIS LINN.
CALLISTEPHUS CHINENSIS (LINN.)
CANNA SP.
CANNABIS SATIVA LINN. (HEMP)
CAPSICUM ANNUM LINN. (PEPPERS)
ÇATALPA SP.
CHENOPODIUM ALBUM LINN.
CHRYSANTHEMYM SP.
CHRYSANTHEMUM LEUCANTHEUM LINN.
CIRSIUM SP.
CIRSIUM ARVENSE (LINN.)
COLLINSONIA CANADENSIS LINN.
CUCUMIS MELO LINN. (MELONS)
CUCURBITA PEPA LINN. (PUMPKIN)
DACTYLIS GLOMERATA LINN.
DAHLIA VARIABILIS DESF.
DELPHINIUM SP.
DIANTHUS CARYOPHYLLUS LINN.
FRAGARIA CHILOENSIS (LINN.) (STRAWBERRY)
FRAXINUS SP. (ASH SHOOTS)
GAILLARDIA SP.
GLADIOLUS SP.
GOSSYPIUM HERBACEUM LINN. (COTTON)
HELIANTHUS SP.
HIBISCUS MOSCHEUTOS LINN.
HORDUM VULGARE LINN.
IRIS SP.
LILIUM SP.
LILIUM CANDIDUM LINN
LILIUM SUPERBUM LINN.
MALUS PUMILA MILL. (APPLE)
MELILOTUS ALBA DESR.
MONARD DIDYMA LINN.
NICOTIANA TABACUM LINN. (TOBACCO)
PAEONIA OFFICINALIS LINN.
PHASEOLUS VULGARIS LINN. (BEANS)
PHLEUM PRATENSE LINN.
PHLOX SP.
PISUM SATIVUM LINN. (PEAS)
POA ANNUA LINN.
POA PRATENSIS LINN.
POLYGONUM HYDROPIPER LINN.
PRUNUS DOMESTICA LINN. (PLUM)
PRUNUS PERSICA (LINN.) (PEACH)
RANUNCULUS ACRIS LINN.
RHEUM RHAPONTICUM LINN. (RHUBARB)
RIBES LACUSTRE (PERS.)
RIBES SATIVUM SYME (CURRANT)
RICINUS COMMUNIS LINN.
ROBINIS PSEUDO-ACACIA LINN.
ROSA SP.
RUBUS ALLEGHENIENSIS PORTER
RUBUS IDAEUS V. STRIGOSUS (MICHX.)
RUDBECKIA SP.
RUDBECKIA HIRTA LINN.
RUMEX CRISPUS LINN.
RUMEX PATIENTA LINN.
SALIX XP. (WILLOWS)
SALVIA SP.
SAMBUCUS CANADENSIS LINN.
SAPONARIA OFFICINALIS LINN.
SECALE CEREALE LINN. (RYE)
SOLANUM LYCOPERSICUM LINN. (TOMATO)

Solanum melongena Linn. (egg plant)
Solanum tuberosum Linn. (potato)
Spinacia oleracea Linn. (spinach)
Spiraea sp.
Trifolium pratense Linn.
Triticum aestivum Linn. (wheat)
Verbascum thapsus Linn.
Vitis sp. (grapes)
Vitis vinifera Linn. (grape)
Xanthium strumarium Linn.
Zea mays Linn. (corn)
Zinnia elegans Jacq. (Giant Bur Elder"

NEBULOSA Edwards, CATOCALA-Noctuidae
syn. PONDEROSA Grote
LIFE HISTORY
FOOD PLANTS
Carya sp. (hickories)
Juglans nigra Linn.

NEBULOSA Hulst
syn. see MISERULATA Grote (Eupithecia)

NEBULOSA Neumoegen
syn. see BISECTA Lintner

NEBULOSARIA Freyer
syn. see CAMBRICA Curtis

NECOPINA (Grote), PAPAIPEMA-Noctuidae
syn. IMPERTURBATA Bird
LIFE HISTORY
Can. Ent. 30:131, 1898 Bird larva, pupa
Loc. cit. 39:274, 1907 Bird larva, pupa
FOOD PLANTS
Arctium lappa Linn.
Helianthus divaricatus Linn.
Helianthus grosseserratus Mart.
Helianthus tuberosus Linn.

NEGLECTA Edwards
ssp. see PSEUDARGIOLUS Boisduval & Leconte

NEGLECTA-MAJOR Tutt
ssp. see PSEUDARGIOLUS Boisduval & Leconte

NELITA (Strecker), PAPAIPEMA-Noctuidae
form LINDA Bird
ab ORBICULARIS Strand
LIFE HISTORY
Can. Ent. 48:13, 1916 Bird larva
FOOD PLANTS
Rudbeckia laciniata Linn.

NELSONI (Boisduval), MITOURA-Lycaenidae
ab EXOLETA Henry Edwards
a MUIRI Henry Edwards
LIFE HISTORY
Southern Cal. Acad. Sci. Bull. 31:88, 1932 Comstock & Dammers life hist., egg* larva* both Pl. 24
FOOD PLANTS
Fragaria chiloensis (Linn.)(?)
Libocedrus decurrens Torr.
Thyya sp. (cedars)

NELSONI Edwards
ssp. see OCCIDENTALIS Reakirt

NEMESIS (Edwards), CALEPHELIS-Riodinidae
syn. AUSTRALIS Edwards
LIFE HISTORY
Southern Cal. Acad. Sci. Bull. 31:12, 1932 Comstock & Dammers life hist. egg* larva* Pls. 5, 6; chrysalis* Pl. 7
FOOD PLANTS
Baccharis glutinosa Pers.
Bebbia juncea (Benth.)

NEMORELLA Hulst
ssp. see MUNITATA Hubner

NEMORIS Edwards
syn. see HEGON Scudder

NEMORUM (Boisduval), OCHLODES-Hesperiidae
syn. PRATINCOLA Boisduval

LIFE HISTORY
SOUTHERN CAL. ACAD. SCI. BULL. 28:56, 1929 COMSTOCK EGG*
FOOD PLANTS

NEOGAMA (ABBOT & SMITH), CATOCALA- NOCTUIDAE
SYN. COMMUNIS GROTE
SYN. SNOWIANA GROTE
A LORETTA BARNES & MCDUNNOUGH
B MILDREDAE FRANCLEMONT
LIFE HISTORY
AMER. MUS. NAT. HIST. BULL. 16:385 1902 BEUTEN. LARVAL HEAD* F. 5
LOC. CIT. 38:156, 1918 BARNES & MCD. EGG, LARVA
AMER. MUS. NAT. HIST. MEM. NEW SERIES 3, PART 1, 1918 BARNS & MCDUNNOUGH LARVA* PL. 10, F. 29; PL. 11, FF. 8, 9 COLOR
BROOKLYN ENT. SOC. BULL. 7:51 1884 HULST LARVA
CAN. ENT. 33:199, 1901 DODGE LARVA
NEW YORK STATE MUS. MEM. 8:719, 1905 FELT LARVA (BRIEF)
ABBOT & SMITH-LEP. INS. GEORGIA 2: 175, 1797 LARVA* PL. 88 COLOR
DUNCAN-NAT. LIBRARY 7:203, 1841 LARVA* PL. 26, F. 1 COLOR
GUENEE-SPEC. GEN. LEP. NOCT. 3: 96, 1852 LARVA
FOOD PLANTS
CARYA SP. (HICKORIES)
JUGLANS NIGRA LINN.
JUGLANS REGIA LINN.
QUERCUS SP. (OAKS)

NEOMEXICANA BARNES & BENJAMIN
SSP. SEE IO FABRICIUS

NEOMEXICANA BREHME
SYN. SEE OSLARI ROTHSCHILD (ANISOTA)

NEOMEXICANA SMITH
SSP. SEE CANDEFACTA HUBNER

NEONARIA WALKER
SYN. SEE AMICARIA HERRICH-SCHAEFFER

NEOTELIS SMITH
SYN. SEE TESSELLATA HARRIS

NEOTERICA SMITH
SSP. SEE DTRACTA WALKER

NEPHELE KIRBY
SSP. SEE ALOPE FABRICIUS

NEPHELEPTENA (DYAR), PAPAIPEMA- NOCTUIDAE
SYN. MOESERI BIRD
LIFE HISTORY
CAN. ENT. 45:120, 1913 BIRD LARVA, PUPA (MOESERI)
LOC. CIT. 48:15, 1916 BIRD LARVA
FOOD PLANTS
CHELONA GLABRA LINN.

NEPI BARNES & BENJAMIN
SSP. SEE INTERIOR SCUDDER

NEPIASARIA WALKER
SSP. SEE AMICARIA HERRICH-SCHAEFFER

NEPTARIA (GUENEE), SEMIOTHISA- GEOMETRIDAE
SYN. FLAVOFASCIATA PACKARD
SYN. CINEREATA BATES
A TRIFASCIATA PACKARD
B SINUATA PACKARD
LIFE HISTORY
NEW YORK STATE AGRI. EXP. STA. (CORNELL), MEM. 274:48, 1948 FORBES LARVA
PSYCHE 12:115, 1905 DYAR EGG, LARVA
UNITED STATES NAT. MUS. PROC. 25:389, 1903 DYAR LARVA (TRIFASCIATA)
FOOD PLANTS
ALNUS SP. (ALDERS)
POPULUS BALSAMIFERA LINN.
SALIX SP. (WILLOWS)

NEPTIS SMITH
    SSP.SEE ALBILINEA HUBNER

NEREA BOISDUVAL
    SSP.SEE DORIS BOISDUVAL

NERISSA HENRY EDWARDS
    SSP.SEE AMICA HUBNER

NERVOSA NEUMOEGEN & DYAR
    SSP.SEE ARGE DRURY

NESILENS SMITH
    SSP.SEE TRISTICULA MORRISON

NESSUS(CRAMER), AMPHION-SPHINGIDAE
    A FLORIDENSIS CLARK
    LIFE HISTORY
        AMER.MUS.NAT.HIST.BULL.7:282 1895 BEUTEN.LARVA
        CAN.ENT.9:20,1877 ANDREWS LARVA(BRIEF),PUPA
        LOC.CIT.29:28,1897 BEUTEN.EGG LARVA
        ENT.SOC.AMER.ANNALS 11:437, 1819 MOSHER PUPA
        MINNESOTA AGRI.EXP.STA.BULL. 61:86,1899 LUGGER LARVA
        NEW YORK STATE AGRI.EXP.STA. (CORNELL),MEM.274:197, 1948 FORBES LARVA
        NEW YORK STATE MUS.MEM.8:740 1905 FELT LARVA(BRIEF)
        NOVITATES ZOOL.(TRING)9 SUPPL. P.606,1903 ROTHSCHILD & JORDAN LARVA,PUPA
        ONTARIO ENT.SOC.ANN.REPT.P 52,1881 REED LARVA,PUPA
        PAPILIO 2:147,1882 SAUNDERS LIFE HISTORY
        PSYCHE 7:212 1895 SOULE LARVA PUPA
        PSYCHE 7:272,1895 SOULE LARVA
        UNITED STATES DEPT.AGRI.MISC. PUB.657:349,;1950 CRAIGHEAD LARVA
        BOISDUVAL-SPEC.GEN.LEP.SPHING P.317,1874 LARVA*PL.1 COLOR
        ELLIOT&SOULE-CATER.&MOTHS P.79,1902 EGG,LARVA*PUPA
        FERNALD-SPHINGIDAE NEW ENGLAND P.21,1886 LARVA
        KELLOGG-AMER.INS.P.438,1908 LARVA
    FOOD PLANTS
        EPILOBIUM COLORATUM BIEHLER
        FUCHSIA SP.
        OENOTHERA BIENNIS LINN.
        PARTHENOCISSUS QUINQUEFOLIA (LINN.)
        PARTHENOCISSUS TRICUSPIDATA (S.&Z.)
        RUBIACEA
        VITIS SP.(GRAPES)
        VITIS VINIFERA LINN.(GRAPE)

NESTOS BOISDUVAL
    SYN.SEE ACQUILO BOISDUVAL

NEUI RUMMEL
    SSP.SEE HYPOPHLAEAS BOISDUVAL

NEUMOEGENI EDWARDS, MEGATHYMUS (MEGATHYMIDAE
    SYN.ARYXNA DYAR
    A STEPHENSI SKINNER
    LIFE HISTORY
        SOUTHERN CAL.ACAD.SCI.BULL. 33:81,1934 COMSTOCK & DAMMERS EGG*PL.28;LARVA* AND TUNNELS*PLS.26,27,29; PUPA*PL.30
    FOOD PLANTS
        AGAVA DESERTI ENGLM.

NEUMOEGENI(HENRY EDWARDS), HEMILEUCA-SATURNIIDAE
    LIFE HISTORY
        NAT.ACAD.SCI.MEM.12:123,1914 PACKARD LARVA*PL.24,F.2 COLOR
        PSYCHE 8:151,1897 DYAR LARVA
    FOOD PLANTS'

NEUMOEGENI SKINNER, MELITAEA-NYMPHALIDAE

AB FRIDAYI GUNDER
AB BOHARTI GUNDER
LIFE HISTORY
SOUTHERN CAL. ACAD. SCI. BULL. 22:69, 1923 COMSTOCK LIFE HIST. PUPA*PL. 15
LOC. CIT. 27:66, 1928 COMSTOCK EGG*
LOC. CIT. 30:18, 1931 COMSTOCK LARVA*
COMSTOCK, J. A. BUTTER. CAL. P. 107, 1927 LARVA, CHRYSALIS* (A 38)
FOOD PLANTS
SERICOCARPUS BIFOLIATUS (WALT.)

NEUMOEGENII SKINNER
SYN. SEE ILAIRE GODART

NEURONA (SKINNER), PLEBEIUS-LYCAENIDAE
LIFE HISTORY
SOUTHERN CAL. ACAD. SCI. BULL. 32:79, 1933 COMSTOCK & DAMMERS LIFE HIST. EGG, LARVA* PL. 23; CHRYSALIS* PL. 24
FOOD PLANTS
ERIOGONUM WRIGHTII TORR.

NEVADA SCUDDER, THORYBES-HESPERIIDAE
SYN. AEMILEA SKINNER
LIFE HISTORY
FOOD PLANTS
TRIFOLIUM MONANTHUM GRAY

NEVADAE BARNES & BENJAMIN
SSP. SEE WEIDEMEYERII EDWARDS

NEVADAE BARNES & MCDUNNOUGH
SSP. SEE RUBERATA FREYER

NEVADAE (GROTE), POLIA-NOCTUIDAE
A CANADENSIS SMITH
LIFE HISTORY
UNITED STATES NAT. MUS. PROC. 27:845, 1904 DYAR EGG, LARVA
FOOD PLANTS

NEVADARIA (HULST), DREPANULATRIS-GEOMETRIDAE
LIFE HISTORY
FOOD PLANTS
CEANOTHUS CORDULATUS KELL.

NEVADARIA HULST
SSP. SEE TITEA CRAMER

NEVADARIA PACKARD, CHLOROSEA-GEOMETRIDAE
LIFE HISTORY
FOOD PLANTS
ALNUS RUBRA BONG
PRUNUS SEROTINA EHRH.

NEVADENSIS BENJAMIN
SSP. SEE BARNESI BENJAMIN

NEVADENSIS BEUTENMULLER, CATOCALA-NOCTUIDAE
FORM MONTANA BEUTENMULLER
LIFE HISTORY
AMER. MUS. NAT. HIST. BULL. 38: 167, 1918 BARNES & MCD. EGG, LARVA
FOOD PLANTS
POPULUS SP. (POPLARS)
SALIX SP. (WILLOWS)

NEVADENSIS (FELDER & FELDER), OENEIS-SATYRIDAE
SYN. GIGAS BUTLER
SYN. CALIFORNICA BOISDUVAL
FORM IDUNA EDWARDS
LIFE HISTORY
CAN. ENT. 11:143, 1879 W. H. EDWARDS EGG
COMSTOCK, J. A. BUTTER CAL. P. 77 1927 EGG*LARVA*(A31)
EDWARDS, W. H. BUTTER. NORTH AMER. 2ND SERIES 1879 EGG*PL. CHIONOBAS NO. 3 COLOR
EDWARDS-LOC. CIT. 3RD SERIES 1897 EGG*LARVA*PL. CHIONOBAS NO. 11, FF. A-D
GIGAS
EDWARDS-LOC. CIT. 3RD SERIES 1897 EGG*LARVA*PL. CHIONOBAS NO. 12, FF. G-K

CALIFORNICA
EDWARDS-LOC.CIT.3RD SERIES
1896 EGG*LARVA*PL.CHIONOBAS
NO.12,FF.A-F IDUNA

FOOD PLANTS
LICHENS SP.

NEVADENSIS(GROTE & ROBINSON)APANTESIS-
ARCTIIDAE
SYN.BEHRII STRETCH
A GENEURA STRECKER
SYN.INCORRUPTA HENRY EDWARDS
B SULPHURICA NEUMOEGEN
SYN.OCHRACEA NEUMOEGEN
C SUPERBA STREGCH

LIFE HISTORY
BROOKLYN MUS.ARTS.&SCI.BULL.
1:194,1906 DYAR LARVA
CAN.ENT.37:344,1905 GIBSON
LARVA,PUPA
LOC.CIT.37:346,1905 GIBSON LARVA
PUPA SUPERBA
LOC.CIT.44:188,1912 BARNES &
MCD.EGG,LARVA INCORRUPTA
NEW YORK ENT.SOC.JOURN.6:250
1898 COQUILLETT LARVA
LOC.CIT.8:44,1900 DYAR LARVA
LOC.CIT.8:45,1900 DYAR LARVA
SUPERBA
SOUTHERN CAL.ACAD.SCI.BULL.
34:150,1935 COMSTOCK LARVA*
PL.42;PUPA*PL.43
LOC.CIT.39:118,1940 MEDLAR
TABLE SHOWING DIFFERENCES
BETWEEN LARVAE OF NEVADENSIS
AND INCORRUPTA. DESCRIBES
LARVA OF INCORRUPTA ALSO.
UNITED STATES NAT.MUS.PROC.25:
374,1903 DYAR LARVA

FOOD PLANTS
ARTEMISIA SP.
ASTRAGALUS SP.
CASTILLEJA SESSILIFLORA PURSH
JUNIPERUS SP.(JUMIPERS)
LUPINUS SP.(LUPINES)
MALVA ROTUNDIFOLIA LINN.
PURSHIA SP.
TRIFOLIM SP.(CLOVERS)
GENERAL FEEDER,LOW PLANTS

NEVADENSIS STRETCH,HEMILEUCA-
SATURNIIDAE
SYN.ARTEMIS PACKARD
A CALIFORNICA WRIGHT

LIFE HISTORY
AMER.PHIL.SOC.PROC.31:72,
1893 PACKARD LARVA*PL.9,
F.13 ARTEMIS
CAL.ACAD.SCI.PROC.7:23,1876
EDWARDS LARVA
CAN.ENT.20:34,1888 WRIGHT
LARVA CALIFORNICA
LOC.CIT.26:293,1894 FRENCH
EGG,LARVA CALIFORNICA
LOC.CIT.27:325,1895 DYAR LARVA
ENT.AMER.3:191,1888 HULST
LARVA
ENT.AMER.3:191,1888 HULST
LARVA
NAT.ACAD.SCI.MEM.12:121,1914
PACKARD LARVA*PL.23,FF.
1-5 COLOR
SOUTHERN CAL.ACAD.SCI.BULL.
37:110,1938 COMSTOCK &
DAMMERS LIFE HIST.EGG*PL.
34;LARVA*PL.35,35A;PUPA*
PL.35
UNITED STATES NAT.MUS.PROC.
25:371,1903 DYAR LARVA
ESSIG-INS.WEST.NORTH AMER.P.
671,1926 EGG,LARVA

FOOD PLANTS
AMBROSIA ARTEMISIIFOLIA LINN.
EGG LAID ON THIS PROBABLY
NOT THE FOOD
POPULUS FREMONTI S.WATS.
SALIX INTERIOR ROWLEE
SALIX LASIANDRA BENTH.
SALIX LASIOLEPIS BENTH.

NEWCOMBI COMSTOCK
SSP.SEE GABBII BEHR.

NEWCOMBI GUNDER
SSP.SEE EURYDICE BOISDUVAL

NEXA(BOISDUVAL),CISTHENE-ARCTIIDAE
SYN.GRISEA PACKARD
SYN.DESERTA FELDER & FELDER
LIFE HISTORY
CAL.ACAD.SCI.PROC.1 NEW SERIES P.104,JUNE 1888 EDWARDS EGG,LARVA,PUPA
FOOD PLANTS
LICHENS

NEZAHUALCOTYL STRECKER
SYN.SEE PHILENOR LINNAEUS

NI(HUBNER),AUTOGRAPHA-NOCTUIDAE
BRASSICAE RILEY(AUCT)
SYN.ECHINOCYSTIDIS STRECKER
SYN.U-NOTATA STRAND
LIFE HISTORY
ILLINOIS AGRI.EXP.STA.BULL.60: 500,1900 FORBES LARVA*PUPA*
ILLINOIS DEPT.AGRI.TRANS.15:230, 1877 FRENCH LARVA
LOC.CIT.17:41,1879 THOMAS LARVA
LOC.CIT.18,APPEND.P.140,1880 MARTEN LARVA
MICHIGAN AGRI.EXP.STA.BULL. 134:52,1924 MCDANIEL EGG, LARVA
LOC.CIT.314:86 ,1941 MCDANIEL LARVA*PUPA*
MISSOURI,REPT.INS.OF 2:110,1870 RILEY LARVA*PUPA*
NEW YORK STATE GENEVA EXP.STA. BUL.83,NEW SERIES P.668, 1894 EGG*PL.I
NEW YORK STATE,REPT.INS.OF 2: 90,1885 LINTNER LARVA*PUPA*
ONTARIO ENT.SOC.ANN.REPT.P.87 1871 GETHUNE LARVA*PUPA*
LOC.CIT.P.61,1886 BETHUNE LARVA* COCOON*
LOC.CIT.PP.58,1890 FYLES LARVA* PUPA
UNITED STATES DEPT.AGRI.BUR. ENT.BULL.33:60,1902 NEW SERIES CHITTENDENEGG* F.13B LARVA*F.13C,F.14;PUPA*F.13D
UNITED STATES DEPT.AGRI.CIRC. 615,1942 REED SMITH LARVA* F.1B NO SESCR.
UNITED STATES DEPT.AGRI. FAMER'S BULL.120:24,1900 HOWARD LARVA*F.19B;PUPA* F.19C
LOC.CIT.856:33,1917 ORTON CHITTENDEN LARVA*F.37; PUPA&COCOON* F.37
UNITED STATES DEPT.AGRI. FAMER'SBULL.1362:26,1922 WEIGEL & SASSCER LARVA* F.18;PUPA*F.18
LOC.CIT.1371:18,1924 GILBERT LARVA* PUPA* COCOON* ALL F.24
LOC.CIT.1495:11,1926 WEIGEL & MIDDLETON LARVA* F.6
UNITED STATES DEPT.AGRI.REPT. ENT.P.119,1883 LIFE HIST* RILEY PL.11,FF.2-5
UNITED STATES DEPT.AGRI.TECH BULL.782:2,1941 SMITH, THOMAS LARVA* F.1B
LOC.CIT.846:11,1944 MCKENNY EGG* F.8A;LARVA*F.8B; PUPA*F.8D
UNITED STATES DEPT.AGRI.YEAR-BOOK P.142 FOR 1898 CHITTENDEN LARVA* F.25B;PUPA*F.25C NO DESCR.
CHITTENDEN-IN.SINJ.VEG.P.140, 1907 EGG*LARVA*PUPA*ALL F.9
CRAGEN-OUR INS.FRIENDS&FOES P.131,1899 LARVA
CROSBY&LEONAED-MANUAL VEG. GARD.INS.P.9,1918 EGG*F.8; LARVA*F.9;PUPA*F.11; COCOON* F.10
GLOVER-ILLUST.NORTH AMER.ENT. 1878 LARVA*PL.18,F.1
HART-INS.COLL.ILLINOIS HIGH SCHOOLS P.39,1903 LARVA* PUPA* F.53
HOLLAND-MOTH BOOK.P.239,1905 LARVA*PUPA*F.148
METCALF & FLINT-DESTR.&USE-

FUL INS.P.533,1939 EGG, LARVA* F.333;PUPA
O'KANE-INJ.INS.P.178,1912 LARVA* F.207
PACKARD-INS.INJ.WEST P.752, 1877 LARVA*PUPA*COCOON*
PEAIRS-PESTS FARM,GARD.ORCHARD P.21,1941 EGG*F.27;LARVA* FF.27,336;PUPA*F.27
INDEX AMER.ECON.ENT.
VOL.1:45;VOL.2:46;
VOL.3:51;VOL.4:57;
VOL.5:68;VOL.6:78/
VOL.7:1032 AS NI ALL OTHERS AS BRASSICAE INCLUDING ALL PREVIOUS LISTINGS

FOOD PLANTS

AGERATUM CONYZOIDES LINN.
ALTHAEA ROSEA CAV.
ANTIRRHINUM SP.
APIUM GRAVEOLENS LINN. (CELERY)
ARGEMONE PLATYCERAS V.HISPIDA GRAY
ASPARAGUS OFFICINALIS LINN. (ASPARAGUS)
BETA VULGARIS LINN. (BEET)
BRASSICA HIRTA MOENCH(MUSTARD)
BRASSICA NIGRA(LINN.)(MUSTARD)
BRASSICA OLERACEA LINN. (CABBAGE)
BRASSICA OLERACES V.BOTYRTIS DC. (CAULIFLOWER)
BRASSICA OLERACEA VIRDIS LAM. (KALE)
BRASSICA RAPA LINN. (TURNIP)
CALENDULA OFFICINALIS LINN.
CHAENACTIS STEVIOIDES H.&A.
CHENOPODIUM ALBUM LINN.
CHRYSANTHEMUM SP.
CITRULLUS VULGARIS SCHAD. (WATERMELON)
CREPIS SP.
CUCUMIS MELO LINN. (MELON)
CUCUMIS SATIVUS LINN. (CUCUMBER)
CUCURBITA MAXIMA DUSCH. (SQUASH)
CYDONIA JAPONICA(THUNB.)
DIANTHUS CARYOPHYLLUS LINN.
ERODIUM SP.
ERODIUM CICUTARIUM(LINN.)
ESCHSCHOLTZIA CALIFORNICA CHAMB.
GARANIUM SP.
GOSSYPIUM HERBACEUM LINN.
LACTUCA CANADENSIS LINN.
LACUTCA SATIVA LINN. (LETTUCE)
LATHYRUS ODORATUS LINN.
MALVA SP.
MENTHA SP. (MINTS)
NICOTIANA GLAUCA GRAHAM
NICOTIANA TABACUM LINN. (TOBACCO)
NICOTIANA TRIGONOPHYLLA DUNAL
PECTIS PAPPOSA HAW.&GRAY
PHASEOLUS VULGARIS LINN. (BEAN)
PISUM SATIVUM LINN. *(PEAS)
PLANTAGO SP.
RAPHANUS SATIVUS LINN. (RADISH)
RESEDA ODORATA LINN.
RUMEX SP.
SENECIO CRUENTIS DC.
SENECIA MIKANIODIDES OTTO.
SMILAX ROTUNDIFOLIA LINN.
SOLANUM LYCOPERSICUM LINN. (TOMATO)
SOLANUM TUBEROSUM LINN. (POTATO)
SPINACIA OLERACES MILL. (SPINACH)
TARAXACUM OFFICINALE WEBBER
TRIFOLIUM SP. (CLOVERS)
TROPAELUM MAJUS LINN.
VIGNA SINENSIS LINN.

NICALIS SMITH
SSP.SEE CURIALIS GROTE

NICIPPE(CRAMER), EUREMA-PIERIDAE
SYN. CALLAE FIELD
FORM FEMALE FLAVA STRECKER
AB DAMMERSI GUNDER

LIFE HISTORY

AMER.MUS.NAT.HIST.BULL.5: 252,1893 BEUTEN.LARVA, CHRYSALIS
CAL.ACAD.SCI.PROC.5:166,1873 EDWARDS LARVA,CHRYSALIS
CAN.ENT.13:61,1881 EDWARDS EGG,LARVA,CHRYSALIS
LOC.CIT.76:17,1944 TINKHAM

CHRYSALIS
ILLINOIS DEPT. AGRI. TRANS. 18:79 APPEND. 1880 MIDDLETON LARVA
NAT. CANADIEN P. 91, 1927 HUARD LARVA, CHRYSALIS
NEW YORK STATE AGRI. SOC. TRANS. P. 105, 1855 GLOVER LARVA, CHRYSALIS
SMITHSONIAN MISC. COLL. 4, ART. 2:33, 1862 MORRIS LARVA, CHRYSALIS
SOUTHERN CAL. ACAD. SCI. BULL. 25:63, 65, 1926 COMSTOCK EGG*CHRYSALIST
WASHINGTON ENT. SOC. PROC. 19: 78, 1917 AINSLIE COLOR VARIATION IN CHRYSALIS
BOISDUVAL-SPEC. GEN. LEP. DIURNAE 1:654, 1836 LARVA, CHRYSALIS
BOISDUVAL&LECONTE-HIST. LEP. D NORTH AMER. P. 57, 1833 LARVA* PL. 20
CHENU LUCAS-ENCYCL. NAT. HIST. PAPILLONS P. 60, 1856 LARVA, CHRYSALIS
COMSTOCK, J. A. BUTTER. CAL. P. 56, 1927 EGG*PL. 24A; LARVA* PL. 24A; CHRYSALIS*PL. 25A AND PL. 63, F. 11 IN COLOR
COMSTOCK J. H. HOW KNOW BUTTER P. 97, 1904 LARVA
FRENCH-BUTTER. EAST. UNITED STATES P. 137, 1886 CHRYSALIS* LARVA, EGG
GLOVER-ILLUST. NORTH AMER. ENT. 1878 LARVA, PUPA*PL. 1, F. 5
GUNDLACH-ENT. CUBANA P. 83, 1881 LARVA, CHRYSALIS
HOLLAND-BUTTERFLY BOOK 1905 LARVA* PL. 2, F. 6COLOR: CHRYSALIS*PL. 5, FF. 51, 52
MAYNARD-BUTTER. NEW ENGLAND P. 46, 1886 LARVA, CHRYSALIS
PEALE-LEP. NORTH AMER. 1833 LARVA*PL. 8COLOR
SCUDDER-BUTTER. EAST. UNITED STATES & CANADA PP. 1069-1072, 1889 LIFE HIST*PL. 65, F. 31; PL. 76, F. 6; PL. 79, F. 68; PL. 84, FF. 51, 52 COLOR
SCUDDER-BUTTER. THEIR STRUCTIRE P. 184, 1881 CHRYSALIS
SCUDDER-EVERYDAY BUTTER. P. 65 1899LLARVA, CHRYSALIS
SCUDDER-GUIDE TO COMMON BUTTER P. 137, 1893 LARVA, CHRYSALIS

FOOD PLANTS
CASSIA MARILANDICA LINN.
LEGUMINOSAE
TRIFOLIUM SP. (CLOVERS)

NICOTIANAE MENETRIES
SYN. SEE SEXTA JOHNANSSON

NIG STRECKER
SSP. SEE CLYTON BOISDUVAL & LECONTE

NIG STRECKER
SSP. SEE ARCHIPPUS CRAMER

NIG STRECKER
SSP. SEE PHILODICE GODART

NIG STRECKER
SSP. SEE PSEUDARGIOLUS BOISDUVAL & LECONTE

NIGER SMITH
SSP. SEE EXSERTISTIGMA MORRISON

NIGER WRIGHT
SSP. SEE SMINTHEUS DOUBLEDAY & HEWITSON

NIGERRIMA COCKERELL
SSP. SEE WEIDEMEYERII EDWARDS

NIGERRIMA VERITY
SSP. SEE SMINTHEUS DOUBLEDAY & HEWITSON

NIGRA CARY
SSP. SEE GORGONE HUBNER

NIGRA CHERMOCK

SSP. SEE MARTHESIA CRAMER

NIGRA CHERMOCK
SSP. SEE UMBRATA WALKER

NIGRA COCKERELL
SYN. SEE LACINIA GEYER

NIGRA EASTMAN
SSP. SEE PURA HULST

N IGRA EDWARDS
SYN. SEE PSEUDARGIOLUS BOISDUVAL & LECONTE

NIGRA EDWARDS
SSP. SEE MELISSA FABRICIUS

NIGRA STRETCH
SSP. SEE FULVA STREGCH

NIGRACOSTA COMSTOCK
SSP. SEE MANAPIA FELDER & FELDER

NIGRARIA FORBES
SSP. SEE DIVISATA WALKER

NIGRESCARIA HULST, PLAGODIS-GEOMETRIDAE
SYN. KEUTZINGI GROTE
LIFE HISTORY
FOOD PLANTS
FRAXINUS SP. (ASHES)

NIGRESCENS CASSINO
SSP. SEE ULTRONIA HUBNER

NIGRESCENS CLARK
SSP. SEE CERISYI KIRBY

NIGRESCENS COCKERELL
SSP. SEE LACINIA GEYER

NIGRESCENS FLETCHER
SSP. SEE PSEUDARGIOLUS BOISDUVAL & LECONTE

NIGRESCENS GROTE & ROBINSON
SYN. SEE INDOMITA WALKER

NIGRESCENS HALL
SSP. SEE THAROS DRURY

NIGRICANS BETHUNE
SYN. SEE UNDULARIS DRURY

NIGRICANS REAKIRT
SSP. SEE OPELLA GROTE

NIGRICANS STRECKER
SYN. SEE ARCHIPPUS CRAMER

NIGRICARIA ROTKE
SSP. SEE PENDULINARIA GUENEE

NIGRICEPS WALKER
SYN. SEE CLANDESTINA HARRIS

NIGRICOSTA CHERMOCK
SSP. SEE EURYTHEME BOISDUVAL

NIGRIDICE SCUDDER
SYN. SEE PHILODICE GODART

NIGRIFERA WALKER
SYN. SEE OPELLA GROTE

NIGRINA STRECKER
SYN. SEE PHILODICE GODART

NIGRIOR STRAND
SYN. SEE AERUGINOSA GUENEE

NIGRIRENA HAWORTH
SYN. SEE NUNDINA DRUTY

NIGRISUPERNIPENNIS GUNDER
SSP. SEE PERIDICCAS EDWARDS

NIGROANALIS BRYK & EISNER
SYN. SEE CLODIUS MENETRIES

NIGROANGULATA STRECKER
SSP. SEE LIMITARIA WALKER

NIGROCAERULEA COCKERELL
SYN. NITOCRIS EDWARDS

NIGROCAPUT MORRISON
SYN. SEE RIDINGSI RILEY

NIGROCOMMA WARREN, SEMIOTHISA-GEOMETRIDAE
HELIOTHIDATA AUCT.
LIFE HISTORY
NEW YORK STATE MUS. MEM. 8:730 1905 FELT LARVA(BRIEF)
PSYCHE 9:203, 1901 DYAR EGG, LARVA, PUPA HELIOTHIDATA
FOOD PLANTS
ROBINIA PSEUDO-ACACIA LINN.

NIGROFASCIA HULST
SYN. SEE ERIDANIA CRAMER

NIGROFASCIATA REIFF
SYN. SEE PHILODICE GODART

NIGROFIMBRIA GUENEE, XANTHOPTERA-NOCTUIDAE
LIFE HISTORY
INDEX AMER. ECON. ENT. VOL. 2:386; VOL. 4:513
FOOD PLANTS
IPOMOEA SAGITATA CAV.

NIGROINITA GUNDER
SSP. SEE BEHRII EDWARDS(CALLIPSYCHE)

NIGROSIGNATA WALKER
SYN. SEE IPOMOEAE DOUBLEDAY

NIGROSUFFUSA BARNES & MCDUNNOUGH
SSP. SEE COENIA HUBNER

NIGROSUFFUSA FIELD
SSP. SEE MARCELLUS CRAMER

NIGROSUFFUSANA STRAND
SYN. SEE DERIDENS GUENEE

NIGROVENARIA PACKARD
SSP. SEE UMBROSARIA PACKARD

NIGROVENOSA WATSON
SSP. SEE BURNSI WATSON

NIKIAS EHRMANN
SSP. SEE ATLANTIS EDWARDS

NINONIA BOISDUVAL
SYN. SEE MANAPIA FELDER & FELDER

NINUS EDWARDS
SYN. SEE SPINETORUM HEWITSON

NIOBE STRECKER
SYN. ZEE ECHO ABBOT & SMITH

NIPHADOPHILATA DYAR
SSP. SEE SORBINATA HUBNER

NIPHON(HUBNER), INCISALIS-LYCAENIDAE
LIFE HISTORY
AMER. MUS. NAT. HIST. BULL. 5: 281, 1893 BEUTEN. LARVA CHRYSALIS
CAN. ENT. 1:96, 1868 SAUNDERS LARVA
LOC. CIT. 16:94, 1884 FLETCHER EGG
LOC. CIT. 39:257, 194, 1907 COOK EGG*PL. 6; LARVA*PL. 6, 9; CHRYSALIS*PL. 9
ILLINOIS DEPT. AGRI. TRANS. 18. APPEND:94, 1880 MIDDLETON LARVA, CHRYSALIS
MONTANA UNIV. BULL. 30:229, 1906 ELROD LARVA, CHRYSALIS
NEW YORK STATE AGRI. SCO. REPT. 4:57, 743, 1858 FITCH LARVA, CHRYSALIS
NEW YORK STATE MUS. MEM. 8:752 1905 FELT LARVA (BRIEF)
NEW YORK STATE REPT. NOX. INS. OF 4:57, SEC. 272, 1858 FITCH LARVA, PUPA
SMITHSONIAN MISC. COLL. 4, ART. 2:98, 1862 MORRIS LARVA, CHRYSALIS
BOISDUVAL & LECONTE-HIST. LEP. NORTH AMER. P. 106, 1833 LARVA* CHRYSALIS* PL. 33
COMSTOCK, J. H. HOW KNOW BUTTER. P. 234, 1904 LARVA

Fernald-Butter. Maine p. 83, 1884 larva, chrysalis
French-Butter. East. United States p. 276, 1886 larva, chrysalis
Holland-Butterfly Book 1005 chrysalis* Pl. 5, ff. 38, 40 color.
Maynard-Butter. New England p. 36, 1886 larva, chrysalis
Morris-Synop. Lep. North Amer. p. 97, 1862 larva, chrysalis
Packard-Ins. Inj. Forest Trees p. 201, 1881 larva
Scudder-Butter. East. United States & Canada pp. 831-833 1889 life hist*Pl. 65, f. 7; Pl. 68, f. 7; Pl. 75, f. 24; Pl. 84, ff. 38, 40
Scudder-Guide to Common Butter p. 114, 1893 larva, chrysalis
Index Amer. Econ. Ent. Vol. 4:232

FOOD PLANTS
Juniperus sp. (junipers)
Lupinus perennis Linn.
Pinus rigida Mill.

NITELA Comstock
ssp. see DYMAS Edwards

NITELA Guenee
ssp. see NEBRIS Greene

NITENS Dyar
ssp. see MIRANDA Grote

NITENS Grote
syn. see VECORS Guenee (Orthodes)

NITIDA Packard
syn. see CONCINNA Abbot & Smith

NITOCRIS (Cramer), AMETRIS-Geometridae
syn. NITOCRITARIA Hubner
syn. PENINSULARIA Grote
syn. SCHAUSARIA Henry Edwards
LIFE HISTORY
Psyche 9:69, 1900 Dyar egg, larva, pupa
FOOD PLANTS
Coccoloba diversifolia Jacq.

NITOCRIS Edwards, ARGYNNIS-Nymphalidae
syn. NIGROCAERULEA Cockerell
a CAERULESCENS Holland
LIFE HISTORY
Ent. News 18:318, 1907 Skinner egg, two larval stages NIGROCAERULEA
Ent. Rec. 22:72, 1910 Cockerell oviposition
FOOD PLANTS

NITOCRITIARIA Hubner
syn. see NITOCRIS Cramer

NITTANYENSIS Chermock
ssp. see LYGDAMUS Doubleday

NIVALIS (Boisduval), LYCAENA-Lycaenidae
syn. IANTHE Edwards
LIFE HISTORY
FOOD PLANTS
Ribes sp.?

NIVEA Gunder
ssp. see MYRINA Cramer

NIVEA Herrich-Schaeffer
syn. see INVOLUTUM Sepp.

NIVEA Neumoegen, CERURA-Notodontidae
a MERIDIONALIS Dyar
b NIVEATA Barnes & Benjamin
LIFE HISTORY
New York Ent. Soc. Journ. 6:189 190, 1898 Kunze egg, larva
FOOD PLANTS
Populus fremonti S. Wats.
Salix sp. (willows)

NIVEATA Barnes & Benjamin
ssp. see NIVEA Neumoegen

NIVEIFASCIATA Hulst

SSP. SEE VASILIATA GUENEE

NIVEIGUTTA WALKER
SYN. SEE ANNULOSA WALKER

NIVEILINEA (GROTE), EUXOA-NOCTUIDAE
A RABIATA SMITH
LIFE HISTORY
NEBRASKA AGRI. EXP. STA. RESEARCH BULL. 81:12,1935 WHELAN TECH. DESCR. LARVA; LARVAL CLAW* F. 9B
FOOD PLANTS
MELILOTUS ALBA DESR.
TRITICUM AESTIVUM LINN. (WHEAT)
ZEA MAYS LINN. (CORN)

NIVEIVENOSA (GROTE), PROTAGROTIS-NOCTUIDAE
SYN. VIRALIS GROTE
LIFE HISTORY
INDEX AMER. ECON. ENT. VOL. 7:817
FOOD PLANTS
TARAXACUM OFFICINALE WEBER

NIVEOCILIARIA (HERRICH-SCHAEFFER), EUEANA-GEOMETRIDAE
SYN. SALTUSARIA HULST
LIFE HISTORY
PSYCHE 9:118,1900 DYAR EGG, LARVA, PUPA, COCOON
WASHINGTON ENT. SOC. PROC. 5:456 1900 DYAR LARVAL NOTE (BRIEF) SALTUSARIA
FOOD PLANTS
CONDALIA FERREA GRIS.

NIVEOCOSTA FRANCLEMONT
SSP. SEE PATEFACTA WALKER

NIVEOSERICAETUS JONES
SYN. SEE SUBSIGNARIUS HUBNER

NIVEUM GROTE
SEE INVOLUTUM SEPP.

NIVOSA GUNDER
SSP. SEE CARYE HUBNER

NIVOSUS GUNDER
SSP. SEE ARCHIPPUS CRAMER

NIVOSUS GUNDER
SSP. SEE PLEXIPPUS LINNAEUS

NOBILIS NEUMOEGEN
SSP. SEE IMPERIALIS DRURY

NOCTIVAGA GROTE, ACRONICTA-NOCTUIDAE
LIFE HISTORY
AMER. MUS. NAT. HIST. BULL. 14: 258,1901 BEUTEN. LARVA
NEW YORK STATE MUS. MEM. 8: 747,1905 FELT LARVA
PAPILIO 3:15,1883 THAXTER EGG, LARVA
UNITED STATES NAT. MUS. PROC. 21:156,1899 SMITH&DYAR LARVA*PL. 8, FF. 27,28 COLOR OR UNCOLORED; COCOON
FOOD PLANTS
POPULUS SP. (POPLARS)
GENERAL FEEDER LOW PLANTS

NOKOMIS BRODIE
SSP. SEE COLUBIA SMITH (PLATYSAMIA)

NOKOMIS EDWARDS, ARGYNNIS-NYMPHALIDAE
SYN. APACHEANA SKINNER
AB HERMOSA COMSTOCK
LIFE HISTORY
SOUTHERN CAL. ACAD. SCI. BULL. 27:89,1928 COMSTOCK LIFE HIST EGG*APACHEANA
LOC. CIT. 39:75,1940 COMSTOCK EGG*PL. 8B
FOOD PLANTS
VIOLA SP. (VIOLETS)

NORDA SMITH
SSP. SEE MINEREA GUENEE

NORDEGGENSIS CASSINO & SWETT
SYN. SEE GELIDATA MOESCHLER

NORDICA SMITH
SSP. SEE TESSELLATA HARRIS

NORMA MORRISON
SYN. SEE INCLUDENS WALKER

NORMALIS DYAR
SSP. SEE LINTNERI GROTE(GLUPHISIA)

NORMALIS DYAR
SYN. SEE EGLANTERINA BOISDUVAL

NORMALIS GROTE, ACERRA-NOCTUIDAE
LIFE HISTORY
FOOD PLANTS
ACER MACROPHYLLUM PURSH
ALNUS RUBRA BONG.
AMELANCHIER FLORIDA LINDL.
HOLODISCUS DISCOLOR(PURSH)
PRUNUS EMARGINATA(DOUGL.)
RIBES SANGUINEUM PURSH
SALIX HOOKERIANA BARR.
SALIX SCOULERIANA BARR.

NORMANI BARTSCH
SSP. SEE ILIA CRAMER

NORMANI(GROTE), CROCIGRAPHA-NOCTUIDAE
LIFE HISTORY
CAN. ENT. 35:17, 1903 GIBSON EGG, LARVA, PUPA
ENT. NEWS 19:321, 1908 PATCH LARVA* PL. 13; PUPA
INDEX AMER. ECON. ENT. VOL. 1:95; VOL. 2:101
FOOD PLANTS
CAULOPHYLLUM THALICTROIDES(LINN.)
CORYLUS CORNUTA MARSH.
MALUS PUMILA MILL. (APPLE)
PRUNUS SEROTINA EHRH.
QUERCUS SP. (OAKS)
QUERCUS ALBA LINN.

NORMANIANA(GROTE), GRAPHIPHORA-NOCTUIDAE
SYN. OBTUSA SPEYER
LIFE HISTORY
ENT. SOC. AMER. ANNALS 23:396, 1930 HAWKINS LARVAL CLAWS
MAINE AGRI. EXP. STA. BULL. 356:159 1930 PHIPPS LARVA, PUPA
INDEX AMER. ECON. ENT. VOL. 5:277
FOOD PLANTS
COMPTONIA PEREGRINA(LINN.)
PRUNUS PENSYLVANICA LINN.
RUBUS IDAEUS V. STRIGOSUS(MICHX.)
SPIRAEA LATIFOLIA(AIT.)
VACCINIUM ANGUSTIFOLIUM AIT.
VACCINIUM MYRTILLOIDES MICHX.

NORNA(THUNBERG), EOENEIS-SATYRIDAE
LIFE HISTORY
FOOD PLANTS
CAREX CONCOLOR R. BR.

NORTONII EDWARDS
SYN. SEE ACCIUS ABBOT & SMITH

NORUS PLOETZ
XYN. SEE MACULATA EDWARDS

NOSTRA(SMITH), EUXOA-NOCTUIDAE
LIFE HISTORY
FOOD PLANTS
CONRINGIA ORIENTALIS(LINN.)
VIOLA PEDATIFIDA DON.

NOTARIA HENRY EDWARDS
SYN. SEE ELEGANS STRECKER

NOTATA STRECKER, TRICHOLITA-NOCTUIDAE
SYN. SYRISSA STRECKER
FORM OTA BARNES & BENJAMIN
LIFE HISTORY
BROOKLYN ENT. SOC. BULL. 33:135 1938 HERZ LARVA
FOOD PLANTS
SILPHIUM TEREBINTHINACEUM JACQ.
SOLIDAGO RIDIDA LINN.

NOTATARIA HULST
SYN. SEE TRIANGULIFERATA PACKARD

NOTATARIA(WALKER), EUFIDONIA-GEOMETRIDAE
SYN. MALE FIDONIATA WALKER
SYN. QUADRIPUNCTATA MORRISON
AB FAMULATA HULST
A CONVERGARAI WALKER
SYN. BICOLORARIA MINOT

LIFE HISTORY
CAN. ENT. 14:199, 1882 GOODELL
EGG, LARVA, PUPA
LOC. CIT. 77:198, 1945 MCGUFFIN
LARVA* FF. 1, 2
NEW YORK STATE AGRI. EXP. STA. (CORNELL), MEM. 274:63, 1948 FORBES LARVA
NEW YORK STATE MUS. MEM. 8:753 1905 FELT LARVA
UNITED STATES DEPT. AGRI. MISC. PUB. 657:361, 429, 1950 CRAIGHEAD LARVA
INDEX AMER. ECON. ENT. VOL. 7:365
FOOD PLANTS
ABIES BALSAMEA (LINN.)
LARIX LARICINA (DUROI)
PICEA GLAUCA (MOENCH)
PICEA MARIANA (MILL.)
PINUS BANKSIANA LAMB.
PINUS STROBUS LINN.
TSUGA CANADENSIS (LINN.)

NOVA FITCH
SYN. SEE ANTIQUA LINNAEUS

NOVA SMITH
SSP. SEE BELLA LINNAEUS

NOVANDA (GUENEE), MELIPOTIS-NOCTUIDAE
LIFE HISTORY
FOOD PLANTS
ACACIA GREGGII A. GRAY

NOVANGLIAE REIFF
SYN. SEE AMICA HUBNER

NOVANGLIAE SCUDDER
SSP. SEE RAPAE LINNAEUS

NOVASCOTIAE MCDUNNOUGH
SSP. SEE CYBELE FABRICIUS

NOVA-SCOTIAE MCDUNNOUGH
SSP. SEE TARQUINIUS FABRICIUS

NOX GUNDER
SSP. SEE GORGONE HUBNER

NUBECULARIA GUENEE, PARAPHIA-GEOMETRIDAE
SYN. TRIPLIPUNCTATA FITCH
LIFE HISTORY
NEW YORK STATE AGRI. EXP. STA. (CORNELL), MEM. 274:62, 1948 FORBES LARVA TRIPLIPUNCTATA
NEW YORK STATE REPT. NOX. INX. OF 5:45, SEC. 32, 1858 FITCH LARVA
FOOD PLANTS
QUERCUS SP. (OAKS)

NUBES SCUDDER
SSP. SEE MYRINA CRAMER

NUMBIFACTA DYAR
SYN. SEE ANNULOSA WALKER

NUBIFERARIA SWETT
SSP. SEE PAMPINARIA GUENEE

NUBIGENS (BEHR.) EUPHYDRYAS-NYMHPALIDAE
FORM RUBROSUFFUSA COMSTOCK
AB TIOGAENSIS GUNDER
AB BOHARTI GUNDER
A AURILACUS GUNDER
AB ALBIRADIATA GUNDER
AB FOXI GUNDER
B MONOENSIS GUNDER
SYN. FRIDAYI GUNDER
C LEHMANI GUNDER
AB CAVERNA GUNDER
LIFE HISTORY
MEAD-REPT. LEP WHEELER SURVEY 5:758, 1875 LARVA, CHRYSALIS
FOOD PLANTS
PLANTAGO SP?

NUBILA COMSTOCK
SSP. SEE AROTA BOISDUVAL

NUBILATA (PACKARD), EUSTOMA-GEOMETRIDAE
SYN. SEMIATRATA HULST
FORM MACDUNNOUGHI BLACKMORE
LIFE HISTORY
NEW YORK STAAT AGRI EXP. ST.

(CORNELL), MEM. 274:130, 1948
FORBES LARVA
PSYCHE 11:29, 1904 DYAR EGG, LARVA
FOOD PLANTS
EPILOBIUM SP.

NUBILATA PACKARD
SYN. SEE FORFICARIA GUENEE

NUBILATA SMITH
SSP. SEE HIBISCI GUENEE

NUBILATA WARREN
SYN. SEE LORICARIA EVERSMANN

NUBILIS GUERIN-NEMEVILLE
SYN. SEE DISSTRIA HUBNER

NUBILIS (HUBNER), EUPARTHENOS-NOCTUIDAE
SYN. MALE FASCIATA BEUTENMULLER
A APACHE POLING
B OSIRIS BARNES & BENJAMIN
LIFE HISTORY
NEW YORK STATE MUS. MEM. 8:730 1905 FELT LARVA (BRIEF)
HARRIS-ENT. CORRESP. P. 319, 1869 LARVA
FOOD PLANTS
ROBINIA PSEUDO-ACACIA LINN.

NUBILOFASCIATA (PACKARD), HYDRIOMENA-GEOMETRIDAE
?SYN. VANAVAHARATA STRECKER
FORM RAPTATA SWETT
FORM CUPIDATA SWETT
FORM CUMULATA SWETT
FORM VULNERATA SWETT
FORM SPARSIMACULA HULST
?A SCALATA WARREN
LIFE HISTORY
KIRBY-EUROPEAN BUTTER. & MOTHS. P. 381, 1882 LARVA SORDITATA FABRICIUS
NEWMAN-BRITISH MOTHS. P. 153, 1869 LIFE HISTORY
PACKARD-MONOGRAPH GEOMETRIDAE P. 99, 1876 LIFE HISTORY
FOOD PLANTS
QUERCUS SP. (OAKS)
SALIX CAPREA LINN.
SALIX CINEREA LINN.

NUBS SCUDDER
SSP. SEE MYSTIC SCUDDER

NUCICOLORA (GUENEE), ELAPHRIA-NOCTUIDAE
SYN. UNISIGNATA WALKER
SYN. PAGINATA MORRISON
FORM CLARA HARVEY
LIFE HISTORY
GUENEE-SPEC. GEN. LEP. NOC. 1: 241, 1852 LARVA
FOOD PLANTS
GENERAL FEEDER LOW PLANTS

NUMBIFACTA DYAR
SYN. SEE ANNULOSA WALKER

NUMITOR (FABRICIUS), ANCYLOXYPHA-HESPERIIDAE
SYN. PUER HUBNER
SYN. MARGINATUS HARRIS
FORM LONGLEYI FRENCH
LIFE HISTORY
CAN. ENT. 21:160, 1889 BEUTEN. EGG
LOC. CIT. 61:118, 1939 DETHIER EGG, LARVA
ILLINOIS DEPT. AGRI. TRANS. 18, APPEND: 96, 1880 MIDDLETON CHRYSALIS
MUS. COMP. ZOOL. BULL. 87, NO. 6, 1941 DETHIER LARVAL ANTENNA* F. 46
SOUTHERN CAL. ACAD. SCI. BULL. 37:74, 1938 DETHIER LARVA* PL. 20; LARVAL HEAD* PL. 18; CHRYDSALIS PL. 19
FRENCH-BUTTER. EAST. UNITED STATES P. 301, 1886 CHRYSALIS
HARRIS-INS. INJ. VEG. FLINT EDIT. P. 308, 1862 CHRYSALIS
KELLOGG-AMER. INS. P. 443, 1908 LARVA
MAYNARD-BUTTER. NEW ENGLAND P. 57, 1886 CHRYSALIS
S

Scudder-Butter. East. United States & Canada pp. 1560-1562 1889 life hist* Pl. 66, ff. 22 25; Pl. 69, f. 8; Pl. 73, f. 4; Pl. 80, f. 51
Scudder-Everyday Butter. p. 366, 1899 egg, larva, chrysalis
Scudder-Guide to Common Butter p. 167, 1893 larva, chrysalis
Vance-Live Amont Butter. p. 54 1925 egg, larva, chrysalis
Index Amer. Econ. Ent. Vol. 5:30

FOOD PLANTS

Setaria sp.
Zea mays Linn. (corn)
Grasses

NUNDINA (Drury), SCHINIA-Noctuidae
syn. NIGRIRENA Haworth

LIFE HISTORY

Ent. News 34:196, 1923 Bird larva, pupa

FOOD PLANTS

Solidago altissima Linn. (flowers and seeds)

NUNENMACHERI Strand
ssp. see PSEUDARGIOLUS Boisduval & Leconte

NUPERA (Lintner), XYLENA-Noctuidae
syn. ?VETUSTA Walker

LIFE HISTORY

Brooklyn Ent. Soc. Bull. 27:96 1932 Crumb larva
Massachusetts Agri. Exp. Sta. Bull. 239:23, 1928 Franklin larva* Pl. 2, ff. 7 A-D COLOR
Psyche 2:122, 1878 Thaxter larva
O'Kane-Inj. Ins. p. 275, 1912 larva
Slingerland & Crosby-Manual Fruit Ins. p. 470, 1914 egg, larva, pupa
Index Amer. Econ. Ent. Vol. 1:58; Vol. 2:386

FOOD PLANTS

Acer macrophyllum Pursh
Alnus rubra Bong.
Amelanchier florida Lindl.
Commandra pallida A. Dc.
Holodiscus discolor (Pursh)
Iris sp.
Malus pumila Mill. (apple)
Prunus emarginata (Dougl.)
Ribes sanguineum Pursh
Rosa gymnocarpa Nutt. ex. T. G.
Rosa nutkana Presl
Rubus sp.
Salix hookeriana Barr.
Salix scouleriana Barr.
Vaccinium angustifolium Ait.
Vaccinium macrocarpon Ait.
General feeder low plants. grass, weeds

NUPTULA Walker
syn. see GRYNEA Cramer

NURUS Walker
syn. see AMATRIX Hubner

NUTTALLI Strecker
ssp. see EGLANTERINA Boisduval

NYCTEIS (Doubleday & Hewitson) PHYCIODES-Nymphalidae
syn. OENONE Scudder
ab female MILBURNI Rummel
ab LACETEUS Gunder
ab BREYI Field
a DRUSIUS Edwards
ab HEWITSONI Field

LIFE HISTORY

Amer. Assoc, Advancement Sci. Proc. 23:B:108, 1874 Riley larva, chrysalis
Amer. Mus. Nat. Hist. Bull. 5:261 1893 Beuten. larva, chrysalis
Brooklyn Ent. Soc. Bull. 14:47 1919 Rummel larva (brief)
Buffalo Soc. Nat. Sci. Bull. 19: 24, 1939 Wild larva
Can. Ent. 5:224, 1873 Edwards larva

Loc. cit. 7:151, 1875 Edwards chrysalis
Loc. cit. 11:101, 1879 Edwards egg, larva, chrysalis
Illinois Dept. Agri. Trans. 18, Append: 83, 165, 1880 Coquillett & Martin larva
New York State Cabin. Nat. Hist. Rept. No. 23:158, 1873 Lintner egg, larva, chrysalis Pl. 8, f. 14
Comstock-How Know Butter p. 128, 1904 larva
Fernald-Butter. Maine p. 46, 1884 life history
French-Butter. East. United States p. 172, 1886 life hist.
Holland-Butterfly Book p. 152 1905 chrysalis* Pl. 5, f. 19
Holland-loc. cit. 3rd edit. p. 136, 1931 life hist* Pl. 5. f. 9
Lintner-Ent. Contrib. No. 1:26, 1872 larva
Maynard-Butter. New England p. 28, 1886 larva, chrysalis
Scudder-Butter. East. United States & Canada pp. 661-666 1889 life hist* Pl. 64, f. 32; Pl. 75, ff. 12, 15; Pl. 79, ff. 16-19; Pl. 84, f. 19
Scudder-Everyday Butter. p. 232, 1899 egg, larva, chrysalis
Scudder-Guide to Common Butter p. 69, 1873 larva, chrysalis
Index Amer. Econ. Ent. Vol. 5:125; Vol. 6:584

FOOD PLANTS

Actinomeris alternifolia (Linn.)
Aster sp.
Aster umbellatus Mill.
Conyza canadensis Linn.
Helianthus spl
Helianthus divarcatus Linn.
Plantago sp?
Rudbeckia laciniata Linn.
Verbesina helianthoides Michx.

nympha Edwards
ssp. see minuta Edwards (Melitaea)

nyssaria Guenee
ssp. see amicaria Herrich-Schaeffer

O

obductata Moeschler
ssp. see luctuata Schiffenmuller

obesalis Stephens
syn. see scabra Fabricius

obfirmaria (Hubner), Metarranthis- Geometridae

LIFE HISTORY

Can. Ent. 29:258, 1897 Fyles egg, larva
New York Ent. Soc. Jour. 51:136 1943 Rupert egg, larva
Loc. cit. 53:200, 1945 Forbes pupa
New York State Agri. Exp. Sta. (Cornell), Mem. 274:93, 1948 Forbes larva

FOOD PLANTS

Calyculata sp.
Prunus virginiana Linn.
Quercus sp. (oaks)
Vaccinium sp.

objurgata Smith
syn. see tessellata Harris

oblata (Morrison), Graphiphora- Noctuidae
syn. hilliana Harvey
a streckeri Barnes & Benjamin

LIFE HISTORY

maine Agri. Exp. Sta. Bull. 356: 160, 1930 Phipps larva, pupa
United States Nat. Mus. Proc. 27:827, 1904 Dyar egg, larva

FOOD PLANTS

Salis sp. (willows)
Spiraea latifolia Borkl.
Vaccinium angustifolium Ait.
Vaccinium myrtilloides Michx.

OBLINITA(ABBOT & SMITH), ACRONICTA-
NOCTUIDAE

SYN. SALICIS HARRIS
FORM INSOLITA GROTE

LIFE HISTORY

AMER. ENT. 2:341, 1870 LARVA*COCOON
AMER. MUS. NAT. HIST. BULL. 14: 261, 1901 BEUTEN. LARVA
CAN. ENT. 3:226, 1871 SAUNDERS LARVA*COCOON*
LOC. CIT. 10:66, 1878 GODELL LARVA
ILLINOIS DEPT. AGRI. TRANS. 15:201 1877 FRENCH LARVA
LOC. CIT. 18, APPEND:131, 170, 1880 COQUILLETT & MARTEN LARVA* COCOON*
KENTUCKY AGRI. EXP. STA. BULL. 227: 80, 1910 GARMAN LARVAL CUTICLE* F. 10
MINNESOTA AGRI. EXP. STA. BULL. 61:209, 1899 LUGGER LARVA* COCOON* BOTH F. 156
MISSOURI, REPT. NOX. INS. OF 3:70, 1871 RILEY LARVA*PUPA*
NEW YORK STATE CABIN. NAT. HIST. ANN. REPT. NO. 26:159, 1874 LINTNER LARVA
NEW YORK STATE MUS. MEM. 8:560 743, 1905 FELT LARVA
OHIO AGRI. EXP. STA. BULL. 45:175 1893 WEBSTER LARVA
ONTARIO ENT. SOC. ANN. REPT. P. 23, 1872 SAUNDERS LARVA*PUPA*
LOC. CIT. P. 49, 1899 SAUNDERS LARVA
UNITED STATES DEPT. AGRI. MISC. PUB. 657:355, 393, 1950 CRAIGHEAD LARVA
UNITED STATES NAT. MUS. PROC. 21:170, 1899 SMITH, DYAR LARVA*PL. 8, FF. 29, 30 COLOR OR UNCOLORED; PUPA, COCOON
VERMONT BOARD AGRI. REPT. 5:275 1878 PERKINS LARVA*PUPA*
WISCONSIN AGRI. SOC. TRANS. 21:7, 1883 FORBES LARVA*PUPA* COCOON
ABBOT & SMITH-LEP. INS. GEORGIA 2:187, 1797 LARVA*PL. 94 COLOR
FERNALD KINGSLEY'S STAND. NAT. HIST. 2:452, 1885 LARVA
GLOVER-ILLUST. NORTH AMER. ENT. 1878 LARVA*PL. 9, F. 20; PL. 16, F. 8
GUENEE-SPEC. GEN. LEPID. NOCT. 1:49, 1852 LARVA
HART-INS. COLLECTIONS ILLINOIS HIGH SCHOOLS P. 25, 1903 LARVA*COCOON*
HOLLAND-MOTH BOOK P. 158, 1905 LARVA*COCOON*F. 87
O'KANE-INJ. INS. P. 275, 1912 LARVA* F. 391
PA; CKARC-GUIDE STUDY INS. P. 304 1869 LARVA*
SAUNDERS-INS. INJ. FRUITS P. 325, 1883 LARVA*COCOON*
SAUNDERS-LOC. CIT. P. 325, 1889 LARVA*COCOON*F. 337
INDEX AMER. ECON. ENT.
VOL. 1:20 APATELA
VOL. 3:4 APATELA

FOOD PLANTS

ACER SACCHARINUM LINN.
ALNUS RUGOSA(DUROI)
ALNUS SERRULATA(AIT.)
ASPARAGUS OFFICINALIS LINN.
CANNA SP.
CEPHALANTHUS OCCIDENTALIS LINN.
FAGOPYRUM SAGITTATUM GILIB.
FRAGARIA CHILOENSIS(LINN.)
GOSSYPIUM HERBACEUM LINN (COTTON)
JUGLANS CINEREA LINN.
LONCIERA SP.
MALUS PUMILA MILL. (APPLE)
PHASEOLUS VULGARIS LINN. (BEAN)
PINUS SP. (PINES)
POLYGONUM LYDROPIPER LINN.
POLYGONUM PUNCTATUM ELL.
POPULUS NIGRA V. ITALICA MUENCH.
PRUNUS PERSICA(LINN. ((PEACH)
PYRUS COMMUNIS LINN. (PEAR)
QUERCUS SP. (OAKS)
RUBUS ALLEGHENIENSIS PORTER

RUBUS IDAEUS V. STRIGOSUS(MICHX.)
SALIX SP. (WILLOWS)
SYRINGA SP.
TRIFOLIUM SP. (CLOVERS)
TRITICUM AESTIVUM LINN. (WHEAT)
TYPHA LATIFOLIA LINN.
ULMUS SP. (ELMS)
VITIS SP. (GRAPES)
ZEA MAYS LINN. (CORN)
GRASSES

OBLIQUA PACKARD, HETEROCAMPA-NOTODONTIDAE
SYN. TROUVELOTI PACKARD
SYN. BRUNNEA GROTE & ROBINSON
LIFE HISTORY
AMER. MUS. NAT. HIST. BULL. 10: 426, 1898 BEUTEN. LARVA
NAT. ACAD. SCI. MEM. 7:244, 1895 PACKARD EGG, LARVA*PL. 30, FF. 1-2 COLOR AND PL. 85 UNCOLORED; PUPA*
NEW YORK ENT. SOC. JOURN. 3:27 1895 PACKARD EGG, LARVA* PL. 1, FF. 1, 3
NEW YORK STATE AGRI. EXP. STA. (CORNELL), MEM. 274:233, 1948 FORBES LARVA
NEW YORK STATE MUS. BULL. 8:705 1905 FELT LARVA(BRIEF)
GLOVER- ILLUST. NORTH AMER. ENT. 1879 LARVA*PL. 13, FF. 22, 26. 27
FOOD PLANTS
QUERCUS SP. (OAKS)
QUERCUS MACROCARPA MICHX.

OBLIQUA(WALKER), ARZAMA-NOCTUIDAE
SYN. OBLIQUATA GROTE
A PALLIDA BARNES & MCDUNNOUGH
LIFE HISTORY
AMER. MUS. NAT. HIST. BULL. 16: 439, 1902 BEUTEN. LARVA
BRITISH COL. ENT. SOC. PROC. 44: 17, 1948 GUPPY LARVA(BRIEF)
CAN. ENT. 10:15, 1878 WORTHINGTON LARVA, PUPA
LOC. CIT. 15:174, 1883 RILEY LARVAL HABITS
LOC. CIT. 20:119, 1888 BREHME LIFE HISTORY(BRIEF)
LOC. CIT. 21:160, 1889 BEUTEN. LARVA OBLIQUATA
ENT. SOC. AMER. ANNALS 16:377, 1923 ROBERTSON-MILLER EGG*LARVA*
INSECUTOR INSCITIAE MENSTRUUS 1:19, 1913 DYAR
NEW YORK STATE AGRI. EXP. STA. (CORNELL), MEM. 47:477, 1921 CLAASEN EGG*PL. 41, FF. 21, 24, 25; PL. 46, F. 65; LARVA* PL. 40, FF. 22, 27, 28, 29, 20; PL. 46, F. 66; PUPA*PL. 41, F. 23
OHIO AGRI. EXP. STA. TECH. BULL. 1:6, 1889 WEED LARVA*PL. 1 FF. 1, 2A, 2B; PUPA*PL. 1, F. 2C
OHIO STATE ACAD. SCI. ANN. REPT. P. 37, 1896 (FOURTH REPT.) HINE NOTE ON LIFE HIST. NO FF.
ONTARIO ENT. SOC. ANN. REPT. P. 18, 1883 RILEY EGG, LARVA
LOC. CIT. P. 20, 1883 KELLICOTT LARVAL HABITS
OTTAWA NATURALIST 17:37, 1903 FLETCHER LARVAL SPIRACLES
INDEX AMER. ECON. ENT. VOL. 2:41; VOL. 3:45
FOOD PLANTS
NELUMBO SP.
PONTEDERIA CORDATA LINN. (CAPTIVITY)
SAGITTARIA
SPARGANIUM SP.
SYMPLOCARPUS FOETIDUS(LINN.)
TYPHA LATIFOLIA LINN.

OBLIQUATA GROTE
SYN. SEE OBLIQUA WALKER

OBLIQUATA GROTE & ROBINSON
SYN. SEE ATOMARIA WALKER(OLENE)

OBLIQUIFERA WALKER
SYN. SEE MALANA FITCH

OBLITA HARRIS
SSP. SEE AUTUMNATA GUENEE

OBLITERATA GROTE
SSP. SEE DIPHTEROIDES GUENEE

OBLITERARA SCHWARZ
SSP. SEE MINUTA EDWARDS (CATOCALA)

OBLITERATA SCUDDER
SSP. SEE HYPOPHLAEAS BOISDUVAL

OBLITERATA STRECH
SSP. SEE ORNATA PACKARD

OBLITERATUS SCUDDER
SSP. SEE CLARUS CRAMER

OBLIVIOSA HENRY EDWARDS
SYN. SEE LEUCOSTIGMA ABBOT & SMITH

OBLONGA (GROTE), ARCHANARA-NOCTUIDAE
SYN. PERMAGNA GROTE
SYN. SUBCARNEA KELLICOTT
LIFE HISTORY
AMER. NAT. 17:1172, 1883 LARVA (BRIEF)
BUFF. SOC. NAT. SCI. BULL. 5:40, 1886 KELLICOTT EGG, LARVA, PUPA SUBCARNEA
CAN. ENT. 15:174, 1883 RILEY LARVA (BRIEF)
ENT. NEWS 19:295, 1908 WALTON LIFE HIST. LARVA* PL. 12; PUPA* PL. 12
NEW YORK STATE AGRI. EXP. STA. (CORNELL), MEM. 47:482, 1921 CLAASEN LARVA* PL. 47, F. 67; PUPA* PL. 42, F. 33; PL. 47, F. 77
ONTARIO ENT. SOC. ANN. REPT. P. 20, 1883 KELLICOTT LARVA
INDEX AMER. ECON. ENT.
VOL. 2:252 NONAGRIA
VOL. 3:279 NONAGRIA
VOL. 5:534 NONAGRIA
FOOD PLANTS
SCIRPUS SP. (RARELY)
TYPHA ANGUSTIFOLIA LINN.
TYPHA LATIFOLIA LINN.

OBNIGRA SMITH
SSP. SEE COMIS GROTE

OBRUPTATA WALKER
SYN. SEE OBSTIPATA FABRICIUS

OBRUSSATA GROTE
SYN. SEE ARGENTATA DRURY

OBSCURA BUTLER
SYN. SEE TETRIO LINNAEUS

OBSCURA COCKERELL
SSP. SEE ANTIOPA LINNAEUS

OBSCURA DYAR
SSP. SEE CERINTHA TREITSCHKE

OBSCUAR (FABRICIUS), ERINNYIS-SPHINGIDAE
SYN. STHENO HUBNER
SYN. PALLIDA GROTE
SYN. CINEROSA GROTE & ROBINSON
SYN. RHAEBUS BOISDUVAL
LIFE HISTORY
NEW YORK ENT. SOC. JOURN. 8:32, 1900 SWAINSON LARVA, PUPA
SOUTHERN CAL. ACAD. SCI. BULL. 34:128, 1935 COMSTOCK & DAMMERS LARVA* PLS. 23, 24; PUPA* PL. 24
WASHINGTON ENT. SOC. PROC. 4:450 1900 DYAR LARVA, PUPA
FOOD PLANTS
CYNANCHUM PALUSTRE (GRAY)
SARCOSTEMMA CYNANCHOIDES V. HARTWEGII (VAIL)

OBSCURA HENRY EDWARDS
SSP. SEE AMERICANA HARRIS (ACRONICTA)

OBSCURA HILL
SYN. SEE SPONSA SMITH

OBSCURA RILEY
SSP. SEE FRUGIPERDA ABBOT & SMITH

OBSCURA SAGEFER
SSP. SEE CECROPIA LINNAEUS

OBSCURA(SMITH), POLIA-NOCTUIDAE
LIFE HISTORY
FOOD PLANTS

OBSCUAR STRECKER, CATOCALA-NOCTUIDAE
SYN. SIMULATILIS GROTE
FORM OBVIA SCHWARZ
LIFE HISTORY
CAN. ENT. 36:115, 1904 DODGE LARVA
LOC. CIT. 41:227, 1919 ROWLEY LARVA
ENT. AMER. 2:45, 1866 KELLICOTT LARVA
ENT. NEWS 20:134, 1909 ROWLEY LARVA
NEW YORK STATE MUS. MEM. 8:716 1905 FELT LARVA
FOOD PLANTS
CARYA SP.
CARYA ILLINOENSIS(WANG.)
CARYA OVATA(MILL.)

OBSCURA STRETCH
SYN. SEE OPELLA GROTE

OBSCURATA VERITY
SSP. SEE NASTES BOISDUVAL

OBSCURIOR SMITH
SYN. SEE OLIVACEA MORRISON

OBSCURIOR SMITH
SSP. SEE ALBILINEA HUBNER

OBSCURIPENNIS(GROTE), ZANCLOGNATHA-NOCTUIDAE
LIFE HISTORY
ENT. AMER. 14:181, 1934 CRUMB LARVA
WASHINGTON ENT. SOC. PROC. 4:329 1898 DYAR LARVA
FOOD PLANTS
DEAD LEAVES

OBSCURIPENNIS GUNDER
SSP. SEE EPITHORE EDWARDS

OBSERVABILIS GROTE
SSP. SEE EXSERTISTIGMA MORRISON

OBSOLESCENS STRAND
SSP. SEE SUPLICATA BIRD

OBSOLETA EDWARDS
SSP. SEE ARCHIPPUS CRAMER

OBSOLETA EHRMANN
SSP. SEE PHILENOR LINNEUS

OBSOLETA(FABRICIUS)HELIOTHIS-NOCTUIDAE
NOTE-THE AMERICAN SPECIES IS NOW KNOWN AS HELIOTHIS ZEA BRODIE. SINCE THE LITERATURE CITED HERE LISTS THE SPECIAS AS OBSOLETA THIS NAME IS RETAINED.
SYN. ARMIGER HUBNER
FORM UMBROSA GROTE
LIFE HISTORY
AGRI. OF PA. REPT. P. 238, 1883 RATHVON LARVA*PUPA* COCOON
AMER. ENT. 1:212, 1869 LIFE HIST* FF. 150, 151 ALL STAGES
LOC. CIT. 2:42, 1869 LARVA*PUPA*
AMER. NAT. 7:214, 1873 LARVA*F. 30 NO DESCR.
ARKANSAS AGRI. EXP. STA. BULL. 310:13, 1935 ISLEY EGG*F. 4; LARVA*FF. 1, 5; PUPA*F. 6
CAL. AGRI. EXP. STA. BULL. 660;30 1942 SMITH LARVA*F. 23A; PUPA*F. 23 B, C
LOC. CIT. 707:5, 1948 MICHELBACHER & ETC. LARVA*F. 1A; PUPA*F. 1 C
FLORIDA AGRI. EXP. STA. BULL. 151; 171, 1919 WATSON LARVA*F. 97
ILLINOIS AGRI. EXP. STA. BULL. 95: 397, 1904 FORBES EGG*LARVA* PUPA*ALL ON A COLOR PLATE
ILLINOIS DEPT. AGRI. TRANS. 15: 103, 232, 1877 FRENCH, THOMAS LARVA

LOC.CIT.18,APPEND:150,1880 COQUILLETT EGG*LARVA*PUPA* COCOON

JOURN.AGRI.RESEARCH 30:790,1925 ELLIS LARVA*PL.2,FF.H,M. SETAL MAPS

JOURN.ECON.ENT.8:215,1915 HASEMAN EGG

LOC.CIT.12:268,1919 MOSHER LARVAL SETAL MAP* & CLAWS.*

LOC.CIT.34:351,1941 BARBER EGG SIZES AND HATCHING

KENTUCKY AGRI.EXP.STA.BULL. 187,1914 GARMAN & JEWETT LARVA*COLOR PLATES II, III SHOW VARIATIONS EXCELLENT; PUPA*PL.3 COLOR; EGG

LOC.CIT.227,1920 GARMAN LARVA* FF.12,13,14;STRUCTURE PATERN

MISSOURI,REPT.NOX.INS.3:105, 1871 RILEY EGG*LARVA*

NEBRASKA UNIV.AGRI.EXP.STA. RESEARCH BULL.81:11,1935 WHELAN LARVA TECH.DESCR.

NEW YORK STATE AGRI.SOC.TRANS. 99,1855 GLOVER EGG,LARVA, PUPA

ONTARIO ENT.SOC.ANN.REPT.P. 75,1900 LOCKHEAD LARVA*

LOC.CIT.P.100,1919 ROSS LARVA*

PAN PACIFIC ENT.15:123,1939 STAHLER COMPARISON OF LARVAE OF OBSOLETA ON PHLOXIPHAGA G. & R.

TENNESSEE AGRI.EXP.STA.BULL. 186:41,1943 STANLEY LARVA* F.30;PUPA*F.32 NO DESCR.

UNITED STATES DEPT.AGRI.BUR. ENT.BULL.29,OLD SERIES P. 18,1893 MALLY LARVA,PUPA

LOC.CIT.BULL.50 NEW SERIES P. 41,1904 QUAINTANCE,BRUES EGG*PL.4,F.2;PL.3,F.1; LARVA*PL.3,FF.2-7;PL.6, F.5;PUPA*PL.12,F.4

UNITED STATES DEPT.AGRI.DIV. ENT.CIRC.133,133,1911 WILDERMUTH LARVA*F.8 NO DESCR.

UNITED STATES DEPT.AGRI.CIRC. 615:2,1942 REED,SMITH,& BARE LARVA*F.1D NO DESCR.

UNITED STATES DEPT.AGRI. FARMER'S BULL.47:10,1897 HOWARD EGG*F.6 NO.1;LARVA* F.6 NOS.2,4,5;PUPA*F.6 NO.6

LOC.CIT.120:4,1900 HOWARD LARVA*F.8 B,C;PUPA*F.8D

LOC.CIT.191:6,1904 QUAINTANCE EGG*LARVA*PUPA*F.1

LOC.CIT.290:8,1907 BISHOPP & JONES EGG*F.1;LARVA* PUPA*F.2

LOC.CIT.856:42,1917 ORTON & CHITTENDEN LARVA

LOC.CIT.872:6,1917 BISHOPP EGG* F.5;LARVA*PUPA*F.6

LOC.CIT.890:18,1917 PIERCE LARVA*F.19B;PUPA*FF.19,20

LOC.CIT.1206:9,1921 LUNGINBILL EGG*F.5;LARVA*FF.1,6;PUPA* FF.7,8

LOC.CIT.1310:8,1923 PHILLIPS & KING EGG* F.8;LARVA*F.9; PUPA*FF.10,11

LOC.CIT.1371:39,1924 GILBERT LARVA* F.59B

LOC.CIT.1595:4,1929 BISHOPP EGG*F.9;LARVA*FF.3,5;PUPA* F.10

LOC.CIT. 1651:6,1931 PHILLIPS & BARBER EGG*F.8;LARVA*F.9; PUPA* FF.10-13

LOC.CIT.1688:6,1932 FOLSOM LARVA* F.5B;PUPA*F.5C

UNITED STATES DEPT.AGRI.TECH. BULL.782:2,1941 REED, SMITH,THOMAS LARVA*F.1D

UNITED STATES DEPT.AGRI. YEARBOOD FOR 1898 P.132 HOWARD LARVA* FF.14B,C; PUPA*F.14D NO DESCR.

LOC.CIT. FOR 1898 P.256 CHITTENDEN LARVA*F.79 B,C;

PUPA*F.79D
LOC.CIT. FOR 1910 P.289 MORGAN
LARVA*F.12 B,D;PUPA*F.12D
UNITED STATES DEPT.AGRI.ENT.
CONM.REPT.4,1885 RILEY
LIFE HISTORY* COLOR
UNITED STATES NAT.MUS.PROC.
68:2,1927 CRUMB LARVA
CHITTENDEN-INS.INJ.VEG.P.208
1907 LARVA*FF.2,132,133,149
COMSTOCK-REPT.COTTON INS.
UNITED STATES DEPT.AGRI.P.
297,1879 LIFE HIST* COLOR
CROSBY & LEONARD-MANUAL VEG.
GARD.INS.P.213,1918 EGG*
F.134;LARVA*FF.135,136;
PUPA
ESSIG-INS.WEST.NORTH AMERICA
P.680,1926 EGG,LARVA*F.554
FERNALD-APPLIED ENT.P.275,
1926 LARVA*PUPA* BOTH F.278
FERNALD & SHEPARD-LOC.CIT. P.
270,1942 LARVA*PUPA*BOTH
F.273
GLOVER-ILLUST.NORTH AMER.ENT.
1878 EGG,LARVA,PUPA PL.
19,F.21
GUNDLACH-ENT.CUBANA P.309,
1886 LARVA,PUPA
HART-INS.COLL.ILLINOIS HIGH
SCHOOLS P.38,1903 EGG,
LARVA*PUPA*F.51
HOLLAND-MOTH BOOK P.223,1905
EGG*LARVA*PUPA*FF.130,131
KIRBY-EUROPEAN BUTTER.&MOTHS
P.225,1882 LARVA
METCALF & FLINT-DESTR.&USEFUL
INS.PP.369,370,1939 EGG,
LARVA,F.221
O'KANE-INJ.INS.P.211,1912 LARVA*
F.263
PACKARD-HALF HOURS WITH INS.
P.224,1877 LARVA
PACKARD INS.INJ.WEST P.778,
1877 LARVA
PEAIRS-PESTS FARM,GAR.ORCHARD
P.137,1941 LARVA*FF.154,
198
SMITH-INS.NEW JERSEY P.426,
1900 EGG*LARVA*PUPA*F.198
AND UNNUMBERED
SMITH-OUR INS.FRIENDS &
ENEMIES P.264,1909 EGG*
LARVA*PUPA*ALL F.121
STAINTON-BRITISH BUTTER.MOTHS
1:292,1857 LARVA
INDEX AMER.ECON.ENT.
VOL.1:150;VOL.2:81
CHLORIDEA
VOL.3:83 CHLORIDAE
VOL.4:101 CHLORIDEA
VOL.5:292;VOL.6:342
ARMIGERA
VOL.7:450

FOOD PLANTS

ABUTILON SP.
AGERATUM CONYZOIDES LINN.
AMARANTHUS SP.
APIUM GRAVEOLENS LINN.(CELERY(
ARACHIS HYPOGAEA LINN.(PEANUT)
ASPARAGUS OFFICINALIS LINN
(ASPARAGUS)
ATROPA BELLADONNA LINN.
AVENA SATIVA LINN.(OATS)
BRASSICA OLERACEA LINN.(CABBAGE)
BRASSICA OLERACES V.BOTYRTIS
D.C.(CAULIFLOWER)
CANNA SP.
CANNABIS SATIVA LINN.(HEMP)
CAPSICUM ANNUUM LINN.(PEPPERS)
CARICA PAPAYA LINN.
CHENOPODIUM ALBUM LINN.
CHRYSANTHEMUM SP.
CICER ARIETINUM LINN.
CITRULLUS VULGARIS SCHRAD.
(WATERMELON)
CLEOME GYNANDRA LINN.
CROTALARIA SAGITTALIS LINN.
CUCUMIS MELO LINN.(MELON)
CUCUMIS SATIVUS LINN.(CUCUMBER)
CUCURBITA MAXIMA DUSCH.(SQUASH)
CUCURBITA PEPA LINN.(PUMPKIN)
DAHLIA VARIABILIS DESF.
DATURA STRAMONIUM LINN.
DESMODIUM TORTUOSUM(SW.)
DIANTHUS CARYOPHULLUS LINN.

DIGITARIA ISCHAEMUM(SCHREB.)
FRAGARIA CHILOENSIS(LINN.) (STRAWBERRY)
GERANIUM MACULATUM LINN.
GLADIOLUS SP.
GLYCINE MAX(LINN.)(SOY BEAN)
GOSSYPIUM HERBACEUM LINN. (COTTON)
HAPLOPAPPUS DIVARICATUS(NUTT.)
HELIANTHUS DECAPETALUS LINN.
HIBISCUS SP.
HIBISCUS ESCULENTUS LINN. (OKRA)
HIBISDUS LASIOCARPOS CAV.
HIBISCUS MILITARIS CAV.
HIBISCUS TRIONUM LINN.
HYOSCYAMUS NIGER LINN.
IPOMOEA BATATAS(LINN.)(SWEET POTATO)
IPOMOEA PURPUREA(LINN.)
LACTUCA SATIVA LINN. (LETTUCE)
LATHYRUS ODORATUS LINN.
LESPEDEZA CUNEATA(DUM.)
LINARIA CANADENSIS(LINN.)
LINUM SULCATUM RIDD. (FLAX)
MALUS PUMILA MILL.
MALVA ROTUNDIFOLIA LINN.
MEDICAGO SATIVA LINN.
MELILOTUS ALBA DESR.
MELILOTUS SAUVEOLENS LEDEB.
MENTHA SP. (MINTS)
NASTURTIUM SP.
NICOTIANA TABACUM LINN. (TOBACCO)
ORYZA SATIVA LINN. (RICE)
PANICUM FASCICULATUM SWARTZ
PANICUM OLIGSANTHES V. SCRIBNERIANUM(NASH)
PAPAVE SP.
PERSEA AMERICANA MILL. (AVOCADO)
PHASEOLUS LIMENSIS MACFAD. (LIMA BEAN)
PHASEOLUS VULGARIS LINN. (BEANS)
PHLOX SP.
PHYSALIS SP.
PHYSALIS PUBESCENS LINN.
PHYSALIS VISCOSA LINN.
PISUM SATIVUM LINN. (PEAS)
POLYGONUM PENSYLVANICUM LINN.
PRUNUS PERSICA(LINN.)(PEACH)
PYRUS COMMUNIS LINN. (PEAR)
RESEDA SP.
RESEDA ODORATA LINN.
RICINUS COMMUNIS LINN.
ROSA SP.
SACCHARUM OFFICANARUM LINN. (SUGAR)
SASSAFRAS ALBIDUM NUTT. V. MOLLE(RAF.)
SECALE CEREALE LINN. (RYE)
SETARIA ITALICA(LINN.)
SIDA RHOMBIFOLIA LINN.
SIDA SPINOSA LINN.
SOLANUM LYCOPERSICUM LINN. (TOMATO)
SOLANUM MELONGENA LINN. (EGG PLANT)
SOLANUM TUBEROSUM LINN. (POTATO)
SOLIDAGO LEAVENWORTHII T. & G.
SORGHUM SP.
SORGHUM BICOLOR(LINN.)
SPINACIA OLERACEA MILL. (SPINASH)
STACHYS SIEBOLDII MIQ. (ARTICHOKE)
STIZOLOBIUM DEERINGIANUM BORT.
TRIFOLIUM PRATENSE LINN.
TRITICUM AESTIVUM LINN. (WHEAT)
VICIA SATIVA LINN.
VICIA VILLOSA ROTH
VIGNA SINENSIS(LINN.)
VITIS SP. (GRAPES)
XANTHIUM SP.
XANTHIUM STRUMARIUM LINN.
ZEA EVERTA STURT.
ZEA MAYS LINN. (CORN)
VELVET-LEAF FOXTAIL

<u>OBSOLETA</u> HENRY EDWARDS
SSP. SEE <u>LEANIRA</u> BOISDUVAL

<u>OBSOLETA</u> NIEPELT
SSP. SEE <u>PHILEA</u> LINNAEUS

<u>OBSOLETA</u> REIFF
SSP. SEE <u>LUCINA</u> HENRY EDWARDS

OBSOLETA WORTHINGTON
SSP. SEE ILIA CRAMER

OBSOLETA-LUNULATA TUTT
SSP. SEE PSEUDARGIOLUS BOISDUVAL & LECONTE

OBSTIPATA(FABRICIUS), PERCNOPTILOTA-GEOMETRIDAE
SYN. FLUVIATA HUBNER
SYN. PERACUTATA WALKER
SYN. OBRUPTATA WALKER
SYN. ALTERNATA WALKER
SYN. PIGRATA WALKER
SYN. BRUNNEIPENNIS HULST
LIFE HISTORY
ENT. INTELLIGENCE 1858 ANONYMOUS LIFE HIST.
LYONS(FRANCE)LINN. SOC. ANNALS. P.50,1868 MILLIERE LIFE HIST* PL.90
NEW YORK STATE AGIR. EXP. STA. (CORNELL), MEM. 274:154, 1948 FORBES LARVA
PSYCHE 8:429,1899 DYAR EGG, LARVA FLUVIATA
PSYCHE 19:16,1912 BARNES & MCD. EGG, LARVA FLUVIATA
GUMPPENBERG-SYSTEMA GEOMETRARUM 3:402,1890 LARVA, PUPA
NEWMAN-BRITISH MOTHS. 1869 LIFE HISTORY
PACKARD-MONOGRAPH GEOMETRIDAE PP.78,564,1876 LARVA, PUPA
FOOD PLANTS
ANTHEMIS COTULA LINN.
CHRYSANTHEMUM SP.
POLYGONUM SP. (IN CAPTIVITY)
POLYGONUM HYDROPIPER LINN.
POLYGONUM PERSICARIA LINN.
RUMEX SP.
SENECIO SP.
ULMUS SP. (ELMS)
GENERAL FEEDER LOW PLANTS

OBTUSA SPEYER
SYN. SEE NORMANIANA GROTE

OBTUSARIA(HUBNER), EUCHLAENA-GEOMETRIDAE
SYN. DECISARIA WALKER
SYN. MUZARIA WALKER
SYN. INCISARIA WALKER
LIFE HISTORY
NEW YORK ENT. SOC. JOURN. 53: 199,1945 FORBES LARVA, PUPA
NEW YORK STATE AGRI. EXP. STA. (CORNELL), M3M. 274:76,1948 FORBES LARVA
NEW YORK STATE MUS. MEM. 8:721 1905 FELT LARVA
GUMPPENBERG-SYSTEMA GEOMETRARUM 7:389,1895 LARVA, PUPA
PACKARD-MONOGRAPH GEOMETRIDAE P.517,1876 LARVA*PL.13,F. 5
FOOD PLANTS
CASTANEA DENTATA(MARSH.)
IMPATIENS SP.
PRUNUS SEROTINA EHRH.
ROSA SP. (ROSES)

OBUMBRATA TAYLOR
SSP. SEE PYGMAEATA HUBNER

OBVIA SCHWARZ
SSP. SEE OBSCURA STRECKER

OCCANTARIA HULST
SYN. SEE MILISARIA HULST

OCCIDENTA GROTE
SYN. SEE CONFIGURATA WALKER

OCCIDENTALIS CLARK
SSP. SEE MYOPS ABBOT & SMITH

OCCIDENTALIS FRENCH
SYN. SEE CALIFORNIAE WALKER

OCCIDENTALIS GROTE & ROBINSON
SYN. SEE INTERRUPTA GUENEE

OCCIDENTALIS(HULST), OPOROPHTERA-GEOMETRIDAE

FORM LATIPENNIS HULST
LIFE HISTORY
FOOD PLANTS
ACER MACROPHYLLUM PURSH
AMELANCHIER FLORIDA LINDL.
QUERCUS SP. (OAKS)
SYMPHORICARPOS ALBUS(LINN.)

OCCIDENTALIS(HULST), PERO-GEOMETRIDAE
A BLACKMOREI CASSINO & SWETT
LIFE HISTORY
UNITED STATES NAT. MUS. PROC. 27:914, 1904 DYAR EGG, LARVA
FOOD PLANTS

OCCIDENTALIS HENRY EDWARDS
SYN. SEE MODESTA HARRIS

OCCIDENTALIS LINTNER, CERURA-NOTODONTIDAE
A GIGANS McDUNNOUGH
SYN. DEORUM DYAR
LIFE HISTORY
AMER. MUS. NAT. HIST. BULL. 10: 430, 1898 BEUTEN. LARVA
BOSTON SOC. NAT. HIST. PROC. 24: 555, 1890 PACKARD LARVA
CAN. ENT. 13:144, 1881 FRENCH LARVA
LOC. CIT. 28:307, 1896 DYAR EGG, LARVA
LOC. CIT. 54:140, 1922 McD. EGG, LARVA GIGANS
NAT. ACAD. SCI. MEM. 7:269, 1895 PACKARD LARVA
NEW YORK STATE AGRI. EXP. STA. (CORNELL), MEM. 274:236, 1948 FORBES LARVA
FOOD PLANTS
POPULUS SP. (POPLARS)
SALIX SP. (WILLOWS)
SALIX LASIANDRA BENTH.

OCCIDENTALIS REAKIRT, PIERIS-PIERIDAE
A CALYCE EDWARDS
B NELSONI EDWARDS
LIFE HISTORY
ENT. AMER. 3:162, 1887 HY. EDW. CHRYSALIS
PSYCHE 2:184, 1878 MEAD LARVA, PROBABLY THIS SPECIES
FOOD PLANTS
CLEOME SERRULATA PURSH

OCCIDENTALIS SCUDDER, COLIAS-PIERIDAE
A CHRYSOMELAS HENRY EDWARDS
FORM FEMALE SHASTAE BARNES & BENJAMIN
B BARBARA HENRY EDWARDS
LIFE HISTORY
FOOD PLANTS
CLEOME SERRULATA PURSH
FABACEAE?
TRIFOLIUM SP. (CLOVERS)

OCCIDENTALIS SKINNER
SYN. SEE COMMUNIS GROTE(PYRGUS)

OCCIDENTALIS SMITH
SSP. SEE LUBRICALIS GEYER

OCCIDENTIS WALKER
SYN. SEE AMERICANA HARRIS (EPICNAPTERA)

OCCIDUARIA GUENEE
SYN. SEE CREPUSCULARIA SCHIFFENMULLER

OCCIDUARIA(PACKARD), ITAME-GEOMETRIDAE
A ANDERSONI SWETT
LIFE HISTORY
NEW YORK ENT. SOC. JOURN. 53: 190, 1945 FORBES PUPA
NEW YORK STATE AGRI. EXP. STA. (CORNELL), MEM. 274:40, 1948 FORBES LARVA
UNITED STATES NAT. MUS. PROC. 25:390, 1903 DYAR LARVA
FOOD PLANTS
AMELANCHIER SP.
RIBES AUREUM PURSH
RIBES SATIVUM SYME (CURRANT)
SALIX SP. (WILLOWS)

OCCIDUARIA WALKER
SSP. SEE PULVERARIA LINNAEUS

occulta(Henry Edwards),Drasteria-Noctuidae
LIFE HISTORY
Brooklyn Ent.Soc.Bull.27:80, 1932 Crumb larva graphica
FOOD PLANTS
Vaccinium sp.(blueberry)

occulat(Linnaeus),Eurois-Noctuidae
LIFE HISTORY
Amer.Mus.Nat.Hist.Bull.14: 271,1901 Beuten.larva
Ent.Soc.Amer.Annals 23:396, 1940 Hawkins larval claws
Maine Agri.Exp.Sta.Bull.356: 167,1930 Phipps larva*f.24, c;pupa
United States Nat.Mus.Proc. 27:821,1904 Dyar egg,larva
Curtis-British Ent.6:248,1862 larva*pl.248 COLOR
Humphrey's-Genera British Moths 1:52,1859 larva*pl.21 COLOR
Kirby-European Butter.&Moths p.207,1882 larva
Stainton-British Butter.&Moths 1:271,1882 larva
Stephens-Illust.British Ent. Haust.3:29,1829 larva
Index Amer.Econ.Ent. Vol.5:393 Lycophotia

FOOD PLANTS
Acer sp.(maples)
Populus sp.(poplars)
Vaccinium sp.
Vaccinium angustifolium Ait.
Vaccinium myrtilloides Michx.

ocellata Edwards
syn.see clyton Boisduval & Leconte

ocellifera Grote
syn.see columella Fabricius

ocellinata(Guenee),Semiothisa-Geometridae
syn.duplicata Packard
LIFE HISTORY
New York State Agri.Exp.Sta. (Cornell),Mem.274:48,1948 Forbes larva
United States Dept.Agri.Misc. Pub.657:361,429,1950 Craighead larva
Gumppenberg-Systema Geometarum 7:441,1895 larva, pupa
FOOD PLANTS
Robinia pseudo-acacia Linn.

ochracea Cassino
see radians Walker

ochracea Lindsey
ssp.see comma Linnaeus

ochracea Neumoegen
syn.see navadensis Grote & Robinson

ochracea(Riley),Lasionycta-Noctuidae
LIFE HISTORY
Washington Ent.Soc.Proc.4: 325,1898 Dyar larva
FOOD PLANTS
Artemisia californica Less.

ochracea Stretch
ssp.see ornata Packard

ochracea Stretch
ssp.see guttata Boisduval

ochracea-rivulosa Stretch
ssp.see ornata Packard

ochrae Grote
ssp.see crassiuscula Haworth

ochreata Barnes & MacDunnough
ssp.see stragula Grote

ochreata Butler
syn.see nais Drury

OCHREATA PACKARD
SSP. SEE SULPHURARIA PACKARD (HESPERUMIA)

OCHREATUS HULST
SSP. SEE MAGNARIUS GUENEE

OCHREIPENNIS HARVEY
SYN. SEE INDOMITA WALKER

OCHREOUS MEADOWS
SSP. SEE EDWARDSII PACKARD

OCHROGASTER (GUENEE), EUXOA-NOCTUIDAE
SYN. TURRIS GROTE
FORM INSIGNATA WALKER
SYN. CINEROMACULA MORRISON
FORM GULARIS GROTE
LIFE HISTORY
ENT. SOC. AMER. ANNALS 23:396, 1930 HAWKINS LARVAL CLAWS
ILLINOIS AGRI. EXP. STA. BULL. 93:364, 1904 FORBES LARVA
MAINE AGRI. EXP. STA. BULL. 340: 205, 1927 PHIPPS LARVA*F.30 G,H
NEBRASKA AGRI. EXP. STA. RESEARCH BULL. 81:15, 1935 WHELAN LARVA, TECH. DESCR.
CROSBY&LEONARD-MANUAL VEG. GARD INS. P. 267, 1918 LARVA, PUPA
ESSIG-INS. WEST. NORTH AMERICA P. 682, 1926 LARVA
INDEX AMER. ECON. ENT. VOL. 1:133; VOL. 2:144; VOL. 3:161; VOL. 4:186; VOL. 5:245; VOL. 6:283 VOL. 7:372
FOOD PLANTS
AVENA SATIVA LINN. (OATS)
BETA VULGARIS LINN. (BEETS)
BRASSICA OLERACEA LINN. (CABBAGE)
BRASSICA OLERACES V. BOTYRTUS DC (CAULIFLOWER)
HORDEUM VULGARE LINN. (BARLEY)
PHASEOLUS VULGARIS LINN. (BEANS)
PISUM SATIVUM LINN. (PEAS)
RAPHANUS SATIVUS LINN. (RADISH)
SOLANUM TUBEROSUM LINN. (POTATO)
TRITICUM AESTIVUM LINN. (WHEAT)
ZEA MAYS LINN. (CORN)
GENERAL FEEDER FLOWERS

OCHROLEUCA STRAND
SYN. SEE LECONTEI GUERIN-MENEVILLE

OCHROPTENA (DYAR), PAPAIPEMA-NOCTUIDAE
A HUMULI BIRD
LIFE HISTORY
CAN. ENT. 39:138, 1907 BIRD EGG, LARVA, PUPA AS CIRCUMLUCENS
LOC. CIT. 47:114, 1915 BIRD LARVA* PL. 1, FF. 4,5; GALL*PL. 1, F. 3
LOC. CIT. 57:304, 1925 BIRD LARVA
FOOD PLANTS
HUMULUS LUPULUS LINN. (HOPS)

OCHROSACPUS GROTE & ROBINSON
SSP. SEE RUBROSCAPUS MENETRIES

OCTOMACULATA DEAN
SSP. SEE HYPOPHLAEAS BOISDUVAL

OCTOMACULATA (FABRICIUS), ALYPIA-AGARISTIDAE
SYN. BIMACULATA GMELIN
SYN. QUADRIGUTTALIS HUBNER
AB MALE ALBOMACULATA STOLL
LIFE HISTORY
AMER. ENT. 2:151, 1870 LARVA*
AMER. JOURN. SCI. 36:314, 1839 HARRIS LARVA, PUPA
AMER. MUS. NAT. HIST. BULL. 10: 357, 1898 BEUTEN. LARVA
AMER. NAT. 4:613, 1871 LARVA* F. 613 NO DESCR.
AMER. POMOLOGICAL SOC. REPT. P. 217, 1854 HARRIS LARVA
ILLINOIS DEPT. AGRI. TRANS. 15:176 1877 FRENCH LARVA
LOC. CIT. 18, APPEND: 110, 172, 1880 COQUILLETT MARTEN LARVA*
JOURN. ECON. ENT. 10:47, 1917

Lowry larva
Kentucky Bur. Agri. Rept. p. 225, 188[illegible] Goldsmith larva(brief)
Minnesota Agri. Exp. Sta. Bull. 61:121, 1899 Lugger larva* f. 66
Loc. cit. Bull. 84, 1903 Washburn larva* f. 7 COLOR
Missouri Rept. Ins. of 1:136, 1869 Riley larva*
Loc. cit. 2:80, 1870 Riley life hist* f. 55
Loc. cit. 6:94, 1874 Riley larva*
Mus. Comp. Zool. Bull. 87, No. 6 1941 Dethier larval antenna* f. 73
New York Acad. Sci. Annals 8: 226, 1894 Dyar larval setae
New York State Cabin. Nat. Hist. Ann. Rept. No. 26:119, 1874 Lintner larva
Loc. cit. No. 42:181, 1874 Lintner larva
Loc. cit. No. 42:181, 1889 Lintner larva* f. 8
New York State Mus. Bull. 37:16 1900 Felt larva*(brief)
New York State Mus. Mem. 8:575 1905 Felt larva
New York State Rept. Ins. of 3:399, also p. 81, sec. 122, 1856 Fitch larva
Ontario Ent. Soc. Ann. Rept. p. 78, 1891 Southwick larva*
Loc. cit. p. 108, 1894 Southwick larva*
Smithsonian Misc. Coll. 4, art. 2:132, 1862 Morris larva
United States Dept. Agri. Farmer's Bull. 1220:24, 1921 Quaintance larva* f. 23; pupa* f. 24
Loc. cit. 1893:21, 1942 Demaree & Runner larva* f. 19
United States Dept. Agri. Misc. Pub. 657:364, 390, 1950 Craighead larva
Vermont Bd. Agri. Rept. 5:276, 1878 Perkins larva*
Abbot & Smith–Lepid. Ins. Georgia 1:87, 1797 larva* pl. 44 COLOR
Cragen–Our Ins. Friends & Foes p. 146, 1899 larva
Emmons–Nat. Hist. New York 5: 128, 1854 larva* pl. 38 COLOR
Essit–Ins. West. North Amer. p. 678, 1926 larva
Fernald–Kingsley's Stand. Nat. Hist. 2:463, 1885 larva
Harris–Ent. Correstp. p. 285, 1869 larva
Hart–Ins. Coll. Illinois High Schools pl 24, 1903 larva*
Herrick–Ins. Enemies Shade Trees p. 381, 1935 larva* f. 310
Holland–Moth Book p. 144, 1905 larva* f. 83
Kellogg–Amer. Ins. p. 409, 1908 larva
Morris–Synop. Lep. North Amer. p. 132, 1862 larva
O'Kane–Inj. Ins. p. 275, 1912 larva* f. 393
Packard–Inj. Ins. West p. 785, 1877 larva*
Saunders0Ins. Inj. Fruits p. 263 1883 larva* pupa*
Saunders–Loc. cit. p. 263, 1889 larva* f. 272
Slingerland & Crosby–Manual Fruit Ins. p. 420, 1914 larva
Smith–Ins. New Jersey p. 402 1900 larva* f. 189
Stretch–Zygaen. Bombyc. North Amer. p. 7, 1873 larva* pl. 10, f. 10 COLOR
Wood–Ins. Abroad p. 651, 1874 larva*
Index Amer. Econ. Ent. Vol. 1:8; Vol. 2:14; Vol. 3:17, Vol. 4:17; Vol. 5:21; Vol. 6:23; Vol. 7:31

FOOD PLANTS
Berberis vulgaris Linn.

PARTHENOCISSUS QUINQUEFOLIA (LINN.)
PARTHENOCISSUS TRICUSPIDATA (S.Z.)
ROSA (ROSES)
VITIS SP.
VITIS VINIFERA LINN. (GRAPE)

OCULANA FABRICIUS
SYN. SEE DEFLORATA FABRICIUS

OCULATISSIMA ABBOT & SMITH
SYN. SEE DEFLORATA FABRICIUS

OCULATRIX (GUENEE), PAECTES-NOCTUIDAE
LIFE HISTORY
CAN. ENT. 59:99, 1927 CRIDDLE LARVA
FOOD PLANTS
RHUS TOXICODENDRON LINN.

OCULEA NEUMOEGEN
SSP. SEE POLYPHEMUS CRAMER

ODIUS (FABRICIUS), HISTORIS-NYMPHALIDAE
SYN. ORION FABRICIUS
SYN. DANAE CRAMER
LIFE HISTORY
AMER. ACAD. SCI. PROC. 27:245, 1892 SCUDDER LIFE HIST. ORION
NEW YORK ENT. SOC. JOURN. 9:78 1901 SWAINSON LARVA, CHRYSALIS
FOOD PLANTS
CECROPIA SP.
TABEBUIA SP. (TRUMPET TREE WEST IND.)

ODOAR (LINNAEUS), EREBUS-NOCTUIDAE
SYN. AGARISTA CRAMER
LIFE HISTORY
ENT. AMER. 4:36, 1888 FERNALD EGG
ENT. NEWS 18:454, 1907 HOLLAND LARVA (BRIEF)
ENT. NEWS 28:79, 1917 HISER LARVA, EGG
SOUTHERN CAL. ACAD. SCI. BULL. 35: 95, 1936 COMSTOCK LARVA* PLS. 21, 22; PUPA* PL. 23
WASHINGTON ENT. SOC. PROC. 13: 231, 1911 DYAR LARVA
GUNDLACH-ENT. CUBANA P. 367, 1886 LARVA PUPA
INDEX AMER. ECON. ENT. VOL. 4:171; VOL. 5:227; VOL. 6:270; VOL. 7:360
FOOD PLANTS
ACACIA DECURRENS WILLD
CASSIA FISTULA LINN.
FICUS CARICA LINN. (FIG)
GYMNOCLADUS DIOICA (LINN.)
PITHECELLOBIUM UNGUISCATE (LINN.)
SAMANEA SAMAN (JACQ.)

OECOGENES DYAR
SYN. SEE DENSA WALKER

OENO BOISDUVAL
SYN. SEE MELISSA FABRICIUS

OENONE SCUDDER
SYN. SEE NYCTEIS DOUBLEDAY & HEWITSON

OETUS (BOISDUVAL), MINOIS-SATYRIDAE
SYN. CHARON EDWARDS
A PHOCUS EDWARDS
LIFE HISTORY
CAN. ENT. 18:88, 1886 EDWARDS EGG, LARVA, CHRYSALIS
MONTANA UNIV. BULL. 30:107, 1906 ELROD LARV, CHRYSALIS
COMSTOCK, J. A. BUTTER. CAL. P. 75 1927 EGG*LARVA*CHRYSALIS* A29
EDWARDS-BUTTER. NORTH AMERICA 3RD SERIES 1897 LIFE HIST* PL. SATYRUS S. FF. A-D
HOLLAND-BUTTER. BOOK PP. 217-128 1905 3RD EDIT. P. 192, 1931 LIFE HISTORY CHARON
MEAD-REPT. LEP. WHEELER EXPED. 5:773, 1875 EGG (BRIEF)
FOOD PLANTS
GRASSES

OGDENENSIS HOLLAND
SSP. SEE JUBA SCUDDER

OGEECHENSIS SCUDDER
SYN. SEE LOGAN EDWARDS

OITHONA(STRECKER),APANTESIS-ARCTIIDAE
AB CONSPICUA STRETCH
FORM RECTILINEA FRENCH
LIFE HISTORY
CAN. ENT. 23:130,1891 FRENCH EGG,LARVA,PUPA
CAN. ENT. 35:117,1903 GIBSON LARVA
LOC. CIT. 37:343,1905 GIBSON PUPA
FOOD PLANTS
CASTILLEJA SESSILIFLORA PURSH
GALIUM BOREALE LINN.
LATHYRUS VENOSUS MUHL.
TRIFOLIUM SP. (CLOVER PREFERRED) GENERAL FEEDER LOW PLANTS

OKANAGANA MCDUNNOUGH
SSP. SEE SAEPIUM BOISDUVAL

OKEECHOBEE WORTHINGTON
SYN. SEE BATABANO LUCAS

OKIUS OBERTHUR
SYN. SEE SILVESTRIS EDWARDS

OLANCHA WRIGHT
SSP. SEE CHALCEDONA DOUBLEDAY & HEWITSON

OLENUSARIA WALKER
SYN. SEE COLORARIA FABRICIUS

OLERACEA HARRIS
SSP. SEE NAPI LINNAEUS

OLIVACEA COCKERELL
SSP. SEE POLYPHEMUS CRAMER

OLIVACEA GROTE
SSP. SEE HUMULI HARRIS(HYPENA)

OLIVACEA(MORRISON),LACINIPOLIA-NOCTUIDAE
SYN. FEMALE OBSCURIOR SMITH
A LUCINA SMITH
B MEGARENA SMITH
C ALTUA SMITH
D DAVENA SMITH
E PETITA SMITH
F VAU-MEDIA SMITH
LIFE HISTORY
UNITED STATES NAT. MUS. PROC. 27:855,1904 DYAR EGG, LARVA PETITA
FOOD PLANTS
PHLOX SP.
PLANTAGO LANCEOLATA LINN.
PLANTAGO MAJOR LINN.

OLIVALIS HULST
SYN. SEE ARGILLACEARIA PACKARD

OLIVATA(HARVEY),PSEUDOGLAEA-NOCTUIDAE
SYN. BLANDA GROTE
SYN. TAEDATA GROTE
SYN. DECEPTA GROTE
LIFE HISTORY
WASHINGTON ENT. SOC. PROC. 4:324 1898 DYAR LARVA
FOOD PLANTS
PRUNUS SEROTINA EHRH.
QUERCUS SP. (OAKS)
QUERCUS ALBA LINN.

OLIVATA PACKARD
SYN. SEE BIUNDATA WALKER

OLIVEATA CASSINO
SSP. SEE INDUCTATA GUENEE

OLIVIAE COCKERELL,HEMILEUCA-SATURNIIDAE
AB MALE GRISEA COCKERELL
AB FEMALE SUFFUSA COCKERELL
LIFE HISTORY
ENT. SOC. AMER. ANNALS 7:295, 1914 MOSHER PUPA
JOURN. ECON. ENT. 11:363,1918 CAFFREY LARVAL SPINES
NAT. ACAD. SCI. MEM. 12:127,1914 PACKARD LARVA*PL. 24,F. 4;

COCOON, PUPA
UNITED STATES DEPT. AGRI. BULL. 443:1, 1916 WILDERMUTH, CAFFREY EGG*PL. 2, NO. 5; LARVA*FF. 1, 3; PUPA*F. 7
UNITED STATES DEPT. AGRI. BUR. ENT. BULL. 85 NEW SERIES PART 5. 65, 1910 AINSLIE EGG*F. 32; LARVA*FF. 33-37, 43; PL. 4, F. 2
ESSIG-INS. WEST. NORTH AMER. P. 672, 1926 LARVA
INDEX. AMER. ECON. ENT. VOL. 1:154; VOL. 2:174; VOL. 3:192; VOL. 5:297; VOL. 6:349; VOL. 7:459

FOOD PLANTS
AGROPYRON SMITHII RYDB.
ANDROPOGON HALLII HACK.
BOUTELOUA CURTIPENDULA(MICHX.)
BOUTELOUS GRACILIS H. BK
BOUTELOUS HIRSUTA LAG.
BROMUS POLYANTHUS SCRIB.
BUCHLOE DACTYLOIDES(NUTT.)
DISTICHLIS SPICATA(LINN.)
HILARIA CENCHROIDES H. BK
HORDEUM JUBATUM V. CAESPITOSUM (SCRIBN.)
LYCURUS PHELOIDES H. BK
MUHLENBERGIA PORTERI SCRIB.
SPOROBOLUS AIROIDES TORR.

OLYMPIA BUTLER
SYN. SEE FIDELISSIMA HERRICH-SCHAEFFER

OLYMPIA(EDWARDS), EUCHLOE-PIERIDAE
A ROSA EDWARDS
LIFE HISTORY
ENT. NEWS 18:73, 1907 SHULL EGG* LARVA*PUPA*ALL PL. 1
MISSOURI ACAD. SCI. PROC. 4:154 1938 MEINERS EGG
MONTANA UNIV. BULL. 30:35, 1906 ELROD LARVA
COMSTOCK-HOW KNOW BUTTER. P. 83, 1904 LARVA
EDWARDS-BUTTER. NORTH AMERICA 3RD SERIES SUPPL. NOTES 1897 EGG, LARVA
FOOD PLANTS
ARABIS LYRATA LINN.
ARABIS MISSOURIENSIS GREENE
CRUCIFERAE

OLYMPIANNA BURCIK
SSP. SEE SMINTHEUS DOUBLEDAY& HEWITSON

OLYMPUS EDWARDS
SSP. SEE ALOPE FABRICIUS

OLNYTHUS BOISDUVAL & LECONTE
SYN. SEE ETHLIUS CRAMER/

OMISSA HARRIS
SSP. SEE AUTUMNATA GUENEE

OMNILUTEOFUSCUS GUNDER
SSP. SEE CHALCEDONIA DOUBLDDAY & HIEWTSON

OMPHALE HUBNER
SYN. SEE AUGE LINNAEUS

ONONIS(SCHIFFERMULLER), HELIOTHIS-NOCTUIDAE
SYN. SEPTENTRIONALIS HENRY EDWARDS
LIFE HISTORY
INDEX AMER. ECON. ENT. VOL. 7:456
FOOD PLANTS
LINUM SULCATUM RIDDELL

ONTARIO(EDWARDS), STRYMON-LYCAENIDAE
A VIOLAE STALLINGS TURNER
LIFE HISTORY
ENT. NEWS 46:123, 1935 CLARK LARVAL NOTE
FOOD PLANTS
AMPHICARPA BRACTEATA(LINN.)
AMPHICARPA COMOSA(LINN.)
QUERCUS STELLATA WANG.
QUERCUS VIRGINIANA MILL.

ONUSTA GROTE, MACRONOCTUA-NOCTUIDAE
LIFE HISTORY

CAN. ENT. 46:296, 1914 WINN. EGG
ENT. SOC. AMER. ANNALS 27:548 1934 GRISWOLD EGG*GOOD FIGURE
NEW YORK STATE MUS. BULL. 155: 53, 1912 FELT LARVA, PUPA
LOC. CIT. 180:71, 1915 FELT EGG
ONTARIO ENT. SOC. ANN. REPT. P. 49, 1903 GIBSON LARVA
UNITED STATES DEPT. AGRI. FARMER'SBULL. 1495:26, 1926 WEIGEL-MIDDLETON LARVA* F. 38 B
WASHINGTON ENT. SOC. PROC. 4:321 1898 DYAR LARVA
INDEX AMER. ECON. ENT. VOL. 1:194; VOL. 2:221; VOL. 3:246; VOL. 4:285; VOL. 5:399; VOL. 6:464; VOL. 7:628

FOOD PLANTS
BELAMCANDA CHINENSIS (LINN.)
IRIS SP.
ZEA MAYS LINN. (CORN) "GERMAN LILY"

OO (CRAMER), AUTOGRAPHA-NOCTUIDAE
SYN. ROGATIONIS GUENEE
SYN. DYAUS GROTE
SYN. CULTA LINTNER
AB OONANA STRAND
LIFE HISTORY
JOURN. AGRI. RESEARCH 33:800, 1926 BOTTIMER LARVA
NEW YORK STATE REPT. ENT. 2:94 1885 LINTNER LARVA
INDEX AMER. ECON. ENT. VOL. 1:45 ROGATIONIS VOL. 2:289 ROGATIONIS VOL. 6:79
FOOD PLANTS
BRASSICA OLERACEA LINN. (CABBAGE)
EUPATORIUM SP.
GERANIUM SP.
HIBISCUS ESCULENTUS LINN. (OKRA)
PELARGONIUM SP.
ZEBRINA PENDULA SCHNIZ

OONANA STRAND
SSP. SEE OO CRAMER

OPACA BARNES & BENJAMIN
SSP. SEE SEPTENTRIONALIS WALKER, GLUPHISIA

OPELLA (GROTE), EUBAPHE-ARCTIIDAE
FORM FLAVA BARNES & BENJAMIN
FORM NIGRICANS REAKIRT
SYN. NIGRIFERA WALKER (MALE)
SYN. OBSCURA STRETCH
FORM BELMARIA EHRMANN
SYN. RUBRICOSTA EHRMANN (FEMALE)
LIFE HISTORY
NEW YORK ENT. SOC. JOURN. 18: 163, 1910 FORBES LARVA* NIGRICANS
PSYCHE 8:119, 1887 DYAR EGG, LARVA, COCOON
FOOD PLANTS
GENERAL FEEDER LOW PLANTS

OPHELIA HENRY EDWARDS, CATOCALA-NOCTUIDAE
A DOLLI BEUTENMULLER
LIFE HISTORY
AMER. MUS. NAT. HIST. MEM. NEW SERIES 3, PART 1, 1918 BARNES & MCDUNNOUGH LARVA* PL. 14, F. 8; PL. 15, FF. 26, 34; PL. 16, F. 8 COLOR
PSYCHE 20:194, 1913 B. & MCD. EGG LARVA
FOOD PLANTS
QUERCUS SP. (OAKS)
QUERCUS MACROCARPA MICHX. BUR AND CATKINS

OPHIR BARNES & MCDUNNOUGH
SSP. SEE WILLIAMSII DODGE

OPHIS EDWARDS
SYN. SEE PANOQUIN SCUDDER

OPHTHALMICUS BOISDUVAL
SSP. SEE CERISYI KIRBY

OPIPARA HENRY EDWARDS
SSP. SEE DIVIDUA GROTE

OPIS EDWARDS
SSP. SEE EURYNOME EDWARDS

OPONEARIA WALKER
SSP. SEE ASTYLUSARIA WALKER

OPORARIA ZELLER
SYN. SEE INCERTATA WALKER

OPPOSITA BARNES & BENJAMIN
SSP. SEE PERSPICUA GROTE & ROBINSON

OPPOSITARIA GUENEE
SYN. SEE DISTRIBUARIA HUBNER

OPS DRURY
SYN. SEE ANDRIA SCUDDER

OPTIMATA STRECKER
SYN. SEE MEADI PACKARD

OPULENTA HENRY EDWARDS
SSP. SEE CAJA LINNAEUS

ORBICULARIS STRAND
SSP. SEE NELITA STRECKER

ORCUS EDWARDS
SSP. SEE PHERES BOISDUVAL

ORDINARIA WALKER
SYN. SEE DEVASTATOR BROCE

ORDINATA WALKER
SYN. SEE MESSORIA HARRIS

ORDINATA (WALKER), SCOPULA-GEOMETRIDAE
SYN. PURARIA WALKER
SYN. CANDIDARIA PACKARD
LIFE HISTORY
PACKARD-MONOGRAPH GEOMETRIDAE P.565, 1876 LARVA*PUPA*PL. 13, FF. 16, 16A
FOOD PLANTS
TRILLIUM STYLOSUM NUTT.

OREAS (EDWARDS), POLYGONIA-NYMPHALIDAE
A SILENUS EDWARDS
LIFE HISTORY
FOOD PLANTS
ALNUS SP. (ALDERS)
RIBES DIVARICATUM DOUGL.

OREGONENSIS BARNES & MCDUNNOUGH
SSP. SEE BATTOIDES BEHR

OREGONENSIS BUTLER
SSP. SEE GIBBOSA ABBOT & SMITH

OREGONENSIS HARVEY
SSP. SEE GEORGII GROTE

OREGONENSIS (STRETCH), EUCHAETIAS-ARCTIIDAE
LIFE HISTORY
NAT. CAN. P. 17, 1930 HUARD LARVA
ONTARIO ENT. SOC. ANN. REPT. P. 145 1908 LYMAN EGG, LARVA
FOOD PLANTS
APOCYNUM ANDROSAEMIFOLIUM LINN.

OREGONENSIS SWETT
SSP. SEE INFANS MOESCHLER

OREGONIA EDWARDS
SSP. SEE BAIRDII EDWARDS

OREODAPHNE HENRY EDWARDS
SSP. SEE CHERSIS HUBNER

ORFORDENSIS CASSINO & SWETT
SSP. SEE LIMBATA HAWORTH

ORICUS EDWARDS
SYN. SEE ALPHEUS EDWARDS

ORIENTALIDES VERITY
SSP. SEE CREUSA DOUBLEDAY & HEWITSON

ORIGINES MORRISON
SYN. SEE THEMISTOCLES LATREILLE

ORILLATA (WALKER), SEMIOTHISA-GEOMETRIDAE
LIFE HISTORY
NEW YORK STATE AGRI. EXP. STA.

(CORNELL), MEM. 274:88, 1948 FORBES LARVA?

SOUTHERN CAL. ACAD. SCI. BULL. 34:224, 1935 COMSTOCK & DAMMERS LARVA*PL. 65; PUPA* PL. 65

LOC. CIT. P. 69, 1927 CHERMOCK LARVA*PUPA*

FOOD PLANTS

CUPRESSUS GUADELUPENSIS S. WATS.

JUNIPERUS SP. POSSIBLY

ORINA GUENEE
SYN. SEE GRATA HUBNER

ORINELLA STRAND
SSP. SEE CANESCENS BEHR

ORINULA STRAND
SSP. SEE CANESCENS BEHR

ORION FABRICIUS
SYN. SEE ODIUS FABRICIUS

ORION MCDUNNOUGH
SSP. SEE MESKEI GROTE

ORIZABA (WESTWOOD), ROTHSCHILDIA-SATURNIIDAE

LIFE HISTORY

AMER. ACAD. SCI. PROC. 39:555, 1904 PACKARD LARVA

AMER. MUS. NAT. HIST. BULL. 5:93, 1893 BEUTEN. COCOON

ENT. SOC. AMER. ANNALS 9:148, 1916 MOSHER PUPA

NAT. ACAD. SCI. MEM. 12:253, 1914 PACKARD LIFE HISTORY*LARVA* PL. 4, FF. 1-3 COLOR

INDEX AMER. ECON. ENT. VOL. 2:320

FOOD PLANTS

FRAXINUS AMERICANA LINN. (IN CAPTIVITY)

ORNATA (BARNES & MCDUNNOUGH
SSP. SEE ALCIPHERIA WALKER

ORNATA GROTE & ROBINSON
SSP. SEE APICALIS WALKER

ORNATA (PACKARD), APANTESIS-ARCTIIDAE
FOR SYNS. AND SSP. SEE MCDUNNOUGH'S LIST

LIFE HISTORY

CAN. ENT. 34:120, 1903 GIBSON LARVA

LOC. CIT. 37:344, 1905 GIBSON LARVA

NEW YORK ENT. SOC. JOURN. 8:43 1900 DYAR LARVA

PSYCHE 6:380, 1893 DYAR EGG, LARVA (INCORRECTLY AS BLAKEI- CORRECTION ON P. 556 SAME JOURNAL)

STRETCH-ZYGAEN. & BOMBYC. NORTH AMERICA P. 125, 1873 LARVA, PUPA ACHAIA

FOOD PLANTS

CHENOPODIUM ALBUM LINN.

ERODIUM SP.

LACUTCA SATIVA LINN. (LETTUCE)

PLANTAGO SP.

PLANTAGO MAJOR LINN.

POLYGONUM SP.

POLYGONUM AVICULARE LINN.

SALIX SP. (WILLOWS)

TARAXACUM OFFICINALE WEBER

TRIFOLIUM SP. (CLOVERS)

GENERAL FEEDER LOW PLANTS

ORNATRIX (LINNAEUS), UTETHEISA ARCTIIDAE
FORM STRETCHII BUTLER
SYN. PURA BUTLER
SYN. BUTLERI DYAR

LIFE HISTORY

INSECUTOR INSCITIAE MENSTRUUS 2:115, 130, 1914 DYAR LARVA

INDEX AMER. ECON. ENT. VOL. 5:678; VOL. 7:1042

FOOD PLANTS

LEGUMINOSAE

ORNITHOGALLI GUENEE, PRODENIA-NOCTUIDAE
SYN. LINEATELLA HARVEY
FORM EUDIOPTA GUENEE

SYN. FLAVIMEDIA HARVEY

LIFE HISTORY

AMER. MUS. NAT. HIST. BULL. 16: 423, 1902 BEUTEN. LARVA

BROOKLYN ENT. SOC. BULL. 22:51, 1927 CRUMB MATURE LARVA* PL. 4B; PL. 5 D

LOC. CIT. 27:94, 1932 CRUMB LARVA

CAN. ENT. 13:24, 1881 FRENCH LARVA

ILLINOIS AGRI. EXP. STA. BULL. 60: 496, 1900 FORBES LARVA

ILLINOIS DEPT. AGRI. TRANS. 18, APPEND: 139, 1880 MARTEN LARVA

INSECT LIFE 7:325, 1895 ASHMEAD LARVA LINEATELLA

JOURN. ECON. ENT. 8:393, 1915 CRUMB KEY TO LARVAE LARVAL HEAD* PL. 20, F. 12

NEBRASKA AGRI. EXP. STA. RESEARCH BULL. 81:26, 1935 WHELAN LARVA TECH. DESCR* PL. 7B, 8A STRUCTURE

UNITED STATES DEPT. AGRI. BUR. ENT. BULL. 27:65, 71, 1901 NEW SERIES CHITTENDEN EGG, LARVA* PL. 4, F. 2 B, C; F. 3 B, C EUDIOPTA; PUPA

LOC. CIT. 43:33, 1903 NEW SERIES CHITTENDEN LARVA* F. 30

LOC. CIT. 57:43, 1906 NEW SERIES SANDERSON LARVA* F. 25

UNITED STATES DEPT. AGRI. CIRC. 639:48, 1942 CHAMBERLIN, MADDEN LARVA* F. 53 NO DESCR.

UNITED STATES DEPT. AGRI. FARMER'S BULL. 223:15, 1905 SANDERSON LARVA* F. 15

LOC. CIT. 890:19, 1917 PIERCE LARVA* F. 23

LOC. CIT. 1688:17, 1932 FOLSOM LARVA* F. 18

UNITED STATES DEPT. AGRI. TECH. BULL. 88:150, 1929 CRUMB LARVA* PL. 8 D, K; PUPA* F. 19D

CROSBY & LEONARD-MANUAL VEG. GARD. INS. P. 296, 1918 LARVA* F. 181

ESSIG-INS. WEST. NORTH AMERICA P. 685, 1926 LARVA

FERNALD-APPLIED ENT. P. 278 1926 LARVA* F. 284

FERNALD & SHEPARD-APPLIED ENT. P. 276, 1942 LARVA* F. 279

METCALF & FLINT-DESTR. & USEFUL INS. P. 349, 1939 LARVA

O'KANE-INJ. INS. P. 123, 1912 LARVA

FOOD PLANTS

ALLIUM CEPA LINN. (ONION)

ALLIUM CERNUUM ROTH.

AMARANTHUS RETROFLEXUS LINN.

AMARANTHUS SPINOSUS LINN.

ASPARAGUS OFFICINALIS LINN.

ASTER SP.

BETA VULGARIS LINN. (BEET)

BRASSICA OLERACEA LINN. (CABBAGE)

BRASSICA RAPA LINN. (TURNIP)

CHENOPODIUM ALBUM LINN.

CITRULLUS VULGARIS SCHRAD. (WATERMELON)

CONYZA CANADENSIS (LINN.)

COSMOS BIPINNATUS CAV.

CUCUMIS ATIVUS LINN. (CUCUMBER)

DATURA STRAMONIUM LINN.

GLADIOLUS SP.

GLYCINE MAX (LINN.)

GOSSYPIUM HERBACEUM LINN. (COTTON)

IPOMOEA ALBA LINN.

IPOMOEA BATATUS (LINN.) (SWEET POTATO)

IPOMOEA PURPUREA (LINN.)

LACUTCA SATIVA LINN. (LETTUCE)

MEDICAGO SATIVA LINN.

NICANDRA PHYSALODES (LINN.)

NICOTIANA TABACUM LINN. (TOBACCO)

PETUNA AXILLARIS (LAM.)

PHASEOLUS VULGARIS LINN. (BEANS)

PHYTOLACCA AMERICANA LINN.

PISUM SATIVUM LINN. (PEAS)

PLANTAGO LANCEOLATA LINN.

POPULUS SELTOIDES MARSH.

PRUNUS PERSICA (LINN.) (PEACH)

RHEUM RHAPONTICUM LINN. (RHUBARB)

RICINUS COMMUNIS LINN.
RUBUS ALLEGHENIENSIS PORTER
RUBUS IDAEUS V.STRIGOSUS(MICHX.)
RUMEX SP.
SIDA SPINOSA LINN.
SOLANUM CAROLINENSE LINN.
SOLANUM LYCOPERSICUM LINN. (TOMATO)
SOLANUM TUBEROSA LINN. (POTATO)
SOLIDAGO LEAVENWORTHII T.&G.
TRAGOPOGON PORRIFOLUS LINN.
VERNONIA NOVEBORACENSIS(LINN.)
FIELD CROPS, GRASSES, LOW PLANTS, VEGETABLES

ORO SCUDDER
SSP. SEE LYGDAMUS DOUBLEDAY

ORONO SCUDDER
SYN. SEE CONSPICUA EDWARDS

ORSA BOISDUVAL
SYN. SEE MONTANA BEHR

ORTHOGONIA MORRISON, AGROTIS-NOCTUIDAE
A DELORATA SMITH
B DUAE BARNES & BENJAMIN

LIFE HISTORY

CAN. ENT. 82:181, 1950 JACOBSON BLAKELEY, FARSTADT LARVAL FEEDING HABITS
LOURN. AGRORESEARCH 22:314, 1921 PARKER, STRAND, SEAMANS EGG*F.1 AND PL.C. F.1COLOR;LARVA*PL.C,FF. 2,3 COLOR;PUPA*PL.C,F.4 COLOR
JOURN. ECON. ENT. 7:203, 1914 GIBSON LARVA DELORATA
MONTANA AGRI. EXP. STA. BULL. 225, 1930 COOK EGG*LARVA* PUPA*COLOR
NEBRASKA AGRI. EXP. STA. RESEARCH BULL. 81:11, 1935 WHELAN LARVA TECH. DESCR.
OKLAHOMA AGRI. EXP. STA. BULL. 275:21, 1948 FENTON, WHITEHEAD LARVA
UTAH AGRI. EXP. STA. BULL. 297:7 1941 SORENSON, THORNLEY EGG*F.3 B,D;LARVA*F.3 D, F,G;PUPA*F.3 F
ESSIG-INS. WEST. NORTH AMERICA P.681, 1926 LARVA
METCALF & FLINT-DESTR. &USEFUL INS. P.349, 1939 LARVA
INDEX AMER. ECON. ENT.
VOL.1:253 POROSAGROTIS DELORATA
VOL.2:295 POROSAGROTIS
VOL.3:330 POROSAGROTIS
VOL.4:386 POROSAGROTIS
VOL.5:17;VOL.6:18; VOL.7:24

FOOD PLANTS

AGROPYRON SMITHII RYDB.
ALLIUM CEPA LINN. (ONION)
AVENA SATIVA LINN. (OATS)
BETA VULGARIS LINN. (BEET)
BRASSICA HIRTA MOENCH(MUSTARD)
BRASSICA OLERACEA LINN. (CABBAGE)
BROMUS TECTORUM LINN.
CHENOPODIUM ALBUM LINN.
CIRSIUM ARVENSE(LINN.)
DAUCUS CAROTA V.SATIVA HOFF. (CARROT)
HELIANTHUS SP.
HORDEUM VULGARE LINN. (BARLEY)
LACTUCA SCARIOLA LINN.
LINUS SULCATUM RIDD.
MALVA ROTUNDIFOLIA LINN.
MEDICAGO SATIVA LINN. (ALFALFA)
MELILOTUS ALBA DESR.
SALSOLA KALIV. TENUFILIS TAUSCH
SECALE CEREALE LINN.
SISYMBRIUM ALTISSIMUM LINN.
SOLANUM LYCOPERSICUM LINN. (TOMATO)
SOLANUM TUBEROSUM LINN. (POTATO)
SORGHYM SUDANENSIS(PIPER)
TARAXACUM OFFICINALE WEBER
THLASPI ARVENSE LINN.
TRITICUM AESTIVUM LINN. (WHEAT)
ZEA MAYS LINN. (CORN)
ZEBRINA PENDULA SCHNZ.
GRASSES

ORTHOMENES SCUDDER
SYN. SEE ZABULON BOISDUVAL &LECONTE

ASCEOLA LINTNER
SYN. SEE RURICOLA BOISDUVAL

OSCULATA HULST
SSP. SEE ILIA CRAMER

OSIRIS BARNES & BENJAMIN
SSP. SEE NUBILIS HUBNER

OSLARI(BARNES), HEMEROCAMPA-LIPARIDAE
LIFE HISTORY
GRAHAM-PRINCIP. FOREST ENT. P. 52,1939 LARVA* F. 21
INDEX AMER. ECON. ENT. VOL. 1:154
FOOD PLANTS
ABIES CONCOLOR(GORD & GLAND)

OSLARI GUNDER
SSP. SEE ANICIA DOUBLEDAY&HEWITSON

OSLARI ROTHSCHILD & JORDAN
SSP. SEE JUANITA STRECKER

OSLARI ROTHSCHILD & JORDAN
SSP. SEE GORDIUS CRAMER

OSLARI ROTHSCHILD, ANISOTA-CITHERONIIDAE
SYN, SKINNERI BIEDERMANN
SYN. NEOMEXICANA BREHME
LIFE HISTORY
CAN. ENT. 42:401,1910 BARNES & McD. EGG, LARVA, SKINNERI
ENT. NEWS 19:77,1908 BIEDERMANN LARVA SKINNERI
ENT. SOC. AMER. ANNALS 7:290, 1914 MOSHER PUPA STRUCTS. SKINNERI
FOOD PLANTS
ACER NEGUNDO LINN.
ACER RUBRUM LINN.
ACER SACCHARINUM LINN.
QUERCUS SP. (OAKS)
QUERCUS VELUTINA LAM.

OSLARI ROTHSCHILD
SSP. SEE IMPERIALIS DRURY

OSLARI ROTHSCHILD
SSP. SEE MURINA STRETCH

OSLARI SKINNER
SSP. SEE DAURA STRECKER

OSSULARIA(HUBNER), METASIOPSIS-GEOMETRIDAE
SYN. TENNARIA GUENEE
SYN. SUBLATARIA GUENEE
SYN. MAGNIFERARIA WALKER
SYN. FAVILLIFERA WALKER
SYN. REPLETARIA WALKER
A PIMACOATA CASSINO
LIFE HISTORY
FOOD PLANTS
FRAGARIA SHILOENSIS(LINN.) (CAPTIVITY)
GALIUM SP.
STELLARIA MEDIA(LINN.)

OSYKA EDWARDS
SYN. SEE FEMALE BRETTUS BOISDUVAL & LECONTE

OSYKA EDWARDS
SYN. SEE MALE RURICOLA BOISDUVAL

OTA BARNES & BENJAMIN
SSP. SEE NOTATA STRECKER

OTERO BARNES, KODIOSOMA-ARCTIIDAE
LIFE HISTORY
FOOD PLANTS
ESCHSCHOLTIZIA CAESPITOSA V. HYPECOIDES GRAY
FESTUCA MYUROS LINN.
GILIA ACHILLAEAFOLIA BENTH.
TRIFOLIUM CILIOLATUM BENTH.

OTHO(ABBOT & SMITH), CATIA-HESPERIIDAE
SYN. DRURY LATREILLE
SYN. PUSTULA GEYER
A EGEREMET SCUDDER
SYN. URSA WORTHINGTON

SYN. CINNA PLOETZ
SYN. AETNA SCUDDER
LIFE HISTORY
AMER. MUS. NAT. HIST. BULL. 5:292, 1893 BEUTEN. LARVA
BUFF. SOC. NAT. SCI. BULL. 19:33, 1939 WILD LARVA
ENT. NEWS 19:412, 1908 LARU ENT EGG, EARLY STAGES
MUS. COMP. ZOOL. BULL. 87, NO. 6, 1941 DETHIER LARVAL ANTENNA* F. 62
ABBOT & SMITH-LEP. INS. GEORGIA 1:31, 1797 LARVA*PUPA*PL. 16 COLOR
BOISDUVAL & LECONTE-HIST. LEP. NORTH AMER. 1833 LARVA* CHRYSALIS* PL. 77 COLOR
HOLLAND-BUTTERFLY BOOK 1905 CHRYSALIS* PL. 6, F. 42
SCUDDER-BUTTER. EAST. UNITED STATES & CANADA PP. 1698 1700, 1889 LIFE HIST*PL. 66, F. 27; PL. 73, F. 11; PL. 77, F. 34; PL. 85, F. 42 AETNA
FOOD PLANTS
DIGITARIA SANGUINALIS (LINN.)
SABBATIA CAMPANULATA (LINN.)
SABBATIA ELLIOTTII STEUD.

OTIOSA NEUMOEGEN & DYAR
SSP. SEE VIRGUNCULA KIRBY

OU (GUENEE), AUTOGRAPHA-NOCTUIDAE
SYN. FRATELLA GROTE
AB PEDALIS GROTE
SYN. OUELLS STRAND
AB OUANA STRAND
LIFE HISTORY
FOOD PLANTS
MENTHA SP. (MINTS)

OUANA STRAND
SSP. SEE OU GUENEE

OUELLA STRAND
SYN. SEE OU GUENEE

OUWAH POLING
SSP. SEE MICRONYMPHA GUENEE

OVATA GROTE, ACRONICTA-NOCTUIDAE
LIFE HISTORY
NEW YORK STATE MUS. MEM. *;704, 1905 FELT LARVA (BRIEF)
UNITED STATES NAT. MUS. PROC. 21:137, 1898 SMITH&DYAR EGG, LARVA*PL. 5, F. 3 COLOR OR UNCOLORED; PUPA; COCOON
FOOD PLANTS
BETULA SP. (BIRCHES)
BETULA ALLEGHENENSIS BRITT.
CASTANEA DENTATA (MARSH.)
FAGUS SP. (BEECHES)
QUERCUS SP. (OAKS)

OVIDIUS SCUDDER
SYN. SEE ZARUCCO LUCAS

OVIDUCA (GUENEE), ORTHODES-NOCTUIDAE
SYN. CAPSELLA GROTE
LIFE HISTORY
WASHINGTON ENT. SOC. PROC. 5:293 1903 DYAR EGG, LARVA
FOOD PLANTS
GENERAL FEEDER LOW PLANTS

OVILLA (GROTE), CELAMA-NOLIDAE
LIFE HISTORY
AMER. MUS. NAT. HIST. BULL. 10: 359, 1898 BEUTEN. LARVA
AMER. NAT. 18:726, 1884 PACKARD LARVA, COCOON
NEW YORK STATE MUS. MEM. 8:707 1905 FELT LARVA (BRIEF)
PSYCHE 6:110, 1891 DYAR LARVA, COCOON
PSYCHE 7:137, 1894 DYAR LARVA, COCOON
FOOD PLANTS
QUERCUS SP. (OAKS)
QUERCUS MACROCARPA MICHX.

OWENI EDWARDS
SSP. SEE MONTIVAGA BEHR

o

OXILUS HUBNER
SYN. SEE CRESPHONTES CRAMER

OXYBAPHI CLEMENS
SYN. SEE GALLI ROTTENBURG

OXYGALA (GROTE), LEUCANIA-NOCTUIDAE
LIFE HISTORY
UNITED STATES NAT. MUS. PROC. 27: 862, 1904 DYAR EGG, LARVA
FOOD PLANTS

P

PACIFICA (HARVEY), ORTHOSIA-NOCTUIDAE
AB INFRAPICTA STRAND
LIFE HISTORY
CAN. ENT. 26:66, 1894 DYAR LARVA
WASHINGTON ENT. SOC. PROC. 4:323, 1898 DYAR LARVA
FOOD PLANTS
ACER SP. (MAPLES)
AESCULUS CALIFORNICA (SPACH)
PRUNUS SEROTINA EHRH. (?)
QUERCUS GARRYANA DOUGL.
SALIX SCOULERIANA BARR.

PACIFICA SMITH
SYN. SEE LEPUSCULINA GUENEE

PACIFICA SMITH
SSP. SEE AMERICANA SPEYER

PACKARDARIA GROTE
SYN. SEE RUBRIFRONTARIA PACKARD (NEMORIA)

PACKARDARIA (HULST), AETHALOIDA-GEOMETRIDAE
SYN. LACHRYMOSA HULST
SYN. HOMOPTEROIDES HULST
LIFE HISTORY
SOUTHERN CAL. ACAD. SCI. BULL. 36:114, 1937 COMSTOCK LARVA* PL. 47
FOOD PLANTS
ADENOSTOMA SP.
CEANOTHUS SP.

PACKARDARIA MOESCHLER
SSP. SEE BITACTATA WALKER

PACKARDATA LINTNER
SYN. SEE PROPULSATA WALKER

PACKARDII (GROTE), CISTHENE-ARCTIIDAE
LIFE HISTORY
PSYCHE 3:243, 1881 MURTFELDT LIFE HIST. (*?SUBJECTA WALKER)
FOOD PLANTS
LICHENS

PACKARDII SAUNDERS
SSP. SEE THAROS DRURY

PACKARDII SCHAUPP
SYN. SEE ACREA DRURY

PAENULATARIA GROTE
SYN. SEE QUERNARIA ABBOT & SMITH

PAGINATA MORRISON
SYN. SEE NUCICOLORA GUENEE

PALAEMON (PALLAS), CARTEROCEPHALUS-HESPERIIDAE
SYN. MANDAN EDWARDS
SYN. MESAPANO SCUDDER
SYN. SKADA EDWARDS
SYN. PANISCUS FABRICIUS
LIFE HISTORY
CAN. ENT. 21:113, 1889 FLETCHER EGG, LARVA, CHRYSALIS
ENT. 25:225, OCT. 1892 FROHAWK LIFE HISTORY
ONTARIO ENT. SOC. ANN. REPT. P. 85, 1888 FLETCHER EGG, LARVA
COMSTOCK, J. A. BUTTER. CAL. P. 216, 1927 EGG, LARVA
FLETCHER-A TRIP TO NEPIGON P. 12, 1888 EGG, LARVA
SCUDDER-BUTTER. EAST. UNITED STATES & CANADA PP. 1571-1574, 1889 EGG, LARVA MANDAN
FOOD PLANTS
AGROPYRON REPENS (LINN.)

Bromus ciliatus v. asper B. B
Digitaria sanguinalis (Linn.)
Echinochloa crus-galli (Linn.)
Poa pratensis Linn.

palaeogama Guenee, Catocala-Noctuidae
form annida Fager
form phalanga Grote
form denussa Ehrmann
LIFE HISTORY
Amer. Mus. Nat. Hist. Bull. 16:388 1902 Beuten. larval heads* f. 12
Amer. mus. Nat. Hist. Mem. new series 3, part 1, 1918 Barnes & McD. larva* Pl. 13, f. 4; Pl. 15, ff. 6, 30; Pl. 17, f. 5 COLOR
Can. Ent. 20:108, 1888 French egg larva, pupa
Ent. News 21:108, 1910 Rowley-Berry larva
New York State Mus. Mem. 8:716 1905 Felt larva (brief)
Hart-Ins. Coll. Illinois High Schools p. 42, 1903 cocoon
FOOD PLANTS
Carya sp. (hickories)
Carya illinoensis (Wang.)
Castanea dentata (Marsh.)
Juglans nigra Linn.
Malus pumila Mill. (apple)

palamedes Drury, Papilio, Papilionidae
syn. chalcas Fabricius
LIFE HISTORY
Can. Ent. 13:119, 1881 Edwards egg, larva, chrysalis
Maryland Nat. Hist. Soc. Proc. p. 13, 1933 Haydon egg, larva, chrysalis
Smithsonian Misc. Coll. 4, art. 2:7, 1862 Morris larva, chrysalis chalcas
Boisduval-Spec. Gen. Lep. Diurnae 1:338, 1836 larva, chrysalis
Boisduval & Leconte-Hist. Lep. North Amer. p. 19, 1833 larva* Pl. 5
Comstock-How Know Butter. p. 55, 1904 larva
French-Butter. East. United States p. 95, 1886 egg, larva chrysalis
Jaeger-Life North Amer. Ins. p. 210, 1864 larva
Morris-Synop. Lep. North Amer. p. 7, 1862 larva
Scudder-Butter. East. United States & Canada pp. 1848-1851, 1889 life history
Westwood-In Drury's Illust. Exot. Ent. 1:36, 1837 larva
Index Amer. Econ. Ent. Vol. 7:741
FOOD PLANTS
Citrus sp.
Kalmia sp.
Magnolia sp.
Persea borbonia (Linn.)
Sassafras albidum v. molle (Raf.)

palatka (Edwards), Atrytone-Hesperiidae
syn. floridensis Ploetz
LIFE HISTORY
Can. Ent. 11:192, 1879 Edwards larva
Illinois Dept. Agri. Trans. 18, Append: 154, 1880 Coquillett larva
Boisduval & Leconte-Hist. Lep. North Amer. 1833 larva* Pl. 67, f. 4; chrysalis* Pl. 67, f. 5 COLOR
French-Butter. East. United States p. 341, 1886 larva
FOOD PLANTS
Cladium jamaicenses Crantz

palla Boisduval, Melitaea-Nymphalidae
ab wardi Oberthur
ab blackmorei Gunder
form female eremita Wright
ab stygiana Comstock
ab hemifusa Gunder
a whitneyi Behr

AB VANDUZEEI GUNDER
B CALYDON MEAD
C VALLISMORTIS JOHNSON
LIFE HISTORY
CAL. ACAD. SCI. PROC. 5:167, 1873
HY. EDW. LARVA, CHRYSALIS
COMSTOCK, J. A. BUTTER. CAL. P. 109, 1927 LARVA, CHRYSALIS
MEAD-REPT. LEP. WHEELER EXPED. WEST 100 MERID, P. 759, 1875 LARVA, CHRYSALIS
FOOD PLANTS
CASTILLEJA FLAVA WATSON
PLANTAGO SP.
SCROPHULARIACEAE
GENERAL FEEDER FLOWERS

PALLA FRENCH
SYN. SEE INCLUSA HUBNER

PALLENS HENRY EDWARDS
SSP. SEE FULVICOLLIS HUBNER

PALLENS STRECKER
SSP. SEE JUGLANDIS ABBOT & SMITH

PALLENS WHITAKER
SSP. SEE NICIPPE CRAMER

PALLESCENS MCDUNNOUGH
SSP. SEE FURCILLA PACKARD

PALLESCENS ROTHSCHILD & JORDAN
SSP. SEE CHERSIS HUBNER

PALLIDA BARNES & BENJAMIN
SSP. SEE OBLIQUA WALKER

PALLIDA BARNES & MCDUNNOUGH
SSPL SEE AMATRIX HUBNER

PALLIDA BOWLES
SYN. SEE RUBICUNDA FABRICIUS

PALLIDA COCKERELL
SYN. SEE RUTYTHEME BOISDUVAL

PALLIDA COCKERELL
SSP. SEE CHRISTINA EDWARDS

PALLIDA COCKERELL
SYN. SEE ALEXANDRA EDWARDS

PALLIDA COCKERELL
SSP. SEE SENNAE LINNAEUS

PALLIDA EDWARDS
SSP. SEE MYLITTA EDWARDS

PALLIDA FIELD
SYN. SEE IOLE BOISDUVAL

PALLIDA FRUHST
SSP. SEE STELENES LINNAEUS

PALLIDA GROTE
SYN. SEE OBSCURA FABRICIUS

PALLIDA MCDUNNOUGH
SSP. SEE RUBERATA FREYER

PALLIDA PACKARD, CRAMBIDIA-ARCTIIDAE
LIFE HISTORY
ENT. NEWS 3:245, 1892 DYAR EGG, LARVA
FOOD PLANTS

PALLIDA PACKARD
SSP. SEE CUNEA DRURY

PALLIDA SCUDDER
SSP. SEE NAPI LINNAEUS

PALLIDA SKINNER
SSP. SEE HECLA LEFEBRE

PALLIDA SKINNER
SYN. SEE MYSTIC SCUDDER

PALLIDA (STRECKER), DICENTRIA-NOTODONTIDAE
LIFE HISTORY
FOOD PLANTS
ACER MACROPHYLLUM PURSH
ALNUS SP. (ALDERS)
AMELANCHIER FLORIDA LINDL.
POPULUS SP. (POPLARS)

ROSA SP. (ROSES)
SALIX SP. (WILLOWS)

PALLIDA WATSON
SSP. SEE HOBOMOK HARRIS

PALLIDICE SCUDDER
SYN. SEE PHILODICE GODART

PALLIDICOMA GROTE
SYN. SEE LONGA GUENEE

PALLIDIOR SMITH
SYN. SEE SIMPLEX WALKER (XYLOMYGES)

PALLIDISSIMA BARNES & MCDUNNOUGH
SSP. SEE NAPI LINNAEUS

PALLIDULUS HENRY EDWARDS
SSP. SEE CERISYI KIRBY

PALLIFLAVA MCDUNNOUGH
SSP. SEE NASTES BOISDUVAL

PALLIGERA (SMITH), TETANOLITA-NOCTUIDAE
SYN. GRETA SMITH
LIFE HISTORY
ENT. AMER. 14:164, 1834 CRUMB
LARVA
FOOD PLANTS
DEAD LEAVES (APPLE)

PALLIPENNIS (SMITH), EUXOA-NOCTUIDAE
SYN. ALCESTA SMITH
LIFE HISTORY
INDEX AMER. ECON. ENT.
VOL. 3:161
FOOD PLANTS
SALSOLA KALI V. TENUIFOLIA TAUSCH
TRITICUM AESTIVUM LINN. (WHEAT)

PALMERI (EDWARDS), APODEMIA-RIODINIDAE
FORM MARGINALIS SKINNER
A HEPBURNI GODMAN & SALVIN
LIFE HISTORY
SOUTHERN CAL. ACAD. SCI. BULL.
31:37, 1932 COMSTOCK, DAMMERS
EGG*PL. 17; LARVA*PL. 18;
CHRYSALIS*PL. 19 MARGINALIS
FOOD PLANTS
BELOPERONE CALIFORNICA BENTH.
PROSOPIS GLANDULOSA TORR.

PALMI BEUTENMULLER, DATANA-NOTODONTIDAE
LIFE HISTORY
AMER. MUS. NAT. HIST. BULL. 10:
407, 1898 BEUTEN. LARVA
ENT. AMER. 6:130, 181, 1890 DYAR
EGG, LARVA, PUPA
NAT. ACAD. SCI. MEM. 7:116, 1895
PACKARD EGG, LARVA*PL. 14,
FF. 2, 3 COLOR; PUPA
NEW YORK STATE MUS. MEM. 8:739
1905 FELT LARVA (BRIEF)
PSYCHE 6:573, 1893 DYAR KEY TO
DATANA LARVAE
FOOD PLANTS
ANDROMEDA SP.
VACCINIUM SP.
VACCINIUM STAMINEUM LINN.

PALPALIS GROTE
SYN. SEE DIFFINIS BOISDUVAL

PALPALIS HAWORTH
SYN. SEE SCABRA FABRICIUS

PALPARIA (WALKER), BOMOLOCHA-NOCTUIDAE
SYN. SCUTELLARIS GROTE
LIFE HISTORY
ENT. AMER. 14:143, 1934 CRUMB
LARVA
FOOD PLANTS
QUERCUS SP. (OAKS)

PALPATA PACKARD, EUPITHECIA-GEOMETRIDAE
LIFE HISTORY
CAN. ENT. 74:150, 1942 MCGUFFIN
LARVA* PL. 12, F. A ALSO,
LARVAL MANDIBLE*
LOC. CIT. 77:55, 1945 MCGUFFIN
LARVA
INDEX AMER. ECON. ENT.
VOL. 7:367
FOOD PLANTS
ABIES BALSAMEA (LINN.)

LARIX LARICINA(DUROI)
PICEA SP. (SPRUCES)
PICEA ENGELMANNI (PARRY)
PICEA GLAUCA(MOENCH)
PICEA MARIANA(MILL.)
PINUS SP. (PINES)
PINUS BANKSIANA LAMB.
PINUS RESINOSA AIT.
PINUS STROBUS LINN.

PALUDATA(THUNBERG), CARSIA-GEOMETRIDAE
A ALPINATA PACKARD
SYN. BOREATA PACKARD
B LABRADORIENSIS SOMMER
C THAXTERI SWETT
D COLUMBIATA MCDUNNOUGH
LIFE HISTORY
NEW YORK STATE AGRI. EXP. STA. (CORNELL), MEM. 274:172, 1948 FORBES LARVA THAXTERI
FOOD PLANTS
ACER SP. (MAPLES)
FRAXINUS SP. (ASHES)
MALUS PUMILA MILL. (APPLE)
PRUNUS CERASUS LINN.
PRUNUS SEROTINA EHRH.
ULMUS SP. (ELMS)
VACCINIUM SP. (THAXTERI)

PALUMBES FRANCLEMONT
SSP. SEE COMPTARIA WALKER

PAMINA(NEUMOEGEN), AUTOMERIS-SATURINIIDAE
FORM AUROSEA NEUMOEGEN
LIFE HISTORY
ENT. AMER. 4:62, 1888 HY. EDW. LARVA
ENT. NEWS 18:87, 1908 BUCHHOLZ NOTE ON EGG AND COCOON
ENT. SOC. AMER. ANNALS 7:298 1914 MOSHER PUPA
NAT. ACAD. SCI. MEM. 12:104, 1914 PACKARD EGG, LARVA*PL. 19, FF. 1-6 COLOR
NEW YORK ENT. SOC. JOURN. 8:201 1900 KUNZE EGG, LARVA
FOOD PLANTS
CEANOTHUS FENDLERI GRAY
CEROCARPUS BETULOIDES V. BLANCHERI(SCHNEID.)
OPUNTIA SP.
QUERCUS EMORYII TORR.
QUERCUS MARILANDICA MUENCH
QUERCUS UNDULATA TORR.

PAMPHILIUS STOLL
SYN. SEE BRONTES DRURY

PAMPINARIA(GUENEE), ANAVITRINELLA-GEOMETRIDAE
SYN. FRUGALIARIA GUENEE
SYN. COLLECTA WALKER
SYN. FRAUDULENTARIA ZELLER
AB NUBIFERARIA SWETT
A EROSIATA WALKER
B STYGIA FRENCLEMONT
LIFE HISTORY
IOWA ACAD. SCI. BIO. SURVEY NO. 1, 1936 JAQUES LARVA P. 95
MAINE AGRI. EXP. STA. BULL. 259: 50, 1917 MOSHER LARVA, PUPA* F. 2, F, I
MINN. AGRI. EXP. STA. BULL. 61:242, 1899 LUGGER LARVA
NEW YORK ENT. SOC. JOURN. 53:189 1945 FORBES PUPA
NEW YORK STATE AGRI. EXP. STA. (CORNELL), MEM. 274:54, 1948 FORBES LARVA
NEW YORK STATE MUS. MEM. 8:743 1905 FELT LARVA(BRIEF)
PAPILIO 1:82, 1881 FRENCH LARVA
UNITED STATES DEPT. AGRI. BUR. ENT. BULL. 66, PART 3, P. 22, 1907 NEW SERIES CHITTENDEN LARVA*F. 6 B, C; PUPA*F. 6A
UNITED STATES DEPT. AGRI. FARMER'S BULL. 178:20, 1903 SMITH LARVA, PUPA
GLOVER-ILLUST. NORTH AMER. ENT. 1878 LARVA*PL. 9, FF. 4, 9, 13, 14; PL. 19, F. 15
GUMPPENBERG-SYSTEMA GEOMETRARUM 5:318, 1892 LARVA, PUPA
O'KANE-INJ. INS. P. 283, 1912 LARVA

PACKARD-MONOGRAPH GEOMETRIDAE
P.435,1876 LARVA
SLINGERLAND&CROSBY-MANUAL
FRUIT INS.P.464,1914 LARVA
INDEX AMER.ECON.ENT.
VOL.6:803 VITRINELLA
FOOD PLANTS
ACER SP.(MAPLES)
ASPARAGUS OFFICINALIS LINN.
CARYA SP.(HICKORIES)
CARY ILLINOENSIS(WANG.)
CITRUS SP.
CITRUS SINENSIS(LINN.)(ORANGE)
FRAGARIA CHILOENSIS(LINN.)
FRAXINUS SP.(ASHES)
GERANIUM SP.
GLEDITSIA TRIACANTHOS LINN.
GOSSYPIUM HERBACEUM LINN.
(COTTON)
MALUS PUMILA MILL.(APPLE)
PELARGONIUM SP.
POPULUS SP.(POPLARS)
PSIDIUM GUAJAVA LINN.
PYRUS SP.
PYRUS COMMUNIS LINN.(PEAR)
RUBUS ALLEGHENIENSIS PORTER
RUMEX CRISPUS LINN.
SALIX SP.(WILLOWS)
TRIFOLIUM SP.(CLOVERS)
ULMUS SP.(ELMS)
VACCINIUM MACROCARPON AIT.
GENERAL FEEDER FRUIT TREES
MOST ANYTHING

PAMPINATRIX ABBOT & SMITH
SYN.SEE MYRON CRAMER

PANDARIA WALKER
SYN.SEE CONFUSARIA HUBNER

PANDORA BLAKE, COLORADIA-SATURNIIDAE
SYN.LOIPERDA DYAR
LIFE HISTORY
ENT.AMER.4:61,1888 HENRY
EDWARDS EGG
JOURN.ECON.ENT.34:698,1941
WYGANT EGG*F.1D,E;LARVA*
F.1F;PUPA*F.1G
NAT.ACAD.SCI.MEM.12:112,1914
PACKARD EGG,LARVA*PL.20,
F.8;PL.21,F.1 COLOR
NEW YORK ENT.SOC.JOURN.30:69,
1922 CHAMBERLIN EGG,LARVA
UNITED STATES DEPT.AGRI.TECH.
BULL.137,1929 PATTERSON
EGG*F.1C,D;LARVA*F.2 A,
B,C;PUPA* F.2D
DOANE ETC-FOREST INS.P.276,
1936 EGG* LARVA*PUPA*F.157
ESSIG-INS.WEST.NORTH AMER.
P.670,1926 EGG,LARVA
INDEX AMER.ECON.ENT.
VOL.3:97;VOL.4:115;
VOL.5:150;VOL.6:172;
VOL.7:207
FOOD PLANTS
PINUS SP.(PINES)
PINUS ECHINATA MILL.
PINUS CONTORTA V.MURRAYANA
(GREV. & BALF.)
PINUS PONDEROSA LAWS
POPULUS TREMULOIDES MICHX.

PANDORUS HUBNER
SSP.SEE SATELLITIA LINNAEUS

PANISARIA WALKER
SYN.SEE DUARIA GUENEE

PANISCUS FABRICIUS
SYN.SEE PALAEMON PALLAS

PANNARIA(GUENEE), SCELOPHIA-GEOMETRIDAE
SYN.TREMULARIA WALKER
SYN.PURPURISSATA GROTE
SYN.FORMOSA HULST
LIFE HISTORY
FOOD PLANTS
VIGNA SINENSIS(LINN.)

PANOQUIN(SCUDDER0, PANOQUINA-
HESPERIIDAE
SYN.OPHIS EDWARDS
SYN.COCHLES SCUDDER
SYN.WIMICO PLOETZ
LIFE HISTORY

ENT. NEWS 19:415,1908 LAURENT
EGG, EARLY STAGES
LOC. CIT. 1:8,1890 SKINNER EARLY
STAGES
FOOD PLANTS
GRASSES

PAPAIPEMOIDES BARNES & BENJAMIN
SSP. SEE SAUZALITAE GROTE

PAPKA BARNES & BENJAMIN
SSP. SEE STRICTA WALKER

PARADOXA BEHR
SSP. SEE CINEREA WALKER

PARADOXA CHERMOCK
SSP. SEE MELISSA EDWARDS

PARADOXA MCDUNNOUGH
SSP. SEE PERIDICCAS EDWARDS

PARADOXA WATSON
SSP. SEE BURNSI WATSON

PARALLELA GROTE & BORINSON
SSP. SEE ATOMARIA WALKER (OLENE)

PARDOPSIS HOLLAND
SSP. SEE BELLONA FABRICIUS

PARTA GUENEE, CATOCALA-NOCTUIDAE
AB PETULANS HULST
FORM PERPLEXA STRECKER
FORM FORBESI FRANCLEMONT
LIFE HISTORY
AMER. MUS. NAT. HIST. BULL. 16:
387,1902 BEUTEN. LARVAL
HEAD*F. 9
LOC. CIT. 38:157,1918 BARNES &
MCD. EGG, LARVA
AMER. MUS. NAT. HIST. MEM. NEW
SERIES 3, PART 1,1918 BARNES
&MCD. LARVA*PL. 10, F. 34;
PL. 11, F. 4; PL. 13, F. 10
COLOR
BROOKLYN ENT. SOC. BULL. 7:44,
1884 HULST LARVA
ENT. NEWS 20:12,1909 ROWLEY
EGG, LARVA
LOC. CIT. 21:446,1910 ROWLEY
LARVA
NEW YORK STATE MUS. MEM. 8:
743,1905 FELT LARVA (BRIEF)
PAPILIO 3:24,1883 EDWARDS LARVA
FOOD PLANTS
POPULUS SP. (POPLARS)
POPULUS TREMULOIDES MICHX.
SALIX SP. (WILLOWS)
SALIX BABYLONICA LINNAEUS

PARTHENICE (KIRBY), APANTESIS-ARCTIIDAE
SYN. SAUNDERSII GROTE
SYN. CIRCA STRETCH
AB APPROXIMATA STRETCH
LIFE HISTORY
CAN. ENT. 37:339,1905 GIBSON
EGG, LARVA, PUPA
CAN. JOURN. IND. ART. SCI. 8:354
1863 SAUNDERS LARVA
NEW YORK ENT. SOC. JOURN. 8:40
1900 DYAR LARVA (BRIEF)
ONTARIO ENT. SOC. ANN. REPT. P.
23,1873 SAUNDERS LARVA, PUPA
SAUNDERSII
PHIL. ENT. SOC. PROC. 2:28,1863
SAUNDERS LARVA
LOC. CIT. 3:76,1864 GROTE LARVA
SAUNDERSII
FOOD PLANTS
CHENOPODIUM ALBUM LINN.
PLANTAGO SP.
TARAXACUM OFFICINALE WEBER
VERONIA NOVEBORACENSIS (LINN.)
GRASS

PARTHENOS (HARRIS), PARASEMIA-ARCTIIDAE
SYN. BOREALIS MOESCHLER
LIFE HISTORY
CAN. ENT. 1:5,1868 SAUNDERS
LARVA
LOC. CIT. 3:225,1871 SAUNDERS
LARVA
NEW YORK STATE CABIN. NAT. HIST.
ANN. REPT. NO. 24:132,1871
LINTNER EGG, LARVA

UNITED STATES NAT. MUS. PROC. 27:796, 1904 DYAR LARVA (BRIEF)
FOOD PLANTS
LACTUCA SP.
LACTUCA SATIVA LINN. (LETTUCE)

PARTITA McDUNNOUGH
SSP. SEE PERQUIRITATA MORRISON

PARTITALIS SMITH
SSP. SEE LUBRICALIS GEYER

PARUMPUNCTATA HERRICH-SCHAEFFER
SYN. ACCIUS ABBOT & SMITH

PARVA BLACKMORE
SSP. SEE ERECHTEA CRAMER

PARVA ROTHSCHILD
SSP. SEE CAJA LINNAEUS

PARVIMACULA GROTE
SSP. SEE EURYALUS BOISDUVAL

PARVULA EDWARDS
SSP. SEE MINUTA EDWARDS (CATOCALA)

PASADENAE GUNDER
SSP. SEE GABBII BEHR

PASCOENSIS WRIGHT
SSP. SEE THAROS DRURY

PASSER (GUENEE), LUPERINA-NOCTUIDAE
SYN. INCALLIDA WALKER
SYN. LOCULATA MORRISON
FORM CONSPICUA MORRISON
A BIRNATA SMITH
LIFE HISTORY
BROOKLYN ENT. SOC. BULL. 16:86, 1921 ENGELHARDT LARVA, PUPA
FOOD PLANTS
RUMEX OBTUSIFOLIUS LINN. (BORES IN ROOTS)
RUMES VERTICILLATUS LINN. (BORES IN ROOTS)

PASSIFLORAE FABRICIUS
SYN. SEE VANILLAE LINNAEUS

PASTORALIS GROTE
SYN. SEE PUNCTIGERA WALKER

PASTORALIS GROTE
SYN. SEE FLAVIPUNCTALIS GEYER

PATALIS (GROTE), LACINIPOLIA-NOCTUIDAE
A FLETCHERI GROTE
LIFE HISTORY
WASHINGTON ENT. SOC. PROC. 4: 325, 1898 DYAR LARVA
FOOD PLANTS

PATEFACTA (WALKER), LITHOPNANE-NOCTUIDAE
LIFE HISTORY
FOOD PLANTS
AMELANCHIER FLORIDA LINDL.

PATERSONIA BREHME
SYN. SEE DAMON CRAMER

PATIBILIS WALKER
SYN. SEE ERECHTEA CRAMER

PATRICIA CASSINO
SSP. SEE UNIJUGA WALKER

PATTENII SCUDDER
SYN. SEE ACCIUS ABBOT & SMITH

PATULARIA D'URBAN
SYN. SEE SIGNATARIA WALKER

PAUCIMACULA BARNES & McDUNNOUGH
SSP. SEE ELEGANTARIA HENRY EDWARDS

PAULINA HENRY EDWARDS
SSP. SEE LACRYMOSA GUENEE

PAULUS EDWARDS
SSP. SEE SILVESTRIS EDWARDS

PAVITENSIS MORRISON
SYN. SEE DEDUCTA MORRISON

PAVONIA GEYER
SYN. SEE EXCAECATA ABBOT & SMITH

PECCATARIA SWETT
SSP. SEE FISCELLARIA GUENEE

PACKIUS (KIRBY), POLITES-HESPERIIDAE
CORAS AUCT.
?SYN. AESCULAPIUS FABRICIUS
SYN. WAMSUTTA HARRIS
SYN. ENYS SCUDDER
LIFE HISTORY
CAN. ENT. 1:66, 1869 SAUNDERS EGG, LARVA
ILL. DEPT. AGRI. TRANS. 18, APPEND 178, 1880 COQUILLETT LARVA
SOUTHERN CAL. ACAD. SCI. BULL. 38:188, 1939 DETHIER LIFE HIST., LARVAL HEAD* PL. 50; PUPA* PL. 50
FERNALD-BUTTER. MAINE P. 98, 1884 EGG, LARVA
FRENCH-BUTTER. EAST. UNITED STATES P. 318, 1886 EGG, LARVA
MAYNARD-BUTTER. NEW ENGLAND P. 61, 1886 LARVA
SCUDDER-BUTTER. EAST. UNITED STATES & CANADA PP. 1686-1688, 1889 EGG* LARVA* PL. 66, F. 28; PL. 80, F. 57
SCUDDER-EVERYDAY BUTTER. P. 354 1889 EGG, LARVA
SCUDDER-GUIDE TO COMMON BUTTER P. 171, 1893 LARVA, CHRYSALIS
FOOD PLANTS
GRASSES

PECOSENSIS COCKERELL
SSP. SEE EXCAECATA ABBOT & SMITH

PECTINARIA (DENNIS & SCHIFFERMULLER), EUCHLAENA-GEOMETRIDAE
SYN. DEDUCTARIA WALKER
LIFE HISTORY
NEW YORK STATE AGRI. EXP. STA. (CORNELL), MEM. 274:75, 1948 FORBES LARVA
NEW YORK STATE MUS. MEM. 8: 706, 1905 FELT LARVA (BRIEF)
PSYCHE 9:344, 1902 DYAR EGG, LARVA
GUMPPENBERG-SYSTEMA GEOMETRARUM 7:388, 1895 LARVA, PUPA
PACKARD-INS. INJ. FORREST TREES P. 50, 1881 LARVA
PACKARD-MONOGRAPH GEOMETRIDAE P. 514, 1876 LARVA
FOOD PLANTS
POPULUS SP. (POPLARS)
PRUNUS SEROTINA EHRH.
QUERCUS SP. (OAKS)
GENERAL FEEDER ON DECIDUIOUS FOREST TREES

PECTINATA SMITH
SSP. SEE EMMEDONIA CRAMER

PEDALIS GROTE
SSP. SEE OU GUENEE

PEDIPILALIS (GUENEE), ZANCLOGNATHA-NOCTUIDAE
LIFE HISTORY
ENT. AMER. 14:182, 1934 CRUMB LARVA* PL. 24J
WASH. ENT. SOC. PROC. 4:328, 1898 DYAR LARVA
FOOD PLANTS
DEAD LEAVES

PEGALA FABRICIUS
SEE ALOPE FRABRICIUS

PELASGUS CRAMER
SYN. SEE THYSBE FABRICIUS

PELEUS SULZER
SEE PETREUS CRAMER

PELIDNEIDES STAUDINGER
SYN. SEE GIGANTEA STRECKER

PELLENIS GODART
SYN. SEE ELEUCHEA HUBNER

PELLUCIDA ABBOT & SMITH
SYN. SEE VIRGINIENSIS DRURY(ANISOTA)

PELLUCIDARIA GROTE & ROBINSON, ELLOPIA
GEOMETRIDAE
LIFE HISTORY
NEW YORK STATE AGRI. EXP. STA. (CORNELL), MEM. 274:101, 1948 FORBES LARVA
NEW YORK STATE MUS. MEM. 8:753 1905 FELT LARVA(BRIEF)
PSYCHE 9:21, 1900 DYAR EGG, LARVA
UNITED STATES DEPT. AGRI. MISC. PUB. 657:362, 435, 1950 CRAIGHEAD LARVA* F. 911
FOOD PLANTS
PINUS SP. (PINES)
PINUS ECHINATA MILL.
PINUS RESINOSA AIT.
PINUS RIGIDA MILL.

PELLUCIDARIA PACKARD
SSP. SEE SEMICLUSARIA WALKER

PELONA GUNDER
SSP. SEE LEANIRA BOISDUVAL

PEMBINA EDWARDS
SSP. SEE ICARIOIDES BOISDUVAL

PENDULINARIA(GUENEE), COSYMBIA-
GEOMETRIDAE
SYN. QUADRANNULATA WALKER
SYN. DILUCIDARIA ROTHKE
SYN. LUMENARIA HUBNER
FORM NIGRICARIA ROTHKE
A GRISEOER MCDUNNOUGH
LIFE HISTORY
CAN. ENT. 74:225, 1942 DETHIER LARVA*PUPA*ALSO NOTE ON THE LARVA OF NIGRICARIA PL. 14, FF. 1, 2, 5, 7; PL. 17, F. 3
MAINE AGRI. EXP. STA. BULL. 259: 39, 1917 MOSHER LARVA, PUPA* F. 2C
NEW YORK STATE AGRI. EXP. STA. (CORNELL), MEM. 274:120, 1948 FORBES LARVA
NEW YORK STATE MUS. MEM. 8: 749, 1905 FELT LARVA LUMENARIA (BRIEF)
PSYCHE 8:103, 1897 SCUDDER LARVA
PSYCHE 13:93, 1906 DYAR EGG, LARVA LUMENARIA HUBNER
UNITED STATES DEPT. AGRI. MISC. PUB. 657:361, 426, 1950 CRAIGHEAD LARVA
UNITED STATES NAT. MUS. PROC. 27: 901, 1904 DYAR EGG, LARVA LUMENARIA HUBNER
GUMPPENBERG-SYSTEMA GEOMETRARUM 5:292, 1892 LARVA, PUPA
PACKARD-MONOGRAPH GEOMETRIDAE P. 363, 1876 LARVA(NOT FIGURED) PUPA* PL. 13, FF. 14, 14 A, B
PACKARD-MONOGRAPH GEOMETRIDAE P. 365, 1876 LARVA*PUPA* PL. 13, FF. 22, 22 A LUMENARIA HUBNER
INDEX AMER. ECON. ENT. VOL. 3:102
FOOD PLANTS
ALNUS SP. (ALDERS)
BETULA SP. (BIRCHES)
COMPTONIA PEREGRINA(LINN.)
PSORALEA PSORALIOIDES (WALT)
VACCINIUM SP.

PENINSULARIA GROTE
SYN. SEE NITOCRIS CRAMER

PENITA MORRISON
SYN. SEE INCLUDENS WALKER

PENNARIA(LINNAEUS), HIMERA-GEOMETRIAE
LIFE HISTORY
NEW YORK ENT. SOC. JOURN. 53:201 1945 FORBES LARVA, PUPA
NEW YORK STATE AGRI. EXP. STA. (CORNELL), MEM. 274:85, 1948 FORBES LARVA
FOOD PLANTS
QUERCUS SP. (OAKS)

GENERAL FEEDER FOREST TREES

PENSYLVANICA SMITH
SSP. SEE PUDENS GUENEE

PENNSYLBANICUS CHERMOCK
SSP. SEE CRESPHONTES CRAMER

PENROSEAE FIELD
SSP. SEE MARIPOSA REAKIRT

PENSACOLA REIFF
SYN. SEE CONSORS ABBOT & SMITH

PENSILIS(GROTE), LACINIPOLIA-NOCTUIDAE
AB INDISTINCTA STRAND
LIFE HISTORY
UNITED STATES NAT. MUS. PROC. 27: 859, 1904 DYAR EGG, LARVA
FOOD PLANTS
DEAD LEAVES

PENSYLVANICA GUERIN-MENEVILLE
SYN. AMERICANA FABRICIUS

PEPITA GUENEE, BASILODES-NOCTUIDAE
SYN. CHRYSOPASA WALKER
LIFE HISTORY
INSECUTOR INSCITIAE MENSTRUUS 9:196, 1921 DYAR LARVA
FOOD PLANTS
ACTINOMERIS ALTERNIFOLIA(LINN.)

PEPLARIOIDES(HULST), PERO-GEOMETRIDAE
A MACDUNNOUGHI CASSINO & SWETT
LIFE HISTORY
ENT. NEWS 11:583, 1900 CAUDELL LARVA
SOUTHERN CAL. ACAD. SCI. BULL. 29:29, 1930 COMSTOCK EGG, LARVA*PUPA*
FOOD PLANTS
LIGUSTRUM SP. (IN CAPTIVITY)
PRUNUS SEROTINA EHRH.

PERACUTATA WALKER
SYN. SEE OBSTIPATA FABRICIUS

PERAMANS HULST
SSP. SEE MULIERCULA GUENEE

PERANGULATA HULST, ENYPIA-GEOMETRIDAE
LIFE HISTORY
UNITED STATES NAT. MUS. PROC. 27:908 1904 DYAR EGG, LARVA
FOOD PLANTS

PERANGULATA HULST
SYN. SEE ALCIPHEARIA WALKER

PERARCUATA WALKER
SYN. SEE SUBCESSARIA WALKER

PERATTENTA(GROTE), EUERETAGROTIS-NOCTUIDAE
A INATTENTA SMITH
LIFE HISTORY
MAINE AGRI. EXP. STA. BULL. 256:169, 1930 PHIPPS LARVA
FOOD PLANTS
PRUNUS PENSYLVANICA LINN.
VACCINIUM ANGUSTIFOLIUM AIT.

PERCOLENS FRANCLEMONT
SSP. SEE VINNULA GROTE

PERDICCAS(EDWARDS), EUPHYDRYAS-NYMPHALIDAE
SYN. PERDICEAS EDWARDS
AB NIGRISUPERNIPENNIS GUNDER
AB SVIHLAE GUNDER
A PARADOXA MCDUNNOUGH
LIFE HISTORY
FOOD PLANTS
PLANTAGO LANCEOLATA LINN.
PLANAGO MAJOR LINN.

PERDITA GROTE, ACRONICTA-NOCTUIDAE
LIFE HISTORY
UNITED STATES NAT. MUS. PROC. 27:805, 1904 DYAR EGG, LARVA
FOOD PLANTS
AMELANCHIER ALNIFOLIA NUTT.
MALUS PUMILA MILL. (APPLE)

PERDITALIS (WALKER), LEGNA-NOCTUIDAE
SYN. SEMILINEATA WALKER
SYN. UMBRIFASCIA GROTE
LIFE HISTORY
WASH. ENT. SOC. PROC. 4:328, 1898 DYAR LARVA
FOOD PLANTS
CEPHALANTHUS OCCIDENTALIS LINN.

PERELEGANS HENRY EDWARDS, SPHINX-SPHINGIDAE
A VANCOUVERENSIS HENRY EDWARDS
B ASELLUS ROTHSCHILD & JORDAN
LIFE HISTORY
NOVITATES ZOOL. (TRING) 9 SUPPL. P. 131, 1903 ROTHSCHILD & JORDAN LARVA, PUPA VANCOUVERENSIS
FOOD PLANTS
ARCTOSTAPHYLOS GLAUCA LINDL.
MALUS PUMILA MILL. (APPLE)
PRUNUS DOMESTICA LINN. (PLUM)
PRUNUS DOMESTICA V. GALATENSIS EX HOOK (PRUNE)
SUMPHORICARPUS ALBUS (LINN.)

PEREXCELLENS GROTE
SYN. SEE EXCELLENS GROTE

PERFUMOSA (HAMPSON), ASEPTIS-NOCTUIDAE
LIFE HISTORY
SOUTHERN CAL. ACAD. SCI. BULL. 39:104, 1940 COMSTOCK LIFE HIST. LARVA*PL. 15; PUPA*PL. 16
FOOD PLANTS
ARCTOSTAPHYLOS GLAUCA LINDL.

PERFUSA DYAR
SSP. SEE PERSPICUA GROTE & ROBINSON

PERAGAMUS HENRY EDWARDS
SSP. SEE INDRA REAKIRT

PERICULOSA GUENEE, PHLOGOPHORA-NOCTUIDAE
FORM V-BRUNNEUM GROTE
LIFE HISTORY
CAN. ENT. 31:47, 1899 FYLES EGG, LARVA
UNITED STATES NAT. MUS. PROC. 27:871, 1904 DYAR EGG, LARVA
FOOD PLANTS
ABIES BALSAMEA (LINN.)
PLANTAGO MAJOR LINN.
URTICA LYALLII S. WATS.

PERITALIS SMITH
SSP. SEE CRUCIALIS HARVEY

PERLATA (GUENEE), CAMPAEA-GEOMETRIDAE
SYN. VIRIDOPERLATA PACKARD
SYN. PRAEGRANDIA GUENEE
LIFE HISTORY
NEW YORK STATE AGRI. EXP. STA. (CORNELL), MEM. 274:85, 1948 FORBES EGG, LARVA
PSYCHE 10:190, 1903 DYAR EGG, LARVA PRAEGRANDARIA
FOOD PLANTS
ACER SP. (MAPLES)
ALNUS SP. (ALDERS)
AMELANCHIER FLORIDA LINDL.
BETULA PAPYRIFERA MARSH.
POPULUS SP. (POPLARS)
QUERCUS SP. (OAKS)
SALIX SP. (WILLOWS)
TILIA SP. (LINDENS)
ULMUS SP. (ELMS)

PERLENTANS WALKER
SYN. SEE TESSELLATA HARRIS

PERLINEATA PACKARD
SYN. SEE COMPTARIA WALKER

PERLUTEA NEUMOEGEN & DYAR
SSP. SEE FRAGILIS STRETCH

PERMACULATA (PACKARD), TURUPTIANA-ARCTIIDAE
SYN. REDUCTA GROTE
SYN. CAECA STRECKER
LIFE HISTORY
ENT. AMER. 3:14, 1887 BRUCE EGG, LARVA
NEW YORK ACAD. SCI. ANNALS 8:

81,1893 PACKARD LARVA
FOOD PLANTS
PLANTAGO SP.
POLYGONUM SP
STELLARIA MEDIA(LINN.)
TARAXACUM OFFICINALE WEBER
GENERAL FEEDER LOW PLANTS

PERMAGNA GROTE
SYN.SEE OBLONGA GROTE

PERNIGRA GRINNELL
SSP.SEE PERSIUS SCUDDER

PERNOTATA HULST
SSP.SEE MAGNOLIATA GUENEE

PEROBSOLETA LYMAN
SYN.SEE FRIGIDA SMITH(PAPAIPEMA)

PEROLIVATA PACKARD
SSP.SEE ASTARTE DOUBLEDAY

PEROPHOROIDES(STRECKER),HYPARPAX-NOTODONTIDAE
SYN.TYRIA SLOSSON
A AUROSTRIATA GRAEF
LIFE HISTORY
WASH.ENT.SOC.PROC.11:171,1909 DYAR DESCRIBES A LARVA WHICH HE THINKS IS THIS SPECIES
FOOD PLANTS
PERPALLIDA HENRY EDWARDS
SSP.SEE MIRIFICA HENRY EDWARDS

PERPALLIDARIA GROTE
SYN.SEE BIFILATA HULST, DREPANULATRIX

PERPICTA DYAR
SYN.SEE ORNATA PACKARD

PERPLEXA BARNES & BENJAMIN
SSP.SEE DUMETORUM BOISDUVAL

PERPLEXA GUENEE
SYN.SEE FLAVISTRIARIA HUBNER

PERPLEXA STRECKER
SSP.SEE PARTA GUENEE

PERQUIRITATA(MORRISON),ANOMOGYNA-NOCTUIDAE
SYN.BAILEYANA GROTE
A PARTITA MCDUNNOUGH
B BEDDEKI HAMPSON
C CLARKEI BENJAMIN
LIFE HISTORY
CAN.ENT.74:23,1942 MCGUFFIN LARVA*
INDEX AMER.ECON.ENT. VOL.7:43
FOOD PLANTS
ABIES BALSAMEA(LINN.)
PICEA GLAUCA(MUENCH)
PICEA MARIANA(MILL.)

PERSCRIPTA GUENEE,LEPIPOLYS-NOCTUIDAE
LIFE HISTORY
WASH.ENT.SOC.PROC.5:292, 1903 DYAR LARVA
FOOD PLANTS
LINARIA CANADENSIS(LINN.)

PERSEPHONE GROTE
SSP.SEE ANNA GROTE(APANTESIS)

PERSIMILATA GROTE
SYN.SEE INSULSARIA GUENEE

PERSINUARIA GUENEE
SYN.SEE AMICARIA HERRICH-SCHAEFFER

PERSIUS(SCUDDER0,ERYNNIS-HESPERIIDAE
A PERNIGRA GRINNELL
B AVINOFFI HOLLAND
LIFE HISTORY
AMER.MUS.NAT.HIST.BULL.5: 300,1893 BEUTEN.LARVA
BUFF.SOC.NAT.SCI.BULL.19:31 1939 WILD LARVA
DENNISON UNIV.SCI.LAB.JOURN. 22:111,1927 LINDSEY LARVAL HEAD* F.1 A
NEW YORK STATE MUS.MEM.8: 742,1905 FELT LARVA(BRIEF)
HOLLAND-BUTTERFLY BOOK1905 CHRYSALIS* PL.6,F.34
SCUDDER-BUTTER.EAST.UNITED

STATES&CANADA PP.1471-1475,1889 LIFE HIST*PL.66, F.11;PL.69,F.1;PL.77,FF.7, 15;PL.80,FF.37-41;PL.82, F.7;PL.85,F.34
SCUDDER-BUTTER.THEIR STRUCTURE ETC-P.108,1881 NEST*
SCUDDER-EVERYDAY BUTTER.P. 324,1899 EGG,LARVA
SCUDDER-GUIDE TO COMMON BUTTER P.160,1893 LARVA,CHRYSALIS

FOOD PLANTS
BAPTISIA TINCTORIA(LINN.)
CEANOTHUS DIVARICATUS NUTT.
GALACTIA SP.
LESPEDEZA SP.
LOTUS AMERICANUS(NUTT.)
POPULUS BALSAMIFERA LINN.
POPULUS GRANDIDENTATA MICHX.
POPULUS TREMULOIDES MICHX.
QUERCUS ILICIFOLIA WANG.
SALIX HUMILIS MARSH
TRIFOLIUM CILIOLATUM BENTH.

PERSONATA MORRISON
SSP.SEE DETERSA WALKER

PERSONATA WALKER
SYN.SEE FRATER GROTE

PERSPICUA GROTE &ROBINSON,DATANA-NOTODONTIDE
SYN.MESILLAE COCKERELL
A DISCALIS DYAR
SYN.EILEENA DYAR
FORM INFUSA DYAR
FORM PERFUSA DYAR
B OPOSITA BARNES & BENJAMIN

LIFE HISTORY
AMER.MUS.NAT.HIST.BULL.10: 409,1898 BEUTEN.LARVA
CAN.ENT.20:135,1888 BEUTEN. KEY TO DANANA LARVAE
LOC.CIT.23:82,1891 DYAR EGG, LARVA,PUPA
ENT.AMER.3:170,1887 HY.EDW. EGG,LARVA
INSECUTOR INSCITIAE MENSTRUUS 11:10,1923 DYAR LARVA (BRIEF)
MUS.COMP.ZOOLBULL.87,NO.6, 1941 DETHIER LARVAL ANTENNA*F.104
NAT.ACAD.SCI.MEM.7:118,1895 PACKARD EGG,LARVA*PL.14, F.4 COLOR; PUPA*
NEW YORK ENT.SOC.JOURN.1:61, 1893 PACKARD LARVA
NEW YORK STATE AGRI.EXP.STA. (CORNELL)MMEM.274:215,1948 FORBES LARVA
NEW YORK STATE MUS.MEM.8:737 1905 FELT LARVA(BRIEF
PSYCHE 6:573,1893 DYAR KEY TO DANATA LARVAE
PSYCHE 8:41,1897 DYAR LARVA
SOUTHERN CAL.ACAD.SCI.BULL. 37;123,1938 COMSTOCK & DAMMERS EGG*LARVA*PUPA*
UNITED STATES DEPT.AGRI.MISC. PUB.PP.355,399,1950 CRAIGHEAD LARVA* F.80 A

FOOD PLANTS
RHUS AROMATICA AIT.
RHUS COPALLINA LINN.
RHUS TYPHINA LINN.

PERSPICUA WALKER
SYN.SEE FLAVISTRIARIA HUBNER

PERSUASA HARVEY
SSP.SEE BRUMOSA GUENEE

PERTINCTA DYAR,PAPAIPEMA-NOCTUIDAE

LIFE HISTORY
CAN.ENT.58:250,1926 BIRD LARVA

FOOD PLANTS
LUPINUS POLYPHYLLUS LINDL.

PERVELATA WALKER
SYN.SEE SCOLOPACINARIUS GUENEE

PERVERSA NEUMOEGEN & DYAR
SSP.SEE DISSTRIA HUBNER

PERVERSA STRAND

SYN. SEE RUBIFERA GROTE

PERVOLATA(HULST), SEMITHISA-GEOMETRIDAE
LIFE HISTORY
NEW YORK ENT. SOC. JOURN. 53:
190,1945 FORBES PUPA
UNITED STATES NAT. MUS. PROC.
25:389,1903 DYAR LARVA
FOOD PLANTS
RIBES CYNOSBATI LINN.

PETERSII CLARK
SSP. SEE MACHAON LINNAEUS

PETITA SMITH
SSP. SEE OLIVACEA MORRISON

PETRAEA WALKER
SYN. SEE LINEOLATA WALKER

PETREIUS EDWARDS
SYN. SEE RURALIS BOISDUVAL

PETREUS(CRAMER), ATHENA-NYMPHALIDAE
PELEUS SULZER
THETYS FABRICIUS
LIFE HISTORY
AMER. ACAD. SCI. PROX. 27:247
1892 SCUDDER LARVA,
CHRYSALIS
FLA. ENT. 15:34,35,1931 ROBERTSON-MILLER LARVA, CHRYSALIS
LOC. CIT. 18:29,30,1934 ROBERTSON-MILLER LARVA, CHRYSALIS
INSECUTOR INSCITIAE MENSTRUUS
2:113,1914 DYAR LARVA
OHIO JOURN. SCI. 38:294,1938
STROHECKER LARVA*CHRYSALIS*
PA. ACAD. SCI. PROX. 19:42,1945
CHERMOCK CHRYSALIS*F. 1 E,F
INDEX AMER. ECON. ENT.
VOL. 5:66
FOOD PLANTS
ANACARDIUM OCCIDENTALE LINN.
FICUS SP.
FICUS CARICA LINN.

PETRONIUS LINTNER
SYN. SEE HORATIUS SCUDDER & BURGESS

PETROSA WALKER
SYN. SEE PLANTAGINIS LINNAEUS

PETULANS HULST
SSP. SEE PARTA GUENEE

PHAEALIS GUENEE
SYN. SEE LUBRICALIS GEYER

PHAEDON HERRICH-SCHAEFFER
SYN. SEE PHAETON DRURY

PHAEDRUS HALL
SYN. SEE EPIXANTHE BOISDUVAL &
LECONTE

PHAEORRHOEA(DONOVAN), NYGMIA-LIPARIDAE
CHRYSORRHOEA AUCT.
LIFE HISTORY
BOSTON SOC. NAT. HIST. PROC. 27:
143,1896 DYAR LARVAL
SETAE
ILL. AGRI. EXP. STA. BULL. 151:
478,1911 FORBES LARVA*
MAINE AGRI. EXP. STA. BULL.
134:210,1906 PATCH EGG,
LARVA, PUPA
NEW YORK STATE AGRI. EXP. STA.
(CORNELL), MEM. 274:240,1948
FORBES LARVAE OF PHAEORRHOEA
AND CHRYSORRHOEA LINN.
NEW YORK STATE MUS. BULL. NO. 64,
P. 94,1903 FELT LARVA*PUPA*
COLOR
LOC. CIT. NO. 97:385,1905 FELT
LIFE HISTORY
LOC. CIT. NO. 103:16,1906 FELT
LIFE HISTORY*ALL STAGES*
COLOR
LOC. CIT. NO. 141:19,1910 FELT
EGG, LARVA
LOC. CIT. NO. 180:31,1915 FELT
EGG, LARVA, PUPA
NEW YORK STATE MUS. MEM. 8:
164,1905 FELT LIFE HIST*
NOVA SCOTIA ENT. SOC. PROC. 1:47

1915 Saunders life history
Ohio Agri. Exp. Sta. Bull. 332: 277,1918 Houser egg*larva* Pl. XLV
Ontario Ent. Soc. Ann. Rept. p.72 1898 Fletcher life history
Loc. cit. p.28,1908 Felt larva
United States Dept. Agri. Bur. Ent. Bull. 17:26,1898 new series C. H. Fernald life hist.
Loc. cit. 34:31,1902 new series Banks egg*larva*pupa*all f.21
Loc. cit. 87:21,1910 new series Rogers-Burgess egg*Pl.4, f.4;larva*Pl.4,f.5;pupa* Pl.4,f.1
United States Dept. Agri. Circ. 464:31,1938 Burgess,Baker egg*f.16 G,H;larva*f.16 E,G;Pupa*f.16 B,C
United States Dept. Agri. Farmer's Bull. 274: 10,1906 Howard egg,larva*f.1,2
Loc. cit. 453:19,1911 Marlatt larva*f.6;nest*f.7
Loc. cit. 564:7,1914 Burgess egg*larva*pupa*all f.4
Loc. cit. 1270:44,1922 Quaintance egg,larva*f.85;pupa,nest* f.89
Loc. cit. 1335:11,1923 Burgess egg *larva*pupa*all f.6
Loc. cit. 1623:24,1930 Burgess egg*f.18h;larva*f.18 e,g; pupa*f.18 b,c
United States Dept. Agri. Misc. Pub. 657:355,412,1950 Craighead all stages* f.83
United States Dept. Agri. Yearbook for year 1905 p.126 Howard egg*larva*ff.3,4
Doane etc-Forest Ins. p.270, 1936 larva,pupa
Fernald-Applied Ent. p.269, 1926 egg*larva*both f.269
Fernald & Shepard0Applied Ent. p.265,1942 egg*larva*f.226
Herrick-Ins. Enemies Shade Trees p.84,1935 larva* f.51;nest*f.53
Metcalf&Flint-Destr. &Useful Ins. p.706,1939 egg*larva* pupa* all f.467
O'Kane-Inj. Ins. p.293,1912 egg* f.435;larva*f.433
Peairs-Pests FarmGard. Orchard p.370,1941 egg*larva*both f.437
Slingerland,Crosby-Manual Fruit Ins. p.136,1914 egg, larva*f.149
Index Amer. Econ. Ent. Vol.1:127;Vol.2:141; Vol.3:157;Vol.4:181; Vol.5:451;Vol.6:536; Vol.7:710

FOOD PLANTS

Acacia sp.
Acer sp. (maples)
Acer pseudoplatanus Linn.
Acer saccharinum Linn.
Amelanchier sp.
Arctium lappa Linn.
Betula sp. (n.p.)
Betul papyrifera Marsh.
Carya sp. (N. P.)
Castanea dentata(Marsh.)(n.p.)
Crataegus sp.
Cydonia japonica(Thunb.)
Cydonia oblonga Mill. (quince)
Fagus sp.
Fragaria chiloensis(Linn.)
Fraxinus sp. (n.p.)
Fraxinus americana Linn.
Geranium sanguinem Linn.
Juglans nigra Linn.
Malus coronaria(Linn.)
Malus pumila Mill. (apple)
Myrica pensylvanica Loisel
Parthenocissus quinquefolia (Linn.)
Plantago sp.
Populus sp. (poplars)
Prunus avium(Linn.)

PRUNUS CERASUS LINN. (CHERRY)
PRUNUS DOMESTICA LINN. (PLUM)
PRUNUS DOMESTICA V. GALATENSIS EX HOOK (PRUNE)
PRUNUS PENSYLVANICA LINN.
PRUNUS PERSICA(LINN.)(PEACH)
PRUNUS WEROTINA EHRH.
PYRUS COMMUNIS LINN.
QUERCUS SP. (OAKS)
RHEUM RHAPONTICUM LINN. (RHUBARB)
RIBES GROSSULARIA LINN.
RIBES SATIVUM SYME(CURRANT)
ROBINIA PSEUDO-ACACIA LINN.
ROSA SP. (N.P.)
ROSA CAROLINA LINN.
RUBUS ALLEGHENIENSIS PORTER
RUBUS IDAEUS V. STRIGOSUS(MICHX.)
RUMEX CRISPUS LINN.
RUMEX VERTICILLATUS LINN.
SALIX SP. (WILLOWS)
SALIX BABYLONICA LINN.
SAMBUCUX CANADENSIS LINN.
SAMBUCUX COERULEA RAF.
SORBUS AMERICANA MARCH.
SORBUS ARIA(LINN.)
SPIRAEA SP.
SPIRAEA THUNBERGII SIEB.
TILIA SP.
TRIFOLIUM PRATENSE LINN.
ULMUS SP.
VIBURNUM CASSINOIDES LINN.
VITIS VULPINA LINN.
WEIGELA FLORIDA (S.&Z.)
WISTERIA SINENSIS SWETT

PHAETAENA HUBNER
SYN. SEE PHAETON DRURY

PHAETHUSA HULST
SSP. SEE PHAETON DRURY

PHAETON(DRURY)EUPHYDRYAS-NYMPHALIDAE
SYN. PHAEDON HERRICH-SCHAEFFER
SYN. PHAETAENA HUBNER
SYN. PHAETONTEA GODART
SYN. SCHAUSI CLARK
AB PHAETHUSA HULST
AB SUPERBA STRECKER
SYN. MAGNIFICA CLARK
AB MOJAVE COMSTOCK

LIFE HISTORY

AMER. MUS. NAT. HIST. BULL. 5: 259, 1893 BEUTEN. LARVA, CHRYSALIS
AMER. NAT. 2:219, 1868 LARVA* PUPA*
BUFF. SOC. NAT. SCI. BULL. 19:23, 1939 WILD LARVA
CAN. ENT. 7:150, 1875 EDWARDS EGG, LARVA
ILL. DEPT. AGRI. TRANS. 18, APPEND:' 83, 1880 MIDDLETON LARVA
NEW YORK STATE CABIN. NAT. HIST ANN. REPT. NO. 23:154, 1873 LINTNER EGG, LARVA
PAPILIO 4:69, 1884 GRUBER LARVA
PEABODY ACAD. SCI. REPT. P. 48, 1871 SCUDDER EGG
SOUTHERN CAL. ACAD. SCI. BULL. 37:38, 1938 COMSTOCK LARVA* PL. 8; CHRYSALIS* PL. 9) MOJAVE
UNITED STATES NAT. MUS. PROC. 71: 9, 1928 CLARK EGG, LARVA, PUPA GOOD ARTICLE ON FEEDING HABITS MAGNIFICA
COMSTOCK-HOW KNOW BUTTER. P. 125, 1904 LARVA
CRAGEN-OUR INS. FRIENDS&FOES P. 85, 1899 LARVA* F. 22
EDWARDS-BUTTER. NORTH AMER. 2ND SERIES 1875 LIFE HIST* PL. MELITEA NO. 1 COLOR
FERNALD-BUTTER. MAINE P. 44, 1884 EGG, LARVA, CHRYSALIS
FRENCH-BUTTER. EAST. UNITED STATES P. 169, 1886 LIFE HISTORY*
GLOVER-ILLUST. NORTH AMER. ENT. 1878 LARVA, PUPA PL. 22, F. 6
HOLLAND-BUTTERFLY BOOK PP. 138 139, 1905 CHRYSALIS* FF. 15, 16
LOC. CIT. 3RD EDIT. P. 115, 1931 1931 LIFE HISTORY
KELLOGG-AMER. INS. P. 457 1908

LARVA
MAYNARD–BUTTER. NEW ENGLAND
P.30,1886 LARVA,CHRYSALIS
PACKARD–HALF HOURS WITH INS.
1877 LARVA*CHRYSALIS*
COLOR
SAUNDERS–BUTTER. ALLEGHANY
STATE PARK(NEW YORK STATE
HANDBOOK)NO.13,1932 LARVA*
NESTS* F.19
SCUDDER–BUTTER. EAST. UNITED
STATES&CANADA PP.693–701
1889 LIFE HIST*PL.64,FF.37,
43;PL.75,F.11;PL.79,FF.
29–31;PL.82,F.3;PL.84,
FF.15,16
SCUDDER–BUTTER. THEIR STRUCTURE
P.9,1881 EGG WEBB*NEST*P.
106
SCUDDER–EVERYDAY BUTTER. P.198,
1899 EGG,LARVA,PUPA
SCUDDER–GUIDE TO COMMON BUTTER.
P.67,1893 LARVA,CHRYSALIS
INDEX AMER. ECON. ENT.
VOL.4:298 MELITEA
VOL.5:239;VOL.7:367

FOOD PLANTS
ASTER SP.
AUREOLARIA GRANDIFLORA(BENTH.)
CHELONE GLABRA LINN. (IT IS SAID
THAT THE NEWLY HATCHED AND
YOUNG LARVAE MUST START WITH
THIS PLANT,BUT THE OLDER
LARVAE CAN MATURE ON THE
OTHER PLANTS LISTED HERE.)
CORYLUS SP. (HAZELS)
FRAXINUS SP. (YOUNG ASH SHOOTS)
GERARDIA SP.
LONICERA CANADENSIS BARTR.
LONICERA JAPONICA THUNB.
LONICERA TATARICA LINN.
MIMULUS RINGENS LINN.
OENOTHERA DENTATA V. PARISHII
(ABRAMS)
PLANTAGO SP.
RIBES SP.
RIBES NIGRUM LINN.
SCROPHULARIA MARYLANDICA LINN.
TYPHA LATIFOLIA LINN.
VIBURNUM SP.
VIBURNUM DENTATUM LINN.

PHAETON GROTE & ROBINSON,
EUPROSERPINUS–SPHINGIDAE
SYN. ERATO BOISDUVAL
LIFE HISTORY
SOUTHERN CAL. ACAD. SCI. BULL.
33:141,1934 COMSTOCK &
DAMMERS LIFE HIST. EGG,
LARVA*PLS.46,47;PUPA*
PL.47
FOOD PLANTS
OENOTHERA BISTORTA NUTT.

PHAETONTEA GODART
SYN. SEE PHAETON DRURY

PHALANGA GROTE
SSP. SEE PALAEOGAMA GUENEE

PHALETATA (HARRIS), APANTESIS–ARCTIIDAE
SYN. RHODA BUTLER
SYNPULCHERRIMA STRETCH(FEMALE)
AB INCARNATA STRETCH
LIFE HISTORY
AMER. MUS. NAT. HIST. BULL. 10:
382,1898 BEUTEN. LARVA
CAN. ENT. 32:369,1900 GIBSON
LIFE HIST. EGG,LARVA,PUPA
LOC. CIT. 34:50,1902 GIBSON LARVA
LOC. CIT. 44:189,1912 BARNES &
MCD. LARVA
ILL. DEPT. AGRI. TRANS. 15:182,
1877 FRENCH LARVA
LOC. CIT. 18,APPEND:115,171,1880
COQUILLETT & MARTIN LARVA
NEW YORK ENT. SOC. JOURN. 8:38
1900 DYAR LARVA(BRIEF)
INDEX AMER. ECON. ENT.
VOL.1:20
FOOD PLANTS
PLANTAGO SP.
SPARTINA PATENS(AIT.)
TARAXACUM OFFICINALE WEBER
TRIFOLIUM SP. (CLOVERS)
ZEA MAYS LINN. (CORN)

PHALERATULS STRAND
SSP. SEE VITTATA FABRICIUS

PHALEROSALIS WALKER
SYN. SEE FLAVIPUNCTALIS GEYER

PHANTAMARIA (STRECKER), NEPYTIA-GEOMETRIDAE
LIFE HISTORY
BRITISH COL. ENT. SOC. PROC. NO. 18 SUPPLEMATIC SERIES P. 30,1921 G.O. DAY LARVA
NEW YORK ENT. SOC. JOURN. 53: 195,1945 FORBES PUPA
FOOD PLANTS
PICEA SP. (SPRUCES)
PSEUDOTSUGA MENZIESII (MIRB.)
TSUGA HETEROPHYLLA (RAF.)

PHAON (EDWARDS), PHYCIODES, NYMPHALIDAE
SYN. AESTIVA EDWARDS
FORM HIEMALIS EDWARDS
AB THORNEI GUNDER
LIFE HISTORY
SOUTHERN CAL. ACAD. SCI. BULL. 28:22,1929 COMSTOCK LIFE HIST. LARVA*CHRYSALIS*
FOOD PLANTS
LIPPIA LANCEOLATA MICHX.
LIPPIA NODIFLORA (LINN.)

PHAREUS PANZER
SYN. SEE PHYLEUS DRURY

PHASIANARIA GUENEE
SYN. SEE CONFUSARIA HUBNER

PHERES (BOISDUVAL), PLEBEIUS-LYCAENIDAE
AB ORCUS EDWARDS
LIFE HISTORY
FOOD PLANTS
LUPINUS CHAMISSONIS ESCHSCH.

PHIGALIARIA GUENEE
SYN. SEE QUERNARIA ABBOT & SMITH

PHILA SMITH
SYN. SEE ALFARATA STRECKER

PHILEA (LINNAEUS), PHOEBIS-PIERIDAE
SYN. ARICYE CRAMER
SYN. CORDAY HUBNER
GEN. VERN. MELANIPPE CRAMER
FORM FEMALE OBSOLETA NIEPELT
LIFE HISTORY
ENT. AMERICANA 3:162,1887 HY. EDW. CHRYSALIS
PA. ACAD. SCI. PROC. 19:41,1945 CHERMOCK CHRYSALIS*F.1C NO DESCR.
BUTLER-LEP. EXOTICA P.93,1874 LARVA (BRIEF)
BUTLER-MONOGRAPH CALLIDRYAS P.22,1873 LARVA, CHRYSALIS
INDEX AMER. ECON. ENT. VOL.5:116; VOL.7:769
FOOD PLANTS
ACACIA SP. (?)
CASSIA MARILANDICA LINN.

PHILEMON BOISDUVAL
SYN. SEE ANNA EDWARDS

PHILENOR LINNAEUS. PAPILIO. PAPILIONIDAE
SYN. ASTINOUS DRURY
AB WASMUTHII WEEKS
AB OBSOLETA EHRMANN
A ACAUDA OBERTHUR
SYN. NEZAHAULCOYOTL STRECKER
SYN. CORBIS GODMAN & SALVIN
B HIRSUTA SKINNER
AB INGHMAI GUNDER
LIFE HISTORY
AMER. MUS. NAT. HIST. BULL. 5: 242,1893 BEUTEN. LARVA, CHRYSALIS
AMER. NAT. 15:327,1881 RILEY LARVA*CHRYSALIS*
BUFF. SOC. NAT. SCI. BULL. 19:17, 1939 WILD LARVA
CAL. ACAD. SCI. PROC. 5:162,1873 HY. EDW. LARVA, CHRYSALIS
LOC. CIT. 7:19,1876 HY. EDW. CHRYSALIS
CAN. ENT. 13:9,1881 EDWARDS 3GG, LARVA, CHRYSALIS
LOC. CIT. 39:209,1907 GIRAULT

LARVA
ENT. REC. 21:146-148 AND 157,158
1909 FLORSHEIM LARVAL HABITS
ILL. DEPT. AGRI. TRANS. 15:136,
1877 FRENCH LARVA
LOC. CIT. 18, APPEND:164, 1880
COQUILLETT LARVA*
LOC. CIT. 18, APPEND:73, 1880
MIDDLETON LARVA*
MARYLAND ACAD. SCI. BULL. 3(2)
8, 1923 MOORE EGG, LARVA,
CHRYSALIS
MARYLAND NAT. HIST. SOC. PROC.
P. 3, 1933 HAYDON EGG, LARVA,
CHRYSALIS
MO. REPT. INS. OF 2:116, 1870
RILEY LIFE HISTORY*
MUS. COMPZOOL. BULL. 87, NO. 6,
1941 DETHIER LARVAL
ANTENNA* FF. 17, 18, 20
ONTARIO ENT. SOC. ANN. REPT. P. 39,
1880 SAUNDERS LARVA*
CHRYSALIS
LOC. CIT. P. 41, 1894 BETHUNE
LARVA* CHRYSALIS
LOC. CIT. P. 77, 1901 LOCKHEAD
CHRYSALIS
PAPILIO 4:88, 1884 GRUBER LARVA*
PSYCHE 4:238, 1892 SCUDDER NOTE
ON GREEN CHRYSALIS
PSYCHE 8:207, 1898 SCUDDER LARVA
SMITHSONIAN MIS. COLL. 4, ART. 2:6
1862 MORRIS LARVA, CHRYSALIS
SOUTHERN CAL. ACAD. SCIL. BULL.
25:84, 1926 COMSTOCK LARVA*
CHRYSALIA*
UNITED STATES DEPT. AGRI. MISC.
PUB. 657:360, 367, 1950
CRAIGHEAD LARVA* F. 70A
UNITED STATES NAT. MUS. BULL. 157:
177, 1931 AUSTIN-CLARK LARVA,
CHRYSALIS
ABBOT & SMITH-LEP. INS. GA. 1:5,
1797 LARVA* PL. 3
BOISDUVAL-SPEC. GEN. LEP. DIURN.
1:325, 1836 LARVA, CHRYSALIS
BOISDUVAL&LECONTE-HIST. LEP.
NORTH AMER. P. 31, 1833 LARVA*
PL. 11 COLOR
COMSTOCK, J. A. BUTTER. CAL.
P. 18, 1927 LARVA* CHRYSALIS*
PL. 63, F. 4 COLOR
COMSTOCK, J. H. HOW KNOW
BUTTER. P. 66, 1904 LARVA
ESSIG-INS. WEST. NORTH AMER. P.
633, 1926 EGG* F. 509; LARVA*
F. 509
FRENCH-BUTTER EAST. UNITED
STATES PP. 87-89, 1886 LARAV *
CHRYSALIS *
GLOVER-ILLUST. NORTH AMER. ENT.
1878 LARVA* PL. 9, F. 18
HARRIS-ENT. CORRESP. P. 273,
1869 LARVA* CHRYSALIS*
HART-INS. COLL. ILL. HIGH
SCHOOLS P. 7, 1903 LARVA*
F. 1
HOLLAND-BUTTERFLY BOOK 1905
LARVA* PL. 2, FF. 13, 20, 21;
CHRYSALIS* PL. 6, FF. 14, 17,
20
JAEGER-LIFE NORTH AMER. INS.
P. 209, 1864 LARVA
MAYNARD-BUTTER. NEW ENGLAND
P. 49, 1886 LARVA, CHRYSALIS
MORRIS-SYNOP. LEP. NORTH AMER.
P. 6, 1862 LARVA
SCUDDER-BUTTER. EAST. UNITED
STATES&CANADA PP. 1244-
1245; 1248-1250, 1889 LIFE
HIST* PL. 66, FF. 4, 8; PL. 72,
F. 7; PL. 76, FF. 13, 20, 21;
PL. 80, FF. 1-5; PL. 85, FF. 14,
17, 20
SCUDDER-BUTTER. THEIR STRUCTURE
P. 22, 1881 LARVA* CHRYSALIS*
SCUDDER EVERYDAY BUTTER. P. 382
1899 EGG, LARVA, CHRYSALIS
SCUDDER-GEOLOGY NEW HAMPSHIRE
P. 380, 1874 CHRYSALIS* PL. A
SCUDDER-GUIDE TO COMMON BUTTER.
P. 145, 1893 LARVA, CHRYSLIS
WESTEOOD-IN DRURY'SEXOTIC. ENT.
1:21, 1837 LARVA, CHRYSALIS
INDEX AMER. ECON. ENT.
VOL. 3:295; VOL. 4:343;
VOL. 5:469; VOL. 7:741

FOOD PLANTS
ARISTOLOCHIA CALIFORNICA TORR.
ARISTOLOCHIA DURIOR HILL
ARISTOLOCHIA SERPENTARIA LINN.
ASARUM CANADENSE LINN.
IPOMOEA ALBA LINN.
POLYGONUM CONVOLVULUS LINN.
POLYGONUM SCANDENS LINN. (FEMALE SEEN OVIPOSTIGING ON THIS PLANT. ENT. NEWS 6:281)

PHILEROS BOISDUVAL
SYN. SEE ICARIOIDES BOISDUVAL

PHILETA FABRICIUS
SSP. SEE MONUSTE LINNAEUS

PHILODICE (GODART), COLIAS-PIERIDAE
FOR THE MANY SYNS. ANDSSPS. SEE McDUNNOUGH'S LIST
LIFE HISTORY
AMER. MUS. NAT. HIST. BULL. 5:250 1893 BEUTEN. LARVA, CHRYSALIS
BJFF. SOC. NAT. SCI. BULL. 19:20, 1939 WILD LARVA
CAN. ENT. 1:54, 1869 SAUNDERS EGG, LARVA, CHRYSALIS
LOC. CIT. 5:221, 1873 SAUNDERS NOTE ON LIFE HISTORY
CAN. NAT. &GEOL. 2:318, 1857 LARVA
ILL. DEPT. AGRI. TRANS. 18:78, APPEND. 1880 MIDDLETON LARVA
MONTANA UNIV. BULL. 30:37, 1906 ELROD EGG, LARVA
MUS. COMP. ZOOLBULL. 87, NO. 6, 1941 DETHIER LARVAL ANTENNAE* F. 21
NAT. CAN. P. 70, 1927 HUARD LARVA
ONTARIO ENT. SOC. ANN. REPT. P. 27 1874 SAUNDERS EGG, LARVA, CHRYSALIS
LOC. CIT. P. 48, 1881 SAUNDERS EGG, LARVA, CHRYSALIS
PSYCHE 3:308, 1882 SCUDDER LARVA (INTERNAL STRUCTURE)
COMSTOCK-HOW KNOW BUTTER. P. 93, 1904 LARVA
DICKERSON-MOTHS & BUTTER. P. 113, 1901 LIFE HISTORY
EDWARDS-BUTTER. NORTH AMER. 2ND SERIES 1876 LIFE HIST* COLIAS NO. 3 COLOR
FERNALD-BUTTER. MAINE P. 31, 1884 EGG, LARVA, CHRYSALIS
FRENCH-BUTTEREEAST. UNITED STATES P. 134, 1886 LIFE HISTORY
GLOVER-ILLUST. NORTH AMER. ENT. 1878 LARVA, PUPAPL. 1, F. 7
HARRIS-INS. INJ. VEG. FLINT EDIT. P. 272, 1862 LARVA*CHRYSALIS*
HOLLAND-BUTTERFLY BOOK. P. 291 1905 LARVA*PL. 2, F. 10; CHRYSALIS PL. 5, FF. 54, 55
SCUDDER-BUTTER. EAST UNITED STATES&CANADA PP. 1115-1117, 1119-1123, 1889 LIFE HIST*PL. 65, F. 25; PL. 68, F. 14; PL. 72, F. 3; PL. 76, F. 10; PL. 79, FF. 46-48; PL. 86. FF. 13, 42, PL. 84, FF. 54, 55; PL.
87 87, F. 4
SCUDDER-BUTTER. THEIR STRUCTURE PP. 8, 42, 116, 1881 EGG*LARVA* CHRYSALIS*
SCUDDER-EVERYDAY BUTTERP. 27, 1899 EGG, LARVA, CHRYSALIS
SCUDDER-GUIDE TO COMMON BUTTER. P. 134, 1893 LARVA, CHRYSALIS
VANCE-LIFE AMONG BUTTRE. P. 48, 1925 EGG, LARVA, CHRYSALIS
INDEX AMER. ECON. ENT. VOL. 1:130; VOL. 2:142; VOL. 3:159; VOL. 4:115; VOL. 5:149; VOL. 7;206
FOOD PLANTS
ASTRAGALUS SP.
BAPTISIA TINCTORIA(LINN.)
CARAGANA SP.
CYTISUS SP.
LUPINUS PERENNIS LINN.
MEDICAGO SATIVA LINN.
PISUM SATIVUM LINN. (PEAS)
ROBINIA PSEUDO-ACACIA LINN.
TRIFOLIUM SP. (CLOVERS)
VICIA CRACCA LINN.

PHLAEAS BOISDUVAL & LECONTE
SYN. SEE HYPOPHLAEAS BOISDUVAL

PHLOGOPHAGUS GROTE & ROBINSON
SYN. SEE PHLOXIPHAGA GROTE & ROBINSON

PHLOGOSARIA(GUENEE), PLAGODIS-GEOMETRIDAE
LIFE HISTORY
CAN. ENT. 82:205, 1950 MCGUFFIN LARVA*F. 1A APPROXIMARIA DYAR
NEW YORK STATE AGRI. EXP. STA. (CORNELL), MEM. 274:90, 1948 FORBES LARVA
FOOD PLANTS
ALNUS SP.
BETULA PAPYRIFERA MARSH.
MALUS PUMILA MILL. (APPLE)
PRUNUS SEROTINA EHRH.
PRUNUS VIRGINIANA LINN.
SALIX SP. (WILLOWS)
TILIA SP. (LINDENS)

PHLOXIPHAGA GROTE & ROBINSON, HELIOTHIS NOCTUIDAE
SYN. PHLOGOPHAGUS GROTE & ROBINSON
SYN. ACESIAS FELDER
SYN. INTERJACENS GROTE
FORM LUTEITINCTUS GROTE
LIFE HISTORY
ILL. DEPT. AGRI. TRANS. 15:230, 1877 FRENCH LARVA
KENT. AGRI. EXP. STA. BULL. 227: 82, 1920 GARMAN LARVA
PAN PACIFIC ENT. 15:123, 1939 STAHLER LARVAE OF PHLOXIPHAGA & OBSOLETA FABRIDIUS COMPARED
PAPILIO 1:56, 1881 COQUILLETT LARVA LUTEITINCTUS
UNITED STATES NAT. MUS. PROC. 25:374, 1903 DYAR LARVA
LOC. CIT. 68:5, 1927 CRUMB LARVA
INDEX AMER. ECON. ENT.
VOL. 3:84 CHLORIDEA
VOL. 6:346:VOL. 7:456
FOOD PLANTS
ANTIRRHINUM SP. (SEED PODS)
ASTER SP.
GRINDELIA CAMPORUM GREENE
GRINDELIA SQUARROSA(PURSH)
HEMIZONIA CONGESTA DC.
LACUTC SATIVA LINN. (LETTUCE)
LATHYRUS ODORATUS LINN.
MEDICAGO SATIVA LINN. (ALFALFA)
PARTHENIUM ARGENTATUM GRAY
PHLOX SP.
SCHIZANTHUS SP. (PODS)
SOLANUM LYCOPERSICUM LINN. (TOMATO)

GRASSES

PHOCION(FABRICIUS), NEONYMPHA-SATYRIDAE
SYN. AREOLATUS ABBOT & SMITH
SYN. HELICTA HUBNER
A SEPTENTRIONALIS DAVIS
LIFE HISTORY
CAN. ENT. 14:163, 1882, EDWARDS EGG, LARVA CHRYSALIS
ABBOT&SMITH-LEP. INS. GA. 1:25 1797 LARVA* PL. 13
BOISDUVAL&LECONTE-HISTLLEP. NORTH AMER. 1833 LARVA* CHRYSALIS*PL. 63 COLOR
COMSTOCK-HOW TO KNOW BUTTER. P. 196, 1904 LARVA
EDWARDS-BUTTER. NORTH AMER. 3RD SERIES 1897 LIFE HIST* PL. NEONYMPHA 2, FF. A-H
FRENCH-BUTTER. EAST. UNITED STATEA P. 237, 1886 LIFE HISTORY AEREOLATUS
HOLLAND-BUTTERFLY BOOK 1905 LARVA*CHRYSALIS*PL. 3, F. 8; PL. 4, FF. 10, 11
SCUDDER-BUTTER. EAST. UNITED STATES & CANADA PP. 205-207 1889 LIFE HIST*PL. 64, F. 9; PL. 70, F. 11; PL. 74, FF. 8, 12; PL. 78, FF. 15, 16; PL. 83, FF. 10, 11
FOOD PLANTS
DACTYLOCTENIUM AEGYPTIUM(LINN.)
DIGITARIA SANGUINALIS(LINN.)
SORGHASTRUM NUTANS(LINN.)

COARSE GRASSES

PHOCION FABRICIUS
SEE THEMISTOCLES LATREILLE

PHOCUS EDWARDS
SSP. SEE OETUS BOISDUVAL

PHOEBE HENRY EDWARDS
SSP. SEE COELEBS GROTE

PHOLUS (CRAMER), DARAPSA-SPHINGIDAE
SYN. CHOERILUS CRAMER
SYN. AZALEAE ABBOT & SMITH
SYN. CHLORINDA MARTYN
A BRODIEI CLARK
LIFE HISTORY
AMER. JOURNSCI. 36:302, 1839 HARRIS LARVA, PUPA
AMER. MUS. NAT. HIST. BULL. 7:292, 1895 BEUTEN. LARVA, PUPA
ENT. SOC. AMER. ANNALS 11:435. 1918 MOSHER PUPA
NEW YORK STATE AGRI. EXP. STA. (CORNELL), MEM. 274:199, 1948 FORBES LARVA
NEW YORK STATE MUS. MEM. 8:738, 1905 FELT LARVA
NOVITATES ZOOL. (TRING) 9 SUPPL. P. 526, 1903 ROTHSCHILD & JORDAN LARVA, PUPA
ONTARIO ENT. SOC. ANN. REPT. P. 57, 1881 REED LARVA
PAPILIO 3:126, 1883 EDWARDS LARVA, PUPA CHOERILUS
PHIL. ACAD. NAT. SCI. JOURN. 5:147, 1859 CLEMENS LARVA
SMITHSONIAN MISC. COLL. 4, ART. 2:168, 1862 MORRIS LARVA, PUPA
ABBOT&SMITH-LEP. INS. GA. 1:53 1797 LARVA*PUPA*PL. 27 AZALEAE
BOISDUVAL-SPEC. GEN. LEP. SPHING P. 211, 1874 LARVA
ELIOT&SOULE-CATER. &THEIR MOTHS P. 112, 1902 EGG, LARVA*PUPA
FERNALD-SPHING. NEW ENGLAND P. 65, 1886 LARVA, PUPA
GLOVER-ILLUST. NORTH AMER. ENT. 1878 LARVA*PL. 22, F. 14
MORRIS-SYNOP. LEP. NORTH AMER. P. 168, 1862 LARVA, PUPA
FOOD PLANTS
KALMIA ANGUSTIFOLIA LINN.
NYSSA SYLVATICA MARSH.
RHODODENDRON NUDIFLORA LINN.
RHODODENDRON VISCOSA LINN.
TRADESCANTIA VIRGINIANA LINN.
VIBURNUM DENTATUM LINN.
VIBURNUM LENTAGO LINN.
VIBURNUM OPULUS LINN.

PHOLUS (DRURY), LYCOMORPHA-AMATIDAE
A MINIATA PACKARD
LIFE HISTORY
AMER. JOURN. SCIL 36:318, 1839 HARRIS LARVA, PUPA
AMER. MUS. NAT. HIST. BULL. 10: 355, 1898 BENTEN. LARVA
BOSTON SOC. NAT. HIST. PROC. 27: 137, 1896 DYAR LARVAL SETAE
NAT. CAN. P. 257, 1929 HUARD LARVA
PAPILIO 2:181, 1882 PACKARD LARVA, PUPA
PSYCHE 3:82, 1897 DYAR EGG, LARVA, PUPA, COCOON
SMITHSONIAN MISC, COLLECTIONS R, ART. 2:135, 1862 MORRIS LARVA
FERNALD-IN KINGSLEY'S STAND. NAT. HIST. 2:462, 1895 LARVA
HARRIS-INS. INJ. VEG. FLINT EDIT. P. 341, 1862 LARVA (BRIEF)
MORRIS=SYNOP. LEP. NORTH AMER. P. 135, 1862 LARVA
STRETCH-ZYGAEN. ET BOMBYC. NORTH AMER. P. 42, 1873 LARVA
INDEX AMER. ECON. ENT. VOL. 7:621
FOOD PLANTS
LICHENS

PHRAGMATIDICOLA GUENEE, LEUCANIA-NOCTUIDAE
LIFE HISTORY
AMER. MUS. NAT. HIST. BULL. 16:

447,1902 BEUTEN. LARVA
BROOKLYN ENT. SOC. BULL. 22:46, 1927 CRUMB MATURE LARVA
ILL. DEPT. AGRI. TRANS. 18, APPEND. 186,1880 DOQUILLETT LARVA
INDEX AMER. ECON. ENT. VOL. 7:193
FOOD PLANTS
ELYMUS VIRGINICUS LINN.
POA PRATENSIS LINN.
GENERAL FEEDER LOW PLANTS

PHRYNIA HENRY EDWARDS
SSP. SEE RELICTA WALKER

PHYLEUS(DRURY), HYLEPHILA-HESPERIIDAE
SYN. BUCEPHALUS STEPHENS
LIFE HISTORY
AMER. ENT. SOC. TRANS. 50:319 1925 COOLIDGE LARVA, PUPA
AMER. MUS. NAT. HIST. BULL. 5:291 1893 BEUTEN, LARVA, CHRYSALYS
CAN. ENT. 11:190,1879 EDWARDS LARVA, CHRYSALIS
ILL. DEPT. AGRI. TRANS. 18, APPEND. 96,176,1880 COQUILLETT & MIDDLETON LARVA
NEW YORK ENT. SOC. JOURN. 7:211 1899 COQUILLETT LARVA
SOUTHERN CAL. ACAD. SCI. BULL. 32:82,1933 COMSTOCK= DAMMERS EGG*LARVA*CHRYSALIS* ALL PL. 25
COMSTOCK-BUTTER. CAL. P. 220, 1927 EGG, LARVA, CHRYSALIS
FRENCH-BUTTER. EAST. UNITED STATES P. 314,1886 LARVA, CHRYSALIS
HOLLAND-BUTTERFLY BOOK 1905 CHRYSALIS*PL. 5, F. 39
SCUDDER-BUTTER. EAST. UNITED STATES & CANADA P. 1633, 1889 LARVA*CHRYSALIS*PL. 85, F. 39; PL. 77, F. 19
FOOD PLANTS
DIGITARIA SANGUINALIS(LINN.)
POA PRATENSIS LINN.
ROSA SP. ?(ROSES)
GRASSES

PHYLLA DYAR
SSP. SEE MINUSCULA ZELLER

PHYLLIRA(DRURY), APANTESIS-ARCTIIDAE
SYN. B-ATA GOEZE
SYN. DODGEI BUTLER
LIFE HISTORY
AMER. MUS. NAT. HIST. BULL. 10: 381,1898 BEUTEN. LARVAL NOTE
CAN. ENT. 43:259,1911 BARNES & MCD. LARVA
LOC. CIT. 44:132,1912 BARNES & MCD. EGG, LARVA
NEW YORK ENT. SOC. JOURN. 3:178, 1895 PACKARD LARVA
LOC. CIT. 8:41,1900 DYAR LARVA
ABBOT & SMITH-LEP. INS. GA. 2:127,1797 LARVA*PL. 64 COLOR
EMMONS-NAT. HIST. NEW YORK 5: 228,1854 LARVA
GLOVER-ILLUST. NORTH AMER. ENT. 1878 LARVA*PL. 8, F. 6
INDEX AMER. ECON. ENT. VOL. 1:20
FOOD PLANTS
LUPINUS SP.
ZEA MAYS LINN. (CORN)
GENERAL FEEDER LOW PLANTS

PHYLLOPHORA(GROTE), HEPTAGROTIS-NOCTUIDAE
LIFE HISTORY
BKLYN. ENT. SOC. BULL. 27:85, 1932 CRUMB LARVA
ENT. SOC. AMER. ANNALS23:396, 1930 HAWKINS LARVAL CLAWS
MAINE AGRI. EXP. STA. BULL. 356:160,1930 PHIPPS LARVA& F. 23 D; PUPA
INDEX AMER. ECON. ENT. VOL. 5:297
FOOD PLANTS
ALNUS SP. (ALDERS)
BETULS SP. (BIRCHES)
PRUNUS SEROTINA EHRH.

SALIX SP. (WILLOWS)
SPIREAE SP.
VACCINIUM SP.
VIBURNUM SP.

PHYTOLACCAE ABBOT & SMITH
SYN. SEE ERIDANIA CRAMER

PIASUS (BOISDUVAL), PHAEDROTES-LYCAENIDAE
FOR THE MANY SSP.S&SYNS. OF THIS SPECIES SEE MCDUNNOUGH'S LIST
LIFE HISTORY
CAN. ENT. 20:97, 1888 WRIGHT EGG
LOC. CIT. 40:347, 1908 COOLIDGE LARVA, CHRYSALIS
ENT. NEWS 34:297, 1923 COOLIDGE LIFE HISTORY
COMSTOCK, J. A. BUTTER CA. P. 196 1927 EGG* A64
FOOD PLANTS
ADENOSTOMA FASCICULATUM H. & A.
AESCULUS CALIFORNICA SPACH
AESCULUS PAVIA LINN.
LUPINUS CHAMISSONIS ESCHSCH.
LUPINUS DENSIFLORUS BENTH.

PIATRIX GROTE, CATOCALA-NOCTUIDAE
A DIONYZA HENRY EDWARDS
LIFE HISTORY
AMER. MUS. NAT. HIST. BULL. 16: 389, 1902 BEUTEN. LARVAL HEAD* F. 13
LOC. CIT. 38:148, 1918 BARNES & MCD. EGG, LARVA
AMER. MUS. NAT. HIST. MEM. NEW SERIES 3, PART 1, 1918 BARNES & MCD. EGG, LARVA* PL. 10, F. 27; PL. 11, F. 6; PL. 13, F. 5; PL. 15, F. 8 COLOR
CAN. ENT. 33:299, 1901 DODGE ROWLEY LARVA
NEW YORK STATE MUS. MEM. 8: 719 1905 FELT LARVA
INDEX AMER. ECON. ENT. VOL. 1:67; VOL. 3:74; VOL. 7:171
FOOD PLANTS
CARYA SP. (HICKORIES)
CARYA ILLINOENSIS (WANG.)
DIOSPYROS VIRGINIANA LINN.
JUGLANS CINEREA LINN.
JUGLANS NIGRA LINN.
JUGLANS REGIA LINN.
SALIX SP. (WILLOWS)

PIAZZATA CASSINO & SWETT
SYN. SEE SCOLOPACINARIUS GUENEE

PICA WALKER
SYN. SEE HERA HARRIS

PICINA GROTE
SSP. SEE FORTIS GROTE

PICTA (EDWARDS), PHYCIODES-NYMPHALIDAE
SYN. CANACE EDWARDS
AB JEMEZENSIS BREHME
LIFE HISTORY
CAN. ENT. 16:163, 1884 EDWARDS EGG, LARVA, CHRYSALIS
HOLLAND-BUTTERFLY BOOK. P. 156, 1905 LIFE HISTORY
LOC. CIT. 3RD EDIT. P. 139, 1931 LIFE HISTORY
FOOD PLANTS
ASTER SP.
MEDICAGO SATIVA LINN.

PICTA FELDER
SYN. SEE ASTARTE DOUBLEDAY

PICTA (HARRIS), CARAMICA-NOCTUIDAE
SYN. EXUSTA GUENEE
SYN. CONTRARIA WALKER
LIFE HISTORY
AMER. MUS. NAT. HIST. BULL. 16: 294, 1901 BEUTEN. LARVA
AMER. NAT. 18:1266, 1884 PACKARD LARVAL STAGES
ILL. AGRI. EXP. STA. BULL. 60:502 1900 FORBES LARVA*
ILL. DEPT. AGRI. TRANS. 14:60, 1876 THOMAS LARVA
LOC. CIT. 15:226, 1877 FRENCH LARVA
LOC. CIT. 17, APPEND: 51, 1870 THOMAS LARVA

Loc. cit. 18, Append: 185, 1880 Coquillett larva*
Kentucky Agri. Exp. Sta. Bull. 227:69, 1920 Garman larval structure
Maine Agri. Exp. Sta. Bull. 356: 174, 1930 Phipps larva* f. 24A
Minn. Agri. Exp. Sta. Bull. 61:126 1899 Lugger larva* f. 161
Loc. cit. 84, 6, 1903 Washburn larva* f. 3 COLOR
Loc. cit. 123:67, 1911 Washburn larva* f. 10 COLOR
Mo. Rept. Nox. Ins. of 2:112, 1870 Riley life history*
Mus. Comp. Zool. Bull. 87, No. 6, 1941 Dethier larval antenna* f. 82
New York State Cabin. Nat. Hist. Ann. Rept. no. 26:137, 1874 Lintner larva
Loc. cit. No. 42:207, 1889 Lintner larva* f. 18
New York State Mus. Bull. 6:21, 1888 Lintner larva*
Loc. cit. 37:29, 1900 Felt larva*
Nova Scotia Ent. Soc. Proc. 3:44 1917 Payne life hist. egg* larva* pupa* Pl. 10
Ontario Ent. Soc. Ann. Rept. p. 86 1871 larva* pupa
Loc. cit. p. 61, 1886 Bethune larva*
Loc. cit. 8: 1891 Bethune larva*
Loc. cit. p. 64, 1896 Bethune larva*
Loc. cit. p. 37, 1902 Young larva*
Ontario Ent. Soc. Ann. Rept. p. 10 1905 Williams larva*
Loc. cit. p. 112, 1905 Gibson larva
Loc. cit. p. 16, 1906 Young larva
Loc. cit. p. 13, 1912 Gibson larva*
Loc. cit. p. 39, 1917 Caesar larva*
Loc. cit. p. 97, 1917 Brittain egg, larva* pupa
Practical Ent. 2:21, 1866 larva
Southern Cal. Acad. Sci. Bull. 41:172, 1942 Comstock-Dammers egg, larva* Pl. 33: pupa* Pl. 33
United States Dept. Agri. Bur. Ent. Bull. 43:42, 1903 new series Chittenden larva* f. 39, c, d, e; f. 40: pupa* f. 39e
United States Dept. Agri. Rept. Ent. p. 124, 1883 life hist. Riley Pl. 1, f. 3; Pl. 12, f. 2
Cragen-Our Ins. Friends & Foes p. 128, 1899 larva* f. 63
Crosby & Leonard-Manual Veg. Gard. Ins. p. 21, 1918 egg, larva* f. 17; pupa
Doane etc-Forest Ins. p. 284, 1936 larva (brief)
Essig-Ins. West North Amer. p. 685, 1926 larva
Fernald-In Kingsley's Stand Nat. Hist. 2:452, 1885 larva
Glover-Illust. North Amer. Ent. 1878 larva* Pl. 4, f. 13; Pl. 10, f. 14; Pl. 18, f. 8
Harris-Ins. Inj. Veg. 1st edit. p. 329, 1841 larva
Harris-loc. cit. 2nd edit. p. 350 1852 larva
Harris-loc. cit. Flint edit. p. 451 1862 larva* pupa
Hart-Ins. Coll. Ill. High Schools p. 33, 1903 larva* f. 46
Holland-Moth Book p. 194, 1905 larva* f. 107; Pl. 1, f. 11 COLOR
O'Kane-Inj. Ins. p. 171, 1912 larva* f. 194
Packard-Inj. Ins. West. Hayden's Rept. p. 753, 1877 larva*
Peairs-Pests Farm. Gard. Orchard p. 251, 1941 Larva
Smith-Ins. New Jersey p. 418, 1900 larva* f. 194
Stauffer-Agri. of Pa. p. 239, 1878 larva*
Index Amer. Econ. Ent. Vol. 1:198 Mamestra Vol. 2:73; Vol. 4:91; Vol. 5:119; Vol. 6;136 Vol. 7:15

FOOD PLANTS

ALLIUM CEPA LINN. (ONION)
APIUM GRAVEOLENS LINN. (CELERY)
APOCYNUM CANABINUM LINN.
ARCTIUM LAPPA LINN.
ASPARAGUS OFFICINALIS LINN.
ASTER SP.
AVENA SATIVA LINN. (OATS)
BETA VULGARIS LINN. (BEET)
BRASSICA NAPOBRASSICA MILL. (RUTA BAGA)
BRASSICA OLERACEA LINN. (CABBAGE)
BRASSICA OLERACEA V. BOTYRITS D.C. (CAULIFLOWER)
BRASSICA RAPA LINN. (TURNIP)
CHENOPODIUM ALBUM LINN.
CITRUS SINENSIS(LINN.)(ORANGE)
DAUCUS CAROTA LINN.
DAUCUS CAROTA V. SATIVUS HOFF. (CORROT)
DYSSODIA PAPPOSA(VENT.)
FAGOPYRUM SAGITTATUM GILIB.
FRAGARIA CHILOENSIS(LINN.)
GLADIOLUS SP.
HUMULUS LUPULUS LINN. (HOPS)
HYDRANGEA SP.
HYMENOCALLIS OCCIDENTALIS (LECONTE)
LATHYRUS JAPONICUS V. GLABER(SER.)
LATHYRUS ODORATUS LINN.
LONICERA SP.
MALUS PUMILA MILL. (APPLED)
MEDICAGO SATIVA LINN. (ALFALFA)
PARTHENIUM ARGENTATUM GRAY
PHASEOLUS VULGARIS LINN. (BEANS)
PHLEUM PRATENSE LINN.
PICEA PUNGENS ENGELM.
PICEA SITCHENSIS(BONG.)
PISUM SATIVUM LINN. (PEAS)
PLANTAGO SP.
POLYGONUM HYDROPIPER LINN.
PRUNUS DOMESTICA LINN. (PLUM)
PRUNUS SEROTINA EHRH.
PSEUDOTSUGA MENZIESII(MIRB.)
RAPHANUS SATIVA LINN. (RADISH)
RESEDA ODORATA LINN.
RIBES SATIVUM SYME(CURRANT)
RUBUS IDAEUS V. STRIGOSUS(MICHX.) (RASPBERRY)
SALIX SP. (WILLOWS)
SAMBUCUS CANADENSIS LINN.
SOLANUM TUBEROSUM LINN. (POTATO)
SPINACIA OLERACEA MILL. (SPINACH)
SYMPHORICARPOS ALBUS(LINN.)
TAMARIX CHINENSIS LOUR.
TRIFOLIUM SP. (CLOVERS)
VACCINIUM CORYMBUSUM LINN.
VACCINUM MACROCARPON AIT. (CRANBERRY)
ZEA MAYS LINN. (CORN)

PICTA PACKARD, ARACHNIS-ARCTIIDAE
A MAIA OTTOLENGUI
B CITRA NEUMOEGEN & DYAR
C VERNA BARNES & MCDUNNOUGH
D HAMPSONI DYAR
E INSULARIS CLARK
F. MEADOWSI COMSTOCK

LIFE HISTORY
ENT. AMER. 6:73, 1890 DYAR EGG, LARVA, PUPA
PAN PACIFIC ENT. 24:31, 1948 TILDEN LARVA, LARVAL HABITS
SOUTHERN CAL. ACAD. SCI. BULL. 39:189, 1940 COMSTOCK HENNE LARVAE OF PICTA V. INSULARIS AND PICTA MAIA COMPARED
ESSIG-INS. WEST. NORTH AMER. P. 678, 1926 LARVA
STRETCH-ZYGAEN. &BOMBYC. NORTH AMER. P. 84, 1873 LARVA*PL. 10, F. 3 COLOR

FOOD PLANTS
ARTEMISIA SP.
EUCALYPTUS SP.
GERANIUM SP.
LUPINUS CHAMISSONIS ESCH.
MALVA SP.
MEDICAGO SATIVA LINN..
PELARGONIUM SP.
PLANTAGO SP.
ROSA SP.
TRIFOLIUM SP. (CLOVERS)

PIGRATA WALKER
SYN. SEE OBSTIPATA FABRICIUS

PILOSARIA(PACKARD), METARRANTHIS-GEOMETRIDAE
LIFE HISTORY
NEW YORK ENT. SOC. JOURN. 51:146 1943 RUPERT LARVA
NEW YORK STATE AGRI. EXP. STA. (CORNELL), MEM. 274:94, 1948 FORBES LARVA
FOOD PLANTS
BETULA PAPYRIFERA MARSH.

PILUMNUS BOISDUVAL, PAPILIO-PAPILIONIDAE
LIFE HISTORY
PAPILIO 4:100, 1884 SCHAUS, JR. LARVA, CHRYSALIS
FOOD PLANTS
KALMIA SP. (LAURELS)

PIMA EDWARDS, ANTOCHARIS-PIERIDAE
LIFE HISTORY
PSYCHE 15:81, 1908 COOLIDGE CHRYDALIS
COMSTOCK, J. A. BUTTER CAL. P. 42 1927 CHRYSALIS
FOOD PLANTS
ARABIS GLABRA(LINN.)(?)

PIMACOATA CASSINO
SSP. SEE OSSULARIA HUBNER

PINASTRI LINNAEUS, SPHINX-SPHINGIDAE
SYN. ASIATICUS BUTLER
SYN. SANIPTRI STRECKER
LIFE HISTORY
AMER. ENT. SOC. TRANS. 15:187, 1888 SMITH LARVA
NOVITATES ZOOL. (TRING) 9 SUPPL: 146, 1903 ROTHSCHILD & JORDAN LARVA, PUPA
STRECKER-LEP. RHOP. ET HETER. 13;118, 1876 LARVA OF THIS SP. (?) SANIPTRI
FOOD PLANTS

PINDAR SMITH
SYN. SEE MESSORIA HARRIS

PINEUM LINTNER
SSP. SEE BOMBYCOIDES WALKER

PINI DYAR
SYN. SEE PLAGIATA WALKER (OLENE)

PINI MICHNER
SSP. SEE IMPERIALIS DRURY

PINI RETZIUS
SYN. SEE TOGATA HUBNER

PINIARIA PACKARD
SSP. SEE CANOSARIA WALKER

PINIATA(PACKARD), CARIPETA-GEOMETRIDAE
SYN. SEDUCTARIA STRECKER
LIFE HISTORY
NEW YORK STATE AGRI. EXP. STA. (CORNELL), MEM. 274:105, 1948 FORBES LARVA LIKE THAT OF ANGUSTIORATA WALKER
FOOD PLANTS
PINUS SP. (PINES)

PINIATA PACKARD, PARAPHIA-GEOMETRIDAE
SYN. GUTTATA HULST
LIFE HISTORY
CAN. ENT. 75:136, 1943 MCGUFFIN LIFE HIST. LARVA* PL. 12, F. B; TEXT FIG. LARVAL MANDIBLE
INDEX AMER. ECON. ENT. VOL. 7:742
FOOD PLANTS
ABIES BALSAMEA(LINN.)
BETULA PAPYRIFERA MARSH.
CHAMECYPARIS THYOIDES(LINN.)
LARIX LARICINA(DUROI)
PICEA ENGELMANNI (PARRY)
PICEA GLAUCA(MOENCH)
PICEA MARIANA(MILL.)
PINUS SP.
PINUS BANKSIANA LAMB.
PINUS CONTORTA V. LATIFOLIA ENGELM.
PINUS STROBUS LINN.
PINUS SYLVESTRIS LINN.
TSUGA CANADENSIS(LINN.)

PINICOLA DYAR

SSP. SEE PLAGIATA WALKER(OLENE)

PISTACIARIA(PACKARD), NEMORIA-GEOMETRIDAE
SYN. UNISTRIGATA GUMPPENBERG
LIFE HISTORY
NEW YORK STATE AGRI. EXP. STA. (CORNELL)MEM. 274:116, 1948 FORBES LARVA
SOUTHERN CAL. ACAD. SCI. BULL. 39: 78, 1940 COMSTOCK EGG*PL. 9; LARVA*PL. 10; PUPA*PL. 11. THIS REFERENCE IS AN ERROR IN IDENTIFICATION. THE SPECIES IS REALLY PUNCTULARIA BARNES & MCDUNNOUGH(SEE SOUTHERN CAL. ACAD. SCI. BULL. 44:20, 1945)
FOOD PLANTS
QUERCUS SP.

PITYCHOROUS GROTE
SYN. SEE DETERSA WALKER

PITYOCHROMUS GROTE, PLAGIMIMICUS-NOCTUIDAE
SYN. MEDIA MORRISON
LIFE HISTORY
FOOD PLANTS
AMBROSIA TRIFIDA LINN.

PLACENTIA(ABBOT & SMITH), APANTESIS-ARCTIIDAE
FORM FLAMMEA NEUMOEGEN
LIFE HISTORY
CAN. ENT. 43:259, 1911 BARNES & MCD. LARVA
LOC. CIT. 44:135, 1912 BARNES & MCD. EGG, LARVA
ABBOT & SMITH-LEP. INS. GA. 2:128 1797 LARVA*PL. 65 COLOR
FOOD PLANTS
LUPINUS SP.

PLACIDA DYAR
SSP. SEE CINEREA WALKER

PLACIDA(GROTE), ABAGROTIS-NOCTUIDAE
FORM MINIMALIS GROTE
LIFE HISTORY
BKLYN. ENT. SOC. BULL. 27:89, 1932 CRUMB LARVA
MAINE AGRI. EXP. STA. BULL. 356: 168, 1930 PHIPPS LARVA
INDEX AMER. ECON. ENT. VOL. 3:357 RHYNCHAGROTIS
FOOD PLANTS
CRATAEGUS SP. (THORNS)
MALUS PUMILA MILL. (APPLE)
POPULUS SP. (POPLARS)
POPULUS FREMONTI S. WATS.
SARCOBATUS VERMICULATUS(HOOK)
VACCINIM ANGUSTIFOLIUM AIT.

PLACIDATA TAYLOR, EUPITHECIA-GEOMETRIDAE
LIFE HISTORY
FOOD PLANTS
LIBOCEDRUS DECURRENS TORR.

PLACIDATA WALKER
SYN. SEE MAGNOLIATA GUENEE

PLAGIATA AUCT.
SEE VAGANS BARNES & MCDUNNOUGH

PLAGIATA WALKER
SYN. SEE BICARNEA GUENEE

PLAGIATA(WALKER), HEMEROCAMPA-LIPERIDAE
SYN. BEFINITA PACKARD
LIFE HISTORY
AMER. MUS. NAT. HIST. BULL. 10: 382, 1898 BEUTEN. LARVA DEFINITA
INS. LIFE 3:390, 1891 DYAR EGG, LARVA COMPARES EGGS OF THIS SPECIES WITH THOSE OF LEUCOSTIGMA ABBOT & SMITH AND ANTIQUA LINN.
NEW YORK ACAD SCI. ANNALS 8:65 1893 PACKARD YOUNG LARVA
NEW YORK STATE AGRI. EXP. STA. (CORNELL), MEM. 274:246, 1948 FORBES LARVA
NEW YORK STATE MUS. MEM. 8:522 1905 FELT LARVA DEFINITA
NEW YORK STATE, REPT. OF ENTOM.

4:50,1888 LINTNER LARVA
PSYCHE 6:111,145,1891 DYAR EGG, LARVA,PUPA,COCOON
PSYCHE 7:341,1896 DYAR KEY TO LARVAE
UNITED STATES DEPT.AGRI.MISC. PUB.657:356,407,1950 CRAIGHEAD LARVA
INDEX AMER.ECON.ENT. VOL.1:153

FOOD PLANTS
ACER SP.(MAPLES)
HAMAMELIS VIRGINIANA LINN.
PINUS SP.(PINES)
QUERCUS SP.(OAKS)
SALIX SP.(WILLOWS)
TILIA SP.(LINDENS)
FOREST TREES,FRUIT TREES, SHRUBS

PLAGIATA(WALKER),OLENE-LIPARIDAE
SYN.PINI DYAR
A PINICOLA DYAR

LIFE HISTORY
AMER.MUS.NAT.HIDT.BULL.10: 386,1898 BEUTEN.LARVA
NEW YORK STATE AGRI.EXP.STA. (CORNELL),MEM.274:244,1948 FORBES LARVA PINI
WASH.ENT.SOC.PROC.13:19,1911 DYAR LARVA PINI
INDEX AMER.ECON.ENT. VOL.2:198 LASIOCAMPA PINI VOL.7:721

FOOD PLANTS
ABIES SP.(FIRS)
CHAMAECYPARIS THYOIDES(LINN.)
JUNIPERUS SP.
LARIX LARICINA(DUROI)
PICEA SP.
PINUS SP.(PINES)
PINUS BANKSIANA LAMB.
PINUS RESINOSA AIT.
PINUS STROBUS LINN.
PRUNUS SEROTINA EHRH.
QUERCUS SP.(OAKS)
TAXODIUM DISTICHUM(LINN.)
THUJA SP.(CEDARS)
TILIA SP.(LINDENS)
TSUGA CANADENSIS(LINN.)
ULMUS SP.(ELMS)
CONIGERS

PLAGIFASCIATA WALKER
SYN.SEE PULVERARIA LINNAEUS

PLANATA TAYLOR
SYN.SEE IDUATA GUENEE

PLANTAGINIS(LINNAEUS),NEMEOPHILA-ARCTIIDAE
SYN.PETROSA WALKER
SYN.CAESPITIS GROTE & ROBINSON
SYN.CHICHORII GROTE & ROBINSON
FORM HOSPITA DENNIS & SCHIFFERMULLER
FORM MODESTA PACKARD
SYN.ALASCENSIS STRETCH
FORM GEOMETRICA GROTE
FORM GEDDESI NEUMOEGEN
FORM SCUDDERI PACKARD
SYN.SELWYNII HENRY EDWARDS

LIFE HISTORY
CAN.ENT.25:248,1893 LYMAN EGG,LARVA,PUPA SCUDDERI
ENT.AMER.5:112,1889 BRUCE, EGG,LARVA

FOOD PLANTS
ASTER SP.
PLANTAGO SP.

PLAUTUS SCUDDER & BURGESS
SYN.SEE JUVENALIS FABRICIUS

PLEBEIS(FABRICIUS),ATREIDES-SPHINGIDAE

LIFE HISTORY
AMER.MUS.NAT.HIST.BULL.7:305 1895 BEUTEN.LARVA,PUPA
LOC.CIT.8:292,1896 BEUTEN.EGG, LARVA,PUPA
ENT.SOC.AMER.ANNALS11:418, 1918 MOSHER PUPA
NEW YORK STATE AGRI.EXP.STA (CORNELL),MEM.274:188,1948 FORBES LARVA
NEW YORK STATE MUS.MEM.8:741 1905 FELT LARVQ

PSYCHE 2:78,1877 SCUDDER LARVA
PSYCHE 6:437,1893 SCUDDER PUPA
BOISDUVAL-SPEC.GEN.LEP.SPHING.
P.100,1874 LARVA*PL.1
COLOR;PUPA
FERNALD-SPHINGIDAE NEW ENGLAND
P.53,1886 LARVA

FOOD PLANTS
ANISOSTICHUS SP.
ANISOSTICHUS CARREOLATAUS (LINN.)
CAMPSIS RADICANS(LINN.)
PASSIFLORA SP.
SYRINGA VULGARIS LINN.

PLEBEIA(SMITH),HEMIGRAPHIPHORA-NOCTUIDAE
A BAJOIDES BARNES & BENJAMIN
LIFE HISTORY
BKLYN.ENT.SOC.BULL.27:88,1932
CRUMB LARVA
FOOD PLANTS
CORYLUS SP.(HAZELS)
MALUS PUMILA MILL.(APPLE)
PRUNUS CERASUS LINN.(CHERRY)
RHODODENDRON MACROPHYLLUM DON
RUBUS IDAEUS V.STRIGOSUS(MICHX.)
RUBUS PARVIFLORUS NUTT.
RUBUS VITIFOLIUS C. & S.
SALIX SP.(WILLOWS)
TARAXACUM OFFICINALE WEBER

PLECTA(LINNAEUS),OCHROPLEURA-NOCTUIDAE
SYN.VICARIA WALKER
A PLECTELLA STRAND
LIFE HISTORY
AMER.MUS.NAT.HIST.BULL.4:275
1901 BEUTEN.LARVA
KIRBY-EUROPEAN BUTTER.&MOTHS.
P.194,1882 LARVA*PL.36
COLOR
STAINTON-BRITISH BUTTER.&MOTHS
1:234,1857 LARVA
STEPHENS-ILLUST.BRITISH ENT.
HAUST 2:137,1829 LARVA
INDEX AMER.ECON.ENT.
VOL.7:717
FOOD PLANTS
BETA VULGARIS LINN.(BEETS)

PLECTELLA STRAND
SSP.SEE PLECTA LINNAEUS

OLENILINEALIS GROTE
SYN.SEE FACTIOSALIS WALKER

PLENIPENNIS WALKER
SYN.SEE AERUGINOSA GUENEE

PLEONECTUSA GROTE,IPIMORPHA-NOCTUIDAE
SYN.AEQUILINEA SMITH
A MANITOBAE STRAND
LIFE HISTORY
UNITED STATES NAT.MUS.PROC.
25:381,1903 DYAR LARVA
LOC.CIT.27:874,1904 DYAR LARVA
FOOD PLANTS
POPULUS DELTOIDES MARSH.
POPULUS FREMONTI S.WATS.
POPULUS TREMULOIDES MICHX.

PLEURITICA(GROTE),EUXOA-NOCTUIDAE
LIFE HISTORY
FOOD PLANTS
SALSOLA KALI V.TENUFOLIA TAUSCH

PLEXIPPUS(LINNAEUS),DANAUS-DANAIDAE
SYN.MENIPPE HUBNER
SYN.ARCHIPPUS FABRICIUS
SYN.PULCHRA STRECKER
AB FUMOSUS HULST
AB AMERICANUS GUNDER
AB NIVOSUS GUNDER
LIFE HISTORY
AMER.ASSOC,ADVANCE.SCI.PROC.
28 B:455,1879 RILEY LARVA*
CHRYSALIS*
AMER.ENT.1:28,1868 RILEY LARVA*
F.20;CHRYSALIS
AMER.ENT.2:307,1870 RILEY PUPA*
AMER.MUS.NAT.HIST.BULL.5:253
1893 BEUTEN.LARVA,CHRYSALIS
AMER.MUS.JOURN.P.530,1916
LUTZ LARVA*CHRYSALIS*
ARCHIVES NATURGESCH.44:5,1878
DEWITZ LARVA*CHRYSALIS*PL.1
BUFF.SOC.NAT.SCI.BULL. 19:20
1939 WILD LARVA

Cal. Acad. Sci. Proc. 5:166, 1873 Edwards larva, chrysalis
Can. Ent. 1:74, 1869 Saunders larva, chrysalis
Loc. cit. 2:148, 1870 Saunders egg
Loc. cit. 5:4, 1873 Saunders life hist. egg*larva*chrysalis*
Loc. cit. 8:119, 1876 Edwards note on life history
Can. Nat. Sci. News 1:15, 1897 Ware note on life history
Can. Nat. &Geol. 2:351, 1857 larva*chrysalis*
Ent. News 23:407, 1912 Girault length of pupal stage
Ent. 44:377-382, 1911 Frohawk life history
Loc. cit. 45:148, 147, 1912 Aiken larva, chrysalis
Everhart Mus. Bull. (Lacka-wanna Co. Penn.), 1:7, 1915 Davis larva
Ill. Dept. Agri. Trans. 18, Append: 79, 161, 1880 Coquillett & Middleton larva*
Minnesota Agri. Exp. Sta. Bull. 61:67, 1899 Lugger egg8 larva*chrysalis* all f. 16
Missouri, Rept. Ins. of 3:144, 1871 Riley life history*
Montana Univ. Bull. 30:45, 1906 Elrod egg, larva, chrysalis
Mus. Comp. Zool. Bull. 87, No. 6, 1941 Dethier larval antenna* ff. 23, 30
Nat. Can. p. 93, 1927 Huard larva chrysalis
New York State Cabin. Nat. Hist. Ann. Rept. No. 41:255, 1888 Lintner larva*chrysalis*
New York State Mus. Handbook #13:19, 1932 Saunders(Butter. Allegheny State Park)Larva* chrysalis* ff. 4, 5
Ontario Ent. Soc. Ann. Rept. p. 57, 1898 Lockhead larva*chrysalis*
Psyche 2:169, 1878 Edwards life history
Psyche 3:261, 1881 Scudder larva, chrysalis anatomy
Psyche 5:61, 1888 Scudder life history in New England
Smithsonian Misc. Coll. 4, art. 2:38, 1862 Morris larva, chrysalis
Zeitschr. Gesament. Nat. 1878, 1879 Dewitz larva, chrysalis
Abbot & Smith-Lep. Ins. Ga. 1:11 1797 larva*chrysalis*Pl. 6 COLOR
Ballard-Ins. Lives p. 34, 1879 larva*chrysalis*
Boisduval&Leconte-Hist. Lep. North Amer. p. 137, 1833 larva*chrysalis*Pl. 40 COLOR
Comstock, J. A. Butter. Cal. p. 59 1927 egg*larva*chrysalis* Pl. A27, 80
Comstock, J. H.-How Know Butter p. 204, 1904 larva*chrysalis* Pl. I COLOR
Cragen-Our Ins. Friends & Foes p. 83, 1899 larva*f. 19
Dickerson-Moths & Butter. p. 13 1901 life hist*larva*ff. 6, 7, 8, 12-15, 83; chrysalis*ff. 16-23
Edwards-In Kingsley's Stand. Nat. Hist. 2:492, 1885 larva, chrysalis
Emmons-Nat. Hist. New York 5: 202, 1854 larva* chrysalis* Pl. 38 COLOR
Essit-Ins. West. North Amer. p. 639, 1926 egg, larva, chrysalis
Fernald, C. H. Butter. of Maine p. 35, 1884 egg*larva*chrysalis*
Fernald, H. T. Applied Ent. p. 295 1926 larva, chrysalis
Fernald&Shepard-Applied Ent. p. 290 1942 larva, chrysalis
French-ButtereEast. United States p. 144, 1886 life hist* _archippus_
Gosse-Letters from Ala. p. 185

1859 LARVA, CHRYSALIS
GUNDLACH-ENT. CUBANA P. 23, 1881
LARVA, CHRYSALIS
HARRIS-INS. INJ. VEG. FLINT EDIT.
P. 280, 1862 LARVA, CHRYSALIS
HART-INS. COLL. ILL. HIGH SCHOOLS
P. 12, 1903 LARVA* F. 12
KELLOGG-AMER. INS. P. 452, 1908
LARVA* F. 791
LUTZ-A LOT OF INS. P. 9, 1941 EGG*
LARVA* CHRYSALIS*
MAYNARD-BUTTER. NEW ENGLAND P. 8
1886 LARVA
MORRIS-SYNOP. LEP. NORTH AMER.
P. 38, 1862 LARVA, CHRYSALIS
PEALE0LEP. AMER. 1833 LARVA* PL.
7 COLOR
SAY-AMER. ENT. 3, 1828 LARVA*
CHRYSALIS* PL. 54
SCUDDER-BUTTER EAST. UNITED
STATES & CANADA PP. 724-726
737-744, 1889 LIFE HIST* PL. 64
F. 1; PL. 76, F. 4; PL. 62, F. 6; PL.
70, F. 3; PL. 74, F. 5; PL. 78, FF.
1-5; PL. 86, FF. 1, 37; PL. 78,
FF. 10, 16; PL. 62, F. 5; PL. 83,
FF. 1-3, PL. 83, F. 17; PL. 87,
F. 20
SCUDDER-BUTTER. THEIR STRUCTURE
PP. 10, 19, 44, 1881 EGG* LARVA*
CHRYSALIS* FF. 17, 18, 22, 61
SCUDDER-EVERYDAY BUTTER. P. 96
1899 EGG, LARVA, CHRYSALIS
SCUDDER-GUIDE TO COMMON BUTTER
P. 63, 1893 LARVA, CHRYSALIS
VANCE-LIFE AMONT BUTTER. P. 39
1925 EGG, LARVA, CHRYSALIS
WEED-LIFE HIST. AMER. INS. 1897
EGG, LARVA* FF. 51, 52;
CHRYSALIS* F. 53
WESTWOOD-INTROD. MOD. CLASS. INS.
2:352, 1840 LARVA* CHRYSALIS*
INDEX AMER. ECON. ENT.
VOL. 1:14; ANOSIA; VOL. 3:114;
VOL. 4:132; VOL. 5:174;
VOL. 6:197; VOL. 7:245

FOOD PLANTS

ACERATES SP.
APOCYNUM ANDROSAEMIFOLIUM LINN.
ASCLEPIAS AMPLEXICAULIS SM.
ASCLEPIAS CORDIFOLIA (BENTH.)
ASCLEPIAS CURASSAVICA LINN.
ASCLEPIAS MEXICANA CA.
ASCLEPIAS NIVEA LINN.
ASLEPIAS PURPURASCENS LINN.
ASCLEPIAS SPECIOSA TORR.
ASCLEPIAS SYRIACA LINN.
ASCLEPIAS TOMENTOSA ELL.
ASCLEPIAS TUBEROSA LINN.

PLICADUTA NAKAHARA
SSP. SEE PHILODICE GODART

PLICATUS GROTE, UFEUS-NOCTUIDAE
SYN. BAROMETRICUS GOSSENENS
LIFE HISTORY
INDEX AMER. ECON. ENT.
VOL. 1:317
FOOD PLANTS
POPULUS DELOIDES MARSH.

PLOTA STRECKER
SYN. SEE CANADENSIS BOISDUVAL

PLUMBARIA HULST
SYN. SEE MISERULATA GROTE (EUPITHECIA)

PLUMBEA HENRY EDWARDS
SSP. SEE FUCOSA HUBNER

PLUMOGERARIA (HULST), CONIODES-GEOMETRIDAE
FORM IMMACULARIA CASSINO & SWETT
LIFE HISTORY
JOURN. ECON. ENT. 17:418, 1924
FLANDERS EGG, LARVA
NEW YORK ENT. SOC. JOURN. 53:184
1945 FORBES LARVA
PSYCHE 10:168, 1903 DYAR EGG,
LARVA, PUPA
UNITED STATES DEPT. AGRI. BUR.
ENT. BULL. 30 OLD SERIES P.
28, 1893 COQUILLETT LARVA,
PUPA
LOC. CIT. BULL. 7 NEW SERIES P. 64
1897 COQUILLETT LIFE HIST.
EGG, LARVA* F. 40; PUPA

ESSIG-INS. WEST. NORTH AMER. P. 700, 1926 LARVA
HERRICK-INS. ENEMIES SHADE TREES P. 303, 1935 LARVA
HOLLAND-MOTH BOOK. P. 346, 1905 LARVA*F. 206
QUAYLE-INS. OF CITRUS &SUB. TROP. FRUITS P. 387, 1938 EGG, LARVA

FOOD PLANTS
JUGLANS NIGRA LINN.
JUGLANS REGIA LINN.
MALUS PUMILA MILL. (APPLE)
PRUNUS DOMESTICUS V. GALATENSIS EX HOOK(PRUNE)
QUERCUS GARRYANA DOUGL.
QUERCUS VIRGINANA MILL.

PLUMOSA HENRY EDWARDS
SYN. LUNATA HENRY EDWARDS

PLUSIAEFORMIA, HENRY EDWARDS, STRETCHIA-NOCTUIDAE
A COLORADICOLA STRAND
LIFE HISTORY
JOURN. ECON. ENT. 30:900, 1937 HOERNER EGG*F. 2A; LARVA* F. 1; PUPA*F. 2B
SOUTHERN CAL. ACAD. SCI. BULL. 38: 183, 1939 COMSTOCK DAMMERS LARVA*PL. 47
UNITED STATES NAT. MUS. PROC. 25: 377, 1903 DYAR EGG, LARVA
INDEX AMER. ECON. ENT. VOL. 5:627; VOL. 6:744

1 FOOD PLANTS
RIBES SP.
RIBES AMERICANUM MILL.
RIBES LACUSTRE(PERS.)

PLUTO(FABRICIUS), XYLOPHANES-SPHINGIDAE
SYN. CROESUS DALMAN
SYN. THORATES HUBNER
LIFE HISTORY
NOVITATES ZOOL. (TRING.)9 SUPPL. 682, 1903 ROTHSCHILD & JORDAN LARVA, PUPA
DEWITZ-BESCHREIBUNGER VON JUNGENSTADIEN EXOTISCHER LEPIDOPTERAN, HALLE P. 249 1882 LARVA*F. 9 COLOR
FOOD PLANTS

PLUVIALIS DYAR
SSP. SEE SCOLOPENDRINA BOISDUVAL

PLUVIALIS(DYAR), MALACOSOMA-LASIOCAMPIDAE
A RECENSED DYAR
LIFE HISTORY
CAN. ENT. 25:42, 1893 DYAR LARVA
LOC. CIT. 74:204, 1943 ATWOOD LARVA*
NEW YORK ACAD. SCI. ANNALS 8:229 1894 DYAR LARVAL SETAE
PSYCHE 7:259, 1894 DYAR EGG, LARVA, PUPA, COCOON
UNITED STATES NAT. MUS. PROC. 27:887, 1904 DYAR LARVA
DOANE ETC.-FOREST INS. P. 281 1936 LARVA
ESSIG-INJ. &BENEF. INS. P. 411, 1915 LARVA
ESSIG-INS. OF WEST. NORTH AMER. P. 697, 1926 LARVA
HERRICK-INS. ENEMIES SHADE TREES' P. 34, 1935 LARVA
FOOD PLANTS
ALNUS SP.
ALNUS RUBRA BONG.
BETULA SP. (BIRCHES)
CRATAEGUS DOUGLASII LINDL.
MALUS DIVERSIFOLIA(BONG.)
MALUS PUMILA MILL. (APPLE)
POPULUS NIGRA V. ITALICA MUENCH.
POPULUS TREMULOIDES MICHX.
PRUNUS CERASUS LINN. (CHERRY)
PRUNUS EMARGINATA(DOUGL.)
PRUNUS VIRGINIANA LINN.
QUERCUS COCCINEA MUENCH.
QUERCUS GARRYANA DOUGL.
RIBES DIVARICATUM DOUGL.
RIBES LACUSTRE(PERS.)
RIBES SATIVUM SYME(CURRANT)
ROBINIA PSEUDO-ACACIA LINN.
ROSA CANINA LINN.
ROSA GYMOCARPA NUTT. EX T. & G.
ROSA NUTKANA PRESL.
SALIS SP. (WILLOWS)

SORBUS AUCUPARIA LINN.

PLUVIATA(GUENEE),HYDRIOMENA-GEOMETRIDAE
SYN.(?) TRIFASCIATA BORKHAUSEN
SYN.TRANSFIGURATA SWETT
LIFE HISTORY
UNITED STATES NAT.MUS.PROC. 25:389,1890 LARVA,PUPA
FOOD PLANTS
ALNUS SP.(ALDERS)

PLUVIOSA WALKER
SYN.SEE ARCTICA BOISDUVAL

POCOHONTAS SCUDDER
SYN.SEE J-ALBUM BOISDUVAL & LECONTE

POCAHONTAS SCUDDER
SSP.SEE HOBOMOK HARRIS

PODARCE FELDER & FELDER
SSP.SEE AQUILO BOISDUVAL

POEAS HUBNER
SYN.SEE CECROPS FABRICIUS

POECILA STEPHENS
SYN.SEE GORDIUS CRAMER

POEYI BUTLER
SSP.SEE ILAIRE GODART

POLA BOISDUVAL,MELITAEA-NYMPHALIDAE
A ARACHNE EDWARDS
AB POLINGI GUNDER
AB GUNDERIAE HOLLAND
B MONACHE COMSTOCK
LIFE HISTORY
SOUTHERN CAL.ACAD.SCI.BULL. 31:8,1932 SPERRY MATURE LARVA
FOOD PLANTS
PENSTEMON ACUMINATUS DOUGL.

POLARIS HULST
SSP.SEE SABINI KIRBY

POLINGI GUNDER
SSP.SEE POLA BOISDUVAL

POLINGI SKINNER,MEGATHYMUS-MEGTHYMIDAE
FORM MARIAE BARNES & BENJAMIN
LIFE HISTORY
CARNEGIE MUS.ANNALS20:264, 1931 BONNIWELL LIFE HIST. MARIAE
FOOD PLANTS
AGAVE LECHEGUILLA TORR.

POLIOS COOK & WATSON,INCISALIS-LYCAENIDAE
AB DAVISI WATSON & COMSTOCK
LIFE HISTORY
CAN.ENT.40:37,1908 COOK EGG*
FOOD PLANTS
ARCTOSTAPHYLOS UVA-URSI(LINN.)

POLIXENES(FABRICIUS),OENEIS-SATYRIDAE
SYN.CRAMBIS FREYER
A SUBHYALINA CURTIN
B KATAHDIN NEWCOMB
LIFE HISTORY
EDWARDS-BUTTER.NORTH AMER. 3RD SERIES 1897 EGG*PL. CHIONOBAS 7,F.A CRAMBIS
FOOD PLANTS

POLUS EHRMANN
SYN.SEE SMINTHEUS DOUBLEDAY & HEWITSON

POLYDAMAS LINNAEUS,PAPILIO PAPILIONADAE
A LUCAYUS ROTHSCHILD & JORDAN
LIFE HISTORY
ARCHIVES NAT.44:2,1878 DEWITZ LARVA*F.1
JAMAICA INS.JOURN.1:27,1891 COCKERELL ARVA,CHRYSALIS V POLYCRATEZ
SMITHSONIAN MISC.COLL.4,ART. 2:13,1862 MORRIS LARVA
SOUTHERN CAL.ACAD.SCI.BULL. 34:76,1935 COMSTOCK-GRIMSHAW LIFE HIST.EGG*PL. 3;LARVA*PL.4,5; CHRYSALIS* PL.6

WASH. ENT. SOC. PROC. 13:227, 1911
DYAR LARVA
ZEITSCHRIFT GESAMENT. NATUR. 1878
DEWITZ LARVA, CHRYSALIS
LOC. CIT. 1879, DEWITZ LARVA, CHRYSALIS
BOISDUVAL-SPEC. GEN. LEP. DIURNAE 1:321, 1836 LARVA
COMSTOCK, J. A.-BUTTER. CAL. P. 18 1927 LARVA, CHRYSALIS
GUNDLACH-ENT. CUBANA P. 121, 1881 LARVA, CHRYSALIS
JAEGER-LIFE NORTH AMER. INS. P. 210, 1864 LARVA
MORRIS-SYNOP. LEP. NORTH AMER. P. 13, 1862 LARVA
INDEX AMER. ECON. ENT. VOL. 3:295; VOL. 7:741

FOOD PLANTS
ARISTOLOCHIA DURIOR HILL
ARISTOLOCHIS GRANDIFLORA SWARTZ
ARISTOLOCHIA PENTANDRA JACQ.
ARISTOLOCHIA RINGENS VAHL.

POLYGAMA GUENEE
SYN. SEE GRYNEA CRAMER

POLYGRAMMARIA (PACKARD), STENOPORPIA-GEOMETRIDAE
LIFE HISTORY
FOOD PLANTS
GALIUM SP.

POLYGRAMMATA HULST
SSP. SEE CUSTODIATA GUENEE

POLYMNIAE BIRD, PAPAIPEMA-NOCTUIDAE
LIFE HISTORY
CAN. ENT. 49:123, 1917 BIRD LARVA* PL. 8, FF. 3, 4, 5; PUPA
FOOD PLANTS
POLYMIA UVEDALIA LINN.

POLYOCHROA HAMPSON
SYN. SEE MURINA STRETCH

POLYOMMATA TEPPER
SYN. SEE CALLETA WESTWOOD

POLYPHEME HUBNER
SYN. SEE POLYPHEMUS CRAMER

POLYPHEMUS BOISDUVAL
SSE SEE XERCES BOISDUVAL

POLYPHEMUS (CRAMER), TELEA-SATURNIIDAE
SYN. POLYPHEME HUBNER
SYN. RUBRA DEAN
AB FLAVA GROTE
AB FUMOSUS WURSTER
A OCULEA NEUMOEGEN
SYN. AURELIA DRURY
B OLIVACEA COCKERELL
C BRUNNEA SAGED
LIFE HISTORY
AMER. ACAD. ARTS. SCI. PROX. 28: 80, 1893 PACKARD, EGG, LARVA
AMER. ENT. 1:121, 1869 LARVA
LOC. CIT. 1:252, 1869 LARVA (BRIEF)
AMER. MUS. NAT. HIST. BULL. 10: 436, 1898 BEUTEN. LARVA
AMER. NAT. 1:30, 85, 145, 1867 TROUVELOT LIFE HIST. LARVA* PUPA* COCOON*
CAN. ENT. 6:86, 1874 GENTRY LARVA NORMAL AND VARIATIONS
LOC. CIT. 7:63, 1875 BUNKER COCOON
LOC. CIT. 14:41, 1882 SAUNDERS EGG, LARVA* PUPA* COCOON*
LOC. CIT. 28:107, 1896 HARRINGTON EGG
CAN. NAT. P. 309, 1840, GOSSE LARVA*
ENT. NEWS 9:150, 1898 SOULE DURATION OF EGG STAGE
LOC. CIT. 19:13, 141, 1908 COOLIDGE & RUSSEL ATTACHED COCOONS
ENT. SOC. AMER. ANNALS 9:140 1916 MOSHER PUPAL STRUCTURE* PL. 6, FF. 11, 12, 13
ILL. DEPT. AGRI. TRANS. 18, APPEND. 124, 176, 1880 COQUILLETE, MARTEN LARVA
MARYLAND NAT. HIST. SOC. BULL. 8: 50, 1938 MASTERS COCOON*
MARYLAND NAT. HIST. SOC. PUB. NO PAGE 1930 HAYDEN EGG, LARVA, PUPA